普通高等教育“十三五”规划教材

山东省精品课程教材

化工原理

双语

王许云　王晓红　田文德　主编

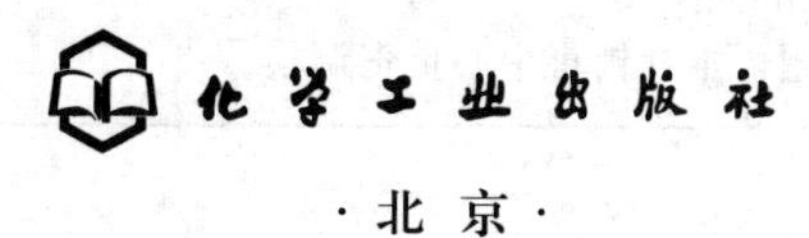

·北京·

《化工原理》（双语）以动量、热量、质量传递理论为主线，系统简明地阐述了典型过程工程单元操作的基本原理、工艺计算、主要设备的结构特点及性能、过程或设备的强化途径等。内容包括流体流动原理及应用、传热及传热设备、传质原理及应用、固体颗粒流体力学基础与机械分离以及其他单元操作共五章。教材用中英文双语进行表达，教材与教学内容始终与国际接轨，使学生及时了解国内外相关学科发展前沿，同时为化工原理课程双语教学奠定基础。为保证教学效果，每章均配有习题、思考题、工程案例分析，可以引导读者应用所学知识，培养对复杂工程问题的深入分析和解决能力。

《化工原理》（双语）可作为高等学校化工及高分子、化学、材料、化工自动化、环境、化工安全、食品、制药、冶金等专业学生的教材，也可供从事科研、设计和实际生产的科技人员参考。

图书在版编目(CIP)数据

化工原理：汉、英/王许云，王晓红，田文德主编.—北京：化学工业出版社，2019.12

普通高等教育“十三五”规划教材　山东省精品课程教材

ISBN 978-7-122-35822-6

Ⅰ.①化…　Ⅱ.①王…②王…③田…　Ⅲ.①化工原理-高等学校-教材-汉、英　Ⅳ.①TQ02

中国版本图书馆 CIP 数据核字（2019）第 266311 号

责任编辑：刘俊之　　文字编辑：刘志茹

责任校对：王素芹　　装帧设计：韩　飞

出版发行：化学工业出版社（北京市东城区青年湖南街 13 号　邮政编码 100011）

印　　刷：三河市航远印刷有限公司

装　　订：三河市宇新装订厂

787mm×1092mm　1/16　印张 29¾　字数 747 千字　2019 年 12 月北京第 1 版第 1 次印刷

购书咨询：010-64518888　　售后服务：010-64518899

网　　址：http://www.cip.com.cn

凡购买本书，如有缺损质量问题，本社销售中心负责调换。

定　　价：69.00 元

前 言

化工原理是化学工程与技术学科最重要的核心课程之一，是从基础理论课程到工程专业课程过渡的桥梁。本书是青岛科技大学化工原理教研室结合山东省精品课程建设编写的系列教材之一。《化工原理》（双语）的编写结合化工类相关专业人才培养方案、教学体系要求，并基于教育部工程教育认证“具备一定的国际视野，能够在跨文化背景下进行沟通和交流”的要求进行编写，为化工原理课程双语教学奠定基础。

本书以动量、热量、质量传递理论为主线，突出工程学科的特点，系统简明地阐述了典型过程工程单元操作的基本原理、工艺计算、主要设备的结构特点及性能、过程或设备的强化途径等。内容包括流体流动原理及应用、传热及传热设备、传质原理及应用、固体颗粒流体力学基础与机械分离以及其他单元操作五章内容，其中第3章传质原理及应用包括传质基本概念、蒸馏、吸收、萃取以及传质设备；第5章其他单元操作包括了蒸发、干燥、膜分离、结晶、吸附、离子交换等。

《化工原理》（双语）在编写过程中，注意吸收青岛科技大学化工原理教学团队在教学方面的丰富经验和体会，力争深入浅出、循序渐进、层次分明、论述严谨。同时注意引入国内外过程工业领域不断更新的新理论、新技术、新设备等最新动态，用中英文双语进行表达，教学内容始终与国际接轨，使学生及时了解国内外相关学科发展前沿。同时，教材的编写中注意结合本校化工、高分子、材料、机械等特色专业的最新科研成果，充分体现不断创新的理念。

为保证教学效果，每章章末均配有习题、思考题、工程案例分析，可以激发读者的学习兴趣，引导读者应用所学知识，培养对复杂工程问题的深入分析和解决能力。每章后面均配有名词术语中英文对照，可帮助读者有效进行英文资料的阅读，提升跨文化交流的能力。

本书可作为高等学校化工及高分子、化学、材料、化工自动化、环境、化工安全、食品、制药、冶金等专业的教材，也可供从事科研、设计和实际生产的科技人员参考。

参加本书编写工作的有青岛科技大学化工学院化工原理教研室王许云（英文部分和部分中文部分）、王晓红、田文德（中文部分）。另外，杜蕾、赵丽芳、白立俊、何海波、杜留娟、贾云、綦琪、许春莉、宁文君、巩伟、刘虎等也参与了本书的校改工作，对此一并致以诚挚的谢意！

由于水平有限，书中不妥之处在所难免，恳请读者提出宝贵意见。

编者

2019 年 10 月

目　录

第 4 章 固体颗粒流体力学基础与机械分离
Chapter 4 Solid Particles' Fluid Mechanics Basis and Mechanical Separations / 288

第 5 章 其他单元操作
Chapter 5 Others Unit Operation / 330

0

绪　论

Introduction

0.1　化工原理课程的基本内容及特点

化学工程是通过化工手段将原材料经过物理或化学的改变或分离获得所需要的产品的工业过程。化学工程师必须具备如下能力：对整个过程和相关设备进行开发、设计；选择合适的原料；保证工厂高效、安全、经济地运营；确保产品符合客户的要求。因此，化学工程既是一门艺术，也是一门科学。当工程师需要借助科学的方法解决问题时，就应该使用科学，通常，当科学不能给出完整的答案时，则有必要运用经验和判断。一名工程师的专业地位取决于他能够利用所有信息处理和解决实际问题的能力。

需要化学工程师服务的过程和行业种类繁多。在过去，化学工程师最关心的领域是选矿、石油精炼以及重化学物质和有机物的制造，如硫酸、甲醇和聚乙烯的制造。如今，聚合物平版印刷、电子工业、高强度复合材料、食品加工领域的转基因生化制剂，以及药物制造和药物输送等领域的技术变得越来越重要。大量化学技术和工艺及生化工业相关文献所描述的工艺为化学工程领域贡献出大量新理念。

化工原理则是上述化工、生物、冶金、食品等多个专业的一门技术基础课程，其主要任务是研究化工类型生产中各种物理操作（原料的预处理、产品的分离及精制等）问题的基本原理、典型设备的结构原理、操作性能和涉及的计算。化工原理课程是工程类专业学生在大学学习中的一个转折点，是从基础课程过渡到专业课程的一个桥梁，其目的是帮助学习者建立起工程概念。

为了清晰介绍化工原理课程的任务及特点，首先需要了解以下基本概念。

0.1.1　单元操作（Unit operations）

化工过程是指对原料进行大规模加工处理，使其不仅在状态与物理性质上发生变化，而且在化学性质上也发生变化，最终得到具有特定物理化学性质产品的工业过程。化工过程涉及的范围相当广泛，如石油炼制、化工、冶金、制药等，在国民经济中占有十分重要的地位。

通常，一种产品的生产过程往往需要几个或几十个物理加工过程，而且各种生产过程工艺路线千差万别，但所发生的各种物理变化过程及操作原理基本相同，所用设备也大同小异。例如聚氯乙烯和纯碱生产中最后工序都要脱水、干燥；酿酒和乙烯生产中都要将液体混合物分开；制糖和制盐的过程中都要将水溶液中的水分蒸发等等。这些操作工序是过程工业中共有的，像物料的加热与冷却，即传热操作几乎在所有的工业领域，甚至在日常生活中都是离不开的。正是这些共有的操作工序，引起了人们的研究兴趣。20 世纪 20 年代提出了“单元操作”（unit operations）的概念。

单元操作在国内较多的时候称为“化工原理”（principles of chemical engineering），20 世纪 50～60 年代曾称为“化工过程及设备”，80 年代也曾称为“化学工程”“化学工程基础”。

常用的单元操作已有几十种之多，主要包括：流体流动、流体输送机械、沉降、过滤、蒸发、传热、蒸馏、吸收、萃取、塔设备、干燥、吸附、膜分离、搅拌、冷冻、流态化、结晶、升华等，如表 0-1 所示。

表 0-1　化工常用单元操作

单元操作名称	原理与目的	理论基础
流体输送	输入机械能将一定量流体由一处送到另一处	流体动力过程（动量传递）
沉降	利用密度差，从气体或液体中分离悬浮的固体颗粒、液滴或气泡	
过滤	根据尺寸不同的截留，从气体或液体中分离悬浮的固体颗粒	
搅拌	输入机械能使流体间或与其他物质均匀混合	
流态化	输入机械能使固体颗粒悬浮，得到具有流体状态的特性，用于燃烧、干燥等过程	
换热	利用温差输入或移出热量，使物料升温、降温或改变相态	传热过程（热量传递）
蒸发	加热以气化物料，使之浓缩	
蒸馏	利用各组分间挥发度不同，使液体混合物分离	传质过程（质量传递）
吸收	利用各组分在溶剂中的溶解度不同，分离气体混合物	
萃取	利用各组分在萃取剂中的溶解度不同分离液体混合物	
吸附	利用各组分在吸附剂中的吸附能力不同分离气、液混合物	
膜分离	利用各组分对膜渗透能力的差异，分离气体或液体混合物	
干燥	加热湿固体物料，使之干燥	热、质同时传递过程
增减湿	利用加热或冷却来调节或控制空气或其他气体中的水汽含量	
结晶	利用不同温度下溶质溶解度不同，使溶液中溶质变成晶体析出	
压缩	利用外力做功，提高气体压力	热力过程
冷冻	加入功，使热量从低温物体向高温物体转移	
粉碎	用外力使固体物体破碎	机械过程
颗粒分级	将固体颗粒分成大小不同的部分	

动量传递是研究动量在运动的介质中所发生的变化规律，如流体流动、沉降和混合等操作中的动量传递；热量传递是研究热量由一个地方到另一个地方的传递，如传热、干燥、蒸发、蒸馏等操作中存在的热量传递；质量传递涉及物质由一相转移到另一不同的相，在气相、液相和固相中，其传递机理都是一样的，如蒸馏、吸收、萃取等操作中存在的质量传递。

0.1.2　化工原理与其他课程的关系

虽然化工原理有时也叫作化学工程，实际上前者仅是后者的基础。就化学工程的发展来看，包括如下几个分支：①化工传递过程，在许多单元操作的发展过程中，人们逐渐认识到它们之间存在共同的原则，可进一步归纳为动量传递、热量传递和质量传递，总称为化工传递过程。②化工热力学，是在化学热力学和工程热力学的基础上形成的，主要研究多组分系统的温度、压力、各相组成和各种热力学性质间相互关系的数学模型以及能量（包括低品位能量的有效利用）问题。③化学反应工程，从化学反应设备的实际出发，深入分析其中的过程规律性，找出其共同点，逐渐形成了化学反应工程，其研究的核心问题是反应器中化学反应速率的快慢及其影响因素，从而能够正确选择反应器类型和操作条件，使化学反应实现工业化。④化工系统工程，主要研究化工过程模拟分析、综合和最优化等。过程综合指已知过程的输入和输出，确定过程的结构，即选择适宜的设备类型、流程结构和操作条件等；最优化指要求过程的性能指标达到某些最优数值，包含了过程最优设计、最优控制和最优管理等性能指标。

化学工程是直接支撑化学工业的主要工程技术，是重要的学科支持。化学工程经过近一个世纪的发展，其应用领域不但覆盖了几乎所有的过程工业，而且新的生长点正不断产生，如“生化工程”“环境工程”等，其研究对象广泛而复杂，已远远地超出了化学工程的范畴，遍布于能源、资源、环境、运输、医药卫生、材料、农业以及生物等诸多领域。

0.1.3 化工原理的任务及特点

化工原理课程的主要任务是培养学生运用本学科的基础理论及基本技能，来分析解决化工生产实际问题的能力，包括：①选型，即根据生产工艺要求、物料特性及技术要求，能合理选择恰当的单元操作及设备；②设计，对已选的单元操作进行设备设计和工艺计算；③操作，熟悉该单元的操作原理、操作方法，具备初步的分析及解决操作故障的能力。

化工原理课程突出特点是符号多、公式多、单位换算多。尤其是刚开始学习的时候总觉得没有头绪。尽管如此，只要下点功夫仍然可以把课本学好。然而，要强调指出的是，化工原理面临着真实的、复杂的生产问题，即特定的物料，在特定的设备内，进行特定的过程，这就使问题的复杂性不完全在于过程本身，而首先在于过程工业设备复杂的几何形状和多变的物性。所以，研究工程问题的方法论和解决生产实际问题的能力，在化工原理的学习中上升到了显著的地位。

Chemical Engineering

Chemical engineering has to do with industrial processes in which raw materials are changed or separated into useful products. The chemical engineer must develop, design, and engineer both the complete process and the equipment used; choose the proper raw materials; operate the plants efficiently, safely, and economically; and see to it that products meet the requirements set by the customers. Chemical engineering is both an art and a science. Whenever science helps the engineer to solve a problem, science should be used. When, as is usually the case, science does not give a complete answer, it is necessary to use experience and judgement. The professional stature of an engineer depends on skill in utilizing all sources of information to reach practical solutions to processing problems.

The variety of processes and industries that call for the services of chemical engineers is enormous. In the past, the areas of most concern to chemical engineers were ore beneficiation, petroleum refining, and the manufacture of heavy chemicals and organics such as sulfuric acid, methyl alcohol, and polyethylene. Today items such as polymeric lithographic supports far the electronics industry, high-strength composite materials, genetically modified biochemical agents in areas of food processing, and drug manufacture and drug delivery have become increasingly important. The processes described in standard treatises on chemical technology and the process and biochemical industries give a good idea of the field of chemical engineering.

Because of the variety and complexity of modern processes, it is not practicable to cover the entire subject matter of chemical engineering under a single head. The field is divided into convenient, but arbitrary, sectors. This text covers that portion of chemical engineering known as the unit operations.

Unit Operations

An economical method of organizing much of the subject matter of chemical engineering is based on two facts: (1) Although the number of individual processes is great, each one can be broken down into a series of steps, called operations, each of which in turn appears in process after process; (2) the individual operations have common techniques and are based on . the same scientific principles. For example, in most processes solids and fluids must be moved; heat or other forms of energy must be transferred from one substance to another; and tasks such as drying, size reduction, distillation, and evaporation must be performed. The unit operation concept is this: By studying systematically these operations themselves—operations that clearly cross industry and process lines—the treatment of all processes is unified and simplified.

The strictly chemical aspects of processing are studied in a companion area of chemical engineering called reaction kinetics. The unit operations are largely used to conduct the primarily physical steps of preparing the reactants, separating and purifying the products, recycling unconverted reactants, and controlling the energy transfer into or out of the chemical reactor.

The unit operations are as applicable to many physical processes as to chemical ones. For example, the process used to manufacture common salt consists of the following sequence of unit operations: transportation of solids and liquids, transfer of heat, evaporation, crystallization, drying, and screening. No chemical reaction appears in these steps. On the other hand, the cracking of petroleum, with or without the aid of a catalyst, is a typical chemical reaction conducted on an enormous scale. Here the unit operations-transportation of fluids and solids, distillation, and various mechanical separations-are vital, and the cracking reaction could not be utilized without them. The chemical steps themselves are conducted by controlling the flow of material and energy to and from the reaction zone.

Because the unit operations are a branch of engineering, they are based on both science and experience. Theory and practice must combine to yield designs for equipment that can be fabricated, assembled, operated, and maintained. A balanced discussion of each operation requires that theory and equipment be considered together. This book presents such a balanced treatment.

0.2 单位制与单位换算

0.2.1 单位制（Unit systems）

凡是物理量均有单位，可分为基本单位和导出单位两类。在描述单元操作的众多物理量中，独立的物理量叫基本量，其单位叫基本单位，如时间、长度、质量等。不独立的物理量叫导出量，其单位叫导出单位，如速度、加速度、密度等。基本单位仅有几个，而导出单位由基本单位组成，数量很多。

基本单位加上导出单位称为单位制度（单位制）。由于历史和地区的原因，出现了对基本单位的不同选择，因而产生了不同的单位制度。常用的单位制有绝对单位制（包括物理单位制和米制）、重力单位制［工程单位制，centimeter-gram-second（cgs）and foot-pound-second（fps）］和国际单位制（SI 制）。

长期以来，科技领域存在多种单位制度并用的局面。同一个物理量，有时在不同的单位制中具有不同的单位和数值，给计算和交流带来麻烦，且很容易出错。为了改变这一局面，1960 年 10 月，第十一届国际计量大会通过了一种新的单位制，叫国际单位制。该单位制共有七个基本单位，分别为长度、时间、质量、热力学温度、电磁强度、光强度和物质的量，外加平面角和立体角两个辅助单位。其优点是，所有的物理量都可以用上述七个基本单位导出（有时要借助辅助单位），且任何一个导出量由上述七个基本单位导出时，都不需要引入比例系数。

1984 年，国内确定了统一实行以 SI 制为基础，包括由我国指定的若干非 SI 制在内的法定单位制，并规定，自 1991 年起除个别领域外，不允许使用非法定单位制。本课程主要采用法定单位制，兼顾各单位制之间的换算。

0.2.2 单位换算

单位换算虽然简单，但即使是一个经验丰富的工程师，稍一马虎也会出错，所以必须认真对待。如 SI 制与工程单位制的换算：

① 质量与重力　在 SI 制中 1kg 质量的物体，若用工程单位制表示，该物体的重力为 1kgf。即同一物体用 SI 制表示的质量与用工程单位制表示的重力在数值上相等。所以在有关手册中查得工程单位制的重力，SI 制的质量可直接取其数值。但应注意，两者数值相等，但概念不同。质量是物体所含物质的多少，而重力是物体受地球引力的大小，一般认为地球附近的引力大小近似不变。

② 重力　在工程单位中重力是基本单位，但在 SI 制中重力是导出单位。因此，在工程单位制中重力为 1kgf 的物体，若用 SI 制表示，该物体的重力为 9.81N。

【例 0-1】 实验测得空气垂直流过管子外侧时，管壁对流传热系数计算公式为：

$$\alpha=0.37G^{0.37}[\text{Btu}/(\text{ft}^2\cdot\text{h}\cdot{}^{\circ}\text{F})]$$

式中，G——空气的质量流速，lb/(ft^2·h)。试把空气质量流速的单位换算成 kg/(m^2·s)，对流传热系数的单位为 J/(m^2·s·℃)，写出相应的计算式。

解

$$1\text{lb}/(\text{ft}^2\cdot\text{h})=0.4536\ \text{kg}/(0.3048^2\ \text{m}^2\times3600\text{s})$$

$$=1.356\times10^{-3}\ \text{kg}/(\text{m}^2\cdot\text{s})$$

$$1\text{Btu}/(\text{ft}^2\cdot\text{h}\cdot{}^{\circ}\text{F})=1.055\times10^3\text{J}/\left(0.3048^2\ \text{m}^2\times3600\text{s}\times\frac{1}{1.8}℃\right)$$

$$=5.678\ \text{J}/(\text{m}^2\cdot\text{s}\cdot℃)$$

则：

$$\alpha'\left(\frac{1}{5.678}\right)=0.37\times\left(G'\times\frac{1}{1.356\times10^{-3}}\right)^{0.37}$$

处理得到：

$$\alpha'=24.18(G')^{0.37}\ \text{J}/(\text{m}^2\cdot\text{s}\cdot℃)$$

式中，G'的单位是 kg/(m^2·s)。

Scientific Foundations of Unit Operations

A number of scientific principles and techniques are basic to the treatment of the unit operations. Some are elementary physical and chemical laws such as the conservation of mass and energy, physical equilibria, kinetics, and certain properties of matter. Their general use is described in the remainder of this chapter. Other special techniques important in chemical engineering are considered at the proper places in the text.

Unit Systems

The official international system of units is SI (System International Units). Strong efforts are underway for its universal adoption as the exclusive system for all engineering and science, but older systems, particularly the centimeter-gram-second (cgs) and foot-pound-second (fps) engineering gravitational systems, are still in use and probably will be around for some time. The chemical engineer fords many physiochemical data given in cgs units; that many calculations are most conveniently done in fps units; and that SI units are increasingly encountered in science and engineering. Thus it becomes necessary to be expert in the use of all three systems.

In the following treatment, SI is discussed first, and then the other systems are derived from it. The procedure reverses the historical order, as the SI units evolved from the cgs system. Because of the growing importance of SI, it should logically be given a preference. If, in time, the other systems are phased out, they can be ignored and SI used exclusively.

Physical Quantities

Any physical quantity consists of two parts: a unit, which tells what the quantity is and gives the standard by which it is measured, and a number, which tells how many units are needed to make up the quantity. For example, the statement that the distance between two points is 3 m means all this: A definite length has been measured; to measure it, a standard length, called the meter, has been chosen as a unit; and three 1m units, laid end to end, are needed to cover the distance. If an integral number of units are either too few or too many to cover a given distance, submultiples, which are fractions of the unit, are defined by dividing the unit into fractions, so that a measurement can be made to any degree of precision in terms of the fractional units. No physical quantity is defined until both the number and the unit are given.

SI Units

The SI system covers the entire field of science and engineering, including electromagnetics and illumination. For the purposes of this book, a subset of the SI units covering chemistry, gravity, mechanics, and thermodynamics is sufficient. The units are derivable from (1) four proportionalities of chemistry and physics; (2) arbitrary standards for mass, length, time, temperature, and the mole; and (3) arbitrary choices for the numerical values of two proportionality constants.

0.2.3 量纲分析

量纲与单位不是一个概念。如长度的单位有米、厘米、毫米、英尺和英寸等，为了明确长度的特性，可用量纲 L 表示。人们规定，用一个符号表示一个基本量，这个符号连同它的指数叫基本量纲。而基本量纳的组合叫导出量纲。基本量纲和导出量纲统称量纲。各物理量均可以用量纲表示，如长度用 L，质量用 M，时间用 θ，温度用 T，密度用 $M \cdot L^{-3}$ 等。

在过程工业中，由于一些物理过程十分复杂，建立理论模型颇为困难。如果过程的影响因素已经明了，作为影响因素的物理量的相互关系，可进行某种程度的预测，这种预测方法称为量纲分析。量纲分析的依据是量纲一致性原则或 π 定理。所谓量纲一致性原则是指一个物理量方程的各项量纲必相同。所谓 π 定理是指量纲一致性的方程都可以化为无量纲数群的形式，方法是将方程中的各项同除以其中的任何一项即可，且有：

无量纲数群的个数＝变量数－基本量纲的个数

若能找出过程的影响因素，使用量纲分析的方法将其归纳为无量纲数群表示的经验模型，用实验确定模型的系数和指数，这在化工原理上是可行的。这样的经验模型不仅关联式简单，而且可减少实验的工作量。

Dimensional Analysis

Many important engineering problems cannot be solved completely by theoretical or mathematical methods. Problems of this type are especially common in fluid-flow, heat-flow, and diffusion operations. One method of attacking a problem for which no mathematical equation can be derived is that of empirical experimentation. For example, the pressure loss from friction in a long, round, straight, smooth pipe depends on all these variables: the length and diameter of the pipe, the flow rate of the liquid, and the density and viscosity of the liquid. If any one of these variables is changed, the pressure drop also changes. The empirical method of obtaining an equation relating these factors to pressure drop requires that the effect of each separate variable be determined in turn by systematically varying that variable while keeping all others constant. The procedure is laborious, and it is difficult to organize or correlate the results so obtained into a useful relationship for calculations.

There exists a method intermediate between formal mathematical development and a completely empirical study. It is based on the fact that if a theoretical equation does exist among the variables affecting a physical process, that equation must be dimensionally homogeneous. Because of this requirement, it is possible to group many factors into a smaller number of dimensionless groups of variables. The groups themselves rather than the separate factors appear in the final equation.

This method is called dimensional analysis, which is an algebraic treatment of the symbols for units considered independently of magnitude. It drastically simplifies the task of fitting experimental data to design equations; it is also useful in checking the consistency of the units in equations, in converting units, and in the scale-up of data obtained in model test units to predict the performance of full-scale equipment.

In making a dimensional analysis, the variables thought to be important are chosen and their dimensions tabulated. If the physical laws that would be involved in a mathematical solution are known, the choice of variables is relatively easy. The fundamental differential equations of fluid flow, for example, combined with the Laws of heat conduction and diffusion, suffice to establish the dimensions and dimensionless groups appropriate to a large number of chemical engineering problems. In other situations the choice of variables may be speculative, and testing of the resulting relationships may be needed to establish whether some variables were left out or whether some of those chosen are not needed.

0.3 基本概念与定律

要做一个合格的工程师是不容易的，过程工业的复杂性和影响过程的众多因素，使得问题的解决十分困难，一般的处理方法是：理论分析、实验研究、经验估计、权衡调整。要运用以上各类方法处理问题，首先从掌握以下概念入手。

0.3.1 平衡关系及过程速率

在化学工业的许多单元操作中，例如吸收、蒸馏等，平衡关系具有重要的意义。平衡关系是一种动态平衡，平衡条件可以用热力学法则来描述，而过程进行的方向和所能达到的极限都可以由平衡关系推知。因此，物理化学是化工原理的重要基础。

任何一个物系如果不是处在平衡状态，则必然会发生趋向平衡的过程，而过程变化的速率总是和它所处状态与平衡状态的差距（推动力）成正比，而与阻力成反比，即过程速率＝过程推动力/过程阻力。推动力的性质取决于过程的内容，如传热的推动力是温度差，流体流动的推动力是压力差；与推动力相对应的阻力则与操作条件和物性有关。过程速率指明了过程进行的快慢程度，属于动力学在工程问题中的应用，动力学特征主要取决于过程的机理，而大部分过程的机理与动量、热量和质量传递密切相关，所以，传递过程是化工原理的另一个重要基础。

0.3.2 质量衡算

质量衡算、能量衡算和动量衡算是化工原理课程中分析问题采用的基本方法。质量衡算的依据是质量守恒定律，能量衡算的依据是能量守恒定律。进行衡算的一般步骤如下：首先确定衡算范围，具体包括微分衡算和总衡算，微分衡算取微元体为衡算范围，而总衡算的衡算范围可以是单个装置，也可以是一段流程、一个车间或一个工厂；其次是确定衡算对象和衡算基准；最后按衡算的通式进行计算。上述三种衡算中，质量衡算和能量衡算最为常用。在过程的开发设计、模拟优化、操作控制等工作中，质量衡算和能量衡算是必不可少的依据。

根据质量守恒定律，即物质既不会凭空产生也不会凭空消失，参与任何化工过程的物料物质的量是守恒的。即按照上述衡算步骤可建立质量衡算通式：

输入系统的物质的量＝输出系统的物质的量＋累积的物质的量

当过程为稳态时，在衡算范围内累积的量等于零，即：

输入系统的物质的量＝输出系统的物质的量

【例 0-2】 在硝酸钾的生产过程中，20%（质量分数，下同）的硝酸钾水溶液以 1000 kg/h 的流量送入蒸发器，在 422 K 下蒸发出部分水而得到 50%的浓硝酸钾溶液，然后送入冷却结晶器，在 311K 下结晶，得到含水 4%的硝酸钾结晶和含硝酸钾 37.5%的饱和溶液。前者作为产品取出，后者循环回到蒸发器。过程为稳态操作，试计算硝酸钾结晶产品量、水蒸发量和循环的饱和溶液量。

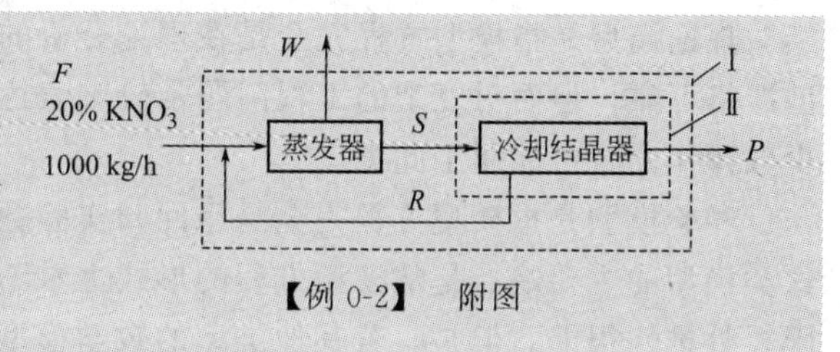

【例 0-2】 附图

解 根据题意画出过程的物料流程图。

(1) 求硝酸钾结晶产品量 P 取包括蒸发器和冷却器的整个过程为衡算系统（虚线框Ⅰ)，取 1h 为衡算基准，以硝酸钾为衡算对象，因系统稳态操作，输入系统的硝酸钾量应等于输出系统的硝酸钾量。则：

$$1000\times0.2=0.96P$$

即：

$$P=1000\times0.2/0.96=208.3(\text{kg/h})$$

(2) 求水蒸发量 W 仍取系统Ⅰ，衡算基准为 1h，以总物料为衡算对象，则：

$$1000=W+P$$

$$W=1000-P=1000-208.3=791.7(\text{kg/h})$$

(3) 求循环的饱和溶液量 R 取冷却结晶器为系统Ⅱ，衡算基准为 1h，以总物料为衡算对象，做总物料衡算，可得：

$$S=208.3+R$$

以硝酸钾为衡算对象，做硝酸钾的物料衡算，得：

$$S\times0.5=208.3\times0.96+R\times0.375$$

上两式联立，可得：

$$R=766.6(\text{kg/h})$$

0.3.3 能量衡算

过程工业中涉及的能量主要是机械能和热能，能量衡算的依据是能量守恒定律，即：

输入系统的总能量＝输出系统的总能量＋系统累积的能量

对于稳态操作过程，系统内无能量累积，即：

输入系统的总能量＝输出系统的总能量

0.4 研究方法

将基础理论研究成果应用到现代大型工业设备中，主要采用数学模型法，其实质是通过数学模型来放大和设计工业过程和设备。该方法的关键在于所建立的数学模型是否能够描述过程的本质问题，而对过程本质的认识又来源于实践，因此，实验仍然是数学模型法的主要依据。

数学模型大体可分为四类，即理论模型、经验模型、半经验半理论模型和人工智能模型。

理论模型是指模型方程完全是在理论分析的基础上建立起来的数学表达式。尽管这类模型严格可靠，但面对复杂的工程问题常较难表达，只有那些过程十分简单，关系极其明确的少数操作才有可能属于此类。

经验模型是指模型方程完全是靠回归实验数据得到的数学表达式，没有任何理论依据。这类模型的主要缺点是从模型方程的形式上看不出所研究问题的内在规律，且受实验范围所限，外推性很差。但是，若其他类型的数学模型难以得到时，使用经验模型仍不失为一种补偿的方法。

半经验半理论模型介于上述两种模型之间，是处理复杂工程问题的最有效、最常用的模型。这种模型的建立是理论与实验的结合，即通过对所研究对象的过程机理进行理论分析，建立模型方程的表达式，进而通过实验确定模型参数。这样，可减少盲目性，增加可信性，外推性也有较大改善。

人工智能模型是模拟或部分模拟人类智能的数学模型。人工神经网络模型是典型的人工智能模型，它是由大量的神经元互连而成的网络，模拟人脑神经系统的学习和记忆功能是人工神经网络的核心任务。近些年来，人工智能模型的理论研究已取得了突破性进展并在解决某些复杂的工程问题中获得了成功。相信随计算机技术的发展和非经典数学方法的研究，人工智能模型必将广泛地应用到各种工程系统和生产系统中去。

本书主要介绍和应用经验模型、半经验半理论模型。

第1章

流体流动原理及应用

Chapter 1 Principles of Fluid Flow and Applications of Fluid

学习目标

通过本章学习，学生应具备下列知识和能力：

1. 掌握流体流动单元操作的基本原理；能够基于流体流动的基本原理，建立复杂工程问题中关于流体输送所涉及的数学模型，分析、解决流体输送过程中的复杂化学工程问题。

2. 掌握流体流动单元操作设备的结构特点、工作原理、操作方法；能够针对复杂化学工程问题所涉及的流体输送问题，提出合理的解决措施。

3. 掌握流体流动操作设备的计算及选型；能够针对复杂化学工程问题，使用恰当的技术、资源、现代工程工具和信息技术工具，通过对复杂工程问题进行建模，选取合适的流体输送机械。

4. 熟悉相关的英文表达，增强国际交流能力，拓宽国际视野。

学习要求

1. 通过本章学习，能够解决管径的选择及管路布置。

2. 能估算输送流体所需要的能量，确定流体输送机械的型式及其所需要的功率。

3. 能够对测定流体流速、流量及压强等仪表设备进行合理选择以及操作。

4. 掌握化工中常用流体输送机械的基本结构、工作原理和操作特性，能够根据生产工艺要求和流体特性，合理选择和正确操作流体输送机械，并使之能高效、安全、可靠运行。

1.1 流体基本概念（Basic Concepts of Fluid）

1.1.1 流体特征（Nature of fluids）

气体和液体总称为流体，流体具有流动性且无固定形状。过程工业（process industry）

中所处理的物料，包括原料、半成品及产品等，大多数是流体。流体的输送（transport）、传热（heat transfer）、传质（mass transfer）或化学反应（chemistry reaction），大多是在流体流动情况下进行的，因而流体流动状态对这些过程有很大影响，它是过程工业的基础。

1.1.2 流体力学基本概念（Basic concepts of fluid mechanics）

流体力学（fluid mechanics）是研究流体在相对静止和运动时所遵循的宏观基本规律，同时研究流体与固体相互作用的学科，流体力学有许多分支，例如“水力学”（hydraulics）及“空气动力学”（aerodynamics）等。“流体流动”（fluid flow）是为“化工原理”课程需要而编写的流体力学最基础的内容，主要是研究流体的宏观运动规律，其介绍范围主要局限在流体静力学（fluid statics）、流体在管道内流动的基本规律及流量计量（metering of fluids）等方面。

运用流体流动基本知识，可以解决管径的选择及管路的布置；估算输送流体所需的能量、确定流体输送机械的型式及其所需的功率；测量流体的流速、流量及压强等；为强化设备操作及设计高效能设备提供最适宜的流体流动条件。

由于讨论流体流动问题时，着眼点不在于研究流体复杂的分子运动，因此本章采用连续性假设（continuity hypothesis），即把流体看成是由大量质点（又称分子集团）组成的连续介质，因为质点的大小与管道或设备的尺寸相比是微不足道的，可认为质点间是没有间隙的，可用连续函数描述。但是，高真空下的气体，连续性假定不能成立。

Fluid Flow

The behavior of fluids is important to process engineering generally and constitutes one of the foundations for the study of unit operations. An understanding of fluids is essential, not only for accurately treating problems in the movement of fluids through pipes, pumps, and all kinds of process equipment, but also for the study of heat flow and the many separation operations that depend on diffusion and mass transfer.

The branch of engineering science that has to do with the behavior of fluids—and fluids are understood to include liquids, gases, and vapors—is called *fluid mechanics*. Fluid mechanics in turn is part of a larger discipline called *continuum mechanics*, which also includes the study of stressed solids.

Fluid mechanics has two branches important to the study of unit operations, *fluid statics*, which treats fluids in the equilibrium state of no shear stress, and *fluid dynamics*, which treats fluids when portions of the fluid are in motion relative to other parts.

Although gases and liquids consist of molecules, it is possible in most cases to treat them as continuous media for the purposes of fluid flow calculations. This treatment as a continuum is valid when the smallest volume of fluid contains a large enough number of molecules so that a statistical average is meaningful and the macroscopic properties of the fluid such as density, pressure, temperature, velocity and so on, vary smoothly or continuously from point to point.

1.1.3 流体密度

(1) 密度(density)ρ

在给定的温度(temperature)和压力(pressure)下,单位体积流体所具有的质量,称为流体的密度,通常以 ρ 示之,单位为 kg/m^3。

$$\rho=\frac{m}{V} \tag{1-1}$$

式中 m——流体的质量,kg;

V——流体的体积,m^3。

不同流体的密度是不同的。对任何一种流体,其密度是压力与温度的函数,即 $\rho=f(p, T)$。其中,压力对液体密度的影响很小,可忽略不计,故液体可视为不可压缩流体(incompressible fluid);而气体是可压缩性流体(compressible fluid),其密度随系统压力明显变化。

温度对气体及液体的密度均有一定的影响,故在平时查取流体密度时应注明温度条件。

① 气体密度计算 因为气体具有可压缩性及膨胀性,其密度随温度、压力的变化较大。当温度不太低,压力不太高时,气体可按理想气体处理,根据理想气体状态方程(state equation of ideal gas):

$$pV=nRT=\frac{m}{M}RT \tag{1-2}$$

则

$$\rho=\frac{pM}{RT} \tag{1-2a}$$

式中 n——气体的物质的量,kmol;

p——气体的绝对压力,kPa;

T——气体的热力学温度,K;

M——气体的千摩尔质量,kg/kmol;

R——气体常数,$R=8.314$kJ/(kmol·K)。

理想气体操作状态(operating state)(压力 p、温度 T)下的密度 ρ 与标准状态(normal temperature pressure)[压力 $p_0=1$atm(1atm=101.325kPa),温度 $T_0=273$K]下的密度 ρ_0 之间的换算可由下式进行:

$$\rho=\rho_0\frac{p}{p_0}\times\frac{T_0}{T} \tag{1-3}$$

若气体按真实气体计算,则需引入压缩系数进行校正。

当计算气体混合物密度时,可假设混合物各组分在混合前后质量不变,取 $1m^3$ 混合气体为基准,则气体混合物密度可由下式计算。

$$\rho_m=\rho_1 y_1+\rho_2 y_2+\rho_3 y_3+\cdots+\rho_n y_n \tag{1-4}$$

式中 ρ_m——气体混合物密度,kg/m^3;

$\rho_1, \rho_2, \cdots, \rho_n$——气体中各组分的密度,$kg/m^3$;

$y_1, y_2, \cdots, y_n$——各组分的体积分数,由于理想气体遵守道尔顿分压定律,所以混合气体中各组分的体积分数等于摩尔分数或分压比。

气体混合物的密度也可由式(1-2a)计算,此时式中的千摩尔质量 M 应由混合气体的平均千摩尔质量 M_m 代替。

$$M_m = M_1 y_1 + M_2 y_2 + \cdots + M_n y_n \tag{1-5}$$

式中　M_1，M_2，…，M_n——气体混合物中各组分的千摩尔质量，kg/kmol。

② 液体密度计算　对于纯组分液体密度，可查取附录或有关的工艺及物化手册。

液体混合物密度的计算，可取1kg混合物为基准，并假定混合前、后总体积不变。液体混合物组成常用组分的质量分数表示，故液体混合物密度 ρ_m 可表示为

$$\frac{1}{\rho_m} = \frac{x_1}{\rho_1} + \frac{x_2}{\rho_2} + \cdots + \frac{x_n}{\rho_n} \tag{1-6}$$

式中　x_1，x_2，…，x_n——液体混合物中各组分的质量分数；

ρ_1，ρ_2，…，ρ_n——液体混合物中各组分的密度，kg/m^3。

【例1-1】 在盐酸制造过程中，氯化氢气体混合物（其中含25%HCl、75%空气，均为体积分数），在50℃及743mmHg（1mmHg=0.1333kPa）（绝压）的条件下进入吸收塔，试计算气体混合物的密度。

解　已知HCl的千摩尔质量为36.5kg/kmol，空气的千摩尔质量为29kg/kmol，混合气体中各组分的体积分数为：HCl=0.25，空气=0.75。T=50K+273K=323K，p=743mmHg。

$$\begin{aligned} M_m &= M_1 y_1 + M_2 y_2 \\ &= 36.5 \times 0.25 + 29 \times 0.75 \\ &= 9.125 + 21.75 \\ &= 30.875 (kg/kmol) \end{aligned}$$

气体混合物的密度，由式(1-3)求得：

$$\begin{aligned} \rho_m &= \rho_0 \frac{p}{p_0} \times \frac{T_0}{T} \\ &= \frac{30.875}{22.4} \times \frac{743}{760} \times \frac{273}{323} \\ &= 1.377 \times 0.978 \times 0.845 \\ &= 1.14 (kg/m^3) \end{aligned}$$

【例1-2】 计算293K时60%（质量分数）的乙酸水溶液的密度。

解　293K时，$\rho_{水} = 998kg/m^3$，$\rho_{乙酸} = 1049kg/m^3$。

在293K时乙酸水溶液的密度为：

$$\frac{1}{\rho_m} = \frac{x_{水}}{\rho_{水}} + \frac{x_{乙酸}}{\rho_{乙酸}} = \frac{0.40}{998} + \frac{0.60}{1049}$$

则 $\rho_m = 1029kg/m^3$

(2) 比容（specific volume）v

比容是密度的倒数，单位为 m^3/kg，即：

$$v = \frac{1}{\rho} \tag{1-7}$$

1.2 流体静力学（Fluid Statics）

1.2.1 压强（Pressure）

流体单位面积上所承受的垂直作用力，称为流体的静压强，简称压强，以符号 p 示之，

而流体的压力 P 称为总压力。

$$p=\frac{P}{A} \tag{1-8}$$

式中 p——流体压强，N/m^2 或 Pa；

P——垂直作用于面积 A 上的总压力，N；

A——作用面的表面积，m^2。

压强的单位除用 Pa 表示外，还可用大气压（atm）、米水柱（mH_2O）、毫米汞柱（mmHg）、巴（bar）等表示，所以熟练掌握压强不同单位间的换算十分重要。

1atm（物理大气压）$=760mmHg=10.33mH_2O=1.033kgf/cm^2$（工程大气压，at）$=1.013\times10^5Pa=1.0133bar$

工程上为计算方便，还引入工程大气压换算系统。

1at(工程大气压)$=735.6mmHg=10mH_2O=1kgf/cm^2=9.807\times10^4Pa=0.9807bar$

此处需要指出：1kgf 指 1kg 物体在 $g=9.81m/s^2$ 重力场中受到的重力，称为“千克力”，则 $1kgf/cm^2$ 作为压强单位有时在有些文献或工程现场中用到。

流体的压强除用不同的单位来表示外，还可以用不同的方法来表示。

设备内流体的真实压强称为绝对压强（absolute pressure），简称绝压。

当设备内流体的绝对压强高于外界大气压时，常在设备上安装压力表，压力表上的读数称为表压强（gage pressure），它反映流体绝压高于外界大气压（atmospheric pressure）的数值，即可表示为：

表压强＝绝对压强－外界大气压(当时当地)

当设备内流体的真实压强低于外界大气压时，工程上视为负压操作，常在设备上安装真空表，真空表上的读数称为真空度（vacuum pressure），它反映流体绝压低于外界大气压的数值，即可表示为：

真空度＝外界大气压(当时当地)－绝对压强

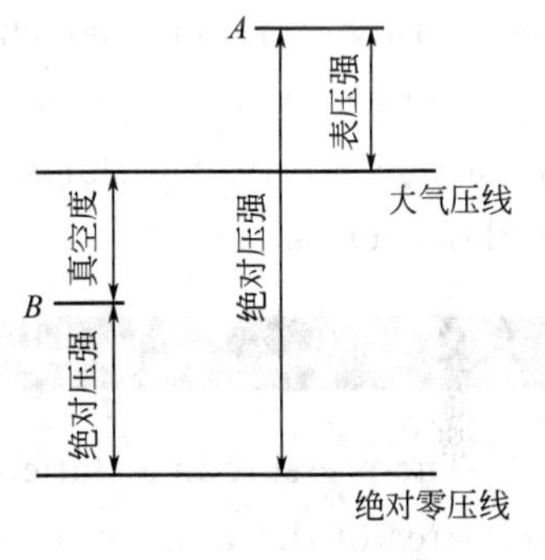

图 1-1 绝对压强、外界大气压、表压和真空度的关系

绝对压强、外界大气压、表压和真空度之间的关系如图 1-1 所示。不难看出，真空度实际上是流体表压的负值（negative gage pressure）。例如，体系的真空度为 3.3×10^3Pa，则其表压为 -3.3×10^3Pa。

为了避免不必要的错误，在工程计算中，必须在压强的单位后加括号或加注脚注明压强的不同表示方法。例如：$p=2.0kgf/cm^2$（表压），$p_{真空度}=300mmHg$，$p=4.9\times10^5Pa$（绝对压强）等。

【例 1-3】 已知当地大气压（绝压）p_0 为 750mmHg，某容器内气体的真空度为 $0.86kgf/cm^2$，试求该气体的绝对压强及表压分别是多少，单位用 Pa 表示。

解
$$p_0=750mmHg=750\times\frac{1.013\times10^5}{760}=9.997\times10^4(Pa)$$

$$p=0.86kgf/cm^2=0.86\times9.81\times10^4=8.437\times10^4(Pa)$$

所以
$$p(绝压)=p_0-p_{真}=(9.997-8.437)\times10^4=1.56\times10^4(Pa)$$

$$p_{表}=p-p_0=-p_{真}=-8.437\times10^4(Pa)$$

Nature of Fluids

A fluid is a substance that does not permanently resist distortion. An attempt to change the shape of a mass of fluid results in layers of fluid sliding over one another until a new shape is attained. During the change in shape, shear stresses (Shear is the lateral displacement of one layer of material relative to another layer by an external force. Shear stress is defined as the ratio of this force to the area of the layer.) exist, the magnitudes of which depend upon the viscosity of the fluid and the rate of sliding, but when a final shape has been reached, all shear stresses will have disappeared. A fluid in equilibrium is free from shear stresses.

Density

At a given temperature and pressure, a fluid possesses a definite density, which in engineering practice is usually measured in kilograms per cubic meter. Although the density of all fluids depends on the temperature and pressure, the variation in density with changes in these variables may be small or large. If the density changes only slightly with moderate changes in temperature and pressure, the fluid is said to be incompressible, if the changes in density are significant, the fluid is said to be compressible. Liquids are generally considered to be incompressible and gases compressible. The terms are relative, however, and the density of a liquid can change appreciably if pressure and temperature are changed over wide limits. Also, gases subjected to small percentage changes in pressure and temperature act as incompressible fluids, and density changes under such conditions may be neglected without serious error.

Pressure

The pressure in a static fluid is familiar as a surface force exerted by the fluid against a unit area of the walls of its container. Pressure also exists at every point within a volume of fluid. It is a scalar quantity, and at any given point its magnitude is the same in all directions.

The pressure at a point within a volume of fluid will be designated as either an *absolute pressure* or a *gage pressure*. Absolute pressure is measured relative to a perfect vacuum (absolute zero pressure), whereas gage pressure is measured relative to the local atmospheric pressure. Thus, a gage pressure of zero corresponds to a pressure that is equal to the local atmospheric pressure. Absolute pressures are always positive, but gage pressures can be either positive or negative depending on whether the pressure is above atmospheric pressure (a positive value) or below atmospheric pressure (a negative value). A negative gage pressure is also referred to as a suction or vacuum pressure. It is to be noted that pressure differences are independent of the reference, so that no special notation is required in this case.

In addition to the reference used for the pressure measurement, the units used to express the value are obviously of importance. Pressure is a force per unit area. In the SI system the units are N/m^2, this combination is called the pascal and written as Pa ($1\ N/m^2 = 1\ Pa$). Pressure can also be expressed as the height of a column of liquid, the units will refer to the height of the column (mm, m, etc.), and in addition, the liquid in the column must be specified (H_2O, Hg, etc.). For example, standard atmospheric pressure can be expressed as 760mmHg (abs).

1.2.2 流体静力学平衡 (Hydrostatic equilibrium)

流体的静止状态是流体运动的一种特殊形式，它之所以能在设备内维持相对静止状态，是它在重力与压力作用下达到平衡的结果。所以，静止流体的规律就是流体在重力场的作用下流体内部压力变化的规律。该变化规律的数学描述，称为流体静力学平衡方程，简称静力学方程（fluid statics equation）。

静力学方程导出的思路是在静止的流体中取微元体作受力分析，建立微分方程，然后在一定的边界条件下积分。

如图 1-2 所示，在面积为 A 的液柱（方形、矩形、圆形均可）上取微元高度 dz，对微元体 $A dz$ 作受力分析：

下底面总压力　　pA

上底面总压力　　$-(p+dp)A$

本身重力　　$\rho g A dz$

流体静止时，上述三力之和等于零，即：

$$dp+\rho g z=0 \tag{1-9}$$

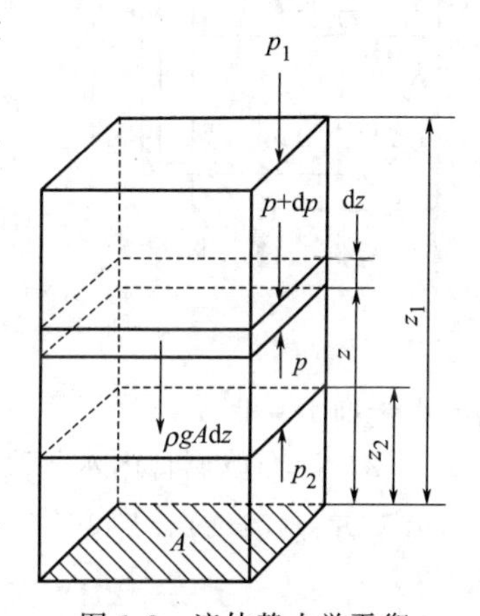

图 1-2　流体静力学平衡

对于不可压缩流体［即 ρ＝常数（constant）］，则上式不定积分求得：

$$\frac{p}{\rho}+gz=常数 \tag{1-9a}$$

若取边界条件为：$z=z_1$，$p=p_1$；$z=z_2$，$p=p_2$，则式(1-9) 定积分得：

$$\frac{p_1}{\rho}+gz_1=\frac{p_2}{\rho}+gz_2 \tag{1-10}$$

对式(1-10) 整理变形还可以得到如下几式：

$$\frac{p_2-p_1}{\rho}=g(z_1-z_2)=gh \tag{1-10a}$$

$$p_2=p_1+\rho g h \tag{1-10b}$$

$$\frac{p_2-p_1}{\rho g}=z_1-z_2=h(\text{m 流体柱高}) \tag{1-10c}$$

式(1-10) 和式(1-10a)、式(1-10b)、式(1-10c) 均称为流体静力学基本方程式，说明了在重力场中，静止流体内部压强的变化规律。

［讨论］：

① 静力学基本方程成立的前提条件：在重力场中，流体是静止的、连续的同一种流体，且流体的密度为常数。若流体的密度不能作为常数处理时，则式(1-10) 和式(1-10a)、

式(1-10b)、式(1-10c) 表示的静力学基本方程式不成立。

② 由式(1-10b) 可知，液体中任一点的压强大小与液面上方压强 p_1 及液体密度 ρ 和该点所处深度 h 有关，所在位置愈低、密度愈大，则其压力愈大。而且，液面上方压力有任何数量的改变，液体内部任一点的压力也将有同样大小和方向的改变，即压力可以同样大小传至液体内各点处。

③ 液体中任意水平面上各点的压强相同，称为等压面（equipressure surface）。等压面可用静止（static）、连续（continuous）、均一（uniform）、水平（horizontal）八个字来体现，等压面的正确选取是流体静力学基本方程应用的关键所在。

④ 因各类常见工业容器中气体密度变化不大，所以上述静力学基本方程式也适用于气体。

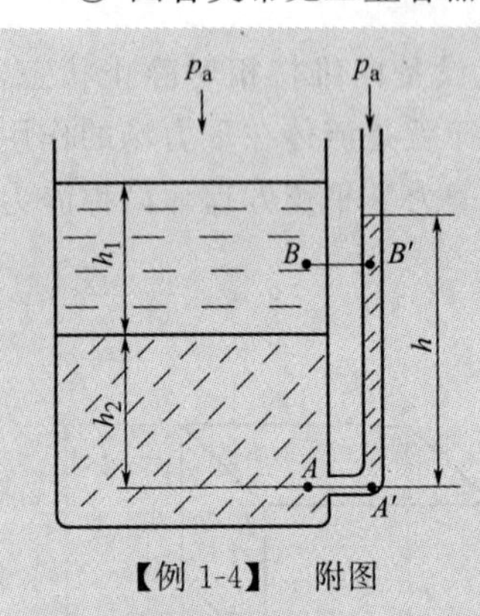

【例 1-4】 附图

【例 1-4】 本题附图所示的开口容器内盛有油和水。油层高度 $h_1=0.8\text{m}$，密度 $\rho_1=800\text{kg/m}^3$，水层高度 $h_2=0.6\text{m}$、密度 $\rho_2=1000\text{kg/m}^3$。(1) 判断下列关系是否成立：$p_A=p'_A$，$p_B=p'_B$。(2) 计算水在玻璃管内的高度 h。

解 本题是静力学基本方程的应用，解题要点是找恰当的等压面。

(1) 判断题给两关系式是否成立 $p_A=p'_A$的关系成立，因 A 及 A'两点在静止的连通着的同一种流体内，并在同一水平面上，所以截面 $A—A'$称为等压面。

$p_B=p'_B$的关系不成立。因 B 及 B'两点虽在静止流体的同一水平面上，但不是连通着的同一种流体，即截面 $B—B'$不是等压面。

(2) 计算玻璃管内水的高度 h 由上面讨论知，$p_A=p'_A$，而 p_A 与 p'_A都可以用流体静力学方程式计算，即：

$$p_A=p_a+\rho_1 gh_1+\rho_2 gh_2 \text{ 及 } p'_A=p_a+\rho_2 gh$$

于是：$h=h_2+\dfrac{\rho_1}{\rho_2}h_1=0.6+\dfrac{800}{1000}\times 0.8=1.24(\text{m})$

1.2.3 流体静力学方程应用（Application of fluid statics equation）

静力学方程的应用十分广泛，如流体在设备或管道内压力变化的测量，液体在储罐内液位的测量，设备的液封高度的确定等，均以静力学方程为依据，以下举例说明。

(1) 压差计的应用（application of manometer）

图 1-3 为 U 形管压差计（U tube manometer）。U 形管内装有指示液 A，U 形管两端连接被测的流体 B，且指示液密度 ρ_A 要大于被测流体的密度 ρ_B。

U 形管两端的流体压力是不相等的（$p_1>p_2$），两端的压差值（$\Delta p=p_1-p_2$）可通过静力学方程的应用来得到。如图 1-3 所示，a，a'二点的静压力是相等。而且，由此向下，在 U 形管内的任意水平线都为等压面。按静力学方程可得到：

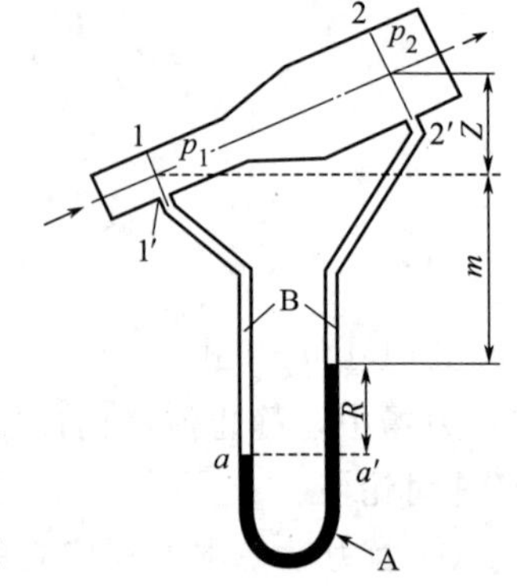

图 1-3 U 形管压差计示意图

$$p_a = p_1 + (m+R)g\rho_B$$

$$p_{a'} = p_2 + \rho_B g(Z+m) + \rho_A gR$$

因为 $p_a = p_{a'}$，故：

$$\Delta p = p_1 - p_2 = Rg(\rho_A - \rho_B) + \rho_B gZ \tag{1-11}$$

当流体输送管段水平放置时，$Z=0$，则上式可简化成：

$$\Delta p = p_1 - p_2 = Rg(\rho_A - \rho_B) \tag{1-11a}$$

常用的指示液有汞、乙醇水溶液、四氯化碳及矿物油等。

当被测流体是气体时，由于气体密度远小于指示液密度，即 $\rho_A \gg \rho_B$，此时上式可简化写成：

$$\Delta p \approx Rg\rho_A \tag{1-11b}$$

【例 1-5】 如本题附图（a）所示，在某输送管路上装一复式 U 形管压差计以测量 A、B 两点间的压差，指示剂为水银，两指示剂之间的流体与管内流体相同。已知管内流体密度 $\rho=900\text{kg/m}^3$，压差计读数 $R_1=0.35\text{m}$，$R_2=0.45\text{m}$，试求：A、B 两点间的压差？

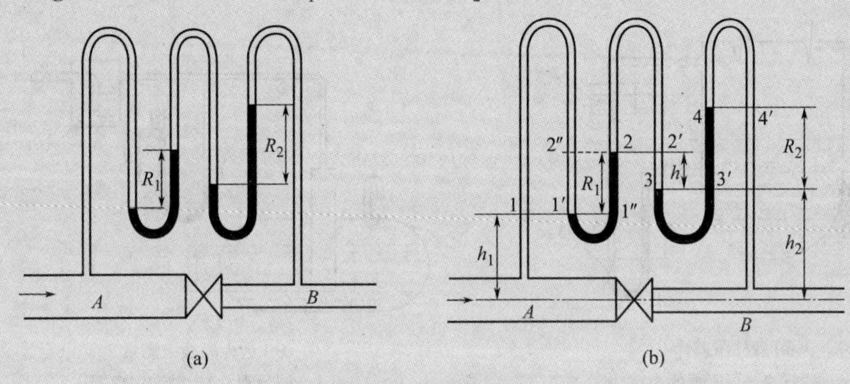

【例 1-5】 附图

解 本题目的是应用静力学方程，而其关键是正确选取等压面。

如附图（b）所示，取水平面 1-1′-1″、2-2′、3-3′和 4-4′，则以上四个水平面均为等压面，于是：

$$p_1 = p_1' = p_1'', p_2 = p_2', p_3 = p_3', p_4 = p_4'$$

又由静力学方程可知：

$$p_A = p_1 + h_1\rho g \tag{1}$$

$$p_1 = p_1'' = p_2 + R_1\rho_0 g \tag{2}$$

$$p_2 = p_2' = p_3 - h\rho g \tag{3}$$

$$p_3 = p_3' = p_4 + R_2\rho_0 g \tag{4}$$

$$p_4 = p_4' = p_B - (h_2 + R_2)\rho g \tag{5}$$

将式(1)～式(5) 相加，得：

$$p_A = p_B + (R_1 + R_2)\rho_0 g - (h_2 + R_2 + h - h_1)\rho g \tag{6}$$

又由几何关系可知：

$$h_2 + R_2 - (R_1 + h_1) = R_2 - h$$

即：

$$h_2 + h - h_1 = R_1 \tag{7}$$

将式(7) 代入式(6)，整理得：

$$p_A-p_B=(R_1+R_2)(\rho_0-\rho)g$$
$$=(0.35+0.45)\times(13600-900)\times9.81$$
$$=99670(\text{Pa})$$

（2）液位测量（level gauging）

过程工业中经常需要了解各类容器的储存量，或要控制设备里的液面，这就要对液面进行测定，液面测定是依据同一流体在同一水平面上的压力相等的原则来设计的。图 1-4（a）为液柱压差计测定液面的示意图，将 U 形管压差计的两端分别接在储槽的顶端和底端，利用 U 形管压差计上 R 的数值，即可得出容器内液面的高度，所测液面高度与液面计玻璃管粗细无关；当容器或设备的位置离操作室较远时，可采用远距离液位测量装置，如图 1-4（b）所示，压缩氮气经调节阀 1 调节后进入鼓泡观察器 2。管路中氮气的流速控制得很小，只要在鼓泡观察器 2 内看到有气泡缓慢逸出即可。因此气体通过吹气管 4 的流动阻力可以忽略不计。吹气管某截面处的压力用 U 形管压差计 3 来计量。压差计读数 R 的大小，即反映储罐 5 内液面的高度。

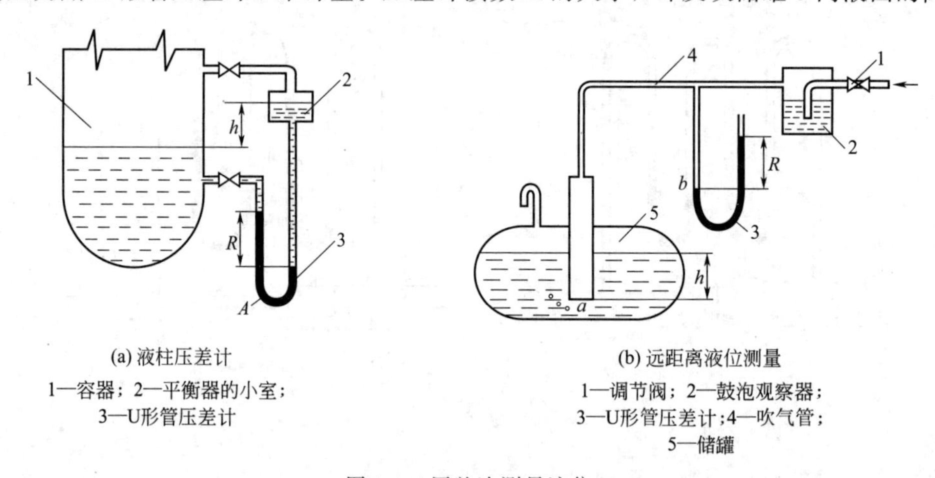

(a) 液柱压差计
1—容器；2—平衡器的小室；
3—U形管压差计

(b) 远距离液位测量
1—调节阀；2—鼓泡观察器；
3—U形管压差计；4—吹气管；
5—储罐

图 1-4　压差法测量液位

【例 1-6】 如图 1-4（b）所示装置，现已知 U 形管压差计的指示液为水银，其读数 $R=130\text{mm}$，罐内有机液体的密度 $\rho=1250\text{kg/m}^3$，储罐上方与大气相通。试求储罐中液面离吹气管出口的距离 h 为多少？

解 由于吹气管内氮气的流速很低，且管内不能存有液体，故可认为管出口 a 处与 U 形管压差计 b 处的压力近似相等，即 $p_a\approx p_b$。若 p_a 与 p_b 均用表压强表示，根据流体静力学平衡方程，得：

$$p_a=\rho gh, p_b=\rho_{Hg}gR$$

故
$$h=R\frac{\rho_{Hg}}{\rho}=0.13\times\frac{13600}{1250}=1.41(\text{m})$$

（3）液封（liquid seal）

液封在化工生产中应用非常广泛。为了防止设备中气体的泄漏，往往将带有压力的气体管路插入液体中，让足够的液层高度阻止气体外泄。这个液层高度可用流体静力学基本方程加以计算而确定。例如真空蒸发操作中产生的水蒸气，往往送入图 1-5（a）所示的混合冷凝器中与冷水直接接触而冷凝。为了维持操作的真空度，冷凝器上方与真空泵相通，随时将冷凝器内的不凝气体（空气）抽走。同时为了防止外界空气由气压管 4 漏入，致使设备内真空度降低，因此，气压管必须插入液封槽 5 中，水即在管内上升一定的高度 h，这种措施即为液封。

图 1-5（b）为乙炔发生炉的液封示意图。若炉内的表压为 p，则水封高度 Z 可由下式确定：

$$Z \geqslant \frac{p}{\rho_{水} g} \tag{1-12}$$

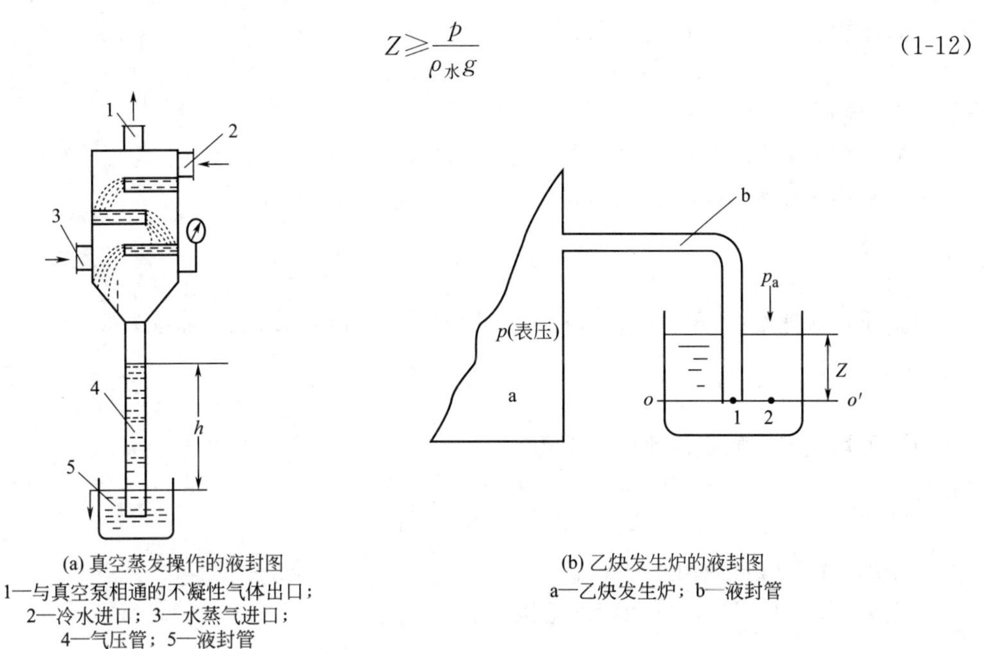

(a) 真空蒸发操作的液封图
1—与真空泵相通的不凝性气体出口；
2—冷水进口；3—水蒸气进口；
4—气压管；5—液封管

(b) 乙炔发生炉的液封图
a—乙炔发生炉；b—液封管

图 1-5　液封

【例 1-7】 对于图 1-5（a）所示装置，若真空表的读数为 86×10^3 Pa，试求气压管中水上升的高度 h。

解　设气压管内水面上方的绝对压强为 p，作用于液封槽内水面的压强为大气压强 p_a，根据流体静力学基本方程式知：

$$p_a = p + \rho g h$$

于是：

$$h = \frac{p_a - p}{\rho g} = \frac{86 \times 10^3}{1000 \times 9.81} = 8.77(\text{m})$$

1.3　流体流动的基本概念 (Basic Concepts of Fluid Flow)

过程工业中，流体往往在密闭的管路中流动。流体在管内流动的规律，可用其流动的基本方程来描述，基本方程包括连续性方程和柏努利方程。流体在管内宏观上的流动是轴向流动，所以，本节所研究的对象为一维流动（one dimensional flow）。

1.3.1　流量（Flow rate）与流速（Velocity）

（1）流量

流体在单位时间内流经管道任一截面的量称为流量，流量常分为体积流量和质量流量两种。

① 体积流量 V_s（volume flow rate）单位为 m^3/s。

$$V_s=V/\theta \tag{1-13}$$

② 质量流量 m_s（mass flow rate）单位为 kg/s。

$$m_s=m/\theta \tag{1-14}$$

③ 体积流量与质量的关系为：

$$m_s=V_s\rho \tag{1-15}$$

（2）流速

流体在单位时间内流经管道单位截面的量称为流速。流速分为平均流速和质量流速。

① 平均流速 u（average velocity）简称为流速，单位为 m/s。

$$u=\frac{V_s}{A} \tag{1-16}$$

② 质量流速 G（mass velocity）又称质量通量（mass flux），单位为 $kg/(m^2 \cdot s)$

$$G=\frac{m_s}{A} \tag{1-17}$$

③ 平均流速与质量流速的关系为：

$$G=\frac{m_s}{A}=\frac{V_s\rho}{A}=u\rho \tag{1-18}$$

式中，A 为管道截面积，m^2。

某些流体在管道中的常用流速范围见表 1-1。

表 1-1　某些流体在管道中的常用流速范围

流体及其流动类别	流速范围/(m/s)	流体及其流动类别	流速范围/(m/s)
自来水(3×10^5 Pa 左右)	1～1.5	一般气体(常压)	10～20
水及低黏度液体(1×10^5～1×10^6 Pa)	1.5～3.0	鼓风机吸入管	10～20
高黏度液体	0.5～1.0	鼓风机排出管	15～20
工业供水(8×10^5 Pa 以下)	1.5～3.0	离心泵吸入管(水类液体)	1.5～2.0
锅炉供水(8×10^5 Pa 以下)	>3.0	离心泵排出管(水类液体)	2.5～3.0
饱和蒸气	20～40	往复泵吸入管(水类液体)	0.75～1.0
过热蒸气	30～50	往复泵排出管(水类液体)	1.0～2.0
蛇管、螺旋管内的冷却水	<1.0	液体自流速度(冷凝水等)	0.5
低压空气	12～15	真空操作下气体流速	<10
高压空气	15～25		

1.3.2 稳态流动（Steady-state flow）及非稳态流动（Unsteady-state flow）

流体在流动过程中，任一截面处的流速、流量和压力等有关物理参数都不随时间变化，只随空间位置而变化，这种流动称为稳态流动（steady-state flow）。

若各流动参数不仅随空间位置变化，还随时间变化，则称为非稳态流动（unsteady-state flow）。

图 1-6（a）为稳态流动，（b）为非稳态流动。在图 1-6（a）中的水槽有进水管补充水，又有溢流装置，使水槽液面维持恒定，则排水管任一截面处的流速、压力等参数均不随时间变化，只随位置而变，即属于稳态流动；而图 1-6（b）图中，由于水槽中的液位不断下降，使得各流动参数都随时间和空间而不断变化，所以属于非稳态流动。

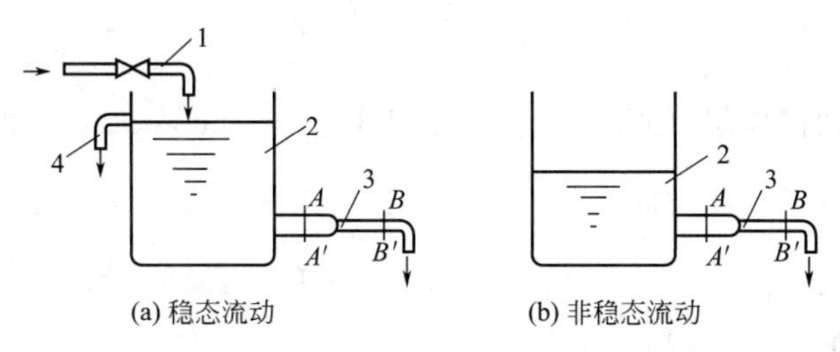

图 1-6　稳态流动和非稳态流动

1—进水管；2—出水管；3—排水管；4—溢水管

连续化生产的过程工业中，正常情况下多数为稳态流动，但开车、停车阶段及间歇操作属于非稳态流动，本章主要介绍稳态流动。

1.3.3　牛顿黏性定律（Newton's law of viscosity）

（1）流体流动中的内摩擦

两个固体之间作相对运动时，必须施加一定的外力以克服接触表面的摩擦力（外摩擦）。与固体间的相对运动类似，假定流动流体可分成许多流体层，由于流体流经固体壁面时存在附着力，所以壁面上黏附一层静止的流体层，其速度 $u_W=0$，该流体层对相邻的流体层有一个向后的曳力，类似地，逐个流体层相互作用，且曳力的作用也逐渐减弱，其结果是与流体流动方向垂直的同一截面上出现了点速度分布。

上述发生在流体层之间的作用力，称为剪切力（黏性力），因为在流体内部产生，故叫作内摩擦力。

（2）黏性（viscosity）

流体流动时，不同速度的流体层之间产生内摩擦，内摩擦力可看作是一层流体抵抗另一层流体引起形变的力。运动一旦停止，这种抵抗力随即消失。通常，这种表明流体受剪切力作用时，本身抵抗形变的物理特性叫黏性。实际流体都有黏性，但各种流体的黏性差别很大，如空气、水等流体的黏性较小，而蜂蜜、油类等流体的黏性较大。应该注意，黏性是流体在运动中表现出来的一种物理属性。

（3）牛顿黏性定律的概念

假设相距很近的两平行大平板间充满黏稠液体，如图 1-7 所示。若下面平板保持不动，上板施加一平行于平板的外力，使其以速度 u 沿 x 方向运动，此时，两板间的液体就分成许许多多的流体层而运动，附在上平板的流体层随上板以速度 u 运动，以下各层流体流速逐渐降低，附在下平板表面上的流体层速度为零。

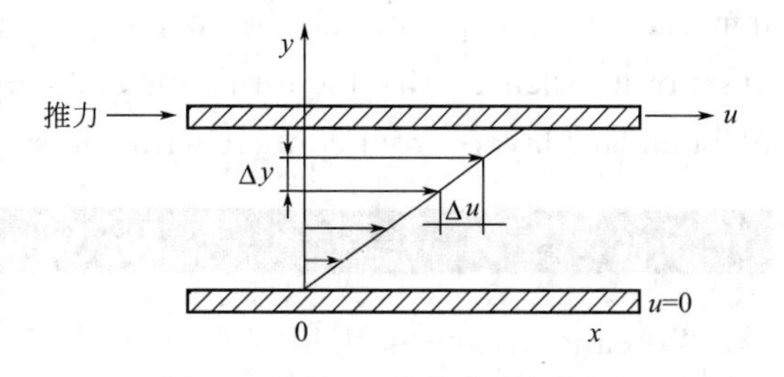

图 1-7　平板间液体速度变化图

在流体层之间存在速度分布（velocity distribution）（速度差异），相邻流体层以大小相等方向相反的剪切力 F（shear force）相互作用。实验证明，对全部气体与大部分液体而言，剪应力（shear stress）τ 服从牛顿黏性定律。

$$\tau=\frac{F}{A}=\mu\frac{du}{dy} \tag{1-19}$$

式中 τ——剪应力，N/m^2；

A——相邻的两流体层（adjacent layers slide past one another）的作用面积，m^2；

$\frac{du}{dy}$——流体速度沿法线方向上的变化率，称为速度梯度（velocity gradient），1/s；

μ——比例系数（proportionality constant），称为黏性系数或动力黏度，简称黏度，$N\cdot s/m^2$（$Pa\cdot s$）。

Fluid Flow Phenomena

The behavior of a flowing fluid depends strongly on whether the fluid is under the influence of solid boundaries. In the region where the influence of the wall is small, the shear stress may be negligible and the fluid behavior may approach that of an ideal fluid, one that is incompressible and has zero viscosity. The flow of such an ideal fluid is called *potential flow* and is completely described by the principles of Newton mechanics and conservation of mass. Potential flow has two important characteristics: (1) neither circulations nor eddies can form within the stream, so that potential flow is also called irrotational flow, and (2) friction cannot develop, so that there is no dissipation of mechanical energy into heat.

Potential flow can exist at distances not far from a solid boundary. A fundamental principle of fluid mechanics, originally stated by Prandtl in 1904, is that, except for fluids moving at low velocities or possessing high viscosities, the effect of the solid boundary on the flow is confined to a layer of the fluid immediately adjacent to the solid wall. This layer is called the boundary layer, and shear and shear forces are confined to this part of the fluid. Outside the boundary layer, potential flow survives. Most technical flow processes are best studied by considering the fluid stream as two parts, the boundary layer and the remaining fluid. In some situations such as flow in a converging nozzle, the boundary layer may be neglected, and in others, such as flow through pipes, the boundary layer fills the entire channel, and there is no potential flow.

Within the current of an incompressible fluid under the influence of solid boundaries, four important effects appear. (1) the coupling of velocity-gradient and shear-stress fields, (2) the onset of turbulence, (3) the formation and growth of boundary layers, and (4) the separation of boundary layers from contact with the solid boundary.

The Velocity Field

When a stream of fluid is flowing in bulk past a solid wall, the fluid adheres to the solid at the actual interface between solid and fluid. The adhesion is a result of the force fields at the boundary, which are also responsible for the interfacial tension between solid and fluid. If, therefore, the wall is at rest in the reference frame chosen for the solid-fluid system, the velocity of the fluid at the interface is zero. Since at distances away from the solid the velocity is not zero, there must be variations in velocity from point to point in the flowing stream. Therefore, the velocity at any point is a function of the space coordinates of that point, and a velocity field exists in the space occupied by the fluid. The velocity at a given location may also vary with time. When the velocity at each location is constant, the field is invariant with time and the flow is said to be steady.

One-dimensional Flow

Velocity is a vector, and in general, the velocity at a point has three components, one for each space coordinate. In many simple situations all velocity vectors in the field are parallel or practically so, and only one velocity component, which may be taken as a scalar, is required. This situation, which obviously is much simpler than the general vector field, is called one-dimensional flow. An example is steady flow through straight pipe. The following discussion is based on the assumptions of steady one-dimensional flow.

(4) 黏度（viscosity）

黏度是用来度量流体黏性大小的物理量。由式(1-19) 可看出，当流体的速度梯度$\frac{du}{dy}$为1时，流体的黏度μ在数值上即等于单位面积上的黏性力（内摩擦力）（skin friction）。因此，在相同的流速下，黏度愈大的流体，所产生的黏性力也愈大，即流体因克服阻力而损耗的能量愈大，所以，对于黏度较大的流体，应选用较小的流速。流体的黏度愈大，表示其流动性愈差，如油的黏度比水为大，则油比水的流动性差，在相同流速下，输送油所消耗的能量要比输送水消耗的能量大得多。

黏度的单位可由式(1-19) 导出：

$$[\mu]=\frac{F}{A\cdot\frac{du}{dy}}=\frac{\mathrm{N}}{\mathrm{m}^2\ \frac{\mathrm{m/s}}{\mathrm{m}}}=\frac{\mathrm{N\cdot s}}{\mathrm{m}^2}=\mathrm{Pa\cdot s}$$

黏度值一般由实验测定，通过手册可查取流体在某一温度下的黏度，手册查得的黏度单位多为物理制单位 泊（P）或厘泊（cP），与法定单位（Pa·s）之间的换算如下：

$$1\text{厘泊(cP)}=10^{-2}\text{泊(P)}=10^{-3}\mathrm{Pa\cdot s}=1\mathrm{mPa\cdot s}$$

液体黏度随温度升高而减小，气体黏度则随温度升高而增大；液体黏度随压力变化基本不变，气体黏度只有当压力较高时（如4×10^6Pa以上）才随压力增大略有增大。

混合物的黏度一般用实验测定，当缺乏实验数据时，可由如下经验式求算。

① 低压混合气体（low pressure mixed gas）：

$$\mu_m=\frac{\sum y_i\mu_iM_i^{1/2}}{\sum y_iM_i^{1/2}} \tag{1-20}$$

式中 μ_m——混合气体黏度，Pa·s；

y_i——混合气体中i组分的摩尔分数；

μ_i——混合气体中i组分的黏度，Pa·s；

M_i——混合气体中 i 组分的分子量。

② 不缔合混合液体（no association mixed liquid）

$$\lg\mu_m=\sum_{i=1}^{n}x_i\lg\mu_i \tag{1-21}$$

式中 μ_m——混合液体黏度，Pa·s；

x_i——混合液体中 i 组分的摩尔分数；

μ_i——混合液体中 i 组分的黏度，Pa·s。

【例 1-8】 甲烷与丙烷组成混合气体，其摩尔分数分别为 0.4 和 0.6，求在常压下及 293K 时混合气的黏度。

解 由附录查得常压下 293K 时纯组分的黏度：

$$\mu_{甲烷}=0.0107\text{mPa·s};\mu_{丙烷}=0.0077\text{mPa·s}$$

各组分的分子量：$M_{甲烷}=16$，$M_{丙烷}=44$

将各数值代入式(1-20)，可计算出混合气黏度：

$$\mu_m=\frac{0.4\times0.0107\times16^{1/2}+0.6\times0.0077\times44^{1/2}}{0.4\times16^{1/2}+0.6\times44^{1/2}}$$

$$=\frac{0.0171+0.0306}{1.6+3.97}$$

$$=0.00857(\text{mPa·s})$$

1.3.4 流动型态（Types of fluid flow）

（1）雷诺实验及流动型态（Reynolds experiment and types of fluid flow）

1883 年英国科学家雷诺（Reynolds）按图 1-8 所示装置进行实验。

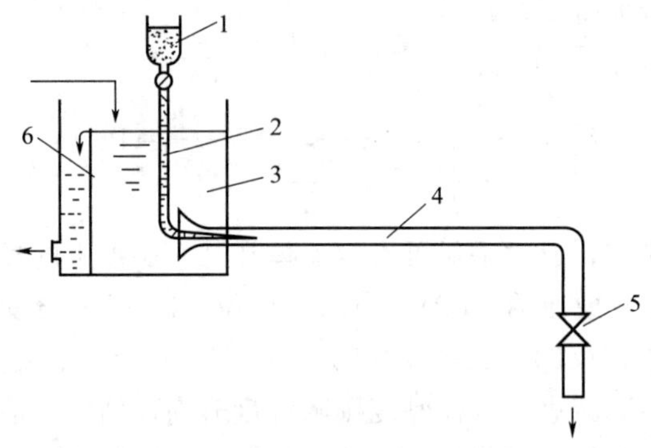

图 1-8 雷诺实验演示图

1—小瓶；2—细管；3—水箱；4—水平玻璃管；5—阀门；6—溢流装置

透明储水槽中的液位由溢流装置维持恒定，水槽的下部插入一带有喇叭口的水平玻璃管，管内水的流速由出口阀门调节。水槽上方设置一个盛有色液体的吊瓶，有色液体通过导管及针形细嘴由玻璃管的轴线引入。

从实验观察到，当水的流速很小时，有色液体沿管轴线作直线运动，与相邻的流体质点无宏观上的混合，如图 1-9（a）所示，这种流动型态称为层流（laminar flow）或滞流（viscous flow）；随着水的流速增大至某个临界值（critical velocity）后，有色液体流动的细线开始抖动、弯曲，呈现波浪形，如图 1-9（b）所示；当流速再增大时，波形起伏加剧，出现强烈的骚扰滑动，全管内水的颜色均匀一致，如图 1-9（c）所示，这种流动型态称为湍流（turbulent flow）或紊流。

流体的流动型态只有层流和湍流两种。

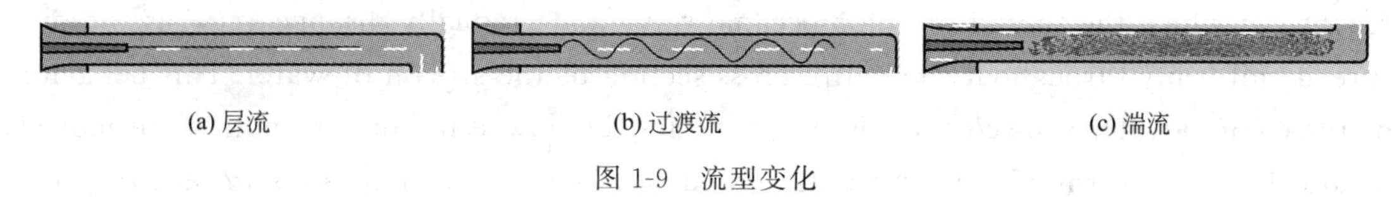

(a) 层流　　(b) 过渡流　　(c) 湍流

图 1-9　流型变化

（2）雷诺数（Reynolds number）

雷诺采用不同的流体和不同的管径多次进行了上述实验，所得结果表明：流体的流动型态除了与流速 u 有关外，还与管径 d、密度 ρ、黏度 μ 这三个因素有关。雷诺将这四个因素组成一个复合数群，以符号 Re 表示，即：

$$Re=\frac{du\rho}{\mu} \tag{1-22}$$

该数群称为雷诺数，是一个无量纲的数群（dimensionless groups）。例如用 SI 制表示：

$$[Re]=\left[\frac{du\rho}{\mu}\right]=\frac{(\mathrm{m})(\mathrm{m/s})(\mathrm{kg/m^3})}{\mathrm{N\cdot s/m^2}}=\mathrm{m^0\cdot kg^0\cdot s^0}$$

不论采用哪种单位制，雷诺数的数值都是一样的。实验结果表明，对于流体在圆管内流动，当 $Re<2000$ 时，流动型态为层流；当 $Re>4000$ 时，流动型态为湍流；当$Re=2000\sim4000$，称为过渡流，但它不是一种流型，实际上是流动的过渡状态，即流动可能是层流，也可能是湍流，受外界条件的干扰而变化（例如在管道入口处，流道弯曲或直径改变，管壁粗糙或有外来震动都易形成湍流）。所以，可用雷诺数的数值大小来判断流体的流动型态，雷诺数愈大说明流体的湍动程度愈剧烈，产生的流体流动阻力愈大。

【例 1-9】 某油品在管内径为 100mm 的管中流动，油的密度为 $900\mathrm{kg/m^3}$，黏度为 $0.072\mathrm{Pa\cdot s}$，流速为 $1.57\mathrm{m/s}$，求 Re 并判断流动型态。

解　已知 $d=0.1\mathrm{m}$，$u=1.57\ \mathrm{m/s}$，$\rho=900\mathrm{kg/m^3}$ 和 $\mu=0.072\ \mathrm{Pa\cdot s}$，可得：

$$Re=\frac{du\rho}{\mu}=\frac{0.1\times1.57\times900}{0.072}=1963<2000$$

该流动型态为层流。

Reynolds Experiment and Reynolds Number

It has long been known that a fluid can flow through a pipe or conduit in two different ways. At low flow rates the pressure drop in the fluid increases directly with the fluid velocity; at high rates it increases much more rapidly, roughly as the square of the velocity. The distinction between the two types of flow was first demonstrated in a classic experiment by Osborne Reynolds, reported in 1883. A horizontal glass tube was immersed in a glass-walled tank filled with water. A controlled flow of water could be drawn through the tube by opening a valve. The entrance to the tube was flared, and provision was made to introduce a fine filament of colored water from the overhead flask into the stream at the tube entrance. Reynolds found that, at low flow rates, the jet of colored water flowed intact along with the mainstream and no cross-mixing occurred. The behavior of the color band showed clearly that the water was flowing in parallel straight lines and that the flow was *laminar*. When the flow rate was increased, a velocity, called the *critical velocity*, was reached at which the thread of color became wavy and gradually disappeared, as the dye spread uniformly throughout the entire cross section of the stream of water. This behavior of the colored water showed that the water no longer flowed in laminar motion but moved erratically in the form of cross-currents and eddies. This type of motion is *turbulent flow*.

Reynolds studied the conditions under which one type of flow changes to the other and found that the critical velocity, at which laminar flow changes to turbulent flow, depends on four quantities: the diameter of the tube and the viscosity, density, and average linear velocity of the liquid. Furthermore, he found that these four factors can be combined into one group and that the change in the kind of flow occurs at a definite value of the group. The grouping of variables so found was:

$$Re = \frac{du\rho}{\mu}$$

Where, d is diameter of tube, u is average velocity of liquid, μ is viscosity of liquid, and ρ is density of liquid.

The dimensionless group of variables denned by (Eq. 1-22) is called the Reynolds number Re. It is one of the named dimensionless groups. Its magnitude is independent of the units used, provided the units are consistent.

Additional observations have shown that the transition from laminar to turbulent flow actually may occur over a wide range of Reynolds numbers. In a pipe, flow is always laminar at Reynolds numbers below 2000, but laminar flow can persist up to Reynolds numbers well above 24000 by eliminating all disturbances at the inlet. If the laminar flow at such high Reynolds numbers is disturbed, however, say by a fluctuation in velocity, the flow quickly becomes turbulent. Disturbances under these conditions are amplified, whereas at Reynolds numbers below 2000 all disturbances are damped and the flow remains laminar. At some flow rates a disturbance may be neither damped nor amplified; the flow is then said to be neutrally stable. Under ordinary conditions, the flow in a pipe or tube is turbulent at Reynolds numbers above about 4000. Between 2000 and 4000 a *transition region* is found where the flow may be either laminar or turbulent, depending upon conditions at the entrance of the tube and on the distance from the entrance.

(3) 管内层流与湍流的比较（comparison of laminar and turbulent flow）

若将管内层流与湍流两种不同的流动型态进行比较，简单地说，两者的本质区别在于流体内部质点的运动方式不同。前者是流体质点沿着与管轴平行方向作有规则的直线运动，是一维流动（one-dimensional flow），流体质点互不干扰，互不碰撞，没有位置交换，是很有规律的分层运动；后者是流体质点除沿轴线方向作主体流动外，还在径向方向上作随机的脉动，湍流质点间发生位置交换，相互剧烈碰撞与混合，使流体内部任一位置上流体质点的速度大小及方向都会随机改变，因而湍流是一个杂乱无章、无规则的运动。

若从速度分布来看，层流与湍流的区别在于与流体流动方向垂直的同一截面上各点速度的变化规律不同。

① 层流速度分布（velocity distribution for laminar flow）　层流时的点速度沿管径按抛物线的规律分布，如图 1-10 所示，其速度分布式为：

$$u_r = u_{max}\left[1-\left(\frac{r}{R}\right)^2\right] \tag{1-23}$$

式中　u_r——与管中轴线垂直距离为 r 处的点速度，m/s；

u_{max}——管中轴线上的最大速度，m/s；

r——与管中轴线的垂直距离，m；

R——管半径，m。

a. 当 $r=0$ 时，即管中轴线处流速最大，$u_{max}=\dfrac{\Delta p_f}{4\mu l}R^2$；

b. 当 $r=R$ 时，即管壁处流体流速为零。

c. 理论和实验结果都表明，层流时各点速度的平均值 u 等于管中心处最大速度 u_{max} 的 0.5 倍。

② 湍流速度分布（velocity distribution for turbulent flow）　湍流时，流体质点运动比较复杂，其速度分布曲线一般由实验测定，如图 1-11 所示。由于流体质点强烈碰撞混合，使截面上靠管中心部分彼此拉平，速度分布比较均匀。管内流体 Re 值愈大，湍动程度愈高，曲线顶端愈平坦，其速度分布式为：

$$u_r = u_{max}\left(\frac{R-r}{R}\right)^{1/7} \tag{1-24}$$

a. 当 $r=0$ 时，管中轴线处流速最大；

b. 当 $r=R$ 时，管壁处流体流速为零。

通常，湍流时的平均速度 u 与管内最大速度 u_{max} 的比值随雷诺数变化，如图 1-12 所示，图中 Re 和 Re_{max} 是分别以平均速度 u 与管内最大速度 u_{max} 计算的雷诺数。

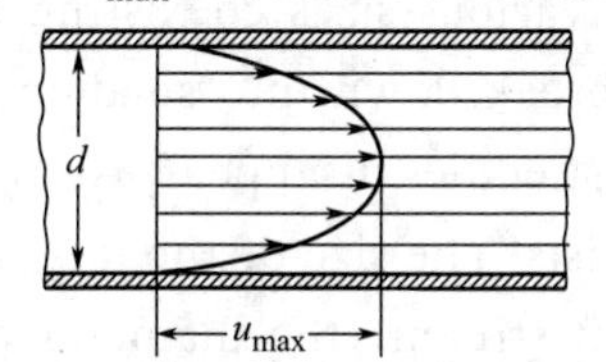

图 1-10　层流时圆管内的速度分布

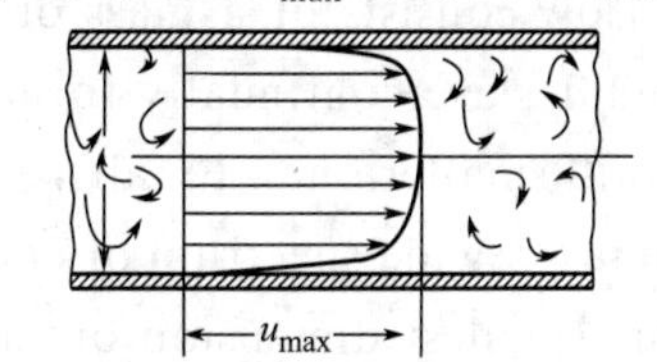

图 1-11　湍流时圆管内的速度分布

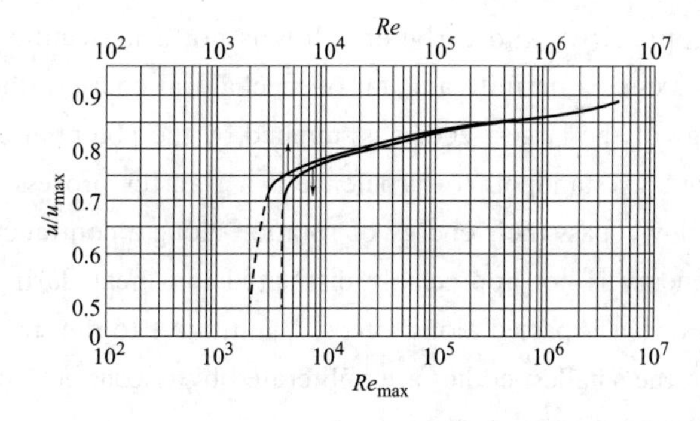

图 1-12　u/u_{max} 与 Re、Re_{max} 的关系

层流与湍流的区别也可以从动量传递的角度更深入地理解。根据牛顿第二定律：

$$F=m\frac{du}{d\theta}=\frac{d(mu)}{d\theta} \tag{1-25}$$

则：

$$\tau=\frac{F}{A}=\frac{1}{A}\frac{d(mu)}{d\theta} \tag{1-25a}$$

式中 θ——时间，s；

mu——动量，kg·m/s。

因此，剪应力意味着相邻的两流体层之间，单位时间、单位面积所传递的动量，即动量通量（momentum flux）。对于层流流动，两流体层间的动量传递是分子交换；对于湍流流动，两流体层间的动量传递是分子交换（molecular exchange）加质点交换，而且是以质点交换为主。

Nature of Turbulence

Because of its importance in many branches of engineering, turbulent flow has been extensively investigated in recent years, and a large literature has accumulated on this subject. Refined methods of measurement have been used to follow in detail the actual velocity fluctuations of the eddies during turbulent flow, and the results of such measurements have shed much qualitative and quantitative light on the nature of turbulence.

Turbulence may be generated in other ways than by flow through a pipe. In general, it can result either from contact of the flowing stream with solid boundaries or from contact between two layers of fluid moving at different velocities. The first kind of turbulence is called wall turbulence and the second kind free turbulence. Wall turbulence appears when the fluid flows through closed or open channels or past solid shapes immersed in the stream. Free turbulence appears in the flow of a jet into a mass of stagnant fluid or when a boundary layer separates from a solid wall and flows through the bulk of the fluid. Free turbulence is especially important in mixing.

Turbulent flow consists of a mass of eddies of various sizes coexisting in the flowing stream. Large eddies are continually formed. They break down into smaller eddies, which in turn evolve still smaller ones. Finally, the smallest eddies disappear. At a given time and in a given volume, a wide spectrum of eddy sizes exists. The size of the largest eddy is comparable with the smallest dimension of the turbulent stream; the diameter of the smallest eddies is 10 to 100μm. Smaller eddies than this are rapidly destroyed by viscous shear. Flow within an eddy is laminar. Since even the smallest eddies contain about 10^{12} molecules, all eddies are of macroscopic size, and turbulent flow is not a molecular phenomenon.

Any given eddy possesses a definite amount of mechanical energy, much like that of a mall spinning top. The energy of the largest eddies is supplied by the potential energy of the bulk flow of the fluid. From an energy standpoint, turbulence is a transfer process in which large eddies, formed from the bulk flow, pass their energy of rotation along a continuous series of smaller eddies. This mechanical energy is not appreciably dissipated into heat during the breakup of large eddies into smaller ones, but is passed along almost quantitative to the smallest eddies. It is finally converted to heat when the smallest eddies are obliterated by viscous action. Energy conversion by viscous action is called *viscous dissipation*.

(4) 边界层（boundary layer）概念

早期的流体力学研究，理论与实验结果差异很大。例如，对于黏度很小的流体，一般的理解是产生的摩擦力也很小，可按理想流体处理，但理论推断结果与实验数据不符。类似的问题一直没能得到圆满的解释。直到20世纪初，普兰特（Prandtl）提出了边界层概念，深刻地揭示了理论与实验结果的差异所在，从此，流体力学得到了迅速的发展。

① 边界层及其形成（boundary layer and its forming）　如图1-13所示，当流体以速度u_0流经固体壁面时，流体与壁面接触，紧贴壁面处的流体速度变为零，受其影响，在垂直于流动方向的截面上，出现了速度分布，且随着离开壁面前缘的距离的增加，流速受影响的区域相应增大。则定义，$u \leqslant 99\% u_0$的区域叫流动边界层（flow boundary layer）。或者说，流动边界层是固体壁面对流体流动的影响所波及的区域。边界层厚度用δ表示，注意$\delta = f(x)$。

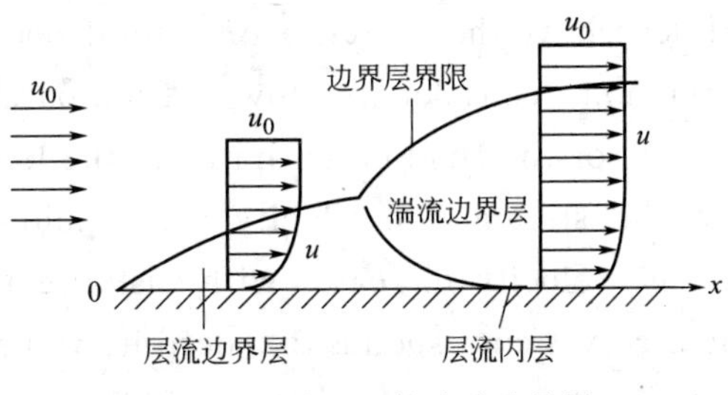

图1-13　壁面上的流动边界层

② 层流边界层与湍流边界层　边界层内也有层流与湍流之分。流体流经固体壁面的前段，若边界层内的流型为层流，叫层流边界层（laminar boundary layer）；当流体离开前沿若干距离后，边界层内的流型转变为湍流，叫湍流边界层（turbulent boundary layer）。

湍流边界层发生处，边界层突然加厚，且其厚度较快地扩展。即使在湍流边界层内，壁面附近仍有一层薄薄的流体层呈层流流动，这个薄层称为层流内层或滞流底层（viscous sublayer）。层流内层到湍流主体间还存在过渡层（buffer layer）。层流内层的厚度随Re值增加而减小，但不论流体湍动得如何剧烈，层流内层的厚度都不会为零。层流内层的厚度对传热和传质过程有很大的影响。

如图1-14所示，对管道而言，仅在流体的进口段，边界层有内外之分，经过某一段距离后，边界层扩展至管中心汇合，边界层厚度即为管道的半径且不再变化，即管壁对流体的影响波及整个管内的流体，把这种流动称为充分发展的流动（fully developed flow）。

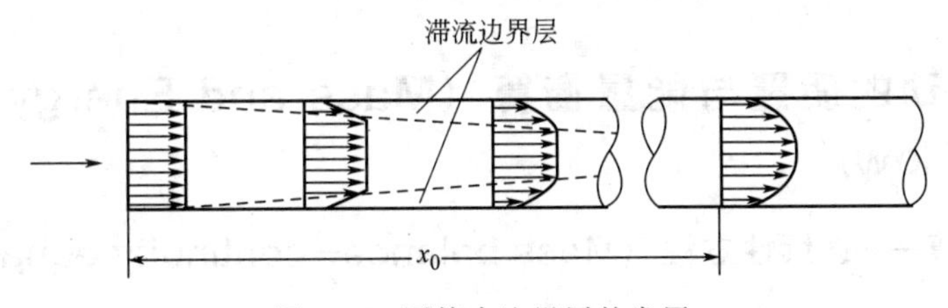

图1-14　圆管中边界层的发展

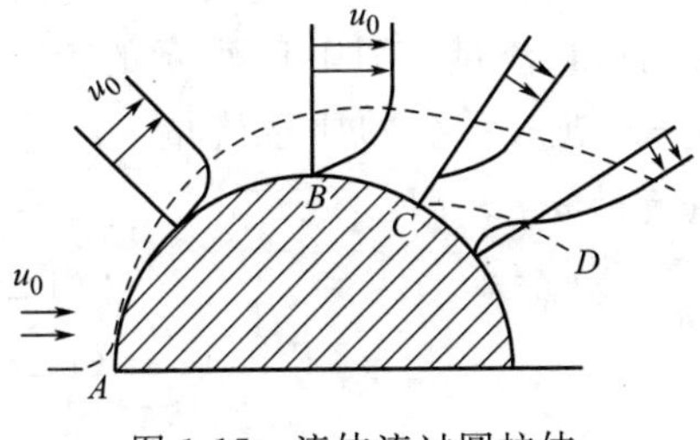

图1-15　流体流过圆柱体表面的边界层分离

③ 边界层分离（boundary layer separation）

边界层的一个重要的特点是，在某些情况下，其内部的流体会发生倒流，引起边界层与固体壁面的分离现象，并同时产生大量的旋涡（eddies，large eddies called vortices），造成流体的机械能损失（mechanical energy loss）（形体阻力），这种现象称为边界层分离（boundary layer separation），如图1-15所示。边界层分离是黏性流体产生能量损失的重要原因之一，这种现象通常在流体绕流物体（流线型物体除外）或流道突然改变（abrupt change in flow channel）时发生。

Boundary Layers

A boundary layer is defined as that part of a moving fluid in which the fluid motion is influenced by the presence of a solid boundary. As a specific example of boundary layer formation, consider the flow of fluid parallel with a thin plate, as shown in Fig. 1-13. The velocity of the fluid upstream from the leading edge of the plate is uniform across the entire fluid stream. The velocity of the fluid at the interface between the solid and fluid is zero. The velocity increases with distance from the plate, as shown in Fig. 1-13. Each of these curves corresponds to a definite value of x, the distance from the leading edge of the plate. The curves change slope rapidly near the plate; they also show that the local velocity approaches asymptotically the velocity of the bulk of the fluid stream.

In Fig. 1-13 the line is so drawn that the velocity changes are confined between this line and the trace of the wall. Because the velocity lines are asymptotic with respect to distance from the plate, it is assumed, in order to locate the dashed line definitely, that the line passes through all points where the velocity is 99 percent of the bulk fluid velocity u_0. The line represents an imaginary surface that separates the fluid stream into two parts: one in which the fluid velocity is constant and the other in which the velocity varies from zero at the wall to a velocity substantially equal to that of the undisturbed fluid. This imaginary surface separates the fluid that is directly affected by the plate from that in which the local velocity is constant and equal to the initial velocity of the approach fluid. The zone, or layer, between the dashed line and the plate constitutes the boundary layer.

The formation and behavior of the boundary layer are important, not only in the flow of fluids but also in the transfer of heat and mass.

1.4 流体流动的质量与能量衡算（Mass and Energy Balance of Fluid Flow）

1.4.1 质量衡算——连续性方程（Mass balance—continuity equation）

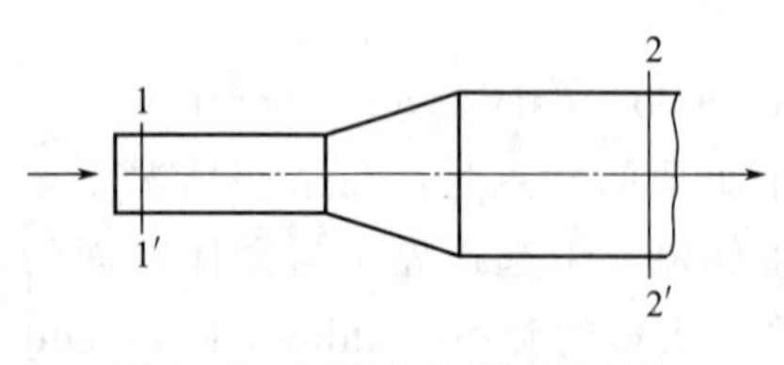

图 1-16 连续性方程式的推导

对于一个稳态流动系统，系统内任意空间位置上均无物料积累，所以物料衡算关系为流入系统的质量流量等于离开系统的质量流量。图 1-16 所示的管路系统，流体从截面 1—1′进入系统的质量流量 m_{s_1} 应等于离开系统的质量流量 m_{s_2}，即：

$$m_{s_1}=m_{s_2} \tag{1-26}$$

或

$$u_1\rho_1A_1=u_2\rho_2A_2 \tag{1-26a}$$

对于不可压缩流体，ρ 可取常数，则：

$$u_1A_1=u_2A_2 \tag{1-26b}$$

即单位时间内通过管路各截面的体积流量也相等，式(1-26)和(1-26a)称为流体在管道内流动时的连续性方程式。在圆形管道中，不可压缩流体稳态流动的连续性方程可以写为：

$$\frac{u_1}{u_2}=\left(\frac{d_2}{d_1}\right)^2 \tag{1-26c}$$

式(1-26c)说明，当流体的体积流量一定时，流速与管径平方成反比。此流动规律与管道的放置方式、管道上是否装有管件、阀门及输送机械的布置情况无关，它只是描述不可压缩流体在圆形管道中的物料衡算关系。

【例 1-10】 水连续由粗管流入细管作稳态流动，粗管的内径为80mm，细管的内径为40mm。水在细管内的流速为3m/s，求水在粗管内的流速。

解 可按式(1-26c)进行计算：

$$\frac{u_1}{u_2}=\left(\frac{d_2}{d_1}\right)^2$$

$$\frac{u_1}{3}=\left(\frac{40}{80}\right)^2$$

所以粗管内流速： $u_1=0.75(\text{m/s})$

1.4.2 总能量衡算（Overall energy balance）

对于流体流动过程，除了掌握流动体系的物料衡算外，还要了解流动体系能量间的相互转化关系。本节介绍能量衡算方法，进而导出可用于解决工程实际问题的柏努利方程（Bernoulli equation）。

流体流动过程必须遵守能量守恒定律。如图1-17所示，流体在系统内作稳态流动，管路中有对流体做功的泵和与流体发生热量交换的换热器。在单位时间内，有质量为 m kg的流体从截面1—1′进入，则同时必有相同量的流体从截面2—2′处排出。这里对1—1′与2—2′两截面间及管路和设备的内表面所共同构成的系统进行能量衡算，并以0—0′为基准水平面。

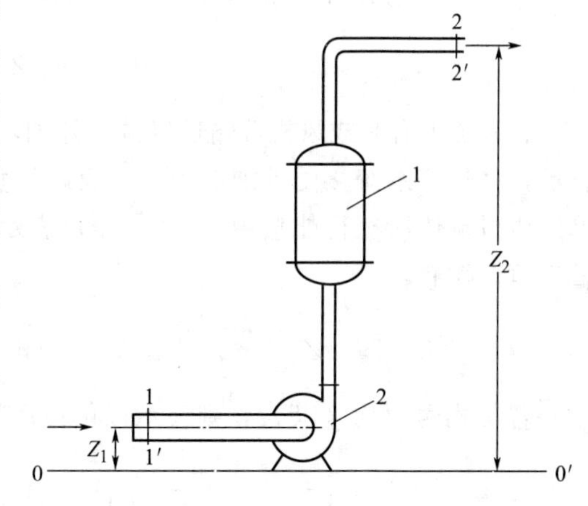

图1-17 流体流动的总能量衡算

1—换热器；2—泵

流体由1—1′截面所输入的能量有以下几种：

(1) 内能（internal energy）U

内能是储存于物质内部的能量，它是由分子运动、分子间作用力及分子振动等产生的能量。从宏观来看，内能是状态函数，它与温度有关，而压力对其影响较小。以 U 表示单位质量流体的内能，对于质量 m kg的流体由1—1′截面带入的内能为：

$$U_1 m,[\text{J/kg}\cdot\text{kg}=\text{J}]$$

(2) 位能（potential energy）gZ

位能是流体在重力作用下，因高出某基准水平面而具有的能量，相当于将质量为 $m$kg 的流体，由基准水平面提高到某一高度克服重力所需的功。位能是个相对值。输入 1—1′截面流体的位能为：

$$mgZ_1, [\text{kg} \cdot (\text{m/s}) \cdot \text{m} = \text{J}]$$

（3）静压能（static pressure energy）

静压能是将流体推进流动体系所需的功或能量。如图 1-17 所示，1—1′截面处的压强为 p_1，则作用于该截面上的总压力为 p_1A_1。现有质量为 $m$kg，体积为 V_1 的流体，要流过 1—1′截面进入体系，必须对其做一定量的功，以克服该截面处的总压力 p_1A_1。换言之，通过 1—1′截面的流体必定携带与所需功相当的能量进入系统，则把这部分能量称为静压能。因为静压能是在流动过程中表现出来的，所以也可叫作流动功（flow work）。

在 p_1A_1 总压力的作用下，$m$kg 流体流经的距离为 $L=V_1/A_1$，则 $m$kg 流体的静压能为：

$$p_1A_1L = p_1A_1\frac{V_1}{A_1} = p_1V_1, [\text{N/m}^2 \cdot \text{m}^3 = \text{J}]$$

（4）动能（kinetic energy）

流体因运动而具有的能量称为动能，它等于将流体由静止状态加速到速度为 u 时所需的功，所以，$m$kg 流体在 1—1′截面的动能为：

$$\frac{mu_1^2}{2}, [\text{kg} \cdot \text{m/s}^2 = \text{J}]$$

以上四种能量的总和为 $m$kg 流体输入 1—1′截面的总能量，即：

$$mU_1 + mgZ_1 + \frac{mu_1^2}{2} + p_1V_1$$

同理，$m$kg 流体离开系统 2—2′截面的总能量为：

$$mU_2 + mgZ_2 + \frac{mu_2^2}{2} + p_2V_2$$

若系统中有泵或风机等输送机械的外加功输入，其单位质量流体所获得能量用 W_e(J/kg) 表示，并规定系统接受外加功为正，反之为负；若利用加热器或冷却器与系统交换能量，其单位质量流体所获得能量用 Q_e（J/kg）表示，并规定系统吸热为正，反之为负。则根据能量守恒定律有：

$$mU_1 + mgZ_1 + \frac{mu_1^2}{2} + p_1V_1 + mW_e + mQ_e = mU_2 + mgZ_2 + \frac{mu_2^2}{2} + p_2V_2 \tag{1-27}$$

若等式两边均除以 m，则表示以单位质量流体为基准的能量衡算式：

$$U_1 + gZ_1 + \frac{u_1^2}{2} + p_1V_1 + W_e + Q_e = U_2 + gZ_2 + \frac{u_2^2}{2} + p_2V_2 \tag{1-28}$$

式(1-28) 称为单位质量流体（per unit mass of fluid）稳态流动过程（steady state-flow process）的总能量衡算式，各项的单位均为 J/kg。

1.4.3 流体流动的总机械能衡算式（Overall mechanical energy balance）——柏努利方程式（Bernoulli equation）

（1）柏努利方程式推导（derivation of bernoulli equation）

上述的总能量衡算式中，各项能量可分成机械能（mechanical energy）和非机械能（non－mechanical energy）两类。其中，动能、位能、静压能及外加功属于机械能；内能和热是非机械能。机械能和非机械能的区别是前者在流动过程中可以相互转化，既可用于流体输送，也可转变成热和内能；而后者不能直接转变成机械能用于流体的输送。因此，为了工程应用的方便，需将总能量衡算式转变为机械能衡算式。

根据热力学第一定律（the first law of thermodynamics），流体内能的变化仅涉及流体获得的热量与流体在该过程的有用功，即：

$$U_1-U_2=Q_e-W=Q_e-\left(\int_1^2 p\,\mathrm{d}V-\sum h_f\right) \tag{1-29}$$

式中，W 为每千克流体的可逆功，它等于流体的膨胀功 $\int_1^2 p\,\mathrm{d}V$ 与流体因克服流动阻力而损耗的能量 $\sum h_f$ 之差。

此外

$$p_2V_2-p_1V_1=\int_1^2 d(pV)=\int_1^2 p dV+\int_1^2 V dp \tag{1-30}$$

将式(1-29) 和式(1-30) 代入式(1-28) 可得：

$$gZ_1+\frac{u_1^2}{2}+W_e=gZ_2+\frac{u_2^2}{2}+\int_1^2 V\mathrm{d}p+\sum h_f \tag{1-31}$$

根据 $V=1/\rho$ 的关系，对于不可压缩流体，ρ 等于常数，则式(1-31) 可简化为：

$$gZ_1+\frac{u_1^2}{2}+\frac{p_1}{\rho}+W_e=gZ_2+\frac{u_2^2}{2}+\frac{p_2}{\rho}+\sum h_f \tag{1-32}$$

式(1-32) 称为不可压缩流体作稳态流动时的机械能衡算式——柏努利方程式。

（2）柏努利方程的讨论

① 对于理想流体（ideal gas）（黏度为零的流体），又无外加功（no mechanical energy added）的情况下，柏努利方程可写成：

$$gZ_1+\frac{u_1^2}{2}+\frac{p_1}{\rho}=gZ_2+\frac{u_2^2}{2}+\frac{p_2}{\rho}=\text{常数}$$

由此可以看出，流体流动过程中，任一截面的总机械能保持不变，而每项机械能不一定相等，能量的形式可相互转化，但必须保证机械能之和为一常数值。

② 柏努利方程通常还采用以单位质量流体为衡算基准，即以压头形式表示，就是将式(1-32) 各项均除以重力加速度 g，可得：

$$Z_1+\frac{u_1^2}{2g}+\frac{p_1}{\rho g}+H_e=Z_2+\frac{u_2^2}{2g}+\frac{p_2}{\rho g}+H_f \tag{1-32a}$$

式中　Z——位压头，m 流体柱；

$\frac{u^2}{2g}$——动压头，m 流体柱；

$\frac{p}{\rho g}$——以压头形式表示的静压能，称为静压头，m 流体柱；

$H_e=\frac{W_e}{g}$——以压头形式表示的外加功，称为有效压头，m 流体柱；

$H_f=\frac{\sum h_f}{g}$——压头损失，m 流体柱。

③ 对于气体流动过程，若$\frac{p_1-p_2}{p_1}<20\%$时，也可用式(1-32）进行计算，此时式中的密度ρ须用气体平均密度ρ_m代替，即$\rho_m=\frac{\rho_1+\rho_2}{2}$。

④ 当速度u为零，则摩擦损失$\sum h_f$不存在，此时体系无须外加功，则柏努利方程演变为流体静力学方程。因而，流体静力学方程可视为柏努利方程的一种特例。

⑤ 输送单位质量流体所需外加功，是选择输送设备的重要依据，若输送流体的质量流量为m_s(kg/s)，则输送流体所需供给的功率（即输送设备有效功率）为：

$$N_e=W_e m_s \tag{1-33}$$

⑥ 摩擦阻力损失（friction loss）$\sum h_f$是流体流动过程的能量消耗，一旦损失能量不可挽回，其值永远为正值。

(3) 柏努利方程式应用（application of Bernoulli equation）

柏努利方程的应用极为广泛，在应用中必须注意下面几个问题：

① 根据题意绘出流程示意图，选择两个截面（stations）构成机械能的衡算范围。选截面时，应考虑流体在衡算范围内必须是连续的，所选截面要与流体流动方向垂直，同时要便于有关物理量的求取。

② 基准水平面（reference plane）可以任意选定，只要求与地面平行即可。但为了计算方便，通常选取基准水平面通过两个截面中相对位置较低的一个，如果该截面与地面平行，则基准水平面与该截面重合。尤其对于水平管道，应使基准水平面与管道的中心线重合。

③ 方程中各项的单位应是同一单位制，尤其应注意流体的压力，方程两边都用绝对压力或都用表压。

④ 衡算范围内所含的外加功及阻力损失不能遗漏。

以下是柏努利方程的应用示例：

① 计算管道内流体的流速及压强。

【例 1-11】 如本题附图所示，水在直径均一的虹吸管内稳态流动，设管路的能量损失可忽略不计。试求：①管内水的流速；②管内截面 2—2′、3—3′、4—4′和 5—5′处的流体压强。已知大气压为 1.0133×10^5Pa，图中所注单位均为 mm。

解 ① 在截面 1—1′与管子出口内侧截面 6—6′之间列柏努利方程式，并以 6—6′为基准水平面。由于管路的能量损失忽略不计，即$\sum h_f=0$，故有：

$$gZ_1+\frac{u_1^2}{2}+\frac{p_1}{\rho}=gZ_6+\frac{u_6^2}{2}+\frac{p_6}{\rho} \tag{1}$$

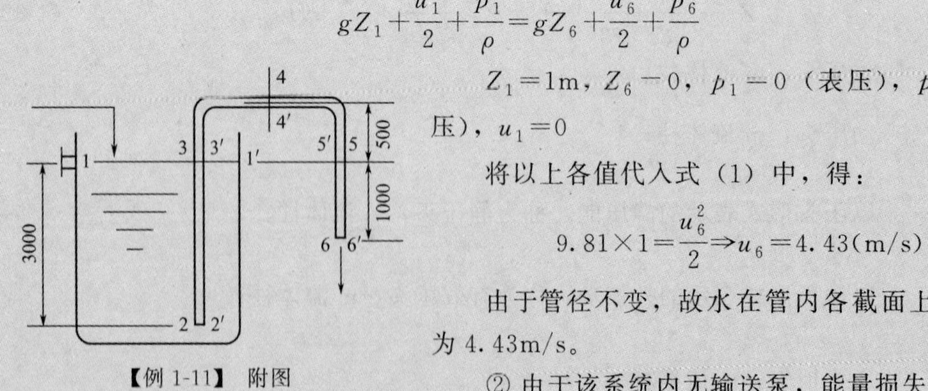

【例 1-11】 附图

$Z_1=1$m，$Z_6=0$，$p_1=0$（表压），$p_6=0$（表压），$u_1=0$

将以上各值代入式（1）中，得：

$$9.81\times1=\frac{u_6^2}{2}\Rightarrow u_6=4.43(\text{m/s})$$

由于管径不变，故水在管内各截面上的流速均为 4.43m/s。

② 由于该系统内无输送泵，能量损失又可不计，

故任一截面上的总机械能相等。按截面1—1′算出其值为（以2—2′为基准水平面）：

$$E=gZ_1+\frac{u_1^2}{2}+\frac{p_1}{\rho}=9.81\times3+\frac{101330}{1000}=130.8(\text{J/kg})$$

因此，可得截面2—2′的压强为：

$$p_2=\left(E-\frac{u_2^2}{2}-gZ_2\right)\rho=(130.8-9.81)\times1000=120990(\text{Pa})$$

截面3—3′的压强为：

$$p_3=\left(E-\frac{u_3^2}{2}-gZ_3\right)\rho=(130.8-9.81-9.81\times3)\times1000=91560(\text{Pa})$$

截面4—4′的压强为：

$$p_4=\left(E-\frac{u_4^2}{2}-gZ_4\right)\rho=(130.8-9.81-9.81\times3.5)\times1000=86650(\text{Pa})$$

截面5—5′的压强为：

$$p_5=\left(E-\frac{u_5^2}{2}-gZ_5\right)\rho=(130.8-9.81-9.81\times3)\times1000=91560(\text{Pa})$$

由以上计算可知，$p_2>p_3>p_4$，而 $p_4<p_5<p_6$，这是由于流体在管内流动时，位能与静压能相互转换的结果。

② 计算输送机械的有效功及功率（actual or theoretical energy and power）。

【例1-12】 如本题附图所示，为 CO_2 水洗塔的供水系统。水洗塔内绝对压强为2100kN/m^2，储槽水面绝对压强为300kN/m^2，塔内水管与喷头连接处高于储槽水面20m，管路为 ϕ57mm×2.5mm钢管，送水量为15m^{3}h。塔内水管与喷头连接处的绝对压强为2250kN/m^2。设能量损失为49J/kg，水的密度取1000kg/m^3，求水泵的有效功率。

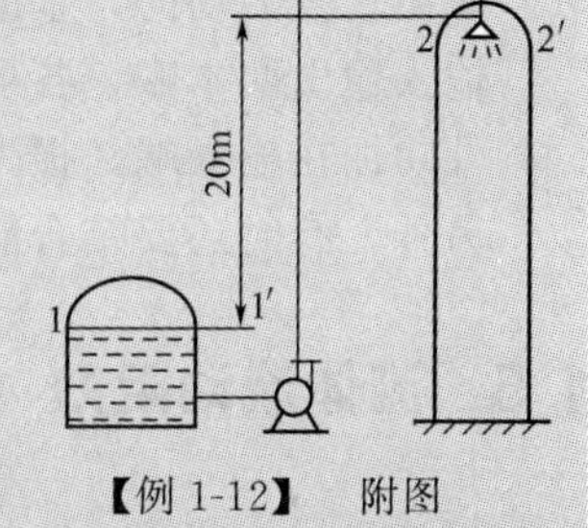

【例1-12】 附图

解 取水槽水面为1—1′截面，塔内水管与喷头连接处为2—2′截面。以1—1′截面为基准水平面，列出1—1′、2—2′截面间柏努利方程式：

$$gZ_1+\frac{u_1^2}{2}+\frac{p_1}{\rho}+W_e=gZ_2+\frac{u_2^2}{2}+\frac{p_2}{\rho}+\sum h_f$$

已知：$Z_1=0$，$Z_2=20\text{m}$，$p_1=300\times10^3\text{N/m}^2$（绝压），$p_2=2250\times10^3\text{N/m}^2$（绝压），$u_1=0$，$\sum h_f=49\text{J/kg}$

将各已知值代入柏努利方程式，可得：

$$\begin{aligned}W_e&=(20-0)\times9.81+(2250\times10^3-300\times10^3)/1000+(1.97^2-0)/2+49\\&=196.2+1950+1.93+49\\&=2197(\text{J/kg})\end{aligned}$$

则水泵的有效功率 N_e

$$N_e=W_e m_s=2197\times15\times10^3/3600=9160(\text{W})=9.16(\text{kW})$$

③ 计算管道及设备间的相对高度（relative height）。

【例1-13】 在如本题附图所示的流程中，容器B液面上方的静压力 p_B 为 1.47×10^5 Pa（绝对压力），储槽A液面上方通大气，静压力 p_A 为 9.81×10^4（绝对压力）。若要求流体

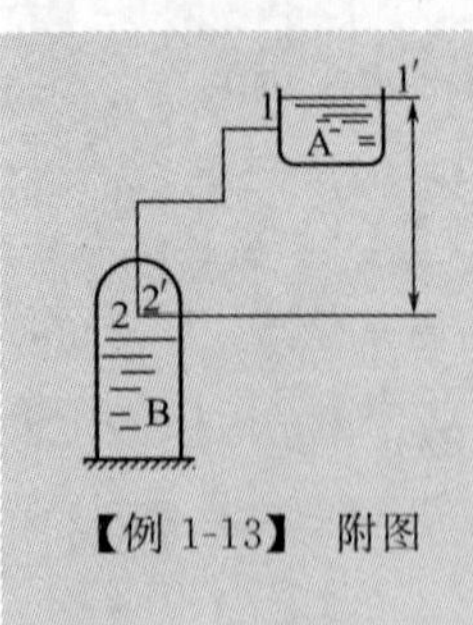

【例 1-13】 附图

以 7.20/m³/h 的流量由 A 流入 B，则储槽 A 的液面应比容器 B 高出多少米？已知该流体的密度为 900kg/m³，管道直径为 100mm，可忽略阻力损失。

解 根据式(1-32) 可知，在 1—1′、2—2′截面间列柏努利方程：

$$\frac{\Delta u^2}{2g}+\frac{\Delta p}{\rho g}+\Delta Z=0$$

因为 $u_1=0$ 且 $u_2=\frac{V}{\frac{\pi}{4}d^2}=\frac{7.20/3600}{0.785\times0.1^2}=0.255(\mathrm{m/s})$

所以：

$$\begin{aligned}Z_1-Z_2&=\frac{u_2^2-u_1^2}{2g}+\frac{p_2-p_1}{\rho g}\\&=\frac{0.255^2-0}{2\times9.81}+\frac{(14.7-9.81)\times10^4}{900\times9.81}\\&=3.3\times10^{-3}+5.55\\&=5.55(\mathrm{m})\end{aligned}$$

综上所述，应用柏努利方程的解题要点可归纳如下：

柏努利方程式，能量衡算是实质；

两个截面划系统，系统以外不考虑；

截面垂直流动向，基准平面选合适；

输入输出两本账，各项单位要统一；

外功加在输入端，损失总是算输出。

另外，有时还要结合静力学方程、连续性方程及阻力公式求解。

1.5 流体流动阻力（Friction Loss of Fluid Flow）计算

讨论机械能衡算式时指出，实际流体流动会产生阻力损失。那么，流体阻力是如何产生的？受哪些因数影响呢？这些问题与流体流动的内部结构密切相关，而流体流动的内部是极其复杂的问题，涉及的知识面较广，因此，本节仅做简要介绍。

流体在管路中流动时的阻力损失可分为直管阻力和局部阻力两部分。

1.5.1 直管阻力（Friction loss in straight pipe）计算

直管阻力是流体流经一定管径的直管时，由于流体内摩擦力的作用而产生的阻力，通常也称为沿程阻力。

如图 1-18 所示，流体在直管内以一定的速度流动时，同时受到静压力的推动和摩擦阻力的阻碍，当两个力达到力平衡时，流体的流动速度才能维持不变，即达到稳态流动。

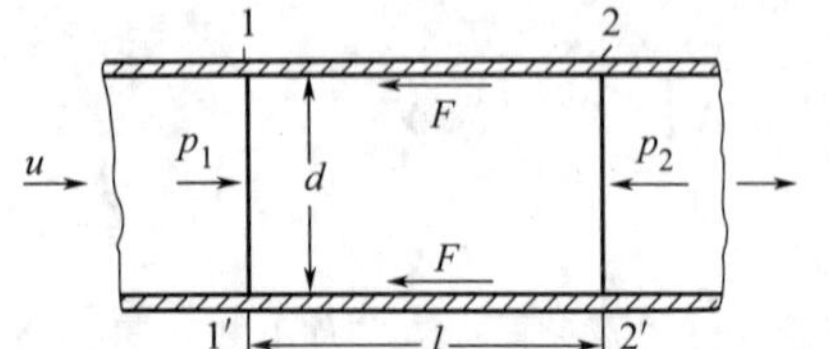

图 1-18 直管阻力通式的推导

可在 1—1′和 2—2′截面间列柏努利方程：

$$gZ_1+\frac{u_1^2}{2}+\frac{p_1}{\rho}+W_e=gZ_2+\frac{u_2^2}{2}+\frac{p_2}{\rho}+\sum h_f$$

这里将 $Z_1=Z_2$；$u_1=u_2$；$W_e=0$ 代入上式中，

可得到：

$$\sum h_f=\frac{p_1-p_2}{\rho}=\frac{-\Delta p}{\rho} \tag{1-34}$$

通过实验可测定流体流过 i 管长后的压降 Δp，进而可得出直管阻力 $\sum h_f$。利用流体在圆管中流动时力的平衡原理，可推导出直管阻力的一般计算式，现分析流体在直径为 d，管长为 l 的水平管内的受力情况：

$$推动力=(p_1-p_2)A=-\Delta p\ \frac{\pi d^2}{4}$$

流体流动阻力（剪切力）$=\tau\pi dl$，τ 为剪应力。

由于流体在圆管内是稳态等速流动，故以上两个力必达到大小相等，方向相反，则平衡方程为：

$$-\Delta p\ \frac{\pi d^2}{4}=\tau\pi dl \tag{1-35}$$

经处理得到：

$$-\Delta p=\frac{8\tau}{\rho u^2}\times\frac{l}{d}\times\frac{\rho u^2}{2} \tag{1-35a}$$

令：

$$\lambda=\frac{8\tau}{\rho u^2} \tag{1-36}$$

则上式可得：

$$-\Delta p=\lambda\ \frac{l}{d}\times\frac{\rho u^2}{2} \tag{1-37}$$

$$\sum h_f=\frac{-\Delta p}{\rho}=\lambda\ \frac{l}{d}\times\frac{u^2}{2} \tag{1-38}$$

上式称为直管阻力计算通式，也称范宁公式（Fanning equation），此式对于层流与湍流，管道水平、垂直、倾斜放置的情况均适用。式中，λ 称为摩擦系数（friction factor），无量纲，它与 Re 及管壁粗糙度 ε 有关，可通过实验测定，也可由相应的关联式计算。

1.5.2　摩擦系数 λ 的确定

1.5.2.1　层流时摩擦系数（friction factor in laminar flow）计算

根据牛顿黏性定律式(1-19）和层流速度分布公式(1-23)，可以导出 $\tau=-8u\mu/d$，代入式(1-36）中可得：

$$\lambda=\frac{64}{Re} \tag{1-39}$$

【例 1-14】 甘油在 20℃下以 $5\times10^{-4}\ m^3/s$ 的流量在内径为 27mm 的钢管内流动，试计算通过每米管长所产生的压降。

解　由附录 4 可查表得 20℃甘油的密度为 $1261kg/m^3$，黏度为 1499 mPa·s，则管内流速为：

$$u=\frac{V_S}{A}=\frac{5\times10^{-4}}{0.785\times0.027^2}=0.874(m/s)$$

雷诺数为：　$Re=\frac{du\rho}{\mu}=\frac{0.027\times 0.874\times 1261}{1499\times 10^{-3}}=19.85<2000$

故流动为层流，摩擦系数为：$\lambda=\frac{64}{Re}=\frac{64}{19.85}=3.224$

则每米管长所产生的压降为：

$$-\Delta p=\rho\sum h_f=\lambda\ \frac{l}{d}\times\frac{\rho u^2}{2}=3.224\times\frac{1}{0.027}\times\frac{1261\times 0.874^2}{2}=57.5(\text{kPa})$$

1.5.2.2 湍流时摩擦系数（friction factor in turbulent flow）计算

对于湍流的 λ，还无法从理论上导出。目前，采用的方法是通过量纲分析（dimensional analysis）和实验确定计算 λ 的关联式（correlation）。所谓量纲分析法（dimensional analysis method）是指当所研究的过程涉及变量较多，且用理论分析或数学模型求解很困难时，要用实验方法进行经验关联。利用量纲分析的方法可将几个变量组合成一个无量纲数群，用无量纲数群代替个别变量做实验，这样因无量纲数群数量必小于变量数，所以实验简化，准确性提高。

量纲分析法的理论基础是量纲一致性原则（dimensional consistency principle）和 π 定理（π theorem）。其中，量纲一致性原则表明，凡是根据基本物理规律导出的物理方程，其中各项的量纲必然相同；π 定理指出任何物理量方程都可转化为无量纲的形式，以无量纲数群的关系式代替原物理量方程，且无量纲数群的个数等于原方程中的变量总数减去所有变量涉及的基本量纲个数。

（1）量纲分析的基本步骤

① 找出影响湍流直管阻力的影响因数：

$$h_f=f(d,l,u,\rho,\mu,\varepsilon)$$

式中　ε——粗糙度，指固体表面凹凸不平的平均高度，m。

② 写出各变量的量纲：

$[h_f]=L^2\theta^{-2},[d]=L,[l]=L,[u]=L\theta^{-1},[\rho]=ML^{-3},[\mu]=ML^{-1}\theta^{-1},[\varepsilon]=L$

所有变量涉及的基本量纲是 3 个，即 M，θ，L。

③ 选择核心物理量　选择核心物理量的依据是：

a. 核心物理量不能是待定的物理量（本例是 h_f）；

b. 核心物理量要涉及全部的基本量纲，且不能形成无量纲数群。本例选择 d，u，ρ 为核心物理量符合要求。

④ 将非核心物理量分别与核心物理量组合成无量纲特征数 π_i　本例的非核心物理量是 μ，l，ε，h_f，分别与核心物理量 d，u，ρ 组合得：

$$\pi_1=d^a u^b \rho^c \mu$$

$$\pi_2=d^i u^f \rho^g l$$

$$\pi_3=d^h u^o \rho^p \varepsilon$$

$$\pi_4=d^r u^s \rho^t h_f$$

以 π_1 为例，按量纲一致性原则，对其展开得：

$$M^0\theta^0L^0=[L]^a[L\theta^{-1}]^b[ML^{-3}]^c[ML^{-1}\theta^{-1}]$$

$$M:0=c+1$$

$$\theta:0=-b-1$$

$$L:0=a+b-3c-1$$

解得：
$$a=-1,b=-1,c=-1$$

故：
$$\pi_1=\frac{\mu}{du\rho}=\frac{1}{Re}$$

同理：
$$\pi_2=\frac{l}{d},\pi_3=\frac{\varepsilon}{d},\pi_4=\frac{h_f}{u^2}$$

那么，湍流阻力式可写为：

$$F(\pi_1,\pi_2,\pi_3,\pi_4)=F\left(\frac{\mu}{du\rho},\frac{l}{d},\frac{\varepsilon}{d},\frac{h_f}{u^2}\right)=0$$

以上内容说明由过程函数式变成无量纲数群式时，变量数减少了3个，使得变量数目明显降低，实验研究更有针对性。

(2) 湍流 λ 的关联式　将上式变化形式：

$$\frac{h_f}{u^2}=f\left(Re,\frac{l}{d},\frac{\varepsilon}{d}\right)$$

将上式与式(1-38) 比较，可知：

$$\lambda=f\left(Re,\frac{\varepsilon}{d}\right)$$

上式的具体函数关系，可由实验确定。

过程工业中的管道可分为光滑管与粗糙管。通常把玻璃管、铝管、铜管、塑料管等称为光滑管；把钢管和铸铁管称为粗糙管。各种管材，在经过一段时间使用后，其粗糙程度都会产生很大差异。管壁粗糙面凸出部分的平均高度称为管壁绝对粗糙度（absolute roughness），以 ε 表示。绝对粗糙度 ε 与管径 d 之比 ε/d 称为管壁相对粗糙度（relative roughness）。表1-2列出某些工业管道的绝对粗糙度。

表1-2　某些工业管道的绝对粗糙度

项目	管道类别	绝对粗糙度 ε/mm
金属管	无缝黄铜管、铜管及铝管	0.01～0.05
	新的无缝钢管或镀锌铁管	0.1～0.2
	新的铸铁管	0.3
	具有轻度腐蚀的无缝钢管	0.2～0.3
	具有显著腐蚀的无缝钢管	0.5以上
	旧的铸铁管	0.85以上
非金属管	干净玻璃管	0.0015～0.01
	橡皮软管	0.01～0.03
	木管道	0.25～1.25
	陶土排水管	0.45～6.0
	很好整平的水泥管	0.33
	石棉水泥管	0.03～0.8

在湍流流动情况下，雷诺数一定，管壁越粗糙，摩擦因子越大。如果对一根粗糙管的内壁进行打磨，同样的流动条件下，摩擦因子会降低，若进一步打磨，以至于在雷诺数一定的条件下，摩擦因子不再进一步降低，则认为该管子水力学光滑（hydraulically smooth）。

典型的几个计算湍流 λ 的关联式如下：

① 光滑管（smooth tube）

a. 柏拉修斯式：

$$\lambda=\frac{0.3164}{Re^{0.25}} \tag{1-40}$$

该式适用于 $Re=3000\sim1\times10^5$ 的水力学光滑管内流动。

b. 顾毓珍式：

$$\lambda=0.0056+\frac{0.5}{Re^{0.32}} \tag{1-41}$$

该式适用于 $Re=3000\sim3\times10^6$ 的情况。

② 粗糙管（rough tube）

a. 顾毓珍式：

$$\lambda=0.0127+0.7543/Re^{0.38} \tag{1-42}$$

此式适用于 $Re=3000\sim3\times10^6$ 范围内，此式所指的粗糙管为内径为 50～200mm 的新钢管和铁管。

b. 尼库拉则与卡门式：

$$\frac{1}{\sqrt{\lambda}}=2\lg\frac{d}{\varepsilon}+1.14 \tag{1-43}$$

该式适用于 $\dfrac{\frac{d}{\varepsilon}}{Re\sqrt{\lambda}}>0.005$ 的情况。

③ 摩擦因子图（the friction factor chart）

为了计算方便，用实验结果将 λ、Re、ε/d 之间的相互关系绘于双对数坐标内，这就是图 1-19 所示的摩擦系数图（the friction factor chart）。图中有四个不同区域：

a. 层流区　$Re\leqslant2000$ 时，因有 $\lambda=64/Re$，则 $\lg\lambda$ 随 $\lg Re$ 的增大呈线性下降，此时 λ 只与 Re 有关，与相对粗糙度 ε 无关。

b. 过渡区　当 $2000<Re<4000$ 时，管内流动属于过渡状态且受外界条件影响，使 λ 值波动较大，为安全起见，工程上一般都按湍流处理，即用湍流时的曲线延伸至过渡区来查取 λ 值。

c. 湍流区　当 $Re>4000$ 且在图 1-19 虚线以下区域时，流体流动进入湍流区。对于一定的管型（即 $\varepsilon/d=$常数），λ 随 Re 的增大而减小；而当 Re 保持恒定时，λ 随 ε/d 的增大而增大。

d. 完全湍流区　图 1-19 虚线以上区域，λ 与 Re 的关系曲线几乎成水平线，说明当 ε/d 一定时，λ 为一定值，几乎与 Re 无关。根据式(1-39) 可知，此时流体流动产生的阻力与速度的平方成正比，故称该区域为阻力平方区，或称完全湍流（very turbulent flow）区。

当查图不便时，层流区采用式(1-39) 计算；在湍流及过渡区内，一般工程计算建议采用 Colebrook 公式：

$$\frac{1}{\sqrt{\lambda}}=1.74-2\lg\left(\frac{2\varepsilon}{d}+\frac{18.7}{Re\sqrt{\lambda}}\right) \tag{1-44}$$

图 1-19 也是按式(1-44) 绘制而成的。

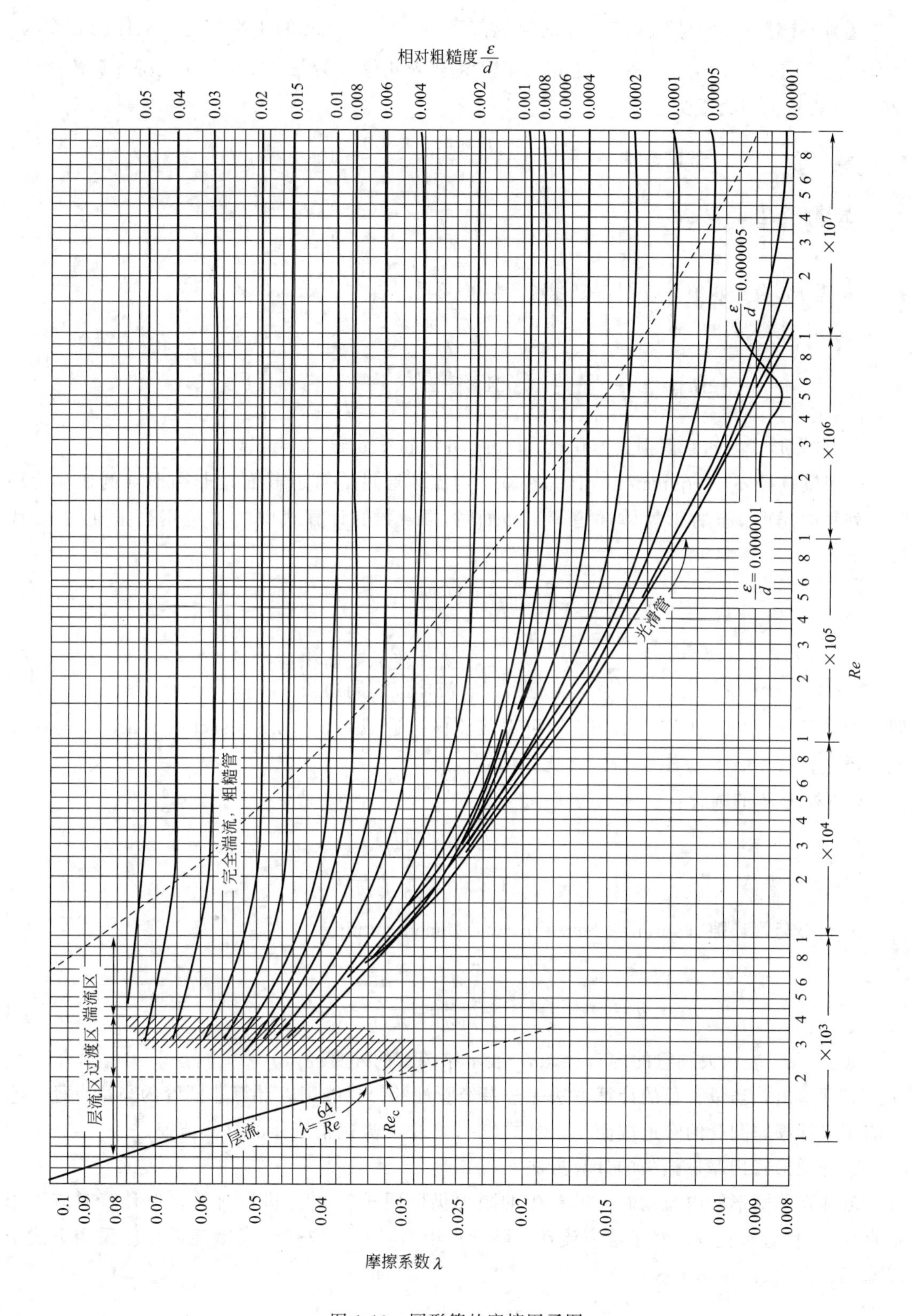

图 1-19　圆形管的摩擦因子图

【例 1-15】 某液体以 4.5m/s 的流速流经内径为 0.05m 的水平钢管，液体的黏度为 4.46×10^{-3}Pa·s，密度为 800kg/m³，钢管的绝对粗糙度为 4.6×10^{-5}m，试计算流体流经 40m 管道的阻力损失。

解 雷诺数 $Re=\dfrac{du\rho}{\mu}=\dfrac{0.05\times4.5\times800}{4.46\times10^{-3}}=40359>4000$

故属于湍流，又：

$$\varepsilon/\mathrm{d}=4.6\times10^{-5}/0.05=0.00092$$

根据 Re 及 ε/d 值，查图 1-19 得：$\lambda=0.024$

故阻力损失：

$$\sum h_{\mathrm{f}}=\lambda\frac{l}{d}\times\frac{u^2}{2}=0.024\times\frac{40}{0.05}\times\frac{4.5^2}{2}=194.4(\mathrm{J/kg})$$

④ 非圆形管内的摩擦损失（friction loss in noncircular channels）

a. 当量直径 d_{e}（equivalent diameter） 当流体通过的管道截面是非圆形（例如套管环隙、列管的壳程、长方形气体通道等）截面时，Re 数的计算式中的 d 应用当量直径 d_{e} 代替，即：

$$Re=\frac{d_{\mathrm{e}}u\rho}{\mu}$$

$$d_{\mathrm{e}}=4r_{水力}=4\times\frac{流体流通截面积}{流体湿润周边长} \tag{1-45}$$

即水力半径 $r_{水力}$（hydraulic radius）可看成流体流通截面积（cross-sectional area）与流体湿润周边长（wetted perimeter）之比。

i. 直径为 d 的圆管：

$$d_{\mathrm{e}}=4r_{水力}=4\times\frac{\frac{\pi}{4}d^2}{\pi d}=d$$

ii. 同心套管环隙（annulus between two concentric pipes）：

$$d_{\mathrm{e}}=4r_{水力}=4\times\frac{\frac{\pi}{4}d_1^2-\frac{\pi}{4}d_2^2}{\pi(d_1+d_1)}=d_1-d_2$$

式中，d_1 表示大圆管的内直径，d_2 表示小圆管的外直径。

必须指出，当量直径的计算方法完全是经验性的，只能用以计算非圆管的当量直径，绝不能用来计算非圆管的管道截面积，且式(1-45) 中的流速指流体的真实流速。

b. 非圆形管内摩擦损失的计算方法

流体在非圆形管内流动时，所产生的阻力仍可用式(1-38) 进行计算，但计算时应以当量直径 d_{e} 代替管径 d。对于层流流动，除管径用当量直径取代外，摩擦系数应采用下式予以修正：

$$\lambda=\frac{C}{Re} \tag{1-46}$$

式中，C 值根据管道截面的形状而定，其值列于表 1-3。

表 1-3　某些非圆形管的常数 C 值

非圆管形的截面形状	正方形	等边三角形	环形	长方形	
				长∶宽=2∶1	长∶宽=4∶1
常数 C	57	53	96	62	73

1.5.3　局部阻力 (Local friction loss)计算

局部阻力又称形体阻力，是指流体通过管路中的管件（如三通、弯头、变径管等）、阀门、管子出入口及流量计等局部障碍处而产生的阻力。由于在局部障碍处，流体流动方向或流速发生突然变化，产生大量旋涡，加剧了流体质点间的内摩擦，因此，局部障碍造成的流体阻力比等长的直管阻力大得多。

局部阻力损失计算一般采用两种方法：阻力系数法和当量长度法。

（1）阻力系数法（loss coefficient method）

将局部阻力损失表示为动能$\frac{u^2}{2}$的倍数，即：

$$h_f'=\xi\frac{u^2}{2} \tag{1-47}$$

式中，ξ 为局部阻力系数，无量纲，由实验测定。当部件发生截面变化时，例如图 1-20 所示的截面突然扩大和突然缩小时，式(1-47) 中的流速均应取小截面处的流速。

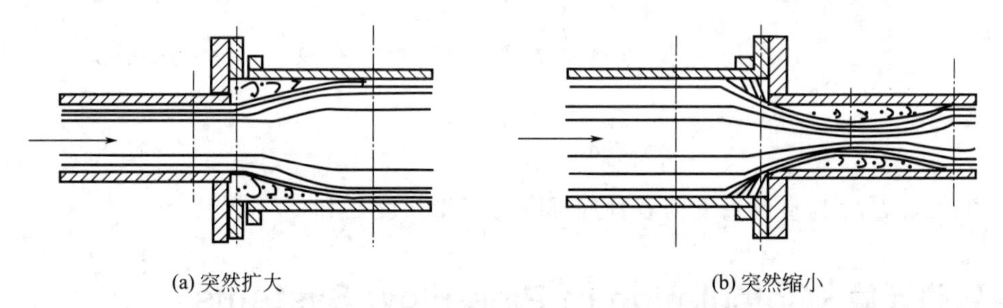

(a) 突然扩大　　(b) 突然缩小

图 1-20　截面突然扩大和突然缩小示意图

其中，流体自大容器进入管内，可看作流体的流道突然缩小（sudden contraction of cross section），由很大截面突然变为很小的截面，此时局部阻力系数 $\xi_c=0.5$，称为进口损失；相反，流体自管子进入大容器或直接排放到管外空间，可看作流体的流道突然扩大（sudden expansion of cross section），此时局部阻力系数 $\xi_e=1.0$，称为出口损失。

（2）当量长度法（equivalent length method）

若将流体局部阻力折合成相当于流体流经同直径管长为 l_e 的直管时所产生的阻力，则局部阻力可表示为：

$$h_f'=\lambda\frac{l_e}{d}\times\frac{u^2}{2} \tag{1-48}$$

式中，l_e 为管件的当量长度，其值由实验测定。

常见的管件（fittings）和阀门（valves）局部阻力系数及当量长度与管径比值见表 1-4。

管路的总阻力损失（overall friction loss）计算为流体流经直管阻力损失（friction loss in straight pipe ）与各局部阻力损失（local friction loss）之和，其表达式可写成：

表 1-4　管件和阀件的局部阻力系数 ξ 及当量长度与管径比值

名称	阻力系数 ξ	当量长度与管径之比 l_e/d	名称	阻力系数 ξ	当量长度与管径之比 l_e/d
弯头(45°)	0.35	17	标准阀(全开)	6.0	300
弯头(90°)	0.75	35	标准阀(半开)	9.5	475
三通	1	50	角阀(全开)	2.0	100
回弯头	1.5	75	止逆阀		
管接头	0.04	2	球阀	70.0	3500
活接头	0.04	2	摇板式	2.0	100
闸阀(全开)	0.17	9	水表(盘式)	7.0	350
闸阀(半开)	4.5	225			

$$\sum h_f = h_f + \sum h_f' = \lambda \frac{l}{d} \times \frac{u^2}{2} + \xi_1 \frac{u^2}{2} + \xi_2 \frac{u^2}{2} + \cdots + \xi_n \frac{u^2}{2} = \left(\lambda \frac{l}{d} + \sum \xi\right)\frac{u^2}{2} \tag{1-49}$$

也可写成：

$$\sum h_f = h_f + \sum h_f' = \lambda \frac{l}{d} \times \frac{u^2}{2} + \lambda \frac{l_{e_1}}{d} \times \frac{u^2}{2} + \lambda \frac{l_{e_2}}{d} \times \frac{u^2}{2} + \cdots + \lambda \frac{l_{e_n}}{d} \times \frac{u^2}{2} = \lambda \frac{l + \sum l_e}{d} \times \frac{u^2}{2} \tag{1-50}$$

式中，$l + \sum l_e$ 为直管长度与各种局部阻力当量长度之和。

1.6　管路计算（Calculation of Pipe Flow Systems）

管路计算是流体流动连续性方程、柏努利方程和流体流动阻力计算式的综合应用，其中管路的安装配置可以分为简单管路和复杂管路两类。

1.6.1　管路组成（Components of pipe flow systems）

1.6.1.1　管子材料和用途（materials of construction and application of pipes）

(1) 铸铁管（cast-iron pipe）

铸铁管常用作埋入地下的给水总管、煤气管及污水管等，也可用来输送碱液及浓硫酸。铸铁管价廉、耐腐蚀性强，但管壁厚较笨重，强度差，故不宜输送蒸汽及在压力下输送爆炸性或有毒性气体。

(2) 有缝钢管（seamed steel pipe）

有缝钢管一般用于压力小于 1.6MPa 的低压管路。小直径的有缝钢管（公称直径 D_g 为 10～150mm）又称水煤气管，是用低碳钢焊制而成，分镀锌管（白铁管）、不镀锌管（黑铁管）两种，常用于水、煤气、空气、低压蒸汽和冷凝液及无腐蚀性的物料管路，其工作温度范围为 0～200℃。

（3）无缝钢管（seamless steel pipe）

无缝钢管分为热轧和冷拔两种，其特点是品质均匀和强度高，可用于输送有压力的物料，如蒸汽、高压水、过热水以及有燃烧性、爆炸性和毒性的物料。

（4）紫铜管（copper pipe）和黄铜管（brass pipe）

铜管质量较轻，导热性能好，低温下冲击韧性高，适宜作热交换器用管及低温输送管。黄铜管可用于海水处理，紫铜管常用于压力输送，适用温度小于 250℃。

（5）铅管（lead pipe）

铅管性软，易于锻制和焊接，但机械强度差，不能承受管子自重，必须铺设在支承托架上，能抗硫酸、60%的氢氟酸、浓度低于 80%的乙酸等，多用于耐酸管道，但硝酸、次氯酸盐和高锰酸盐类介质不宜使用，最高使用温度为 200℃。

（6）铝管（aluminium pipe）

铝管能耐酸腐蚀（不包含盐酸），但不耐碱、盐水及盐酸等含氯离子的化合物，多用于输送浓硝酸、乙酸等，使用温度小于 200℃。

（7）陶瓷管（ceramic pipe）和玻璃管（glass pipe）

陶瓷管、玻璃管耐腐蚀性好，但性脆，强度低，不耐压。陶瓷管多用于排除腐蚀性污水，而玻璃管由于透明，可用于某些特殊介质的输送。

（8）塑料管（plastic tube）

常用的塑料管有聚氯乙烯管、聚乙烯管、玻璃钢管等，其特点是质轻、抗腐蚀性好、易加工，但耐热耐寒性差，强度低，不耐压，一般用于常压、常温下酸、碱液的输送。

（9）橡胶管（rubber tube）

橡胶管能耐酸、碱，抗腐蚀，有弹性，能任意弯曲，但易老化，只能用于临时管路。

（10）铝塑复合管（aluminum-plastic composite tube）

铝塑复合管抗腐蚀性强，可任意弯曲，与管件联接方便，使用寿命较长。

Pipe and Tubing

Fluids are usually transported in pipe or tubing, which is circular in cross section and available in widely varying sizes, wall thicknesses, and materials of construction. There is no clear-cut distinction between the terms pipe and tubing. Generally speaking, pipe is heavy-walled and relatively large in diameter and comes in moderate lengths of 6 or 12 m; tubing is thin-walled and often comes in coils several hundred feet long. Metallic pipe can be threaded; tubing usually cannot. Pipe walls are usually slightly rough; tubing has very smooth walls. Lengths of pipe are joined by screwed, flanged, or welded fittings; pieces of tubing are connected by compression fittings, flare fittings, or soldered fittings. Finally, tubing is usually extruded or cold-drawn, while metallic pipe is made by welding, casting, or piercing a billet in a piercing mill.

Pipe and tubing are made from many materials, including metals and alloys, wood, ceramics, glass, and various plastics, Polyvinyl chloride, or PVC, pipe is extensively used for wastewater lines. In process plants the most common material is low-carbon steel, fabricated into what is sometimes called black-iron pipe. Wrought-iron and cast-iron pipes are also used for a number of special purposes.

1.6.1.2 管路连接（pipes joints and fittings）

（1）承插式连接（bell-and-spigot joints）

铸铁管、耐酸管、水泥管常用承插式连接。管子的一头扩大成钟形，使另一根管子的平头插入钟形口内，环隙先用麻绳、石棉绳填塞，然后用水泥、沥青等胶合剂涂抹。该连接安装方便，对管子中心线的对接允许有较大的偏差，但缺点是难以拆卸，高压时不便使用。

（2）螺纹连接（screwed fitting）

管子两端有螺纹，可用现成的螺纹管件连接而成，通常在小于100mm管道中使用。

（3）法兰连接（flange joint）

法兰连接拆装方便，密封可靠，可适用的压强、温度、管径的范围很大，缺点是费用较高。法兰连接分普通钢管的平焊法兰和高压用的凹凸面对焊法兰两种类型。两法兰间放置垫片，垫片起密封作用，垫片的材料有橡胶石棉板、橡胶、塑料、软金属等，视介质的性质、温度、压力而定。

（4）焊接（welding）

焊接法比上述任何方法都便宜、方便、严密，无论钢管、有色金属管、聚氯乙烯管均可焊接，应用十分广泛。但对经常拆卸的管路和对焊缝有腐蚀性的物料管路，以及不宜动火的车间管路，不便采用焊接。

Joints and Fittings

The methods used to join pieces of pipe or tubing depend in part on the properties of the material but primarily on the thickness of the wall. Thick-walled tubular products are usually connected by screwed fittings, by flanges, or by welding. Pieces of thin-walled tubing are joined by soldering or by compression or flare fittings. Pipe made of brittle materials such as glass or carbon or cast iron is joined by flanges or bell-and-spigot joints.

When screwed fittings are used, the ends of the pipe are threaded externally with a threading tool. The thread is tapered, and the few threads farthest from the end of the pipe are imperfect, so that a tight joint is formed when the pipe is screwed into a fitting. Tape of polytetrafluoroethylene is wrapped around the threaded end to ensure a good seal. Threading weakens the pipe wall, and the fittings are generally weaker than the pipe itself. When screwed fittings are used, therefore, a higher schedule number is needed than with other types of joints. Screwed fittings are standardized for pipe sizes up to 12 in. (300mm), but because of the difficulty of threading and handling large pipe, they are rarely used in the field with pipe larger than 3 in. (75 mm).

Lengths of pipe larger than about 2 in (50 mm) are usually connected by flanges or by welding. Flanges are matching disks or rings of metal bolted together and compressing a gasket between their faces. The flanges themselves are attached to the pipe by screwing them on or by welding or brazing. A flange with no opening, used to close a pipe, is called a *blind flange or a blank flange*. For joining pieces of large steel pipe in process piping, especially for high-pressure service, welding has become the standard method. Welding makes stronger joints than screwed fittings do, and since it does not weaken the pipe wall, lighter pipe can be used for a given pressure. Properly made welded joints are leakproof, whereas other types of joints are not. Environmental protection legislation considers flanged and screwed joints to be sources of emission of volatile materials. Almost the only disadvantage of a welded joint is that it cannot be opened without destroying it.

1.6.1.3 管件与阀门 (fittings and valves)

(1) 常用管件 (common fittings)

常用管件如图1-21所示。用以改变流体流向的管件有90°弯头 (elbow)、45°弯头、180°回弯头等；用以堵截管路的管件有堵头 (丝堵)、管帽、盲板等；用以连接支管的管件有三通、四通；用以改变管径的管件有异径管，内、外螺纹接头等；用以延长管路的管件有管箍 (束节)、外丝接头、活接头、法兰等。

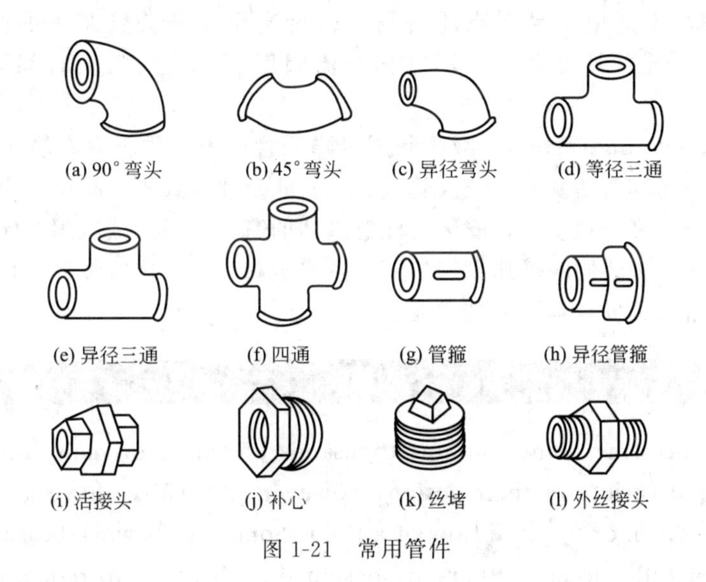

图1-21 常用管件

(2) 常用阀门 (common valves)

① 闸阀 (gate valves) 闸阀的阀体内装有一块闸板，使用时由螺旋升降，其移动方向与管道轴线垂直。它密封性好，流动阻力小，应用广泛，多用于管路作切断或全开之用。这种阀门结构比较复杂，密封面易擦伤，不适用于控制流量的大小及有悬浮物的介质，常用于上水管道和热水供暖管道，但通常不用在蒸汽管道上，因为压力较高时，闸阀会因单面承受压力而难于开启。

② 截止阀 (globe valves) 截止阀是利用圆形阀盘在阀体内的升降来改变阀盘与阀座间的距离，以开关管路和调节流量。该阀门的流体阻力大于闸阀，但较严密可靠，可用于流量调节，不适用于有悬浮物的流体管道。

闸阀和截止阀见图1-22。

③ 止回阀 (check valves) 止回阀又称单向阀或止逆阀，它只允许流体朝一个方向流动，靠流体的压力自动开启，可防止管道或设备中的介质倒流。离心泵吸入管端的底阀

(a) 闸阀　　(b) 截止阀

图 1-22　常用阀门

就属于此类，另外，止回阀多用于给水管路，安装时有严格的方向性，一定不可装反。

④ 旋塞（plug cocks）　旋塞亦称考克，是依靠阀体内带中心孔锥形体来控制启闭的阀门。其特点是结构简单，开闭迅速，流体阻力小，适用于含有固相的液体，但不适用于高温、高压的场合，制造维修费工时。

⑤ 球阀（ball valves）　球阀有一个中间开孔的球体作阀芯，靠旋转球体来开关管路。它的特点是结构简单，体积小，开关迅速，操作方便，流动阻力小，但制造精度要求高。

旋塞和球阀均是快开式阀门，阻力小、流量大。但球阀的密封面易磨损，开关力较大，容易卡住，故不适用于高温高压的情况。旋塞用于开关管路中的介质也可作节流阀门；球阀只用于开关管道介质，不宜作节流阀用，以免阀门长时间受介质冲刷而失去严密性。

⑥ 减压阀（reducing valves）　减压阀用以降低管道内介质压力，使介质压力符合生产的需要，常用的减压阀有活塞式、波纹管式、鼓膜式及弹簧式等。减压阀应直立安装在水平管道上，阀盖要与水平管道垂直，安装时注意阀体的箭头方向。减压阀两侧应安装阀门。高低压管上都设有压力表，同时低压系统还要设置安全阀。安装这些装置的目的是调节和控制压力，对低压系统保证安全运行尤其重要。

Valves

A typical processing plant contains thousands of valves of many different sizes and shapes. Despite the variety in their design, however, All valves have a common primary purpose: to slow down or stop the flow of a fluid. Some valves work best in on-or-off service, fully open or fully closed. Others are designed to throttle, to reduce the pressure and flow rate of a fluid. Still others permit flow in one direction only or only under certain conditions of temperature and pressure. A steam trap, which is a special form of valve, allows water and Inez gas to pass through while holding back the steam. Finally, by using sensors and automatic control systems to adjust the valve position and thus the flow through the valve, the temperature, pressure, liquid level, or other fluid properties can be controlled at points remote from the valve itself.

In all cases, however, the valve initially stops or controls flow. This is done by placing an obstruction in the path of the fluid, an obstruction that can be moved about as desired inside the pipe with little or no leakage of fluid from the pipe to the outside. Where the resistance to flow introduced by an open valve must be Sam, the obstruction and the opening that can be closed by it are large. For precise control of flow rate, usually obtained at the price of a large pressure drop, the cross sectional area of the flow channel is greatly reduced, and a small obstruction is set into the small opening.

Valves containing a bellows seal are often used in processes involving hazardous or toxic materials to ensure against leakage. In these valves an upper stem raises or lowers the top of an expandable bellows, moving a lower stem that is attached inside the bellows. The lower stem raises or lowers the valve disk. The upper stem may rotate; the lower stem does not. The lower end of the bellows is sealed to the valve body by a gasket or by welding. Bellows valves are available in various alloys in sizes from 1/2 in. (12 mm) to 12 in. (300mm).

1.6.1.4 管径的确定 (selection of pipe diameters)

对于圆形管道，$A=(\pi/4)d^2$，代入式(1-16) 可得：

$$d=\sqrt{\frac{4V_s}{\pi u}} \tag{1-51}$$

当 V_s=常数时，随着流速 u 增大，管道直径 d 减小，反之亦然。由于流速 u 的大小体现了操作费用的高低，而管径 d 的大小则体现了设备投资费用的多少。所以，对于较长的管道，两者要权衡考虑，以总费用最低为目标，确定最优流速，如图 1-23 所示。一般的情况下，管道的流速可由经验确定，某些流体在管道内的常用流速范围如表 1-1 所示。

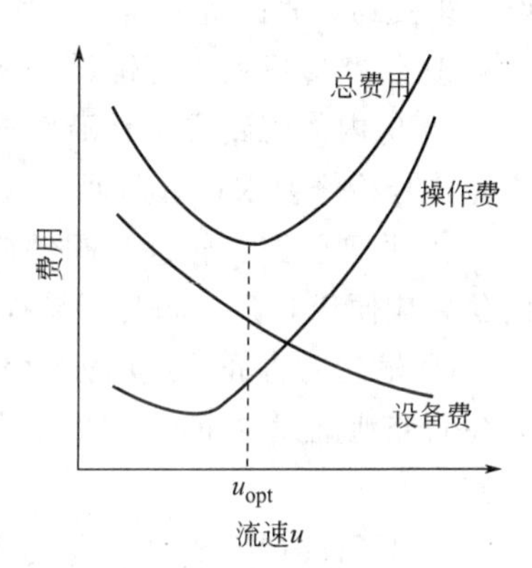

图 1-23 最优流速的确定

由式(1-51) 计算的管径是管子的内径，当管子材料确定后，可查取管子规格。其中，水煤气管（英制）的规格以公称直径表示，公称直径既不是管子外径，也不是管子内径，而是与其相近的整数；无缝钢管（公制）的规格是采用外径×壁厚来表示。各种常用管子的规格见附录 22。

已知流体的体积流量确定管径的基本方法是先查表 1-1 初选流速，以此来计算管道直径，然后按管子的规格圆整管径，最后用圆整后的管径重新核算流速。

【例 1-16】 某厂精馏塔进料量为 10000kg/h，料液的密度为 960kg/m^3，其他性质与水接近，试选择进料管的管径。

解 由题给条件：

$$V_s=\frac{m_s}{\rho}=\frac{10000}{3600\times 960}=0.00289(\mathrm{m^3/s})$$

因料液性质与水相近，参照表 1-1 的经验数据，选取料液在管内流速 $u=1.8$m/s。

再由流体的体积流量 V 和选定的流速 u 按下式求出管道直径 d。

$$d=\sqrt{\frac{4V_s}{\pi u}}=\sqrt{\frac{4\times 0.00289}{3.14\times 1.8}}=0.0452(\mathrm{m})=45.2(\mathrm{mm})$$

查附录 22 水管规格，确定选用 ϕ57mm×3mm 的无缝钢管，其内径为：

$$d=57-3\times 2=51(\mathrm{mm})=0.051\mathrm{m}$$

重新核算流速，即料液在管内的实际流速为：

$$u=\frac{4\times 0.00289}{\pi\times 0.051^2}=1.42(\mathrm{m/s})$$

通过本例计算，应初步熟悉工程上管路流体输送的流速范围以及管路尺寸的表示方法。

1.6.1.5 管子壁厚的选择（selection of pipe wall thickness）

管径决定以后，管子的壁厚应按其承受的压力及管材在操作温度下的许用压力来确定。一般铸铁管的每种内径只有一个厚度，故定出内径，壁厚也就决定。有缝钢管一般有两种壁厚，可根据操作压强先决定选用普通管还是加强管，然后根据算出的内径找出合适的规格。无缝钢管同一种管径有多种壁厚，壁厚是按公称压力 P_g 分级的，即可按 P_g 决定壁厚。

1.6.1.6 管路布置的一般原则（general principles of designing and installing a piping system）

布置管路时，应对车间所有管路（生产系统管路、辅助系统管路、电缆、照明、仪表管路、采暖通风管路等）全盘规划，各安其位，这项工作的一般原则有以下内容。

① 管路应成列平行铺设，尽量走直线，少拐弯，少交叉，力求整齐美观。

② 房内的管路应尽量沿墙或柱子铺设，以便设置支架；各管路之间与建筑物间的距离应能符合检修要求；管路通过人行道时，最低点离地面应在 2m 以上。

③ 并列管路上管件与阀件应错开安装，阀门安装的位置应便于操作，温度计、压力表的位置应便于观察，同时不易损坏。

④ 输送有毒介质或腐蚀性介质的管路，不得在人行道上设置阀件、伸缩器、法兰等，以免管路泄漏发生事故。输送易燃易爆介质的管路，一般应设有防火安全装置和防爆安全装置。

⑤ 长管路要有支承，以免弯曲存液及受震，并要保持适当的坡度。

⑥ 平行管路的排列要遵守一定的原则。如垂直排列时，热介质管路在上，冷介质管路在下；高压管路在上，低压管路在下；输送无腐蚀性介质的管路在上，输送有腐蚀性介质的管路在下。水平排列时，低压管路在外，高压管路靠近墙柱；检修频繁的在外，不常检修的靠近墙柱；质量大的管路要靠近管件支柱或墙。

⑦ 输送必须保持使温度稳定的热流体或冷流体的管路保温或保冷。

⑧ 管路安装完毕，应按规定进行强度及密度试验。未经试验合格，焊缝及连接处不得涂漆及保温。管路在开工前需要压缩空气或用惰性气体吹扫。

对于各种非金属管路及特殊介质管路的布置与安装，还应考虑一些特殊性问题，如聚氯乙烯管应避开热的管路，氧气管路安装前应脱油等。

Recommended Practice

In designing and installing a piping system, many details must be given carefully attention, for the successful operation of the entire plant may turn upon a seemingly insignificant feature of the piping arrangement. Some general principles are important enough to warrant mention. In installing pipe, for example, the lines should be parallel and contain, as far as possible, right-angle bends. In systems where the process lines are likely to become clogged, provision should be made for opening the lines to permit cleaning them out.

Unions or flanged connections should be generously included, and tees or crosses with their extra openings closed with plugs should be substituted for elbows in critical locations. With hazardous materials, especially volatile ones, flanged or screwed fittings should be used sparingly.

In gravity flow systems the pipe should be oversized and contain as few bends as possible. Fouling of the lines is particularly troublesome where flow is by gravity, since the pressure head on the fluid cannot be increased to keep the flow rate up if the pipe becomes restricted.

Leakage through valves should also be expected. Where complete stoppages of flow is essential, therefore, where leakage past a valve would contaminate a valuable product or endanger the operators of the equipment, a valve or check valve is inadequate. In this situation a blind flange set between two ordinary flanges will stop all flow; or the line can be broken at a union or pair of ranges and the open ends capped or plugged.

Valves should be accessible and well supported without strain, with suitable allowance for thermal expansion of the adjacent pipe. Room should be allowed for fully opening the valve and for repacking the stuffing box.

1.6.2　简单管路计算

管路计算按配管情况可分为简单管路（single pipes）和复杂管路（multiple pipes），后者又可分为分支管路（branching system）和并联管路（parallel pipe system）。

简单管路是指流体从入口至出口是在一条管路（管径可以相同，也可以不同）中流动，中间没有出现分支或汇总情况。

简单管路计算常用三个方程联立求解，即：

连续性方程：
$$\frac{u_1}{u_2}=\left(\frac{d_2}{d_1}\right)^2$$

柏努利方程：
$$gZ_1+\frac{u_1^2}{2}+\frac{p_1}{\rho}+W_e=gZ_2+\frac{u_2^2}{2}+\frac{p_2}{\rho}+\left(\lambda\ \frac{\Sigma l+\Sigma l_e}{d}+\Sigma\zeta\right)\frac{u^2}{2}$$

摩擦阻力系数计算式：
$$\lambda=f\left(Re,\frac{\varepsilon}{d}\right)$$

当输送流体确定后，即流体的物性已知时，上述三个方程中仍含有许多变量，用三个方程联立求解，由于给出的已知变量不同，就构成了不同类型的计算问题。例如，求流体输送所需提供的有效压头、设备内的压力、两设备间的相对位置、流体输送的阻力损失、流速及管路直径等。

当进行计算时，经常遇到两种情况，即一种是已知管径 d、管长 i 和允许压降 Δp，求所输送流量 V_s；另一种是已知管长 l、流量 V_s 及允许压降 Δp，求管路直径 d。这两种情况由于管径或流速均为未知，无法求取 Re 数，这样既无法判断流体流动类型，又不能计算摩擦系数 λ。而 λ 与 Re 的关系又是十分复杂的非线性关系，所以计算时需采用试差法。

试差法（trial-and-error）是化工计算中经常采用的方法。试差过程既可采用流速为试差变量，也可选摩擦系数为试差变量。选取流速为试差变量时，应在适宜流速范围内（见表

1-1）选取中间值，采用对等分布方法进行；若选取 λ 为试差变量，由于 λ 值变化不大（通常范围为 0.02～0.03），其值选取可采用流动已进入阻力平方区的 λ 值为初值。

【例 1-17】 10℃的水以 $10m^3/h$ 的流量流经 25m 长的水平管，设管两端的压强差为 $5mH_2O$，求管子的适宜直径。

解 这里需采用试差法来求解。

由附录查得水在 10℃时的物性：$\rho=1000\ kg/m^3$，$\mu=1.3077\times10^{-3}\ Pa\cdot s$。

$$设\ u=2.0m/s\Rightarrow d=\sqrt{\frac{4V_s}{\pi u}}=\sqrt{\frac{4\times10}{\pi\times2\times3600}}=\sqrt{0.00177}=0.0421(m)$$

查附录 22 选 $1\frac{1}{2}''$钢管 $\phi48mm\times3.5mm$，即 $d_{内}=41mm$，查表 1-2 得到新钢管的粗糙度为 0.1mm。

$$校核：\quad u=\frac{V_s}{0.785\times d^2\times3600}=\frac{10}{0.785\times0.041^2\times3600}=2.11(m/s)$$

$$Re=\frac{du\rho}{\mu}=\frac{0.041\times2.11\times1000}{1.3077\times10^{-3}}=6.6\times10^4,\ \frac{\varepsilon}{d}=\frac{0.1}{41}=0.00244$$

查图 1-19 得到：$\lambda=0.023$

$$所需压头\ H_f=\lambda\frac{l}{d}\times\frac{u^2}{2g}=0.023\times\frac{25}{0.041}\times\frac{2.11^2}{2\times9.81}=3.18(m)$$

因为所给 $H_{f_0}=5m>H_f=3.18m$，故所选管径符合要求。

1.6.3 复杂管路计算

（1）分支或汇合管路（branching system）

如图 1-24（a）所示，流体分流后不再汇合称为分支管路，同理也可有汇合管路，如图 1-24（b）所示。分支（或汇合）管路的计算遵循以下两条流动规律：

①总流量等于各支管流量之和，即 $V_s=\sum V_{si}$。②分支点处的总压头为恒定值，即：

$$Z_A+\frac{u_A^2}{2g}+\frac{p_A}{\rho g}=Z_B+\frac{u_B^2}{2g}+\frac{p_B}{\rho g}+H_{f,A-B}$$
$$=Z_C+\frac{u_C^2}{2g}+\frac{p_C}{\rho g}+H_{f,A-C}$$

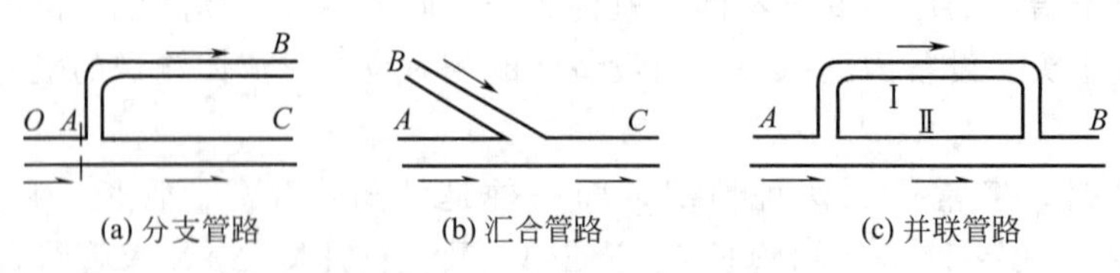

图 1-24 复杂管路示意图

（2）并联管路（parallel pipe system）

如图 1-24（c）所示，流体分流后又汇合称为并联管路。分支管路计算的两条流动规律对并联管路同样适用，由第 2 条规律可导出：$H_{f,A-B}=H_{f,i}$，即各支管的能量损失相等，而 AB 间的总阻力损失应等于其中任意一条支路的阻力，而不是各支管的阻力加和。

Multiple Pipe Systems

Multiple pipe systems contain more than one pipe. The complex system of tubes in our lungs (beginning with the relatively large-diameter branchs and ending in minute branchs after numerous branchings) and the maze of pipes in a city's water distribution system are typical of such systems. The governing mechanisms for the flow in multiple pipe systems are the same as for the single pipe systems. However, because of the numerous unknowns involved, additional complexities may arise in solving for the row in multiple pipe systems. In this section multiple pipe systems containing pipes in parallel or branching configuration will be discussed.

1.7 流体输送机械（Fluid Transportation Machinery）

流体输送机械可按不同的方法分类，若按其工作原理可分为离心式（如离心泵、旋涡泵、轴流泵等）、往复式（如往复泵、柱塞泵、计量泵等）、旋转式（如齿轮泵、螺杆泵等）、流体动力作用式（如空气升液器等）。

1.7.1 离心式输送机械（Centrifugal fluid transportation machinery）

输送机械的种类虽然繁多，但离心式使用得最广，如化学工业使用的离心泵大约占所用泵的80%以上，所以首先重点介绍离心式输送机械。

1.7.1.1 离心泵（centrifugal pumps）

（1）离心泵的基本结构

离心泵的结构如图1-25所示，主要由两部分组成：旋转部件（rotating components）——叶轮和泵轴；静止部件（stationary components）——壳体、密封、轴承等。

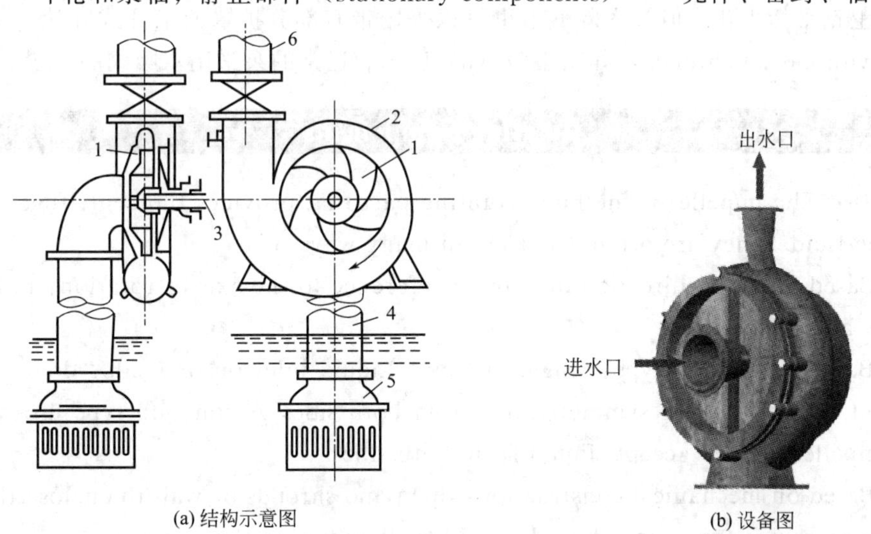

(a) 结构示意图　　(b) 设备图

图1-25 离心泵

1—叶轮；2—泵壳；3—泵轴；4—吸入管；5—底阀；6—压出管

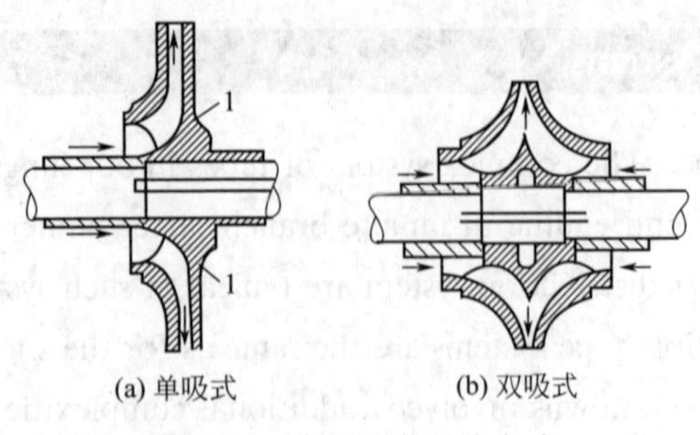

图 1-26　离心泵的吸液方式

① 叶轮（impellers）　叶轮是为流体提供离心加速度的主要旋转部件。叶轮可以按照不同的方式分类：

a. 基于旋转轴中流体的主要流向不同，叶轮可分为径向流叶轮、轴向流叶轮和混合流叶轮；

b. 基于吸液方式的不同，叶轮可分为单吸式和双吸式叶轮。单吸式如图 1-26（a）所示，液体从叶轮的一侧吸入；双吸式如图 1-26（b）所示，液体对称地从叶轮的两侧吸入。

c. 基于结构的不同，叶轮又分为开式、半开式和闭式，如图 1-27 所示。

如图 1-27 所示，叶轮一般有 6～12 片沿旋转方向后弯的叶片，常分为：开式（open impeller），用于输送含有杂质的悬浮液；半闭式（semi-open impeller），用于输送易沉淀或含固体颗粒的物料；闭式（enclosed impeller），用于输送清液。

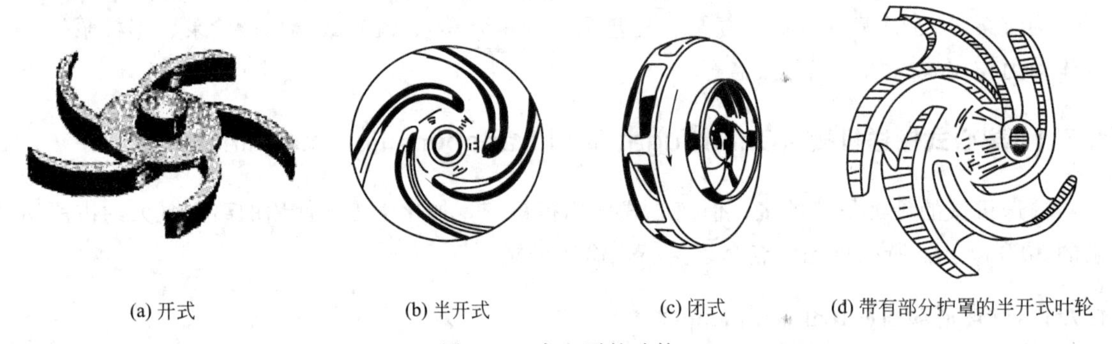

图 1-27　离心泵的叶轮

在效率方面，按开式、半闭式和闭式的顺序依次增加，后两种叶轮由于侧面加了盖板，易产生轴向推力。轴向推力使叶片与壳体接触，引起振动、磨损，增加电机负荷，消除方法是在盖板上钻平衡小孔，但效率降低，也可以采用带有部分护罩的半开式叶轮（semi-open impeller with partial shroud），如图 1-27（d）所示。通常的泵采用双吸式的叶轮。

Rotating Components

Impeller. The impeller is the main rotating part that provides the centrifugal acceleration to the fluid. They are often classified in many ways.

(1) Based on major direction of flow in reference to the axis of rotation：radial flow，axial flow and mixed flow.

(2) Based on suction type：single-suction，Liquid inlet on one side；double-suction，liquid inlet to the impeller symmetrically from both sides. A common type uses a double-suction impeller，which accepts liquid from both sides.

(3) Based on mechanical construction：open，no shrouds or wall to enclose the vanes，semi-open or vortex type，and shrouds or sidewall enclosing the vanes.

An open impeller is characterized by impeller blades that are supported almost entirely by the impeller hub，as shown in Fig. 1-27（a）. This is the simplest impeller style and it is primarily applied to clean, non-abrasive, low horsepower applications. An open impeller is lighter in weight than its shrouded counterpart. Less impeller weight reduces shaft deflection and enables the use of a smaller diameter shaft, at a lower cost, than an equivalent shrouded impeller.

An **open impeller** typically operates at a higher efficiency than a shrouded impeller of the same specific speed. The largest contributor to efficiency loss in an enclosed radial impeller is disc friction caused by the front and back impeller shrouds turning in close proximity to the stationary casing walls. Removing the shrouds eliminates the disc friction.

One drawback of the open impeller is that it is more susceptible to abrasive wear than a shrouded impeller. High velocity fluid on the impeller blades in close proximity to the casing walls establishes rotating vortices that accelerate wear when abrasives are present.

A tight clearance between the impeller and the front and back casing walls is necessary to maximize efficiency. As the impeller wears, these clearances open and efficiency drops rapidly. The tight operating clearances required on both sides of an open impeller for efficient operation precludes adjustment of the impeller axial position to compensate for wear.

A **semi-open impeller** is a compromise between an open and an enclosed impeller. It incorporates a single shroud, usually located on the back of the impeller, as shown in Fig. 1-27 (b). A semi-open impeller has a solids passing capability similar to that found in an open impeller. With only a single shroud a semi-open impeller is easy to manufacture and completely accessible for applying surface hardening treatments. For moderately abrasive slurries, especially if plugging is a concern, a semi-open impeller is a good choice.

A semi-open impeller operates more efficiently than an enclosed impeller because of lower disc friction and tighter axial clearances. It has an advantage over an open impeller in that it can be adjusted axially to compensate for casing wear.

High axial thrust is the primary drawback of a semi-open impeller design. Axial thrust balance is manageable through design for both open and enclosed impellers. On a semi-open impeller, the entire backside surface of the shroud is subject to the full impeller discharge pressure. The front side of the shroud is at suction pressure at the eye of the impeller and increases along the impeller radius due to centrifugal action. The differential between the pressure profiles along the two sides of the shroud creates the axial thrust imbalance. This can be managed somewhat through the use of pump-out vanes on the backside of the shroud, but the vanes will start to lose effectiveness if the impeller is moved forward in the casing to compensate for wear. Some manufacturers have integrated an adjustable wear-plate into the casing design so that clearance adjustments can be made. Combined with hard materials or surface hardening treatments, this option provides a good design in lightly to moderately abrasive applications.

An obvious question is why use a semi-open impeller in a solids application if an open impeller with an adjustable wear plate could be used instead? It might seem logical that an open impeller of hard metal construction, used in conjunction with an adjustable wear liner，would combine good solids handling characteristics，with low thrust imbalance，light weight，and adjustability for wear. Unfortunately，true open impellers lack the structural support to prevent blade collapse or deformation under the demands of most industrial applications. A semi-open impeller is well suited for handling solids in applications where the blades might encounter high impact loads from rocks and the like，or in higher power applications. In both situations the shroud provides additional structural support and reinforcement to protect against blade collapse or deformation.

One improvement that has been made to the semi-open impeller is the use of a partial shroud，as shown in Fig. 1-27（d）. Most of the pressure developed by the impeller，and most of the shroud surface area，is in the outer diameter region of the impeller. Elimination of the shroud in this area reduces the axial thrust in a semi-open impeller without compromise to the structure support provided by the full back shroud.

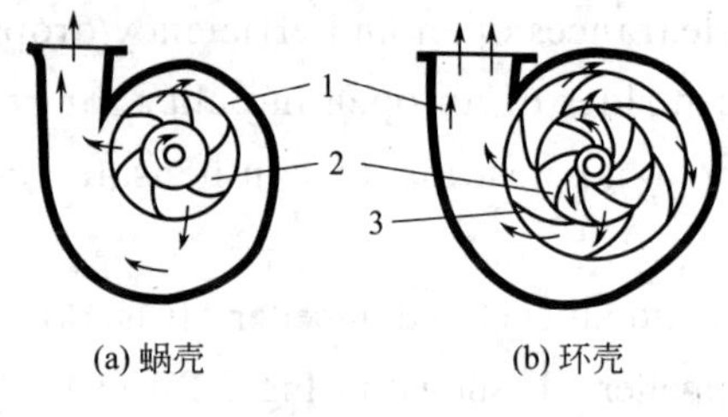

图 1-28　蜗壳与环壳
1—泵壳；2—叶轮；3—导轮

② 泵壳（casings）　泵壳通常分两种：蜗牛形的蜗壳和环壳（图 1-28）。泵壳内装有叶轮。环壳泵通常用于低压头、高流量的情况。

由于液体在蜗壳中流动时流道渐宽，所以动能降低，转化为静压能，所以说泵壳不仅是汇集由叶轮流出的液体的部件，而且是一个能量转化装置（既降低流动阻力损失，又提高流体静压能）。蜗壳泵能提供高压头。

环壳内叶轮的边缘装有固定的导叶轮，可将动能转化成静压能，通常应用于多级泵中。

Stationary Components

Casings are generally of two types：volute and circular. The impellers are fitted inside the casings.

Volute casings build a higher head；circular casings are used for low head and high capacity.

A **volute** is a curved funnel increasing in area to the discharge port as shown in Fig. 1-28. As the area of the cross-section increases，the volute reduces the speed of the liquid and increases the pressure of the liquid. One of the main purposes of a volute casing is to help balance the hydraulic pressure on the shaft of the pump. However，this occurs best at the manufacturer's recommended capacity. Running volute-style pumps at a lower capacity than the manufacturer recommends can put lateral stress on the shaft of the pump，increasing wear-and-tear on the seals and bearings，and on the shaft itself. Double-volute casings are used when the radial thrusts become significant at reduced capacities.

Circular casing have stationary diffusion vanes surrounding the impeller periphery that convert velocity energy to pressure energy. Conventionally the diffusers are applied to multi-stage pumps.

③ 轴（shaft）和轴承（bearings）　泵轴的尺寸和材料应能保证传递驱动机的全部功率。

轴承一般采用标准的滚珠轴承、滚柱轴承或滑动轴承，必要时设推力轴承。当液体温度超过117℃或轴向力较大时，轴承应进行水冷。对低温泵的滑动轴承要注意轴承间隙和材料的选取。

④ 轴封（seal）装置　轴封装置的作用是封住转轴与壳体之间的缝隙，以防止泄漏。轴封可分为填料密封和机械密封两种形式。

填料密封［填料函（stuffing box）或盘根纱］装置中的，填料采用浸油或涂石墨的石棉绳，应注意不能用干填料，不要压得过紧，允许有液体滴漏（1 滴/s），不能用于酸、碱、易燃、易爆液体的输送。

机械密封（mechanical seal）（又称端面密封），由转轴上的动环（合金硬材料）和壳体上的静环（非金属软材料）构成，两环之间形成的薄薄的液膜起密封和润滑作用。其特点是密封性好、功率消耗低，可用于酸、碱、易燃、易爆液体的输送，但价格较高。

Shaft

The basic purpose of a centrifugal pump shaft is to transmit the torques encountered when starting and during operation while supporting the impeller and other rotating parts. It must do this job with a deflection less than the minimum clearance between the rotating and stationary parts.

Seal Chamber and Stuffing Box

Seal chamber and Stuffing box both refer to a chamber, either integral with or separate from the pump case housing that forms the region between the shaft and casing where sealing media are installed. When the sealing is achieved by means of a mechanical seal, the chamber is commonly referred to as a seal chamber. When the sealing is achieved by means of packing, the chamber is referred to as a stuffing box. Both the seal chamber and the stuffing box have the primary function of protecting the pump against leakage at the point where the shaft passes out through the pump pressure casing. When the pressure at the bottom of the chamber is below atmospheric, it prevents air leakage into the pump. When the pressure is above atmospheric, the chambers prevent liquid leakage out of the pump. The seal chambers and stuffing boxes are also provided with cooling or heating arrangement for proper temperature control.

（2）离心泵工作原理

离心泵是依靠叶轮旋转时产生的离心力来输送液体的泵，其工作原理可由两个过程说明。

① 排液过程　启动离心泵以前，应首先向泵内灌满待输送液体，则泵启动后，叶轮带动液体高速旋转并产生离心力，将液体从叶片间甩出并在蜗壳体内汇集。由于壳体内流道渐大，流体的部分动能转化为静压能，则在泵的出口处，液体可获得较高的静压头而排液。

② 吸液过程　离心泵在排液过程中，当液体自叶轮中心被甩向四周后，叶轮中心处（包括泵入口）形成低压区，此时由于外界作用于储槽液面的压强大于泵吸入口处的压强而使泵内外产生足够的压强差，从而保证了液体连续不断地吸入叶轮中心。

③ 气缚现象（aerial binding phenomenon） 若泵内存在空气，由于空气的密度比液体的密度小得多，故产生的离心力不足以在叶轮中心处形成要求的低压区，导致不能吸液，这种现象叫气缚。消除气缚的方法是启动前必须向泵内灌满待输送液体，并保证离心泵的入口底阀不漏，同时防止吸入管路漏气。

(3) 离心泵的主要性能参数

为了正确合理地选择和使用离心泵，必须了解其工作的主要性能参数（the key performance）。

① 流量 Q(m^3/s)（capacity） 离心泵的流量又称送液能力，指离心泵在单位时间内排送到输出管路系统中的液体体积。流量的大小受到泵的转速、结构及尺寸的影响。

② 扬程 H(m 液柱)（head） 离心泵的扬程指泵对单位质量流体所提供的有效压头，扬程的大小受到泵的转速、结构、尺寸及送液量的影响。由于泵内流动情况复杂，目前还不能从理论上导出扬程的计算式，一般采用实验测定。应注意扬程与升扬高度不是同一概念，在特定的管路中，离心泵的扬程等于单位质量流体所需提供的有效压头，即 $H=H_e$。

③ 效率 η(efficiency) 泵轴做功并不能全部用于液体输送，有部分能量损失（$\eta<100\%$），主要包括：

a. 容积损失效率 η_v 由泵的泄漏造成。

b. 水力损失效率 η_w 由泵内的环流、流动阻力和流体冲击损失造成。

c. 机械损失效率 η_c 由机械部件间的摩擦损失造成。

$$\eta=\eta_v\eta_w\eta_c$$

η 由实验测定，一般中小型泵的效率为 50%～70%，大型泵可达到 90%。

④ 轴功率 N(brake horse power，BHP) 离心泵的轴功率是泵轴所需功率，可由功率表直接测定。N 与有效功率的关系是：

$$N=\frac{N_e}{\eta}$$

SI 制：$\left[N=\frac{\rho g Q H_e}{\eta}\right]$ W；工程单位制：$\left[N\approx\frac{\rho Q H_e}{102\eta}\right]$ kW

为了防止电机超负荷，应取（1.1～1.2）N 选电机。

Definition of important terms

The key performance parameters of centrifugal pumps are capacity head, BHP (Brake horse power), BEP (Best efficiency point) and specific speed. The pump curves provide the operating window within which these parameters can be varied for satisfactory pump operation. The following parameters or terms are discussed in detail in this section.

Capacity. Capacity means the flow rate with which liquid is moved or pushed by the pump to the desired point in the process. It is commonly measured in either liter per second (L/s) or cubic meters per hour (m^3/h). The capacity usually changes with the changes in operation of the process. For example, a boiler feed pump is an application that needs a constant pressure with varying capacities to meet a changing steam demand.

The capacity depends on a number of factors, like, process liquid characteristics, i. e. density, viscosity, size of the pump and its inlet and outlet sections, impeller size, size and shape of cavities between the vanes, impeller rotational speed (r/min), pump suction and discharge temperature and pressure conditions.

For a pump with a particular impeller running at a certain speed in a liquid, the only items on the list above that can change the amount flowing through the pump are the pressures at the pump inlet and outlet. The effect on the flow through a pump by changing the outlet pressures is graphed on a pump curve.

Head. The pressure at any point in a liquid can be thought of as being caused by a vertical column of the liquid due to its weight. The height of this column is called the static head and is expressed in terms of meter of liquid.

The same head term is used to measure the kinetic energy created by the pump. In other words, head is a measurement of the height of a liquid column that the pump could create from the kinetic energy impacted to the liquid. Imagine a pipe shooting a jet of water straight up into the air, the height the water goes up would be the head.

The head is not equivalent to press. Head is a term that has units of a length and pressure has units of force per unit area. The main reason for using head instead of pressure to measure a centrifugal pump's energy is that the pressure from a pump will change if the specific gravities (weight) of liquid changes, but the head will not change. Since any given centrifugal pump can move a lot of different fluids with different specific gravities, it is simpler to discuss the pumps head and forget about the pressure. A given pump with a given impeller and speed will raise Q liquid to a certain height regardless of the weight of the liquid.

So a centrifugal pump's performance on any Newtonian fluid, whether it's heavy (sulfuric acid) or light (gasoline) is described by using the term head. The pump performance curves are mostly described in terms of head.

(4) 离心泵的特性曲线（性能曲线）及其应用

① 离心泵特性曲线（characteristic curves）　为了解离心泵的性能，厂方以20℃的清水，在一定的转速下将离心泵的 H-Q，N-Q，η-Q 的关系由实验数据作图，图1-29中的这些曲线称为离心泵的特性曲线。

a. H-Q 曲线（扬程曲线）（head-capacity relation curve）　该曲线表明，离心泵的压头 H 与流量 Q 的对应关系。H 随着 Q 的增加逐渐下降，当 $Q=0$ 时（即指出口阀门关闭），H 也只能达到一个有限值。

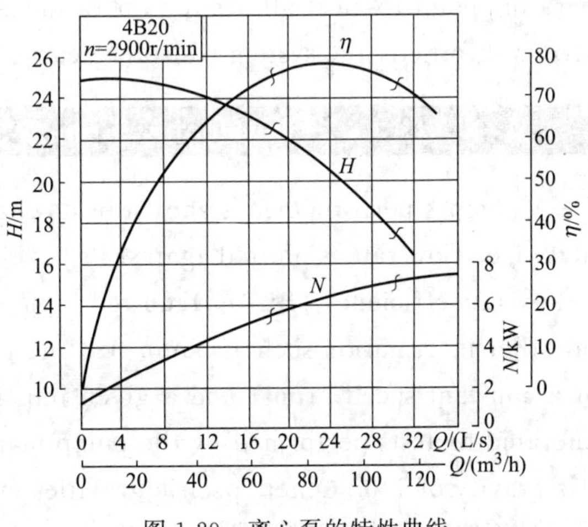

图1-29　离心泵的特性曲线

b. N-Q 曲线（功率曲线）（brake horse power-capacity relation curve）该曲线表明电机传到泵轴上的功率 N 与流量 Q 的关系。当 Q 增大时，N 平缓上升；当 $Q=0$ 时，N 最小。因此，离心泵在零流量下启动（即闭路启动），是为了降低启动功率，保护电机。

c. η-Q 曲线（效率曲线）（efficiency-capacity relation curve） 该曲线反映了离心泵的总效率 η 与流量 Q 的关系。曲线上的每一个点表示泵在某一操作情况下（H，Q 一定）的工作效率。由图 1-29 可知，开始时，η 随着 Q 的增加而增加；达到最大值 η_{max} 后，η 则随 Q 的增加而减小。这说明，离心泵在一定的转速下，有一最高效率点 η_{max}。在最高效率点下操作，泵内的压头损失最小，泵的设计应以此点为设计点。标在泵的铭牌上对应于最高效率点下的流量、压头、功率称为额定值。

对于选泵和操作，应在 $\eta \geqslant 92\% \eta_{max}$ 的左右区域（称为高效率区）考虑，这样比较经济适用。

Characteristic curves, head-capacity relation

Before you select a pump model, examine its performance curve, which is indicated by its head - flow rate or operating curve. The curve shows the pump's capacity plotted against total developed head. It also shows efficiency, required power input, and suction head requirements (net positive suction head requirement) over a range of flow rates. The plots of actual head, total power consumption, and efficiency vs. volumetric flow rate are called the characteristic curves of a pump, shown in Fig. 1-29. For publication purpose, it is more convenient to draw several curves in a single graph. This presentation method shows a number of Head - Capacity curves for one speed and several impeller diameters or one impeller diameter and several different speeds for the same pump.

The capacity and pressure needs of any system can be defined with the help of a graph called a system curve. Similarly the capacity *vs* pressure variation graph for a particular pump defines its characteristic pump performance curve.

The pump suppliers try to match the system curve supplied by the user with a pump curve that satisfies these needs as closely as possible. A pumping system operates where the pump curve and the system resistance curve intersect. The intersection of the two curves defines the operating point of both pump and process, however, it is impossible for one operating point to meet all desired operating conditions. For example, when the discharge valve is throttled, the system resistance curve shift left and so does the operating point.

Developing a pump performance curve

A pump's performance is shown in its characteristics performance curve where its capacity i. e. flow rate is plotted against its developed head. The pump performance curve also shows its efficiency (BEP), required input power (in BHP), NPSH, speed (in RPM), and other information such as pump size and type, impeller size, etc. This curve is plotted for a constant speed (rpm) and a given impeller diameter (or series of diameters). It is generated by tests performed by the pump manufacturer. Pump curves are based on a specific gravity of 1. 0. Other specific gravities must be considered by the user.

② 液体性质对离心泵特性的影响

a. 密度 ρ 的影响 离心泵的流量、压头、效率均与液体密度无关，当被输送液体密度发生变化时，H-Q 与 η-Q 曲线不变，但泵的轴功率与液体密度成正比，轴功率应用下式进行校正。

$$N'=N\frac{\rho'}{\rho} \tag{1-52}$$

式中，ρ 为 20℃清水的密度，kg/m^3；ρ'为工作流体的密度，kg/m^3。

b. 黏度 μ 的影响 若被输送液体的黏度大于常温下清水的黏度，则泵内液体的能量损失增大，因此泵的压头、流量都要减小，效率下降，而轴功率增大，即会导致离心泵的特性曲线发生改变。

但由于黏度变化对离心泵特性参数的影响较复杂，所以无准确的计算方法，通常采用式(1-53) 进行修正。

$$\begin{aligned} Q'&=C_Q Q \\ H'&=C_H H \\ \eta'&=C_\eta \eta \end{aligned} \tag{1-53}$$

式中 C_Q，C_H，C_η——修正系数，可查有关手册获得。对于运动黏度 $v<20$cst（1cst$=10^{-6}m^2/s$）的液体，例如汽油、煤油、柴油等，μ 的影响可以忽略。

③ 转速 n 和叶轮直径 D 对离心泵特性的影响

a. 转速 n 的影响 当 $\left|\frac{n'-n}{n}\right|\leqslant 20\%$时，离心泵效率 η 基本不变。

对同一台泵，设 n 为原转速，n'为新转速，则可以导出：

$$\frac{Q'}{Q}=\frac{n'}{n};\quad \frac{H'}{H}=\left(\frac{n'}{n}\right)^2;\quad \frac{N'}{N}=\left(\frac{n'}{n}\right)^3 \tag{1-54}$$

上述关系称为比例定律。

b. 叶轮直径 D 的影响 对原叶轮直径 D 切割（$D\rightarrow D'$），当 $\left|\frac{D'-D}{D}\right|$ 较小时，离心泵效率 η 基本不变，速度三角形基本相似，且叶轮出口流通截面大致不变，可导出：

$$\frac{Q'}{Q}=\frac{D'}{D};\quad \frac{H'}{H}=\left(\frac{D'}{D}\right)^2;\quad \frac{N'}{N}=\left(\frac{D'}{D}\right)^3 \tag{1-55}$$

上述关系称为切割定律。应注意，切割尺寸太大时，上述关系不成立。

对于几何相似（几何相似指结构尺寸对应成比例）的同一系列泵，且速度三角形相似，可导出：

$$\frac{H'}{H}=\left(\frac{D'}{D}\right)^2;\quad \frac{Q'}{Q}=\left(\frac{D'}{D}\right)^3;\quad \frac{N'}{N}=\left(\frac{D'}{D}\right)^5 \tag{1-56}$$

(5) 离心泵的安装高度（吸上高度）H_g

如图 1-30 所示，H_g 指离心泵入口中心线与吸入槽液面之间的垂直距离。若离心泵在液面之上，H_g 为正值；离心泵在液面之下，H_g 为负值（称为倒灌）。

在工业生产中，离心泵的安装高度不是任意的，是有限制的，而这个限制就是因为离心泵在工作中容易产生汽蚀现象。

① 汽蚀现象（cavitation phenomenons） 离心泵运转时，在叶片入口附近的 K 点（入口叶片的背面）静压头最低，若：

$$p_k\leqslant p_v$$

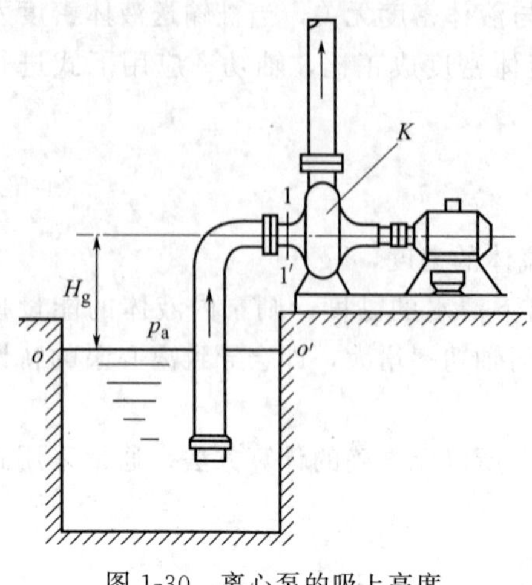

图 1-30　离心泵的吸上高度

式中，p_k 为泵内的最低压强，Pa；p_v 为输送温度下液体的饱和蒸气压，Pa，$p_v=f(t)$。

液体在压强最低点处发生部分汽化，汽泡随同液体从低压区（叶片入口）进入高压区（叶片出口）的过程中，在高压的作用下，汽泡迅速凝结。汽泡的消失产生了局部真空，此时周围的液体以极高的速度和频率冲击原汽泡空间，产生非常大的冲压强，造成对叶轮和壳体的冲击，使其震动并发出噪声，这种现象叫汽蚀。汽蚀现象发生时，传递到叶轮和壳体上的冲击力再加上液体中微量溶解氧释出时对金属的化学腐蚀的共同作用，在一定的时间后，可使其表面出现斑痕和裂缝，甚至呈海绵状逐渐脱落，有的还出现穿孔。

顺便指出，汽蚀现象不仅在离心泵等水利机械中存在，在流量计、阀门、管道及内燃机汽缸的冷却水套壁面上也会发生汽蚀现象。可以毫不夸张地说，凡是与液体流动有关的各种设备中都有可能存在汽蚀问题。

离心泵在汽蚀条件下运转，泵体震动剧烈，发出噪声，流量、压头和效率都明显下降，严重时会吸不上液体。为了避免汽蚀现象发生，保证离心泵正常运转，必要的条件是：

$$p_k > p_v$$

但是，p_k 很难测出，易测定的是泵入口处的绝对压强 p_1，显然应 $p_1 > p_k$。

② 安装高度 H_g 计算　如图 1-30 所示，在吸入槽液面 0—0′与泵入口处 1—1′两截面间列柏努利方程：

$$0+0+\frac{p_0}{\rho g}=H_g+\frac{u_1^2}{2g}+\frac{p_1}{\rho g}+H_{f,0-1}$$

式中　p_0——0—0′截面的绝对压强，Pa

p_1——1—1′截面的绝对压强，Pa。

则安装高度计算式为：

$$H_g=\frac{p_0}{\rho g}-\frac{p_1}{\rho g}-\frac{u_1^2}{2g}-H_{f,0-1} \tag{1-57}$$

$p_k=p_v$ 时，$p_1=(p_1)_{min} \Rightarrow H_g=H_{g,max}$⇒刚发生气蚀；

$p_k>p_v$ 时，$p_1>(p_1)_{min} \Rightarrow H_g<H_{g,max}$⇒泵能正常工作；

$p_k<p_v$ 时，$p_1<(p_1)_{min} \Rightarrow H_g>H_{g,max}$⇒泵可能被损坏。

为了确定离心泵的允许安装高度，需要强调指出的是离心泵的吸上性能（又称抗汽蚀能力）一般不直接用泵吸入口处的最低压强表示，而是将其转换成相应的其他参数表达。

国内离心泵常用两种指标表示泵的吸上性能，它们的数值均可由实验测定。

a. 离心泵的允许汽蚀余量（net positive suction head，NPSH）　为了防止汽蚀现象发生，在离心泵的入口处规定液体的静压头与动压头之和 $\left(\frac{p_1}{\rho g}+\frac{u_1^2}{2g}\right)$ 必须大于操作温度下液体的饱和蒸汽压头 $\left(\frac{p_v}{\rho g}\right)$ 某一最小值，此最小值称为离心泵的允许汽蚀余量，即：

$$NPSH=\frac{p_1}{\rho g}+\frac{u_1^2}{2g}-\frac{p_v}{\rho g} \tag{1-58}$$

式中 NPSH——离心泵的汽蚀余量，对油泵也可用符号 Δh 表示，m 。

则：
$$H_g=\frac{p_0}{\rho g}-\frac{p_v}{\rho g}-NPSH-H_{f,0-1} \tag{1-59}$$

$$NPSH=f(\text{离心泵结构、尺寸、转速、流量})$$

[讨论]：

i. 离心泵的允许汽蚀余量 NPSH 值是在大气压为 $10mH_2O$ 条件下，以 20℃清水为介质由厂方通过实验测定，当输送其他液体时应予修正，但因一般校正系数小于 1，故通常将它作为外加的安全因素，不再校正；

ii. 由于该值随离心泵流量的增大而增大，因此计算 H_g 时应取最高流量下对应的 NPSH 值。

b. 离心泵的允许吸上真空度 H_s 将泵入口处允许的最低绝压 p_1 用相应的真空度表示，即：

$$H_s=\frac{p_a}{\rho g}-\frac{p_1}{\rho g} \tag{1-60}$$

则：
$$H_g=\frac{p_0}{\rho g}-\frac{p_a}{\rho g}+H_s-\frac{u_1^2}{2g}-H_{f,0-1} \tag{1-61}$$

$$H_s=f(\text{泵结构、尺寸、转速、流量、液体的物性、当地大气压})$$

[讨论]：

i. 用 H_s 计算离心泵的安装高度时，要对 H_s 进行选择及修正。由于泵说明书中提供的 H_s 也是在大气压为 $10mH_2O$ 下，以 20℃清水为介质测出的，所以当实际输送清水的操作条件与测试条件不同，或输送其他流体时，应对 H_s 值校正。

$$H_s'=\left[H_s+\left(\frac{p_a}{\rho g}-10\right)-\left(\frac{p_v}{9.81\times10^3}-0.24\right)\right]\frac{1000}{\rho}$$

式中 H_s'——操作条件下输送液体的允许吸上真空度，m 液柱；

H_s——实验条件下输送水时的允许吸上真空度，即在水泵性能表上查得的数值，mH_2O；

p_v——操作条件下输送液体的饱和蒸气压，Pa；

ρ——操作条件下输送液体的密度，kg/m^3。

ii. 因 H_s 随离心泵流量的提高而降低，因此在确定离心泵安装高度时也应使用最大流量下对应的 H_s 数值来计算。

NPSH 与 H_s 之间存在如下关系：

$$NPSH\approx10-H_s \tag{1-62}$$

由于 NPSH 不必修正，故在某离心泵说明书中提供了 H_s 数据后，可按上式把 H_s 转换成 NPSH，再按式(1-59) 计算安装高度。

当储槽液面上方为敞口时，即 $p_0=p_a$ 时，安装高度 H_g 计算可采用如下两式：

$$H_g=\frac{p_a}{\rho g}-\frac{p_v}{\rho g}-NPSH-H_{f,0-1} \tag{1-63}$$

$$H_g = H_s - \frac{u_1^2}{2g} - H_{f,0-1} \tag{1-64}$$

为了确保离心泵正常工作，将实验测定的极限值考虑安全值后作为允许值。

极限值		安全值		允许值
$NPSH_{min}$	+	0.5m	=	$NPSH_{允许}$
$H_{s,max}$	−	0.5m	=	$H_{s,允许}$
$H_{g,max}$	−	0.5m	=	$H_{g,允许}$

【例 1-18】 型号为 IS65-40-200 的离心泵，转速为 2900r/min，流量为 $25m^3/h$，扬程为 50m，允许汽蚀余量为 2.0m，此泵用于将敞口水池中 50℃的水送出，已知吸入管路的总阻力损失为 $2mH_2O$，当地大气压为 100kPa，求泵的允许安装高度。

解 由附录 10 查得 50℃的水的饱和蒸气压为 12.31kPa，水的密度为 998.1 kg/m^3，由式（1-59）可得：

$$H_g = \frac{p_0 - p_v}{\rho g} - NPSH - H_{f,0-1} = \frac{100\times10^3 - 12.31\times10^3}{998.1\times9.81} - 2.0 - 2.0 = 4.96(m)$$

故泵的实际安装高度不应超过液面的 4.96m。

【例 1-19】 用某 Y 型油泵送低位敞口水槽中 35℃的清水，水的流量为 $50.4m^3/h$，当地大气压为 $9.75mH_2O$（绝压）。查得允许吸上真空度为 6.8m，吸入管路阻力为 1.2m，吸入管内径为 80mm，泵安装在比低位水槽液面高 5.4m 处，问该泵能否正常工作？

解 由附录 10 查得 35℃水的饱和蒸气压 p_v=5623.44Pa，密度为 $994kg/m^3$，则：

$$H'_s = \left(H_s + \frac{P_a}{\rho g} - 10 - \frac{p_v}{\rho g} + 0.24\right)\times\frac{1000}{994}$$
$$= \left(6.8 + 9.75 - 10 - \frac{5623.44}{1000\times9.81} + 0.24\right)\times\frac{1000}{994}$$
$$= 6.25\ (m)$$

$$u_1 = \frac{V}{\frac{\pi}{4}d^2} = \frac{\frac{50.4}{3600}}{\frac{\pi\times0.08^2}{4}} = 2.79(m/s)$$

则：
$$H_g = H'_s - \frac{u_1^2}{2g} - \Sigma H_{f,0-1} = 6.25 - \frac{2.79^2}{2\times9.81} - 1.2 = 4.65(m)$$

由计算结果可知，该泵将发生汽蚀现象，不能正常工作，必须将安装高度调至低于 4.65m。

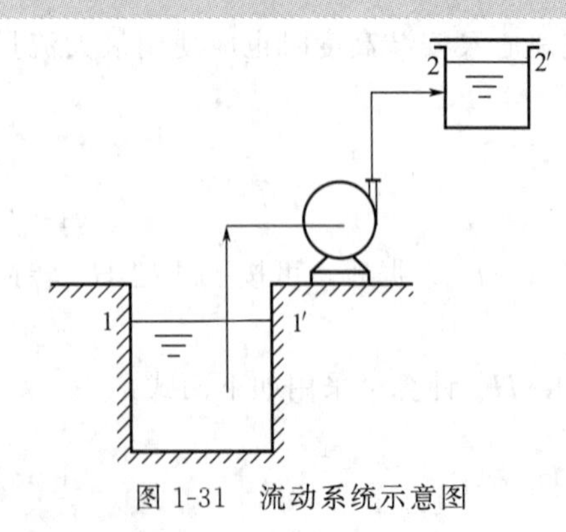

图 1-31 流动系统示意图

（6）离心泵的工作点（operating point）与流量调节（flow rate changing）

① 管路特性曲线（system-head curve） 如图 1-31 所示，在 1—1′与 2—2′截面间列柏努利方程：

$$H_e = (Z_2 - Z_1) + \frac{p_2 - p_1}{\rho g} + H_{f,1-2} = \Delta Z + \frac{\Delta p}{\rho g} + H_{f,1-2}$$

其中，对于特定管路系统而言，ΔZ、Δp 不变，则：

$$\Delta Z+\frac{\Delta p}{\rho g}=K=常数$$

$$H_{f,1-2}=\lambda\frac{l+\sum l_e}{d}\times\frac{(4Q_e/\pi d^2)^2}{2g}$$

$$=\frac{8\lambda}{\pi^2 g}\times\frac{l+\sum l_e}{d^5}Q_e^2$$

上式中除了 λ、Q_e 外均为常数，且：

$$\lambda=\left(\frac{du\rho}{\mu},\frac{\varepsilon}{d}\right)=f(Q_e)$$

$$H_e=K+f(Q_e)$$

在阻力平方区，λ 为常数：

$$H_e=K+BQ_e^2$$

对于上述特定管路系统，驱使流体通过管路所需外加压头 H_e 与流量 Q_e 之间的关系曲线叫管路特性曲线。由以上分析可知，管路特性曲线与泵无关，它是在保持管路系统阀门开度不变的情况下，变化 Q_e 作出的，它与泵的特性无关。

② 离心泵的工作点（operating point）　上述分析可知，离心泵本身有其固有的特性，它与管路特性无关；而管路本身也有其固有的特性，它与泵的特性无关。但是，若将管路特性曲线 H_e-Q_e 与离心泵特性曲线 H-Q 绘在同一个坐标图上，如图 1-32 所示，则两条特性曲线的交点 M 即为离心泵的工作点。

离心泵工作点的含义是：一旦离心泵安装在某一特定的管路上，并在一定的操作条件下工作时，泵所提供压头 H 与管路系统所需要的压头 H_e 应相等；泵所排出的流量 Q 与管路系统输送的流量 Q_e 应相等，这时泵装置处于稳定的工作状态。即离心泵在 M 点工作时，$H=H_e$，$Q=Q_e$。

同时说明，当泵型及其转速、管路特性与操作条件给定时，离心泵稳定运行，只有一个工作点。

泵在操作时，工作点应落在高效区；若泵不正常操作，如发生汽蚀现象，泵的性能迅速恶化，可能为刚发生汽蚀时的工作点。

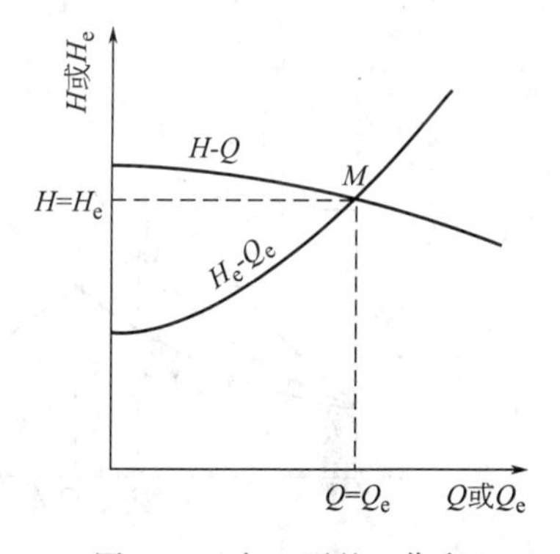

图 1-32　离心泵的工作点

Operating Point

The pump suppliers try to match the system curve supplied by the user with a pump curve that satisfies these needs as closely as possible. A pumping system operates where the pump curve and the system resistance curve intersect. The intersection of the two curves defines the operating point of both pump and process. However, it is impossible for one operating point to meet all desired operating conditions.

A system-head curve can be superimposed on the head-capacity curve of a centrifugal pump. The point of intersection between the characteristic curve of centrifugal pump and system head curve of pump flow system is called operating point. The point M in Fig. 1-32 is the operating point for a given pump at the specific speed.

③ 离心泵的流量调节　离心泵安装在管路上工作，当其工作点对应的流量与生产任务所需的流量不相符合时，就需要进行流量调节。流量调节的实质是通过改变泵的工作点来实现的。由图 1-33 可知，改变管路特性曲线或者改变泵的特性曲线均能使工作点移动，从而达到调节流量的目的。

a. 改变管路特性曲线（change the system-head curve）　改变管路特性曲线最常用的方法是调节离心泵出口阀门的开度。

$$H_e=\Delta Z+\frac{\Delta p}{\rho g}+\lambda\frac{l+\sum l_e}{d}\times\frac{(4Q_e/\pi d^2)^2}{2g}$$

由上式可知：假设 λ = 常数：$H_e=K+BQ_e^2$。

如图 1-33 所示，当改变阀门开度时，由于阀门处局部阻力的改变而导致 $\sum l_e$ 变化，从而使 B 值发生变化。即阀门关小，B 增大，管路特性曲线变陡，工作点由 M 点变化到 M_1 点；阀门开大，B 减小，使管路特性曲线变平坦，工作点由 M 点变化到 M_2 点，即由于工作点沿泵的特性曲线移动位置，从而调节了流量。

该方法的优点简捷方便，适用于经常性调节，广泛使用；缺点是不经济，阀门关小时，压头增大，能量损失增大，且调节幅度大时，工作点易偏离泵的高效率区。

b. 改变泵的特性曲线（change the centrifugal pump characteristic curve）

i. 改变转速 n　设原工况转速为 11，新工况转速为 n_1，由比例定律知：

$$\frac{Q_1}{Q}=\frac{n_1}{n},\ \frac{H_1}{H}=\left(\frac{n_1}{n}\right)^2,\ \frac{N_1}{N}=\left(\frac{n_1}{n}\right)^3$$

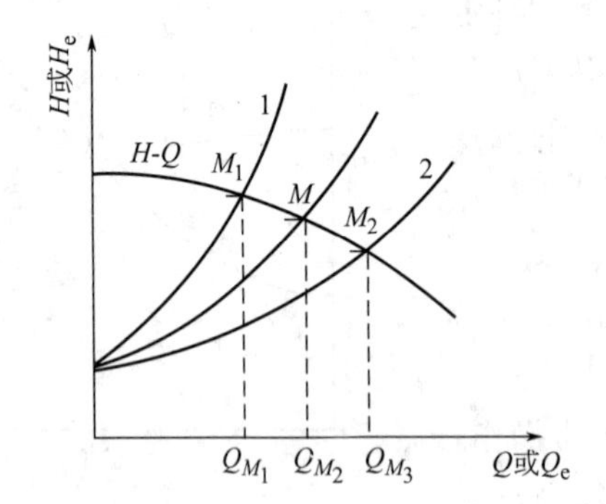

图 1-33　改变管路特性曲线

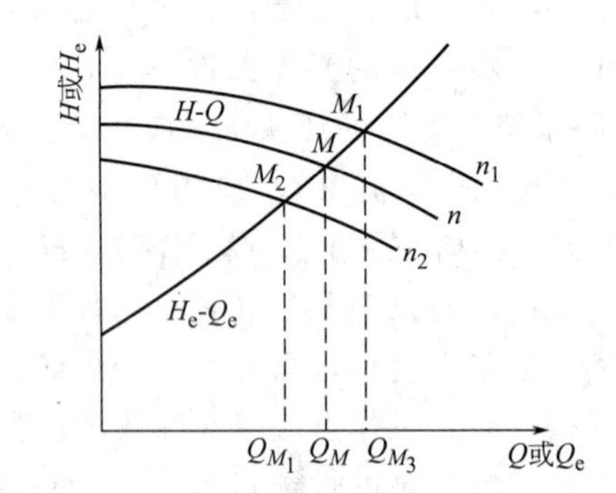

图 1-34　改变转速以改变泵的特性曲线

依据比例定律可作出新工况下泵的性能曲线。由图 1-34 可知，转速变化后，工作点沿着管路特性曲线移动，从而对应于新的流量和压头，其适用条件是离心泵的转速变化不大于 ±20%。

ii. 切削叶轮直径 D　设原工况叶轮直径为 D，新工况叶轮直径为 D_1，由切割定律知：

$$\frac{Q_1}{Q}=\frac{D_1}{D},\ \frac{H_1}{H}=\left(\frac{D_1}{D}\right)^2,\ \frac{N_1}{N}=\left(\frac{D_1}{D}\right)^3$$

与转速变化的情况类似，以切割定律代替比例定律，可作出新工况下的性能曲线，从而看出工作点的变化。其适用条件是固定转速下，叶轮直径的车削不大于 ±5%D。

改变转速或切割叶轮直径，均属于改变泵的特性曲线调节流量。其优点是不额外增加管路阻力，在一定的范围内可保持泵在高效率区工作；缺点是调节不方便，一般在季节性调节使用。调节范围较大时，会降低泵的效率。

Operating Point change

The point at which a pump operates in a given piping system depends on the flow rate and head loss of that system. For a given system, volumetric flow rate is compared to system head loss on a system characteristic curve. By graphing a system characteristic curve and the pump characteristic curve on the same coordinate system, the point at which the pump must operate is identified. For example, in Fig. 1-11, the operating point for the centrifugal pump in the original system is designated by the intersection of the pump curve and the system curve. Several methods can be used to change operating point,

◇ Change the system-head curve

◇ Change the centrifugal pump characteristic curve

It is convenient to change the system-head curve by a valve in discharge. The operating point can be changed as adjusting the opening of valve (Fig. 1-33). For example, when the discharge valve is throttled, the system resistance curve shift left and so does the operating point. The operating point can also be changed as the impeller diameter or revolution speed of centrifugal pump is changed (Fig. 1-34).

Normal Operating Range

A typical performance curve is a plot of Total Head *vs* Flow rate for a specific impeller diameter. The plot starts at zero flow. The head at this point corresponds to the shut-off head point of the pump. The curve then decreases to a point where the flow is in maximum and the head minimum. This point is sometimes called the run-out point. The pump curve is relatively flat and the head decreases gradually as the flow increases. This pattern is common for radial flow pumps. Beyond the run-out point, the pump cannot operate. The pump's range of operation is from the shut-off head point to the run-out point. Trying to run a pump off the right end of the curve will result in pump cavitation and eventually destroy the pump.

(7) 离心泵的联合操作（operation in parallel and in series of centrifugal pump）

在许多情况下，单泵可以送液，只是流量达不到要求，此时，可针对管路的特性选择适当的离心泵组合方式，以增大流量。以两台相同型号的离心泵为例说明。

① 并联（in parallel） 如图1-35所示，两泵性能相同，并联后的合成曲线，可由单泵性能曲线Ⅰ在相等的压头 H 下将流量 Q 加倍，描点做出合成曲线Ⅱ。由图中工作点的位置变化可知，单泵独立工作的送液量＜两泵并联工作的流量＜两泵独立工作的流量之和；并联时的工作点高于单泵独立的工作点；并联后的总效率与单泵的效率相同。

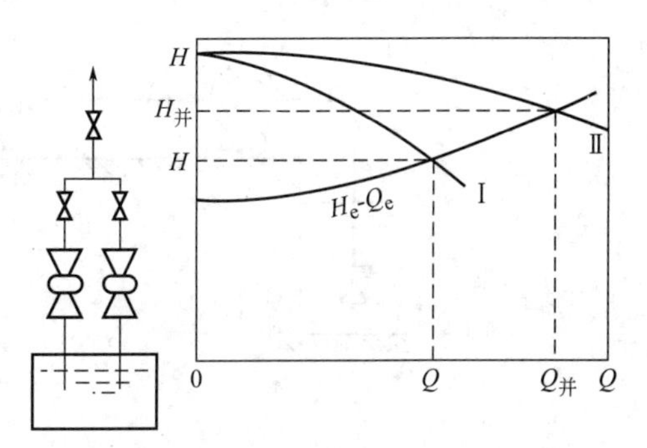

图1-35 两台同型泵的并联操作

Operation in Parallel and in Series of Centrifugal Pump

Centrifugal pumps are typically small in size and can usually be built for a relatively low cost. In addition, centrifugal pumps provide a high volumetric flow rate with a relatively low pressure. In order to increase the volumetric flow rate in a system or to compensate for large flow resistances, centrifugal pumps are often used in parallel or in series.

Two pumps operating in parallel

Two pumps or more discharging into a common line are said to operate in parallel. Fig. 1-35 depicts two identical centrifugal pumps operating at the same speed in parallel. Since the inlet and the outlet of each pump shown in Fig. 1-35 are at identical points in the system, each pump must produce the same pump head. The total flow rate in the system, however, is the sum of the individual flow rates for each pump. For a specific system-head curve, two identical pumps can be selected to discharge, double the capacity of each at the same head. The performance curve for two identical pumps operating in parallel is in Fig. 1-35.

When the system characteristic curve is considered with the curve for pumps in parallel, the operating point at the intersection of the two curves represents a higher volumetric flow rate than for a single pump and a greater system head loss. As shown in Fig. 1-36, a greater system head loss occurs with the increased fluid velocity resulting from the increased volumetric flow rate. Because of the greater system head, the volumetric flow rate is actually less than twice the flow rate achieved by using a single pump.

Two pumps operating in parallel will be more effective with a flatter system head curve. Two pumps or more in parallel should be selected in case one by itself can not perform as requirement for, or certain flexibility is necessary. Otherwise one pump by itself with a spare is, in most cases, the best selection.

② 串联（In Series） 如图 1-36 所示，两泵性能相同，串联后的合成曲线Ⅱ，可由单泵性能曲线Ⅰ在相等的流量 Q 下将压头 H 加倍描点作出。由图中工作点的位置变化可知，单泵独立工作的流量<两泵串联工作的流量<两泵独立工作的流量之和；串联时的工作点高于

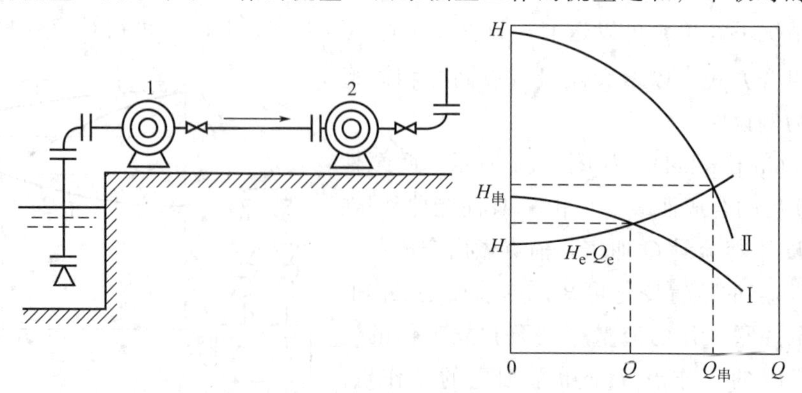

图 1-36 两台同型泵的串联操作

单泵独立的工作点；串联后的总效率与单泵的效率相同。

Two Pumps Operating in Series

Two pumps or more are said to operate in series when the discharge of the first pump serves as suction for the second pump, and the discharge of the second pump as suction for the third one, etc. For a specific system-head curve, two identical pumps can be selected to double the head of the first pump at the same capacity. The performance curve for two identical pumps operating in series is in Fig. 1-36.

Centrifugal pumps are used in series to overcome a larger system head loss than one pump can compensate for individually. As illustrated in Fig. 1-36. Two identical centrifugal pumps operating at the same speed with the same volumetric flow rate contribute the same pump head. Since the inlet to the second pump is the outlet of the first pump, the head produced by both pumps is the sum of the individual heads. The volumetric flow rate from the inlet of the first pump to the outlet of the second remains the same.

As shown in Fig. 1-36, using two pumps in series does not actually double the resistance to flow in the system. The two pumps provide adequate pump head for the new system and also maintain a slightly higher volumetric flow ratc. The operation of two pumps in series a steeper system curve will be more effective than with a flatter system curve.

③ 组合方式的选择　就增大流量而言，可根据管路特性曲线选择组合方式。如图 1-37 所示，对于低阻管路，并联的 H、Q 都高于串联的；而高阻管路，串联的 H、Q 都高于并联的。因此，对低阻管路而言，并联优于串联；对高阻管路而言，串联优于并联。但应注意，若特定管路的 $\Delta Z+\dfrac{\Delta p}{\rho g}=K>$ 大于单泵所能提供的最大压头，则必须采用串联组合操作。

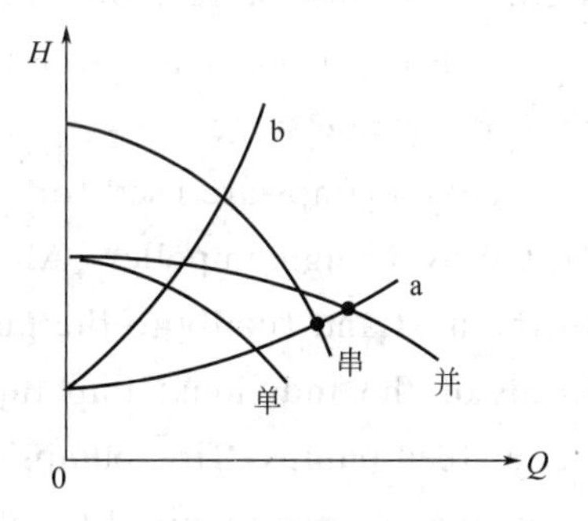

图 1-37　组合方式的选择

(8) 离心泵的类型、型号与选择

① 离心泵的类型　由于过程工业被输送流体的性质、压强、流量差异很大，为了适应各种不同的要求，离心泵的类型是多种多样的。按泵送液体的性质分类：

a. 清水泵（clean water pump）　系列代号为 IS 型、D 型、Sh 型，适用于清水及物化性质类似于清水的液体输送。

i. IS　国际标准单级单吸清水离心泵，全系列扬程范围为 8～98m，流量范围为 4.5～360m^3/h；

ii. D 型　如图 1-38 所示，若所要求的扬程较高而流量不太大时，可采用 D 型多级离心泵。国产多级离心泵的叶轮级数通常为 2～9 级，最多 12 级，全系列扬程范围为 14～351m，流量范围为 10.8～850m^3/h；

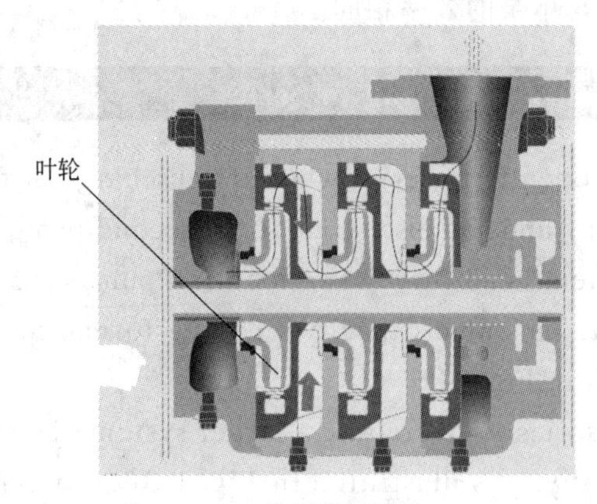

(a) 原理示意图

(b) 设备图

图 1-38　多级离心泵

Multistage Centrifugal Pumps

The maximum head that is practicable to generate in a single impeller is limited by the peripheral speed reasonably attainable. A so-called high-energy centrifugal pump can develop a head of more than 200 m in a single stage, but generally when a head greater than about 60 m is needed, two or more impellers can be mounted in series on a single shaft and a multistage pump so obtained. The discharge from the first stage provides suction for the second, the discharge from the second provides suction for the third, and so forth, as shown in Fig. 1-38. The developed heads of all stages add to give a total head several times that of a single stage.

These pumps are used for services requiring heads (pressures) higher than can be generated by a single impeller. All impellers are in series, the liquid passing from one impeller to the next and finally to the pump discharge. The total head then is the summation of the heads of the individual impellers. Deep-well pumps, high-pressure water-supply pumps, boiler-feed pumps, fire pumps and charge pumps for refinery processes are examples of multistage pumps required for various services.

Multistage pumps may be of the volute type with singe-or double-suction impellers, or of the diffuser type. They may have horizontally split casings or for extremely high pressure, 20 to 40 MPa, vertically split barrel-type exterior casings with inner casings containing diffusers, interstage passages, etc.

iii. Sh　若泵送液体的流量较大而所需扬程并不高时，则可采用双吸离心泵。国产双吸泵系列代号为 Sh，全系列扬程范围为 9～140m，流量范围为 120～12500m^3/h。

在离心泵的产品目录或产品样本中，泵的型号是由字母和数字组合而成，以代表泵的类型、规格等，现举例说明如下：

例如：IS50-32-250

其中　IS——国际标准单级单吸清水离心泵；

50——泵吸入口直径，mm；

32——泵排出口直径，mm；

250——泵叶轮的尺寸，mm。

b. 耐腐蚀泵（corrosion-resistant pump） 当输送酸、碱及浓氨水等腐蚀性液体时应采用防腐蚀泵。该类泵中所有与腐蚀液体接触的部件都用抗腐蚀材料制造，其系列代号为F。F型泵多采用机械密封装置，以保证高度密封要求。F泵全系列扬程范围为15～105m，流量范围为2～400m^3/h。近年来已推出新型号，例如IH型等。

例如：40FM1-26

其中 40——泵吸入口直径，mm；

F——系列代号；

M——与液体接触的材料代号（铬、镍、钼、钛合金）；

1——轴封形式代号（1代表单端面密封）；

26——泵的扬程，m。

c. 油泵（oil pump） 输送石油产品的泵称为油泵。因为油品易燃易爆，因而要求油泵有良好的密封性能。当输送高温油品（200℃以上）时，需采用具有冷却措施的高温泵。油泵有单吸与双吸、单级与多级之分。国产油泵系列代号为Y、双吸式为YS。全系列的扬程范围为60～603m，流量范围为6.25～500m^3/h。近年来已推出新型号，例如SJA型等。

例如：100Y-120×2

其中 100——泵吸入口直径，mm；

Y——系列代号；

120——泵的单级扬程，m；

2——叶轮级数。

d. 杂质泵（trash pump） 用于输送悬浮液及稠厚的浆液时用杂质泵，这类泵的特点是叶轮流道宽、叶片数目少、常采用半闭式或开式叶轮，泵的效率低。

系列代号为P型，分为污水泵（PW）、砂泵（PS）、泥浆泵（PN）。

e. 磁力泵（magnetic driving pump）（图1-39） 磁力泵是高效节能的特种离心泵，采用永磁联轴驱动，无轴封，消除液体渗漏，使用极为安全；在泵运转时无摩擦，故可节能。磁力泵主要用于输送不含固体颗粒的酸、碱、盐溶液和挥发性、剧毒性液体等，特别适用于易燃、易爆液体的输送。磁力泵输送介质是密度不大于1300 kg/m^3，黏度不大于30×10^{-6} Pa·s的不含铁磁性和纤维的液体。常规磁力泵的额定温度对于泵体为金属材质或F46衬里，最

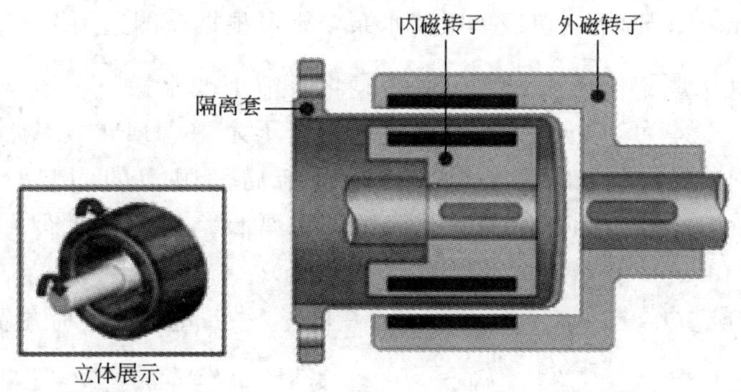

图1-39 磁力泵

高工作温度为 80℃，额定压力为 1.6MPa；高温磁力泵的使用温度≤350℃；泵体为非金属材质，最高温度不超过 60℃，额定压力为 0.6MPa。

对于输送介质是密度大于 1600 kg/m^3 的液体，磁性联轴器需另行设计。磁力泵的轴承采用被输送的介质进行润滑冷却，严禁空载运行。

f. 屏蔽泵（canned motor pump） 近年来，输送易燃、易爆、剧毒及具有放射性的液体时，常采用一种无泄漏的屏蔽泵（图 1-40）。其结构特点是叶轮和电机联为一个整体封在同一泵壳内，不需要轴封装置，又称无密封泵。屏蔽泵在启动时应严格遵守出口阀和入口阀的开启顺序，停泵时先将出口阀关小，当泵运转停止后，先关闭入口阀再关闭出口阀。总之，采用屏蔽泵，完全无泄漏，有效地避免了环境污染和物料损失，只要选型正确，操作条件没有异常变化，在正常运行情况下，几乎没有什么维修工作量。屏蔽泵是输送易燃、易爆、腐蚀、贵重液体的理想用泵。

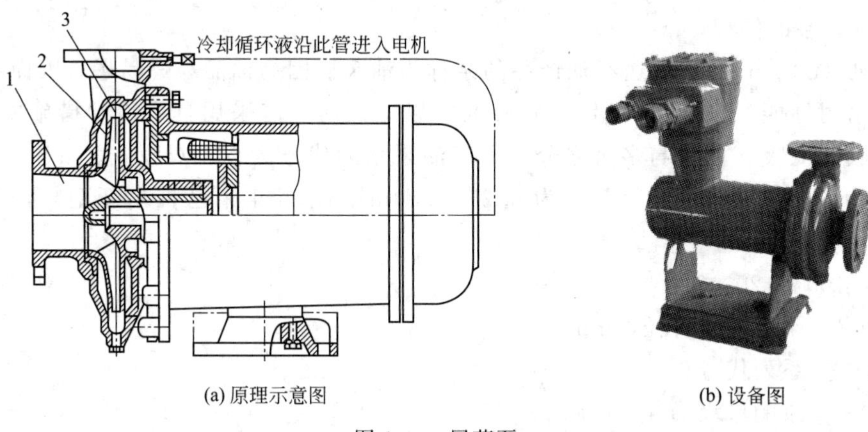

(a) 原理示意图　　(b) 设备图

图 1-40　屏蔽泵

1—吸入口；2—叶轮；3—集液室

② 离心泵的选择　首先，根据被输送液体的性质和操作条件，确定泵的类型；然后，由生产任务 Q_e，根据输送管路系统，计算所需压头 H_e；最后，由 Q_e 和 H_e 查出泵的样本，选择泵的型号。一般要求为：所选泵的 Q、H 应分别比 Q_e、H_e 稍大一些；几台不同型号的泵同时满足要求，应选效率高的；同时也应参考泵的价格。

现以图 1-41 为例，介绍 IS 型水泵系列特性曲线：此图分别以 H 和 Q 为纵、横坐标绘制，图中每一小块面积，表示某型号离心泵的最佳（即最高效率区）的工作范围。利用此图，根据管路要求的流量和压头，可方便地决定泵的具体型号。例如，当输送水时，要求 $H=45$m，$Q=10m^3/h$，选用一单级单吸清水泵，则可根据该图选用 IS50-32-200 离心泵。

【例 1-20】 如本题附图所示，需用离心泵将水池中的水送至密闭高位槽中，高位槽液面与水池液面高度差为 15m，高位槽中的气相表压为 49.1kPa。要求水的流量为 15～25m^3/h，吸入管长 24m，压出管长 60m（均包括局部阻力的当量长度），管子均为 ϕ68mm×4mm，摩擦系数可取 0.021。试选用一台离心泵，并确定安装高度（设水温为 20℃，密度为 1000kg/m^3，当地大气压为 101.3kPa）。

解　以最大流量 $Q=25\ m^3/h$ 计算，在 1—1′与 2—2′间列柏努利方程

$$z_1+\frac{u_1^2}{2g}+\frac{p_1}{\rho g}+H_e=z_2+\frac{u_2^2}{2g}+\frac{p_2}{\rho g}+H_f$$

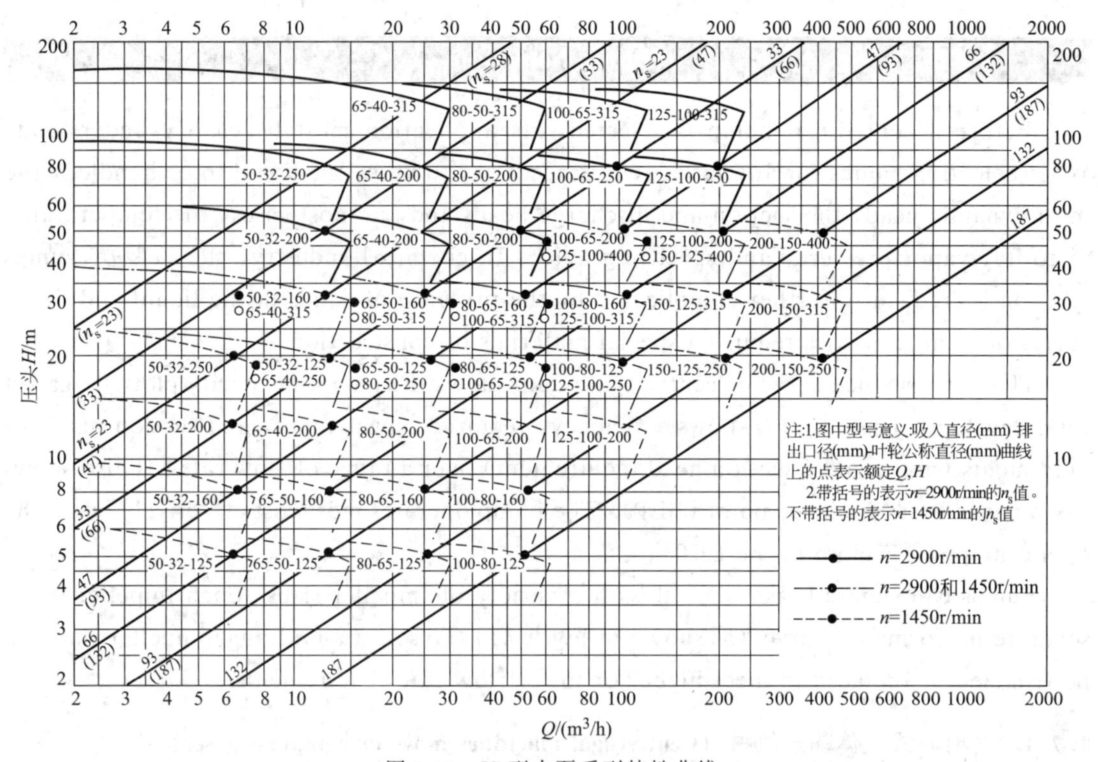

图 1-41　IS 型水泵系列特性曲线

其中：$z_1=0$，$u_1\approx 0$，$p_1=0$（表压）

$z_2=15\text{m}$，$u_2\approx 0$，$p_2=49.1\text{kPa}$（表压）

管中流速：$u=\dfrac{Q}{\dfrac{\pi}{4}d^2}=\dfrac{\dfrac{25}{3600}}{0.785\times 0.06^2}=2.46(\text{m/s})$

总阻力：$H_f=\lambda\dfrac{l+\sum l_e}{d}\times\dfrac{u^2}{2g}=0.021\times\dfrac{24+60}{0.06}\times\dfrac{2.46^2}{2\times 9.81}=9.07(\text{m})$

【例 1-20】 附图

所以：$H_e=z_2+\dfrac{p_2}{\rho g}+H_f=15+\dfrac{49.1\times 10^3}{1000\times 9.81}+9.07=29.07(\text{m})$

根据流量 $Q=25\ \text{m}^3/\text{h}$ 及扬程 $H_e=29.07\text{m}$，查图 1-41 可选型号为 IS65-50-160 的离心泵，再查附录的相应内容，确定其流量 Q 为 25 m^3/h，压头 H 为 32m，转速 n 为 2900r/min，允许汽蚀余量 NPSH 为 2.0m，效率 η 为 65%，轴功率 N 为 3.35 kW。

20℃水的饱和蒸气压 $p_v=2.335\text{kPa}$，吸入管路阻力：

$$H_{f,吸入}=\lambda\frac{l+\sum l_e}{d}\times\frac{u^2}{2g}=0.021\times\frac{24}{0.06}\times\frac{2.46^2}{2\times 9.81}=2.59(\text{m})$$

则离心泵的允许安装高度为：

$$H_g=\frac{p_0-p_v}{\rho g}-\text{NPSH}-H_{f,吸入}=\frac{(101.3-2.335)\times 10^3}{1000\times 9.81}-2.0-2.59=5.5(\text{m})$$

即泵的实际安装高度应低于 5.5m，可取 4.5～5.0m。

Pump Selection

Before a centrifugal pump is selected, its application must be clearly understood. When selecting pumps for any service, it is necessary to know the liquid to be handled, the total dynamic head, the suction and discharge heads and, in most cases, the temperature, viscosity, vapor pressure, and specific gravity. In chemical industry, the task of pumps election is frequently further complicated by the presence of solids in the liquid and liquid corrosion characteristics requiring special materials of construction.

The performance curve of centrifugal pump shows the pump's capacity plotted against total developed head. It also shows efficiency, required power input, and suction head requirements (not positive suction head requirement) over a range of flow rates. It also shows the pump's best efficiency point (BEP). The pump operates most cost effectively when the operating point is close to the BEP.

Pumps can generally be ordered with a variety of impeller sizes. Each impeller has a separate performance curve (as shown in Fig. 1-41). To select a midrange impeller that can be trimmed or replaced to meet higher or lower flow rate requirements.

1.7.1.2 离心式气体输送机械（Centrifugal machines move or compress gases）

气体输送与压缩机械主要用于：气体输送、克服流动阻力及产生高压气体，例如重油加氢、合成氨是在高压下进行，气体要加压送入反应器；产生真空，如许多单元操作，要在低于大气压下进行，所以要从设备中抽出气体产生真空。

但是，与液体相比，气体是可压缩流体，经济流速为 15～25m/s，空气的密度为 1.2kg/m^3；液体是不可压缩，经济流速为 1～3m/s，水的密度为 1000kg/m^3。相同的质量流量下，气体的阻力损失大约是液体的 10 倍，此问题对气体输送尤为突出，因此，压头高、流量大的气体输送比较困难。

正是由于气体本身的一些特点，气体输送机械在结构上与液体输送机械相比出现了一些差异。

若按终压（出口压力）或压缩比（出口压力/入口压力）分类，可分为表 1-5 中的几类。

表 1-5 离心式气体输送机械分类

种类	出口压力（表压）	压缩比
通风机（fans）	14.7×10^3Pa（1500mmH_2O）	1～1.5
鼓风机（blowers）	14.7×10^3～2.94×10^3Pa（0.15～3kgf/cm^2）	<4
压缩机（compressors）	>2.94×10^3Pa（3kgf/cm^2）	>4
真空泵（vacuum pumps）	大气压	视真空度而定

（1）离心式通风机（centrifugal fans）结构和分类

离心式通风机操作原理与离心泵类似，主要靠高速旋转的叶轮产生离心力提高气体的压头。离心式通风机如图 1-42 所示，其特点是叶片数目多、叶片短，有前弯、径向和后弯之分。按前弯、径向和后弯的顺序，效率依次增大，流量依次减小。若按其出口压力分类，则如表 1-6 所示。

表 1-6　离心式通风机按出口压力分类

种类	出口压力(表压)
低压风机	$<0.9807\times10^3$Pa(<100mmH_2O)
中压风机	$0.9807\times10^3\sim2.942\times10^3$Pa($100\sim300mmH_2O$)
高压风机	$2.942\times10^3\sim14.7\times10^3$Pa($300\sim1500mmH_2O$)

(2) 离心式通风机性能参数与特性曲线

① 风量 Q (m^3/h)　指单位时间内从风机出口排出的气体体积，规定以风机进口处的气体状态计算。

$Q=f$(风机的结构、尺寸、转速)

② (全) 风压 p_t(Pa)　p_t 只能由实验测定，方法是在风机的入口与出口之间列柏努利方程：

$$z_1g+\frac{u_1^2}{2}+\frac{p_1}{\rho}+W_e=z_2g+\frac{u_2^2}{2}+\frac{p_2}{\rho}+\sum h_{f,1-2}$$

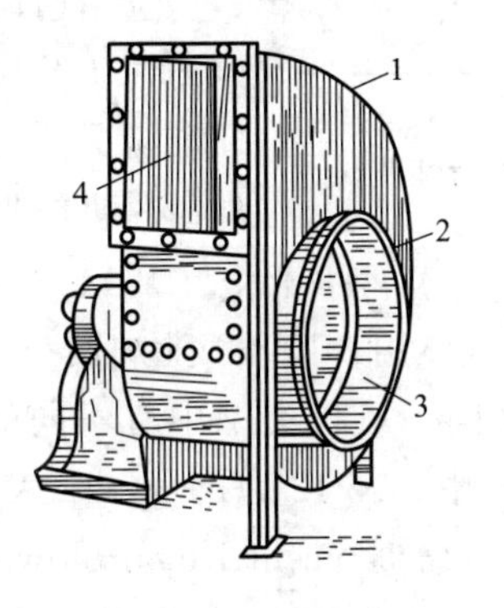

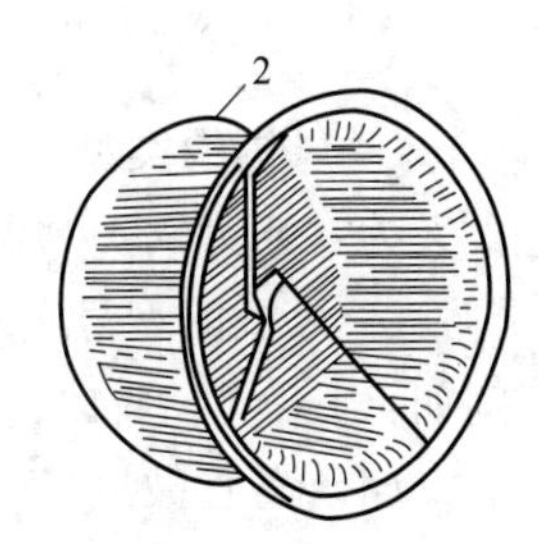

图 1-42　离心式通风机

1—机壳；2—叶轮；3—吸入口；4—排出口

同乘以 ρ 得到：$z_1\rho g+\frac{\rho u_1^2}{2}+p_1+\rho W_e=z_2\rho g+\frac{\rho u_2^2}{2}+p_2+\rho\sum h_{f,1-2}$

令：
$$p_t=\rho W_e=\rho g(z_2-z_1)+\frac{\rho(u_2^2-u_1^2)}{2}+(p_2-p_1)+\rho\sum h_{f,1-2} \tag{1-65}$$

上式说明，全风压是指单位体积的气体流过风机所获得的能量。

一般来说，风机入口与大气相通，所以可取 $u_1\approx0$；并取入口、出口间的位差 $z_2-z_1\approx0$；因入口、出口之间的管路短，则取$\sum h_{f,1-2}\approx0$；而气体经风机后的出口速度 u_2 很大，不能忽略，则上式简化为：

$$p_t=(p_2-p_1)+\frac{\rho u_2^2}{2}=p_{st}+p_{kt} \tag{1-66}$$

式中，p_{st} 称为静风压；p_{kt} 称为动风压。

若 $p_1=0$ (表压)，则测定风机出口压强 p_2 和流量 Q (或 u_2) 就可求出全风压 p_t。

③ 标准全风压 p_{t0}　p_{t0} 是用空气在 20℃、1atm 下测定的全风压，风机的铭牌上或样本上的全风压即指 p_{t0}。

$$p_t\propto\rho,p_{t0}\propto\rho_0=1.2$$

$$\frac{p_{t0}}{p_t}=\frac{1.2}{\rho} \tag{1-67}$$

④ 效率 η 与轴功率 N　η 与密度无关，而轴功率 N 应考虑密度的影响。

操作条件：
$$N=\frac{\rho QW_e}{\eta}=\frac{Qp_t}{\eta}$$

实验条件：
$$N_0=\frac{\rho_0Q_0W_e}{\eta}=\frac{Q_0p_{t0}}{\eta}$$

$$N=\frac{\rho}{\rho_0}N_0 \tag{1-68}$$

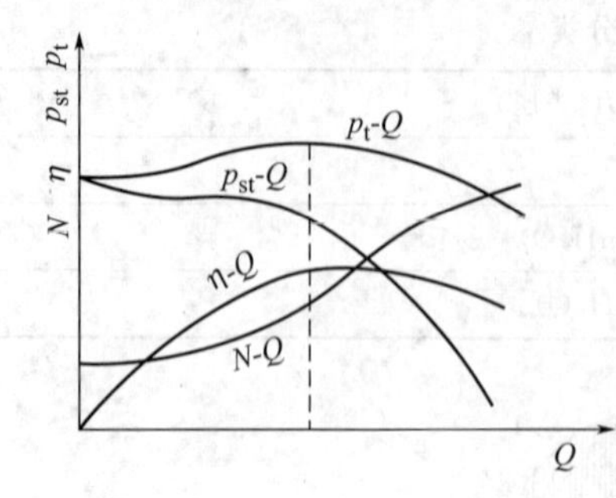

图 1-43　离心式风机的特性曲线

⑤ 特性曲线　在一定的转速下，实验测得 p_t-Q、p_{st}-Q、N-Q、η-Q 的关系曲线叫风机的特性曲线，如图 1-43 所示。η_{max} 对应的 N、p_t、p_{st} 列于风机的铭牌或样本上。

（3）离心式通风机选用

选用通风机的步骤：首先计算全风压 p_t；然后由气体的性质及风压范围确定风机类型；最后由 p_t 换算成标准状况下的 p_{t0} 并结合实际风量 Q 选择风机型号，并列出风机的主要性能参数。

① 风机的类型　有 4-72 型（低、中压）、8-18 型和 9-27 型（高压）。前面加大写字母的含义为：Y 为锅炉高温气体用；L 为工业炉用；F 为防腐用；W 为耐高温用；B 为防爆用。

② 风机的型号　因同一风机类型中有不同的叶轮直径，则在类型的后面加机号区别。例如：4-72 型 $N_0$12。其中，$N_0$12 为机号，12 表示叶轮的直径，单位为 dm。

（4）离心鼓风机与压缩机（centrifugal blowers and compressors）

① 离心鼓风机（centrifugal blowers）　如图 1-44 所示，离心鼓风机外形与离心泵相似，其特点是送气量大，但产生的风压仍不高。一般单级出口压力 < 0.3 kgf/cm^2，多级出口压力可达 3 kgf/cm^2。

图 1-44　离心鼓风机

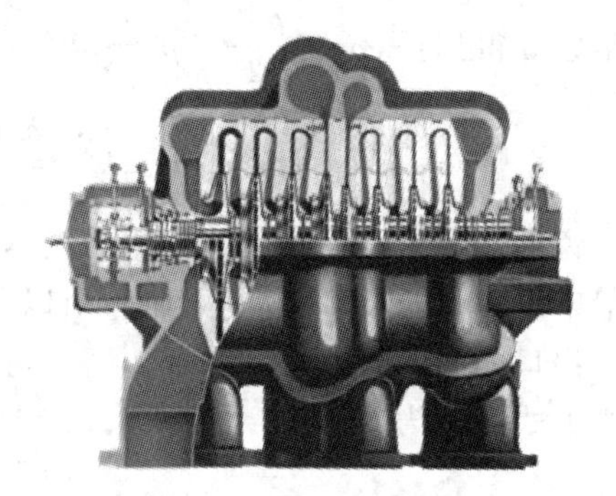
图 1-45　离心压缩机

② 离心压缩机（又称透平压缩机）（centrifugal compressors）　如图 1-45 所示，离心压缩机一律是多级的，外形与多级离心泵类似。其特点是转速高达 5000r/min，压缩比高，出口压力可达 10 kgf/cm^2。压缩时，气体温度升高，需加中间冷却器。与往复压缩机相比，其优点是体积小、质量轻、流量大、供气均匀、运转平稳，应用日趋广泛。目前已有大型透平压缩机，流量大，压头高。

1.7.2　往复式输送机械（Reciprocating conveying machinery）

1.7.2.1　往复泵（reciprocating pump）

（1）往复泵基本结构及工作原理

往复泵的泵头部分主要由泵体（液缸）、活塞（柱塞）、活塞杆、吸液和排液阀门（单向阀）组成，如图 1-46 所示。

通常，把活塞在泵缸内左右移动的两个端点叫死点，把两个死点之间的距离叫行程（冲

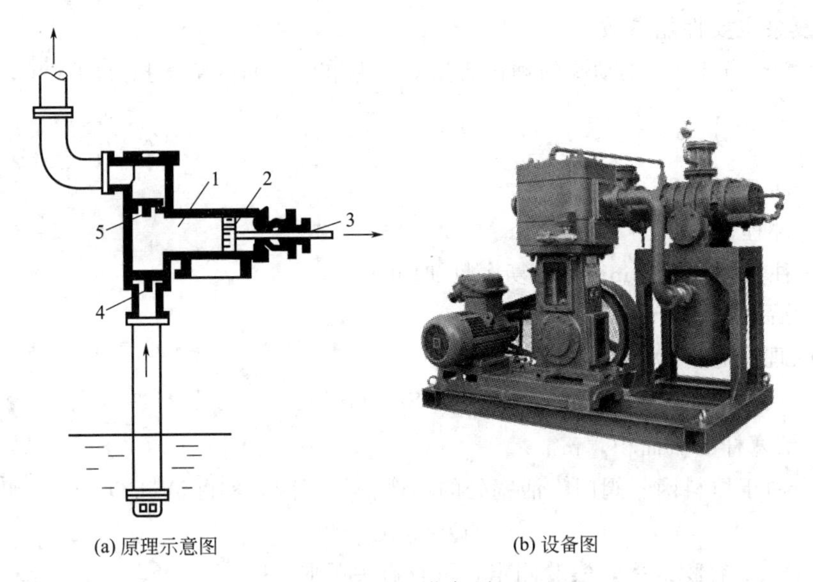

(a) 原理示意图　　(b) 设备图

图 1-46　往复泵

1—泵缸；2—活塞；3—活塞杆；4—吸入阀；5—排出阀

程)；把活塞与阀门之间的空间叫工作室。正常工作时，活塞不断地作往复运动，工作室交替地（不连续）吸液、排液，通过活塞将能量以静压能的形式传递给液体。

由于吸液、排液靠工作室的空间变化，所以往复泵是一种容积式泵。

按活塞作一次往复运动，泵缸排液的次数，可分为以下几类：

① 单动泵（单作用泵）　活塞在两死点间往复运动，速度发生变化，则流量曲线按正弦曲线规律变化，两头小，中间大。其特点是排液间断，流量不均匀，惯性阻力大。

② 双动泵（双作用泵）　如图 1-47（a）所示，双动泵有四个阀门，在活塞的两边各两个，吸液、排液同时进行，即活塞往复运动一次，吸液、排液各一次。其特点是吸液、排液连续，但流量仍不均匀。

③ 三动泵　如图 1-47（b）所示，三动泵的三个曲柄互成 120°，分别推动三个单动泵。三个泵联合操作，当一个泵排液量开始下降时，另外的泵开始排液，三个泵依次进行。其特点是排液连续，流量较均匀。

为了流量均匀，操作平稳，在往复泵排出口上方设有空气室，借助室内气体压缩与膨胀的作用，当往复泵出口流量变化时，能保持管路中的流量大致不变。

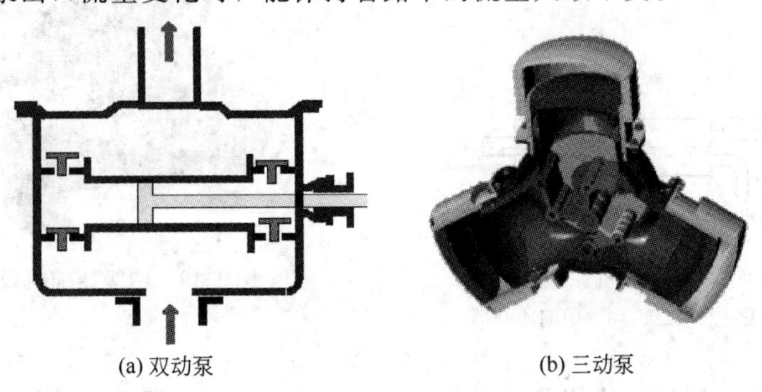
(a) 双动泵　　(b) 三动泵

图 1-47　多动泵

（2）往复泵主要性能参数

① 流量（m^3/min） 单动泵的理论流量 Q_T 为单位时间内活塞扫过的体积，即：

$$Q_T = ASn_r = \frac{\pi}{4}D^2Sn_r \tag{1-69}$$

式中 A——活塞截面积，m^2；

S——冲程，m；

n_r——往复频率，1/min（往复次数/时间）；

D——活塞直径，m。

双动泵的理论流量 Q_T 为：

$$Q_T = 2ASn_r - aSn_r \tag{1-70}$$

式中 a——活塞杆的截面积，m^2。

实际上，由于填料函、阀门、活塞处的不严密等原因，实际流量 $Q < Q_T$，即：

$$Q = \eta_V Q_T \tag{1-71}$$

式中 η_V——容积效率，实验测定，可查有关手册。

② 扬程 H H 的含义与离心泵的一样。但是，对往复泵而言，理论上 H 与 Q_T 无关，定量排液；实际上，H 增加时 η_V 稍有下降，所以实际流量 Q 略有降低。

由于往复泵的压头与流量几乎无关，所以，往复泵常用于压头很大，流量不大的管路输送。

③ 轴功率 N、效率 η 的计算与离心泵相同。

④ 型号 例如，2QS-53/17。其中，2 表示双缸；Q 表示蒸汽驱动；S 表示活塞泵；53 表示最大流量，m^3/min；17 表示最大出口压力。

1.7.2.2 往复压缩机（reciprocating compressor）

往复压缩机与往复泵类似，前者用于气体输送，后者用于液体输送，一般要求，压缩比＞8 时，采用多级压缩，每级压缩比为 3～5。两级压缩机原理示意图见图 1-48。两级压缩机设备图见图 1-49。

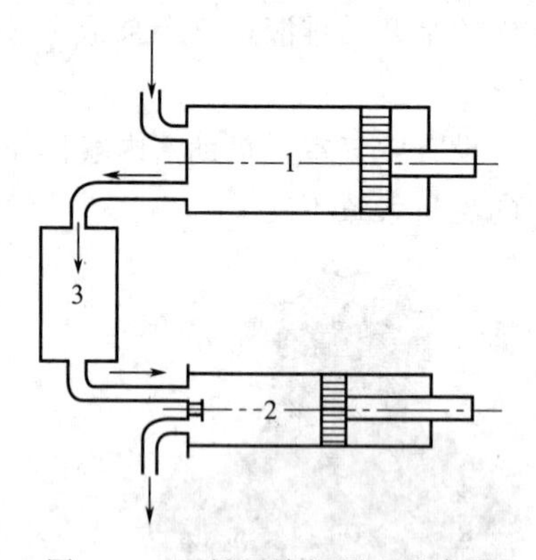

图 1-48 两级压缩机原理示意图

1—一级；2—二级；3—中间冷却器

图 1-49 两级压缩机设备图

多级压缩应使每级压缩比相等，即各级所需外功相等时，所需总功最小；多级压缩通过中间冷却器使过程尽量向等温过程靠近（因为等温过程最省功），即中间冷却器是实现多级压缩的关键。当压缩比较高时，多级压缩比较经济合理。

1.7.3　其他类型输送机械

1.7.3.1　其他类型泵

(1) 计量泵（比例泵）(metering pump)

计量泵就是小流量的往复泵，包括柱塞计量泵和隔膜计量泵两类，工作原理如图 1-50 所示。其中，柱塞计量泵因其结构简单和耐高温高压等优点而被广泛应用于石油化工领域，但因被计量介质和泵内润滑剂之间无法实现完全隔离这一结构性缺点，柱塞计量泵在高防污染要求流体计量应用中受到诸多限制；隔膜计量泵利用特殊设计加工的柔性隔膜取代活塞，在驱动机构作用下实现往复运动，完成吸入-排出过程。由于隔膜的隔离作用，在结构上真正实现了被计量流体与驱动润滑机构之间的隔离，因此该类泵目前已经成为流体计量应用中的主力泵型。

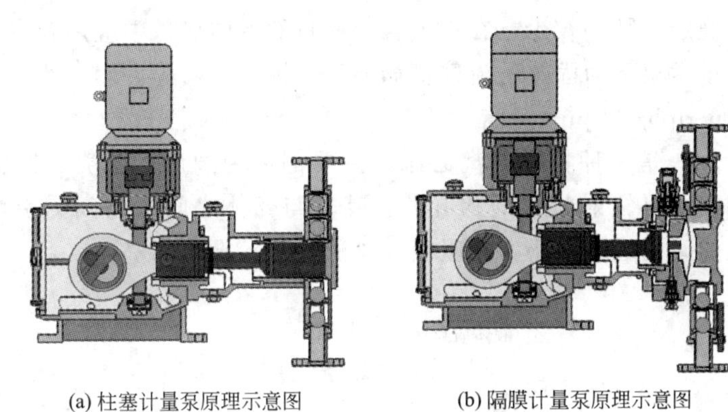

(a) 柱塞计量泵原理示意图　　(b) 隔膜计量泵原理示意图

图 1-50　计量泵

(2) 旋转泵（转子泵）(rotary pump)

旋转泵靠泵内的转子转动而吸液、排液。应注意，其工作原理是靠挤压，而不是离心力。旋转泵按转子的形式分为以下几类。

① 齿轮泵（图 1-51）(gear pump)　两齿轮咬合，反向转动，利用齿轮与泵体之间的间隙挤压液体，转化能量，提高液体的压头而排液。其特点是压头高、流量小，可输送黏稠液体或膏状物料，也可用于给填料函注油封，但不能送悬浮液；缺点是有噪声、震动。

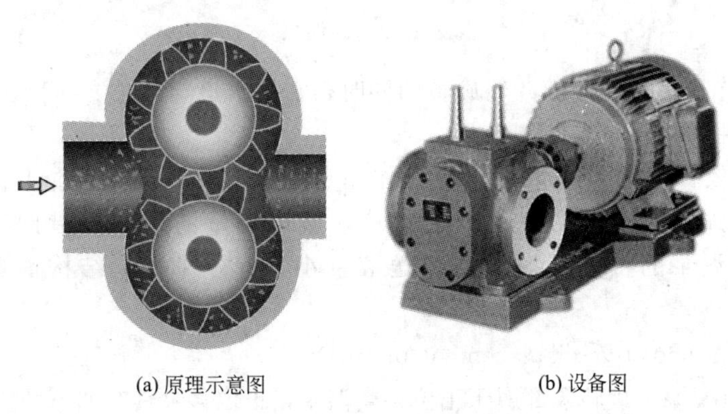

(a) 原理示意图　　(b) 设备图

图 1-51　齿轮泵

② 螺杆泵（screw pump） 螺杆泵（图1-52）工作时，液体被吸入后就进入螺纹与泵壳所围的密封空间，当主动螺杆旋转时，螺杆泵密封容积在螺牙的挤压下提高螺杆泵压力，并沿轴向移动。由于螺杆是等速旋转，所以液体排出流量也是均匀的。

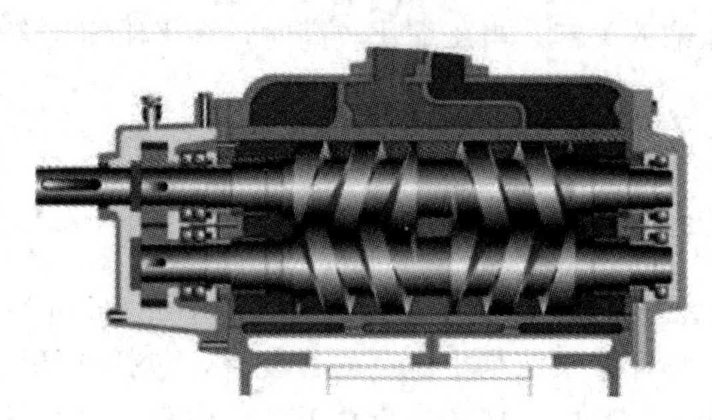

图 1-52 螺杆泵

螺杆泵通常可分为单螺杆泵、双螺杆泵和三螺杆泵，该类泵的特点是结构紧凑、压力高而均匀、流量均匀、转速高、运转平稳、效率高，并能与原动机直联。

螺杆泵可以输送各种油类及高分子聚合物，即特别适用于输送黏稠液体。

③ 凸轮泵（cam pump） 凸轮泵又称凸轮转子泵或旋转活塞泵，其工作原理是依靠两个凸轮相互咬合中工作容积的变化来输送液体的。其特点是能输送介质黏度达100Pa·s以上的液体，且流量稳定，脉冲较少，在三大合成材料的液体输送中应用较多。若在泵体上设加热夹套，则可输送常温下为固体的流体，如沥青、树脂、蜡等。

（3）旋涡泵（peripheral pump）

旋涡泵（图1-53）是一种特殊的离心泵，主要部件为圆形的泵壳、圆盘形的叶轮，叶轮两侧分别铣有凹槽，呈辐射状排列的叶片，叶轮与壳体之间的空间为引水道。在进出口间，为使叶轮与壳体间的缝隙很小，把吸液与排液分开，设有间壁。

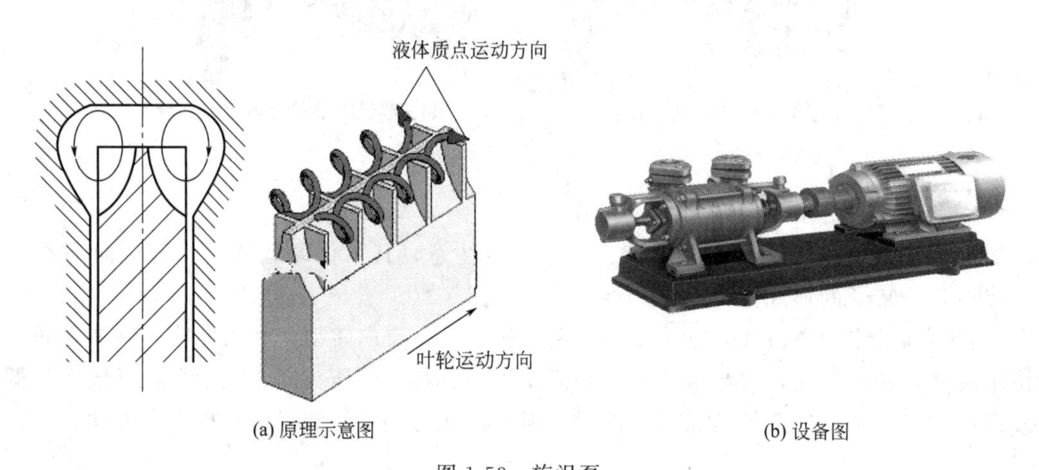

(a) 原理示意图　　(b) 设备图

图 1-53 旋涡泵

旋涡泵的工作原理是液体随叶轮旋转的同时，在引水道与叶片之间反复做旋涡运动，由于叶片的多次作用，获得较高的能量。

与离心泵比较，相同点是依靠离心力作用送液，开车前要灌满液体；不同点是流量下降时，压头增加很快，轴功率也增加很快，所以流量的调节要借助于正位移泵的旁路调节方法。旋涡泵的特点是扬程高但送液量小，且效率低，结构简单，在化工上较多采用。

（4）正位移泵（positive-displacement pumps）

以上介绍的往复泵、旋转泵都属于正位移类型泵，正位移泵具有如下性质：送液能力只取决于活塞或转子的位移，与管路特性无关（而离心泵的送液能力与管路特性有关）；对一定的流量，可提供不同的压头（扬程），提供多大的压头，由管路特性决定。这种特性叫正位移特性，具有这种特性的泵统称正位移泵。

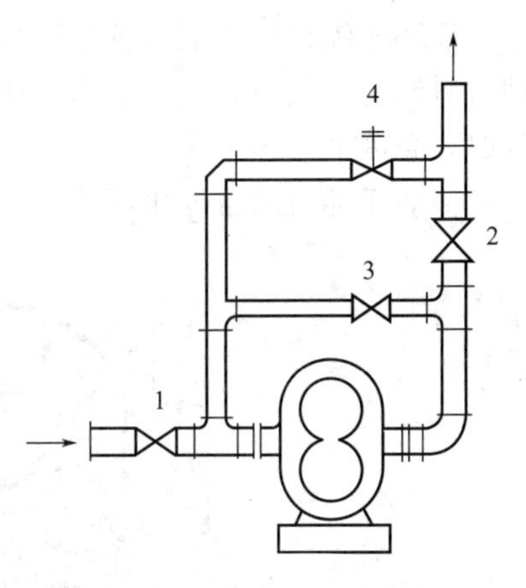

图 1-54　正位移泵的调节流量

1—吸入管路上的调节阀；2—排出管路上的调节阀；3—支路阀；4—安全阀

正位移泵启动前，不用充液体，但必须将出口阀门打开（因泵内液体不倒流，所以必须排出），否则，泵内压头不断升高，设备、管道的强度承受不了；正位移泵安装高度也有一定的限制，以防止汽蚀现象发生。如图1-54所示，调节流量利用旁路（又称回流支路）阀门，并设有安全阀，一旦出口阀关闭或开得太小，泵内压头急剧上升，可能导致机件损坏及电机超负荷时，自动启用安全阀及时泄压；另外，也可以通过改变冲程、往复频率、转子的转速等方法调节流量，但不方便，仅用于定期调节。

1.7.3.2　其他类型气体输送机械

旋转鼓风机及压缩机与旋转泵类似，壳体内有一个或多个转子，适用于压头不大，而流量大的场合。

(1) 罗茨鼓风机（Roots blower）

如图1-55所示，罗茨鼓风机的工作原理与齿轮泵类似，壳体内有两个特殊形状的转子反向运动，以增加气体的压头。其特点是风量正比于转速，在一定的转速下，出口压力提高，风量可保持大致不变，所以也叫定容式鼓风机。应注意，罗茨鼓风机出口要加稳定罐、安全阀，旁路调节流量，操作温度在80～85℃，防止转子受热膨胀而咬死。

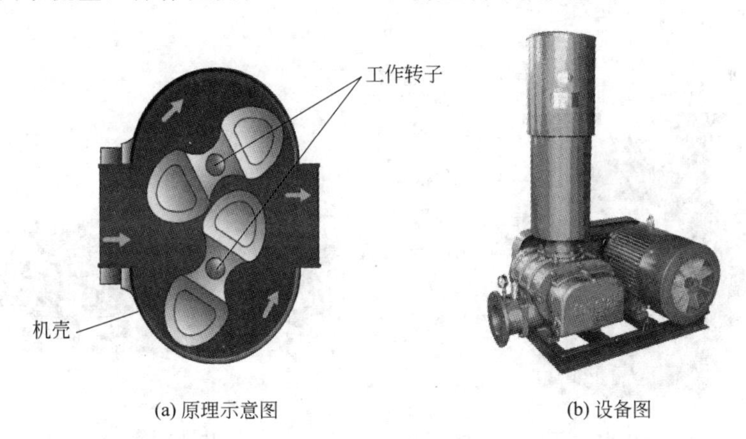

(a) 原理示意图　　(b) 设备图

图 1-55　罗茨鼓风机

(2) 液环压缩机（liquid ring compressor）

液环压缩机（又称纳氏泵）（图1-56）主要由略似椭圆的外壳和旋转叶轮组成，壳中盛有适量的液体。当叶轮旋转时，由于离心力的作用，液体被抛向壳体，形成椭圆形的液环，在椭圆形长轴两端形成两个月牙形空隙。当叶轮回转一周时，叶片和液环间所形成的密闭空间逐渐变大和变小各两次，气体从两个吸入口进入机内，从两个排出口排出。

液环压缩机内的液体将被压缩的气体与机壳隔开，气体仅与叶轮接触，只要叶轮用耐腐蚀材料制造，便适宜于输送腐蚀性气体。壳内的液体应与被输送气体不起作用，例如压送氯气时，壳内的液体可采用硫酸。

液环压缩机的压缩比可达 6～7，但出口表压在 150～180kPa 的范围内效率最高。

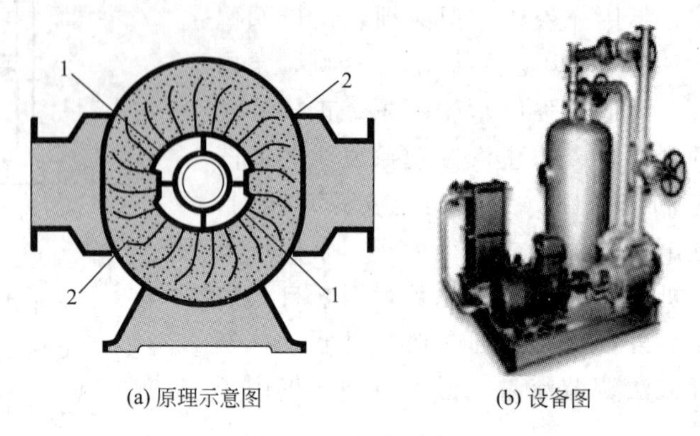

(a) 原理示意图　　(b) 设备图

图 1-56　液环压缩机

1—吸入口；2—排出口

(3) 真空泵（vacuum pump）

原则上讲，真空泵就是在负压下吸气，在大气压下排气的气体输送机械，用于维持工艺系统中要求的真空状态，其主要性能参数为剩余压力和抽气速率。剩余压力（与最大真空度对应）指真空泵能达到的最低绝对压力；抽气速率是在剩余压力下，真空泵单位时间吸入气体的体积。真空泵的选用，主要依据上述两个参数。

① 水环真空泵（water ring vaccum pump）　如图 1-57 所示，水环真空泵为圆形外壳，叶轮偏心安装，充一定量的液体，形成一个新月面，由于液环把气体与壳体隔开，故可用于输送腐蚀性气体。

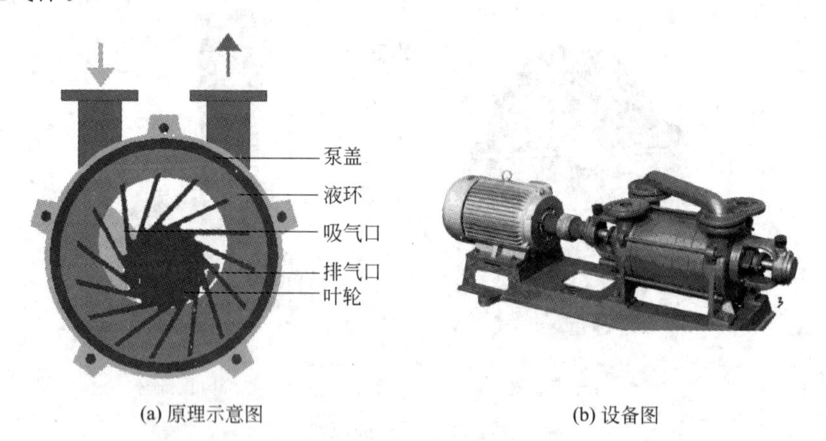

(a) 原理示意图　　(b) 设备图

图 1-57　水环真空泵

Vacuum Pumps

A compressor that takes suction at a pressure below atmospheric and discharges against atmospheric pressure is called a vacuum pump. Any type of blower or compressor-reciprocating, rotary, or centrifugal-can be adapted to vacuum practice by modifying the design to accept very low-density gas at the suction and attain the large compression ratios necessary. As the absolute pressure at the suction decreases, the volumetric efficiency drops and approaches zero at the lowest absolute pressure attainable by the pump. Usually the mechanical efficiency is also lower than that for compressors. The required displacement increases rapidly as the suction pressure falls, so a large machine is needed to move much gas. The compression ratio used in vacuum pumps is higher than that in compressors, ranging up to 100 or more, with a correspondingly high adiabatic discharge temperature. Actually, however, the compression is nearly isothermal because of the low mass flow rate and the effective heat transfer from the relatively large area of exposed metal.

② 喷射真空泵（属于流体动力作用泵）(jet ejectors vacuum pump) 如图1-58所示，喷射真空泵利用一种流体流动时（工作流体），使其静压头转化为动压头，工作流体为水或蒸汽。一般工作流体以高速从喷嘴喷出，部分静压头转化为动压头，形成低压区，可把另一种流体抽进来，混合后经扩散管流速降低，部分静压头恢复。其优点是工作压力范围大，抽气量大，结构简单，适应性强，可抽送含灰尘、腐蚀性、易燃易爆流体；缺点是效率低，一般仅有10%～25%。单级喷射可达到的真空度较低，为了达到更高的真空度，可采用多级喷射。

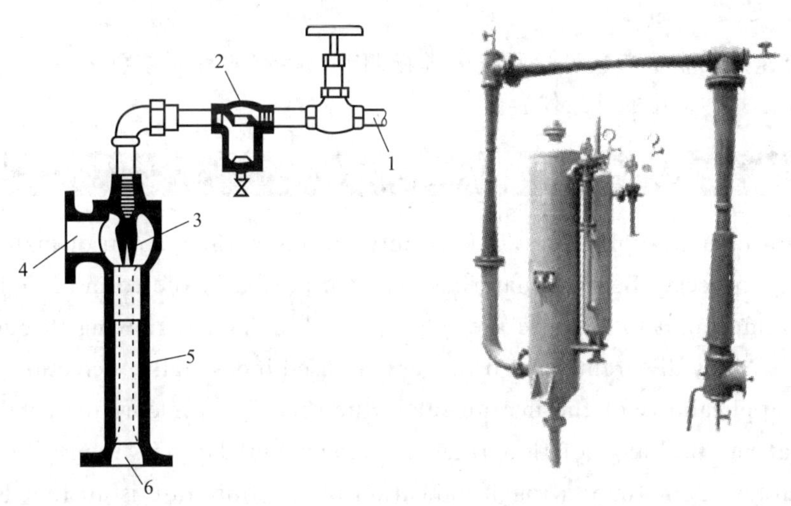

图1-58 喷射真空泵

1—工作蒸汽入口；2—过滤器；3—喷嘴；4—吸入口；5—扩散管；6—压出口

Jet Ejectors

An important kind of vacuum pump that does not use moving parts is the jet ejector, shown in Fig. 1-58, in which the fluid to be moved is entrained in a high-velocity stream of a second fluid. The motive fluid and the fluid to be moved may be the same, such as when compressed air is used to move air, but usually they are not. Industrially greatest use is made of steam-jet ejectors, which are valuable for drawing a fairly high vacuum. As shown in Fig. 1-58, steam at about 7 atm is admitted to a converging-diverging nozzle, from which it issues at supersonic velocity into a diffuser cone. The air or other gas to be moved is mixed with the steam in the first part of the diffuser, lowering the velocity to acoustic velocity or below. In the diverging section of the diffuser the kinetic energy of the mixed gases is converted to pressure energy, so that the mixture can be discharged directly to the atmosphere. Often it is sent to a water-cooled condenser, particularly if more than one stage is used. Otherwise each stage would have to handle all the steam admitted to the preceding stages. As many as five stages are used in industrial processing.

Jet ejectors require very little attention and maintenance and are especially valuable with corrosive gases that would damage mechanical vacuum pumps. For difficult problems the nozzles and diffusers can be made of corrosion-resistant metal, graphite, or other inert material. Ejectors, particularly when multistage, use large quantities of steam and water.

They are rarely used to produce absolute pressures below 1mmHg. Steam jets are no longer as popular as they once were, because of the great increase in the cost of steam. In many instances where corrosion is not a serious consideration, they have been replaced by mechanical vacuum pumps, which use much less energy for the same service.

1.8 流速与流量测量（Metering of Fluid Flow Rate and Velocity）

流量或流速是工业生产和科学研究中进行调节、控制的重要参数之一，其计量方法很多，本节仅介绍几种常用的计量仪表。

Metering of Fluids

To control industrial processes, it is essential to know the amount of material entering and leaving the process. Because materials are transported in the form of fluids wherever possible, it is important to measure the rate at which a fluid is flowing through a pipe or other channel. Many different types of meters are used industrially. Selection of a meter is based on the applicability of the instrument to the specific problem, its installed cost and costs of operation, the range of flow rates it can accommodate (its range ability), and its inherent accuracy. Sometimes a rough indication of the flow rate is all that is needed, at other times a highly accurate measurement, usually of the mass flow rate, is required for such purposes as controlling reactor feeds or transferring custody of the fluid from one owner to another.

A few types of flowmeters measure the mass flow rate directly, but the majority measure the volumetric flow rate or the average fluid velocity from which the volumetric flow rate can be calculated. To convert the volumetric rate to the mass flow rate requires that the fluid density under the operating conditions be known. Most meters operate on all the fluid in the pipe or channel and are known as full-bore meters. Others, called insertion meters, measure the flow rate, or more commonly the fluid velocity at one point only. The total flow rate, however, can often be inferred with considerable accuracy from this single-point measurement.

Detailed descriptions of commercial flowmeters, listing their advantages and limitations, are available in the literature.

1.8.1 变压差恒截面型流量计

变压差恒截面型流量计的基本测量原理是流体流经测量元件时，动压头发生变化转化为静压头，将静压头的变化通过压差计反映出来，即把流速 u 与压差计读数 R 联系起来，由 R 确定 u。

（1）测速管（又称毕托管）(Pitot tube)

如图 1-59 所示，毕托管是测定点速度的流量计，由与流动方向平行位置的两根同心管构成。内管的开口端测定停滞点的冲压头，该冲压头由两部分组成，一是停滞点截面上的静压头，二是停滞点消耗掉的动压头；而外部同心管的一端封死，其曲面上开有许多小孔，用以测定流体的静压头。两根同心管的内管和环隙分别连通 U 形管压差计，停滞点（stagnation point）的冲压头与外部同心管上小孔处静压头之差由压差计读数 R 反映出来，可以导出，停滞点处的点速度（point velocity）为：

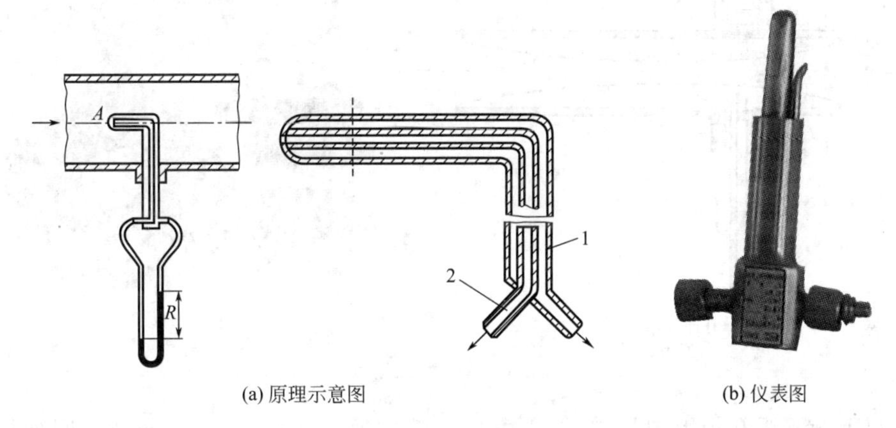

(a) 原理示意图　　(b) 仪表图

图 1-59　测速管

1—静压管；2—冲压管

$$u_A=\sqrt{\frac{2R(\rho'-\rho)g}{\rho}} \tag{1-72}$$

式中，ρ'为指示液的密度，kg/m^3；ρ 为被测流体的密度，kg/m^3。

Pitot Tube

The Pitot tube is a device used to measure the local velocity along a streamline. The velocity measured by an ideal Pitot tube would conform exactly to Eq. 1-86.

Well-designed instruments are in error by not more than l percent of theory, but when precise measurements are to be made, the Pitot tube should be calibrated and an appropriate correction factor applied.

The disadvantages of the Pitot tube are (1) that most designs do not give velocity directly and (2) that its readings for gases are extremely small. When measuring low-pressure gases, some form of multiplying gauge, like at shown must also be used.

Full-Bore Meters

The most common types of full-bore meters are venturi and orifice meters and variable-area meters such as rotameters. Other full-bore measuring devices include V-element, magnetic, vortex-shedding, turbine, and positive-displacement meters, ultrasonic meters, and mass flow devices such as Coriolis flowmeters.

（2）孔板流量计（orifice meter）

孔板流量计的结构如图 1-60 所示，在管道里插入一片带有圆孔的金属板，其圆孔的中心应位于管道中心线上。流体流经孔口，不会马上扩大到整个管截面，而是在惯性作用下，继续收缩到一定距离后，才逐渐扩大到整个管截面。流体流动截面最小处称为缩脉。流体流经孔板前后，流速的变化引起压强的变化，将压强的变化与压差计读数 R 联系起来，可测定流量。

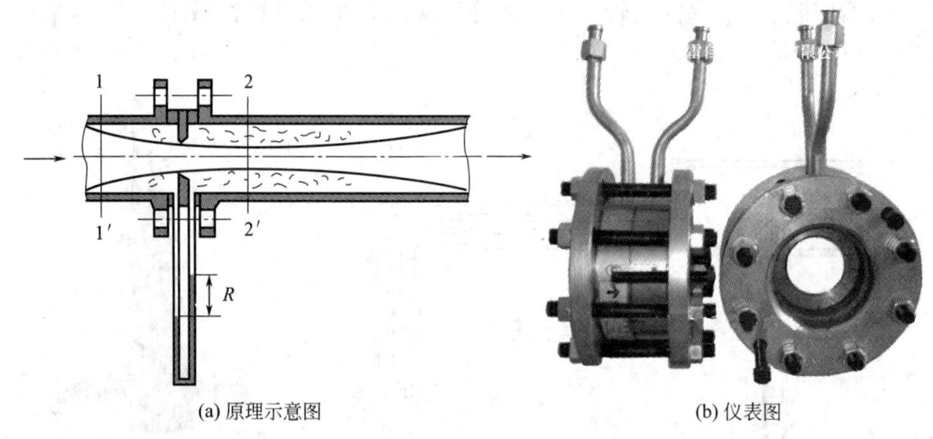

(a) 原理示意图　　(b) 仪表图

图 1-60　孔板流量计

在孔口上游不受孔板影响处取 1—1′截面，缩脉处取 2—2′截面，两截面间列柏努利方程，若忽略阻力损失，可得：

$$\frac{p_1}{\rho}+\frac{u_1^2}{2}=\frac{p_2}{\rho}+\frac{u_2^2}{2}$$

设管道截面积为 A_1，直径为 d_1，孔口截面积为 A_0，直径为 d_0。由于缩脉处的截面积无法测取，所以可用孔口处流速 u_0 代替缩脉处的流速 u_2。再考虑到忽略了阻力损失以及 U 形管压差计的测压位置与所选截面的差异，引进校正系数 C，故有：

$$\sqrt{u_0^2-u_1^2}=C\sqrt{\frac{2(p_1-p_2)}{\rho}} \tag{1-73}$$

根据连续性方程：　$u_1^2=u_0^2\left(\frac{d_0}{d_1}\right)^4$

将该式代入式（1-87）中，可得孔板处的流速为：

$$u_0=\frac{C}{\sqrt{1-\left(\frac{d_0}{d_1}\right)^4}}\sqrt{\frac{2(p_1-p_2)}{\rho}}$$

$$=C_0\sqrt{\frac{2(p_1-p_2)}{\rho}}$$

$$=C_0\sqrt{\frac{2Rg(\rho'-\rho)}{\rho}} \qquad (1\text{-}74)$$

则，管道中的体积流量为：

$$V_s=u_0A_0=C_0A_0\sqrt{\frac{2gR(\rho'-\rho)}{\rho}} \qquad (1\text{-}75)$$

式中，C_0 称为孔流系数，需由实验测定。用角接取压法安装的孔板流量计，其 C_0 与 A_0/A_1 及 Re 有关，如图1-61所示。

当 Re 超过某界限值 Re_c 之后，则 C_0 不再随 Re 而改变，成为定值。流量计所测的流量范围最好落在 C_0 为定值的区域内。此时，流量与压差计的读数 R 的平方根成正比。常用的 C_0 值一般为0.6～0.7之间。

孔板流量计必须安装在管道中的直管段，既可垂直，又可水平或倾斜放置。孔板前后各有一段稳定段，上游不少于 $10d$，下游不少于 $5d$。

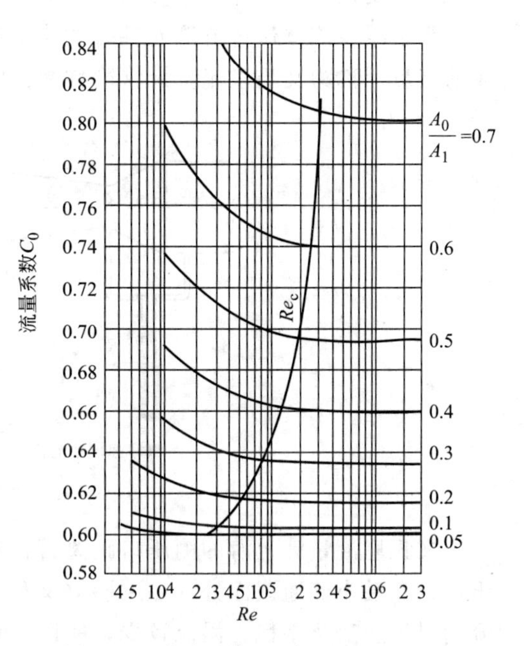

图1-61 孔流系数 C_0 与 Re、A_0/A_1 的关系曲线

【例1-21】 20℃苯在 ϕ133mm×4mm 的钢管中流过，为测量苯的流量，在管路中安装一孔径为75mm的孔板流量计。当孔板前后U形压差计的读数 R 为80mmHg时，求管中苯的流量（m^3/h）。

解 查得20℃苯的物性：$\rho=880\ kg/m^3$，$\mu=0.67\times10^{-3}\ Pa\cdot s$。

面积比：$$\frac{A_0}{A_1}=\left(\frac{d_0}{d_1}\right)^2=\left(\frac{75}{125}\right)^2=0.36$$

设 $Re>Re_c$，由图1-61查得：$C_0=0.648$，$Re_c=1.5\times10^5$。

由式(1-89)计算苯的体积流量：

$$V_s=C_0A_0\sqrt{\frac{2gR(\rho'-\rho)}{\rho}}$$

$$=0.648\times0.785\times0.075^2\times\sqrt{\frac{2\times0.08\times9.81\times(13600-880)}{880}}$$

$$=0.0136(m^3/s)=48.96(m^3/h)$$

管内的流速：$$u=\frac{V_s}{\frac{\pi}{4}d_1^2}=\frac{0.0136}{0.785\times0.125^2}=1.11(m/s)$$

管路的 $$Re=\frac{d_1u\rho}{\mu}=\frac{0.125\times880\times1.11}{0.67\times10^{-3}}=1.82\times10^5>Re_c$$

所以假设成立，以上计算有效，即苯在管路中的流量为 48.96m³/h。

（3）文丘里流量计（Venturi meter）

为了减少流体流过节流元件时的能量损失，可以用一段渐缩、渐扩管代替孔板，这样构成的流量计称为文丘里流量计（简称文氏流量计），如图 1-62 所示。

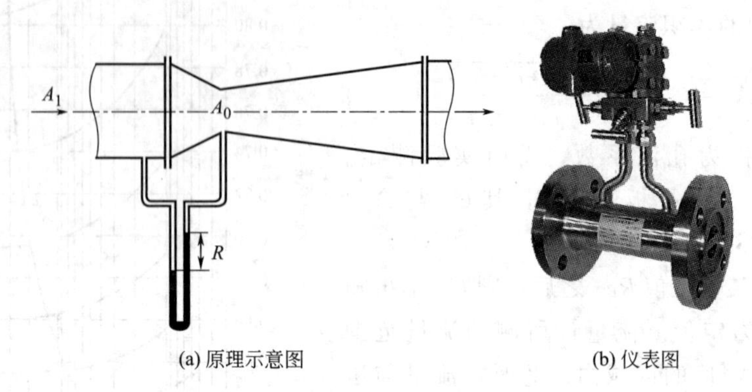

(a) 原理示意图　(b) 仪表图

图 1-62　文丘里流量计

文丘里流量计上游的测压口距离管径开始收缩处的距离至少应为二分之一管径，下游测压口设在最小流通截面处（A_0 处）（又称为文氏喉）。由于有渐缩、渐扩管段，流体在其中流动时流速变化平稳，涡流较少，所以能量损失比孔板流量计大大降低。

文丘里流量计的流量计算式与孔板流量计相似，即：

$$V_s = C_V A_0 \sqrt{\frac{2gR(\rho' - \rho)}{\rho}} \tag{1-76}$$

式中　C_V——流量系数，其值可由实验测定或从仪表手册中查得，无量纲。

文丘里流量计的优点是能量损失小，但各部分尺寸要求严格，加工精细度较高，所以造价较高。

Venturi Meter

A Venturi meter is shown in Fig. 1-62. A short conical inlet section leads to a throat section, then to a long discharge cone. Pressure taps at the start of the inlet section and at the throat are connected to a manometer or differential pressure transmitter.

In the upstream cone the fluid velocity is increased and its pressure decreased. The pressure drop in this cone is used to measure the flow rate. In the discharge cone the velocity is decreased and the original pressure largely recovered. The angle of the discharge cone is made small, between 5° and 15° to prevent boundary layer separation and to minimize friction. Since there is no separation in a contracting cross section, the upstream cone can be made shorter than the downstream cone. Typically 90 percent of the pressure loss in the upstream cone is recovered.

Although Venturi meters can be applied to the measurement of gas flow rates, they are most commonly used with liquids, especially large flow of water where, because of the large pressure recovery, a Venturi requires less power than other types of meters.

1.8.2　恒压差变截面型流量计——转子流量计（Rotameters—area meter）

转子流量计属恒压差变截面型测量装置，其结构如图1-63所示。它是以带有刻度的倒锥形玻璃管为主件，上下通过法兰连接管路，玻璃管内有一个可浮动的转子，转子材料可为金属或其他材质构成。

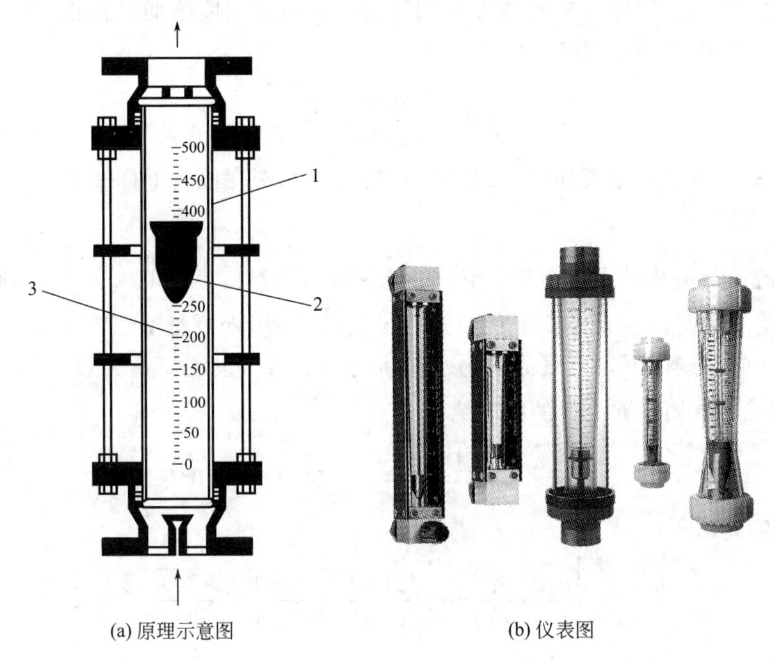

(a) 原理示意图　　(b) 仪表图

图1-63　转子流量计

1—锥形玻璃管；2—转子；3—刻度

当流体自下而上流过转子与玻璃管壁的环隙，由于转子上方截面较大，则环隙的截面较小，此处流速增大，压降减小，使转子上下两端产生压差，在此压差作用下对转子产生一个向上的推力。当该力超过转子的重力与浮力之差时，转子将上移，由于玻璃管是个倒锥形体，故该流体流道截面随之增大，在同一流程下，环隙流速减小，转子两端压差也随之降低。当转子上升到一定高度时，转子两端的压差造成的升力等于转子所受重力与浮力之差时，转子将稳定在这个高度上。由此可见，转子所处的平衡位置与流体流量大小直接相关，流量可由玻璃管上的刻度读出。

转子流量计的流量计算可由转子的力平衡方程导出。如图1-63所示，转子在一定流量下处于平衡状态，即压差产生的升力、转子重力及浮力之间力的平衡方程为：

$$(p_1-p_2)A_f=V_f(\rho_f-\rho)g \tag{1-77}$$

当转子处于某一平衡位置时，转子所受的压差恒定且转子与玻璃管间的环隙面积也固定，因此流体流过环隙通道的流量及压差的关系与孔板流量计相似，即：

$$V_s=C_RA_R\sqrt{\frac{2(p_1-p_2)}{\rho}} \tag{1-78}$$

式中，C_R 为转子流量计的系数；A_R 为玻璃管与转子之间的环隙面积。将式(1-91) 代入式(1-92)，可得流量计算公式：

$$V_s = C_R A_R \sqrt{\frac{2gV_f(\rho_f-\rho)}{A_f\rho}} \tag{1-79}$$

转子流量计的流量系数 C_R 与 Re 和转子形状有关，由实验测定。对于图 1-63 所示转子构型，当 Re 达到 10^4 以后，C_R 值便恒等于 0.98。

转子流量计在出厂时是根据 20℃水或 20℃、101kPa 下的空气进行标定，并将流量数值刻在玻璃管上。当被测流体与标定条件不相符时，应对原刻度值加以校正。

由于在同一刻度下 A_R 相同，则：

$$\frac{V_{s2}}{V_{s1}} = \sqrt{\frac{\rho_1(\rho_f-\rho_2)}{\rho_2(\rho_f-\rho_1)}} \tag{1-80}$$

式中，下标 1 表示标定流体（水或空气）的流量和密度值；下标 2 表示实际操作中所用流体的流量和密度值。

【例 1-22】 某液体转子流量计，转子为硬铅，其密度为 $11000kg/m^3$。现将转子改为形状、大小相同，而密度为 $1150kg/m^3$ 的胶质转子，用于测量空气（50℃、120kPa）的流量。试求在同一刻度下，空气流量为水流量的多少倍（设流量系数 C_R 为常数）？

解 50℃、120kPa 下空气的密度为：

$$\rho_2 = \frac{pM}{RT} = \frac{120\times29}{8.31\times(273+50)} = 1.30(kg/m^3)$$

由式(1-94) 可得：

$$\frac{V_2}{V_1} = \sqrt{\frac{\rho_1(\rho_{f2}-\rho_2)}{\rho_2(\rho_{f1}-\rho_1)}} = \sqrt{\frac{1000\times(1150-1.30)}{1.30\times(11000-1000)}} = 9.4$$

即同刻度下空气的流量为水流量的 9.4 倍。

Rotameters-area Meters

In the orifice, nozzle, or Venturi, the variation of flow rate through a constant area generates a variable pressure drop, which is related to the flow rate. Another class of meters, called area meters, consists of devices in which the pressure drop is constant, or nearly so, and the area through which the fluid flows varies with the flow rate. The area is related, through proper calibration, to the flow rate.

The most important area meter is the rotameter, which is shown in Fig. 1-63. It consists essentially of a gradually tapered glass tube mounted vertically in a frame with the large end up. The fluid flows upward through the tapered tube and suspends freely a float (which actually does not float but is completely submerged in the fluid). The float is the indicating element, and the greater the flow rate, the higher the float rides in the tube. The entire fluid stream must flow through the annular space between the float and the tube wall. The tube is marked in divisions, and the reading of the meter is obtained from the scale reading at the reading edge of the float, which is taken at the largest cross section of the float. A calibration curve must be available to convert the observed scale reading to flow rate. Rotameters can be used for either liquid or gas flow measurement.

Floats may be constructed of metals of various densities from lead to aluminum or from glass or plastic. Stainless－steel floats are common. Float shapes and proportions are also varied for different applications.

Rotameters have a nearly linear relationship between flow and position of the float, compared with a calibration curve for an orifice meter, for which the flow rate is proportional to the square root of the reading. The calibration of a rotameter, unlike that of an orifice meter, is not sensitive to the velocity distribution in the approaching stream, and neither long, straight approaches nor straightening vanes are necessary.

1.8.3 涡轮流量计（Turbine flowmeter）

涡轮流量计是一种速度式流量仪表，它的工作原理如图1-64所示，流体从机壳的进口流入，通过支架将一对轴承固定在管中心轴线上，涡轮安装在轴承上。在涡轮上下游的支架上装有呈辐射形的整流板，对流体起导向作用，以避免流体自旋而改变对涡轮叶片的作用角度。在涡轮上方机壳外部装有传感线圈，接收磁通变化信号。

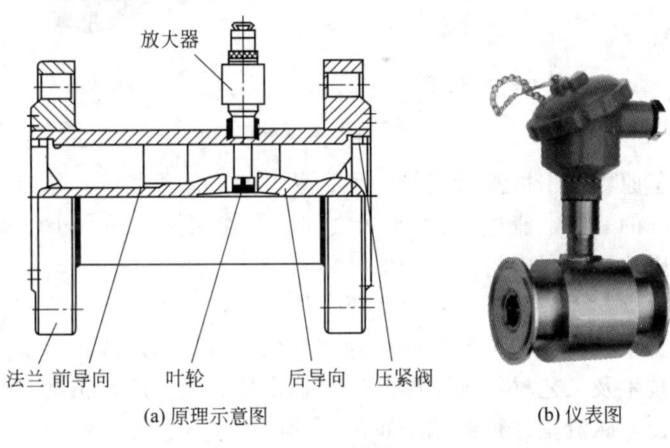

(a) 原理示意图 (b) 仪表图

图1-64 涡轮流量计

使用时，在管道中心安放一个涡轮流量计，两端由轴承支撑。当流体通过管道时，冲击涡轮叶片，对涡轮产生驱动力矩，使涡轮克服摩擦力矩和流体阻力矩而产生旋转。在一定的流量范围内，对于一定的流体介质黏度，涡轮的旋转角速度与流体流速成正比。由此，流体流速可通过涡轮的旋转角速度得到，从而可以计算得到通过管道的流体流量。

使用涡轮流量计的注意事项如下：

① 安装涡轮流量计前，管道要清扫；被测介质不洁净时，要加过滤器，否则涡轮、轴承易被卡住，测不出流量来。

② 拆装流量计时，对磁感应部分不能碰撞。

③ 投运前先进行仪表系数的设定，仔细检查，确定仪表接线无误，接地良好，方可送电。

④ 安装涡轮流量计时，前后管道法兰要水平，否则管道应力对流量计影响很大。

由于涡轮流量计具有测量精度高、反应速度快、测量范围广、价格低廉、安装方便等优点，被广泛应用于化工生产中。

1.8.4 电磁流量计（Electromagnetic flowmeter）

电磁流量计是基于法拉第电磁感应原理研制出的一种测量导电液体体积流量的仪表。根据法拉第电磁感应定律，导电体在磁场中作切割磁力线运动时，导体中产生感应电动势，该电动势的大小与导体在磁场中作垂直于磁场运动的速度成正比，由此再根据管径、介质的不同，转换成流量，如图 1-65 所示。

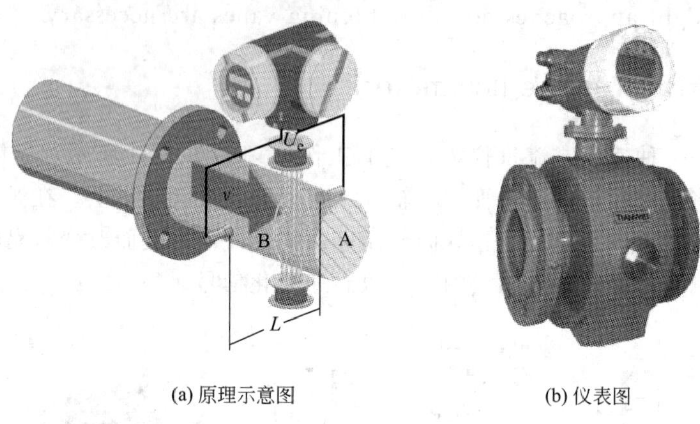

(a) 原理示意图　　(b) 仪表图

图 1-65　电磁流量计

电磁流量计选型原则：①被测量液体必须是导电的液体或浆液；②口径与量程，最好是正常量程超过满量程的一半，流速在 2～4 米之间；③使用压力必须小于流量计耐压；④不同温度及腐蚀性介质选用不同内衬材料和电极材料。

电磁流量计的优点：无节流部件，因此压力损失小，减少能耗，只与被测流体的平均速度有关，测量范围宽；只需经水标定后即可测量其他介质，无须修正，最适合作为结算用计量设备使用。由于技术及工艺材料的不断改进，稳定性、线性度、精度和寿命的不断提高和管径的不断扩大，使电磁流量计得到越来越广泛的应用。

电磁流量计的测量精度建立在液体充满管道的情形下，管道中有空气的测量问题目前尚未得到很好的解决。

1.8.5 超声波流量计（Ultrasonic flowmeter）

超声波流量计（图 1-66）是近代发展起来的一种新型测量流量的仪表，只要能传播声音的流体均可以用此类流量计测量，并可测量高黏度液体、非导电性液体或气体的流量，其测量的原理是：超声波在流体中的传播速度会随被测流体流速而变化。

超声波流量计的种类很多，包括多普勒式、时差式、便携式、管道式及固定式超声波流量计等。

超声波流量计的使用注意事项如下：

① 对于现场安装固定式超声波流量计数量大、范围广的用户，可以配备一台同类型的便携式超声波流量计，用于核校现场仪表的情况；

② 坚持一装一校，即对每一台新装超声波流量计在安装调试时进行核校，确保选位好、安装好、测量准；

③ 在线运行的超声波流量计发生流量突变时，要利用便携式超声波流量计进行及时核校，查清流量突变的原因，弄清楚是仪表发生故障还是流量确实发生了变化。

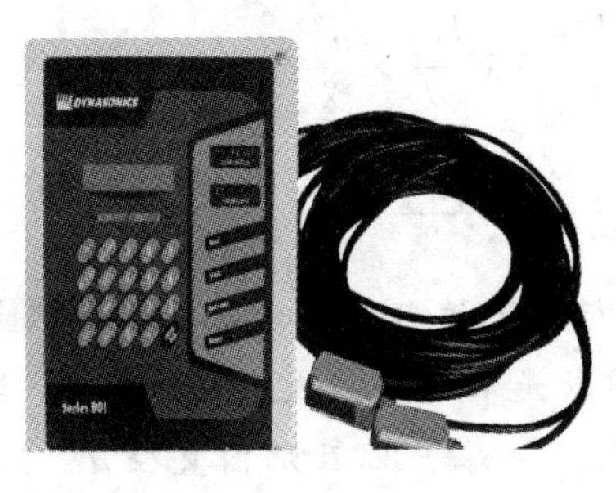

(a) 多普勒式

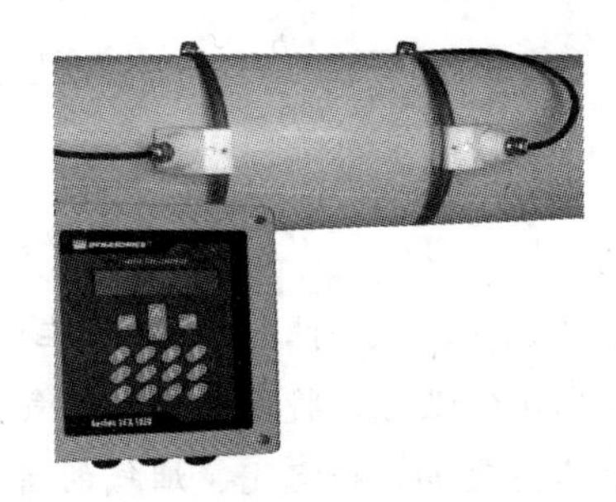

(b) 时差式

(c) 管道式

(d) 固定式

图 1-66 超声波流量计

工程案例分析

烟囱的工作原理

工业中燃烧炉烟囱的设计即遵循流体流动原理。附图为燃烧炉自然排烟系统的示意图，要使烟气从炉内排出，必须克服排烟系统的一系列阻力。烟气之所以能克服这些阻力，是由于烟囱能在其底部（2—2′面）形成吸力（真空）。若炉膛尾部（1—1′面）的压力为大气压，则在两截面间压力差的作用下，高温烟气就可经排烟烟道流进烟囱底部，最后由烟囱排至大气中。

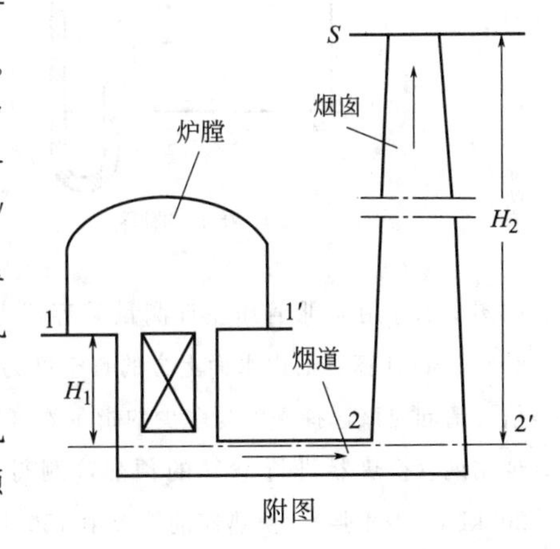

附图

设烟气的密度为 $\rho_{烟}$，空气的密度为 $\rho_{空气}$，烟气在烟囱中的平均流速为 u。在烟囱底部 2—2′截面与顶部 3—3′截面间列柏努利方程：

$$z_2 g+\frac{1}{2}u_2^2+\frac{p_2}{\rho_{烟}}=z_3 g+\frac{1}{2}u_3^2+\frac{p_3}{\rho_{烟}}+\sum W_{f,2-3} \tag{1}$$

若忽略烟囱直径的变化，则 $u_2=u_3$。

且：

$$p_3=p_a-\rho_{空} gH_2 \tag{2}$$

$$\sum W_{f,2\text{-}3}=\lambda\frac{H_2}{d}\times\frac{u^2}{2} \tag{3}$$

将式(2)、式(3) 代入式(1) 中，有：

$$\frac{p_2}{\rho_{烟}}=H_2g+\frac{p_A-\rho_{空}gH_2}{\rho_{烟}}+\lambda\frac{H_2}{d}\times\frac{u^2}{2}$$

$$p_A-p_2=[(\rho_{空}-\rho_{烟})g-\frac{\lambda}{d}\times\frac{\rho_{烟}u^2}{2}]H_2$$

可见，当 $\rho_{烟}<\rho_{空气}$ 时，轻的烟气在烟囱内进行向上的自然运动，从而在烟囱底部造成真空，形成烟囱的吸力，将炉膛内的烟气抽出，这就是烟囱能“拔烟”的原理。烟囱吸力的大小取决于其高度 H_2 和空气、烟气的密度差（$\rho_{空气}-\rho_{烟}$）以及烟囱直径 d。H_2、$\rho_{空气}-\rho_{烟}$ 或 d 越大，则吸力越大。

习　题

1. 在一个容器中，盛有苯和甲苯的混合物。已知苯的质量分数为 0.4，甲苯为 0.6，求 293K 时容器中混合物的密度。
2. 已知某混合气体的组成为 $18\%N_2$、$54\%H_2$ 和 $28\%CO_2$（均为体积分数），试求 $100m^3$ 的混合气体在温度为 300K 和 1MPa 下的质量。
3. 某设备进出口测压仪表的读数分别为 45mmHg（真空度）和 700mmHg（表压），求两处的绝对压强差是多少（kPa）？
4. 如图所示的储槽内盛有密度为 800 kg/m^3 的油，U 形管压差计中的指示液为水银，读数 $R=0.4m$，设容器液面上方的压强 $p_0=15kPa$（表压），求容器内的液面距底面的高度 h。

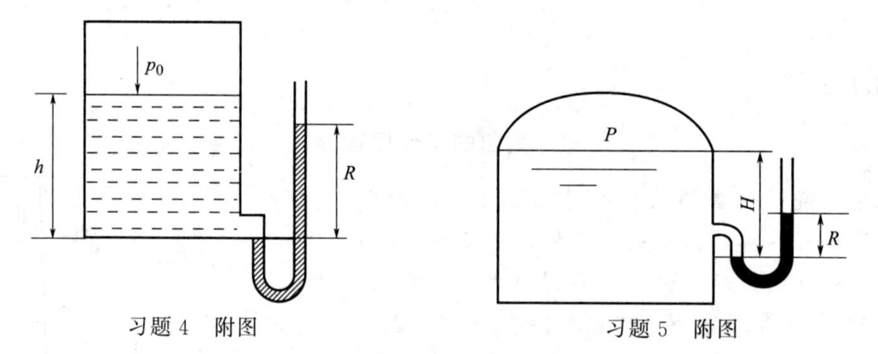

习题 4　附图　　　习题 5　附图

5. 如图所示，用 U 形管压差计测量某容器内水面上方气体的压强，指示液为水银，读数 $R=4cm$，$H=80cm$，试计算容器内水面上方的真空度为多少（kPa）？现在由于精度需要，希望读数增大到原来的 20 倍，请问应该选择密度为多少的指示液（U 形管左端的液位可以认为不变）？
6. 现用列管换热器进行空气的预热，测得空气在管内的流速为 15m/s，其平均温度为 60℃，压强为 200kPa（表压强），换热器的管束由 150 根 $\phi30mm\times2.5mm$ 的钢管组成。请计算：(1) 空气的质量流量；(2) 操作条件下空气的体积流量；(3) 标准状况下空气的体积流量。
7. 有一套管冷却器，内管为 $\phi25mm\times2.5mm$，外管为 $\phi57mm\times3.5mm$，冷冻盐水在套管环隙中流动，已知盐水流量为 3.73t/h，密度为 $1150kg/m^3$，黏度为 1.2mPa·s，试判断其流动类型。
8. 如图所示的高位槽的水面距出水管的垂直距离保持为 6m 不变，水管是采用内径 68mm 的钢管，设总的压头损失为 5.7m H_2O（不包括排水管出口的压头损失），试求每小时可输送的水量。
9. 如图所示为二氧化碳水洗塔的供水系统，水塔内绝对压强为 210kPa，储槽水面绝对压强为 100kPa，塔内水管入口处高于储槽水面 18.5m，管道内径为 52mm，送水量为 $15m^3/h$，塔内水管出口处绝对压强为 225kPa，设系统中全部的能量损失为 5m H_2O，求输水泵所需的外加压头。

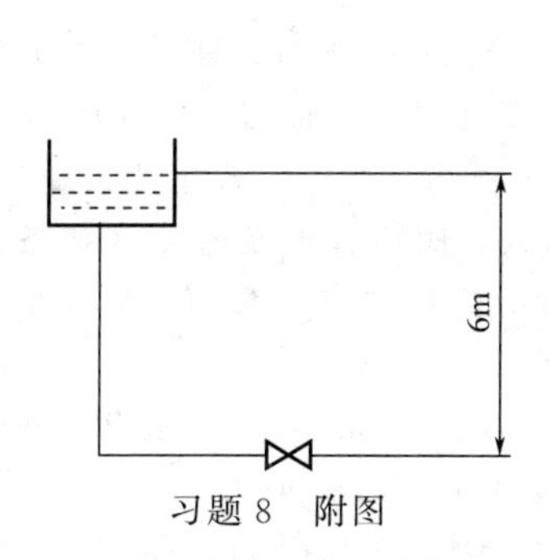

习题 8　附图

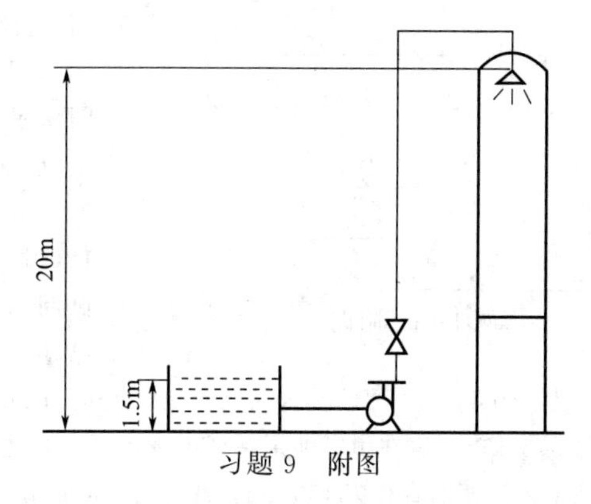

习题 9　附图

10. 如图所示，一管道由两部分组成，一部分管内径为40mm，另一部分管内径则为80mm，流体为水，在管道中流量为13.57m^3/h。两部分管道上均有一个测压点，测压点之间连接一个倒U形管压差计，其中充有一定的空气，若两测压点所在截面之间的摩擦损失为260mmH_2O，求倒U形管压差计中水柱的高度差R为多少（mm）？

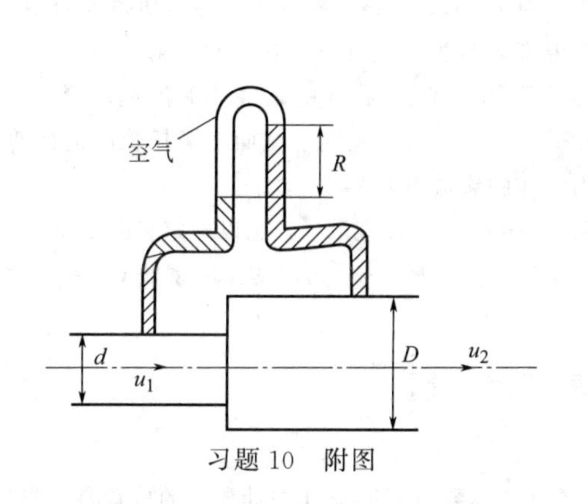

习题 10　附图

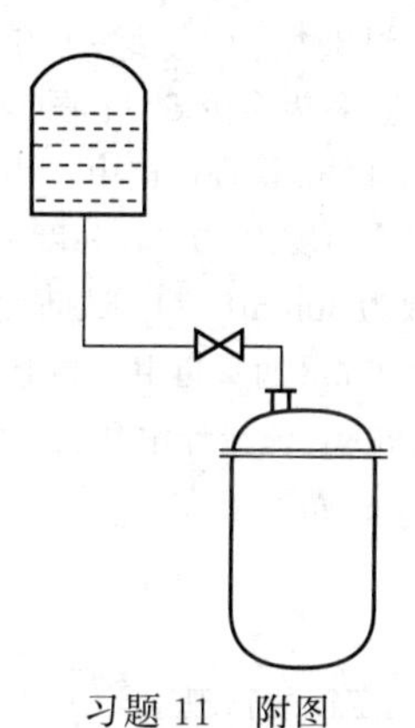
习题 11　附图

11. 如图所示，某料液从高位槽流向反应器，已知该溶液的密度为1100kg/m^3，黏度为1mPa·s，采用ϕ114mm×4mm的钢管，直管全长为20m，管道上有一个全开的闸阀和两个90°的标准弯头，其流量为31.7m^3/h，求总的压头损失。

12. 如图所示，用泵将20℃的苯从地面以下的储槽送到高位槽，流量为300L/min，高位槽的液面比储槽液面高10m。泵的吸入管为ϕ89mm×4mm的无缝钢管，直管长度为15m，并有一个底阀，一个90°弯头。泵排出管为ϕ57mm×3.5mm的无缝钢管，直管长度50m，并有一个闸阀，一个标准阀和三个90°弯头，两个三通。阀门均为全开，泵的效率为75%，试计算泵的轴功率。

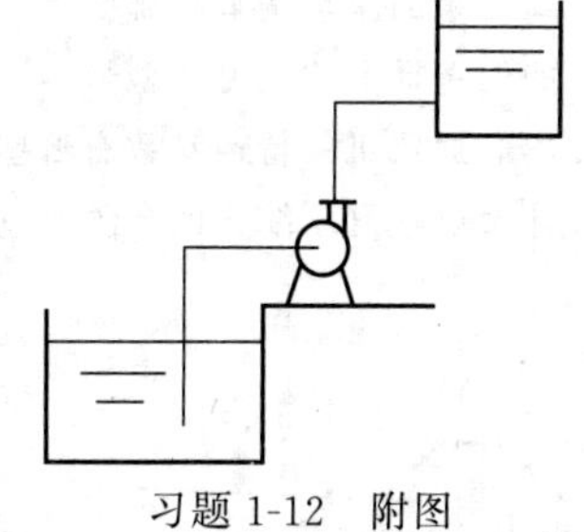
习题 1-12　附图

13. 采用离心泵输送293K的清水，测得离心泵出口处压力表的读数为4.8kgf/cm^2，入口处真空表的读数为147mmHg，两表之间的垂直距离为0.4m。该泵流量为70.37m^3/h，泵的效率为70%，吸入管和出口管的管径相同。试计算该泵的扬程以及泵的轴功率。

14. 某离心泵的铭牌上标有：流量30m^3/h，扬程24m，效率63.5%，电动机功率4kW，最大允许吸上真空度5.7m。现使用该离心泵从池中吸水送入敞口高位槽，高位槽高出地面15m，离心泵高出池面5.0m，要求水流量为6×10^{-3} m^3/s，吸入管长8m，排出管长20m（均包括局部阻力的当量长度），管径均为ϕ57mm×3mm，摩擦系数为0.021。通过计算说明：（1）该泵能否完成任务？（2）该泵安装是否合理？

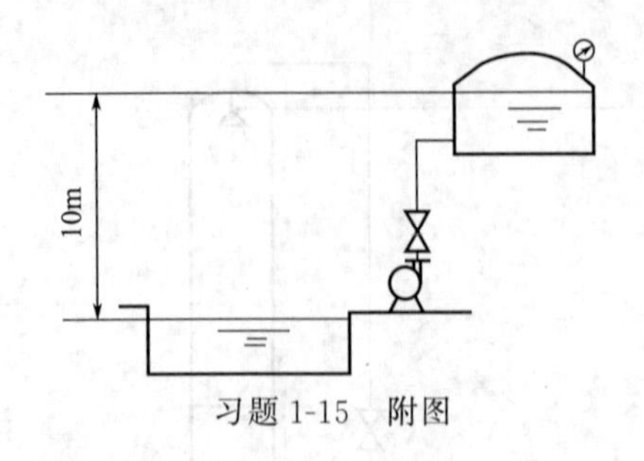

习题 1-15 附图

15. 用一离心泵将敞口水槽中的水输送到表压强为147.2kPa的高位密闭容器中。管路系统尺寸如本题附图所示，当供水量为$36m^3/h$时，管内流动以进入阻力平方区。已知泵的特性曲线方程为$H=45-0.00556Q^2$（Q单位为m^3/h）。若用此泵输送密度为$1200kg/m^3$的碱液，阀门开度及管路其他条件不变，试问输送碱液时的流量和离心泵的有效功率为多少？已知水的密度为$1000kg/m^3$。

16. 将密度为$1500kg/m^3$的硝酸送入反应釜，流量为$8m^3/h$，提升高度为8m，釜内压强为400kPa，管路的压降为30kPa，试在耐腐蚀泵系列中选定一个合适的型号，并估计泵的轴功率。

17. 单动往复泵活塞的直径为160mm，冲程为200mm。现拟用此泵将密度为930 kg/m^3的液体从储槽送至某设备中，流量为$25.8m^3/h$，设备的液体入口比储槽液面高19.5m，设备内液面上方的压力为0.32MPa（表压），储槽为敞口，外界大气压为0.098MPa，管路的总压头损失为10.3m，当有15%的液体漏回和总效率为72%时，试分别计算此泵的活塞每分钟的往复次数及轴功率。

18. 某单级空气压缩机每小时将$360m^3$的空气压缩到0.8MPa（表压），设空气的压缩过程为：(1) 绝热压缩；(2) 等温压缩；(3) 多变压缩（多变指数为1.25）。问该压缩机所消耗的功率各为多少（kW）？压缩后空气温度各为多少℃？已知空气的进口温度为20℃，大气压为0.098MPa。

19. 在一内径为320mm的管道中，用毕托管测定平均分子量为64的某气体的流速。管内气体的压强为101.33 kPa，温度为40℃，黏度为0.022mPa·s。已知在管道同一截面上毕托管所连接的U形管压差计最大读数为$30mmH_2O$。试问此时管道内气体的流量为多大？

20. 在内径为156mm的管道中，装上一块孔径为78mm的孔板流量计，用以测定管路中苯的流量。已知苯的温度为293K，流量计中测压计的指示液为汞，读数$R=30mm$，设孔流系数$C_0=0.625$，试求管路中每小时苯的流量。

思 考 题

1. 说明牛顿黏性定律的物理意义。
2. 采用U形管压差计测某阀门前后的压差，压差计的读数与U形管压差计放置的位置有关吗？
3. 层流与湍流的本质区别是什么？
4. 影响离心泵性能的因素有哪些？
5. 什么是离心泵的汽蚀现象？为避免汽蚀可采取什么措施？
6. 离心泵的流量调节有哪些方法？各种方法的实质及优缺点是什么？
7. 如何选择泵的类型和型号？
8. 离心通风机的特性参数有哪些？若输送空气的温度增加，其性能如何变化？
9. 什么是往复压缩机的余隙？它对压缩过程有何影响？

名 词 术 语

英文/中文

actual head/实际压头
adiabatic compression/绝热压缩
affinity law/相似定律
agitator/搅拌器
air bound/气缚
automatic control valve/自动控制阀
axial flow/轴向流

英文/中文

best efficiency point/最佳效率点
black-iron/黑铁
blowcase/加压箱
brake horse power/制动功率
cam pump/凸轮泵
capacity/流率
carbon steel/碳钢

cast-iron/铸铁
cavitation/汽蚀
centrifugal blower/离心鼓风机
centrifugal compressor/离心压缩机
centrifugal pump/离心泵
characteristic curve/特性曲线
cheek valve/止回阀
circular casing/环壳
circulatory flow/环流
cold-drawn/冷拔
developed head/扬程
diaphragm pump/隔膜泵
diffuser/扩散器
discharge connection/排出管
discharge nozzle/排出口
discharge pressure/排出压力
double-acting/双动
double-suction/双吸
efficiency of the electric motor/电机效率
enclosed impeller/闭式叶轮
end suction/top discharge/轴向吸入/顶部排出
expansion/膨胀
fan/通风机
fitting / 管件
flanged cap/法兰盖
flanged connection/法兰连接
flange/法兰
flare fitting/扩口管件
flexible diaphragm/柔性膜
flexible metal hose/挠性金属管
follower ring/从动环
fouling/污垢
fully close/全关
fully open/全开
gate valve/闸式阀
gear pump/齿轮泵
globe valve/球心阀
graphite/石墨
handwheel/手轮
head-capacity relation/压头-流量关系
head/压头
heat-exchanger/热交换器
ideal pump/理想泵
impeller/叶轮
internal-gear pump/内啮合齿轮泵
isothermal compression/等温压缩
jet ejector/ 喷射器
kinetic energy/动能
lantern gland/液封环
lantern ring/液封环
leakage/泄漏
lobe pump/凸轮旋转泵
low-carbon steel/低碳钢
lubricant/润滑剂
mechanical efficiency/机械效率
mechanical seal/机械密封
multistage pump/多级泵
net positive suction head（NPSH）/允许汽蚀余量
nickel pipe/镀镍管
nominal diameter/公称直径
open impeller/开式叶轮
operate in series/串联操作
operating curve/操作曲线
operating point / 工作点
operation in parallel/并联操作
packing nut/密封螺母
packing / 填料
packless joints/无填料接合
performance characteristics/性能特性
performance curve/性能曲线
peristaltic pump/蠕动泵
pipe size/管子尺寸
plug cock/旋塞
plunger pump/柱塞泵
polytropic compression/多变压缩
polyvinyl chloride（PVC）/聚氯乙烯
positive-displacement blower/正位移鼓风机
positive-displacement pump/正位移泵
pressure head/压头
pumpable/可用泵输送的
pump priming/泵的启动
radial flow/径向流
reciprocating compressor/往复式压缩机
reciprocating piston/往复活塞
required input power/输入功率（制动功率）
rotary positive-displacement compressor/旋转式正位移压缩机
rotary pump/旋转泵

rotating member/回转空腔
saturated vapor pressure/饱和蒸气压
screwed fitting/螺纹接头
screwed pump/螺杆泵
sealed chamber/密封室
semi-open impeller/半开式叶轮
single-acting/单动
single-suction/单吸
soldered fitting/焊接管件
spur-gear pump/正齿轮泵
stationary diffusion vane/静态导叶
steam trap/疏水器
stoppage/中断
stuffing box/填料箱
suction and discharge nozzle/吸入口和排出口
suction connection/吸入连接
suction lift/吸力
suction line/吸入管
system head curve/管路特性曲线
theoretical head/理论压头
top suction /top discharge nozzle/顶部吸入口/顶部排出口
transportation of fluid/流体输送
vacuum pump/真空泵
vane pump/叶片泵
variable-speed driver/变速器
volumetric efficiency/体积效率
volumetric flow rate/体积流量
volute/蜗壳
wastewater line/废水管路
wrought-iron/熟铁

第2章

传热及传热设备

Chapter 2 Heat Transfer and Its Equipments

学习目标

通过本章学习，学生应具备下列知识和能力：

1. 掌握传热单元操作的基本原理；能够基于传热的基本原理，建立复杂工程问题中关于热量传递所涉及的数学模型，分析、解决复杂化学工程问题中的热量传递问题。

2. 掌握传热单元操作设备的结构特点、工作原理、操作方法；能够针对复杂化学工程问题所涉及的热量传递相关问题，进行合理分析并提出合理的解决措施。

3. 掌握传热操作设备的设计计算及选型；能够针对复杂化学工程问题，使用恰当的技术、资源、现代工程工具和信息技术工具，通过对复杂工程问题建模，选取合适的换热设备。

4. 熟悉相关的英文表达，增强国际交流能力，拓宽国际视野。

学习要求

1. 通过本章学习，掌握传热的基本原理和规律。

2. 能够运用热量传递的原理和规律去分析和计算传热过程的有关问题，并能够根据传热基本原理和规律设计和选择合适的换热器。

3. 能够对换热器能否满足换热要求进行核校，并能够采取有效方法强化换热或强化保温效果。

2.1 传热基本概念（Basic Concepts of Heat Transfer）

若系统或物体内存在温度差，必有热量的传递，热量总是自发地由温度较高部分向较低部分传递。传热的应用相当广泛，不仅在过程工业，就是在航空、电子、机电及日常生活等各个方面，都可以遇到许多加热、冷却、蒸发、制冷、凝结、隔热或保温等实际的传热问题。

在过程工业中，往往需要化学反应和单元操作过程的加热或冷却，以维持过程进行所需要的温度。例如在蒸馏操作中，为了使塔釜达到一定温度并产生一定量的上升蒸汽，就需要对塔釜液加热，同时为了使塔顶上升蒸汽冷凝以得到液体产品，还需要对塔顶蒸汽进行冷凝；再如在蒸发、干燥等单元操作中也都要向相应的设备加入或取出热量；此外，化工设备的保温、生产过程中热能的合理应用以及废热的回收等都涉及传热问题。

综上所述，化工生产中对传热过程的要求主要有以下两种情况：其一是强化传热过程，如各种换热设备中的传热；其二是削弱传热过程，如对设备或管道的保温，以减少热损失。显然，研究和掌握传热的基本规律，对探求强化或削弱传热的有效途径及方法具有十分重要的意义。

2.1.1 传热基本方式

根据传热机理不同，传热的基本方式分为热传导、热对流和热辐射三种。

热传导（conduction）简称导热，是指直接接触的系统之间或系统内各部分之间没有宏观的相对运动，仅仅依靠分子、原子及自由电子等微观粒子的热运动而实现热量传递的现象。导热在固体、液体和气体中均可进行，但它们的导热机理各有不同。气体热传导是气体分子作不规则热运动时相互碰撞的结果；液体热传导的机理与气体类似，是依靠分子、原子在其平衡位置附近振动；固体以两种方式传导热能，即自由电子的迁移和晶格振动。

热对流（convection）是指流体中各部分质点之间发生宏观相对运动和混合而引起的热量传递过程，即热对流只能发生在流体内部。热对流分为强制对流和自然对流两种。自然对流是指流体中因各部分温度不同而引起密度的差别，从而使流体质点间产生相对运动而进行的对流传热；因泵或搅拌等外力所产生的质点强制运动而进行的对流传热，称为强制对流。

热辐射（radiation）是能量以电磁波（electromagnetic waves）的形式在空间传递的现象。当任何物体的温度大于热力学零度时，都会以电磁波形式向外界辐射能量，当被另一物体部分或全部接收后，又重新变为热能，这种传热方式称为辐射传热（热辐射），即辐射传热即是物体间相互辐射和吸收能量的总结果。但只有当物体间的温度差别较大时，辐射传热才能成为主要的传热方式。

2.1.2 传热速率

传热速率（又称热流量）Q 是指单位时间内通过一台换热器的传热面或某指定传热面的热量。

$$Q=\frac{\text{传热推动力}}{\text{传热阻力}}$$

传热过程的推动力是指两流体之间的传热温度差，但在传热面的不同位置上，流体的温度差不同，因此在传热计算中，通常采用平均温度差表示；传热阻力则与具体的传热方式、流体物性、壁面材料等多个因素有关，具体将在本章后续相应部分分别介绍。

Heat Transfer and Its Applications

Practically all the operations that are carried out by the chemical engineer involve the production or absorption of energy in the form of heat. The laws governing the transfer of heat and the types of apparatus that have for their main object the control of heat flow are therefore of great importance. This section deals with heat transfer and its applications in process engineering.

Nature of heat flow. When two objects at different temperatures are brought into contact, heat flows from the object at the higher temperature to that at the lower temperature. The net flow is always in the direction of the temperature decrease. The mechanisms by which the heat may flow are three: conduction, convection, and radiation.

Conduction. If a temperature gradient exists in a continuous substance, heat can flow unaccompanied by any observable motion of matter. Heat flow of this kind is called conduction, and according to Fourier's law, the heat flux is proportional to the temperature gradient and opposite to it in sign, for one-dimensional heat flow, Fourier's law is shown as eq. 2-6.

In metals, thermal conduction results from the motion of free electrons, and there is close correspondence between thermal conductivity and electrical conductivity. In solids that are poor conductors of electricity and in most liquids, thermal conduction results from momentum transfer between adjacent vibrating molecules or atoms. In gases, conduction occurs by the random motion of molecules, so that heat is "diffused" from hotter regions to colder ones. The most common example of pure conduction is heat flow in opaque solids such as the brick wall of a furnace or the metal wall of a heat exchanger tube. Conduction of heat in liquids or gases is often influenced by flow of the fluids, and both conductive and convective processes are lumped together under the term convection or convective heat transfer.

Convection. Convection can refer to the flow of heat associated with the movement of a fluid, such as when hot air from a furnace enters a room, or to the transfer of heat from a hot surface to a flowing fluid. The second meaning is more important for unit operations, as it includes heat transfer from metal walls, solid particles, and liquid surfaces. The convective flux is usually proportional to the difference between the surface temperature and the temperature of the fluid. Unlike thermal conductivity, the heat-transfer coefficient is not an intrinsic property of the fluid, but depends on the flow patterns determined by fluid mechanics as well as on the thermal properties of the fluid.

Natural and forced convection. When currents in a fluid result from buoyancy forces created by density differences, and the density differences are caused by temperature gradients in the fluid, the action is called natural convection. When the currents are due to a mechanical device such as a pump or agitator, the flow is independent of density differences and is called forced convection. Buoyancy forces also exist in forced convection, but usually they have only a small effect.

Radiation. Radiation is a term given to the transfer of energy through space by electromagnetic waves. If radiation is passing through empty space, it is not transformed to heat or any other form of energy, nor is it diverted from its path. If, however, matter appears in its path, the radiation will be transmitted, reflected, or absorbed. It is only the absorbed energy that appears as heat, and this transformation is quantitative. For example, fused quartz transmits practically all the radiation that strikes it; a polished opaque surface or mirror will reflect most of the radiation impinging on it; a black or matte surface will absorb most of the radiation received by it and will transform such absorbed energy quantitatively to heat.

Monatomic and most diatomic gases are transparent to thermal radiation, and it is quite common to find that heat is flowing through masses of such gases both by radiation and by conduction-convection. Examples are the loss of heat from a radiator or uninsulated steam pipe to the air of a room and heat transfer in furnaces and other high-temperature gas-heating equipment. The two mechanisms are mutually independent and occur in parallel, so that one type of heat flow can be controlled or varied independently of the other. Conduction-convection and radiation can be studied separately and their separate effects added in cases where both are important. In very general terms, radiation becomes important at high temperatures and is independent of the circumstances of the flow of the fluid. Conduction-convection is sensitive to flow conditions and is relatively unaffected by temperature level.

2.2 热传导（Heat Transfer by Conduction）

2.2.1 热传导基本概念

（1）温度场与等温面

温度差的存在是产生导热的必要条件，而热量的传递与物体内部的温度分布有着密切的关系。所以，首先必须建立起有关温度分布的概念。

如果用直角坐标 x、y、z 来描述物体内各点的位置，以 θ 代表时间，t 代表温度，那么，在某一瞬间，温度在空间各点分布的综合情况称为温度场，其数学描述为：

$$t=f(x,y,z,\theta) \tag{2-1}$$

若温度场不随时间变化，即称为稳态温度场（steady temperature field），在稳态温度场中的导热叫作稳态导热（steady-state conduction），其数学关系如式(2-2) 所示：

$$t=f(x,y,z) \tag{2-2}$$

如只考虑温度仅沿着 x 方向发生变化，则称为稳态一维温度场（one-dimensional steady-state conduction），它具有最简单的数学表达式，即：

$$t=f(x) \tag{2-3}$$

在某一瞬间，温度场内温度相同的各点组成的面叫等温面（isothermal surface），用任意平面与等温面相切可得到等温线，如图 2-1 所示。等温面的特点是：由于某瞬间内空间任一点不可能同时有不同的温度，故温度不同的等温面彼此不相交；由于等温面上温度处处相等，故在等温面上将无热量传递，而沿等温面相交的任何方向，因温度发生变化，则有热量传递。

(2) 温度梯度（temperature gradient）

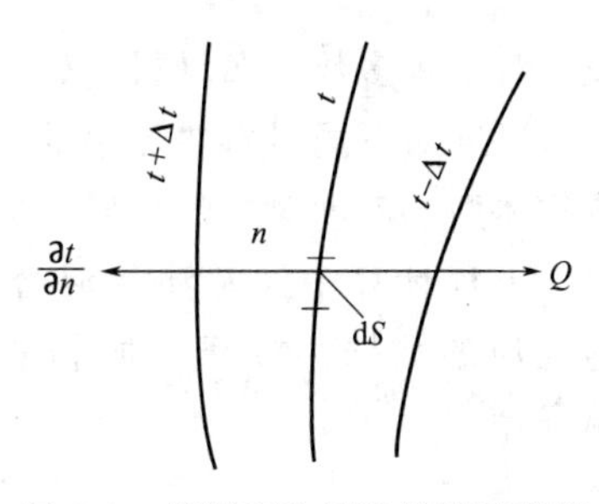

图 2-1　等温面和温度梯度示意图

由图 2-1 可知，不同温度的等温面之间存在温度差。所以，沿着与等温面相交的任何方向都有温度的变化，这种变化在法线方向上距离最短，单位长度的温度变化最大。为了描述这种变化情况，把两相邻等温面之间沿着法线方向的温度差 Δt 与法向距离 Δn 之比叫作温度梯度，即：

$$\mathrm{grad}(t)=\lim_{\Delta n\to 0}\frac{\Delta t}{\Delta n}=\frac{\partial t}{\partial n} \tag{2-4}$$

温度梯度是沿着与等温线垂直方向的矢量，它的方向以温度升高的方向为正，以温度降低的方向为负。

对于一维稳态温度场，等温线全都垂直于 x 方向，其温度梯度为：

$$\mathrm{grad}(t)=\frac{\mathrm{d}t}{\mathrm{d}x} \tag{2-5}$$

2.2.2　傅里叶定律（Fourier's law）

实验研究结果表明，热能总是朝着温度降低的方向传导，其导热速率的大小与温度梯度以及导热面积成正比，数学表达式为

$$\mathrm{d}Q=-\lambda \mathrm{d}S\frac{\partial t}{\partial n} \tag{2-6}$$

式中　Q——热传导速率，即单位时间传导的热量，其方向与温度梯度的方向相反，W；

S——与热传导方向垂直的导热面（等温面）面积，m^2；

λ——物质的热导率，W/(m·℃)。

上式称为傅里叶定律，是热传导的基本定律，式中负号表示热传导的方向与温度梯度的方向相反，即热量朝着温度下降的方向传递。

式(2-6) 可改写为：

$$\lambda=-\frac{\mathrm{d}Q}{\mathrm{d}S\frac{\partial t}{\partial n}} \tag{2-7}$$

上式说明，热导率在数值上等于单位导热面积、单位温度梯度、单位时间内传导的热量。因此，热导率是反映物质导热能力大小的参数，是物质的重要物理性质之一。

热导率一般用实验方法进行测定。通常金属固体的热导率最大，数值在 2.2～420W/(m·℃) 范围内（液态金属除外）；非金属固体的热导率较小，数值在 0.025～3W/(m·℃) 范围内；液体的热导率更小，数值在 0.07～0.7W/(m·℃) 范围内；而气体的热导率最小，数值在 0.006～0.6W/(m·℃) 范围内。热导率受物质的种类、温度、压力、湿度、密度以及物质组成结构型式的影响。作为参考，本书末附录表中摘录了某些工程上常用的 λ 值。

大多数均质固体的热导率与温度成直线关系，即：

$$\lambda=\lambda_0(1+at) \tag{2-8}$$

式中，λ_0 为 0℃时物质的热导率；a 为由实验测定的温度系数，可为正值，也可为负

值。对于大多数金属材料，a 为负值；而对于大多数非金属材料，a 为正值。

温度对不同材料热导率的影响不同，其数值可从有关手册中查到。在一般的情况下，可取物体两端温度对应的热导率的算术平均值，把 λ 当常数处理。

工程计算中经常涉及混合气体或混合溶液的热导率，一般由实验测定，也可以根据纯物质的数据，按摩尔加和法估算混合气体的热导率，按质量加和法估算混合液体的热导率。

2.2.3 固体平壁稳态热传导

（1）单层平壁热传导（conduction of single layer flat wall）

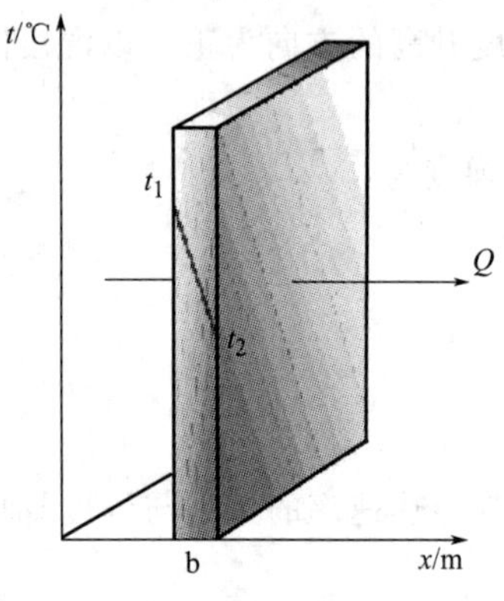

图 2-2 单层平壁的热传导

如图 2-2 所示，设有一厚度为 b 的无限大平壁（其长度和宽度远比厚度大），假设材料均匀，热导率不随温度变化；平壁内的温度仅沿垂直与平壁的方向变化（这里指 x 轴），即等温面垂直于传热方向；平壁面积与平壁厚度相比很大，故可以忽略热损失，现导出通过此平壁的导热速率的计算式。

按照上述问题的描述，此种热传导属于平壁一维稳态热传导（one-dimensional steady-state conduction），若边界条件（the boundary conditions）为：

$$x=0 \text{ 时}, t=t_1$$

$$x=\mathrm{b} \text{ 时}, t=t_2$$

对于一维稳态热传导，傅里叶定律式(2-6) 可表示为：$\mathrm{d}Q=-\lambda S\dfrac{\mathrm{d}t}{\mathrm{d}x}$，根据上述边界条件可推出单层平壁稳态热传导公式，如下：

$$Q=\frac{\lambda S}{b}(t_1-t_2)=\frac{t_1-t_2}{\dfrac{b}{\lambda S}}=\frac{\Delta t}{R} \tag{2-9}$$

式中 b——平壁厚度，m；

R——热传导热阻（conduction resistance），℃/W，$R=\dfrac{b}{\lambda S}$。

Δt——平壁两侧的温度差（temperature drop）（热传导推动力），℃。

式(2-9) 表明导热速率与导热推动力成正比，与导热热阻成反比；还可以看出，导热距离愈大，传热面积和热导率愈小，则导热热阻愈大。

式(2-9) 还可改写为：

$$q=\frac{Q}{S}=\frac{\lambda}{b}(t_1-t_2)=\frac{t_1-t_2}{\dfrac{b}{\lambda}} \tag{2-10}$$

式中，q 为单位时间、单位面积的导热量，称为导热通量或热流密度。

若设壁厚为 x 处的温度为 t，则可得平壁内任意位置的温度分布关系式：

$$t=t_1-\frac{Qx}{\lambda S} \tag{2-11}$$

式(2-11) 即为平壁内的温度分布，它是一条直线，当 $x=b$ 时，$t=t_2$。

【例 2-1】 有一钢板平壁，厚度为 5mm，若高温壁 $T_1=330\text{K}$，低温壁 $T_2=310\text{K}$，热导率 $\lambda=45.4\text{W/(m·K)}$，求该平壁在单位时间内、单位面积上所传递的热量。

解 本题实际上是求导热通量。已知 $T_1=330\text{K}$，$T_2=310\text{K}$，$b=0.005\text{m}$，$\lambda=45.4\text{W/(m·K)}$，将以上数据代入式(2-10) 中得：

$$q=\frac{Q}{S}=\frac{\lambda}{b}(t_1-t_2)$$
$$=\frac{45.4}{0.005}\times(330-310)$$
$$=181.6(\text{kW/m}^2)$$

(2) 多层平壁热传导 (conduction of a series of layers flat wall)

在工业上常见到的是多层平壁的导热问题，例如锅炉炉墙是由耐火砖、保温砖、普通砖等构成的多层平壁，以图 2-3 为例。假设多层平壁层与层之间接触良好，相互接触的表面上温度相等，在稳态导热情况下，边界条件为：

$$x=0\text{ 时}, t=t_1;\ x=b_1\text{ 时}, t=t_2$$
$$x=b_2\text{ 时}, t=t_3;\ x=b_3\text{ 时}, t=t_4$$

对各层分别应用由傅里叶定律推出的单层平壁热传导公式(2-9) 可得：

$$Q_1=\frac{\lambda_1 S}{b_1}(t_1-t_2)=\frac{t_1-t_2}{\dfrac{b_1}{\lambda_1 S}}=\frac{\Delta t_1}{R_1} \tag{2-12}$$

$$Q_2=\frac{\lambda_2 S}{b_2}(t_2-t_3)=\frac{t_2-t_3}{\dfrac{b_2}{\lambda_2 S}}=\frac{\Delta t_2}{R_2} \tag{2-12a}$$

$$Q_3=\frac{\lambda_3 S}{b_3}(t_3-t_4)=\frac{t_3-t_4}{\dfrac{b_3}{\lambda_3 S}}=\frac{\Delta t_3}{R_3} \tag{2-12b}$$

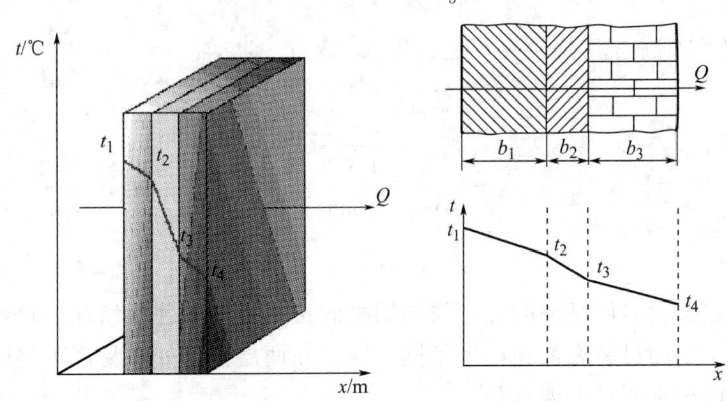

图 2-3 三层平壁的热传导

对于多层平壁热传导，在稳态导热过程中，热量在平壁内没有积累，因而数量相等的热量依次通过各层平壁，即各层平壁的传热速率相等，这是一种典型的串联传热过程，则式(2-12) 及式(2-12a)、式(2-12b) 可联合表示为：

$$Q=\frac{\lambda_1 S(t_1-t_2)}{b_1}=\frac{\lambda_2 S(t_2-t_3)}{b_2}=\frac{\lambda_3 S(t_3-t_4)}{b_3} \tag{2-13}$$

或：

$$Q=\frac{\Delta t_1}{\dfrac{b_1}{\lambda_1 S}}=\frac{\Delta t_2}{\dfrac{b_2}{\lambda_2 S}}=\frac{\Delta t_3}{\dfrac{b_3}{\lambda_3 S}} \tag{2-13a}$$

那么，3 层平壁导热速率的计算式可利用加和性原理，处理上式后整理得：

$$Q=\frac{\Delta t_1+\Delta t_2+\Delta t_3}{\dfrac{b_1}{\lambda_1 S}+\dfrac{b_2}{\lambda_2 S}+\dfrac{b_3}{\lambda_3 S}}=\frac{t_1-t_4}{\sum\limits_{i=1}^{3}\dfrac{b_i}{\lambda_i S}} \tag{2-14}$$

同理，推广至 n 层平壁的传导速率计算式为：

$$Q=\frac{t_1-t_{n+1}}{\sum\limits_{i=1}^{n}\dfrac{b_i}{\lambda_i S}} \tag{2-15}$$

由式(2-15) 可见，多层平壁热传导的总推动力为各层温度差之和，即总温度差；总热阻为各层热阻之和。

【例 2-2】 由耐火砖、硅藻土焙烧板和金属密封护板构成的炉墙，其热导率依次为 $\lambda_1=1.09\text{W/(m·K)}$，$\lambda_2=0.116\text{W/(m·K)}$，$\lambda_3=45\text{W/(m·K)}$。各层厚度为 $b_1=115\text{mm}$，$b_2=185\text{mm}$，$b_3=3\text{mm}$。炉墙的内表面平均温度为 642℃，外表面平均温度为 54℃，试求：(1) 炉墙单位面积的散热速率；(2) 耐火砖和硅藻土焙烧板交界处的温度。

解 已知 $\lambda_1=1/09\text{W/(m·K)}$，$\lambda_2=0.116\text{W/(m·K)}$，$\lambda_3=45\text{W/(m·K)}$；$b_1=115\text{mm}$，$b_2=185\text{mm}$，$b_3=3\text{mm}$；$t_1=642℃$，$t_4=54℃$。

(1) 炉墙单位面积的散热速率

将以上数据代入式(2-15) 得：

$$q=\frac{Q}{S}=\frac{t_1-t_{n+1}}{\sum\limits_{i=1}^{n}\dfrac{b_i}{\lambda_i}}=\frac{642-54}{\dfrac{0.115}{1.09}+\dfrac{0.185}{0.116}+\dfrac{0.003}{45}}=346(\text{W/m}^2)$$

(2) 耐火砖和硅藻土焙烧板交界处温度

$$\frac{t_1-t_2}{\dfrac{b_1}{\lambda_1}}=\frac{642-t_2}{\dfrac{0.115}{1.09}}=346$$

$$t_2=606℃$$

(3) 接触热阻

以上推导是在假设多层平壁的任意层与层之间接触良好的理想情况下得到的结论，但在实际操作中，由于不同材料表面粗糙度不同，所以在两层接触处极易产生接触热阻，因而导致了界面之间可能出现明显的温度降。

接触热阻产生的具体原因：两种材料接触表面间因粗糙不平而留有空穴，空穴中充满了空气，因此，传热过程包括通过实际接触面的热传导和通过空穴的热传导，因气体的热导率很小，所以热阻变大，也就是说接触热阻主要是由空穴造成的。表 2-1 列出了几种常见材料的接触热阻。

表 2-1　几种常见材料的接触热阻

接触面材料	粗糙度/μm	温度/℃	表压强/kPa	接触热阻/(m^2·℃/W)
不锈钢(磨光)、空气	2.54	90～200	300～2500	10^{-3}～0.264
铝(磨光)、空气	2.54	150	1200～2500	10^{-4}～0.88
铝(磨光)、空气	0.25	150	1200～2500	10^{-4}～0.18
铜(磨光)、空气	1.27	20	1200～20000	10^{-5}～0.7

2.2.4　固体圆筒壁稳态热传导（Steady-state conduction of heat flow through solid cylinders）

化工生产中，经常遇到圆筒壁的热传导问题，它与平壁热传导的不同之处在于圆筒壁的传热面积和热通量不再是常量，而是随半径而变，同时温度也随半径而变，但传热速率在稳态时依然是常量。

(1) 单层圆筒壁热传导

设有一长度为 L，内外半径各为 r_1 和 r_2 的单层圆筒壁，如图 2-4 所示。当 L 超过 $10r_2$ 时，在工程计算上可看成是无限长的圆筒壁，或者长度虽短，但两端被绝热，热量仅沿半径 r 方向改变，$t=f(r)$，温度场为一维，等温面是同轴的圆柱面。

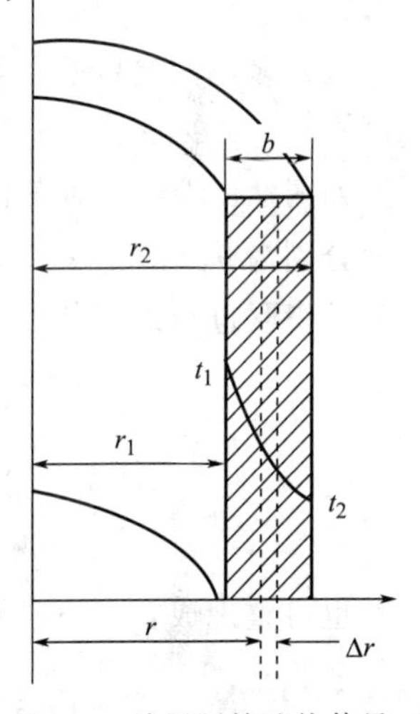

图 2-4　单层圆筒壁热传导

圆筒壁热传导面积可表示为：$S=2\pi rL$，即不是定值，是半径的函数，但传热速率在稳态时是常量。仿照平壁热传导公式，通过该圆筒壁的导热速率可以表示为：

$$Q=-\lambda(2\pi rL)\frac{dt}{dr} \tag{2-16}$$

分离变量后得：

$$dt=-\frac{Q}{2\pi L\lambda}\times\frac{dr}{r} \tag{2-17}$$

根据边界条件：$r=r_1$，$t=t_1$

$r=r_2$，$t=t_2$

将式(2-17) 积分并整理得：

$$Q=\frac{2\pi L\lambda(t_1-t_2)}{\ln\dfrac{r_2}{r_1}} \tag{2-18}$$

式(2-18) 即为单层圆筒壁的热传导速率方程。也可将该式分子、分母同乘 r_2-r_1，化成与平壁热传导速率方程相类似的形式，即：

$$Q=\frac{2\pi L(r_2-r_1)\lambda(t_1-t_2)}{(r_2-r_1)\ln\dfrac{r_2}{r_1}}=\lambda S_m\frac{t_1-t_2}{b}=\frac{t_1-t_2}{\dfrac{b}{\lambda S_m}}=\frac{\Delta t}{R} \tag{2-19}$$

式中　b——圆筒壁的厚度，m，$b=r_2-r_1$。

对比式(2-18)与式(2-19) 可知：

$$S_m = 2\pi\frac{r_2 - r_1}{\ln(r_2/r_1)}L = 2\pi r_m L \tag{2-20}$$

其中：

$$r_m = \frac{r_2 - r_1}{\ln\dfrac{r_2}{r_1}} \tag{2-21}$$

或：

$$S_m = \frac{2\pi L r_2 - 2\pi L r_1}{\ln(2\pi L r_2/2\pi L r_1)} = \frac{S_2 - S_1}{\ln(S_2/S_1)} \tag{2-22}$$

式中 S_m——圆筒壁的对数平均导热面积，m^2；

r_m——圆筒壁的对数平均半径，m。

对于圆筒壁的对数平均半径，当$\frac{r_2}{r_1}\leqslant 2$或对数平均面积$\frac{S_2}{S_1}\leqslant 2$时，可采用算术平均值代替，这时算术平均值与对数平均值相比，计算误差仅为4%，这是工程计算允许的。

设任意壁厚r处的温度为t，则可得出圆筒壁内温度分布的对数曲线关系，即：

$$t = t_1 - \frac{Q}{2\pi L\lambda}\ln\frac{r}{r_1} \tag{2-23}$$

（2）多层圆筒壁热传导（conduction of heat flow through a series of cylinders）（以三层为例）

如图2-5所示，对于各层间接触良好的三层圆筒壁，各层的热导率分别为λ_1，λ_2，λ_3；厚度分别为$b_1 = r_2 - r_1$；$b_2 = r_3 - r_2$和$b_3 = r_4 - r_3$，三层圆筒壁的导热过程可视为各单层圆筒壁串联进行的导热过程，那么与多层平壁热传导的计算类似，得到三层圆筒壁的计算式如下：

$$Q = \frac{2\pi L(t_1 - t_4)}{\frac{1}{\lambda_1}\ln\frac{r_2}{r_1} + \frac{1}{\lambda_2}\ln\frac{r_3}{r_2} + \frac{1}{\lambda_3}\ln\frac{r_4}{r_3}} \tag{2-24}$$

也可整理成：

$$Q = \frac{\Delta t_1 + \Delta t_2 + \Delta t_3}{\frac{b_1}{\lambda_1 S_{m1}} + \frac{b_2}{\lambda_2 S_{m2}} + \frac{b_3}{\lambda_3 S_{m3}}} = \frac{t_1 - t_4}{R_1 + R_2 + R_3} = \frac{\sum\Delta t_i}{\sum R} \tag{2-25}$$

式中：$S_{m1} = \frac{2\pi L(r_2 - r_1)}{\ln\frac{r_2}{r_1}}, S_{m2} = \frac{2\pi L(r_3 - r_2)}{\ln\frac{r_3}{r_2}}, S_{m3} = \frac{2\pi L(r_4 - r_3)}{\ln\frac{r_4}{r_3}}$

那么以此类推，n层圆筒壁热传导速率方程为：

$$Q = \frac{t_1 - t_{n+1}}{\sum\limits_{i=1}^{n}\frac{b_i}{\lambda_i S_{mi}}} = \frac{t_1 - t_{n+1}}{\sum\limits_{i=1}^{n}\frac{1}{2\pi L\lambda_i}\ln\frac{r_{i+1}}{r_i}} \tag{2-26}$$

式中，下标i表示圆筒壁的序号。

上述关于固体壁的一维稳态热传导的计算问题，虽然有平壁与圆筒壁、单层壁与多层壁之分，但就热传导速率计算式的形式而言，则是完全相同的，即都可采用一般的公式表示：

$$Q=\frac{\sum \Delta t}{\sum_{i=1}^{n}\frac{b_i}{\lambda_i S_i}}=\frac{\sum \Delta t}{\sum_{i=1}^{n}R_i} \tag{2-27}$$

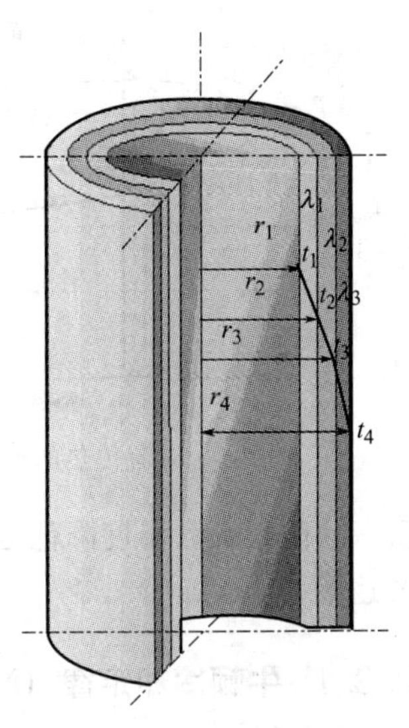

图 2-5 三层圆筒壁热传导

[讨论]：

① 与多层平壁一样，多层圆筒壁导热的总推动力亦为总温度差，总热阻亦为各层圆筒热阻之和，只是计算各层热阻所用的传热面积不相等，而应采用各自的平均面积。其中，对于平壁，任意层的导热面积均相等；对于圆筒壁，导热面积随半径而变化，故计算中常取对数平均面积 S_m；对于空心球壁，有 $S=4\pi r^2$，计算式中传热面积取几何平均面积 $S_m=\sqrt{S_1S_2}$。

② 由于各层圆筒的内外表面积不等，所以在稳态传热时，单位时间通过各层的传热量，即导热效率 Q 虽然相同，但单位时间单位面积通过各层内壁和外壁的热量，即热量通量 q 却不相等，有下面的关系式：

$$Q=2\pi r_1Lq_1=2\pi r_2Lq_2=2\pi r_3Lq_3=2\pi r_4Lq_4 \tag{2-28}$$

可化简为：

$$r_1q_1=r_2q_2=r_3q_3=r_4q_4 \tag{2-28a}$$

式中 Q——导热速率，J/s；

q——导热通量，J/(m^2·s)。

③ 材料的热导率视为常数，是指工程计算时一般取两侧壁面温度下的热导率的算术平均值。

【例 2-3】 某物料管路的管内、外直径分别为 160 mm 和 170 mm。管外包有两层绝热材料，内层绝热材料厚 20 mm，外层厚 40 mm。管子及内、外层绝热材料的 λ 值分别为 58.2W/(m·℃)、0.174W/(m·℃) 及 0.093W/(m·℃)。已知管内壁温度为 300℃，外层绝热层的外表面温度为 50℃，求每米管长的热损失。

解

$$\frac{Q}{L}=\frac{2\pi(t_1-t_4)}{\frac{1}{\lambda_1}\ln\frac{r_2}{r_1}+\frac{1}{\lambda_2}\ln\frac{r_3}{r_2}+\frac{1}{\lambda_3}\ln\frac{r_4}{r_3}}$$

因：

$$\frac{d_2}{d_1}=\frac{170}{160},\frac{d_3}{d_2}=\frac{170+40}{170}=\frac{210}{170},\frac{d_4}{d_3}=\frac{210+80}{210}=\frac{290}{210}$$

所以：

$$\frac{Q}{L}=\frac{2\pi(300-50)}{\frac{1}{58.2}\ln\frac{170}{160}+\frac{1}{0.174}\ln\frac{210}{170}+\frac{1}{0.093}\ln\frac{290}{210}}=335.2(\mathrm{W/m})$$

2.3 对流传热 (Heat transfer by convection)

对流传热在工程技术中非常重要，工业生产中经常遇到两流体之间或流体与壁面之间的换热问题，这类问题需用对流传热理论予以解决。如图 2-6 所示，当流体流经固体壁面时，由于流体黏性的存在，靠近壁面处存在一层很薄的层流内层，其外侧有一个过渡区，然后是湍流主体区。层流内层中流体层之间平行流动，以导热方式传热；而湍流主体内流体质点剧

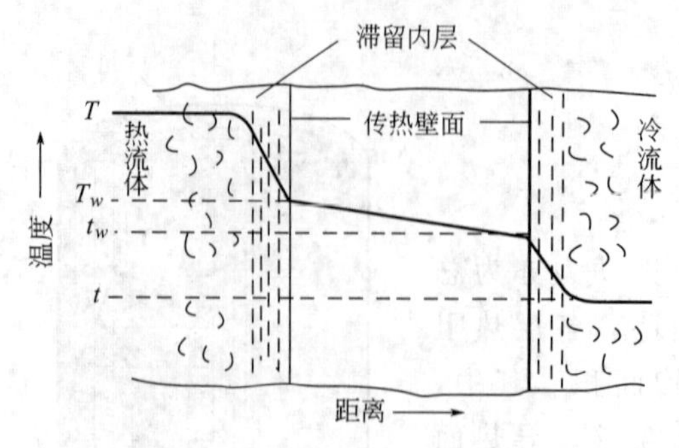

图 2-6　对流传热的温度分布情况

烈湍动，流体以热对流方式传热，由于主体各部分充分混合，使得流速趋于一致，温度也趋于一致，即流体与固体壁面之间进行对流传热时，热对流总是伴随着热传导同时发生。对流传热时，与流体流动方向垂直的同一截面上的温度分布情况也如图 2-6 所示。

由图 2-6 可知，固体壁面两侧均存在层流内层，层流内层的传热以导热方式进行。流体层流内层虽然很薄，但由于流体热导率很小，所以热阻很大，温度降低也主要集中在这里，因此，减薄层流内层厚度是强化对流传热的主要途径。

2.3.1 牛顿冷却定律（Newton's law of cooling）

影响对流传热的因素很多，问题相当复杂，而且不同的对流传热情况又有差别，因此目前的工程计算仍按半经验法处理。根据传递过程普遍关系，壁面与流体间（或反之）的对流传热速率也应该等于推动力和阻力之比，即：

对流传热速率＝对流传热推动力/对流传热阻力＝系数×推动力

上式中推动力是指壁面和流体间的温度差，影响阻力的因素很多，一般将对流传热的全部阻力均集中在厚度为 δ 的有效层流膜（虚拟膜）内。

还应指出，在换热器中，沿流体流动方向上，流体和壁面的温度一般是变化的，在换热器不同位置上的对流传热速率也随之而异，所以对流传热速率方程应该用微分形式表示。

若以流体和壁面间的对流传热为例，对流传热速率方程可以表示为：

$$\mathrm{d}Q=\frac{T-T_{\mathrm{w}}}{\dfrac{1}{\alpha \mathrm{d}s}}=\alpha \mathrm{d}S(T-T_{\mathrm{w}}) \tag{2-29}$$

式中　$\mathrm{d}Q$——局部对流传热速率，W；

$\mathrm{d}S$——微元传热面积，m^2；

T——换热器任一截面上热流体的主体温度，℃；

T_{w}——换热器任一截面上与热流体相接触一侧的壁面温度，℃；

α——局部对流传热系数，W/（m^2·℃）。

上式称为牛顿冷却定律，描述了换热器任一截面上的局部对流传热规律。

在换热器中，局部对流传热系数 α 随管长而变化，但是在工程计算中，常常使用平均对流传热系数 α_{m} 来描述整个换热器内的对流传热情况，此时牛顿冷却定律可以表示为：

$$Q=\alpha_{\mathrm{m}}S\Delta t \tag{2-30}$$

式中　Q——整个换热器内流体与壁面间的总传热速率，W；

α_{m}——平均对流传热系数，W/（m^2·℃）；

S——总传热面积，m^2；

Δt——流体与壁面间温度差的平均值，℃。

还应指出，换热器的传热面积表示方法不同，则牛顿冷却定律就有不同的形式。例如，若热流体在换热器的管间（环隙）流动，冷流体在换热器的管内流动，则与之对应的对流传热速率方程式可分别表示为：

$$dQ=\alpha_o(T-T_w)dS_o=\alpha_i(t_w-t)dS_i \tag{2-31}$$

式中 S_i，S_o——换热器的内、外侧表面积，m^2；

α_i，α_o——换热器的内、外侧流体对流传热系数，W/（m^2·℃）；

T——换热器任一截面上热流体的主体温度，℃；

T_w——换热器任一截面上与热流体相接触一侧的壁面温度，℃；

t——换热器任一截面上冷流体的主体温度，℃；

t_w——换热器任一截面与冷流体相接触一侧的壁温，℃。

从以上表达式可知，对流传热系数必然是和传热面积以及温度差相对应。牛顿冷却定律表达了复杂的对流传热问题，实质上是将矛盾集中到对流传热系数α，因此研究各种对流传热情况下α的大小、影响因素及α的计算式，成为研究对流传热的核心。α的物理意义为单位温度差时，在单位时间内通过单位面积以对流方式所传递的热量，它反映了对流传热的强度。

牛顿冷却定律并非理论导出，而是一种经验推论。因此，用该定律描述对流传热，将一系列影响对流传热过程的因素隐藏在对流传热系数α里，并没有改变问题本身的复杂性。为了揭示对流传热的本质，应该了解各影响因数与α的联系，建立相互之间的函数关系，以便计算α。目前，常用两种方法：解析法和经验公式法。前者只有在少数情况下，才能求得解析解，详细的介绍可参见传热学专著或化工传递过程；后者是通过分析影响α的因数，结合实验建立关联式。

实验表明，影响α的主要因素如下。

（1）流体的种类和相变化的情况

各种液体、气体或蒸汽的α值是不同的，牛顿型流体和非牛顿型流体的α值也有区别，流体有无相变化，对传热也有不同的影响。

（2）流体的物理性质

流体的密度、黏度、比热容、热导率以及容积膨胀系数等物理性质不同，得到的α值不同。例如流体的热导率，对于层流对流传热来说是一个决定性因数，而在湍流对流传热过程中，它直接支配着层流内层的导热性能，也是一个不可忽视的因数。在其他条件相同时，热导率较大的流体，其α值也较大。

（3）流体的流动状况

流体的流动状况直接影响对流传热结果，所以层流和湍流状态对应的α值明显不同；而同为湍流时，因湍动程度影响层流内层的厚度，因而也影响α值。

（4）引起流体流动的原因

按引起流体流动的原因来分类，可分为自然对流和强制对流。自然对流是指由于流体各部分温度不同而导致密度差异所引起的流动，例如利用暖气取暖就是一个典型的实例，此时，暖气片周围受热的那部分气体因密度减小而上升，附近密度较大的空气就流过来补充，这种流体的密度差使流体产生所谓升浮力。升浮力的大小取决于流体的受热情况、物理性质以及流体所在空间的大小和形状。强制对流是由于外力的作用，如泵、风机、搅拌器等迫使流体流动。

实际上进行对流传热时，流体作强制对流的同时，也会有自然对流存在。当强制对流的速度很大时，自然对流的影响可忽略不计。

（5）传热表面的形状、大小及位置情况

在对流传热时，流体沿着壁面流动，壁面的形状（如圆管、平板、管束）、大小（如长度、直径等）及位置（如水平、倾斜或垂直放置等）对流体流动有很大的影响，从而影响对流传热情况。

对流传热系数的确定是一个极其复杂的问题，影响因素很多，一般的处理方法是针对具体情况，用量纲分析方法得出特征数表达式，再用实验确定特征数之间的具体关系，进而得到特征数关联式加以表达。类似于第一章中介绍的量纲分析方法，可得到影响对流传热系数的特征数表达式为：

$$Nu=f(Re,Pr,Gr) \tag{2-32}$$

式中各特征数的名称、符号及意义见表 2-2。

表 2-2　特征数关联式

特征数名称	符号及特征数式	意义
努塞尔特数 (Nusselt number)	$Nu=\frac{\alpha l}{\lambda}$	表示对流传热系数的特征数
雷诺数 (Reynolds number)	$Re=\frac{du\rho}{\mu}$	表示流体流动状态和湍动程度对对流传热的影响
普兰特数 (Prandtl number)	$Pr=\frac{C_p\mu}{\lambda}$	表示流体物性对对流传热的影响
格拉斯霍夫数 (Grashaf number)	$Gr=\frac{\beta g\Delta t l^3\rho^2}{\mu^2}$	表示自然对流对对流传热的影响

注：流体无相变时强制对流传热过程中，$Nu=f(Re,Pr)$。流体无相变时自然对流传热过程中，$Nu=f(Pr,Gr)$。

[讨论]：

使用由实验数据整理得到的关联式应注意的问题：各种不同情况下的对流传热的具体函数关系由实验来决定，在整理实验结果及使用关联式时必须注意以下问题。

（1）应用范围

在使用对流传热的经验公式时，必须注意符合公式的应用条件，例如关联式中 Re、Pr 等特征数的数值范围等。

（2）特征尺寸

Nu、Re 等特征数中所包含的传热相关尺寸 l 称为特征尺寸，该尺寸如何确定要看公式的具体要求，例如流体在圆管内强制对流传热时，特征尺寸取管内径；而对于非圆形管通常取当量直径等。

（3）定性温度

决定特征数中各类流体物性的温度称为定性温度，常用以下几种表示法：

① 取流体的平均温度 $t=\frac{(t_1+t_2)}{2}$ 为定性温度，其中 t_1，t_2 分别为流体的进、出口温度；

② 取壁面的平均温度 t_w 为定性温度；

③ 取膜温，即流体和壁面的平均温度 $t_m=\frac{(t_w+t)}{2}$ 为定性温度。

工程上大多以流体的平均温度为定性温度，使用经验公式时必须按照公式规定的定性温度进行计算。

Regimes of Heat Transfer

A fluid being heated or cooled may be flowing in laminar flow, in turbulent flow, or in the transition range between laminar and turbulent flow. Also, the fluid may be flowing in forced or natural convection. In some instances more than one flow type may occur in the same stream; for instance, in laminar flow at low velocities, natural convection maybe superimposed on forced laminar flow.

The direction of flow of the fluid may be parallel to that of the heating surface, so that boundary layer separation does not occur; or the direction of flow may be perpendicular or at an angle to the heating surface, and then boundary layer separation often occurs.

At ordinary velocities the heat generated from fluid friction is usually negligible in comparison with the heat transferred between the fluids. In most cases friction heating may be neglected. It may be important, however, in operations involving very viscous fluids such as the injection molding of polymers. Friction heating of crude oil in the Alaska pipeline helps keep the oil above the ambient temperature. This decreases the viscosity and lowers the pumping cost. In gas flow at high velocities, at Mach numbers approaching 1.0, friction heat becomes appreciable and cannot be ignored. At very high velocities it may be of controlling importance, as it is for spacecraft reentering the earth's atmosphere.

Because the conditions of flow at the entrance to a tube differ from those well downstream from the entrance, the velocity field and the associated temperature field may depend on the distance from the tube entrance. Also, in some situations the fluid flows through a preliminary length of unheated or uncooled pipe, so that the fully developed velocity field is established before heat is transferred to the fluid, and the temperature field is created within an existing velocity field.

Finally, the properties of the fluid-viscosity, thermal conductivity, specific heat, and density are important parameters in heat transfer. Each of these, especially viscosity, is temperature-dependent. Since the temperature varies from point to point in a flowing stream undergoing heat transfer, a problem appears in the choice of temperature at which the properties should be evaluated. For small temperature differences between fluid and wall and for fluids with weak dependence of viscosity on temperature, the problem is not acute. But for highly viscous fluids such as heavy petroleum oils or where the temperature difference between the tube wall and the fluid is large, the variations in fluid properties within the stream become large, and the difficulty of calculating the heat-transfer rate is increased.

Because of the various effects noted above, the entire subject of heat transfer to fluids without phase change is complex and in practice is treated as a series of special cases rather than as a general theory. All cases considered in this chapter do, however, have a phenomenon in common: In all the formation of a thermal boundary layer, analogous to the hydrodynamic Prandtl boundary layer described before, takes place; it profoundly influences the temperature field and so controls the rate of heat flow.

2.3.2 无相变对流传热（Heat transfer to fluids without phase change）系数计算

2.3.2.1 无相变时流体在管内流动的 α 计算

（1）流体在圆形直管内做强制湍流（heat transfer by force convection in turbulent flow in straight tubes）

① 低黏度流体（约低于 2 倍常温水的黏度）可应用迪特斯-贝尔特关联式（Dittus-Boelter correlation）：

$$Nu=0.023Re^{0.8}Pr^{n} \tag{2-33}$$

$$\alpha=0.023\frac{\lambda}{d_i}\left(\frac{d_i u\rho}{\mu}\right)^{0.8}\left(\frac{C_p\mu}{\lambda}\right)^{n} \tag{2-33a}$$

当流体被加热时，$n=0.4$；当流体被冷却时，$n=0.3$。

应用范围：$Re>10^4$，$0.7<Pr<120$，$\frac{L}{d_i}>60$（L 为管长），若$\frac{L}{d_i}<60$，需考虑传热进口段对 α 的影响，此时可将由式(2-33a) 求得的 α 值乘以 $\left[1+\left(\frac{d_i}{L}\right)^{0.7}\right]$ 进行校正。特征尺寸：管内径 d_i。

定性温度：流体进、出口温度的算术平均值。

② 高黏度流体可应用西德尔（Sieder)-泰特（Tate）关联式：

$$Nu=0.027Re^{0.8}Pr^{1/3}\varphi_w \tag{2-34}$$

式中，$\varphi_w=\left(\frac{\mu}{\mu_w}\right)^{0.14}$ 是考虑热流方向的校正项。其中，液体被加热时，取 $\varphi_w=1.05$，液体被冷却时，$\varphi_w=0.95$；对气体，则不论加热或冷却，均取 $\varphi_w\approx1.0$，μ_w 为壁面温度下流体的黏度。

应用范围：$Re>10^4$，$0.7<Pr<16700$，$\frac{L}{d_i}>60$（L 为管长）。

特征尺寸：管内径 d_i。

定性温度：除 μ_w 取壁温外，均取流体进、出口温度的算术平均值。

（2）流体在圆形直管内呈层流流动（heat transfer by force convection in laminar flow in straight tubes）

流体在圆形直管内层流流动时，应考虑自然对流的影响，情况比较复杂，关联式的误差比湍流时的要大。当管径较小，且流体和壁面的温差不大时，自然对流的影响可以忽略，这时可采用西德尔（Sieder)-泰特（Tate）关联式：

$$Nu=1.86\left(RePr\frac{d_i}{L}\right)^{1/3}\left(\frac{\mu}{\mu_w}\right)^{0.14} \tag{2-35}$$

或：

$$\alpha=1.86\frac{\lambda}{d_i}Re^{1/3}Pr^{1/3}\left(\frac{d_i}{L}\right)^{1/3}\left(\frac{\mu}{\mu_w}\right)^{0.14} \tag{2-35a}$$

应用范围：$Re<2300$，$0.6<Pr<6700$，$RePr\frac{d_i}{L}>100$。

特征尺寸：管内径 d_i。

定性温度：除 μ_w 取壁温外，均取流体进、出口温度的算术平均值。

上式适用于管长较小时 α 的计算，但当管子极长时则不再适用。

必须指出，由于层流时对流传热系数很低，故在换热器设计中，应尽量避免在层流条件下进行换热。

（3）圆形直管内其他情况

对于 $\frac{L}{d_i}<60$ 的短管、圆形弯管、圆形直管内作强制过渡流等情况，均可先用湍流时的公式计算，然后乘以系数进行修正 φ。

① 流体在弯管内作强制对流　圆形弯管的校正系数为：

$$\varphi=1+1.77\frac{d_i}{R} \tag{2-36}$$

式中，d_i 为管内径，m；R 为弯管轴的曲率半径，m。

② 流体在圆形直管中呈过渡流（heat transfer in transition region between laminar and turbulent flow）　过渡流的校正系数为：

$$\varphi=1-\frac{6\times10^5}{Re^{1.8}} \tag{2-37}$$

③ 流体在短管内作强制对流　短管内的校正系数为：

$$\varphi=1+\left(\frac{d}{L}\right)^{0.7} \tag{2-38}$$

式中，L 为短管的长度，m。

对于高黏度液体，也应采用适当的修正式。

表 2-3 中列出空气和水在圆形直管内流动时的对流传热系数，以供参考。由表可见，水的 α 值较空气的大得多。同一种流体，流速愈大，α 也愈大；管径愈大，则 α 愈小。

表 2-3　空气和水的 α 值（16℃和 101.3kPa）

物质	d_i /mm	u /(m/s)	α /[W/(m² · ℃)]	物质	d_i /mm	u /(m/s)	α /[W/(m² · ℃)]
空气	25	6.1	34.1	水	25	0.61	2498
		24.4	101.1			1.22	4372
		42.7	159.9			2.44	7609
		61.0	210.1				
	50	6.1	29.5		50	0.61	2158
		24.4	89.7			1.22	3804
		42.7	137.4			2.44	6586
		61.0	184.0				
	75	6.1	26.1		75	0.61	2044
		24.4	80.6			1.22	3520
		42.7	126.1			2.44	6132
		61.0	169.2				

（4）流体在非圆形管内作强制对流（heat transfer by force convection in turbulent flow in pipe with cross section other than circular）

此时，只要将管内径改为当量直径 d_e，则仍可采用上述各关联式。d_e 的定义式为：

$$d_e=4\times\frac{流体流通截面积}{流体润湿周边长} \tag{2-39}$$

但有些资料中规定某些关联式采用传热当量直径：

$$d_e=4\times\frac{流体流通截面积}{传热周边长} \tag{2-39a}$$

例如对于套管环隙内，传热当量直径为：

$$d_e=\frac{4\times\frac{\pi}{4}(d_1^2-d_2^2)}{\pi d_2}=\frac{d_1^2-d_2^2}{d_2}$$

式中　d_1——套管换热器外管内径，m；

d_2——套管换热器内管外径，m。

至于传热计算中究竟采用哪个当量直径，要由具体的关联式决定。

对于非圆形管内强制对流计算，也可直接通过实验求得相应计算 α 的关联式。

【例 2-4】 1 个绝对大气压的干空气，以 4m/s 的流速通过内径为 60mm，长为 4m 的管子后被预热。干空气入口温度为 45℃，出口温度为 55℃，试求：(1) 管壁对空气的对流传热系数；(2) 若空气的流速增大一倍，而其他条件不变，求此时的对流传热系数。

解　定性温度 $t_m=(45+55)/2=50$℃，查附录知在 1 个绝对大气压下、50℃时干空气的物性常数为：$\rho=1.093\text{kg/m}^3$，$C_p=1.005\text{kJ/(kg·K)}$，$\lambda=0.0283\text{W/(m·K)}$，$\mu=1.96\times10^{-5}\text{Pa·s}$。

已知 $d_i=0.06$m。

(1) 当 $u=4$m/s 时：

$$Re=\frac{du\rho}{\mu}=\frac{0.06\times4\times1.093}{1.96\times10^{-5}}=13384>4000(\text{湍流})$$

又因为：$L/d>60$，$Pr=\frac{C_p\mu}{\lambda}=\frac{1.005\times1000\times1.96\times10^{-5}}{0.0283}=0.7$

因此，可用式(2-33a) 计算对流传热系数：

$$\begin{aligned}\alpha&=0.023\frac{\lambda}{d_i}Re^{0.8}Pr^{0.4}\\&=0.023\times\frac{0.0283}{0.06}\times13384^{0.8}\times0.7^{0.4}\\&=18.82[\text{W/(m}^2\cdot\text{K)}]\end{aligned}$$

(2) 当 $u'=8$m/s 时：

$$Re'=13384\times2=26768$$

所以：

$$\begin{aligned}\alpha'&=0.023\times\frac{0.0283}{0.06}\times26766^{0.8}\times0.7^{0.4}\\&=32.7[\text{W/(m}^2\cdot\text{K)}]\end{aligned}$$

或：

$$\alpha\propto u^{0.8}$$

$$\alpha/\alpha'=(2u/u)^{0.8}=2^{0.8}=1.74$$

$$\alpha'=1.74\alpha=1.74\times18.82=32.7[\text{W/(m}^2\cdot\text{K)}]$$

【例 2-5】 铜氨溶液在一蛇管冷却器中由 38℃冷却至 8℃。蛇管由 ϕ45mm×3.5mm 的管子按4组并联而成，平均圈径约为0.57m。已知铜氨液流量为$2.7m^3/h$，密度$\rho=1.2\times10^3\ kg/m^3$，黏度$\mu=2.2\times10^{-3}Pa\cdot s$，其余物性可按水的0.9倍取。试求：铜氨液在蛇管中的对流传热系数。

解 流体在弯管中流动时，由于不断改变流动方向，增大了流体的湍动程度，使对流传热系数较直管中的大。为此，应按流体在直管的情况计算，然后再乘以弯管的校正系数。

对于流体在直管中作强制对流时的α，应先求Re以判断流动型态，然后确定计算式。

流体的流动截面：

$$S=\frac{\pi}{4}d^2\times4=\frac{3.14}{4}\times(0.038)^2\times4=4.54\times10^{-3}(m^2)$$

流体的流速：

$$u=\frac{V_S}{S}=\frac{2.7}{3600\times4.54\times10^{-3}}=0.165(m/s)$$

$$Re=\frac{du\rho}{\mu}=\frac{0.038\times0.165\times1200}{2.2\times10^{-3}}=2000<3420<4000(过渡流)$$

需按湍流计算后乘以校正系数。定性温度$t_m=(38+8)/2=23℃$，查附录水的物性，再乘以0.9。

$$C_p=0.9\times4.18=3.76[kJ/(kg\cdot℃)],\lambda=0.9\times0.605=0.544[W/(m\cdot℃)]$$

$$Pr=\frac{C_p\mu}{\lambda}=\frac{3.76\times10^3\times2.2\times10^{-3}}{0.544}=15.2$$

湍流时：

$$Nu=0.023Re^{0.8}Pr^n$$

流体被冷却$n=0.3$：

$$\begin{aligned}\alpha&=0.023\frac{\lambda}{d}Re^{0.8}Pr^{0.3}\\&=0.023\times\frac{0.544}{0.038}\times3420^{0.8}\times15.2^{0.3}\\&=500[W/(m^2\cdot K)]\end{aligned}$$

过渡流时：

$$\alpha'=f\alpha$$

$$f=1-\frac{6\times10^5}{Re^{1.8}}=1-\frac{6\times10^5}{3420^{1.8}}=0.739$$

$$\alpha'=0.739\times500=370[W/(m^2\cdot K)]$$

弯管时：

$$\begin{aligned}\alpha''&=\alpha'\left(1+1.77\frac{d_i}{R}\right)\\&=370\times\left(1+1.77\times\frac{0.038}{0.285}\right)\\&=457[W/(m^2\cdot K)]\end{aligned}$$

2.3.2.2 无相变时流体在管外流动的α计算

(1) 流体在管束外强制流动（heating and cooling of fluid in forced convection outside

tubes)

管子的排列方式分为直列和错列两种，错列又可分为正方形排列和等边三角形排列，如图 2-7 所示。

当流体在直列或错列的管束外强制垂直流动时，由于沿管子圆周各点的流动情况不同，所以各点的局部对流传热系数也不同，通过整个圆管的平均对流传热系数为：

① 流体在错列管束外流过时：

$$Nu=0.33Re^{0.6}Pr^{0.33} \tag{2-40}$$

② 流体在直列管束外流过时：

$$Nu=0.26Re^{0.6}Pr^{0.33} \tag{2-41}$$

应用范围：$Re>3000$，管束在 10 排以上。

特征尺寸：管外径 d_o，流速取流体通过每排管子中最狭窄处的速度，其中距最狭窄处的距离应在 x_1-d_o 和 $2(t-d_o)$ 两者中取小者。

定性温度：流体进出口温度的算术平均值。

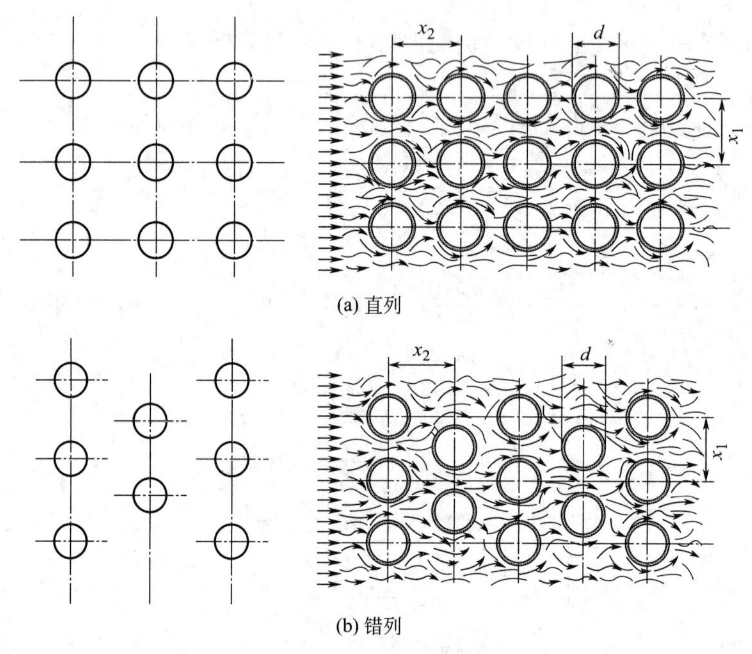

图 2-7 管子排列方式及流体流过管束时的流动形式

（2）流体在换热器的管间流动

对于常用的管壳式换热器，由于壳体是圆筒，管束中各列的管子数目并不相同，而且大都装有折流挡板，使得流体的流向和流速不断地变化，因而在 $Re>100$ 时即可达到湍流。此时对流传热系数的计算，要视具体结构选用相应的计算公式。

管壳式换热器折流挡板的形式较多，如图 2-8 所示，其中以圆缺形（弓形）挡板最为常见，当换热器内装有圆缺形挡板（缺口面积约为 25％的壳体内截面积）时，壳方流体的对流传热系数关联式可采用多诺呼（Donohue）法或凯恩（Kern）法等。此外，若换热器的管间无挡板，则管外流体将沿管束平行流动，此时可采用管内强制对流的公式计算，但需将式中的管内径改为管间的当量直径。

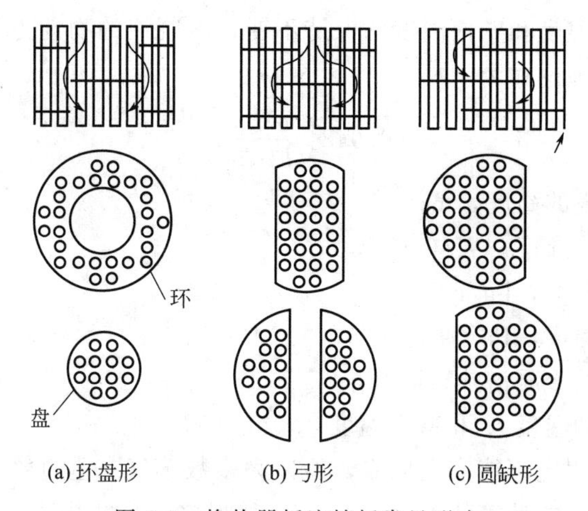

图 2-8 换热器折流挡板常见形式

2.3.3 有相变对流传热系数计算

蒸气冷凝和液体沸腾都是伴有相变化的对流传热过程，这类传热过程的特点是：流体放出或吸收大量的潜热，流体的温度不发生变化，其对流传热系数较无相变化时大很多，例如水的沸腾或水蒸气冷凝时的 α 比水单相流动的 α 要大得多。有相变时流体的对流传热在工业上是很重要的，但是其传热机理至今尚未完全清楚，以下简要介绍蒸气冷凝和液体沸腾的基本机理。

(1) 蒸气冷凝 (heat transfer from condensing vapors)

当饱和蒸气接触比其饱和温度低的冷却壁面时，蒸气放出潜热，在壁面上冷凝成液体，产生有相变化的对流传热。蒸气冷凝方式分为膜状冷凝 (film-type condensation) 和滴状冷凝 (dropwise-type condensation) 两种，如图 2-9 所示。

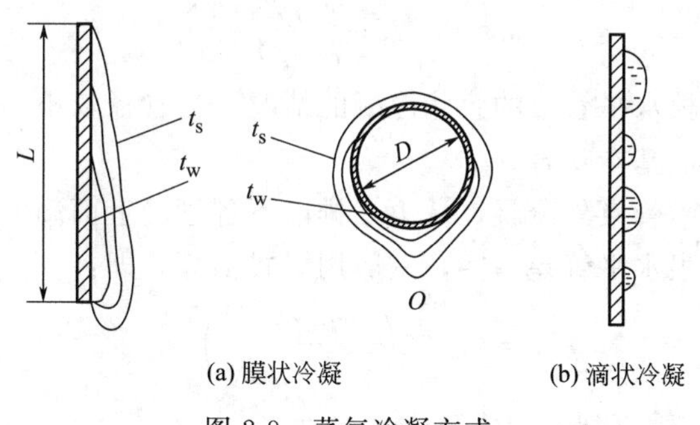

图 2-9 蒸气冷凝方式

通常膜状冷凝发生在易于润湿的冷却表面上，冷凝液在传热表面形成一层连续液膜流下；而滴状冷凝则发生在润湿性不好的表面，蒸气在冷却面上冷凝成液滴，液滴又因进一步冷凝与合并而长大、脱落，然后冷却面上又形成新的液滴。

滴状冷凝时，由于传热面的大部分直接暴露在蒸气中，不存在冷凝液膜引起的附加热阻，所以其对流传热系数比膜状冷凝要大 5～10 倍以上。但到目前为止，描述滴状冷凝的理论和技术仍不成熟，工业冷凝器的设计通常是以膜状冷凝来处理的。

对于蒸气在垂直管外或垂直平板侧的膜状冷凝，若假定冷凝液膜呈层流流动，蒸气与液膜间无摩擦阻力，蒸气温度和壁面温度均保持不变，冷凝液的物性为常数，可推导出计算膜状冷凝对流传热系数的理论式为：

$$\alpha=0.943\left(\frac{r\rho^2 g\lambda^3}{L\mu\Delta t}\right)^{1/4} \tag{2-42}$$

式中 L——垂直管或板的高度，m；

λ——冷凝液的热导率，W/(m·℃)；

ρ——冷凝液的密度，kg/m^3；

μ——冷凝液的黏度，Pa·s；

r——饱和蒸气的冷凝潜热，kJ/kg；

Δt——蒸气的饱和温度 t_s 和壁面温度 t_w 之差，℃。

由于理论推导中的假定不能完全成立，所以大多数蒸气在垂直管外或垂直平板侧的膜状冷凝的实验结果较理论式的计算值差大约 20%，故得修正公式为：

$$\alpha=1.13\left(\frac{r\rho^2 g\lambda^3}{L\mu\Delta t}\right)^{1/4} \tag{2-43}$$

上式也常用无量纲冷凝传热系数 α^* 表示，即：

$$\alpha^*=1.76Re^{-1/3} \tag{2-44}$$

其中：

$$\alpha^*=\alpha\left(\frac{\mu^2}{\lambda^3\rho^2 g}\right)^{1/3} \tag{2-45}$$

$$Re=\frac{Lu\rho}{\mu} \tag{2-46}$$

式(2-42) 和式(2-43) 适用于冷凝液膜为层流（$Re\leqslant1800$），若冷凝液膜为湍流（$Re>1800$）时，可采用如下关联式

$$\alpha^*=0.0077Re^{0.4} \tag{2-47}$$

对于蒸气在单根水平管外的膜状冷凝，可理论推导得：

$$\alpha=0.725\left(\frac{r\rho^2 g\lambda^3}{d_0\mu\Delta t}\right)^{1/4} \tag{2-48}$$

应指出，蒸气在单根水平管上的膜状冷凝的情况，当管径较小，液膜呈层流流动时，实验结果与理论式的计算值基本吻合。

对于蒸气在纵排水平管束外冷凝，从第二排以下各管受上面滴下的冷凝液的影响使液膜增厚，其传热效果较单根水平管差一些，一般用下式估算，即：

$$\alpha=0.725\left(\frac{r\rho^2 g\lambda^3}{n^{2/3}d_0\mu\Delta t}\right)^{1/4} \tag{2-49}$$

式中 n——水平管束在垂直列上的管数。

［讨论］：

① 由于液膜的厚度及其流动状况是影响冷凝传热的关键因素，所以凡是有利于减薄液膜厚度的因素都可提高冷凝传热系数，这些因素包括：加大冷凝液膜两侧的温度差，使蒸气冷凝速率增加，因而液膜层厚度增加，使冷凝传热系数降低。考虑蒸气的流速和流向的影响：若蒸气和液膜同向流动，则蒸气和液膜间的摩擦力，使液膜流动加速，厚度减薄，传热系数增大；若逆向流动，则相反；若液膜被蒸气吹离壁面，则随蒸气流速的增加，对流传热系数急剧增大。

② 由于蒸气冷凝时的 α 值较流体无相变化时大得多，所以在间壁两侧流体进行热交换

时，若热流体为蒸气冷凝，冷流体无相变化，则蒸气冷凝不是过程的主要矛盾，其 α 值可进行估算。例如水蒸气作膜状冷凝，其 α 值可取 12000W/（m^2·K）左右。

③ 应当注意，计算蒸气冷凝的 α 关联式是针对纯净的饱和蒸气在清洁的壁面上冷凝时得到的。若蒸气中含有空气或其他不凝性气体，冷凝的过程中将逐渐在壁面附近形成一层气膜，气膜将使热阻迅速增大，α 值急剧下降。实验证明，蒸气中含有 1%的空气时，α 值将下降 60%以上。因此，冷凝器上方都装有排气阀，以便及时排除不凝性气体。

Heat Transfer to Fluids with Phase Change

Processes of heat transfer accompanied by phase change are more complex than simple heat exchange between fluids. A phase change involves the addition or subtraction of considerable quantities of heat at constant or nearly constant temperature. The rate of phase change may be governed by the rate of heat transfer, but it is often influenced by the rate of nucleation of bubbles, drops, or crystals and by the behavior of the new phase after it is formed. This part covers condensation of vapors and boiling of liquids.

Heat Transfer from Condensing Vapors

The condensation of vapors on the surfaces of tubes cooler than the condensing temperature of the vapor is important when vapors such as those of water, hydrocarbons, and other volatile substances are processed. Some examples will be met later in this text, in discussing the unit operations of evaporation, distillation, and drying.

The condensing vapor may consist of a single substance, a mixture of condensable and noncondensable substances, or a mixture of two or more condensable vapors. Friction losses in a condenser are normally small, so that condensation is essentially a constant-pressure process. The condensing temperature of a single pure substance depends only on the pressure, and therefore the process of condensation of a pure substance is isothermal. Also, the condensate is a pure liquid. Mixed vapors, condensing at constant pressure, condense over a temperature range and yield a condensate of variable composition until the entire vapor stream is condensed, when the composition of the condensate equals that of the original uncondensed vapor.

Common examples of condensation in the presence of an inert gas are the condensation of water from a mixture of steam and air and the recovery of hydrocarbon solvents from air streams leaving extraction or drying operations.

Condensation of mixed vapors and condensation in the presence of noncondensing gases are discussed briefly later in this chapter. The following discussion is limited to the condensation of a single volatile substance on a cold tube.

Dropwise and Film-Type Condensation

A vapor may condense on a cold surface in one of two ways, which are well described by the terms dropwise and film type. In film condensation, which is more common than dropwise condensation, the liquid condensate forms a film, or continuous layer, of liquid that flows over the surface of the tube under the action of gravity. It is the layer of liquid interposed between the vapor and the wall of the tube that provides the resistance to heat flow and therefore fixes the magnitude of the heat transfer coefficient.

In dropwise condensation the condensate begins to form at microscopic nucleation sites. Typical sites are tiny pits, scratches, and dust specks. The drops grow and coalesce with their neighbors to form visible fine drops like those often seen on the outside of a cold-water pitcher in a humid room. The fine drops, in turn, coalesce into rivulets, which flow down the tube under the action of gravity, sweep away condensate, and clear the surface for more droplets. During dropwise condensation, large areas of the tube surface are covered with an extremely thin film of liquid of negligible thermal resistance. Because of this the heat-transfer coefficient at these areas is very high; the average coefficient for dropwise condensation may be 5 to 10 times that for film-type condensation. On long tubes, condensation on some of the surface may be film condensation and the remainder dropwise condensation.

The most important and extensive observations of dropwise condensation have been made on steam, but it has also been observed in ethylene glycol, glycerin, nitrobenzene, isoheptane, and some other organic vapors. Liquid metals usually condense in the dropwise manner. The appearance of dropwise condensation depends upon the wetting or nonwetting of the surface by the liquid, and fundamentally, the phenomenon lies in the field of surface chemistry. Much of the experimental work on the dropwise condensation of steam is summarized in the following paragraphs.

(1) Film-type condensation of water occurs on tubes of the common metals if both the steam and the tube are clean, in the presence or absence of air, on rough or on polished surfaces.

(2) Dropwise condensation is obtainable only when the cooling surface is not wetted by the liquid. In the condensation of steam it is often induced by contamination of the vapor with droplets of oil. It is more easily maintained on a smooth surface than on a rough surface.

(3) The quantity of contaminant or promoter required to cause dropwise condensation is minute, and apparently only a monomolecular film is necessary.

(4) Effective drop promotors are strongly adsorbed by the surface, and substances that merely prevent wetting are ineffective. Some promoters are especially effective on certain metals, for example, mercaptans on copper alloys; other promoters, such as oleic acid, are quite generally effective. Some metals, such as steel and aluminum, are difficult to treat to give dropwise condensation.

(5) The average coefficient obtainable in pure dropwise condensation may be as high as 115 kW/(m^2 · ℃).

Although attempts are sometimes made to realize practical benefits from these large coefficients by artificially inducing dropwise condensation, this type of condensation is so unstable and the difficulty of maintaining it so great that the method is not common. Also the resistance of the layer of steam condensate even in film type condensation is ordinarily small in comparison with the resistance inside the condenser tube, and the increase in the overall coefficient is relatively small when dropwise condensation is achieved. For normal design, therefore, film-type condensation is assumed.

Effect of noncondensables

When a multicomponent mixture contains a noncondensing gas, the rate of condensation is seriously reduced. As in the condensation of a mixture of condensable vapors, there is mass transfer of one or more components in the vapor phase; but here the condensing molecules must diffuse through a film of noncondensing gas which does not move toward the condensate surface. As condensation proceeds, the relative amount of this inert gas in the vapor phase increases significantly.

The partial pressure of the condensing vapor is less than the total pressure, which lowers the equilibrium condensation temperature. In addition, the partial pressure of the condensing vapor at the condensate surface must be less than it is in the bulk vapor-gas phase, to provide the driving force for mass transfer through the gas film. This further lowers the condensing temperature, and usually the change in temperature due to mass transfer is greater than the change in the equilibrium temperature.

Even a small amount of gas can have a large effect on the rate of condensation. Less than 1 percent air in steam can reduce the condensation rate by more than one-half and 5 percent of inert gas can decrease the steam condensation rate by a factor of 5. Whenever air or other noncondensable gas is present in the feed, a fraction of the incoming gas must be vented from the condenser. If the vapor fed to a nonvented condenser, for instance, contains 0.1 percent air and 99 percent of the vapor is condensed, then the remaining vapor will contain 10 percent air and the condensation rate will be quite low in the last part of the condenser.

(2) 液体沸腾（heat transfer to boiling liquids）

液体吸热后，在其内部或表面产生气泡或气膜的过程称为液体沸腾，如图 2-10 所示。

工业上的液体沸腾主要有两种：当加热面浸入比它大的容器里，没有强制对流，加热壁面附近液体的流动仅由自然对流和所产生的气泡的扰动引起时，壁面上的沸腾称为大容积或池内沸腾（pool boiling of saturated liquid）；当液体在管内流动时受热沸腾，称为管内沸腾（calandria boiling）。后者液体的流速对沸腾有很大的影响，产生的气泡与液体一起流动，形成了复杂的气液两相流问题。

按照液体主体温度 t_1、液体饱和温度 t_s 和加热壁面温度 t_w 的差异，上述两类沸腾现象又有过冷沸腾与饱和沸腾之分。当 t_1 小于 t_s，而 t_w 大于 t_s 时，加热壁面上产生的气泡还未脱离壁面或刚脱离壁面就迅速被冷凝成

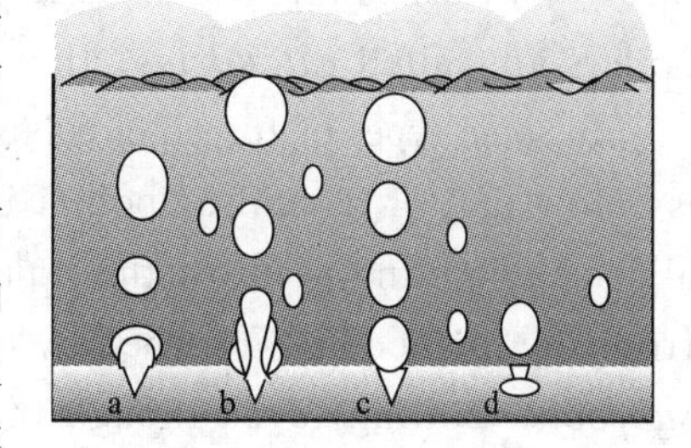

图 2-10　液体沸腾示意图

液体，这种情况称为过冷沸腾（subcooled boiling）；当 t_1 略高于 t_s 时，在加热壁面上产生的气泡进入液相主体，并不断长大、上升，最后从液体表面逸出，这种情况称为饱和沸腾（saturated boiling）。

Heat Transfer to Boiling Liquids

Heat transfer to a boiling liquid is a necessary step in evaporation, distillation, and steam generation, and it may also be used to control the temperature of a chemical reactor. The boiling liquid may be contained in a vessel equipped with a heating surface fabricated from horizontal or vertical tubes, in which steam or other vapor is condensed or hot fluid is circulated to supply the heat needed to boil the liquid. Alternatively, the liquid to be boiled may flow, under either natural or forced convection, inside heated tubes. An important application of boiling in tubes is the evaporation of water from solution, as discussed later.

When boiling is accomplished by a hot immersed surface, the temperature of the mass of the liquid is the same as the boiling point of the liquid under the pressure existing in the equipment. Bubbles of vapor are generated at the heating surface, rise through the mass of liquid, and disengage from the surface of the liquid. Vapor accumulates in a vapor space over the liquid; a vapor outlet from the vapor space removes the vapor as fast as it is formed. This type of boiling can be described as pool boiling of saturated liquid since the vapor leaves the liquid in equilibrium with the liquid at its boiling temperature. When a liquid is boiled under natural circulation inside a vertical tube, relatively cool liquid enters the bottom of the tube and is heated as it flows upward at a low velocity. The liquid temperature rises to the boiling point under the pressure prevailing at that particular level in the tube. Vaporization begins, and the upward velocity of the two-phase liquid-vapor mixture increases enormously. The resulting pressure drop causes the boiling point to fall as the mixture proceeds up the tube and vaporization continues. Liquid and vapor emerge from the top of the tubes at very high velocity.

With forced circulation through horizontal or vertical tubes, the liquid may also enter at a fairly low temperature and be heated to its boiling point, changing into vapor near the discharge end of the tube. Sometimes a flow control valve is placed in the discharge line beyond the tube so that the liquid in the tube may be heated to a temperature considerably above the boiling point corresponding to the downstream pressure. Under these conditions there is no boiling in the tube: The liquid is merely heated, as a liquid, to a high temperature, and flashes into vapor as it passes through the valve. Natural-and forced-circulation boilers are called *calandrias*; they are discussed later.

In some types of forced-circulation equipment, the temperature of the mass of the liquid is below that of its boiling point, but the temperature of the heating surface is considerably above the boiling point of the liquid. Bubbles form on the heating surface, but on release from the surface are absorbed by the mass of the liquid. This type of heat transfer is called *subcooled* boiling, even though the fluid leaving the heat exchanger is entirely liquid.

以下仅简要介绍大容积内饱和液体沸腾的特性。实验表明，大容积内饱和液体沸腾的表

面热通量 q 或对流传热系数 α 可表示为温度差 $\Delta t=t_w-t_s$ 的函数，描述 q 或 α 随 $\Delta t=t_w-t_s$ 变化的曲线称为沸腾曲线。不同工质、不同操作条件下的沸腾曲线是不同的，但基本形式相似。图 2-11 为 1atm 下水的沸腾曲线（the boiling curve of atmospheric water），两条曲线中，一条是 q 随 Δt 的变化，另一条是 α 随 Δt 的变化。由图 2-11 可知，沸腾曲线呈现出不同的变化规律。

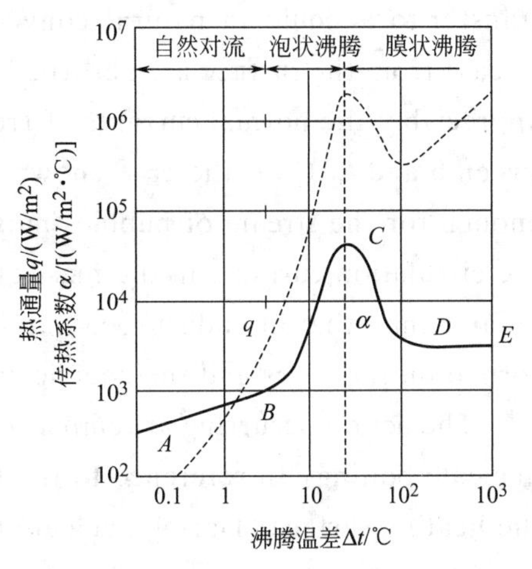

图 2-11 常压下水的沸腾曲线

根据图 2-11 可知，沸腾曲线通常分为三个区域。

① 自然对流（natural convection）区：$\Delta t \leqslant 5℃$，液体稍微过热，液体内产生自然对流，但没有气泡从液体中逸出液面，仅仅是液体表面发生蒸发，q 或 α 都较小。

② 泡状（核状）沸腾（nucleate boiling）区：$5℃<\Delta t<25℃$，加热壁面上局部产生气泡，其产生速度随 Δt 上升而增加。气泡脱离壁面后，不断长大、上升，最后逸出液面。由于气泡的上述作用，使液体受到剧烈的扰动，因此 q 或 α 都随 Δt 上升而急剧增大。

③ 膜状沸腾（film boiling）区：$\Delta t>25℃$ 时，加热壁面上的气泡产生速度大于脱离速度，气泡在加热壁面连接起来形成不稳定的蒸气膜，使液体不能与加热面直接接触，由于蒸气膜的导热性能差，使 q 或 α 都随 Δt 上升而急剧下降。当 Δt 上升到一定数值后，随着 Δt 上升，α 基本上不变，q 又开始上升，这是由于 t_w 较高，传热面几乎全部为气膜所覆盖，形成了较稳定的气膜，辐射传热的影响显著增加。

由沸腾曲线可知，泡状沸腾的 α 较自然对流的大，比膜状沸腾的容易控制，因此工业生产中一般控制在泡状沸腾下操作。由泡状沸腾向膜状沸腾过渡的转折点称为临界点，临界点对应于临界温度差 Δt_c、临界沸腾传热系数 α_c 和临界热通量 q_c，确定不同液体在临界点下的上述参数具有实际意义。

Pool Boiling of Saturated Liquid

Consider a horizontal, electrically heated wire immersed in a vessel containing a boiling liquid. A plot of q versus Δt on logarithmic coordinates will give a curve of the type shown in Fig. 2-11. This curve can be divided into four segments. The first section is line *AB*. The second section, line *BC*, is approximately straight, but its slope is greater than that of line *AB*. The second segment terminates at a definite point of maximum flux, which is point *C* in Fig. 2-11. The temperature drop corresponding to point *C* is called the critical temperature drop, and the flux at point *C* is the peak flux. In the third segment, line *CD* in Fig. 2-11, the flux decreases as the temperature drop rises and reaches a flux again increases with Δt and, at large temperature drops, surpasses the previous maximum reached at point *C*.

Each of the four segments of the graph in Fig. 2-11 corresponds to a deinit mechanism of boiling. In the first section, at low temperature drops, the mechanism is that of heat transfer to a liquid in natural convection. Bubbles form on the heating surface, but on release from the surface are absorbed by the mass of the liquid. They are too few to disturb appreciably the normal currents of free convection. At larger temperature drops, lying between 5 and 25℃ in the case shown in Fig. 2-11, the rate of bubble production is large enough for the stream of bubbles moving up through the liquid to increase the velocity of the circulation currents in the mass of liquid, and the coefficient of heat transfer becomes greater than that in undisturbed natural convection. As Δt is increased, the rate of bubble formation increases and the coefficient increases rapidly.

The action occurring at temperature drops below the critical temperature drop is called nucleate boiling, In reference to the formation of tiny bubbles, or vaporization nuclei, on the heating surface. During nucleate boiling, the bubbles occupy but a small portion of the heating surface at a time, and most of the surface is in direct contact with liquid. The bubbles are generated at localized active sites, usually small pits or scratches on the heating surface. As the temperature drop is raised, more sites become active, improving the agitation of the liquid and increasing the heat flux and the heat-transfer coefficient.

Eventually, however, so many bubbles are present that they tend to coalesce and cover portions of the heating surface with a layer of insulating vapor. This layer has a highly unstable surface, from which miniature ‘explosions’ send jets of vapor away from the heating element into the bulk of the liquid. This type of action is called transition boiling. In this region, corresponding to segment *CD* in Fig. 2-11, increasing the temperature drop increases the thickness of the layer of vapor and reduces the number of explosions that occur in a given time. The heat flux and the heat-transfer coeffcient both fall as the temperature drop is raised.

Near the Leidenfrost point another distinct change in mechanism occurs. The hot surface becomes covered with a quiescent film of vapor, through which heat is transferred by conduction and (at very high temperature drops) by radiation. The random explosions characteristic of transition boiling disappear and are replaced by the slow and orderly formation of bubbles at the interface between the liquid and the film of hot vapor. These bubbles detach themselves from the interface and rise through the liquid. Virtually all the resistance to heat transfer is offered by the vapor sheath covering the heating element. As the temperature drop increases, the heat flux rises, slowly at first and then more rapidly as radiation heat transfer becomes important. The boiling action in this region is known as film boiling.

Film boiling is not usually desired in commercial equipment because the heat transfer rate is low for such a large temperature drop. Heat-transfer apparatus should be so designed and operated that the temperature drop in the film of boiling liquid is smaller than the critical temperature drop, although with cryogenic liquids this is not always feasible.

在管内作强制对流的液体发生沸腾时，还受流体流动状况的影响。除了流速以外，在很大程度上要取决于流体内的蒸气含量，这种含量是沿管程变化的，因此管内沸腾传热现象就更加复杂。

关于沸腾传热α的计算，难以理论求解，工业上常用的经验公式可参见有关专著，这里仅介绍按照对比压强计算泡状沸腾传热系数的莫斯廷凯（Mostinki）公式：

$$\alpha=1.163Z(\Delta t)^{2.33} \tag{2-50}$$

式中 Δt——壁面过热度，$\Delta t=t_w-t_s$，℃；

Z——与操作压强及临界压强有关的参数，其值可按照相关公式计算，$W/(m^2 \cdot ℃^{0.33})$。

综上所述，各类不同情况下α计算式是各不相同的，一般可在化工手册中查到。但须指出，特征数关联式是一种经验公式，选用时应注意以下几个问题：

① 首先分析所处理的问题是属于哪一类，如有相变、无相变、强制对流或自然对流；

② 选用相应的α计算式时，应注意被选用的公式所规定的使用范围，特性尺寸、定性温度等；

③ 公式中物性参数的单位要予以注意，按规定代入；

④ 流体在圆形直管内作强制湍流的无相变计算，在工程上最为常见，但许多情况下，如过渡流、弯管、非圆管等的计算，都是由该式派生而得，应予以充分掌握；

⑤ 蒸气冷凝与液体沸腾要特别注意α的影响因素，如蒸气冷凝中不凝性气体、蒸气流速和流向、蒸气过热、冷凝面的高度及布置方式等对冷凝的影响；

⑥ 液体沸腾中，温度差、操作压力、流体物性、加热面等都对其产生影响。

表2-4列出了一些常见传热情况下流体的对流传热系数α值的大致范围。

表2-4 α值的大致范围

传热情况	$\alpha/[W/(m^2 \cdot K)]$	传热情况	$\alpha/[W/(m^2 \cdot K)]$
空气自然对流	5～25	水蒸气膜状冷凝	5000～15000
气体强制对流	20～100	水蒸气滴状冷凝	400000～120000
水自然对流	200～1000	水沸腾	2500～25000
水强制对流	1000～15000	有机蒸气冷凝	500～2000

2.4 热辐射（Radiation Heat Transfer）

2.4.1 辐射传热基本概念

（1）热辐射及其特点

物体温度大于热力学零度即可向外发射辐射能，辐射能以电磁波形式传递，当与另一物体相遇，则可被吸收、反射、透过，其中吸收的部分又可将电磁波转变为热能。这种仅与物体本身的温度有关而引起的热能传播过程，称为热辐射。热辐射具有以下特点：

① 物体的热能转化为辐射能，只要物体的温度不变，其发射的辐射能亦不变；

② 物体向外发射辐射能的同时，并不断吸收周围物体辐射来的能量，结果是高温物体向低温物体传递了能量，称为辐射传热；

③ 理论上，热辐射的电磁波波长为零到无穷大，实际上，波长仅在0.4～40μm的范围是明显的；

④ 可见光的波长为0.4～0.8μm，红外线的波长为0.8～40μm，可见光与红外线统称为热射线；

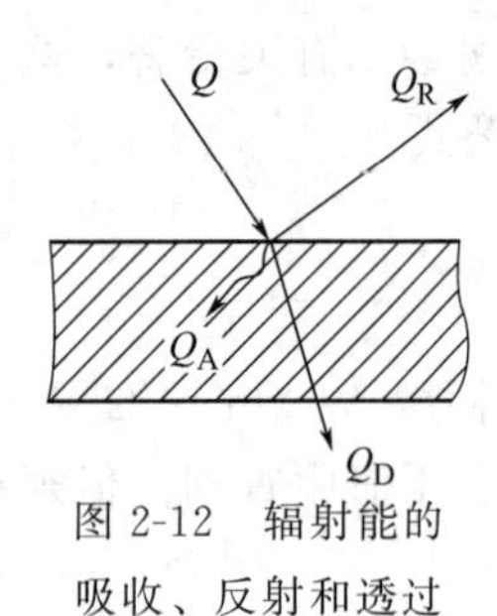

图 2-12　辐射能的吸收、反射和透过

⑤ 虽然热射线与可见光的波长范围不同，但本质一样。

（2）辐射能的吸收、反射和透过

如图 2-12 所示，投射到某一物体表面上的总辐射能为 Q，部分能量 Q_R 被反射，部分能量 Q_A 被吸收，部分能量 Q_D 被透过。

根据能量守恒定律：

$$Q = Q_A + Q_R + Q_D \tag{2-51}$$

令 $\frac{Q_A}{Q}=A$，为吸收率；$\frac{Q_R}{Q}=R$，为反射率；$\frac{Q_D}{Q}=D$，为透过率。则：

$$A + R + D = 1 \tag{2-51a}$$

① $A=1$ 的物体称为绝对黑体或黑体，实际上，$A=1$ 的物体是不存在的，但有些物体接近于黑体，例如无光泽的黑漆表面，$A=0.96\sim0.98$。

② $R=1$ 的物体称为绝对白体或镜体，$R=1$ 的物体也不存在，但有些物体性质接近于白体，例如表面磨光的铜，$R=0.97$。

③ $D=1$ 的物体称为透热体，虽然实际中并不存在透热体，但是单原子或对称双原子气体可近似为透热体。

A、R、$D=f$（物体性质、温度、表面状况、波长）

一般来说，表面粗糙的物体，A 大；固体和液体的 $D\approx0$，为不透热体；气体的 $R\approx0$。

应该注意，黑体、白体不能由颜色区分，例如霜在光学上是白色，但其 $A\approx1$，是黑体。

（3）灰体（gray body）

工程上为了处理问题方便起见，提出了“灰体”的概念。所谓灰体，从辐射的角度来讲，是在相同温度下，能以相同的吸收率吸收各种波长辐射能的物体，是一种理想化物体，但大多数工程材料都可以近似为灰体。

灰体的特点是：①吸收率与波长无关；②灰体是不透热体。

Radiation Heat Transfer

Radiation, which may be considered to be energy streaming through space at the speed of light, may originate in various ways. Some types of material will emit radiation when they are treated by external agencies, such as electron bombardment, electric discharge, or radiation of definite wavelengths. Radiation due to these effects will not be discussed here. All substances at temperatures above absolute zero emit radiation that is independent of external agencies. Radiation that is the result of temperature only is called thermal radiation, and this discussion is restricted to radiation of this type.

Fundamental Facts Concerning Radiation

Radiation moves through space in straight lines, or beams, and only substances in sight of a radiating body can intercept radiation from that body. The fraction of the radiation falling on a body that is reflected is called the reflectivity, R. The fraction that is absorbed is called the absorptivity A. The fraction that is transmitted is called the transmissivity D. The sum of these fractions must be unity, or

$$R + A + D = 1$$

Radiation as such is not heat, and when transformed to heat on absorption, it is no longer radiation. In practice, however, reflected or transmitted radiation usually falls on other absorptive bodies and is eventually converted to heat, perhaps after many successive reflections.

The maximum possible absorptivity is unity, attained only if the body absorbs all radiation incident upon it and reflects or transmits none. A body that absorbs all incident radiation is called a blackbody.

The complex subject of thermal radiation transfer has received much study in recent years and is covered in a number of texts. The following introductory treatment discusses the following topics: emission of radiation, absorption by opaque solids, radiation between surfaces, radiation to and from semitransparent materials.

2.4.2 物体的辐射能力 (Emissive power)

物体的辐射能力指物体在一定温度时，单位时间单位面积发射的能量，以 E 表示，单位是 W/m^2，辐射能力表征物体发射辐射能的本领。

(1) 斯蒂芬-波尔兹曼定律 (Stefan-Boltzmann law)

该定律的描述是：黑体的辐射能力 E_0 与其表面温度（指热力学温度）的四次方成正比，即：

$$E_0 = C_0 \left(\frac{T}{100}\right)^4 \tag{2-52}$$

式中，C_0 为黑体的辐射系数，$C_0 = 5.67W/(m^2 \cdot K^4)$。

(2) 灰体的辐射能力

因为许多工程材料的辐射特性近似于灰体，所以通常将实际物体视为灰体来计算其辐射能力 E，可用下式计算：

$$E = C\left(\frac{T}{100}\right)^4 \tag{2-53}$$

式中，C 为灰体的辐射系数，单位与 C_0 相同。不同物体的辐射系数 C 值不相同，其值与物体的性质、表面状况和温度等有关。

(3) 黑度 ε

在计算辐射传热中，由于相同温度下黑体的辐射能力最强，通常将灰体辐射能力与同温度下黑体辐射能力之比定义为物体的黑度 ε [又称发射率 (emissivity)]，即：

$$\varepsilon = E/E_0 = C/C_0 \tag{2-54}$$

或：

$$E = \varepsilon E_0 = \varepsilon C_0 \left(\frac{T}{100}\right)^4 \tag{2-54a}$$

利用黑度的数值可以判断相同温度下任何物体的辐射能力与黑体辐射能力的差别，黑度的大小在 0～1 范围内变化。

黑度与物体的种类、温度及表面粗糙度、表面氧化程度等因素有关，一般由实验测定，常用工业材料的黑度数值列于表 2-5 中。

表 2-5　常用工业材料的黑度

材料	温度/℃	黑度	材料	温度/℃	黑度
红砖	20	0.93	铝(磨光的)	225～575	0.039～0.057
耐火砖	—	0.8～0.9	铜(氧化的)	200～600	0.57～0.87
钢板(氧化的)	200～600	0.8	铜(磨光的)	—	0.03
钢板(磨光的)	940～1100	0.55～0.61	铸铁(氧化的)	200～600	0.64～0.78
铝(氧化的)	200～600	0.11～0.19	铸铁(磨光的)	330～910	0.6～0.7

(4) 克希霍夫定律（Kirchhoff's law）

克希霍夫定律描述了灰体的辐射能力与其吸收率之间的关系，其表达式为：

$$\frac{E}{A}=\frac{E_i}{A_i}=E_0 \tag{2-55}$$

其中，$i=1,2,3,4,\cdots$。

上式说明，任何物体的辐射能力与其吸收率的比值恒为常数，且等于同温度下黑体的辐射能力，其值仅与物体的温度有关。

由上式可得：

$$\frac{E}{E_0}=A=\varepsilon \tag{2-56}$$

即在同一温度下，物体的吸收率与黑度在数值上相等，但两者的物理意义则完全不同，前者为吸收率，表示由其他物体发射来的辐射能可被该物体吸收的分数；后者为发射率，表示物体的辐射能力占黑体辐射能力的分数。但是，由于物体的吸收率 A 不易测定，但黑度 ε 可测，故工程计算中大都用物体的黑度 ε 代替吸收率 A。

2.4.3　物体间的辐射传热（Radiation between surfaces）

工业上两固体间的相互辐射传热计算是很复杂的，一般都把两物体视为灰体，在两灰体间的辐射传热中，相互进行着辐射能的多次被吸收和多次被反射的过程，应考虑两固体间的吸收率、反射率、形状、大小及两物体间的距离和相互位置的影响，下面给出两固体间相互辐射传热的传热速率计算式：

$$Q_{1-2}=C_{1-2}\varphi S\left[\left(\frac{T_1}{100}\right)^4-\left(\frac{T_2}{100}\right)^4\right] \tag{2-57}$$

上式表明，两灰体间的辐射传热速率正比于二者的热力学温度四次方之差。显然，此结果与另外两种传热方式（热传导和对流传热）完全不同。

式中　C_{1-2}——总辐射系数，$C_{1-2}=\dfrac{C_0}{1/\varepsilon_1+1/\varepsilon_2-1}$ W/(m^2·K^4)；

C_0——黑体辐射系数，$C_0=5.67$ W/(m^2·K^4)；

S——辐射面积，m^2；

T_1、T_2——高温及低温物体的热力学温度，K；

φ——几何因数（角系数）。

角系数φ表示从辐射面积S所发射的能量被另一物体表面所截获的分数。它的数值既与两物体的几何排列有关。又与式(2-57)中的S是用板1的面积S_1，还是用板2的面积S_2作为辐射面积有关，角系数φ符号的定义是：

① φ_{1-2}=灰体1直接辐射在灰体2上的能量/灰体1辐射出的总能量；

② φ_{1-1}=灰体1直接辐射在本身的能量/灰体1辐射出的总能量；

③ φ_{2-1}=灰体2直接辐射在灰体1上的能量/灰体2辐射出的总能量；

④ φ_{2-2}=灰体2直接辐射在本身的能量/灰体2辐射出的总能量。

因此，在计算中，φ必须和选定的辐射面积S相对应。φ值的大小可通过实验测定，其中，对于两个极大平面之间的热辐射，$\varphi_{1-1}=\varphi_{2-2}=0$，而$\varphi_{1-2}=\varphi_{2-1}=1$。几种简单情况下$\varphi$值与总辐射系数$C_{1-2}$的计算式如表2-6所示。当$\varphi$值小于1时，可查图2-13获得。

表2-6 φ值与C_{1-2}的计算式

序号	辐射情况	S	φ	C_{1-2}
1	极大的两平行面	S_1或S_2	1	$C_0/(1/\varepsilon_1+1/\varepsilon_2-1)$
2	面积相等的两平行面	S_1	查图2-13	$\varepsilon_1\varepsilon_2C_0$
3	很大的物体2包住物体1	S_1	1	ε_1C_0
4	物体2恰好包住物体1，$S_1\approx S_2$	S_1	1	$C_0/(1/\varepsilon_1+1/\varepsilon_2-1)$
5	介于3、4两种情况之间	S_1	1	$C_0/[1/\varepsilon_1+(1/\varepsilon_2-1)S_1/S_2]$

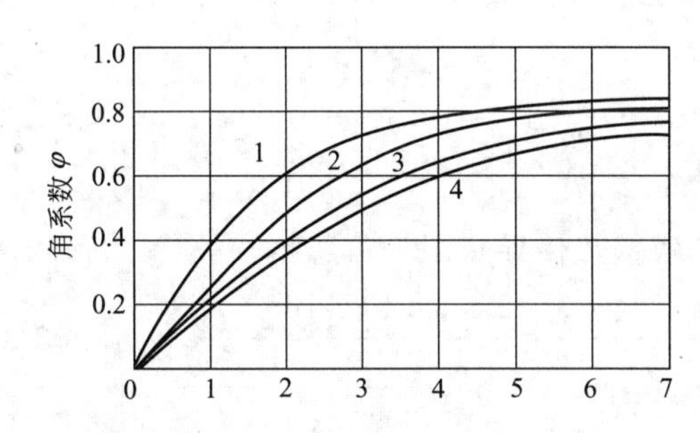

图2-13 平行面间辐射传热的角系数

1—圆盘形；2—正方形；3—长方形（边长比2∶1）；4—长方形（狭长）

2.4.4 对流与辐射联合传热

在化工生产中，许多设备的外壁温度常高于（或低于）环境温度，此时热量将以对流和辐射两种方式自壁面向环境传递而引起热损失（或反向传热而导致冷损失）。为减少热损失或冷损失，许多温度较高或较低的设备，如换热器、塔器、反应器及蒸汽管道等都必需进行保温或隔热处理。

对流热损失：
$$Q_C=\alpha_C S_w(t_w-t) \tag{2-58}$$

辐射热损失：
$$Q_R=\alpha_R S_w(t_w-t) \tag{2-59}$$

式中　α_R——辐射传热系数，$\alpha_R=\dfrac{C_{1-2}[(T_w/100)^4-(T/100)^4]}{t_w-t}$。

因设备向大气辐射传热时角系数 $\varphi=1$，故上式中取消了 φ 项。

那么，总的热损失为：

$$Q=Q_C+Q_R=(\alpha_C+\alpha_R)S_w(t_w-t)=\alpha_T S_w(t_w-t) \tag{2-60}$$

式中　α_T——对流辐射联合传热系数，其值可用近似公式估算，W/（m^2·℃）。

S_w——设备外壁表面，m^2；

t_w——设备外壁温度，℃；

t——环境温度，℃。

通常，对流辐射联合传热系数 α_T 可用如下公式估算：

（1）空气自然对流（t_w<150℃）

平壁：
$$\alpha_T=9.8+0.07(t_w-t) \tag{2-61}$$

管或圆筒壁：
$$\alpha_T=9.4+0.052(t_w-t) \tag{2-62}$$

（2）空气沿粗糙壁面强制对流

空气流速 $u\leqslant 5$m/s：
$$\alpha_T=6.2+4.2u \tag{2-63}$$

空气流速 $u>5$m/s：
$$\alpha_T=7.8u^{0.78} \tag{2-64}$$

【例 2-6】 在 ϕ219mm×8mm 的蒸汽管道外包扎一层厚为 75mm、热导率为 0.1W/(m·℃) 的保温材料，管内饱和蒸汽温度为 160℃，周围环境温度为 20℃，试估算管道外表面的温度及单位长度管道的热损失。假设管内冷凝传热和管壁热传导热阻均可忽略。

解　管道保温层外对流辐射联合传热系数为：

$$\alpha_T=9.4+0.052(t_w-t)=9.4+0.052(t_w-20)$$

单位管长热损失为：

$$\begin{aligned}Q/L&=\alpha_T\pi d_0(t_w-t)=[9.4+0.052(t_w-20)]\pi d_0(t_w-20)\\&=0.06025(t_w-20)^2+10.8914(t_w-20)\end{aligned}$$

由于管内冷凝传热和管壁热传导热阻均可忽略，故：

$$Q/L=\frac{2\pi\lambda(T-t_w)}{\ln\dfrac{d_0}{d}}=\frac{2\pi\times0.1\times(160-t_w)}{\ln\dfrac{0.219+0.075\times2}{0.219}}=1.2037(160-t_w)$$

即：$0.06025\ (t_w-20)^2+10.8914\ (t_w-20)\ =1.2037\ (160-t_w)$

$$t_w=102.5℃$$

则：$Q/L=1.2037\times\ (160-102.5)\ =69.2$（W/m）

2.5　间壁式换热器传热计算（Calculation of Wall Heat-Exchange Equipments）

2.5.1　间壁式换热简介（Introduction of wall heat-exchange equipments）

以套管换热器为例简介间壁式换热过程。套管换热器是典型的间壁式换热器，它是由两种不同直径的直管套在一起组成同心套管，其基本结构如图 2-14 所示。其中的细管称为内管，而内外管之间的空间称为套管环隙。若热流体走内管放出热量，温度从初温降到终温；

则冷流体走套管环隙吸收热量，温度从初温升到终温。由图 2-15 可知，冷、热流体之间的温差也沿换热过程变化。径向的温差是传热的推动力，因此，冷、热流体间热量传递过程的机理是：热量首先由热流体主体以对流的方式传递到间壁内侧；然后以导热的方式穿过间壁；最后由间壁外侧以对流的方式传递至冷流体主体。在垂直于流动方向的同一截面上，温度分布如图 2-15 所示。

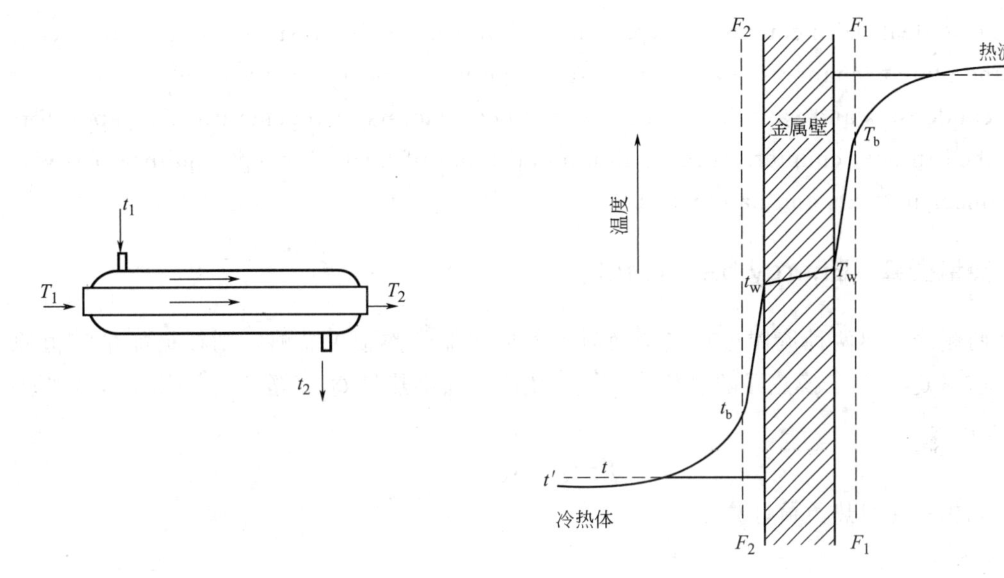

图 2-14　套管换热器基本结构示意

图 2-15　换热截面上温度分布

由温度分布曲线来看，间壁内热传导只有一种分布规律；间壁两侧的对流传热由于流动状况的影响，分别呈现出三种分布规律，在壁面附近为直线，再往外为曲线，在流体主体为比较平坦的曲线。若按上述三种温度分布规律处理流体与壁面间的对流传热问题过于复杂，实际应用上，将流体主体与壁面间的对流虚拟为有效膜内的导热问题，有效膜内温度分布为直线，有效膜外流体的温度取其平均温度（将同一流动截面上的流体绝热混合后测定的温度）。因此，对同一截面而言，热流体的平均温度小于其中心温度，冷流体的平均温度大于其中心温度。

Heat-Exchange Equipment

Heat transfer from a warmer fluid to a cooler fluid, usually through a solid wall separating the two fluids, is common in chemical engineering practice. The heat transferred may be latent heat accompanying a phase change such as condensation or vaporization or it may be sensible heat from the rise or fall in the temperature of a fluid without any phase change. Typical examples are reducing the temperature of a fluid by transfer of sensible heat to a cooler fluid, the temperature of which is increased thereby; condensing steam using cooling water; and vaporizing water from a solution at a given pressure by condensing steam at a higher pressure. All such cases require that heat be transferred by conduction and convection.

In industrial processes heat energy is transferred by a variety of methods, including conduction in electric-resistance heaters; conduction-convection in exchangers, boilers, and condensers; radiation in furnaces and radiant heat dryers; and by special methods such as dielectric heating. Often the equipment operates under steady-state conditions, but in many processes it operates cyclically, as in regenerative furnaces and agitated process vessels.

This part deals with equipment types that are of greatest interest to a process engineer: tubular and plate exchangers; extended-surface equipment mechanically aided heat-transfer devices; condensers and vaporizers; and packed-bed reactors or regenerators. Evaporators are described in next chapter. Information on all types of heat-exchange equipment is given in engineering texts and handbooks.

2.5.2 热量衡算（Energy balances）

间壁两侧冷、热两种流体进行热交换时，若换热器的热损失忽略，则根据能量守恒原理，传热速率 Q 应等于换热器的热负荷，等于热流体放出热量 Q_h，等于冷流体所吸收的热量 Q_c，即：

$$Q=Q_h=Q_c$$

（1）无相变化时热负荷计算

① 热焓法

$$Q=W_h(H_{h_1}-H_{h_2})=W_c(H_{c_2}-H_{c_1}) \tag{2-65}$$

式中 Q——热负荷（rate of heat transfer into stream），J/S 或 W；

H——单位质量流体的焓（enthalpies per unit mass of stream），kJ/kg。

式(2-65）中下标 h 为热流体，c 为冷流体。

② 比热法　若换热器内两流体均无相变化，且流体的比热容 C_p 不随温度而变化（或取流体平均温度下的比热容），则有：

$$Q=W_hC_{ph}(T_1-T_2)=W_cC_{pc}(t_2-t_1) \tag{2-66}$$

式中 W_h、W_c——热、冷流体质量流量（mass flow rate of stream），kg/s；

C_{ph}、C_{pc}——热、冷流体的定压比热容（specific heat），取流体进、出口算术平均温度下的比热值，kJ/(kg·K)；

T_1、T_2——热流体进、出口温度，K(℃)；

t_1、t_2——冷流体进、出口温度，K(℃)。

（2）有相变化时热负荷计算

① 一侧流体发生相变化，且处于饱和状态（例如热流体一侧蒸气饱和冷凝），则：

$$Q=W_hr=W_cC_{pc}(t_2-t_1) \tag{2-67}$$

式中 r——蒸汽冷凝潜热（latent heat of vaporization of vapour），等于饱和蒸汽的焓与同温度下的液体焓之差值，kJ/kg。

② 一侧流体发生相变后，又进一步有显热变化（例如热流体一侧饱和蒸气冷凝后，又进一步冷却），则：

$$Q=W_h[r+C_{ph}(T_d-T_2)]=W_cC_{pc}(t_2-t_1) \tag{2-68}$$

式中　T_d——蒸汽饱和温度（露点温度），K(℃)。

T_2——冷却液实际温度，K(℃)。

应当注意，在热负荷计算时，必须分清属于有相变还是无相变，然后依据不同公式进行计算。对蒸汽的冷凝、冷却过程热负荷要予以分别计算，然后相加。上述热负荷的计算方法，当要考虑热损失时，则有：

$$Q'=Q+Q_{损} \tag{2-69}$$

通常在保温良好的换热器中可取 $Q_{损}=(2\%\sim5\%)\ Q'$。

2.5.3 总传热速率方程（Overrall rate equation of heat transfer）

原则上，根据上述介绍的导热速率方程和对流传热速率方程即可进行换热器的传热计算。但是，采用上述方程计算冷、热流体间的传热速率时，必须知道壁温，而实际上壁温往往是未知的。为便于计算，需避开壁温，而直接用已知的冷、热流体的温度进行计算。为此，需要建立以冷、热流体主体温度差为传热推动力的传热速率方程，该方程称为总传热速率方程。

由上述分析可知，间壁两侧流体进行热交换时，一方面热量在径向上进行传递，另一方面热流体、冷流体、两侧壁温均沿着管长变化。因此，首先应从局部传热面 dS 入手建立总传热速率方程。

当冷、热流体通过间壁换热时，其传热机理如下：①热流体以对流方式将热量传给高温壁面；②热量由高温壁面以导热方式通过间壁传给低温壁面；③热量由低温壁面以对流方式传给冷流体。

由此可见，冷、热流体通过间壁换热是一个“对流—传导—对流”的串联过程，其传热速率方程可分别表示为（假设热流体走内侧）：

内侧热流体至内壁面的对流传热方程：

$$dQ=\alpha_i dS_i(T-T_w)=\frac{T-T_w}{\dfrac{1}{\alpha_i dS_i}} \tag{2-70}$$

内壁面至外壁面的热传导方程：

$$dQ=\frac{\lambda dS_m(T_w-t_w)}{b}=\frac{T_w-t_w}{\dfrac{b}{\lambda dS_m}} \tag{2-71}$$

外壁面至外侧冷流体的对流传热方程：

$$dQ=\alpha_o dS_o(t_w-t)=\frac{t_w-t}{\dfrac{1}{\alpha_o dS_o}} \tag{2-72}$$

对于间壁两侧流体间的稳态传热，上述各串联环节传热速率必然相等，即 dQ 为一个常数，则有：

$$dQ=\frac{T-T_w}{\dfrac{1}{\alpha_i dS_i}}=\frac{T_w-t_w}{\dfrac{b}{\lambda dS_m}}=\frac{t_w-t}{\dfrac{1}{\alpha_o dS_o}} \tag{2-73}$$

将上式加和后可得到：

$$dQ=\frac{T-t}{\frac{1}{\alpha_i dS_i}+\frac{b}{\lambda dS_m}+\frac{1}{\alpha_o dS_o}} \tag{2-74}$$

若令：
$$\frac{1}{K dS}=\frac{1}{\alpha_i dS_i}+\frac{b}{\lambda dS_m}+\frac{1}{\alpha_o dS_o} \tag{2-75}$$

则：
$$dQ=K dS(T-t) \tag{2-76}$$

式中　K——局部总传热系数，W/(m^2·℃)；

T——换热器的任一截面上热流体的主体温度，℃；

t——换热器的任一截面上冷流体的主体温度，℃。

式(2-76) 称为总传热速率微分方程，它是换热器传热计算的基本关系式。由该式可推出局部总传热系数 K 的物理意义，即 K 表示单位传热面积、单位传热温差下的传热速率，它反映了传热过程的强度。

上面的讨论是针对微元传热面积 dS 而言，具有局部性，而对整个换热器计算总传热速率时，可将式(2-76) 积分至整个换热器，即：

$$Q=\int_0^Q dQ=\int_0^S K(T-t)dS \tag{2-77}$$

则有：
$$Q=KS\Delta t_m \tag{2-78}$$

式中　Q——总传热速率，J/S 或 W；

K——总传热系数，W/(m^2·℃)；

S——换热器的总传热面积，m^2；

Δt_m——总平均温度差，℃。

为了顺利应用式(2-78) 求得整个换热器的总传热速率，下面讨论对总传热系数 K 和总平均温度差 Δt_m 的准确计算方法。

2.5.4 总传热系数（Overall heat-transfer coefficient）的确定

总传热系数 K 是评价换热器性能的一个重要参数，也是对换热器进行传热计算的依据。K 的数值取决于流体的物性、传热过程的操作条件及换热器的类型等，因而 K 值变化范围很大。

通常，取得总传热系数 K 值有三种方法：理论计算法、经验选取法及实验测定法。

2.5.4.1 理论计算法

（1）理论公式

当冷、热流体通过管式换热器进行传热时，沿传热方向传热面积是变化的，此时总传热系数 K 必须和所选择的传热面积相对应，选择的传热面积不同，总传热系数的数值也不同。

由式(2-75) 可知，若以传热管的外表面积 S_o（$S_o=\pi d_o L$）为基准，其对应的总传热系数 K_o 为：

$$K_o=\frac{1}{\frac{1}{\alpha_i}\times\frac{S_o}{S_i}+\frac{b}{\lambda}\times\frac{S_o}{S_m}+\frac{1}{\alpha_o}}=\frac{1}{\frac{1}{\alpha_i}\times\frac{d_o}{d_i}+\frac{b}{\lambda}\times\frac{d_o}{d_m}+\frac{1}{\alpha_o}} \tag{2-79}$$

同理，若以传热管内表面积 S_i（$S_i=\pi d_i L$）为基准，其对应的总传热系数 K_i 值为：

$$K_i=\frac{1}{\frac{1}{\alpha_i}+\frac{b}{\lambda}\times\frac{S_i}{S_m}+\frac{1}{\alpha_o}\times\frac{S_i}{S_o}}=\frac{1}{\frac{1}{\alpha_i}+\frac{b}{\lambda}\times\frac{d_i}{d_m}+\frac{1}{\alpha_o}\times\frac{d_i}{d_o}} \tag{2-80}$$

若以传热管的平均面积 $S_m(S_m=\pi d_m L)$ 为基准，其对应的总传热系数 S_m 值为：

$$K_m=\frac{1}{\frac{1}{\alpha_i}\times\frac{S_m}{S_i}+\frac{b}{\lambda}+\frac{1}{\alpha_o}\times\frac{S_m}{S_o}}=\frac{1}{\frac{1}{\alpha_i}\times\frac{d_m}{d_i}+\frac{b}{\lambda}+\frac{1}{\alpha_o}\times\frac{d_m}{d_o}} \tag{2-81}$$

式中 K_o、K_i、K_m——基于管内表面积、外表面积、内外平均表面积的总传热系数，W/(m^2·℃)；

S_o、S_i、S_m——管内表面积、外表面积、内外平均表面积，m^2。

可见，所取基准传热面积不同，K 值也不同，即：$K_o\neq K_i\neq K_m$。

当传热面为平壁时，则 $S_o=S_i=S_m$，此时的总传热系数 K 为：

$$K=\frac{1}{\frac{1}{\alpha_o}+\frac{b}{\lambda}+\frac{1}{\alpha_i}} \tag{2-82}$$

综上所述，总传热系数和传热面积的对应关系十分重要，所选基准面积不同，总传热系数的数值也不相同。手册中所列 K 值，无特殊说明，均视为以管外表面为基准的 K 值。对于管壁薄或管径较大时，可近似取 $S_o=S_i=S_m$，即圆筒壁视为平壁计算。

(2) 污垢热阻（fouling resistances）影响

实际操作的换热器传热表面上常有污垢积存，对传热产生附加热阻，称为污垢热阻。通常污垢热阻比传热壁的热阻大得多，因而设计中应考虑污垢热阻的影响。由于污垢层的厚度及其热导率难以准确地估计，因此通常选用一些经验值，表2-7给出了污垢热阻常见值。

表2-7 污垢热阻常见值

流体	污垢热阻 R_S/(m^2·K/kW)	流体	污垢热阻 R_S/(m^2·K/kW)
水(1m/s, $t>50$℃)		水蒸气	
蒸馏水	0.09	优质(不含油)	0.052
海水	0.09	劣质(不含油)	0.09
清净的河水	0.21	往复机排出	0.176
未处理的凉水塔用水	0.58	液体	
已处理的凉水塔用水	0.26	处理过的盐水	0.264
硬水、井水	0.58	有机物	0.176
气体		燃料油	1.056
空气	0.26～0.53	焦油	1.76
溶剂蒸气	0.14		

设管壁内、外侧表面上的污垢热阻分别为 R_{Si} 及 R_{So}，根据串联热阻叠加原理，式(2-79)可表示为：

$$\frac{1}{K}=\frac{1}{\alpha_o}+R_{S_o}+\frac{b}{\lambda}\times\frac{d_o}{d_m}+R_{S_i}\frac{d_o}{d_i}+\frac{1}{\alpha_i}\times\frac{d_o}{d_i} \tag{2-83}$$

(3) 提高 K 值讨论

若不考虑管壁热阻及污垢热阻的影响，且同时将传热面取为平壁或薄管壁时（即 d_o、d_i、d_m 相等或近于相等），则上式可化简为：

$$K=\frac{1}{\frac{1}{\alpha_o}+\frac{1}{\alpha_i}} \tag{2-84}$$

总传热系数 K 值比两侧流体中 α 值小者还小，其中：

① 当 $\alpha_o \ll \alpha_i$ 时，则 $K \approx \alpha_o$；当 $\alpha_i \ll \alpha_o$ 时，则 $K \approx \alpha_i$，即 K 值总是接近于 α 小的一侧流体（即代表该侧流体的热阻很大）的对流传热系数值。由此可知，总热阻是由热阻大的那一侧流体的对流传热情况所控制，即当两侧流体的对流传热系数相差较大时，要提高 K 值，关键在于提高 α 较小侧流体的对流传热系数。

② 若两侧 α 相差不大时，则必须同时提高两侧 α 的值，才能提高 K 值。

③ 同样，若管壁两侧流体的对流传热系数均很大，即两侧流体的对流传热热阻都很小，而污垢热阻却很大，则称为污垢热阻控制，此时欲提高 K 值，必须设法减慢污垢形成速率或及时清除污垢。

【例 2-7】 现有一管壳式换热器，CO_2 流经管程，壳程为冷却水，管壳由 ϕ25mm×2.5mm 钢管组成。已知管内、外对流传热系数分别为 $\alpha_i=50\mathrm{W/(m^2\cdot K)}$，$\alpha_o=2500\mathrm{W/(m^2\cdot K)}$，钢管的热导率为 $\lambda=45\mathrm{W/(m\cdot K)}$，考虑使用时管内、外都有垢层，其热阻分别为 $R_{Si}=0.0005\mathrm{m^2\cdot K/W}$，$R_{So}=0.00058\mathrm{m^2\cdot K/W}$，试求：

（1）该换热器的总传热系数 K；

（2）在其他条件不变的情况下，分别将管内、外的对流传热系数提高一倍，并计算 K 值增大的百分数。

解 （1）由式(2-79) 并考虑污垢的影响，可知：

$$K_o=\frac{1}{\frac{1}{\alpha_i}\times\frac{d_o}{d_i}+R_{Si}\frac{d_o}{d_i}+\frac{b}{\lambda}\times\frac{d_o}{d_m}+R_{So}+\frac{1}{\alpha_o}}$$

$$=\frac{1}{\frac{1}{50}\times\frac{25}{20}+0.0005\times\frac{25}{20}+\frac{0.0025}{45}\times\frac{25}{22.5}+0.00058+\frac{1}{2500}}$$

$$=37.5\mathrm{W/(m^2\cdot K)}$$

（2）计算 K 值增大的百分数：

当 $\alpha_i'=2\alpha_i=100\mathrm{W/(m^2\cdot K)}$ 时：

$$K'=\frac{1}{\frac{1}{100}\times\frac{25}{20}+0.0005\times\frac{25}{20}+\frac{0.0025}{45}\times\frac{25}{22.5}+0.00058+\frac{1}{2500}}$$

$$=70.6\mathrm{W/(m^2\cdot K)}$$

$$\frac{K'-K}{K}=\frac{70.6-37.5}{37.5}\times100\%=88.2\%$$

当 $\alpha_o'=2\alpha_o=5000\mathrm{W/(m^2\cdot K)}$ 时：

$$K''=\frac{1}{\frac{1}{50}\times\frac{25}{20}+0.0005\times\frac{25}{20}+\frac{0.0025}{45}\times\frac{25}{22.5}+0.00058+\frac{1}{5000}}$$

$$=37.8\mathrm{W/(m^2\cdot K)}$$

$$\frac{K''-K}{K}=\frac{37.8-37.5}{37.5}\times100\%=0.8\%$$

计算结果表明，K 值总是接近热阻大的流体侧的 α 值，因此要提高 K 值，必须对影响 K 值的各项因素进行分析，如在本题条件下，应提高 CO_2 侧的 α 值，才更有利于传热。

2.5.4.2 经验选取法

在实际设计计算中，总传热系数通常采用经验值。由于 K 值的变化范围很大，取值时应注意设备型式相同，工艺条件相仿，表 2-8 给出管壳式换热器中总传系数 K 的经验值。

表 2-8 管壳式换热器中的总传系数 K 的经验值

冷流体	热流体	传热系数 K/[W/(m^2·℃)]
水	水	850～1700
水	气体	17～280
水	有机溶剂	280～850
水	轻油	340～910
水	重油	60～280
有机溶剂	有机溶剂	115～340
水	水蒸气冷凝	1420～4250
气体	水蒸气冷凝	30～300
水	低沸点烃类冷凝	455～1140
水沸腾	水蒸气冷凝	2000～4250

2.5.4.3 实验测定法

对于已有的换热器，可以通过测定有关数据，如设备的尺寸、流体的流量和温度等，然后由传热基本方程式计算 K 值。显然，这样得到的总传热系数 K 值最为可靠。实测 K 值的意义，不仅可以为换热器设计提供依据，而且可以分析了解所用换热器的性能，寻求提高设备传热能力的途径。

2.5.5 平均传热温度差（Mean temperature difference）的计算

在间壁式换热器中，Δt_m 的计算可分为下列几种类型。

（1）恒温传热时的 Δt_m

当间壁换热器两侧流体均发生饱和相变化时，温度差的计算非常简单。例如蒸发器操作，一侧为饱和蒸气冷凝，热流体保持在恒温 T 下放热；另一侧为饱和液体沸腾，冷流体保持在恒温 t 下吸热。因此，换热器间壁两侧流体的温度差处处相等，则：

$$\Delta t_m = T - t = \text{常数} \tag{2-85}$$

（2）变温传热时的 Δt_m

变温传热的情况在生产实际中应用较多，即当间壁两侧流体在传热过程中至少有一侧流体没有发生相变化，则冷、热流体的主体温差必然随换热器位置而变化。在换热器中，冷、热流体若以相同的方向流动，称为并流；两流体若以相反的方向流动，称为逆流。下面举出几种变温传热情况，如图 2-16 及图 2-17 所示。

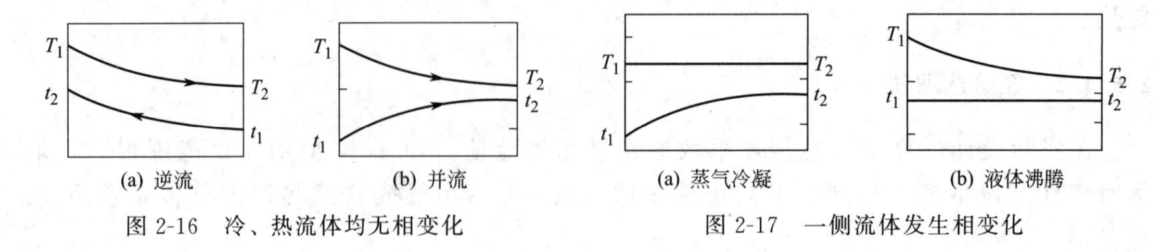

图 2-16　冷、热流体均无相变化　　图 2-17　一侧流体发生相变化

① 并流与逆流时的 Δt_m 计算式导出　现以逆流操作为例，取微元传热面积 dS 分析。对于稳态传热，且热损失可忽略时，由热量衡算得：

$$dQ=-W_h C_{ph} dT=W_c C_{pc} dt \tag{2-86}$$

上式可写为：

$$\frac{dQ}{dT}=-W_h C_{ph}$$

$$\frac{dQ}{dt}=W_c C_{pc}$$

对整个换热器而言，两流体的质量流量不随换热器长度 L 变化。若以定性温度确定两流体的比热容，比热容也可以当常数处理。所以，T-Q 和 t-Q 的关系均为直线，可表示为：

$$T=mQ+k$$

$$t=m'Q+k'$$

上两式相减得：

$$T-t=(m-m')Q+(k-k')$$

即 $(T-t)$-Q 的关系也是直线。

令：
$$\Delta t=T-t$$

则：
$$\Delta t_1=T_1-t_1, \Delta t_2=T_2-t_2$$

故：
$$\frac{d(\Delta t)}{dQ}=\frac{\Delta t_2-\Delta t_1}{Q} \tag{2-87}$$

又：
$$dQ=K dS(T-t)=K dS \Delta t \tag{2-88}$$

将式(2-88) 代入式(2-87) 中，并取 K 为常数，不随 L 变化，则积分后可得：

$$Q=KS\frac{\Delta t_1-\Delta t_2}{\ln\frac{\Delta t_1}{\Delta t_2}} \tag{2-89}$$

上式与式(2-78) 比较知：

$$\Delta t_m=\frac{\Delta t_1-\Delta t_2}{\ln\frac{\Delta t_1}{\Delta t_2}} \tag{2-90}$$

上式为冷、热两种流体进、出口温度 T_1、T_2、t_1 和 t_2 的对数平均值。

对于并流操作或仅一侧流体变温的情况，可采用类似的方法导出同样的表达式，即式(2-90) 是计算逆流和并流时平均温度差的通式。

[讨论]：

① 当 $\Delta t_1/\Delta t_2<2$ 时，Δt_m 可用算术平均值代替对数平均值，其误差不超过4%。

② 利用上式计算对数平均温度差时，取换热器两端的 Δt 中数值大者为 Δt_2，小者为 Δt_1，这样计算 Δt_m 比较简便。

③ 逆流与并流比较：在冷热流体进出口温度相同的前提下，逆流操作的平均温差最大，因此，在换热器的传热量 Q 及总传热系数 K 值相同的条件下，采用逆流操作，若换热介质流量一定时可以节省传热面积，减少设备费；若传热面积一定时，可减少换热介质的流量，降低操作费，因而工业上多采用逆流操作；只有当冷流体被加热或热流体被冷却而不允许超过某一温度时，采用并流更可靠。

【例2-8】 在一套管换热器内用热水来加热某料液，已知热水的进出口温度分别为360K和340K，而冷料液则从290K被加热到330K，试分别计算并流和逆流操作时的平均温度差。

解 并流时，$\Delta t_1=70\text{K}$，$\Delta t_2=10\text{K}$，代入式(2-90)得：

$$\Delta t_m=\frac{70-10}{\ln\frac{70}{10}}=30.8(\text{K})$$

逆流时，$\Delta t_1=50\text{K}$，$\Delta t_2=30\text{K}$，代入式(2-90)得：

$$\Delta t_m=\frac{50-30}{\ln\frac{50}{30}}=39.2(\text{K})$$

当冷、热流体操作温度一定时，$\Delta t_{m逆流}$ 总是大于 $\Delta t_{m并流}$。

② 其他流型的平均温度差计算　在大多数换热器中，冷热流体并非作简单的逆流或并流，而是比较复杂的多程流动。其中，冷热两流体垂直交叉流动，称为错流；一种流体只沿一个方向流动，而另一流体反复改变流向，称为折流，如图2-18所示。

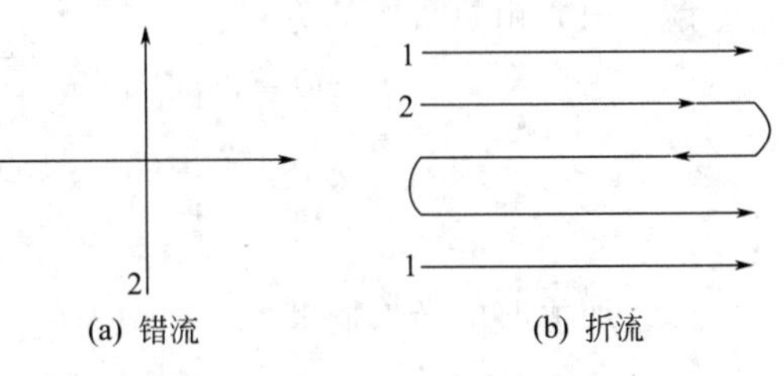

图2-18　错流与折流示意图

当两流体呈错流和折流流动时，平均温度差 Δt_m 的计算较为复杂，通常采用安德伍德（Underwood）和鲍曼（Bowman）提出的图算法，其基本思路是先按逆流计算对数平均温度差，再乘以流动方向的校正因素，即：

$$\Delta t'_m=\Delta t_m\varphi_{\Delta t} \tag{2-91}$$

式中　Δt_m——按逆流操作情况下的平均温度差；

$\varphi_{\Delta t}$——温度校正系数，无量纲，为 P，R 两因数的函数，$\varphi_{\Delta t}=f(P, R)$。

$$R=\frac{T_1-T_2}{t_2-t_1}=\frac{\text{热流体的温降}}{\text{冷流体的温升}}$$

$$P=\frac{t_2-t_1}{T_1-t_1}=\frac{\text{冷流体的温升}}{\text{两流体的最初温度差}} \tag{2-92}$$

对于各种换热情况下的 $\varphi_{\Delta t}$ 值，可在后面图2-39中查到。采用折流或其他流动形式除了是为了满足换热器的结构要求外，还为了提高总传热系数，但是平均温度差较逆流时低。在选择流向时应综合考虑，$\varphi_{\Delta t}$ 值不宜过低，一般设计时应取 $\varphi_{\Delta t}>0.9$，至少不能低于

0.8，否则另选其他流动形式。

通过以上讨论可知，要强化传热过程主要应着眼于增加推动力，减少热阻，也就是设法增大 Δt_m 或增大传热面积 S 和总传热系数 K。

在生产上，无论是选用或设计一个新的换热器还是对已有的换热器进行检定，都是以上述的总传热速率方程为基础，传热计算则主要解决方程中的 Q、S、K、Δt_m 及相关量的计算。因此，总传热速率方程是传热章中最基本的方程式。

2.5.6 换热器的传热计算

传热过程的工艺计算主要有两类：一类是设计型计算，即根据生产要求的热负荷，确定换热器的传热面积；另一类是操作型计算（又称为校核型），即对现有的换热器，判断其对指定的传热任务是否适用，或预测在生产中某些参数变化对传热有影响等，均属于换热器的操作型计算。两类计算都是以上述介绍的换热器热量衡算和传热速率方程为理论基础，但后者的计算较为复杂，往往需要试差或迭代，下面分别举例计算加以说明。

2.5.6.1 计算举例

【例 2-9】 在一钢制套管换热器中，用冷水将 1kg/s 的苯由 65℃冷却至 15℃，冷却水在 ϕ25mm×2.5mm 的内管中逆流流动，其进出口温度为 10℃和 45℃。已知苯和水的对流传热系数分别为 0.82×10^3 W/(m^2·K) 和 1.7×10^3 W/(m^2·K)，在定性温度下水和苯的比热容分别为 4.18×10^3 J/(kg·K) 和 1.88×10^3 J/(kg·K)，钢材热导率为 45W/(m·K)，两侧的污垢热阻可忽略不计。试求：(1) 冷却水消耗量；(2) 所需的总管长。

解 (1) 由热量衡算方程：

$$Q=W_hC_{ph}(T_1-T_2)=W_cC_{pc}(t_2-t_1)$$

$$1\times1.88\times10^3\times(65-15)=W_c\times4.18\times10^3\times(45-10)$$

$$W_c=0.643\text{kg/s}$$

(2) 逆流时的平均温度差：

$$\Delta t_m=\frac{\Delta t_1-\Delta t_2}{\ln(\Delta t_1/\Delta t_2)}=\frac{20-5}{\ln(20/5)}=10.8(℃)$$

以管外面积为计算基准，忽略污垢热阻，则总传热系数：

$$\frac{1}{K}=\frac{d_o}{\alpha_i d_i}+\frac{bd_o}{\lambda d_m}+\frac{1}{\alpha_o}$$

$$=\frac{25}{1.7\times10^3\times20}+\frac{0.0025\times25}{45\times22.5}+\frac{1}{0.82\times10^3}$$

$$K=496\text{W/(m}^2\cdot\text{K)}$$

由传热速率方程可得传热面积：

$$S=\frac{Q}{K\Delta t_m}=\frac{W_hC_{ph}(T_1-T_2)}{K\Delta t_m}$$

$$=\frac{1\times1.88\times10^3\times(65-15)}{496\times10.8}=17.5(\text{m}^2)$$

由于 K 以管外的传热面积为基准，计算 L 时应以 d_o 为基准。故所需总管长

$$L=\frac{S}{\pi d_o}=\frac{17.5}{3.14\times 0.025}=224(\mathrm{m})$$

实际上，上述总管长 L 的实现也可通过管路的并联实现，例如采用列管式换热器。

【例 2-10】 有一套管换热器进行逆流操作，管内通流量为 0.6kg/s 的冷水，冷水的进口温度为 30℃，管间通流量为 2.52kg/s 的空气，进出口温度分别为 130℃和 70℃。已知水侧和空气侧的对流传热系数分别为 2000W/（m^2・K）、50 W/（m^2・K），水和空气的平均比热容分别为 4200 J/（kg・K）、1000 J/（kg・K）。试求水量增加一倍后换热器热流量与原来之比。假设管壁较薄，污垢热阻忽略不计，流体流动 Re 均大于 10^4。

解　因为管壁较薄，污垢热阻忽略不计，则在原工况下：

$$K=\frac{1}{\dfrac{1}{\alpha_o}+\dfrac{1}{\alpha_i}}=\frac{1}{\dfrac{1}{50}+\dfrac{1}{2000}}48.8[\mathrm{W/(m^2\cdot K)}]$$

$$\frac{W_c C_{pc}}{W_h C_{ph}}=\frac{0.6\times 4200}{2.52\times 1000}=1.0$$

由于两端温度差相等，则平均传热推动力：

$$\Delta t_m=T_2-t_1=70-30=40(℃)$$

由热量衡算式得：

$$t_2=t_1+\frac{W_h C_{ph}}{W_c C_{pc}}(T_1-T_2)=30+1\times(130-70)=90(℃)$$

又：

$$\frac{KS}{W_h C_{ph}}=\frac{T_1-T_2}{\Delta t_m}=\frac{130-70}{40}=1.5$$

在新工况下：

$$K'=\frac{1}{\dfrac{1}{\alpha_1}+\dfrac{1}{\alpha'_2}}=\frac{1}{\dfrac{1}{50}+\dfrac{1}{2^{0.8}\times 2000}}=49.3$$

热量衡算式为：

$$t'_2=t_1+\frac{W_h C_{ph}}{W'_c C_{pc}}(T_1-T'_2)=30+0.5\times(130-T'_2) \tag{1}$$

$$W_h C_{ph}(T_1-T'_2)=K'S\frac{(T_1-t'_2)-(T'_2-t_1)}{\ln\dfrac{T_1-t'_2}{T'_2-t_1}} \tag{2}$$

$$=K'S\frac{(T_1-T'_2)-(t'_2-t_1)}{\ln\dfrac{T_1-t'_2}{T'_2-t_1}}$$

由式(1)、式(2) 得到：

$$\ln\frac{T_1-t'_2}{T'_2-t_1}=\frac{K'S}{W_h C_{ph}}\left(1-\frac{W_h C_{ph}}{2W_c C_{pc}}\right)=\frac{K'}{K}\times\frac{KS}{W_h C_{ph}}\left(1-\frac{W_h C_{ph}}{2W_c C_{pc}}\right)$$

$$\ln\frac{130-t'_2}{T'_2-30}=\frac{4.93\times10^{-2}\times1.5}{4.88\times10^{-2}}\times(1-0.5)=0.758$$

$$130-t'_2=2.13\times(T'_2-30) \tag{3}$$

由式(1)、式(3) 得

$$t'_2=64.7℃ \qquad T'_2=60.6℃$$

在新的工况下的传热推动力：

$$\Delta t'_m=\frac{(130-64.7)-(60.6-30)}{\ln\frac{(130-64.7)}{(60.6-30)}}=45.8(℃)$$

两工况下的热流量之比：

$$\frac{Q'}{Q}=\frac{K'S\Delta t'_m}{KS\Delta t_m}=\frac{4.93\times10^{-2}\times45.8}{4.88\times10^{-2}\times40}=1.156\approx\frac{\Delta t'_m}{\Delta t_m}$$

由此例可以看出，增加热阻较小（即对流传热系数较大）侧流体的流量，总传热系数变化很小，热流量的增加主要是由传热推动力增大而引起的，即此时传热过程的调节主要靠 Δt_m 的变化。

2.5.6.2 传热效率 ε 与传热单元数 NTU 法

在传热计算中，当冷、热两种流体的出口温度 T_2 和 t_2 同时未知时，若采用对数平均推动力法求解，必须用试差法，十分麻烦。为避免试差，有人提出了 ε-NTU 法。

（1）传热效率 ε

ε 为实际的传热速率 Q 与最大可能的传热速率 Q_{max} 之比，即：

$$\varepsilon=\frac{Q}{Q_{max}} \tag{2-93}$$

设冷、热流体在一个面积无限大的逆流换热器中换热，无相变化和热损失时，实际传热速率为：

$$Q=W_hC_{ph}(T_1-T_2)=W_cC_{pc}(t_2-t_1) \tag{2-94}$$

因热流体最大可能的温度变化为 T_1-t_1（即 $T_2=t_1$ 时），冷流体最大可能的温度变化也为 T_1-t_1（即 $t_2=T_1$ 时），所以最大可能的传热速率为：

$$Q_{max}=(WC_p)_{min}(T_1-t_1) \tag{2-95}$$

令热容流率 W_hC_{ph} 和 W_cC_{pc} 中较大者为 $(WC_p)_{max}$，较小者为 $(WC_p)_{min}$，则：当 $W_hC_{ph}=(WC_p)_{min}$ 时，联立式(2-93)、式(2-94) 及式(2-95) 可得：

$$\varepsilon_h=\frac{T_1-T_2}{T_1-t_1} \tag{2-96}$$

同理，当 $W_cC_{pc}=(WC_p)_{min}$ 时：

$$\varepsilon_c=\frac{t_2-t_1}{T_1-t_1} \tag{2-97}$$

（2）传热单元数

以逆流操作为例，在换热器中取微元传热面积 dS，由热量衡算式和传热基本方程可得：

$$dQ=-W_hC_{ph}dT=W_cC_{pc}dt=K\,dS(T-t)$$

对于热流体有：

$$NTU_h=-\int_{T_1}^{T_2}\frac{dT}{T-t}=\int_0^S\frac{K\,dS}{W_hC_{ph}} \tag{2-98}$$

对于冷流体：

$$NTU_c=-\int_{t_2}^{t_1}\frac{dt}{T-t}=\int_0^S\frac{K\,dS}{W_cC_{pc}} \tag{2-99}$$

(3) 传热系数与传热单元数的关系

令：

$$C_R=\frac{(WC_p)_{min}}{(WC_p)_{max}} \qquad (NTU)_{min}=\frac{KS}{(WC_p)_{min}}$$

逆流操作时，传热系数与传热单元数间关系经推导得：

$$\varepsilon=\frac{1-\exp[-(NTU)_{min}(1-C_R)]}{1-C_R\exp[-(NTU)_{min}(1-C_R)]} \tag{2-100}$$

并流操作时，传热系数与传热单元数间关系经推导得：

$$\varepsilon=\frac{1-\exp[-(NTU)_{min}(1+C_R)]}{1+C_R} \tag{2-101}$$

两流体之一有相变化时：

$$\varepsilon=1-\exp[-NTU] \tag{2-102}$$

两流体热容流率相等时：

逆流操作：

$$\varepsilon=\frac{NTU}{1+NTU} \tag{2-103}$$

并流操作：

$$\varepsilon=\frac{1-\exp[-2NTU]}{2} \tag{2-104}$$

上述各种情况有相应的关系图可查，如图 2-19 所示。

ε-NTU 法既可对换热器进行校核计算，也可用作新换热器的设计。当传热系数 K 已知时，应用 ε-NTU 法进行换热器计算，具有下述优点：

① 计算简单，查图方便；

② 在两流体的出口温度（T_2，t_2）均未知时，无须试差；

③ 对正在使用的换热器，如果改变进口温度（T_1 或 t_1），可以直接求出相应的出口温度，也无须试算；

④ 对一组串联的换热器，可将 NTU 累计起来计算比常用方法更易估算。

【例 2-11】 在某换热设备中，冷、热流体进行逆流操作。热流体油的流量为 2.85kg/s，进口温度为 110℃，平均比热容为 1.9kJ/(kg·℃)；冷流体水的流量为 0.667kg/s，进口温度为 35℃，平均比热容为 4.18kJ/(kg·℃)。已知换热器面积为 15.8m²，总传热系数为 320W/(m²·℃)。试求冷、热流体的出口温度及传热量。

解 $W_hC_{ph}=2.85\times1.9\times10^3=5415$(W/℃)；$W_cC_{pc}=0.667\times4.18\times10^3=2788$(W/℃)。

因为 $W_cC_{pc}<W_hC_{ph}$，即说明冷流体为最小热容流率流体，所以：

$$C_R=\frac{W_cC_{pc}}{W_hC_{ph}}=\frac{2788}{5415}=0.515$$

$$NTU=\frac{KS}{W_cC_{pc}}=\frac{320\times15.8}{2788}=1.81$$

由图 2-19 查得传热效率：$\varepsilon=0.73$

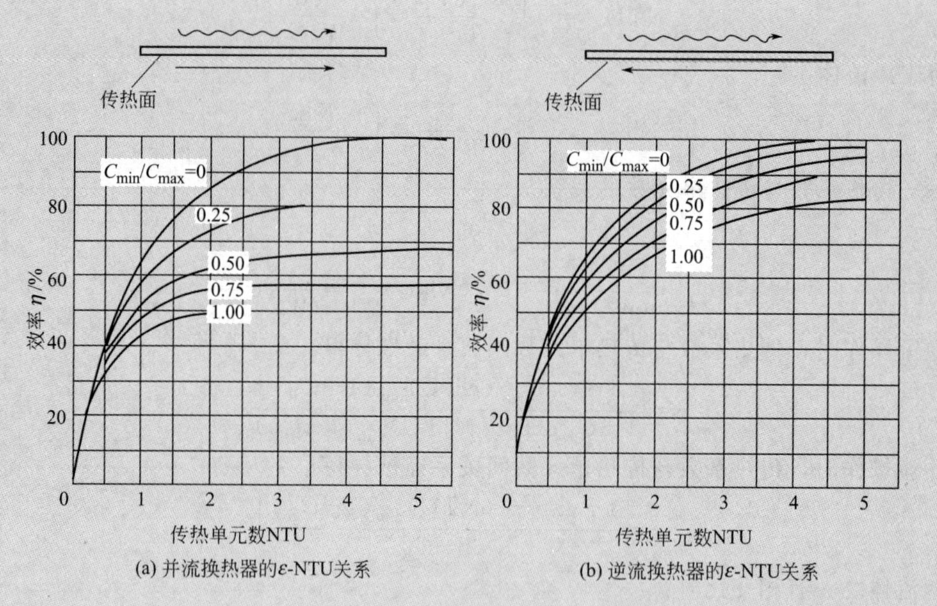

(a) 并流换热器的ε-NTU关系　　(b) 逆流换热器的ε-NTU关系

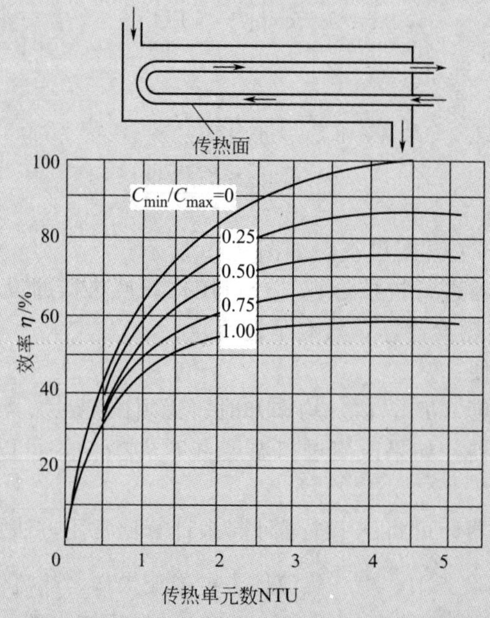

(c) 折流换热器的ε-NTU关系(单壳程，2、4、6管程)

图 2-19　各流型换热器的 ε-NTU 关系

又根据：

$$\varepsilon_c=\frac{t_2-t_1}{T_1-t_1}=0.73$$

所以：

$$t_2=t_1+\varepsilon_c(T_1-t_1)=35+0.73\times(110-35)=89.8(℃)$$

$$Q=W_cC_{pc}(t_2-t_1)=2788\times(89.8-35)=1.53\times10^5\,\text{W}$$

又因为：

$$C_R=\frac{W_cC_{pc}}{W_hC_{ph}}=\frac{T_1-T_2}{t_2-t_1}$$

所以：

$$T_2=T_1-C_R(t_2-t_1)=110-0.515\times(89.8-35)=81.8(℃)$$

2.6 换热设备（Heat Exchange Equipment）

换热器是过程工业及其他许多工业部门的通用设备，在生产中占有重要的地位，其设备投资往往在整个设备总投资中占有很大的比例，如现代石油化学企业中一般可达 30%～40%。换热器的类型多种多样，按用途可分为：加热器、冷却器、冷凝器、蒸发器、再沸器等。本节主要介绍各种换热器的性能、特点以及由工艺条件选择定型设备的方法。

2.6.1 换热器类型

若按传热特征，换热器可分为以下三种。

2.6.1.1 直接接触式（混合式）

直接接触式换热器的特点是冷热两种流体在换热器内直接混合进行热交换。这类换热器主要应用于气体的冷却、除尘、增湿或蒸汽的冷凝，常见的设备有凉水塔、洗涤塔、文氏管及喷射冷凝器等。其优点是传热效果好，设备简单，易于防腐；缺点是仅允许两流体可混合时才能使用。如图 2-20 所示的混合式冷凝器就是一种典型的直接接触式换热器。

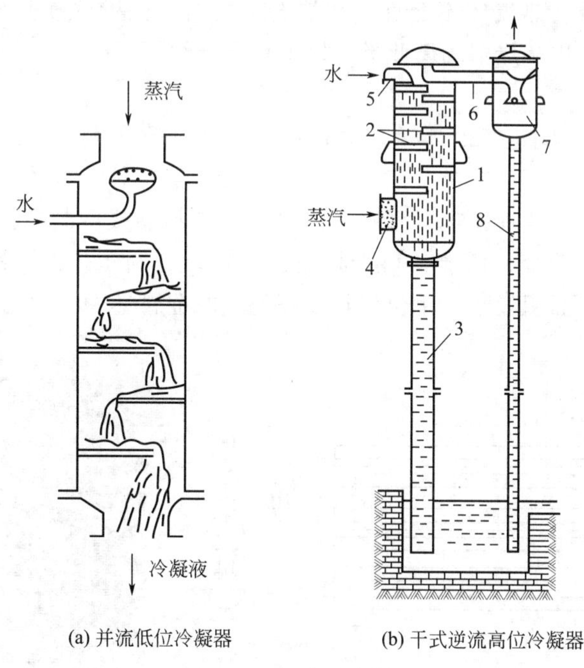

(a) 并流低位冷凝器　　(b) 干式逆流高位冷凝器

图 2-20 混合式冷凝器

1—外壳；2—淋水板；3,8—气压管；4—蒸汽进口；5—进水口；6—不凝气出口；7—分离罐

2.6.1.2 蓄热式换热器

蓄热式换热器如图 2-21 所示，其特点是换热器内装有填充物（如耐火砖），热流体和冷流体交替流过填充物，以填充物交替吸热和放热的方式进行热交换。这类换热器主要应用于高温气体的余热利用。其优点是设备简单，耐高温；缺点是设备体积庞大，不能完全避免两种流体混合。

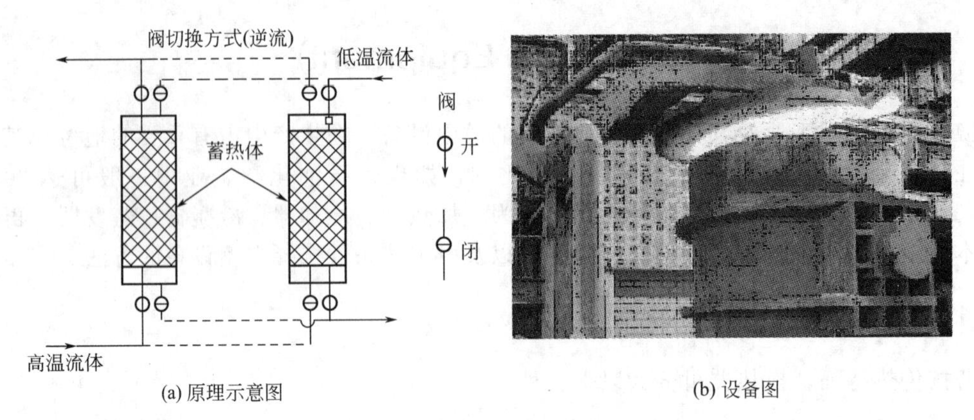

(a) 原理示意图　　(b) 设备图

图 2-21　蓄热式换热器

2.6.1.3　间壁式换热器

在化工生产中遇到的多是间壁两侧流体的热交换，即冷、热流体被固体壁面（传热面）所隔开，互不接触，固体壁面构成间壁式换热器。在此类换热器中，热量由热流体通过壁面传给冷流体，适用于冷、热流体不允许直接混合的场合。间壁式换热器应用广泛，形式多样，各种管式和板式结构换热器均属此类，以下重点介绍。

（1）管式换热器

① 套管式换热器　如图 2-22 所示，两种尺寸的标准管子套在一起（同心安装），各段用 180°的回弯管连接。每程的有效长度为 4～6m。冷、热流体分别流过内管和套管的环隙，并通过间壁进行热交换。这种换热器可用作加热器、冷却器或冷凝器。其优点是结构简单，耐高压，传热面积容易改变，两流体可严格逆流操作，有利于传热；缺点是接头多，易泄漏，单位管长上的传热面积较小，单位传热面积消耗的金属量大。

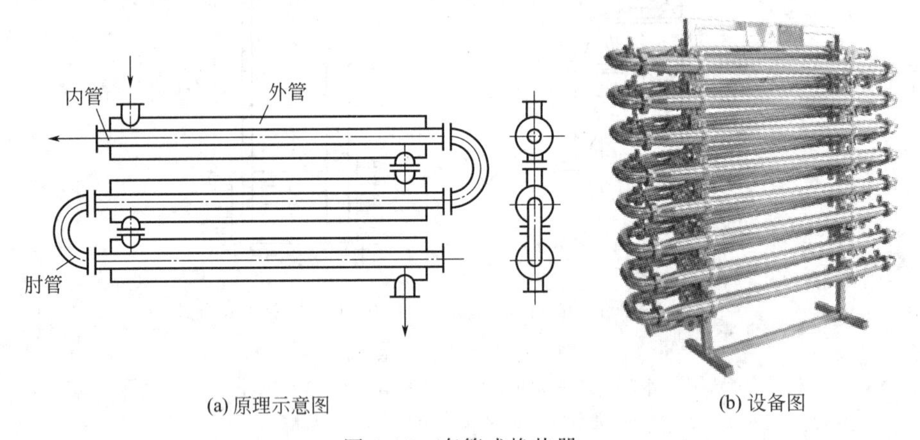

(a) 原理示意图　　(b) 设备图

图 2-22　套管式换热器

② 蛇管式换热器

a. 沉浸式　如图 2-23 所示，管子按容器的形状弯制，并沉浸在容器中，使容器内的流体和蛇管内的流体通过管壁进行换热。这种换热器主要用于反应器或容器内的加热或冷却。其优点是结构简单，价格低廉，易防腐，耐高压；缺点是蛇管外的对流传热系数小，所以总传热系数较小。

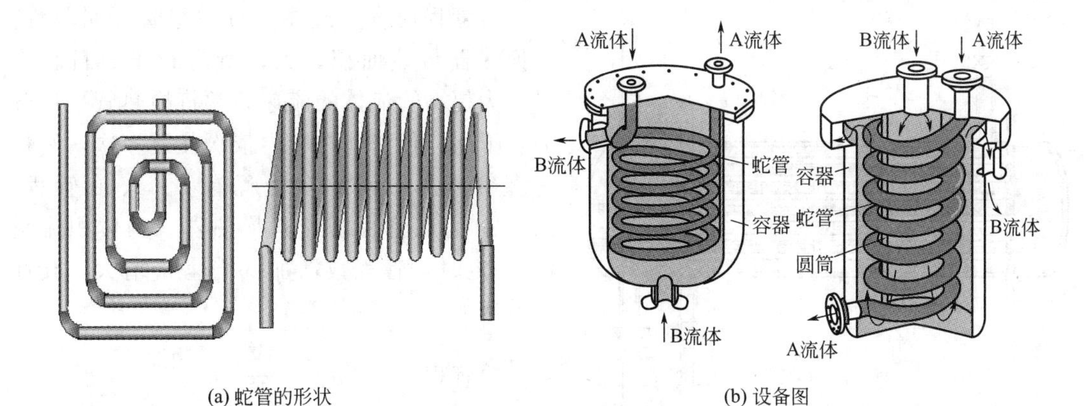

图 2-23 沉浸式蛇管换热器

b. 喷淋式 如图 2-24 所示，将蛇管成排固定在支架上，管束上面装有喷淋装置，冷却水淋洒在管排上形成下流的液膜与管内流体热交换，部分冷却水在空气中汽化，带走部分热量。这种换热器多用作冷却器，室外放置（空气流通处）。其优点是结构简单，检修清洗方便，传热效果比沉浸式好；缺点是喷淋不易均匀。

③ 管壳式换热器 管壳式换热器又称列管式换热器，是目前化工上应用最广泛的一种换热设备。与前面介绍的换热器相比，其优点是单位体积的传热面积大，可用多种材料加工制造，耐高温、高压，适应性强，传热效果较好；缺点是还算不上高效换热器。

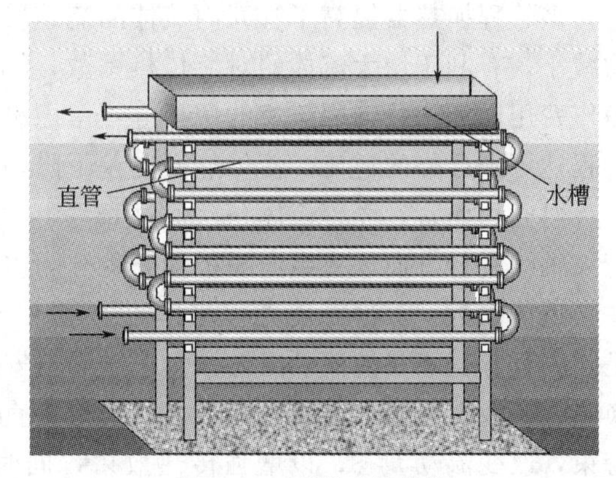

图 2-24 喷淋式换热器

管壳式换热器的基本形式如图 2-25 所示，其主要部件为：壳体、管束（管壳）、管板（花板）、封头（顶盖）。管束安装在壳体内，两端固定在花板上。封头用法兰与壳体连接。进行热交换时，一种流体由封头的进口接管进入，然后分配到平行管束，从另一端封头的出口管流出，这种流动方式称为管程流动。另一种流体则由靠近花板处的连接管进入壳体，在壳体内从管束的空隙流过，由壳体的另一接管流出，这种流动方式称为壳程流动。

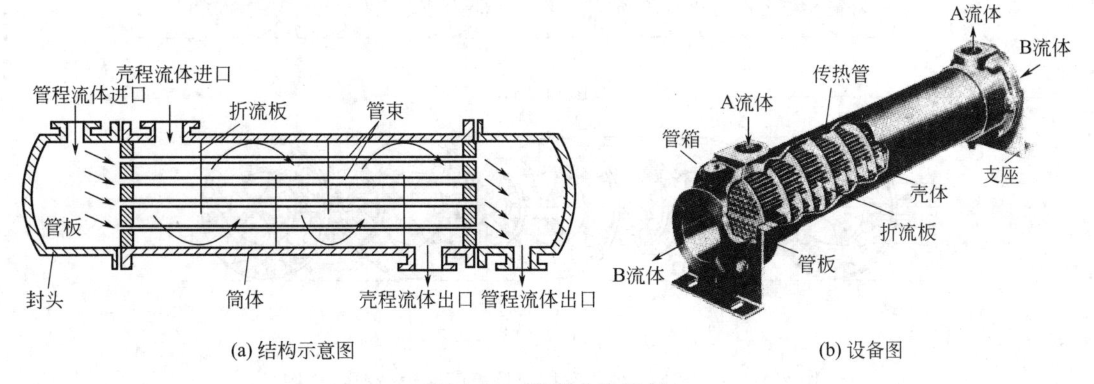

图 2-25 管壳式换热器

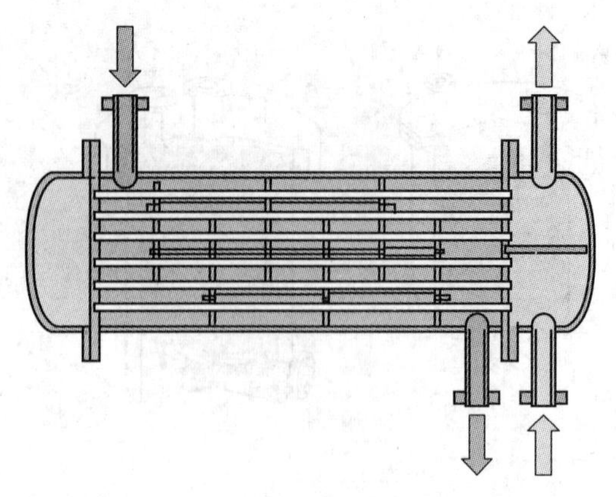
图 2-26　双管程管壳式换热器示意图

对管程而言，流体流过一组管子叫单程。当换热器传热面积较大，所需管子数目较多时，为提高管流体的流速，常将换热管平均分为若干组，使流体在管内依次往返多次，则称为多管程。常用的多管程有 2、4、6 管程，图 2-26 为双管程管壳式换热器。当体积流量 V_S 一定时，管程数增加，流速 u 增大。其计算方法如下：

单管程：
$$u_1=\frac{4V_S}{\pi d_i^2 n}$$

2 管程：
$$u_2=\frac{8V_S}{\pi d_i^2 n}=2u_1$$

4 管程：
$$u_4=\frac{16V_S}{\pi d_i^2 n}=4u_1$$

式中，d_i 为管壳内径；n 为每管程的管壳数。

多管程换热器提高了管壳内流体的流速。一方面，流体的湍动程度加剧，管子表面沉积物减小，有利于增大流体的对流传热系数，减小传热面积；另一方面，流动阻力增大，导致操作费用增加。故管程流速的选择应该考虑设备费用和投资费用之间的权衡问题。另外，多管程达不到严格的逆流，使传热温度差下降，且两端封头内装有隔板，占去了部分管面积。

对壳程而言，壳程流体一次通过壳程，称为单壳程。为提高壳程流体的流速，也可在与管束轴线平行方向放置纵向隔板使壳程分为多程。壳程数即为壳程流体在壳程内沿壳体轴向往、返的次数。分程可使壳程流体流速增大，流程增长，扰动加剧，有助于强化传热。但是，壳程分程不仅使流动阻力增大，且制造安装较为困难，故工程上应用较少。为改善壳程换热，一般在壳体内安装一定数目与管束垂直的折流挡板。折流挡板用以引导流体横向流过管束，改变流动型态，以增强传热效果，同时也起到支撑管束、防止管束振动和弯曲的作用。折流挡板的形式有圆缺形（弓形）、环盘形和孔流形等，其间的流体流动情况如图 2-27 所示。

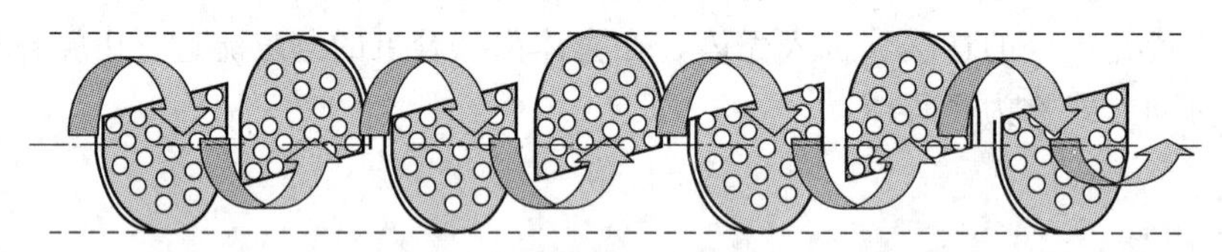
(a) 圆缺形

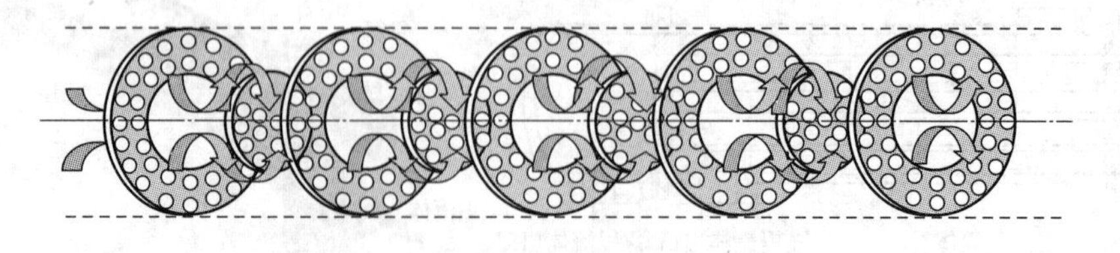
(b) 环盘形

图 2-27　壳程流体在各类折流挡板间流动情况示意图

管壳式换热器根据结构特点分为以下三种常见形式。

a. 固定管板式 如图2-28所示，固定管板式管壳换热器的特点是两端管板与壳体连成一体。其优点是结构简单，造价低廉，每根换热管都可以进行更换，且管内清洗方便；缺点是壳程不易检修、清洗。这种换热器的热补偿方式是加补偿圈，适用范围为：两流体温度<70℃；壳程流体压力<6atm。因而壳程压力受膨胀节强度的限制不能太高。固定管板式换热器适用于壳方流体清洁且不易结垢，两流体温差不大或温差较大但壳程压力不高的场合。

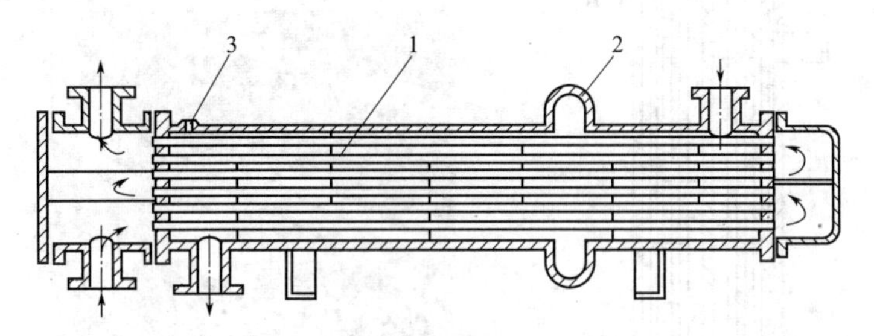

图2-28 固定管板式换热器

1—挡板；2—补偿圈；3—放气嘴

b. 浮头式 如图2-29所示，浮头式管壳换热器的特点是两端花板之一不与壳体固定连接，可在壳体内沿轴向自由伸缩。其优点是当管束受热或受冷时，管束连同浮头可自由伸缩，互不约束，不会产生热应力，且壳程便于清洗、检修；缺点是结构复杂，金属耗量多，造价高。浮头盖与浮动管板之间若密封不严，发生内漏，会造成两种介质的混合。浮头式换热器适用于壳体和管束壁温差较大或壳程介质易结垢的场合。

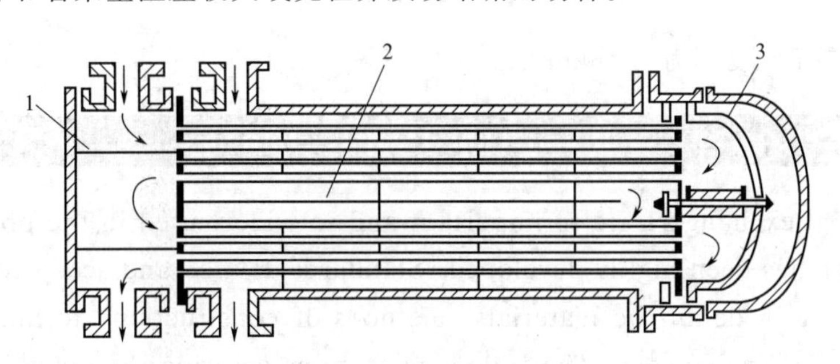

图2-29 浮头式换热器

1—管程隔板；2—壳程隔板；3—浮头

c. U形管式 如图2-30所示，U形管式换热器的特点是每根换热管都弯成U形，管子两端分别固定在同一管板上，管束可以自由伸缩，当壳体与U形换热管有温差时，不会产生热应力。其优点是结构简单，质量轻，耐高温、高压；缺点是管内清洗困难，管板上布管利用率低（因管壳弯成U形需一定的弯曲半径），内层管子坏了不能更换，因而报废率较高。U形管式换热器适用于管、壳壁温差较大或壳程介质易结垢，而管程介质清洁不易结垢以及高温、高压、腐蚀性强的场合。一般高温、高压、腐蚀性强的介质走管内，可使高压空间减小，密封易解决，并可节约材料和减少热损失。

上述几种管壳式换热器都有国标系列标准，例如，F_A600-130-16-2。其中，F表示浮头式；下标A表示A型管壳，ϕ19mm×2mm为正三角形排列；下标B表示B型管壳，ϕ25mm×2.5mm为正方形排列；600表示壳体公称直径为600mm；130表示公称传热面积

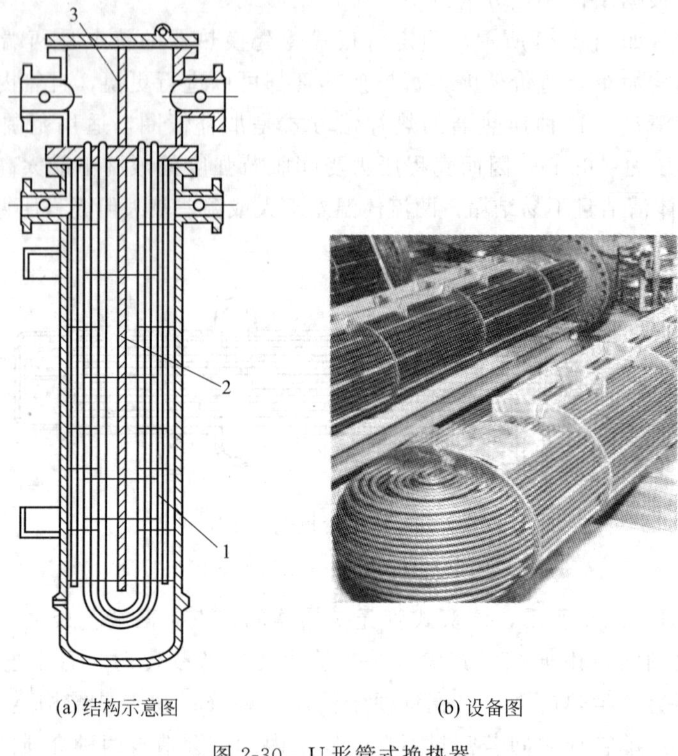

(a) 结构示意图　　(b) 设备图

图 2-30　U 形管式换热器

1—U 形管；2—壳程隔板；3—管程隔板

为 130m^2；16 表示承受压力为 16kgf/cm^2。

Shell-and-Tube Heat Exchangers

Tubular heat exchangers are so important and so widely used in the process industries that their design has been highly developed. Standards devised and accepted by TEMA are available covering in detail the materials, methods of construction, technique of design, and dimensions for exchangers. The following sections describe the more important types of exchanger and cover the fundamentals of their engineering, design, and operation.

Single-pass 1-1 exchanger

The simple double-pipe exchanger shown in Fig. 2-25 is inadequate for flow rates that cannot readily be handled in a few tubes. If several double pipes are used in parallel, the weight of metal required for the outer tubes becomes so large that the shell-and-tube construction, such as that shown in Fig. 2-25, where one shell serves for many tubes, is more economical. This exchanger, because it has one shell-side pass and one tube-side pass, is a 1-1 exchanger.

In an exchanger the shell-side and tube-side heat-transfer coefficients are of comparable importance, and both must be large if a satisfactory overall coefficient is to be attained. The velocity and turbulence of the shell-side liquid are as important as those of the tube-side fluid.

To promote crossflow and raise the average velocity of the shell-side fluid, baffles are installed in the shell. In the construction shown in Fig. 2-25, baffles 1 consist of circular disks of sheet metal with one side cut away, common practice is to cut away a segment having a height equal to one-fourth the inside diameter of the shell. Such baffles are called 25 percent baffles. The cut edges of the baffles may be horizontal for up-and-down flow or rotated 90° to provide side-to-side flow. The baffles are perforated to receive the tubes. To minimize leakage, the clearances between baffles and shell and tubes should be small. The baffles are supported by one or more guide rods C, which are fastened between the tube sheets by setscrews.

(2) 板式换热器

① 夹套式换热器　如图 2-31 所示，夹套式换热器是在反应器或容器的外壁上安装夹套制成。夹套与器壁之间形成加热介质或冷却介质的通道。这种换热器主要用于反应过程的加热或冷却。其优点是结构简单，造价低，可衬耐腐蚀材料；缺点是总传热系数较小，传热面积受容器筒体大小的限制，且夹套内清洗困难。为了提高总传热系数，可在夹套内装挡板，容器内加搅拌装置。为了增加传热面积，可在容器内装蛇管换热器。

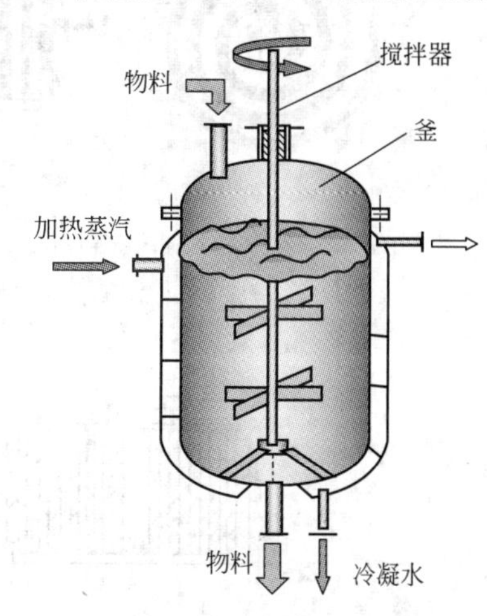

图 2-31　夹套式换热器

② 螺旋板换热器　螺旋板换热器如图 2-32 所示，有两块薄金属板，每块的一端分别与分割挡板焊在一起，卷成螺旋形。两板之间焊有定距柱以维持通道间距，在螺旋板两侧焊有盖板。进行热交换时，冷、热两种流体分别进入两条通道，在换热器内作严格的逆流流动。这种换热器的优点是结构紧凑，单位体积的传热面积约为管壳式换热器的 3 倍，流速可达 20m/s（指气体），$Re=1400\sim1800$，可达完全湍流，总传热系数高，不易结垢、堵塞；同时由于流体的流程长和两流体可进行完全逆流，故可在较小的温差下操作，能充分利用低温热源，可用于低温差传热和精密控制温度。其缺点是操作压力和温度不能太高（$p<2000$kPa，$T<300\sim400$℃），不易检修，流动阻力大（比管壳式的大 2～3 倍）。

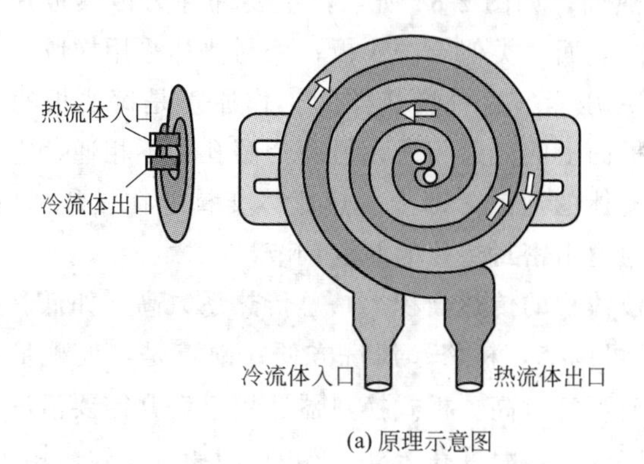

(a) 原理示意图

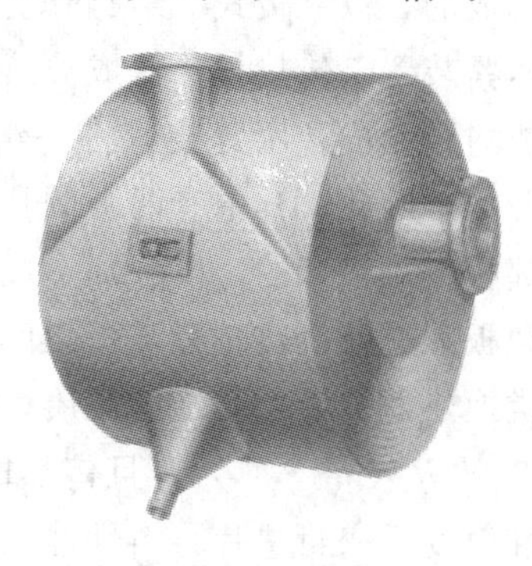
(b) 设备图

图 2-32　螺旋板换热器

常用的螺旋板换热器根据流动方式不同，分为Ⅰ型、Ⅱ型、Ⅲ型及G型四种。其中，Ⅰ型的两个螺旋通道的两侧完全焊接密封，为不可拆结构，如图2-33（a）所示。换热器中，两流体均做螺旋流动，通常冷流体由外周流向中心，热流体由中心流向外周，呈完全逆流流动，此类换热器主要用于液体与液体间的传热。Ⅱ型换热器的一个螺旋通道的两侧为焊接密封，另一通道的两侧是敞开的，如图2-33(b）所示。换热器中，一流体沿螺旋通道流动，而另一流体沿换热器的轴向流动。此类换热器适用于两流体流量差别很大的场合，常用作冷凝器、气体冷却器等。Ⅲ型换热器的结构如图2-33(c）所示，一种流体做螺旋流动，另一流体做兼有轴向和螺旋向两者组合的流动，该类换热器适用于汽体冷凝。G型换热器的结构如图2-33(d）所示，该结构又称塔上型，常被安装在塔顶作为冷凝器。

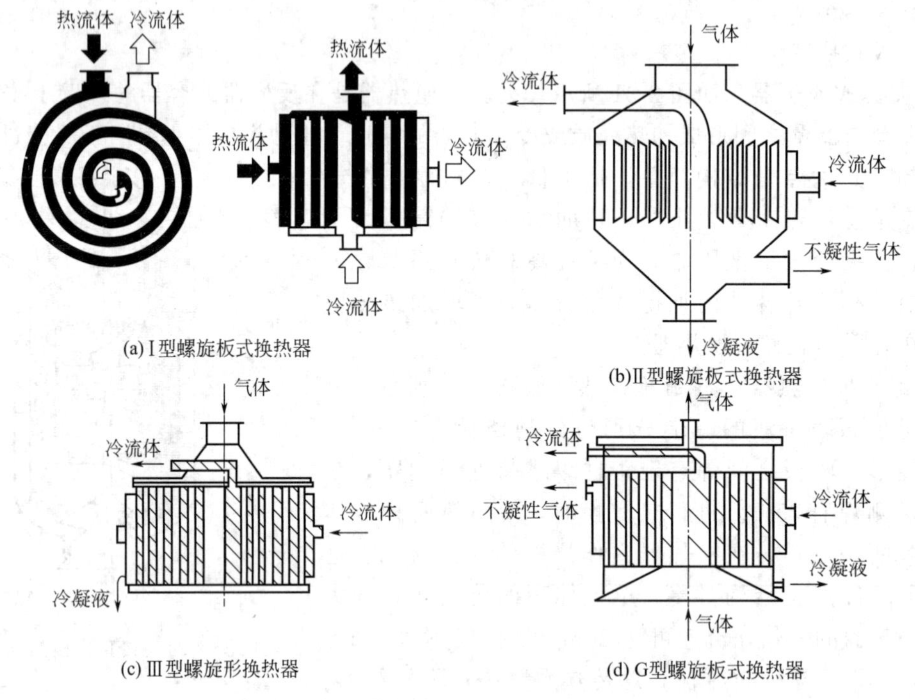

图2-33　螺旋板换热器类型

③ 平板式换热器　平板式换热器是由一组长方形的薄金属板平行排列，夹紧组装于支架上面构成的，以“板式”结构作为换热面，如图2-34所示，主要部件为传热板片、密封垫片和压紧装置。传热板片压成波纹形的表面，类似于洗衣板；密封垫片采用橡胶、压缩石棉或合成树脂制成；压紧装置将一组板片压紧。这种换热器的工作原理是每块板的四个角上，各开一个圆孔，其中有两个圆孔和板面上的流道相通，另两个圆孔则不相通。它们的位置在相邻板上是错开的，以分别形成两流体的通道，即冷、热流体交替地在板片两侧流动，通过金属板片进行换热，板与板之间的通道由密封垫片的厚度调节。

该类换热器的优点是结构紧凑，单位体积的传热面积大，总传热系数高［如低黏度的液体可达7000W/(m^2·℃)］，且传热面积可调节，检修、清洗方便；缺点是，处理量小，操作压力和温度受密封垫片材料性能限制而不宜过高。板式换热器适用于经常需要清洗、工作环境要求十分紧凑，工作压力在2.5MPa以下，温度在−35～200℃的场合，较多用于化工、食品、医药工业。

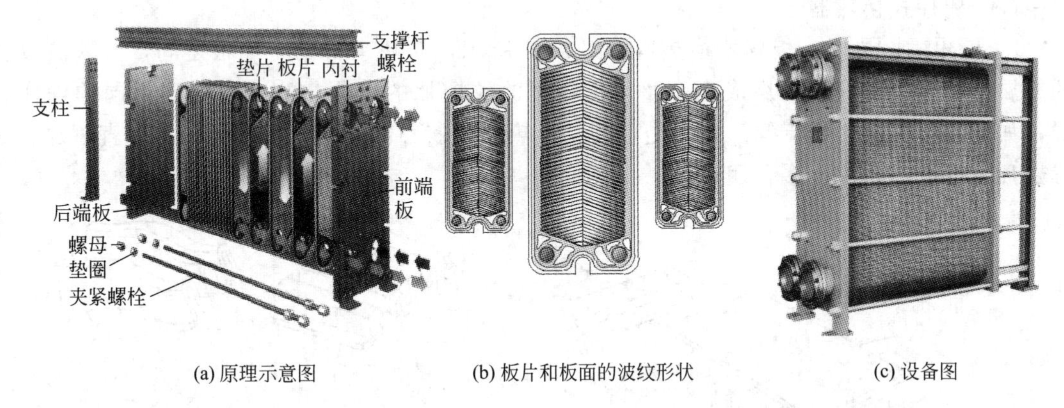

(a) 原理示意图　(b) 板片和板面的波纹形状　(c) 设备图

图 2-34 平板式换热器

Plate-Type Exchangers

For many applications at moderate temperature and pressure, an alternative to the shell-and-tube exchanger is the gasketed plate exchanger, which consists of many corrugated stainless-steel sheets separated by polymer gaskets and clamped in a steel frame. Inlet portals and slots in the gaskets direct the hot and cold fluid to alternate spaces between the plates. The corrugations induce turbulence for improved heat transfer, and each plate is supported by multiple contacts with adjoining plates, which have a different pattern or angle of corrugation.

The space between plates is equal to the depth of the corrugations and is usually 2 to 5mm. A typical plate design is shown in Fig. 2-34.

For a liquid-liquid exchanger, the usual fluid velocity is 0.2 to 1.0m/s, and because of the small spacing, the Reynolds number is often less than 2100. However, the corrugations give the flow turbulent characteristics at Reynolds numbers of 100 to 400, depending on the plate design. Evidence for turbulent flow is that the heat transfer coefficient varies with the 0.6 to 0.8 power of the flow rate, and the pressure drop depends on the 1.7 to 2.0 power of the flow rate.

With water or aqueous solutions on both sides, the overall coefficient for a clean plate-type exchanger may be 3000 to 6000W ($m^2 \cdot K$), several times the normal value for l shell-and-tube exchanger. Because of high shear rates, the fouling factors experienced are much lower than those for shell-and-tube exchangers, and the designer may just add 10 percent to the calculated area to allow for fouling. The units can easily be taken apart for thorough cleaning.

Plate exchangers are widely used in the dairy and food processing industries because they have high overall coefficients and are easily cleaned or sanitized. Several exchangers with different heat duties can be grouped in a single unit. For example, the high-temperature short-time (HTST) process for pasteurizing milk uses a plate exchanger with three or four sections.

（3）翅片式换热器

① 板翅式换热器　板翅式换热器如图2-35所示，两块平行的金属板之间，夹入波纹状的金属翅片，边侧密封，组成一个整体。根据波纹翅片的不同排列，可得到不同的操作方式，如逆流、并流、错流等。翅片是板翅式换热器的核心部件，其常用形式有平直翅片、波形翅片、锯齿形翅片、多孔翅片等。

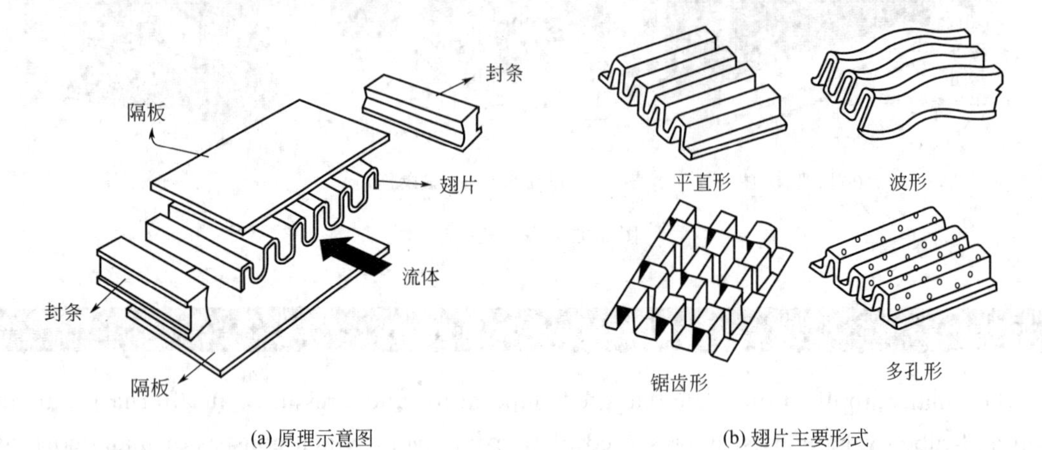

(a) 原理示意图　　(b) 翅片主要形式

图2-35　板翅式换热器

这种换热器的优点是：翅片可增加湍动程度，破坏层流内层，总传热系数高；冷热流体不仅通过平板换热，而且主要通过翅片换热，单位体积的传热面可达2500～4300W/(m^2·℃)；轻巧牢固（翅片不仅是传热面，而且是两平板的支撑）；适应性强（气-气、气-液、液-液、冷凝、蒸发均可）；可由多种介质在同一设备内换热。其缺点是结构复杂，清洗、检修困难，隔板、翅片一般用薄铝片材料，要求介质对铝不发生腐蚀。

板翅式换热器因轻巧牢固，常用于飞机、舰船和车辆的动力设备以及在电子、电器设备中，作为散热器和油冷却器等；也适用于气体的低温分离装置，如空气分离装置中作为蒸发冷凝器、液氮过冷器以及用于乙烯厂、天然气液化厂的低温装置中。

② 翅片管式换热器　翅片管式换热器是指在管子内表面或外表面上装有很多径向或轴向翅片，其中常见的翅片形式如图2-36所示。这种换热器主要用于两流体对流传热系数相

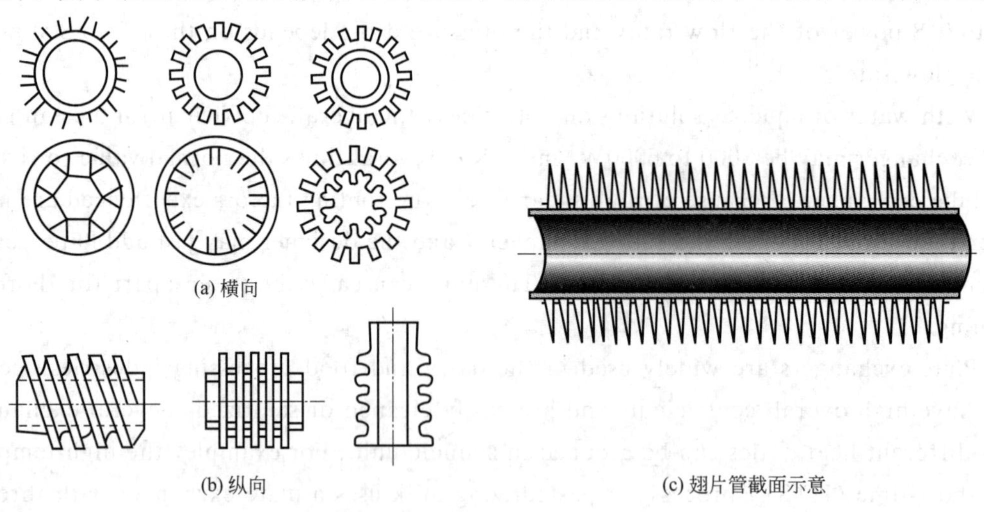

(a) 横向

(b) 纵向

(c) 翅片管截面示意

图2-36　翅片管式换热器的翅片形状

差较大的情况，例如工业上常见的气体加热和冷却问题，因气体的对流传热系数很小，所以当与气体换热的另一流体是水蒸气冷凝或是冷却水时，则气体侧热阻成为传热控制因素。此时要强化传热，就必须增加气体侧的对流传热面积，若在管外装上翅片，既可增加传热面积，又可增加空气湍动程度，减少了气体侧的热阻，使气体传热系数提高，从而明显提高换热器的传热效率。一般来说，当两流体的对流传热系数比大于3时，易采用翅片换热器。翅片管式换热器作为空气冷却器，在工业上应用很广。用空冷代替水冷，不仅可在缺水地区使用，在水源充足的地方，采用空冷也取得了较大的经济效益。

Extended-Surface Equipment

Difficult heat-exchange problems arise when one of two fluid streams has a much lower heat transfer coefficient than the other. A typical case is heating a fixed gas, such as air, by means of condensing steam. The individual coefficient for the steam is typically 100 to 200 times that for the air stream; consequently, the overall coefficient is essentially equal to the individual coefficient for the air, the capacity of a unit area of heating surface will be low, and many meters of tube will be required to provide reasonable capacity. Other variations of the same problem are found in heating or cooling viscous liquids or in treating a stream of fluid at low flow rate, because of the low rate of heat transfer in laminar flow.

To conserve space and to reduce the cost of the equipment in these cases, certain types of heat exchange surfaces, called *extended surfaces*, have been developed in which the outside area of the tube is multiplied, or extended, by fins, pegs, disks, and other appendages and the outside area in contact with the fluid thereby made much larger than the inside area.

The fluid stream having the lower coefficient is brought into contact with the extended surface and flows outside the tubes, while the other fluid, having the high coefficient, flows through the tubes.

Types of extended surface

Two common types of extended surfaces are available, examples of which are shown in Fig. 2-36. Longitudinal fins are used when the direction of flow of the fluid is parallel to the axis of the tube; transverse fins are used when the direction of flow of the fluid is across the tubes. Spikes, pins, studs, or spines are also used to extend surfaces and tubes carrying these can be used for either direction of flow.

In all types, it is important that the fins be in tight contact with the tube, both for structural reasons and to ensure good thermal contact between the base of the fin and the wall.

(4) 热管换热器（heat pipe exchanger）

热管换热器是一种新型高效换热装置，由壳体、热管和隔板组成的。其中，热管作为主要的传热元件，圆管内抽除不凝性气体并充以某种定量的可凝性液体（工作流体）。工作流体在吸热蒸发端沸腾，产生蒸汽流至冷却端凝结放出潜热，由冷却端回流至加热端再次沸腾，如此连续进行，热量则由加热端传递到冷却端。

热管按冷凝液循环方式分为毛细管热管、重力热管和离心热管三种。如图 2-37(a) 所示，毛细管热管的冷凝液依靠毛细管的作用回到热端，这种热管可以在失重情况下工作；重力热管的冷凝液是依靠重力流回热端，它的传热具有单向性，一般为垂直放置，如图 2-37(b) 所示；离心热管是靠离心力使冷凝液回到热端，通常用于旋转部件的冷却。

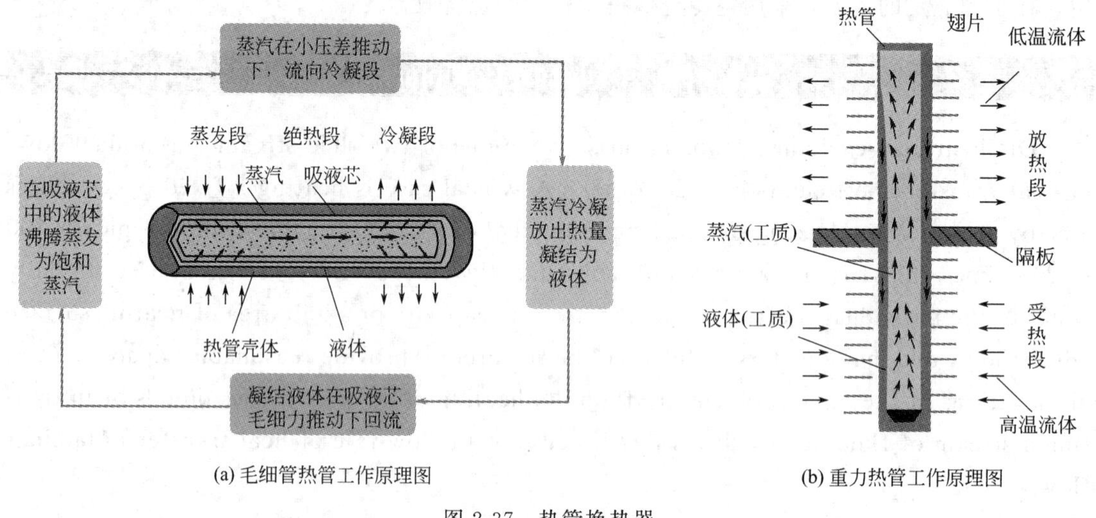

(a) 毛细管热管工作原理图

(b) 重力热管工作原理图

图 2-37 热管换热器

图 2-38 是青岛科技大学化工学院引进的重力热管换热器，换热器主要由壳体和热管元件组成。壳体是一个钢结构件，一侧为热流体通道，另一侧为冷流体通道，中间由管板分隔。热管管壳为无缝钢管，上、下两端焊有封头，内部灌装有工质。

热管传导热量的能力很强，为导热性能最优的金属的 $10^3 \sim 10^4$ 倍。由于热管的沸腾和冷凝对流传热强度都很大，通过管外翅片增加传热面，可巧妙地把管内、外流体间的传热转变为两侧管外的传热，使热管成为高效而结构简单、投资少的传热设备，特别适用于低温差传热及某些等温性要求较高的场合，目前，热管换热器已被广泛应用于烟道气废热的回收利用，并取得了很好的节能效果。

图 2-38 重力热管换热器设备图

2.6.2 强化传热途径

热能是过程工业的主要能源，表 2-9 为美国六大工业（冶金、化工、造纸、石油煤炭、食品、建材）不同单元操作的用能情况。由表可见，热能的消耗大大超过电能，因此，在传热过程中如何节能，对过程工业的生产具有普遍的指导意义。

表 2-9 美国六大工业不同单元操作的用能情况

操作名称	能量/$\times10^{12}$kJ	电能或热能	操作名称	能量/$\times10^{12}$kJ	电能或热能
工艺物料直接加热	7440±1670	主要为热能	蒸煮、消毒等	770±125	热能与电能
压缩	1420±630	电能或机械能	原料	2050±210	热能
电解	1250±420	热能	其他	2920±210	
蒸发	690±125	电能	合计	19090	
干燥	1130±210	热能			

节能是采取技术上可行、经济上合理以及环境和社会可接受的一切措施，来更有效地利用能源。节能的一般途径有：燃料燃烧的合理化，加热、冷却等传热过程的合理化，压缩空气的有效运行，废热（余热）回收，另外还有热能向动力转换，电力向动力和热转换等途径。以下简要介绍有关的节能措施。

（1）利用大量低品位热能

过程工业中所消耗的总热能的80%左右最终以低品位热能形式向环境排放，造成能量的大量流失。因此，有效地利用低品位热能具有十分重要的现实意义。如蒸馏操作中可利用塔顶产品的潜热或显热来预热物料，也可以将塔顶蒸汽经压缩机压缩后作为再沸器的热源。

（2）利用化学反应热

在过程工业中，经常会遇到放热的化学反应。如甲醇氧化制甲醛生产中，反应物甲醛气的温度高达600多摄氏度，用工业冷却水冷却，造成了极大的浪费，若利用该反应热副产蒸汽，则将反应气体从640℃降至240℃，每吨甲醛可副产1t蒸汽，同时还可节约冷却水120t。

（3）设备及管道的热损失

不保温或保温较差的设备及管道将对环境产生很大的热损失。例如一根1m长裸露的4in（1in=2.54cm）蒸汽管道，每小时将冷凝2～5kg的蒸汽，每年要多耗1000～2000kg煤。对输送冷冻盐水的管道，要注意保冷，保冷的材料要保持干燥，否则会影响保冷效果。

（4）降低传热温差，减少有效能损失

传热温差是传热不可逆过程损耗功的重要来源。传热温差愈小，有效能损失就愈小。由传热速率方程 $Q=KS\Delta t_m$ 可知，对于一定的热负荷，传热温差减小，传热面积必然增加，从而使摩擦损耗功增加。若要做到既不增加传热面积，又保持较小的传热温差，就必须强化传热过程，提高总传热系数。由式(2-83) 分析可知，为了增大 K，应尽量减少各项热阻。

（5）强化传热过程，采用高效换热器

强化传热过程，采用高效换热器是传热节能的最有力措施。一般来说，高效换热器具备以下三个功能：①结构紧凑，单位体积内提供较大的传热面积；②总传热系数大；③传热效果好。因此，高效换热器在过程工业已得到广泛应用。

2.6.3 管壳式换热器设计

2.6.3.1 管壳式换热器设计中应考虑的问题

（1）流体流动通道的选择

一般来说，不洁净的、易结垢的、腐蚀性的及压强高的流体走管程为好；饱和蒸汽、被冷却的、对流传热系数大的、流量小的及黏度大的流体走壳程为好。但应注意，这些原则要综合考虑，合理决定。

（2）流速 u 的选择

增加流体在换热器中的流速，将加大对流传热系数，减少污垢在管子表面上沉积的可能，即降低了污垢热阻，使总传热系数增大，从而可减小换热器的传热面积。但是流速增加，又使流动阻力增大，动力消耗就增加。综上所述，在管壳式换热器中，流速的选择要进行经济权衡。此外，还应在流体的物性和设备的结构上加以考虑。例如，体积流量和传热面积一定时，要增加流速，则应增加管长或管程数。但管子太长不宜清洗，而一般管子都有出厂的规格标准。

综合考虑上述各因素，管壳式换热器中常用的流速范围及易燃、易爆液体的安全允许速度如表 2-10 及表 2-11 所示。

表 2-10　管壳式换热器中常用的流速范围

流体的种类		一般流体	易结垢液体	气体
流速/(m/s)	管程	0.5～3	>1	5～30
	壳程	0.2～1.5	>0.5	3～15

表 2-11　管壳式换热器中易燃、易爆液体的安全允许速度

液体种类	乙醚、二硫化碳、苯	甲醇、乙醇、汽油	丙酮
安全允许速度/(m/s)	<1	<2～3	<10

（3）管壳式换热器中流体温度的确定及 Δt_m 的计算

若换热器中冷、热流体之一仅已知进口温度，则出口温度应由设计者来确定。例如用冷水冷却某热流体，冷水的进口温度可以根据当地的气温条件进行估计，而流出换热器的冷水出口温度，则需要根据经济衡算来确定。对于水源缺乏地区，可设计提高冷水的出口温度，即采用较大的进出口温度差，这样可以节省用水量，但由于降低了传热推动力，则需增加传热面积；反之，为了减小传热面积，则要增加用水量。通常，冷却水的两端温度差设计为 5～10℃，缺水地区则可选用更大的温差。

计算 Δt_m 的基本步骤是：

① 逆流和并流　逆流和并流的 Δt_m 均可按式(2-90) 计算。

② 错流和折流　前面 2.5.5 节中介绍了错流和折流的 $\Delta t'_m$ 计算方法，即可先按逆流的 Δt_m 计算，然后用温度差校正系数 $\varphi_{\Delta t}$ 加以修正，$\varphi_{\Delta t}=f(P,R)$，可根据 P、R 查图求得，如图 2-39 所示。可查取多管程和多壳程的 $\varphi_{\Delta t}$，设计中要求 $\varphi_{\Delta t}$ 不低于 0.8。

（4）管子规格及排列方法

① 管子规格　目前，我国试行的管壳式换热器系列标准中仅有两种规格的管子：ϕ25mm×2.5mm 和 ϕ19mm×2mm。管长出厂标准 $L=6$m，合理换热器的管长常取为 1.5m、2m、3m 和 6m，一般取 $L/D=4$～6，D 为壳体直径（单位为 m）。

② 管子排列方法　管子排列方法如图 2-40 所示，可分为正三角形排列、正方形排列和正方形错列三种。其中正三角形排列的优点是结构紧凑，管板的强度高，管子排列密度大，管程流体不易短路，且壳程流体扰动较大，因此传热效果好；缺点是清洗困难。正方形排列的优点是管外清洗方便，适用于壳程流体易产生污垢的场合；缺点是传热效果比正三角形排列差。正方形错列的优缺点介于正三角形排列和正方形排列之间，对流传热系数可适当提高。

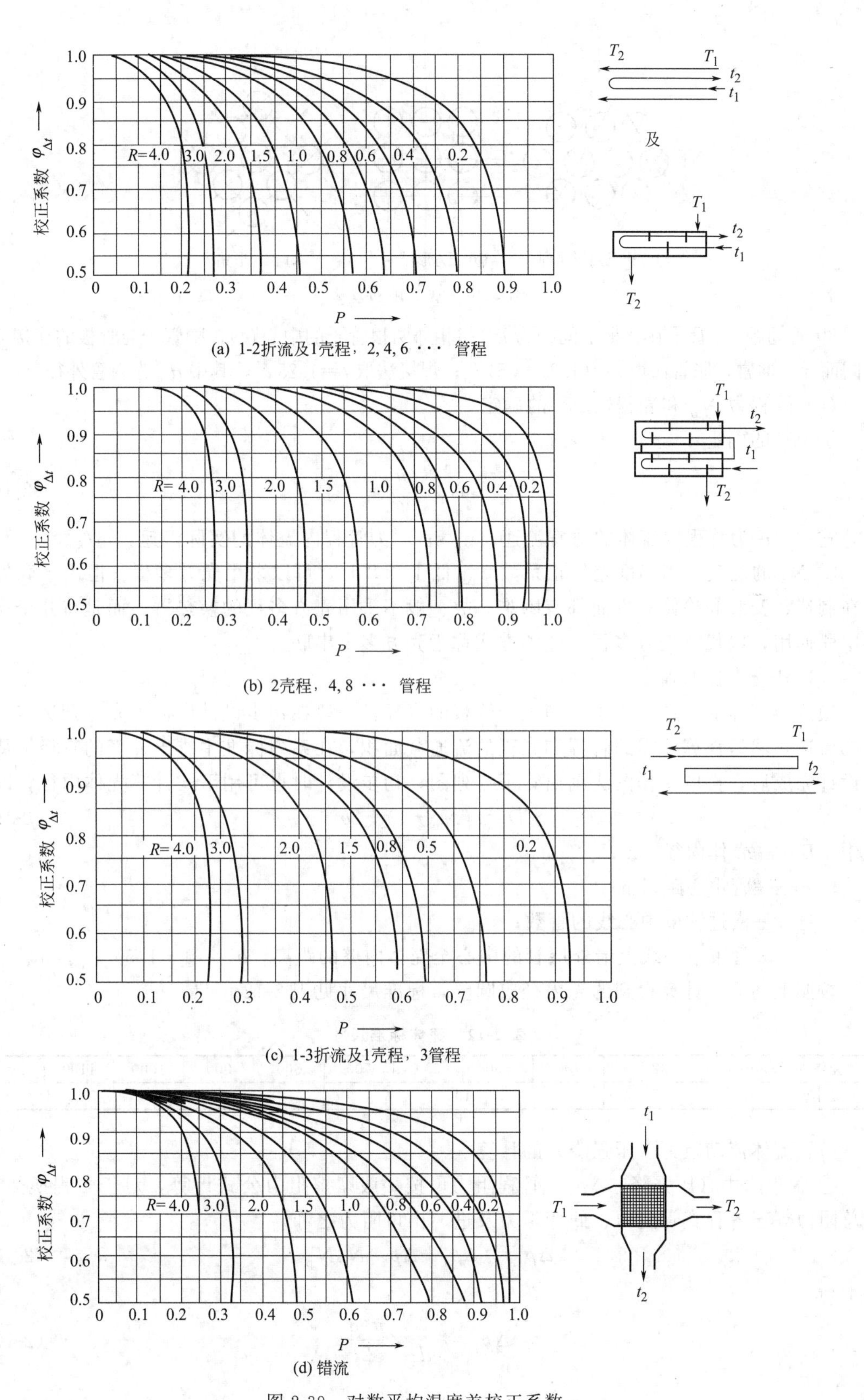

图 2-39　对数平均温度差校正系数 $\varphi_{\Delta t}$

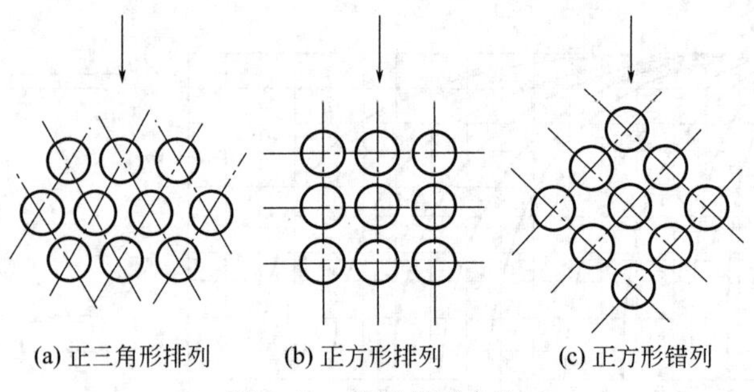

图 2-40 管子排列方法

③ 管间距 t 管子在管板上间距为 t（指相邻两根管子的中心距），随管子与管板的连接方法不同而异。通常，胀管法取 $t=(1.3\sim1.5)d_o$；焊接法取 $t=1.25d_o$，其中 d_o 为内管外径。

（5）管程数 N_p 和壳程数 N_s 的确定

① N_p 的确定：

$$N_p=\frac{u}{u'} \tag{2-105}$$

式中，u 为管程内流体的适宜流速，m/s；u'为管程内流体的实际流速，m/s。

② N_s 的确定 当温度差校正系数 $\varphi_{\Delta t}$ 低于 0.8 时，可以采用壳方多程。但由于壳方隔板在制造、安装和检修等方面都有困难，故一般不采用壳方多程的换热器，而是将几个换热器串联使用，以代替壳方多程，此类方式称为壳方多个串联。

（6）外壳直径的确定

换热器壳体的内径应等于或稍大于管板的直径，一般在初步设计中，可先分别选定两流体的流速，然后计算所需的管程和壳程的流通截面积，在系列标准中查出外壳的直径。待全部设计完成后，再应用作图法画出管子排列图；初步设计时也可用下式计算壳体内径，即：

$$D=t(N_c-1)+2b' \tag{2-106}$$

式中 D——壳体内径，m；

t——管中心距，m；

N_c——横过管束中心线的管数；

b'——管束中心线上最外层管的中心至壳体内壁的距离，$b'=(1-1.5)\ d_o$，m。

按照上述方法计算得到的壳内径应圆整，标准尺寸见表 2-12。

表 2-12 壳体标准尺寸

壳体外径/mm	325	400	500	600	700	800	900	1000	1100	1200
最小壁厚/mm	8	10				12			14	

（7）流体流动阻力（压强降）的计算

① 管程阻力（压强降）Δp_t 管程阻力可按一般摩擦阻力公式计算，对于多程换热器，其总阻力等于各程直管阻力、回弯阻力及进、出口阻力之和。

$$\Delta p_t=(\Delta p_i+\Delta p_r)N_sN_p \tag{2-107}$$

其中：

$$\Delta p_i=\lambda\frac{l}{d_i}\times\frac{u^2\rho}{2} \tag{2-108}$$

$$\Delta p_r=\sum\xi\frac{u^2\rho}{2} \tag{2-109}$$

式中 Δp_i——每程直管阻力（压强降）；

Δp_r——每程局部阻力（压强降）（包括回弯管，进、出口阻力）；

N_s——壳程数；

N_p——管程数。

注意：Δp_t 应该按一根管子计算。

② 壳程阻力（压强降）Δp_s 计算 Δp_s 的经验公式很多，以下式为例说明。

$$\Delta p_s=\lambda_s \frac{D(N_B+1)}{d_e}\times\frac{\rho u_o^2}{2} \tag{2-110}$$

其中：

$$\lambda_s=1.72\left(\frac{d_e u_o \rho}{\mu}\right)^{-0.19} \tag{2-111}$$

$$S_o=hD\left(1-\frac{d_o}{t}\right) \tag{2-112}$$

式中 u_o——壳程流速，以流通面积 S_o 计，m/s；

N_B——折流板数；

h——折流板间距，m；

D——壳体直径，m；

t——管间距，m；

d_o——管子外径；

d_e——壳程当量直径。

一般液体经换热器壳程压强降为 10～100kPa，气体的为 1～10kPa。设计时换热器的工艺尺寸应在压强降与传热面积之间予以权衡，使之既能满足工艺要求，又经济合理。

2.6.3.2 管壳式换热器的选用基本步骤

(1) 估算传热面积，初选换热器型号

① 根据换热任务，计算传热量；

② 确定流体在换热器中的流动途径；

③ 确定流体在换热器中两端的温度，计算定性温度，确定在定性温度下的流体物性；

④ 计算平均温度差，并根据温度差校正系数不应小于 0.8 的原则，确定壳程数或调整加热介质或冷却介质的终温；

⑤ 根据两流体的温差和设计要求，确定换热器的形式；

⑥ 依据换热流体的性质及设计经验，选取总传热系数值 $K_{选}$；

⑦ 依据总传热速率方程，初步算出传热面积 S，并确定换热器的基本尺寸或按系列标准选择设备规格。

(2) 计算管、壳程压降

根据初选的设备规格，计算管、壳程的流速和压降，检查计算结果是否合理或满足工艺要求。若压降不符合要求，要调整流速，再确定管程和折流挡板间距，或选择其他型号的换热器，重新计算压降直至满足要求为止。

(3) 核算

计算管、壳程对流传热系数，确定污垢热阻 R_{Si} 和 R_{So}，再计算总传热系数 $K_{计}$，然后与 $K_{选}$ 值比较，若 $K_{计}/K_{选}=1.15\sim1.25$，则初选的换热器合适，否则需另选 $K_{选}$ 值，重复上述计算步骤。

上述步骤为一般原则，应视具体情况灵活变动，下面给出一个设计举例。

【例 2-12】 某化工厂在生产过程中，需将纯苯液体从 80℃冷却到 55℃，其流量为 20000kg/h；冷却介质采用 35℃的循环水，要求换热器的管程和壳程压降不大于 10kPa，试选用合适型号的换热器。

解：(1) 估算传热面积，初选换热器型号

① 基本物性数据的查取：

$$苯的定性温度=\frac{80+55}{2}=67.5(℃)$$

查得苯在定性温度下的物性数据：

$\rho=828.6\text{kg/m}^3$，$C_p=1.841\text{kJ/(kg}\cdot℃)$，$\lambda=0.129\text{W/(m}^2\cdot℃)$，$\mu=0.352\times10^{-3}\text{Pa}\cdot\text{s}$。

根据设计经验，选择冷却水的温升为 8℃，则水的出口温度为 $t_2=35+8=43(℃)$

$$水的定性温度=\frac{35+43}{2}=39(℃)$$

查得水在定性温度下的物性数据：$\rho=992.3\text{kg/m}^3$，$C_p=4.174\text{kJ/(kg}\cdot℃)$，$\lambda=0.633\text{W/(m}^2\cdot℃)$，$\mu=0.67\times10^{-3}\text{Pa}\cdot\text{s}$。

② 热负荷计算：

$$Q=W_\text{h}C_{p\text{h}}(T_1-T_2)=\frac{20000}{3600}\times1.841\times10^3\times(80-55)=2.56\times10^5(\text{W})$$

冷却水耗量：

$$W_\text{c}=\frac{Q}{C_{p\text{c}}\ (t_2-t_1)}=\frac{2.56\times10^5}{4.174\times10^3\times\ (43-35)}=7.67(\text{kg/s})$$

③ 确定流体的流径　该设计任务的热流体为苯，冷流体为水，为使苯通过壳壁面向空气中散热，提高冷却效果，令苯走壳程，水走管程。

④ 计算平均温度差　暂按单壳程、双管程考虑，先求逆流时平均温度差：

苯	80℃	→	55℃
冷却水	43℃	←	35℃
Δt	37℃		20℃

所以：$\Delta t_\text{m}=\dfrac{\Delta t_2-\Delta t_1}{\ln\dfrac{\Delta t_2}{\Delta t_1}}=\dfrac{37-20}{\ln\dfrac{37}{20}}=27.6\ (℃)$

计算 R 和 P：

$$R=\frac{T_1-T_2}{t_2-t_1}=\frac{80-55}{43-35}=3.125$$

$$P=\frac{t_2-t_1}{T_1-t_1}=\frac{43-35}{80-35}=0.178$$

由 R、P 值，查图 2-39(a)，$\varphi_{\Delta t}=0.94$，因 $\varphi_{\Delta t}>0.8$，选用单壳程可行。

$$\Delta t'_\text{m}=\varphi_{\Delta t}\Delta t_\text{m}=0.94\times27.6=25.9(℃)$$

⑤ 选 K 值，估算传热面积　参照附录，取 $K=450\text{W/(m}^2\cdot℃)$

$$S=\frac{Q}{K\Delta t_\text{m}}=\frac{2.56\times10^5}{450\times25.9}=22(\text{m}^2)$$

⑥ 初选换热器型号　由于两流体温差<50℃，可选用固定管板式换热器。由固定管板式换热器的系列标准，初选换热器型号为：G400Ⅱ-1.6-22。主要参数如下：外壳直径400mm，公称压力1.6MPa，公称面积$22m^2$，管子尺寸$\phi 25mm \times 2.5mm$，管子数102，管长3000mm，管中心距32mm，管程数N_p为2，管子排列方式为正三角形，管程流通面积$0.016m^2$，实际换热面积$S_o = n\pi d_o(L-0.1) = 102 \times 3.14 \times 0.025 \times (3-0.1) = 23.2(m^2)$。

采用此换热面积的换热器，则要求过程的总传热系数为：

$$K_o = \frac{Q}{S_o \Delta t_m} = \frac{2.56 \times 10^5}{23.2 \times 25.9} = 426 W/(m^2 \cdot ℃)$$

（2）核算压降

① 管程压降：

$$\sum \Delta p_i = (\Delta p_1 + \Delta p_2) F_t N_s N_p$$
$$F_t = 1.4, N_s = 1, N_p = 2$$

管程流速：
$$u_i = \frac{V_s}{A_i} = \frac{7.67}{992.3 \times 0.016} = 0.483(m/s)$$

$$Re_i = \frac{d_i u_i \rho}{\mu} = \frac{0.02 \times 0.483 \times 992.3}{0.67 \times 10^{-3}} = 1.43 \times 10^4 (\text{湍流})$$

对于碳钢管，取管壁粗糙度$\varepsilon = 0.1mm$

$$\frac{\varepsilon}{d_i} = \frac{0.1}{20} = 0.005$$

由λ-Re关系图中查得：$\lambda = 0.037$

$$\Delta p_1 = \lambda \frac{L}{d_i} \times \frac{\rho u_i^2}{2} = 0.037 \times \frac{3}{0.02} \times \frac{992.3 \times 0.483^2}{2} = 642.4(Pa)$$

$$\Delta p_2 = 3\left(\frac{\rho u_i^2}{2}\right) = 3 \times \left(\frac{992.3 \times 0.483^2}{2}\right) = 347.2(Pa)$$

$$\sum \Delta p_i = (642.4 + 347.2) \times 1.4 \times 2 = 2771Pa < 10kPa$$

② 壳程压降

$$\sum \Delta p_o = (\Delta p_1' + \Delta p_2') F_s N_s$$
$$F_s = 1.15,\ N_s = 1$$
$$\Delta p_1' = F f_o n_c\ (N_B + 1)\ \frac{\rho u_o^2}{2}$$

管子为正三角形排列：$F = 0.5$

$n_c = 1.1\sqrt{n} = 1.1\sqrt{102} = 11.1$

取折流挡板间距$z = 0.15m\left(\frac{1}{5}D < z < D\right)$

$$N_B = \frac{L}{z} - 1 = \frac{3}{0.15} - 1 = 19$$

壳程流通面积$A_o = z(D - n_c d_o) = 0.15 \times (0.4 - 11.1 \times 0.025) = 0.0184(m^2)$

壳程流速
$$u_o = \frac{V_s}{A_o} = \frac{20000}{3600 \times 828.6 \times 0.0184} = 0.364(m/s)$$

$$Re_o = \frac{d_o u_o \rho}{\mu} = \frac{0.025 \times 0.364 \times 828.6}{0.352 \times 10^{-3}} = 2.14 \times 10^4 > 500$$

$$f_o=5.0Re_o^{-0.228}=5.0\times(2.14\times10^4)^{-0.228}=0.515$$

所以 $\Delta p'_1=0.5\times0.515\times11.1\times(19+1)\times\dfrac{828.6\times0.364^2}{2}=3138(\mathrm{Pa})$

$$\Delta p'_2=N_B\left(3.5-\frac{2z}{D}\right)\frac{\rho u_o^2}{2}=19\times\left(3.5-\frac{2\times0.15}{0.4}\right)\times\frac{828.6\times0.364^2}{2}=2868\ (\mathrm{Pa})$$

$$\sum\Delta p_o=(3138+2868)\times1.15\times1=6907\mathrm{Pa}<10\mathrm{kPa}$$

计算结果表明，管程和壳程的压降均能满足设计条件。

（3）核算总传热系数

① 管程对流传热系数 α_i

$$Re_i=1.43\times10^4>1\times10^4$$

$$Pr_i=\frac{C_p\mu}{\lambda}=\frac{4.174\times10^3\times0.67\times10^{-3}}{0.633}=4.42$$

$$\alpha_i=0.023\frac{\lambda}{d_i}Re_i^{0.8}Pr^{0.4}$$

$$=0.023\times\frac{0.633}{0.02}\times(1.43\times10^4)^{0.8}\times4.42^{0.4}=2783[\mathrm{W/(m^2\cdot℃)}]$$

② 壳程对流传热系数 α_o（Kern 法）

由 $\alpha_o=0.36\left(\dfrac{\lambda}{d_e}\right)\left(\dfrac{d_e u_o\rho}{\mu}\right)^{0.55}\left(\dfrac{C_p\mu}{\lambda}\right)^{1/3}\left(\dfrac{\mu}{\mu_w}\right)^{0.14}$

管子为正三角形排列，则：

$$d_e=\frac{4\times\left(\frac{\sqrt{3}}{2}t^2-\frac{\pi}{4}d_o^2\right)}{\pi d_o}=\frac{4\times\left(\frac{\sqrt{3}}{2}\times0.032^2-\frac{\pi}{4}\times0.025^2\right)}{\pi\times0.025}=0.02(\mathrm{m})$$

$$A=zD\left(1-\frac{d_o}{t}\right)=0.15\times0.4\times\left(1-\frac{0.025}{0.032}\right)=0.0131(\mathrm{m^2})$$

$$u_o=\frac{V_s}{A}=\frac{20000}{3600\times828.6\times0.0131}=0.512(\mathrm{m/s})$$

壳程中苯被冷却，取 $\left(\dfrac{\mu}{\mu_w}\right)^{0.14}=0.95$

$$\alpha_o=0.36\times\left(\frac{0.129}{0.02}\right)\times\left(\frac{0.02\times0.512\times828.6}{0.352\times10^{-3}}\right)^{0.55}\times\left(\frac{1.841\times10^3\times0.352\times10^{-3}}{0.129}\right)^{1/3}\times0.95$$

$$=971.4[\mathrm{W/(m^2\cdot℃)}]$$

③ 污垢热阻

根据表 2-7，管内、外侧污垢热阻分别取为：

$$R_{Si}=2.00\times10^{-4}\mathrm{m^2\cdot℃/W},\ R_{So}=1.72\times10^{-4}\mathrm{m^2\cdot℃/W}$$

④ 总传热系数 K

管壁热阻可忽略时，总传热系数 K 为

$$\frac{1}{K}=\frac{1}{\alpha_o}+R_{So}+R_{Si}\frac{d_o}{d_i}+\frac{1}{\alpha_i}\times\frac{d_o}{d_i}$$

$$K=\frac{1}{\frac{1}{971.4}+0.000172+0.0002\times\frac{0.025}{0.02}+\frac{0.025}{2783\times0.02}}=526.2[\mathrm{W/(m^2\cdot℃)}]$$

$$\frac{K_{计}}{K_{过}}=\frac{526.2}{426}=1.24$$

故所选择的换热器是合适的，安全系数为：

$$\frac{526.2-426}{426}\times 100\%=23.52\%$$

设计结果为：选用固定管板式换热器，型号：G400Ⅱ-1.6-22。

工程案例分析

换热器以小替大改善换热效果

在某化工产品的生产装置中，混合液在分解塔中进行反应时，放出大量的热量，若不及时移走，分解塔内温度将持续上升，会产生过量焦油，这不但会使产品质量下降，甚至堵塞管道造成事故。国内该类生产装置大都采用蒸发冷却的方式来移走反应热（即塔内温度靠液体自身的蒸发来维持，一般维持在88℃左右）。只有北方某厂采用外循环冷却的方式，即将塔内液体用泵抽出，经塔外一双管程列管换热器用冷却水（走管程）冷却后循环流入分解塔，所用的换热器A的主要参数为：壳径1m，双管程，换热管ϕ38mm×2.5mm、长2.5m，换热管数370根，总传热面积100m^2。

后来，该厂欲将塔内温度降至60℃，这一改变要求冷却器热负荷增至4×10^5kJ/h。更换一个传热面积更大的换热器是最容易想到的办法，但这无疑要增加一大笔设备投资。技术人员又想到了仍使用原换热器但增大混合液（走壳程）循环量的办法，因为这样可以提高冷却器的传热系数，总传热速率当然能够获得提高。于是该厂实施了使用原换热器A、更换大泵、将混合液循环提高至原来3倍的技改措施。结果发现，换热效果并未得到明显改善，换热器A的热负荷仅略有提高。鉴于这种情况，厂方又实施了第二种技改措施，即将原换热器A更换为一个传热面积更大的换热器B，其主要参数为；壳径1m，双管程，换热管ϕ25mm×2.5mm、长3m，换热管数1234根，总传热面积213.6m^2。结果发现，采用该换热器的换热效果还不如使用换热器A。于是厂方就此问题向有关专家寻求解决办法。该专家通过在现场收集数据，并进行了大量的技术指标核算，终于找到了问题的症结所在。现将该专家的分析过程简述如下。

虽然换热器A和B的换热面积不小，但这是以直径很大的壳内安置过多的换热管来获得的，这使得该换热器管程和壳程的流通截面都很大，故换热管两侧的对流传热系数都很低。据测算，对换热器A，管内冷却水和管外混合液雷诺数分别只有700和350，即都处于层流流动状态，用层流流动的公式计算对流传热系数，加上污垢热阻后的总传热系数仅为75W/(m^2·K)。即使将混合液循环量提高至原来的3倍，壳侧的雷诺数也仅为1000左右，对应的总传热系数也仅为87W/(m^2·K)，提高不大。

原换热器A和换热面积更大的换热器B都因流通截面积过大，导致传热系数很低而不可用。于是该专家从厂家废品库内找出了壳径为270mm、内装48根ϕ25mm×2.5mm换热管、总传热面积仅为37.5m^{2}的换热器两台（C）。通过计算发现，虽然换热器C的传热面积只有换热器A的37.5%，但由于壳径小，管数少，流通截面积小，因而流速很高，管内、外的对流传热系数分别可达450W/(m^2·K)和600W/(m^2·K)，加上污垢热阻和足够的安全系数，总传热系数可达250W/(m^2·K)以上，其KS值远较换热器A和B的大。于是该专家提出了用C替代A和B的方案。厂家抱着试试看的心理实施了该方案，结果果然如该专家所料，用面积仅为37.5m^{2}的C取代面积为100m^{2}的A后，换热效果不但没有下降，反而有了大幅度的提高，生产能力相应地提高了75%，完全达到了改造的目标。

此案例表明，考察一台换热器的工作能力不能单纯只考虑其传热面积，传热面积和总传热系数的乘积 KS 才能真正代表一台换热器的工作能力。因为存在如下的关系：S-管数-流通截面积-流速-传热系数，因此列管换热器的 K 值和 S 值之间往往存在着“此涨彼消”的关系，过大的传热面积往往由于流体流量的“不匹配”而导致过低的 K 值，结果是换热效果大大低于主观预期。

习　题

1. 锅炉钢板壁厚 $\delta_1=20\text{mm}$，其热导率 $\lambda_1=46.5\text{W/(m·K)}$。若黏附在锅炉内壁上的水垢层厚度为 $\delta_2=1\text{mm}$，其热导率 $\lambda_2=1.162\text{W/(m·K)}$。已知锅炉钢板外表面温度为 $t_1=523\text{K}$，水垢内表面温度为 $t_3=473\text{K}$，求锅炉每平方米表面积的传热速率，并求钢板内表面的温度 t_2。
2. 在一 ϕ60mm×3.5mm 的钢管外包有两层绝热材料，里层为 40mm 的氧化镁粉，平均热导率 $\lambda=0.07\text{W/(m·℃)}$；外层为 20mm 的石棉层，平均热导率 $\lambda=0.15\text{W/(m·℃)}$。现用热电偶测得管内壁温度为 500℃，最外层表面温度为 80℃。已知钢管的平均热导率 $\lambda=45\text{W/(m·℃)}$，试求每米管长的热损失及两层保温层界面的温度。
3. 有一列管式换热器，由 38 根 ϕ25mm×2.5mm 的无缝钢管组成，苯在管程流动，由 20℃ 加热到 80℃，苯的流量为 10.2kg/s，饱和蒸汽在壳程冷凝。试求：(1) 管壁对苯的对流传热系数？(2) 若苯的流量提高一倍，对流传热系数将有何变化？
4. 苯流过一套管换热器的环隙，自 20℃ 升高至 80℃，该换热器的内管规格为 ϕ19mm×2.5mm，外管规格为 ϕ38mm×3mm。苯的流量为 1800kg/h，求苯对内管壁的对流传热系数。
5. 常压下温度为 120℃ 的甲烷以 10m/s 的平均速度在管壳式换热器的管间沿轴向流动。离开换热器时甲烷温度为 30℃，换热器外壳内径为 190mm，管束由 37 根 ϕ19mm×2mm 的钢管组成，试求甲烷对管壁的对流传热系数。
6. 温度为 90℃ 的甲苯以 1500kg/h 的流量通过蛇管而被冷却至 30℃。蛇管的直径为 ϕ57mm×3.5mm，弯曲半径为 0.6m，试求甲苯对蛇管的对流传热系数。
7. 120℃ 的饱和水蒸气在一根 ϕ25mm×2.5mm、长 1m 的管外冷凝，已知管外壁温度为 80℃。分别求该管垂直和水平放置时的蒸气冷凝传热系数。
8. 实验室内有一高为 1m，宽为 0.5m 的铸铁炉门，其表面温度为 600℃，室温为 20℃。试计算每小时通过炉门辐射而散失的热量。如果在炉门前 50mm 处放置一块同等大小同样材料的平板作为隔热板，则散热量为多少？如果将隔热板更换为同等大小材料的已经氧化了的铝板，则散热量有何变化？
9. 现测定一传热面积为 2m^2 的列管式换热器的总传热系数 K 值。已知热水走管程，测得其流量为 1500kg/h，进口温度为 80℃，出口温度为 50℃；冷水走壳程，测得进口温度为 15℃，出口温度为 30℃，逆流流动。试计算该换热器的 K 值。
10. 热气体在套管式换热器中用冷却水冷却，内管为 ϕ25mm×2.5mm 的钢管，冷水在管内湍流流动，对流传热系数为 $2000\text{W/(m}^2\text{·K)}$。热气在套管环隙湍流流动，对流传热系数为 $50\text{W/(m}^2\text{·K)}$。已知钢的热导率为 45.4W/(m·K)。试计算：(1) 管壁热阻占总阻力的百分数；(2) 冷水流速提高一倍，则总传热系数有何变化？(3) 气体流速提高一倍，则总传热系数有何变化？
11. 实验测定管壳式换热器的总传热系数时，水在换热器的列管内作湍流流动，管外为饱和水蒸气冷凝。列管由直径为 ϕ25mm×2.5mm 的钢管组成。当水的流速为 1m/s 时，测得基于管外表面积的总传热系数 K_o 为 $2115\text{W/(m}^2\text{·℃)}$；若其他条件不变，而水的速度变为 1.5m/s 时，测得 K_o 为 $2660\text{W/(m}^2\text{·℃)}$。试求蒸汽冷凝传热系数（假设污垢热阻可以忽略不计）。
12. 某套管式换热器用于机油和原油的换热。原油在套管环隙流动，进口温度为 120℃，出口温度上升到 160℃；机油在内管流动，进口温度为 245℃，出口温度下降到 175℃。

    (1) 试分别计算并流和逆流时的平均温度差；

    (2) 若已知机油质量流量为 0.5kg/s，比热容为 3kJ/(kg·℃)，并流和逆流时的总传热系数 K 均为 $100\text{W/(m}^2\text{·K)}$，求单位时间内传过相同热量分别所需要的传热面积。
13. 某列管换热器由多根 ϕ25mm×2.5mm 的钢管组成，将流量为 15t/h 的苯由 20℃ 加热到 55℃，苯在管中流速为 0.5m/s，加热剂 130℃ 的饱和水蒸气在管外冷凝，其汽化潜热为 2178kJ/kg，苯的比热容

C_p=1.76kJ/(kg·K)，密度 ρ=858kg/m³，黏度 μ=0.52×10⁻³Pa·s，热导率 λ=0.148W/(m·K)，热损失、管壁及污垢热阻均忽略不计，蒸汽冷凝的 $\alpha=10^4$W/(m²·K)。试求：(1) 水蒸气用量（kg/h)；(2) 传热系数 K（以管外表面积为基准)；(3) 换热器所需管数 n 及单根管长度 l。

14. 在一套管式换热器中用水逆流冷却热油，已知换热器的传热面积为 3.5m²。冷却水走管内，流量为 5000kg/h，流动为强制湍流，入口温度为 20℃；热油走套管环隙，流量为 3800kg/h，入口温度为 80℃，其比热容 C_p=2.45kJ/(kg·K)。已知两流体的对流传热系数均为 2000W/(m²·K)。管壁厚度、管壁热阻、污垢热阻均可以忽略。试计算冷热流体的出口温度。如果由于工艺改进，热油的出口温度需要控制在 35℃以下，当通过提高冷却水流量的方法来实现时，冷却水的流量应控制为多少？
15. 某列管换热器由多根 ϕ25mm×2.5mm 的钢管组成，管束的有效长度为 3.0m。通过管程的流体冷凝壳程流量为 5000kg/h、温度为 75℃的饱和有机蒸气，蒸气的冷凝潜热为 310kJ/kg，蒸气冷凝的对流传热系数为 800W/(m²·K)。冷却剂在管内的流速为 0.7m/s，温度由 20℃升高到 50℃，其比热容为 C_p=2.5kJ/(kg·K)，密度为 ρ=860kg/m³，对流传热系数为 2500W/(m²·K)。蒸气侧污垢热阻和管壁热阻忽略不计，冷却剂侧的污垢热阻为 0.00055m²·K/W。试计算该换热器的传热面积并确定该换热器中换热管的总根数及管程数。
16. 在一台管壳式换热器中，壳方 120℃的水蒸气冷凝将一定流量的空气由 20℃加热到 80℃，空气的对流传热系数 α_i=100W/(m²·℃)。换热器由 ϕ25mm×2.5mm，L=6m 的 100 根钢管组成。现因为工艺需要，空气流量增加 50%，要维持对空气的加热要求，有人建议采用如下四种措施：(1) 将管方改为双程，(2) 增加管长到 6m，(3) 将管束改为 ϕ19mm×2mm 的规格，(4) 将蒸汽温度提高到 124℃。已知蒸汽的量足够，试通过定量计算，必要时做合理简化，说明上述措施能否满足需要。
17. 一列管换热器（管径 ϕ25mm×2.5mm)，传热面积为 12m²（按管外径计，下同)。今拟使 80℃的饱和苯蒸气冷凝、冷却到 45℃。苯走管外，流量为 1kg/s；冷却水走管内与苯逆流，其进口温度为 10℃，流量为 5kg/s。已估算出苯冷凝、冷却时的对流传热系数分别为 1500W/(m²·K)、870W/(m²·K)；水的对流传热系数为 2400W/(m²·K)。取水侧的污垢热阻为 0.26×10⁻³m²·K/W，并忽略苯侧的污垢热阻及管壁热阻，问此换热器是否可用？已知水、液体苯的比热容分别为 4.18kJ/(kg·K)、1.76kJ/(kg·K)；苯蒸气的冷凝潜热 r=395kJ/kg。

思 考 题

1. 气体与固体壁面之间、液体与固体壁面之间、有相变流体与固体壁面之间的对流传热系数的数量级分别为多大？
2. 试说明热导率、对流传热系数和总传热系数的物理意义、单位及彼此之间的区别。
3. 物体的吸收率与辐射能力之间存在什么关系？黑度与吸收率之间有何联系？
4. 有一间壁式换热器，管程内空气被加热，壳程为饱和水蒸气，总传热系数 K 接近于哪一侧的对流传热系数？壁温接近于哪一侧流体的温度？
5. 有一套管换热器，用饱和水蒸气加热管内湍流流动的空气，若总传热系数 K 近似等于空气的对流传热系数，今空气流量增加 1 倍，要求空气进出口温度及饱和水蒸气温度仍不变，问该换热器长度应增加为原长度的多少倍？
6. 管壳式换热器在什么情况下要考虑热补偿，热补偿的形式有哪些？
7. 管壳式换热器为何采用多管程和多壳程？
8. 从强化传热角度来比较管式换热器、板式换热器及翅片式换热器的优缺点。

名 词 术 语

英文/中文

absorptivity/吸收率
air-cooled exchanger/空冷器
baffle pitch /挡板间距
baffle /挡板
barometric leg/大气腿
black body/黑体

英文/中文

boiling point/沸点
boiling temperature/沸腾温度
boiling/沸腾
condensate loading/冷凝液负荷
condensate/冷凝液
condensation/冷凝

conduction/传导
conductor/导体
contact condenser/接触式冷凝器
controlling heat resistance/控制热阻
convection/对流
countercurrent flow/逆流
critical temperature drop/临界温度降
crossflow/错流
dehumidifying condenser/除湿冷凝器
dimensional analysis/量纲分析
dimensionless geometric factor/无量纲几何因子
double-pipe heat exchanger/套管换热器
dropwise condensation/滴状冷凝
electromagnetic radiation/电磁辐射
emission/发射
emissive power/发射能力
emissivity/发射率
enthalpy balance/焓衡算（平衡）
evaporator/蒸发器
expansion joint/膨胀节
extended-surface/扩展表面
film boiling/膜沸腾
film-type condensation/膜状冷凝
fin tube exchanger/翅片式换热器
fixed-sheet exchanger/固定板式换热器
floating-head exchanger/浮头式换热器
forced convection/强制对流
fouling factor/污垢因子
Fourier's law/傅里叶定律
Grashof number/格拉晓夫数
gray body/灰体
guide rod/定距杆
heat flow/热流
heat flux/热通量
heating effectiveness/热效率
heat loss/热损失
heat-transfer coefficient/传热系数
heat transfer equipment/换热设备
heat transfer/传热
individual heat-transfer coefficient/传热分（膜）系数
insulator/绝缘体
Kirchhoff's law/柯斯克夫定律
latent heat of vaporization/汽化潜热
logarithmic mean/对数平均
logarithmic mean temperature difference（LMTD）/对数平均温度差
multipass/多程
natural（free）convection/自然对流
Newton's law of cooling/牛顿冷却定律
noncondensed gas/不凝性气体
nucleate boiling/核沸腾
numerical integration/数值积分
Nusselt number/努塞尔数
one-dimensional heat transfer/一维传热
opaque solid/不透射体
overall heat-transfer coefficient/总传热系数
overall temperature drop/总温降
parallel flow/并流
partial pressure/分压
phase change/相变
plate heat exchanger/板式换热器
pool boiling/池沸腾
Prandtl number/普兰德数
provision for thermal expansion/热膨胀补偿
quantitative calculation/定量计算
radiation heat transfer/辐射传热
radiation/辐射
rate of heat flow/热流率
reflectivity/反射率
roughness/粗糙度
saturated liquid/饱和液体
sensible heat/ 显热
shell-and-tube exchanger/列管换热器
single-pass/单程
single-pass 1-1 exchanger/单管程单壳程换热器
specific heat/比热
steady-state conduction/稳态热传导
Stefan-Boltzmann constant/斯蒂芬-波尔兹曼常数
subcooled liquid/过冷液体
subheated liquid/过热液体
temperature difference/温差
temperature gradient/温度梯度
thermal boundary layer/热边界层
thermal conductivity/热导率
thermal radiation/热辐射
thermal resistance/热阻
transition boiling/过渡沸腾
transmissivity/透射率
tube pitch/管间距
tube sheet/管板
wavelength/波长
white body/白体

第3章

传质原理及应用

Chapter 3 Principles of Mass Transfer and Applications

学习目标

通过本章学习，学生应具备下列知识和能力：

1. 掌握质量传递相关单元操作的基本原理；能够基于质量传递的基本原理，建立复杂工程问题中关于质量传递所涉及的数学模型，分析、解决质量传递过程中的复杂化学工程问题。

2. 掌握质量传递单元操作设备的结构特点、工作原理、操作方法；能够针对复杂化学工程问题所涉及的质量传递问题，提出合理的解决措施。

3. 掌握质量传递操作设备的设计计算及设计选型；能够针对复杂化学工程问题，使用恰当的技术、资源、现代工程工具和信息技术工具，通过对复杂工程问题建模，设计合适的分离设备。

4. 熟悉相关的英文表达，增强国际交流能力，拓宽国际视野。

学习要求

1. 通过本章学习，掌握传质与分离过程的基本概念和传质过程的基本计算方法，能够基于本章的传质过程基本原理和概念，为化学工程问题中相关的传质过程选择合适的传质单元操作。

2. 能够基于精馏的基本原理和气液平衡关系，对双组分连续精馏过程进行工程计算，并能够对精馏塔进行设计和操作优化。

3. 能够基于吸收过程的平衡关系与速率关系，对低组成气体吸收过程进行工程计算。

4. 能够基于萃取的基本原理，用三角形相图对单级萃取过程进行分析和计算。

5. 能够针对实际复杂化学工程问题独立设计和优化填料塔和板式塔，分析和解决相关的复杂工程问题。

6. 具备相关领域的国际交流能力。

过程工业中，经常需要将混合物加以分离。例如，原料常需经过分离提纯或净化，以符合工艺的需求；生产中的废气、废液在排放之前，应将其中所含的有害物质尽量除去，以减少对环境的污染，并有可能将其中有用的物质提取出来；反应器出来的混合物含有多种未反应的原料及反应的副产品，必须对其另行处理，以从中分离出纯度合格的产品，并将未反应的物料送回反应器或他处。以上种种情况，都要采用适当的分离方法与设备，并消耗一定的物料和能量。显然，为了实现上述的不同分离目的，必须根据混合物性质的不同而采用不同的方法。

一般将混合物分为两大类，均相混合物及非均相混合物。均相混合物的分离，如吸收、精馏、萃取等，基本特征是物质由一相转移到另一相或生成新相，过程取决于两相之间的平衡关系，称为相际传质（mass transfer between phases）过程。

当要在相界面上利用平衡关系进行分离操作时，从技术角度来说，一方面，应设法尽可能经济地增大两相界面的表面积，以及怎样在界面上充分利用平衡关系；另一方面，平衡状态是过程的极限，分离不能持续进行，还必须研究如何既偏离平衡状态，又进行所期望的分离。

本章主要介绍气液及液液传质原理，蒸馏操作、吸收操作及萃取操作的主要工艺计算，以及板式塔、填料塔及萃取设备的设备特点及流体力学特性。

Principles of Diffusion and Mass Transfer Between Phases

Diffusion is the movement, under the influence of a physical stimulus, of an individual component through a mixture. The most common cause of diffusion is a concentration gradient of the diffusing component. A concentration gradient tends to move the component in such a direction as to equalize concentrations and destroy the gradient. When the gradient is maintained by constantly supplying the diffusing component to the high-concentration end of the gradient and removing it at the low-concentration end, there is a steady-state flux of the diffusing component. This is characteristic of many mass-transfer operations. For example, when ammonia is removed from a gas by absorption in water in a packed column, at each point in the column a concentration gradient in the gas phase causes ammonia to diffuse to the gas-liquid interface, where it dissolves, and a gradient in the liquid phase causes it to diffuse into the bulk liquid. In stripping a solute from a liquid the gradients are reversed; here diffusion brings solute from the bulk liquid to the interface and from there into the gas phase. In some other mass-transfer operations such as leaching and adsorption, unsteady-state diffusion takes place, and the gradients and fluxes decrease with time as equilibrium is approached.

Although the usual cause of diffusion is a concentration gradient, diffusion can also be caused by an activity gradient, as in reverse osmosis, by a pressure gradient, by a temperature gradient, or by the application of an external force field, as in a centrifuge. Molecular diffusion induced by temperature is thermal diffusion, and that from an external field is forced diffusion. Both are uncommon in chemical engineering. Only diffusion under a concentration gradient is considered in this chapter.

Diffusion is not restricted to molecular transfer through stagnant layers of solid or fluid. It also takes place when fluids of different compositions are mixed. The first step in mixing is often mass transfer caused by the eddy motion characteristic of turbulent flow. This is called eddy diffusion. The second step is molecular diffusion between and inside the very small eddies. Sometimes the diffusion process is accompanied by bulk flow of the mixture in a direction parallel to the direction of diffusion.

Role of diffusion in mass transfer

In all the mass-transfer operations, diffusion occurs in at least one phase and often in both phases. In distillation, the low boiler diffuses through the liquid phase to the interface and away from the interface into the vapor. The high boiler diffuses in the reverse direction and passes through the vapor into the liquid. In leaching, diffusion of solute through the solid phase is followed by diffusion into the liquid. In liquid extraction, the solute diffuses through the raffinate phase to the interface and then into the extract phase. In crystallization, solute diffuses through the mother liquor to the crystals and deposits on the solid surfaces. In humidification or dehumidification there is no diffusion through the liquid phase because the liquid phase is pure and no concentration gradient through it can exist; But the vapor diffuses to or from the liquid-gas interface into or out of the gas phase. In membrane separations diffusion occurs in all there phases: in the fluids either side of the membrane and in the membrane itself.

3.1 传质基本概念（Basic Conceptions of Mass Transfer）

质量传递现象出现在诸如蒸馏、吸收、干燥及萃取等单元操作中。当物质由一相转移到另一相，或者在一个均相中，无论是气相、液相还是固相中传递，其基本机理都相同。

其中，不论汽相还是液相，在单相内物质传递的原理有两种：分子扩散传质和涡流扩散传质。由于涡流扩散时也伴有分子扩散，所以这种现象称为对流传质。

3.1.1 分子扩散（Molecular diffusion）

（1）分子扩散与费克定律（Fick's first law of diffusion）

① 费克定律　如图3-1所示，当流体内部存在某一组分的浓度差或浓度梯度时，则因分子的微观运动使该组分由高浓度处向低浓度处转移，这种现象叫分子扩散。分子扩散可以用分子运动论解释，即随机运动，道路曲折，碰撞频繁。因此，分子扩散的速度是很慢的。

分子扩散可以用费克定律作定量描述，即：

$$J_A = -D_{AB}\frac{dc_A}{dZ} \tag{3-1}$$

式中　J_A——组分A在z方向上的扩散速率（扩散通量），$kmolA/(m^2 \cdot s)$；

D_{AB}——组分A在介质B中的分子扩散系数，m^2/s；

Z——扩散方向上的距离，m；

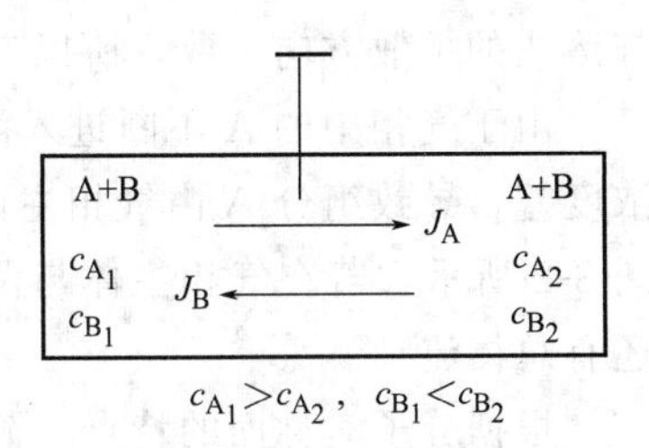

图3-1　分子扩散现象

$\frac{dc_A}{dZ}$——组分A的浓度梯度，$kmol/m^4$。

式中负号表示扩散是沿着物质A浓度降低的方向进行的。

费克定律对B组分同样适用，即只要浓度梯度存在，必然产生分子扩散传质。对于气体混合物，费克定律常以组分的分压表示。费克定律与描述热传导规律的傅里叶定律在形式上类似，但两者有重要的区别，主要在于：热传导传递的是能量，而分子扩散传递的是物质。对于分子扩散而言，一个分子转移后，留下了相应的空间，必由其他分子补充，即介质中的一个或多个组分是运动的，因此，扩散通量（扩散速率）存在一个相对于什么截面的问题；而在热传导中，介质通常是静止的而只有能量以热能的方式进行传递。

② 分子对称面　双组分混合物在总浓度C（对气相是总压P）各处相等的情况下：

因为
$$C=c_A+c_B=\text{常数} \tag{3-2}$$

则
$$\frac{dc_A}{dZ}=-\frac{dc_B}{dZ} \tag{3-3}$$

又因为
$$D_{AB}=D_{BA}=D \tag{3-4}$$

得
$$J_A=-J_B \tag{3-5}$$

则组分A和组分B等量反方向扩散通过的截面叫分子对称面。其特征是仅对分子扩散而言，该截面上净通量等于零，且该截面既可以是固定的截面，也可以是运动的截面。

（2）一维稳态分子扩散（one-dimensional steady-state molecular diffusion）

一维指物质只沿一个方向扩散，其他方向无扩散或扩散量可以忽略，稳态则指扩散速率的大小与时间无关，它只随空间位置而变化。

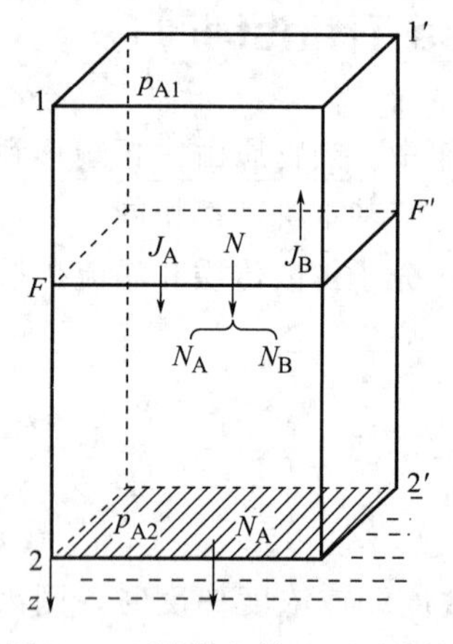

图3-2　可溶性气体A通过惰性气体B的单向扩散（向下）

① 单向扩散（one-way diffusion）　单向扩散适合于吸收操作的分析。现说明如下，如图3-2所示，平面1-1′代表气相主体，平面2-2′代表两相界面，z表示A组分的扩散方向。假设只有气相溶质A不断由气相主体通过两相界面进入液相中，而惰性组分B不溶解且吸收剂S不汽化。当A进入S后，在两相界面的气相一侧留出空间需要补充，由于S不汽化，则只能由气相中的A、B同时补充。也就是说，由于气相中的A不断进入液相，气相主体与两相界面间出现了微小的压差。此压差使气相主体中的A、B同时向两相界面移动，即产生了宏观上的相对运动，这种移动叫对流总体流动（convective bulk flow），总体流动也叫摩尔扩散（指分子群）。

总体流动是由分子扩散（这里指溶质A穿过两相界面）引起的，而不是外力驱动的，这是它的显著特点。因为总体流动与溶质的扩散方向一致，所以有利于传质。

由于气相中的A不断进入液相，使气相主体中的A、B与两相界面上的A、B分别出现浓度差，导致组分A由气相主体向两相界面扩散，组分B由两相界面向气相主体扩散。如图3-2所示，若在气相主体与两相界面间任取一固定截面F，则该截面上不仅有分子扩散，还有总体流动。

根据分子对称面的特点，截面F不是分子对称面；若将分子扩散相对应的分子对称面看成一个与总体流动速度相同的运动截面，则一系列运动的分子对称面与固定的截面F重合。则截面 F 上的净物流通量 N' 为：

$$N' = N + J_A + J_B = N \tag{3-6}$$

固定截面 F 上包括运动的分子对称面，则有：$J_A = -J_B$

N 为总体流动通量，由两部分组成：

$$N = \frac{c_A}{C}N + \frac{c_B}{C}N \tag{3-7}$$

式中　$\frac{c_A}{C}N$——总体流动中携带的组分 A，kmol/(m^2·s)；

$\frac{c_B}{C}N$——总体流动中携带的组分 B，kmol/(m^2·s)。

应注意，式(3-6) 中 N'与 N 数值相等，但两者含义不同。

若在截面 F 与两相界面间做物料衡算：

对组分 A：

$$J_A + \frac{c_A}{C}N = N_A \tag{3-8}$$

对组分 B：

$$\frac{c_B}{C}N = -J_B \tag{3-9}$$

上两式相加得：

$$N_A = N = N' \tag{3-10}$$

式中　N_A——组分 A 通过两相界面的通量，kmol/(m^2·s)。注意，上式中 N_A、N 和 N' 虽然数值相等，但三者含义不同。

这里，组分 B 由两相界面向气相主体的分子扩散与由气相主体向两相界面的总体流动所带的组分 B 数值相等，方向相反，因此宏观上看，组分 B 是不动的或停滞的。所以说，由气相主体到两相界面，只有组分 A 在扩散，称为单向扩散或组分 A 通过静止组分 B 的扩散。由上式(3-8) 可导出单向扩散速率计算式：

$$N_A = \frac{D}{RTZ} \times \frac{P}{p_{Bm}}(p_{A_1} - p_{A_2}) \tag{3-11}$$

其中：

$$p_{Bm} = \frac{p_{B_2} - p_{B_1}}{\ln(p_{B_2}/p_{B_1})} \tag{3-12}$$

式中　p_{Bm}——1、2 两截面上组分 B 分压 p_{B_1} 和 p_{B_2} 的对数平均值，kPa。

对于液体的分子运动规律远不及气体研究得充分，因此只能仿照气相中的扩散速率方程写出液相中的响应关系式。

$$N'_A = \frac{D'C}{Z'c_{Sm}}(c_{A_1} - c_{A_2}) \tag{3-13}$$

式中　N'_A——溶质 A 在液相中的传质速率，kmol/(m^2·s)；

D'——溶质 A 在溶剂 S 中的扩散系数，m^2/s；

c_{A_1}、c_{A_2}——1、2 两截面上的溶质浓度，kmol/m^3。

② 等分子反向扩散（equimolal counter diffusion）　理想精馏操作是等分子反向扩散的

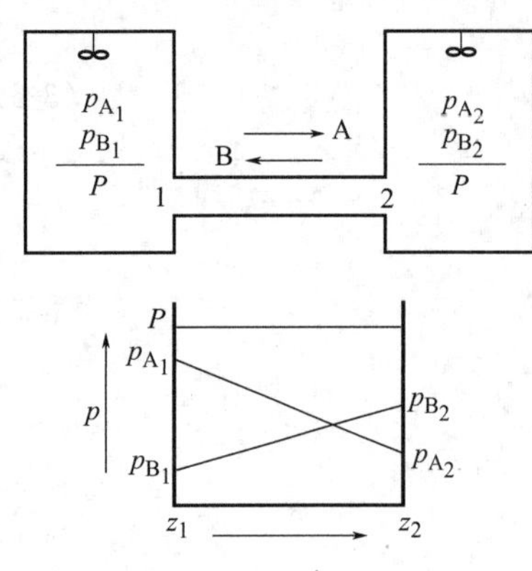

图 3-3　等分子反方向扩散示意图

最好应用实例。根据分子对称面的概念，过程中产生组分 A 的扩散流时，必伴有方向相反的组分 B 的扩散流，即$J_A=-J_B$如图 3-3 所示。

对于等分子反向扩散，因为无总体流动现象，此时，分子对称面就是一个固定的截面，则：

$$J_A=N_A \tag{3-14}$$

$$J_B=N_B \tag{3-14a}$$

由上两式可导出等分子反向扩散速率计算式：

$$N_A=\frac{D}{RTZ}(p_{A_1}-p_{A_2}) \tag{3-15}$$

比较式(3-11）和式(3-15）可以看出，前者比后者多了一个因子 P/p_{Bm}。因为组分 B 分压的对数平均值 p_{Bm} 总小于总压 P，所以 P/p_{Bm} 恒大于 1。显然，这是总体流动的贡献，如同顺水行舟，水流增加了船速，故称 P/p_{Bm} 为“漂流因子”。因为 $P=p_A+p_B=$常数，所以，当 p_A 增加时，p_B 和 p_{Bm} 就会下降，使 P/p_{Bm} 增大，反之亦然。当 P/p_{Bm} 趋近于 1 时，总体流动的因素可忽略，单向扩散与等分子反向扩散无差别。

而液相中发生等分子反向扩散的机会很少，而一个组分通过另一停滞组分的扩散则较多见。

③ 扩散方向上的浓度分布　当发生等分子反向扩散时，组分 A 的分压 p_A 与 Z 的关系为直线关系；而当发生单向扩散时，p_A 与 Z 的关系为对数关系。

Theory of Diffusion

The quantitative relationships for diffusion are discussed in this section, focusing on diffusion in a direction perpendicular to the interface between phases and at a definite location in the equipment. The first topic considered is steady-state diffusion, where the concentrations at any point do not change with time. Equations for transient diffusion are presented later. This discussion is restricted to binary or pseudo-binary mixtures.

Fick's first law of diffusion

In the following general equation for one-dimensional diffusion, the molar flux J is similar to the heat flux q/A, and the concentration gradient dc/dz is similar to the temperature gradient dT/dx. Thus

$$J_A=-D_{AB}\frac{dc_A}{dz} \tag{eq. 3-1}$$

where J_A = molar flux of component A, kmol / (m^2 · s);

D_{AB} = volumetric diffusivity, m^2/s;

c_A = concentration, kmol/m^3;

z = distance in direction of diffusion, m.

Eq.（3-1）is a statement of Fick's first law of diffusion，which may be several different forms. As discussed below，it applies whether or not the total stationary or is in motion with respect to a stationary plane.

（3）分子扩散系数 D（molecular diffusion）

D 简称扩散系数，在传质计算中，D 就像导热计算中的热导率 λ 一样是不可缺少的重要物性参数。由费克定律可知，D 是表征单位浓度梯度下的扩散通量，它反映了某一组分在介质中分子扩散的快慢程度，是物质的一种传递属性。

D 是温度、压力、物质的种类、浓度等因素的函数，比 λ 复杂。对于气体中的扩散，浓度的影响可以忽略；对于液体中的扩散，浓度的影响不能忽略，而压力的影响可以忽略。应用时可由以下三种方法获得：直接查表（不全）；用经验公式或半经验公式估算；实验测定。

① 气体中的扩散系数（diffusivity in gases）　估算气体中扩散系数的经验公式或半经验公式很多，如福勒（Fuller）等人提出的半经验公式为：

$$D=\frac{1.00\times10^{-7}T^{1.75}(1/M_A+1/M_B)^{1/2}}{P[(\Sigma V_A)^{1/3}+(\Sigma V_B)^{1/3}]^2} \tag{3-16}$$

式中　P——总压，atm；

T——热力学温度，K；

M_A、M_B——A、B组分的摩尔质量，g/mol；

ΣV_A、ΣV_B——A、B组分的摩尔扩散体积，一般有机化合物按化学分子式查原子扩散体积相加得到，cm^3/mol。注意，芳烃环或杂环化合物的原子体积要再加上 -20.2，简单物质可直接查得分子扩散体积，此半经验公式误差小于10%。

② 液体中的扩散系数（diffusivity in liquids）　对于很稀的非电解质溶液，威尔盖（Wilke）等人提出：

$$D_{AS}=7.4\times10^{-8}\frac{(aM_S)^{1/2}T}{\mu_S V_A^{0.6}} \tag{3-17}$$

式中　T——溶液的热力学温度，K；

μ_S——溶剂S的黏度，mPa·s；

M_S——S的摩尔质量，g/mol；

a——S的缔合参数，水为2.6，甲醇为1.9，乙醇为1.5，苯、乙醚等不缔合溶剂为1.0；

V_A——溶质A在正常沸点下的摩尔体积 cm^3/mol。

一般来说，液相中扩散速度远远小于气相中的扩散速度，就数量级而论，物质在气相中的扩散系数比在液相中的扩散系数约大 10^5 倍，液相扩散系数的估算式不如气体的可靠。但是，液体的密度往往比气体大得多，因此液相中的物质浓度及浓度梯度可远远高于气相中的值。所以在一定条件下，气液两相中可达到相同的扩散通量。

表3-1、表3-2分别列举了一些物质在空气及水中的扩散系数，供计算时参考。

表 3-1　一些物质在空气中的扩散系数（0℃，101.3kPa）

扩散物质	扩散系数 D_{AB}/(cm²/s)	扩散物质	扩散系数 D_{AB}/(cm²/s)
H_2	0.611	H_2O	0.220
N_2	0.132	C_6H_6	0.077
O_2	0.178	C_7H_8	0.076
CO_2	0.138	CH_3OH	0.132
HCl	0.130	C_2H_5OH	0.102
SO_2	0.103	CS_2	0.089
SO_3	0.095	$C_2H_5OC_2H_5$	0.078
NH_3	0.170		

表 3-2　一些物质在水中的扩散系数（20℃，稀溶液）

扩散物质	扩散系数 $D'_{AB}\times10^{-9}$/(m²/s)	扩散物质	扩散系数 $D'_{AB}\times10^{-9}$/(m²/s)
O_2	1.80	HNO_3	2.60
CO_2	1.50	NaCl	1.35
N_2O	1.51	NaOH	1.51
NH_3	1.76	C_2H_2	1.56
Cl_2	1.22	CH_3COOH	0.88
Br_2	1.20	CH_3OH	1.28
H_2	5.13	C_2H_5OH	1.00
N_2	1.64	C_3H_7OH	0.87
HCl	2.64	C_4H_9OH	0.77
H_2S	1.41	C_6H_5OH	0.84
H_2SO_4	1.73	$C_{12}H_{22}O_{11}$（蔗糖）	0.45

3.1.2　对流传质（Convective mass transfer）

（1）涡流扩散（eddy diffusion）

在传质设备中，流体的流动型态多为湍流。湍流与层流的本质区别是：层流流动时，质点仅沿着主流方向运动；湍流流动时，质点除了沿着主流方向运动外，在其他方向上还存在脉动（或出现涡流、旋涡）。

由于涡流的存在，使流体内部的质点强烈地混合，其结果不仅使流体的质点产生动量、热量传递，而且产生了质量传递。这种由于涡流产生的质量传递过程，称为涡流扩散。

涡流扩散速率要比分子扩散速率大得多，强化了相内的物质传递。与对流传热的同时存在热传导类似，涡流扩散的同时也伴随着分子扩散。因为涡流扩散现象相当复杂，所以至今没有严格的理论予以描述。

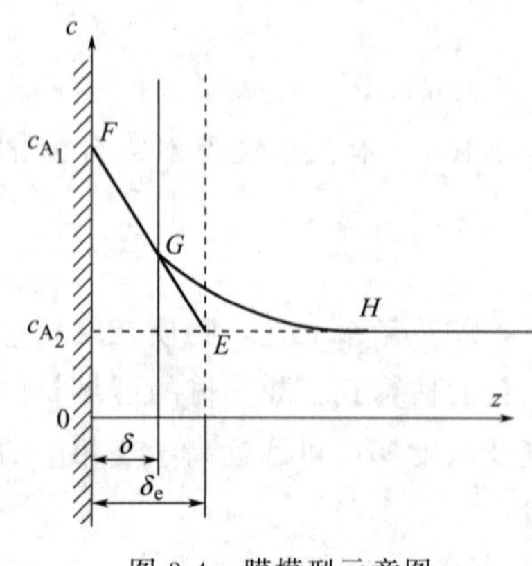

图 3-4　膜模型示意图

（2）对流传质

通常把涡流扩散与分子扩散同时发生的过程称为对流传质，对流传质可用膜模型、渗透理论和表面更新理论描述。

① 膜模型（film theory）　如图 3-4 所示，流体沿固体壁面作湍流流动时，在流体主体与固体壁面之间可分为三个区域，其相应的传质情况为：层流内层（紧靠固

体壁面的很薄流体层）中，流体呈层流流动，以分子扩散方式进行传质，物质的浓度分布是一条直线或近似直线；过渡区中，对涡流扩散和分子扩散都要考虑，物质沿轴向的浓度分布是曲线；而在湍流区中，涡流扩散比分子扩散的速度大得多，则后者可忽略，所以浓度分布曲线近似为一条水平线。

将流体主体与固体壁面之间的传质阻力折合为与其阻力相当的 δ_e 厚的层流膜（又称虚拟膜、当量膜、有效膜）内，这样一来，对流传质的作用就折合成了相当于物质通过 δ_e 距离的分子扩散过程，这种简化处理的理论称为膜模型。

② 对流传质速率方程　应用膜模型，可分别得到气相和液相的对流传质速率方程：

对于气膜：
$$N_A=\frac{D_G}{RT\delta_G}\times\frac{P}{p_{Bm}}(p_{A_1}-p_{A_2}) \tag{3-18}$$

对于液膜：
$$N_A=\frac{D_L}{\delta_L}\times\frac{C}{c_{Sm}}(c_{A_1}-c_{A_2}) \tag{3-19}$$

令：
$$k_G=\frac{D_G}{RT\delta_G}\times\frac{p}{p_{Bm}} \tag{3-20}$$

$$k_L=\frac{D_L}{\delta_L}\times\frac{C}{c_{Sm}} \tag{3-21}$$

则式(3-18) 和式(3-19) 可分别写成：

气相：
$$N_A=k_G(p_{A_1}-p_{A_2}) \tag{3-22}$$

液相：
$$N_A=k_L(c_{A_1}-c_{A_2}) \tag{3-23}$$

式中　δ_G、δ_L——气膜、液膜厚度，m；

k_G、k_L——气相、液相的对流传质系数。

上述处理方法避开难以测定的膜厚 δ_G、δ_L，将所有影响对流传质的因素都集中于 k_G、k_L 中，这样便于实验测定和工程计算。

由理论分析和初步实验可知，影响对流传质系数的主要因素有：物性参数，包括分子扩散系数 D、黏度 μ、密度 ρ；操作参数，包括流速 u、温度 T、压力 P；传质设备特性参数即几何定性尺寸 d。则：

$$k=f(D,\mu,\rho,u,d,T,P) \tag{3-24}$$

由量纲分析得特征数关联式：

$$Sh=f(Re,Sc) \tag{3-25}$$

式中　Sh——施伍德数，包括待求的未知量，它表示分子扩散阻力/对流扩散阻力（它与传热中的努塞尔特数的作用相似），$Sh=\frac{kl}{D}=\frac{l/D}{1/k}$；

l——特征尺寸，可以指填料直径或塔径等，可根据不同的关联式而定；

Re——雷诺数，反映了流动状态的影响，式中的 d 通常采用当量直径，$Re=\frac{du\rho}{\mu}$；

Sc——施密特数，表征物性的影响，$Sc=\frac{\mu}{\rho D}$。

再结合实验，可确定 Sh、Re、Sc 之间的具体函数关系。

③ 其他传质理论简介　膜模型为传质过程的分析提供了一个简化的物理模型。对于具有固定相界面的系统或相界面无明显扰动的相间传质过程，如湿壁塔的吸收操作，按膜模型推导出的传质结果与实际情况基本符合。但是，对于具有自由界面的系统，尤其是剧烈湍动的相间传质过程，如填料塔的吸收操作，则膜模型与实际情况不符。因此，随之出现了新的传质理论，如溶质渗透理论（penetration theory）和表面更新理论（surface renewal theory）。

溶质渗透理论认为，在很多过程工业的传质设备中，例如填料塔内的气液接触，由于接触时间短且湍动剧烈，所以，在任一个微元液体与气体的界面上，所溶解的气体中的组分向微元液体内部进行非定态分子扩散，经过一个很短的接触时间 τ 以后，这个微元液体又与液相主体混合。假设所有微元液体在界面上和气体相接触的时间都是相同的，可以导出液膜传质系数的表达式为：

$$k_L=2\sqrt{\frac{D_L}{\pi\tau}} \tag{3-26}$$

式中　τ——渗透模型参数，是气、液相每次接触的时间，s。

渗透理论比膜模型更符合实际，它指明了缩短 τ 可提高液膜传质系数。因此，增加液相的湍动程度，增加液相的表面积均为强化传质的有效途径。

而表面更新理论认为渗透理论所假定的每个液体微元在表面和气体相接触的时间都保持相同是不可能的。实际的情况应是各个液体微元在相界面上的停留时间不同，而是符合随机的“寿命”分布规律，且不同寿命的表面液体微元被新鲜的液体微元置换的概率是相同的。因此，这一理论将液相分成界面和主体两个区域，在界面区里，质量传递是按照渗透模型进行的，不同的是这里的微元不是固定的，而是不断地与主体区的微元发生交换进行表面的更新；在主体区内，全部液体达到均匀一致的浓度。

由表面更新理论推导出的液相对流传质系数的表达式为：

$$k_L=\sqrt{D_L S} \tag{3-27}$$

式中　k_L——按整个液面的平均值计的液相传质系数，m/s；

　　S——表面更新频率，即单位时间内液体表面被更新的百分率，s^{-1}。

上式表明，液相传质系数与表面更新频率的平方根成正比，因而给传质过程的强化提供了依据。

上述的三个传质模型都有可以接受的物理概念，但每个模型中都包含一个难以测定的模型参数，如膜模型中的虚拟膜厚 δ，渗透模型中的气、液相每次接触的时间 τ 及表面更新模型中的表面更新频率 S。不过，正因为这些模型分别包含一个解决问题的思想方法，为从理论上进一步研究传质过程奠定了基础。

实际上，影响传质系数的因素相当复杂，只有在少数的情况下，传质系数可以由理论推算而得，在绝大多数情况下是采用量纲分析结合实验测定的方法，确定用无量纲数群表示的对流传质系数关联式，并应用到实际的工程设计中去。

3.2 蒸馏（Distillation）

蒸馏操作历史悠久，技术成熟，规模不限，应用广泛，其主要目的是分离液体混合物或液化了的气体混合物，来提纯或回收有用组分。例如在石油炼制中的原油精炼最初阶段，可将混合物分为汽油、煤油、柴油和润滑油等；某原料经过化学反应后，可能产生一个既有反应物又有生成物及副产物的液体混合物，则为了获得纯的生成物，若该混合物是均相的，通常采用精馏的方法予以分离等等。

蒸馏是目前应用最广的一类液体混合物分离方法，具有如下特点。

① 通过蒸馏分离可以直接获得所需要的产品，而吸收、萃取等分离方法，由于有外加的溶剂，需进一步使所提取的组分与外加组分再行分离，因而蒸馏操作流程通常较为简单。

② 蒸馏分离的适用范围广，它不仅可以分离液体混合物，而且可用于气态或固态混合物的分离。例如，可将空气加压液化，再用精馏方法获得氧、氮等产品；再如，脂肪酸的混合物，可用加热使其熔化，并在减压下建立气液两相系统，用蒸馏方法进行分离。

③ 蒸馏过程适用于各种浓度混合物的分离，而吸收、萃取等操作只有当被提取组分浓度较低时才比较经济。

④ 蒸馏操作是通过对混合液加热建立气液两相体系的，所得到的气相还需要再冷凝液化。因此，蒸馏操作耗能较大。蒸馏过程中的节能是个值得重视的问题。

3.2.1 气液相平衡原理

蒸馏操作是气液两相间的传质过程，气液两相达到平衡状态是传质过程的极限。因此，气液平衡关系是分析精馏原理、解决精馏计算的基础。

理想物系气液相平衡关系（gas-liquid equilibrium relations of ideal system）中，若理想物系的液相为理想溶液，平衡关系服从拉乌尔（Raoult）定律；若气相为理想气体，则服从理想气体定律及道尔顿分压定律。

① 自由度分析（degrees of freedom analysis） 相律（phase rule）表示平衡物系中的自由度 F、相数 Φ 及独立组分数 C 之间的关系，即：

$$F=C-\Phi+2 \tag{3-28}$$

式中 F——自由度数，即指该系统的独立变量数；

C——独立组分数；

Φ——相数；

2——外界只有温度和压力两个条件影响物系平衡状态。

由相律可知，双组分气液相平衡系统的自由度数为2。又知，对气相而言，两组分的摩尔分数 y_A 与 y_B 是不独立的；同理液相中的 x_A 与 x_B 也是不独立的。所以，双组分气液相平衡状态涉及的四个独立变量为 p、T、y 和 x，其中的任何两个都可用其他两个表示。换言之，上述的四个变量任意指定其中两个，平衡状态就唯一确定。一般来说，蒸馏常采用恒压操作，则平衡关系可用 $t\text{-}x(y)$ 或 $x\text{-}y$ 的函数关系及其相图表示。

Phase Rule

Generally in a liquid-liquid system we have three components, A, B, and S, and two phases in equilibrium. Substituting into the phase rule,

$$F = C - P + 2 \quad \text{(eq. 3-2)}$$

Where, P is the number of phases at equilibrium, C are the number of total components in the two phases when no chemical reactions are occurring, and F is the number of variants or degrees of freedom of the system.

The number of degrees of freedom is 3. The variables are temperature, pressure, and four concentrations. Four concentrations occur because only two of the three mass-fraction concentrations in a phase can be specified. The third must make the total mass fractions equal to 1.0: $x_A + x_B + x_S = 1.0$. If pressure and temperature are set, which is the usual case, then, at equilibrium, setting one concentration in either phase fixes the system.

② 相平衡组成计算　对于理想溶液，拉乌尔定律（Raoult's law）为：

$$p_A = p_A^0 x_A \tag{3-29}$$

$$p_B = p_B^0 x_B = p_B^0 (1 - x_A) \tag{3-29a}$$

式中　p_A、p_B——相平衡状态下 A、B 组分的蒸气分压，Pa；

p_A^0、p_B^0——同温度下 A、B 组分的饱和蒸气压，Pa；

x_A、x_B——A、B 组分的摩尔分数。

对于理想气体，道尔顿定律为：

$$P = p_A + p_B \tag{3-30}$$

联立式(3-29)、式(3-29a) 和式(3-30) 可得：

$$x_A = \frac{P - p_B^0}{p_A^0 - p_B^0} \tag{3-31}$$

上式称为泡点方程，即在一定的总压 P 下，对于某一指定的平衡温度 t，可求得饱和蒸气压 $p_A^0 = f(t)$ 和 $p_B^0 = f(t)$，由上式即可计算相平衡时的液相组成 x_A。纯组分的饱和蒸气压是温度的非线性函数，可由实验测定或经验公式计算。

纯组分的饱和蒸气压和温度的关系可用安托因（Antoine）方程表示：

$$\lg p_i^0 = a - \frac{b}{t + c} \tag{3-32}$$

式中，a、b、c 为组分的安托因常数，可由有关手册查得，其值由压力和温度的单位而定。

根据道尔顿分压定律：

$$y_A = \frac{p_A}{P} \tag{3-33}$$

将式(3-29) 和式(3-31) 代入上式整理得：

$$y_A = \frac{p_A^0}{P} \times \frac{P - p_B^0}{p_A^0 - p_B^0} \tag{3-34}$$

上式称为露点方程。在一定的总压 P 下，对于某一指定的相平衡温度 t，由上式可计算平衡时的气相组成 y_A。

【例 3-1】 苯（A）-甲苯（B）理想物系，二者的安托因方程分别为：

$$\lg p_A^0 = 6.906 - \frac{1211}{t + 220.8}$$

$$\lg p_B^0 = 6.955 - \frac{1345}{t + 219.5}$$

式中，p^0 的单位是 mmHg；t 的单位是℃。求：$t = 105$℃、$P = 850$mmHg 时的平衡气液相组成 x_A 与 y_A。

解：$\lg p_A^0 = 6.906 - \dfrac{1211}{t+220.8}$，$t=105℃$，$p_A^0 = 1545\text{mmHg}$

$\lg p_B^0 = 6.955 - \dfrac{1345}{t+219.5}$，$t=105℃$，$p_B^0 = 645.9\text{mmHg}$

$$x_A = \frac{P - p_B^0}{p_A^0 - p_B^0} = \frac{850-645.9}{1545-645.9} = 0.227$$

$$y_A = \frac{p_A^0 x_A}{P} = \frac{1545\times 0.227}{850} = 0.413$$

Equilibrium Relations

To evaluate driving forces relating to the mass transfer, a knowledge of equilibria between phases is therefore of basic importance. Several kinds of equilibria are important in mass transfer. In all situations, two phases are involved, and all combinations are found except two gas phases or two solid phases. The controlling variables are temperature, pressure, and concentration. Equilibrium data can be shown in tables, equations, or graphs. For most processes considered in this text, the pertinent equilibrium relationship can be shown in graphically.

Gas-liquid equilibrium

As in the gas-liquid systems, the equilibrium in vapor-liquid systems is restricted by the phase rule which we learned from the textbook of Physical Chemistry. As an example we shall use the ammonia-water, vapor-liquid system. For two components and two phases, this means that there are 2 degrees of freedom. The four variables are temperature, pressure, and the composition y_A of NH_3 in the vapor phase and x_A in the liquid phase. The composition of water (B) is fixed if y_A or x_A is specified, since $y_A + y_B = 1.0$ and $x_A + x_B = 1.0$. If the pressure is fixed, only one more variable can be set. If we set the liquid composition, the temperature and vapor composition are automatically set.

Henry's law

Often the equilibrium relation between p_A in the gas phase and x_A in the liquid phase can be expressed by a straight-line. Henry's law equation at low concentrations:

$$p_A = E x_A \qquad \text{(eq. 3-3)}$$

Where, E is the Henry's law constant in atm/mole fraction for the given system. If both sides of eq. (3-3) are divided by total pressure P in atm:

$$y_A = \frac{E}{P} x_A \qquad \text{(eq. 3-4)}$$

and let

$$m=\frac{E}{P} \qquad \text{(eq. 3-5)}$$

Where, m is the Henry's law constant in mole frac gas/mole frac liquid. Note that m depends on total pressure, whereas E does not.

Vapor-liquid equilibrium relations

The quantity and reliability of data available for equilibrium systems vary widely. Perhaps the most widely and completely investigated type of system is the vapor-liquid system containing paraffin hydrocarbons. For other vapor-liquid systems, such as oxygenated or chlorinated hydrocarbons, the information available is less complete. Generalizations in this area are difficult and the designer is wise to insist on experimental data for the particular system on hand.

Raoult's law

At low pressure the vapor of mixture approaches ideal behavior and follows the ideal gas law. In some mixtures, Raoult's law applies to each component over entire concentration range from 0 to 1.0, such mixtures are called ideal. The partial pressure of component A in gas is proportional to its concentration in liquid.

$$p_A = P_A x_A \qquad \text{(eq. 3-6)}$$

Where, P_A is the vapor pressure of pure A in Pa (or atm), and x_A is the mole fraction of A in the liquid. The same rule, written for component B is:

$$p_B = P_B x_B \qquad \text{(eq. 3-7)}$$

This law holds only for ideal solutions. Actually, few solutions are ideal. The assumption of ideal liquid behavior is generally correct only for members of a homologous series that are close together in molecular weight, such as benzene-toluene, hexane-heptane, and methyl alcohol-ethyl alcohol, which are usually substances very similar to each other. Mixtures of polar substances such as water, alcohol, and electrolytes depart greatly from ideal-solution law. If one component of a two-component mixture follows the ideal-solution law, so does the other. However, the assumption of ideal liquid behavior is good from a practically engineering standpoint.

③ t-$x(y)$ 相图　当压力一定时，在相平衡状态下，分别以泡点方程和露点方程作图，称为温度组成图，如图 3-5 所示。

由图 3-5 可知，曲线两端点分别代表纯组分的沸点，左端点代表纯轻组分沸点，右端点代表纯重组分沸点。上方曲线代表饱和蒸气线，也称为露点线，该线上方的区域代表过热蒸气区；下方曲线代表饱和液体线，也称为泡点线，该线下方的区域代表过冷液体区。两条曲线之间的区域为气液共存区，平衡关系用两曲线间的水平线段表示。显然，只有在气液共存区，才能起到一定的分离作用。混合液的沸点不是一个定值，而是随组成不断变化，在同样组成下，泡点（开始产生第一个气泡时对应的温度）与露点（即开始产生第一滴液滴时对应的温度）并不相等。

④ x-y 相图 x-y 相图可以通过 t-x（y）关系作出，如图3-6所示。

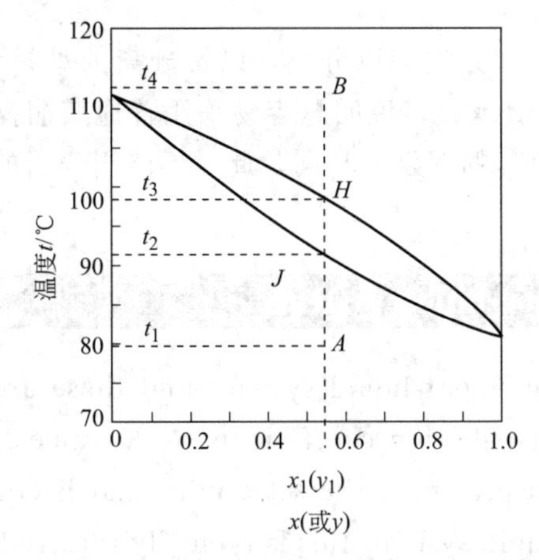

图3-5 常压下苯-甲苯混合液的 t-x（y）相图

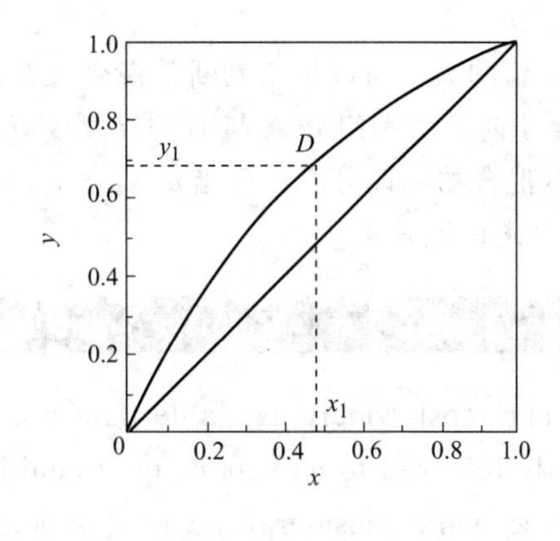

图3-6 常压下苯-甲苯混合液的 x-y 相图

由图3-6可知，图中对角线上方的曲线称为平衡曲线，该曲线上任意点 D 表示组成为 x_1 的液相与组成为 y_1 的气相互成平衡，且表示点 D 有一确定的状态；图中对角线（即 $x=y$ 直线）作查图时参考用。当 $y>x$ 时，平衡线位于对角线上方，平衡线离对角线越远，表示该溶液越容易分离。在总压变化不大时，外压对平衡线的影响可忽略，但 t-x-y 图随压力变化较大，因此常采用 x-y 相平衡曲线较为方便。

⑤ 挥发度和相对挥发度（volatility and relative volatility）

a. 挥发度 ν 对于纯液体而言，其挥发度指该液体在一定温度下的饱和蒸气压，而溶液中各组分的挥发度可用它在蒸气压中的分压和与之平衡的液相中的摩尔分数之比表示，即：

$$\nu_A = p_A / x_A \tag{3-35}$$

$$\nu_B = p_B / x_B \tag{3-35a}$$

对于理想溶液：

$$\nu_A = p_A^0 = f(t) \tag{3-36}$$

$$\nu_B = p_B^0 = f(t) \tag{3-36a}$$

显然，挥发度能反映出混合液中各组分挥发的难易程度，但随温度变化，在使用上不甚方便，故引出相对挥发度的概念。

b. 相对挥发度 α 习惯上将溶液中易挥发组分A的挥发度与难挥发组分B的挥发度之比，称为相对挥发度 α，即：

$$\alpha = \frac{\nu_A}{\nu_B} = \frac{p_A / x_A}{p_B / x_B} = \frac{y_A(1-x_A)}{x_A(1-y_A)} \tag{3-37}$$

由上式可得：

$$y_A = \frac{\alpha x_A}{1+(\alpha-1)x_A} \tag{3-38}$$

当 α 已知时，由上式可求得相平衡时气液相组成 x-y 的关系，故称上式为气液平衡方程或相平衡方程。由于 ν_A 与 ν_B 均随温度同时增大或缩小，因而 α 可取平均值视为常数，使用上非常方便。

α 值的大小可以用来判断某混合液是否能用蒸馏方法加以分离，以及分离的难易程度。当 $\alpha \neq 1$ 时，表示组分 A 与 B 的挥发度有差别，α 值越大，说明越容易采用普通蒸馏操作对 A、B 混合物加以分离；而当 $\alpha = 1$ 时，即 A、B 组分的挥发度相同，此时不能用普通的蒸馏方法分离该混合液。

Relative volatility of vapor-liquid systems

The most widely available data are those for vapor-liquid systems and these are frequently referred to as vapor-liquid equilibrium distribution coefficients or K values. Since K values vary considerably with temperature and pressure, the selectivity that is equal to the ratio of the K values is used. For the vapor-liquid systems this is typically referred to as the relative volatility α. For a binary system, this is defined as the ratio of the concentration of A in the vapor to the concentration of A in the liquid divided by the ratio of the concentration of B in the vapor to the concentration of B in the liquid:

$$\alpha_{AB} = \frac{y_A/x_A}{y_A/x_A} = \frac{y_A/x_A}{(1-y_A)/(1-x_A)} \quad \text{(eq. 3-8)}$$

where α_{AB} is the relative volatility of A with respect to B in the binary system. Eq. (3-8) can be rearranged to give:

$$y_A = \frac{\alpha x_A}{1+(\alpha-1)x_A} \quad \text{(eq. 3-9)}$$

Where, $\alpha = \alpha_{AB}$. For a separation process in which $\alpha = 1$, the compositions of component A would be the same in both phases, so do the component B. obviously, separation is not possible when this occurs since the driving force for mass transfer is zero. When the value of α is above 1.0, a separation is possible. The value of α may change as concentration and pressure change.

If the system obeys Raoult's law, as does the benzene-toluene system:

$$y_A = \frac{P_A}{P}x_A, y_B = \frac{P_B}{P}x_B \quad \text{(eq. 3-10)}$$

Substituting eq. (3-10) into eq. (3-9) for an ideal system:

$$\alpha = \frac{P_A}{P_B} \quad \text{(eq. 3-11)}$$

At a given temperature P_A and P_B are fixed, so that for constant-temperature operations with ideal system α is constant. When binary systems follow Raoult's law, the relative volatility often varies only slightly over a large concentration range at constant total pressure.

⑥ 非理想溶液（non-ideal liquids） 不服从拉乌尔定律的溶液是非理想溶液，实际生产中遇到的大多数溶液都属于非理想溶液。

溶液的非理想性来源于异分子间的作用力与同分子间的作用力不等，表现为其平衡蒸气压偏离拉乌尔定律，偏差可正可负，其中尤以正偏差居多。

a. 正偏差系统　该系统的异分子间作用力小于同分子间作用力时，排斥力占主要地位，分子较易离开液面而进入气相，所以泡点较理想溶液低，混合时，体积变化 $\Delta V>0$，容易出现最低恒沸点。例如乙醇-水系统为正偏差系统，其相图如图 3-7 所示。

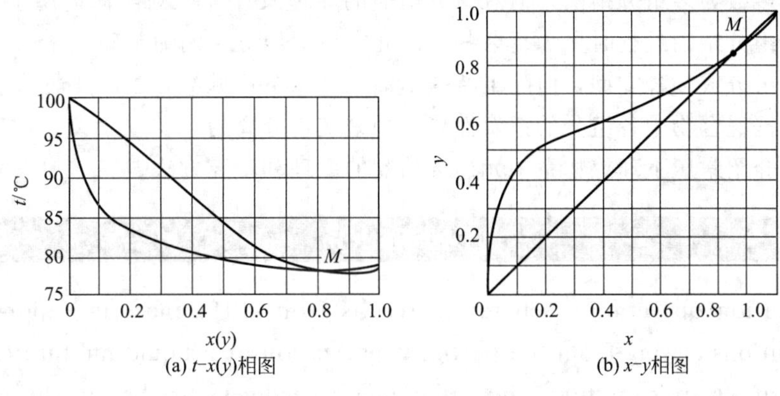

(a) t-x(y)相图　(b) x-y相图

图 3-7　常压下乙醇-水系统相图

b. 负偏差系统　该系统的异分子间作用力大于同分子间作用力，吸引力占主要地位，分子较难离开液相进入气相，所以泡点较理想溶液低，混合时，$\Delta V<0$，容易出现最高恒沸点。例如硝酸-水系统为负偏差系统，如图 3-8 所示。

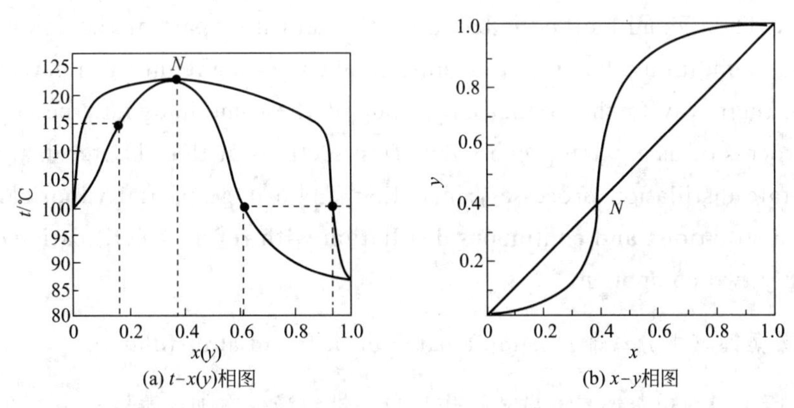

(a) t-x(y)相图　(b) x-y相图

图 3-8　常压下硝酸-水系统相图

应当注意：

ⅰ. 分离恒沸物应采用特殊蒸馏方式；

ⅱ. 恒沸点随总压变化，若采用变化压力来分离恒沸物，要进行经济权衡以做最后处理；

ⅲ. 非理想溶液不一定都有恒沸点，例如甲醇-水系统无恒沸点。

3.2.2　蒸馏方式

蒸馏可按不同的方法分类。按操作方式可分为简单蒸馏、平衡蒸馏、精馏和特殊精馏等，其中，简单蒸馏和平衡蒸馏常用于混合物中各组分的挥发度相差较大，对分离要求又不高的场合。精馏是借助回流技术来实现高纯度和高回收率的分离操作，它是应用最广泛的蒸馏方式。精馏按操作压力可分为常压、减压和加压蒸馏；按物料组分数可分为双组分（二元）精馏和多组分（多元）精馏；按操作流程可分为间歇精馏和连续精馏。例如，常压下为气态的混合物，常采用加压蒸馏的方法分离；而对于泡点较高或热敏性混合物，则宜采用真空精馏操作，以降低操作温度。间歇操作主要应用于小规模、多品种或某些有特殊要求的场合，而工业中以连续蒸馏为主。间歇蒸馏为非稳态操作，而连续精馏一般为稳态操作。工业生产中，绝大多数为多组分精馏，但双组分精馏的原理及计算原则同样适用于多组分精馏，只是在处理多组分精馏过程时更为复杂些，因此常以双组分精馏为基础。

蒸馏尽管可分成多种类型，但除了特殊精馏以外，不论哪类方法，其最基本的原理是相同的，都是依据混合物中各组分的挥发性能不同来实现分离的。

本节重点介绍常压下理想物系二元连续精馏的原理和计算方法。

Distillation

In distillation operations, separation results from differences in vapor- and liquid-phase compositions arising from the partial vaporization of a liquid mixture or the partial condensation of a vapor mixture. The vapor phase becomes enriched in the more volatile components while the liquid phase is depleted of those same components.

In practice, distillation may be carried out by either of two principal methods. The first method is based on the production of a vapor by boiling the liquid mixture to be separated and condensing the vapors without allowing any liquid to return to the still. There is then no reflux. The second method is based on the return of part of the condensate to the still under such conditions that this returning liquid is brought into intimate contact with the vapors on their way to the condenser. Either of these methods may be conducted as a continuous process or as a batch process. The first sections of this chapter deal with continuous steady-state distillation processes, including single-stage partial vaporization without reflux (flash distillation) and continuous distillation with reflux (rectification) for systems containing only two components.

3.2.2.1 简单蒸馏（微分蒸馏）(simple batch or differential distillation)

如图 3-9 所示，简单蒸馏是间歇式操作。将一批料液一次加入蒸馏釜中，在外压恒定下加热到沸腾，生成的蒸气及时引入到冷凝器中冷凝后，冷凝液作为产品分批进入储槽，其中易挥发组分相对富集。过程中釜内液体的易挥发组分浓度不断下降，蒸气中易挥发组分浓度也相应降低，因此釜顶部分批收集流出液，最终将釜内残液一次排出。显然，简单蒸馏得不到大量的高纯度产品。

Simple batch or differential distillation

In simple batch or differential distillation, liquid is first charged to a heated kettle, the liquid charge is boiled slowly and the vapors are withdrawn as rapidly as they form to a condenser, where the condensed vapor (distillate) is collected. The first portion of vapor condensed will be richest in the more volatile component A. As vaporization proceeds, the vaporized product becomes leaner in A.

3.2.2.2 平衡蒸馏（又称闪蒸）(flash distillation)

闪蒸是一种单级连续蒸馏操作，流程如图 3-10 所示。原料液连续加入加热器中，预热至一定温度后，经减压阀减至预定压强，由于压力突然降低，过热液体发生自蒸发，液体部分汽化，气液两相在分离器中分开，塔顶产品 A 得到提浓，底部产品 B 增浓。因为闪蒸操作只经历了一次气液平衡过程，所以对物料的分离十分有限。

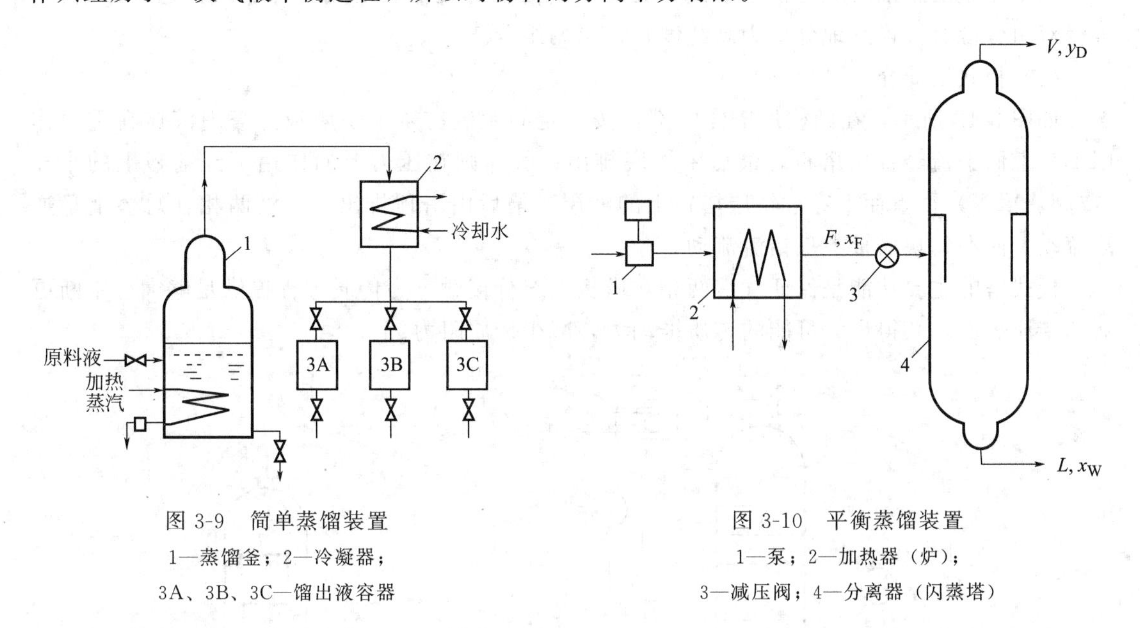

图 3-9 简单蒸馏装置
1—蒸馏釜；2—冷凝器；
3A、3B、3C—馏出液容器

图 3-10 平衡蒸馏装置
1—泵；2—加热器（炉）；
3—减压阀；4—分离器（闪蒸塔）

Flash distillation

Flash distillation consists of vaporizing a definite fraction of the liquid in such a way that the evolved vapor is in equilibrium with the residual liquid, separating the vapor from the liquid, and condensing the vapor. Fig. 3-10 shows the elements of a flash distillation plant. Feed is pumped by pump a through heater 2, and the pressure is reduced through valve 3. An intimate mixture of vapor and liquid enters the vapor separator 4, in which sufficient time is allowed for the vapor and liquid portions to separate. Because of the intimacy of contact of liquid and vapor before separation, the separated streams are in equilibrium. Vapor leaves from the top and liquid through the bottom.

3.2.3 精馏原理 (Principles of distillation)

上述的简单蒸馏及平衡蒸馏都只能使液体混合物得到有限的分离，通常不能满足工业上高产量连续化高纯度分离的要求，但它可以在精馏塔中实现。

(1) 连续精馏操作流程

连续精馏操作流程如图 3-11 所示。料液自塔的中部某适当位置连续加入塔内，塔顶设有冷凝器将塔顶蒸气冷凝为液体，一部分冷凝液回流入塔顶，该过程称为回流，另一部分冷凝液作为塔顶产品（馏出液）连续排出，塔底部设有再沸器（蒸馏釜）以加热液体产生部分蒸气返回塔底，另一部分液体作为塔底产品排出。塔内气、液两相逆流接触进行热量、质量交换。塔内上半部分（加料位置以上），对气相而言，随着气相上升，气相中难挥发组分不断减少，易挥发组分不断增多，完成了气相的精制任务，故称精馏段。塔内下半部分，对液相而言，随着液相不断下流，液相中易挥发组分不断减少，难挥发组分不断增多，完成了液相的提纯任务，故称提馏段。

一个完整的精馏塔应包括精馏段、提馏段、塔顶冷凝器和塔底再沸器。在这样的塔内，可将双组分混合液连续地分离为高纯度轻、重两组分。

（2）板式塔简介

如图 3-12 所示，在圆柱形塔体内部，按一定间距装有若干块塔板。塔内液体在重力作用下自上而下流经各层塔板，最后由塔底排出；气体则在压力差的作用下经塔板上的小孔（或阀、罩等）由下而上穿过每层塔板上的液层，最后由塔顶排出。气液两相在总体上呈逆流流动，而在每块塔板上呈错流流动。

板式塔的主要功能是保证气液两相在塔板上充分接触，为传质过程提供足够大且不断更新的传质表面，获得最大可能的传质推动力，减少传质阻力。

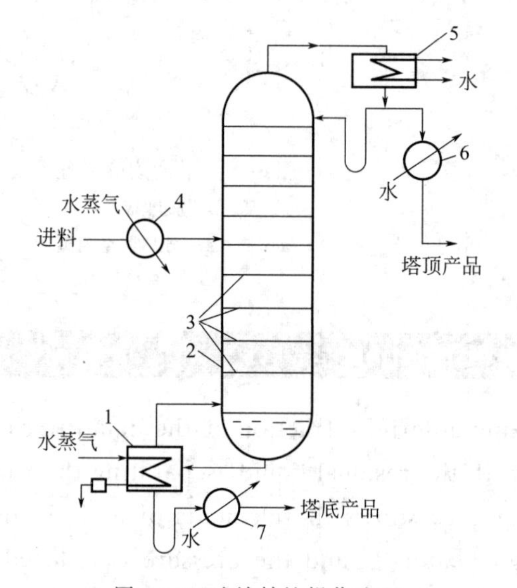

图 3-11　连续精馏操作流程

1—再沸器；2—精馏塔；3—塔板；
4—进料预热器；5—冷凝器；
6—塔顶产品冷却器；7—塔底产品冷却器

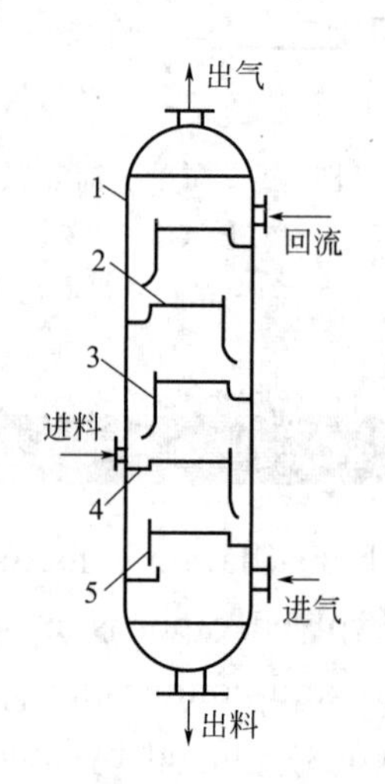

图 3-12　板式塔示意图

1—塔壳体；2—塔板；
3—溢流堰；4—受液盘；
5—降液管

（3）精馏原理

精馏原理可以板式塔操作为例来说明。如图 3-12 所示，相互接触的气液两相在塔板上形成泡沫层，液体为连续相，气泡为分散相，相际界面处于非平衡状态，两相浓度偏离平衡浓度的程度就是传质的推动力。在推动力的作用下，非平衡状态向平衡状态移动，结果是由于 A、B 两组分挥发性的差异，气相中的难挥发组分 B 部分冷凝，转移到液相，并放出潜热，与此同时，液相中的易挥发组分 A 部分汽化，转移到气相，并吸收潜热。若 A 组分的摩尔汽化热与 B 组分的摩尔冷凝热相等，则气液两相间进行等摩尔反向扩散，以使气相中的 A 和液相中的 B 同时增浓。

（4）精馏操作的必要条件

回流是精馏操作能否顺利进行的必要条件，包括塔顶的液相回流与塔底的气相回流。回流能提供气液两相的正常流动，使两相在每块塔板上密切接触，进行传质和传热。精馏区别于蒸馏就在于回流，回流使精馏塔内气液两相充分接触，最终可达到高纯度的分离要求。

Typical distillation equipment

Equipment for continuous distillation is shown in Fig. 3. 1. Column C is fed continuously with the liquid mixture to be distilled, and the liquid in reboiler A is partially converted to vapor by heat transferred from the heating element B. The vapor stream from the still is brought into intimate countercurrent contact with a descending stream of boiling liquid in the column, or tower, C. This liquid must be rich enough in the low boiler that there is mass transfer of the low boiler from the liquid to the vapor at each stage of the column. Such a liquid can be obtained simply by condensing the overhead vapors and returning some of the liquid to the top of the column. This return liquid is called reflux. The use of reflux increases the purity of the overhead product, but not without some cost, since the vapor generated in the reboiler must provide both reflux and overhead product, and this energy cost is a large part of the total cost of separation by distillation.

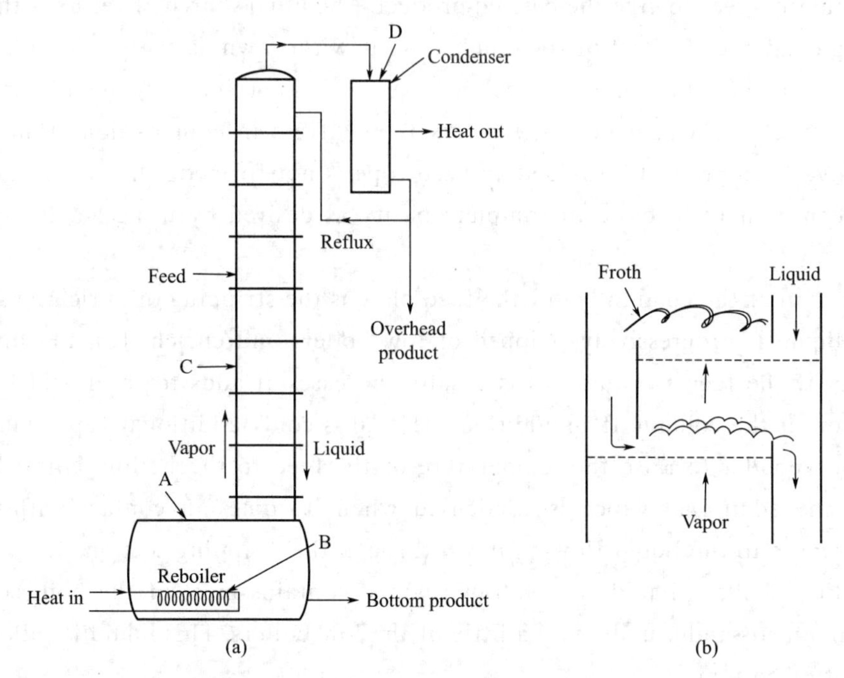

Fig. 3. 1 (a) Reboiler with fractionating column: A, reboiler; B, heating element; C, column; D, condenser. (b) Detail of sieve plate

The reflux entering the top of the column is often at the boiling point; but if it is cold, it is almost immediately heated to its boiling point by the vapor. Throughout the rest of the column, the vapor at any stage is at the same temperature as the liquid, which is at its boiling point. The temperature increases on going down the column because of the increase in pressure and the increasing concentration of high-boiling components.

The vapor is enriched at each stage because the vapor coming to a stage contains less low boilers than the vapor that would be in equilibrium with the liquid fed to that stage. If, as is usual, the overhead vapor is totally condensed, it has the same composition as the product and the reflux.

The reflux, however, has an equilibrium vapor composition which is richer than the vapor coming up to the top stage. This vapor is therefore enriched in low boilers at the expense of the reflux liquid. This partially depletes the reflux of low boilers, but if the flow rates have been set correctly, the liquid passing down to the second stage is still able to enrich the lower-quality vapor coming up to the second stage. At all stages in the column some low boilers diffuse from the liquid into the vapor, and a corresponding amount of high boilers diffuse from the vapor into the liquid. The heat of vaporization of the low boilers is supplied by the heat of condensation of the high boilers, and the total molar flow of vapor up the column is nearly constant.

The upper section of the column, above the feed plate, is known as the rectifying section. Here the vapor stream is enriched in low boilers, as it is brought into contact with the reflux. It is immaterial where the reflux originates, provided its concentration in low boilers is sufficiently great to give the desired product. The usual source of reflux is the condensate leaving condenser D. Part of the condensate is withdrawn as the product, and the remainder is returned to the top of the column. Reflux is sometimes provided by partial condensation of the overhead vapor; the reflux then differs in composition from the vapor leaving as overhead product. Provided an azeotrope is not formed, the vapor reaching the condenser can brought as close to complete purity as desired by using a tall tower and a large reflux.

The section of the column below the feed plate is the stripping or enriching section, in which the liquid is progressively stripped of low boilers and enriched in the high boiling components. If the feed is liquid, as is usually the case, it adds to the liquid flow in the lower section of the column. If in addition the feed is cold, additional vapor must be provided by the reboiler to raise the temperature of the feed to the boiling point. To accomplish this, the additional vapor is condensed when it comes in contact with the feed, adding still more to the liquid flowing down through the stripping section.

From the reboiler, liquid is withdrawn which contains most of the high components boiling components and usually only a little of the low boilers. This liquid is called the bottom product or bottoms.

The column shown in Fig. 3.1(a) contains a number of plates, or trays, stacked one above the other. Often these plates are perforated and are known as sieve plates, details of which are shown in Fig. 3.1(b). They consist of horizontal trays carrying a number of holes and a vertical plate which acts as a down-comer and a segmental weir. The holes contain valves or plugs which are lifted as vapor passes through them. The down-comer from a given plate reaches nearly to the tray below. Liquid flows over the weirs from plate to plate down the column, passing across the plates where the rising vapor causes it to froth. The vapor space above the froth contains a mist of fine droplets formed by the collapsing bubbles. Most of the drops fall back into the liquid, but some are entrained by the vapor and are carried to the plate above.

3.2.4 精馏物料衡算与操作线方程（Material-balance and operating lines）

3.2.4.1 理想传质型塔板——理论塔板（ideal stage or ideal plate）

对塔板上气、液两相传质过程进行完整的数学描述，需要物料衡算、热量衡算、传质速率方程和传热速率方程。由于传质和传热速率方程不仅与物性和操作条件有关，而且与塔板结构有关，所以难于用简单方法描述。为了避免上述困难，引入理论板的概念。

如图3-13所示，所谓理论塔板（理论板），是指一块气、液两相能够充分混合，最终两相间传质、传热阻力为零的理想化塔板。或者说，不论进入该塔板的气液两相组成如何，当两相在该板上经充分接触后，离开塔板的气液两相达相平衡状态，则这样的塔板称为理论板。一块理论塔板即为一个理论级或平衡级。

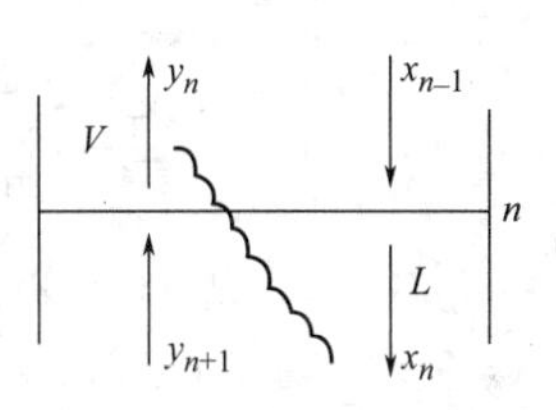

图3-13 理论板示意图

理论板概念的引入，将严格的相平衡及传质、传热理论与复杂的实际精馏操作联系起来，可将复杂的精馏问题分解为两个子问题，分别求精馏塔的理论板数和效率。前者仅取决于分离任务、相平衡关系和两相流率；而后者取决于物性、两相接触情况、操作条件及塔板结构等许多复杂因数。

3.2.4.2 恒摩尔流假定（constant molal overflow assume）

一般情况下，若被分离的A、B两种组分的摩尔汽化潜热相近，且能够忽略显热变化和热损失的影响，则精馏段的气、液摩尔流率分别保持不变，提馏段的气、液摩尔流率也分别保持不变，称为恒摩尔流假定。在少数情况下，若被分离组分的摩尔汽化热相差不大，而单位质量汽化热相近，则取精、提馏段内气液相质量流率分别相等，称为恒质量流假定。但由于加料的缘故，精馏段和提馏段间的气、液流率不一定分别相等。

Constant molal overflow

For most distillations, the molar flow rates of vapor and liquid are nearly constant in each section of the column, and the operating lines are almost straight. This results from nearly equal molar heats of vaporization, so that each mole of high boiler that condenses as the vapor moves up the column provides energy to vaporize about 1 mol of low boiler. For example, the molar heats of vaporization of toluene and benzene are 7960 and 7360 cal/mol, respectively, so that 0.92 mol of toluene corresponds to 1.0 mol of benzene. The changes in enthalpy of the liquid and vapor streams and heat losses from the column often require slightly more vapor to be formed at the bottom, so the molar ratio of vapor flow at the bottom of a column section to that at the top is even closer to 1.0. In designing columns or interpreting plant performance the concept of constant molar overflow is often used. In this simplified model the material-balance equations are linear and the operating lines straight.

3.2.4.3 物料衡算和操作线方程

（1）全塔物料衡算

如图 3-14 所示，连续稳态操作的精馏塔，对全塔物料衡算，可得：

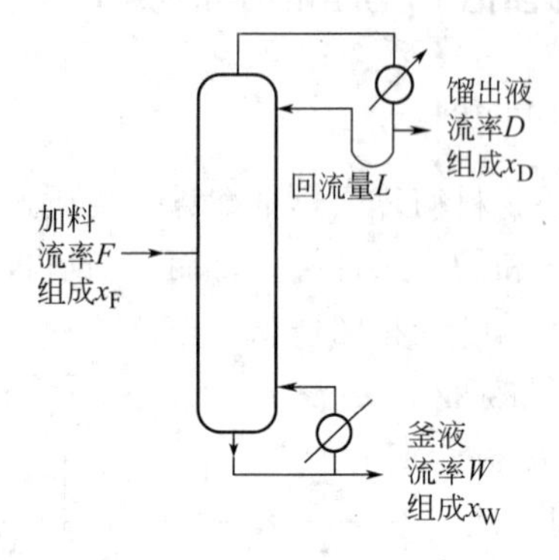

图 3-14 精馏塔的物料衡算示意图

$$F=D+W \tag{3-39}$$

$$Fx_F=Dx_D+Wx_W \tag{3-39a}$$

式中 F——原料液的流量，kmol/h；

D——塔顶产品（馏出液）流量，kmol/h；

W——塔底产品（釜残液）流量，kmol/h；

x_F——原料液中易挥发组分的摩尔分数；

x_D——馏出液中易挥发组分的摩尔分数；

x_W——釜残液中易挥发组分的摩尔分数。

全塔物料衡算式表示了进、出精馏塔的物料流率与组成的平衡关系。

在精馏计算中，分离程度除用轻、重两产品的分离纯度（常以摩尔分数表示）外，有时还用回收率表示，即塔顶 A 的回收率为：

$$\eta_A=\frac{Dx_D}{Fx_F}\times100\% \tag{3-40}$$

塔底 B 的回收率为：

$$\eta_B=\frac{W(1-x_W)}{F(1-x_F)}\times100\% \tag{3-41}$$

【例 3-2】 某常压连续精馏装置，分离苯-甲苯溶液，料液含苯 41%（质量分数，下同），要求塔顶产品中含苯不低于 97.5%，塔底产品中含甲苯不低于 98.2%。每小时处理原料量 8570kg。试求塔顶、塔底产品量（kmol/h）（苯的分子量为 78，甲苯的分子量为 92）。

解 将质量分数换算成摩尔分数：

$$x_D=\frac{97.5/78}{97.5/78+2.5/92}=0.98$$

$$x_W=\frac{(100-98.2)/78}{(100-98.2)/78+98.2/92}=0.02$$

则：

$$M_D=0.98\times78+0.02\times92=78.3(\text{kg/kmol})$$

$$M_W=0.02\times78+0.98\times92=91.7(\text{kg/kmol})$$

由全塔物料衡算：

$$D=F\left(\frac{x_F-x_W}{x_D-x_W}\right)=8570\times\frac{41-1.8}{97.5-1.8}=3510(\text{kg/h})=\frac{3510}{78.3}=44.83(\text{kmol/h})$$

$$W=8570-3510=5260(\text{kg/h})=\frac{5260}{91.7}=57.36(\text{kmol/h})$$

(2) 操作线方程（operating line equations）

对于板式精馏塔，基于上述理论板的概念，若已知离开任意理论板的一相组成，即可利用相平衡关系求得同时离开该理论板的另一相组成。为了进一步确定整个塔内气液两相组成的分布情况，还应设法找出任意相邻的两塔板之间下降的液相组成 x_n 与上升的气相组成 y_{n+1} 之间的关系。y_{n+1} 与 x_n 称为操作关系，其数学描述称为操作线方程。

操作线方程可利用物料衡算导出。由于进料的影响，将使精馏段和提馏段的物流情况有所不同，所以，操作线方程应分段建立。

①精馏段操作线方程（operating line equation of rectifying section） 如图 3-15 所示，在精馏段任意两板之间与塔顶间作物料衡算：

$$V=L+D$$

$$Vy_{n-1}=Lx_n+Dx_D$$

上两式合并得：

$$y_{n+1}=\frac{L}{V}x_n+\frac{D}{V}x_D \tag{3-42}$$

或：

$$y_{n+1}=\frac{R}{R+1}x_n+\frac{1}{R+1}x_D \tag{3-42a}$$

式中，R 称为回流比，是精馏塔重要的操作参数，$R=L/D$。后面还要对其进行详细讨论。

式(3-42a) 即为精馏段操作线方程，稳态操作时，R、x_D 均为常数，则该方程为直线。

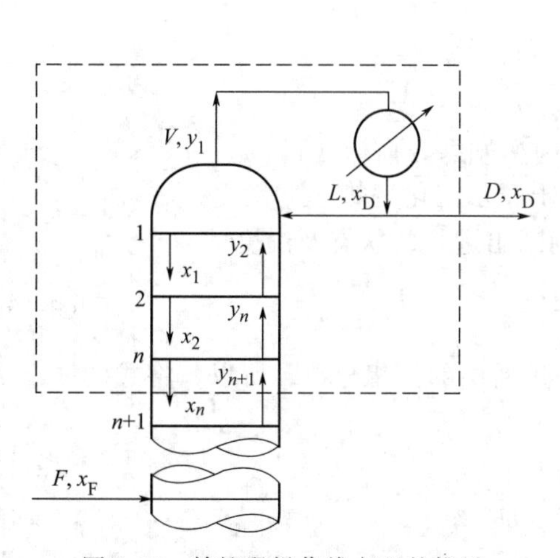

图 3-15 精馏段操作线方程的推导

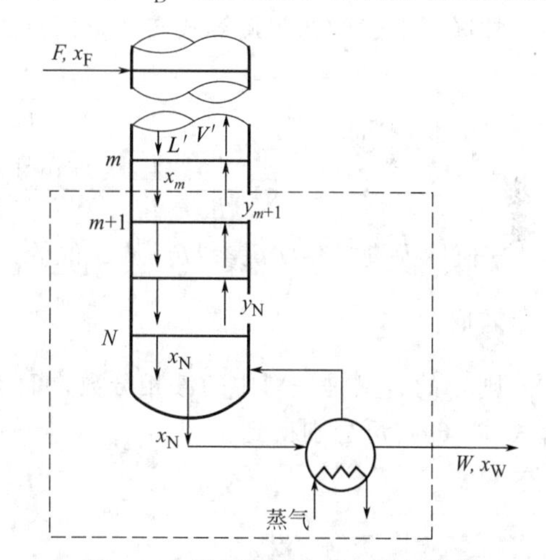

图 3-16 提馏段操作线方程的推导

② 提馏段操作线方程（operating line equation of stripping section） 如图 3-16 所示，在提馏段任意两板之间与塔底间作物料衡算：

$$L'=V'+W$$

$$L'x_m=V'y_{m+1}+Wx_W$$

上两式联立，整理得：

$$y_{m+1}=\frac{L'}{L'-W}x_m-\frac{W}{L'-W}x_W \tag{3-43}$$

式(3-43) 称为提馏段操作线方程，稳态操作时，L'，W，x_W 均为定值，则该方程也为直线。

③ 进料方程（q 线方程）(feed line)　进料板是精馏段和提馏段的连接处，由于有物料自塔外引入，所以其物料、热量关系与普通板不同，必须加以单独讨论。

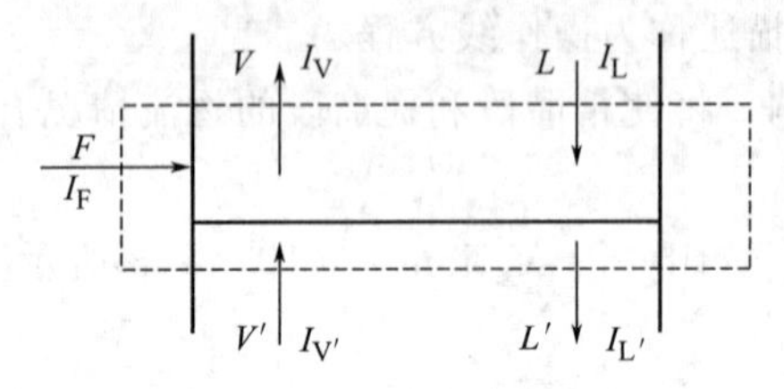

图 3-17　进料板上的物料衡算和热量衡算

a. 进料热状况（condition of feed）　组成一定的物料，可以五种不同的热状况加入塔内，即冷液体进料（料液温度低于泡点）、饱和液体进料（泡点温度进料）、气液混合进料（料液温度介于泡点和露点之间）、温度为露点的饱和蒸气进料、过热蒸气进料。

b. 理论进料板　进料板上的情况非常复杂，但仍可引入理论板的概念，即不管进入加料板的各物流情况如何，假定同时离开加料板的气液两相达平衡状态。

c. 精馏段与提馏段气液流率的关系　如图 3-17 所示，对进料板分别作物料和热量衡算，则：

物料衡算式：
$$F+V'+L=V+L' \tag{3-44}$$

热量衡算式：
$$FI_F+V'I_{V'}+LI_L=VI_V+L'I_{L'} \tag{3-45}$$

式中　I_F——原料液的焓，kJ/kmol；

I_V，$I_{V'}$——加料板上、下方饱和蒸气的焓，kJ/kmol；

I_L，$I_{L'}$——加料板上、下方饱和液体的焓，kJ/kmol。

可近似取：$I_V \approx I_{V'}$，$I_L \approx I_{L'}$

故联立上述的物料衡算式与热量衡算式得：

$$\frac{L'-L}{F}=\frac{I_V-I_F}{I_V-I_L}$$

令：
$$q=\frac{I_V-I_F}{I_V-I_L}=\frac{1\text{kmol 原料变成饱和蒸气所需热量}}{\text{原料液的千摩尔汽化潜热}} \tag{3-46}$$

q 值称为进料热状况参数，从 q 值的大小可判断出进料的状态及温度。

若取：
$$q=\frac{L'-L}{F} \tag{3-46a}$$

则 q 的含义是进料中的液相分数，而 $1-q$ 是进料中的气相分数。所以，将式(3-44) 结合式(3-46a) 可分别得到：

$$L'=L+qF \tag{3-47}$$

$$V=V'+(1-q)F \tag{3-48}$$

【例 3-3】　分离苯-甲苯溶液的常压精馏塔，料液处理量为 175kmol/h，料液含苯 44%（摩尔分数，下同），釜残液含苯小于 2%，馏出液含苯 93.5%。回流比为 2.12，泡点进料，试写出精馏段、提馏段操作线方程，并指出斜率、截距的数值。

解　全塔物料衡算：

$$F=D+W$$

$$Fx_F=Dx_D+Wx_W$$

可求出：$D=80\text{kmol/h}$，$W=95\text{kmol/h}$

则精馏段操作线方程：

$$y=\frac{R}{R+1}x+\frac{x_D}{R+1}=\frac{2.12}{2.12+1}x+\frac{0.935}{2.12+1}=0.68x+0.299$$

操作线的斜率为0.68，在y轴上的截距为0.299。

因泡点进料，即$q=1$，则精馏段的回流量

$$L=RD=2.12\times80=169.6(\text{kmol/h})$$

提馏段的回流量

$$L'=L+F=169.6+175=344.6(\text{kmol/h})$$

故提馏段操作线方程：

$$y'=\frac{L'}{L'-W}x'-\frac{W}{L'-W}x_W=\frac{344.6}{344.6-95}x'-\frac{95}{344.6-95}\times0.02=1.38x'-0.0076$$

操作线的斜率为1.38，在y轴上的截距为-0.0076。

d. q线方程的导出　由上述的分析可知，进料板的物料衡算应同时满足精馏段和提馏段两个方程，即：

精馏段物料衡算式：　$Vy=Lx+Dx_D$

提馏段物料衡算式：　$V'y=L'x-Wx_W$

联立上两式并整理得到：　$$y=\frac{q}{q-1}x-\frac{1}{q-1}x_F \tag{3-49}$$

稳态操作时，q、x_F均为定值，因此，进料板上相互接触的气液两相组成y-x的关系也是直线。该直线是精馏段操作线与提馏段操作线交点的轨迹方程，称为进料方程或q线方程。

e. 进料热状况的特点　不同进料热状况的特点可由式(3-46)分析。由于加料板上饱和蒸气的焓I_V总是大于饱和液体的焓I_L，所以该式中的分母是不为零的正数，则由分子I_V-I_F的大小可知，进料热状况影响精馏段，提馏段的气、液相流率及q线的位置，如表3-3所示，其对应的物流情况如图3-18所示。

表3-3　五种进料状况的特点对照表

进料热状况	进料焓I_F	q值	气、液相流率变化	q线斜率	q线位置
过冷液体	$I_F<I_L$	$q>1$	$L'>L+F, V<V'$	+	第一象限
饱和液体	$I_F=I_L$	$q=1$	$L'=L+F, V=V'$	∞	垂直x轴
气液混合物	$I_L<I_F<I_V$	$0<q<1$	$L'=L+qF$, $V=V'+(1-q)F$	−	第二象限
饱和蒸气	$I_F=I_V$	$q=0$	$L'=L, V=V'+F$	0	水平线
过热蒸气	$I_F>I_V$	$q<0$	$L'<L, V>V'+F$	+	第三象限

表3-3中的q值均可用式(3-46)计算，其中当计算冷液进料的热状况参数时，也可采用下式：

$$q=[r_m+C_{pm}(t_s-t)]/r_m \tag{3-50}$$

式中　t_s——进料液的泡点温度，℃；

t——进料液的实际温度，℃；

r_m——泡点下混合液的平均汽化潜热，kJ/kmol；

C_{pm}——定性温度$\left(\frac{t_s+t}{2}\right)$下混合液的平均定压比热容，kJ/(kg·℃)。

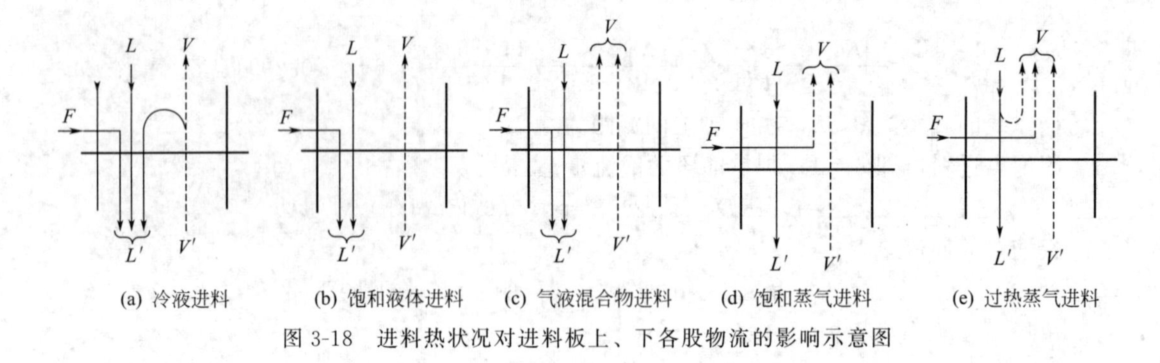

(a) 冷液进料　(b) 饱和液体进料　(c) 气液混合物进料　(d) 饱和蒸气进料　(e) 过热蒸气进料

图 3-18　进料热状况对进料板上、下各股物流的影响示意图

3.2.5　精馏热量衡算

通过精馏装置的热量衡算，可求得冷凝器和再沸器的热负荷以及冷却介质和加热介质的消耗量，并为设计这些换热设备提供基本数据。

（1）全凝器热量衡算

全凝器热量衡算见图 3-19，衡算式如下：

$$Q_c = VI_{VD} - (LI_{LD} + DI_{LD}) = (R+1)D(I_{VD} - I_{LD}) \tag{3-51}$$

式中　Q_c——全凝器热负荷，kJ/h；

I_{VD}——塔顶上升蒸气热焓值，kJ/kmol；

I_{LD}——塔顶流出全凝器的饱和液的热焓值，kJ/kmol。

冷却介质消耗量：

$$W_c = \frac{Q_c}{C_{pc}(t_2 - t_1)} \tag{3-52}$$

式中　W_c——冷却剂的耗用量，kg/h；

t_1，t_2——冷却介质的进、出口温度，℃；

C_{pc}——冷却介质的比热，kJ/(kg·℃)。

常用的冷却剂为冷却水，t_1 为当时当地自来水的温度，t_2 可由设计者给定，通常要求 $t_2 - t_1 = 10$℃左右。

（2）再沸器热量衡算

精馏的加热方式分为直接蒸汽加热与间接蒸汽加热两种方式。直接蒸汽加热时加热蒸汽的消耗量可通过精馏塔的物料衡算求得，而间接蒸汽加热时加热蒸汽消耗量可通过全塔或再沸器的热量衡算求得（图 3-20）。

$$Q_B = V'I_{V'W} + WI_{L'W} - L'I_{L'm} + Q_L$$

设：$I_{L'm} \approx I_{L'W}$，且因 $V' = L' - W$，则：

$$Q_B = V'(I_{V'W} - I_{L'W}) + Q_L \tag{3-53}$$

则加热介质消耗量：

$$W_h = \frac{Q_B}{r} \tag{3-54}$$

式中　Q_B——再沸器的热负荷，kJ/h；

Q_L——再沸器热损失，kJ/h；

$I_{V'W}$——再沸器中上升蒸气焓，kJ/kmol；

$I_{L'W}$——釜残液焓，kJ/kmol；

r——加热蒸汽的汽化潜热，kJ/kmol。

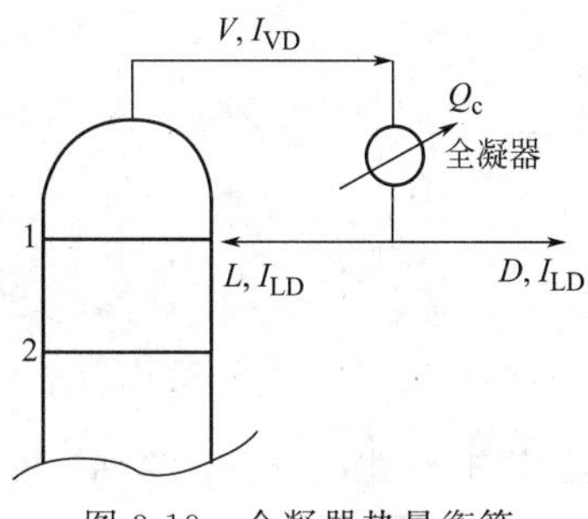

图3-19　全凝器热量衡算

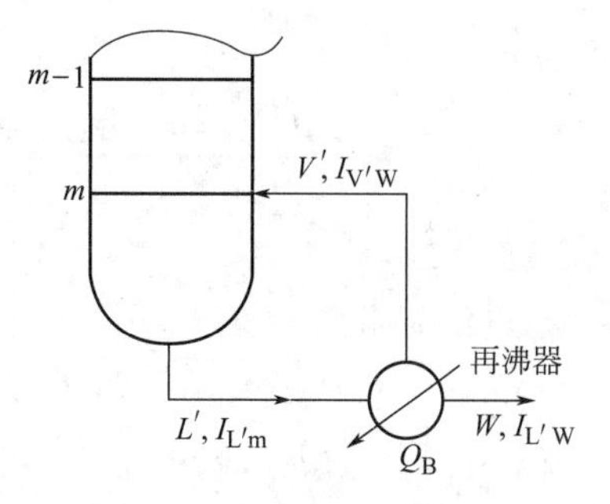

图3-20　再沸器热量衡算

【例3-4】 用一常压连续精馏塔分离含苯0.4的苯-甲苯溶液，进料流量为15000kg/h，进料温度为25℃，回流比为3.5，得到馏出液与釜残液组成分别为0.97和0.02（均为质量分数）。已知再沸器加热蒸气压力为137kPa（表压），塔顶回流液为饱和液体，塔的热损失可以不计，求：（1）再沸器的热负荷及加热蒸汽消耗量；（2）冷却水进、出冷凝器的温度分别为27℃和37℃时，冷凝器的热负荷及冷却水用量。

解　首先将质量分数换算成摩尔分数，得到：

$$x_F=\frac{0.4/78}{0.4/78+0.6/92}=0.44$$

$$x_D=\frac{0.97/78}{0.97/78+0.03/92}=0.975$$

$$x_W=\frac{0.02/78}{0.02/78+0.98/92}=0.0235$$

$$F=\frac{15000\times0.4}{78}+\frac{15000\times0.6}{92}=174.75(\text{kmol/h})$$

根据物料衡算：

$$F=D+W$$

$$Fx_F=Dx_D+Wx_W$$

可以求得：　$D=76.57\text{kmol/h}$，$W=98.15\text{kmol/h}$

（1）为了求出再沸器的加热功率，需要求出提馏段上升蒸气量V'：

$$V'=V-(1-q)F=(R+1)D+(q-1)F$$

于是需要求出q，进料的汽化潜热近似为：

$$r_m=(0.44\times93\times78+0.56\times86\times92)\times4.187=31900(\text{kJ/kmol})$$

当$x_F=0.44$时，溶液的泡点为93℃。计算显热时使用平均温度为$\frac{93+25}{2}=59(℃)$此温度下的比热容［查59℃下的苯和甲苯的质量定压比热容均近似为1.84kJ/(kg·℃)］，故原料液的摩尔定压比热为：

$$C_{p,m}=1.84\times0.44\times78+1.84\times0.56\times92=158[\text{kJ/(kmol·℃)}]$$

则可算出：

$$q=\frac{C_{p,\mathrm{m}}\Delta t+r_{\mathrm{m}}}{r_{\mathrm{m}}}=\frac{158\times(93-25)+31900}{31900}=1.337$$

于是得：　$V'=404\mathrm{kmol/h}$

再沸器热负荷为：

$$Q_{\mathrm{R}}=V'\Delta H_{甲苯}=404\times3.31\times10^{4}(按纯甲苯计算)=1.33\times10^{7}(\mathrm{kJ/h})$$

查得137kPa（表压）或238kPa（绝压）的加热蒸汽的汽化潜热为2240kJ/kg，于是：

$$蒸气用量=\frac{1.33\times10^{7}}{2240}=5955(\mathrm{kg/h})$$

(2) 冷凝器热负荷为：

$$Q_{\mathrm{c}}=(R+1)D\Delta H_{苯}=4.5\times6000\times393.9(近似按纯苯计算)=1.06\times10^{7}(\mathrm{kJ/h})$$

冷却水用量：

$$W_{\mathrm{c}}=\frac{Q_{\mathrm{c}}}{C_{p}\Delta t}=\frac{1.06\times10^{7}}{4.18\times10}=2.54\times10^{5}(\mathrm{kg/h})$$

因精馏过程中，进入再沸器的95%热量需要在塔顶冷凝器中取出，所以精馏过程是能量消耗很大的单元操作之一，其消耗通常占整个单元总能耗的40%～50%左右，如何降低精馏过程的能耗，是一个重要课题。由精馏过程的热力学分析知，减少有效能损失，是精馏过程节能的基本途径，其具体做法是：

① 热节减型　因回流必然消耗大量能量，所以选择经济合理的回流比是精馏过程节能的首要因素。一些新型板式塔和高效填料塔的应用，有可能使回流比大为降低；或减小再沸器与冷凝器的温度差，例如，采用压降低的塔设备，可减少向再沸器提供的热量，从而可提高有效能效率；如果塔底和塔顶的温度差较大，则可在精馏段中间设置冷凝器，在提馏段中间设置再沸器，可有效利用温位合理、价格低廉的换热介质，从而降低精馏的操作费用。

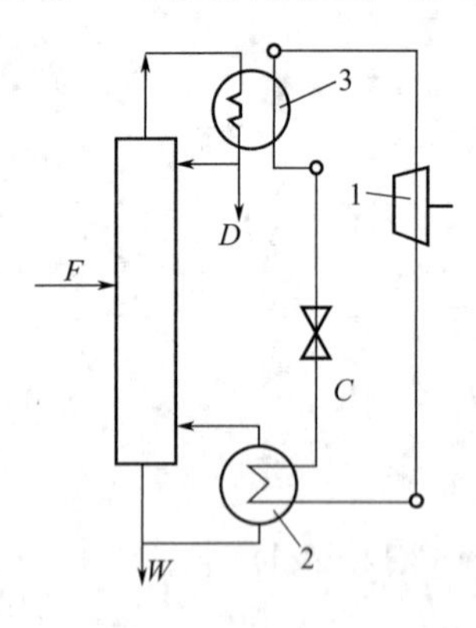

图 3-21　闭式循环热泵精馏

1—压缩机；2—再沸器；3—冷凝器

也可采用精馏节能新技术，如图3-21所示的热泵精馏流程，将塔顶蒸气绝热压缩后升温，重新作为再沸器的热源，把再沸器中的液体部分汽化；而压缩气体本身冷凝成液体，经节流阀后一部分作为塔顶产品抽出，另一部分作为塔顶回流液。这样，除开工阶段以外，可基本上不向再沸器提供另外的热源，节能效果十分显著。应用此法虽然要增加热泵系统的设备费，但可用节能省下的费用很快收回增加的投资。

还可采用多效精馏，其原理如多效蒸发，即采用压力依次降低的若干个精馏塔串联流程，前一精馏塔塔顶蒸气用作后一精馏塔再沸器的加热介质。这样，除两端精馏塔外，中间精馏装置可不必从外界引入加热剂和冷却剂。

② 热能的综合利用（热回收型）　回收精馏装置的余热，作为本系统或其他装置的加热热源，也是精馏操作节能的有效途径。其中包括用塔顶蒸气的潜热直接预热原料或将其用作其他热源；回收馏出液和釜残液的显热用作其他热源等。

对精馏装置进行优化控制，使其在最佳工况下运行，减少操作裕度，确保过程的能耗为最低。多组分精馏中，合理选择流程，也可达到降低能耗的目的。

3.2.6 精馏设计计算

二元连续精馏的工程计算主要涉及两种类型：一类是设计型，主要是根据分离任务的要求确定设备的主要工艺尺寸；另一类是操作型，而操作型主要是在已知设备条件下确定操作时的工况。对于板式塔而言，前者是根据规定的分离要求，选择适应的操作条件，计算所需理论板数，进而求出实际板数；而后者主要是针对已有的设备情况，由已知的操作条件预计分离结果。

设计型命题是本节的重点，连续精馏塔设计型计算的基本步骤是在规定分离要求后（例如 D、x_D 及回收率等)，确定操作条件（选定操作压力、进料热状况参数 q 及回流比 R 等)，再利用相平衡方程和操作线方程计算所需的理论板数。计算理论板数有三种方法：逐板计算法、图解法及简捷法。

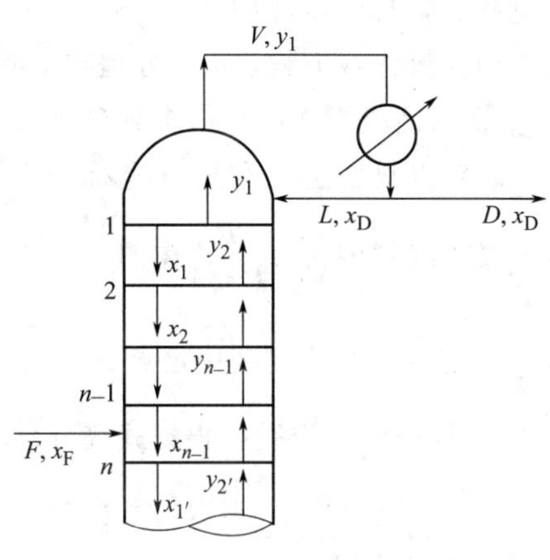

图 3-22 逐板计算法示意图

3.2.6.1 逐板计算法

如图 3-22 所示，假设，塔顶为全凝器，泡点液体回流；塔底为再沸器，间接蒸汽加热；R、q、α 已知，泡点进料。

从塔顶最上一层塔板（序号为 1）上升的蒸气经全凝器全部冷凝成饱和温度下的液体，因此馏出液和回流液的组成均为 y_1，即 $y_1=x_D$。

根据理论板的概念，自第一层板下降的液相组成 x_1 与 y_1 互成平衡，由平衡方程得 $x_1=\dfrac{y_1}{y_1+\alpha(1-y_1)}$。

从第二层塔板上升的蒸气组成 y_2 与 x_1 符合操作关系，故可用精馏段操作线方程由 x_1 求得 y_2，即 $y_2=\dfrac{R}{R+1}x_1+\dfrac{x_D}{R+1}$，按以上方法交替进行，计算步骤如下所示：

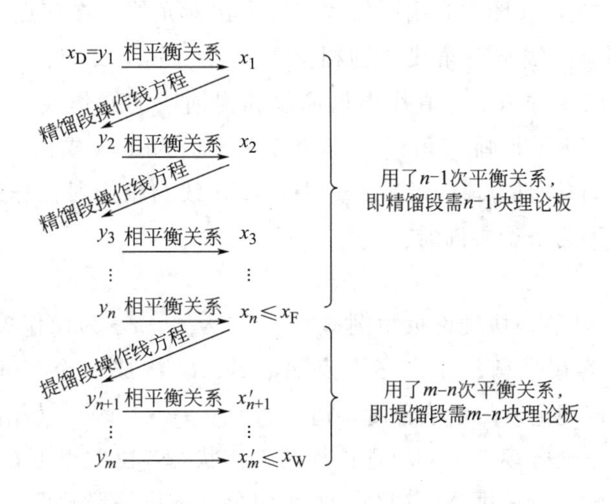

因在计算过程中，每使用一次相平衡关系，即表示需要一块理论板，所以经上述计算得到全塔总理论板数为 $m-1$ 块（塔底再沸器部分汽化，气液两相达平衡状态，起到一定的分离作用，相当于一块理论板），其中精馏段为 $n-1$ 块，提馏段为 $(m-1)-(n-1)=m-n$ 块，进料板位于第 n 板上。

该法计算准确，但手算过程烦琐，当理论板数较多时可用计算机完成。

3.2.6.2 图解法

用图解法求理论板数与逐板计算法原理相同，只是用图线代替方程，以图形的形式求取理论板数，具体步骤如下。

（1）操作线方程、q 线方程在 x-y 相图上的标绘

① 精馏段操作线　根据式(3-42a) 作出精馏段操作线直线图：

a. 定点斜率法　利用点坐标 $x_n=x_D$，$y_{n+1}=x_D$ ［点（x_D，x_D）即为对角线上的一点］及直线斜率 $\frac{L}{V}=\frac{R}{R+1}$ 作图。

b. 定点截距法　利用定点 $x_n=x_D$，$y_{n+1}=x_D$，直线截距为 $\frac{1}{R+1}x_D$。

② 提馏段操作线　根据式(3-43) 作出提馏段操作线直线图：

a. 定点斜率法　利用点坐标 $x_m=x_W$，$y_{m+1}=x_W$ ［点（x_W，x_W）即为对角线上的一点］，直线斜率 L'/V'。

b. 定点截距法　点坐标 $x_n=x_W$，$y_{n+1}=x_W$，直线截距为 $-\frac{W}{V'}x_W$。

c. q 线法　先作出精馏段操作线，并确定提馏段操作线的定点（x_W，x_W），而另一定点（x_e，x_e），由 q 线与精馏段操作线的交点确定，连接两定点（x_W，x_W）与（x_e，x_e），即得提馏段操作线。

③ q 线　q 线由式(3-49) 作图，定点（x_F，x_F）［点（x_F，x_F）即为对角线上的一点］，斜率 $q/(q-1)$。

（2）M-T 图解法基本步骤

如图 3-23 所示，在 x-y 图上画出平衡线，并画出对角线；在三点（$x=x_D$、$x=x_W$ 及 $x=x_F$ 点）处分别引垂直线与对角线分别相交于 a 点（x_D，x_D）、b 点（x_W，x_W）、c 点（x_F，x_F）三点；由已知的 R、q 值作出精馏段和提馏段的操作线；在平衡线与操作线之间，从 a 点开始画梯级（即画三角形）直至 $x \leqslant x_W$ 为止（或由 b 点反向画梯级直至 $x \leqslant x_D$），所画梯级的个数即为理论板数。其中，过 q 线与精馏段操作线交点的三角形为加料板，最后一个三角形为塔釜再沸器。

（3）梯级的含义

如图 3-24 所示，以第 n 块理论板为例 y_n-x_{n-1}、y_{n+1}-x_n 为操作关系，落在操作线上，y_n-x_n 为平衡关系，落在平衡线上。三个点 A、B、C 构成一个三角形，其中边 BA 为 x_{n-1}-x_n 表示液相经该理论板的增浓程度，边 CB 为 y_n-y_{n+1} 表示气相经该理论板的增浓程度。所以，这个三角形充分表达了一块理论板的工作状态，由此也可看出在塔内不论气相还是液相都是自下而上轻组分浓度逐渐增高，而重组分的浓度逐渐减低。

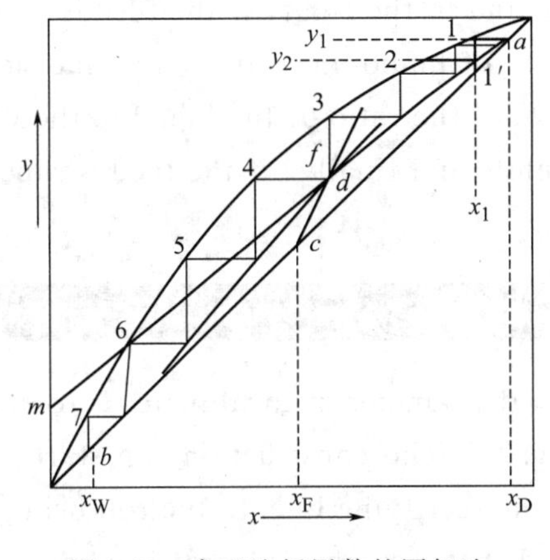

图 3-23　求理论板层数的图解法

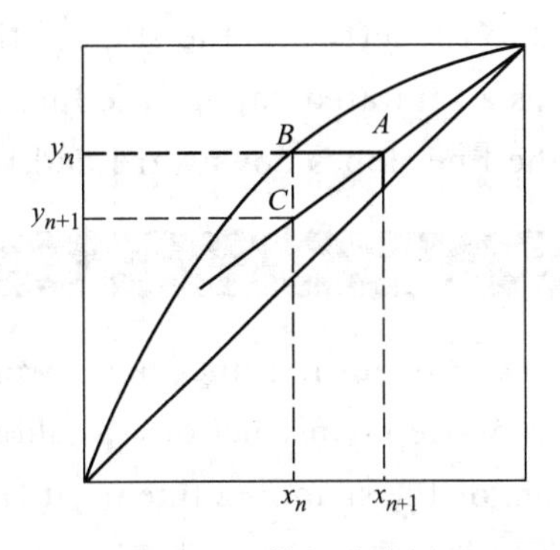

图 3-24　梯级示意图

Construction of operating lines

The simplest method of plotting the operating lines is to (1) locate the feed line; (2) calculate the y-axis intercept $x_D/(R+1)$ of the rectifying line and plot that line through the intercept and the point (x_D, x_D); (3) draw the stripping line through point (x_B, x_B) and the intersection of the rectifying line with the feed line. The operating lines in Fig. 3. 2 show the result of this procedure.

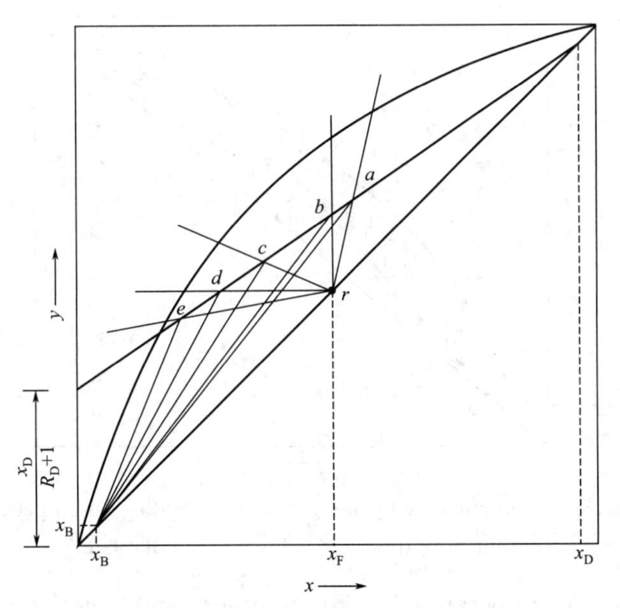

Fig. 3. 2　Effect of feed condition on feed line; *ra*, feed cold liquid; *rb*, feed saturated liquid; *rc*, feed partially vaporized; *rd*, feed saturated vapor; *re*, feed superheated vapor

In Fig. 3. 2 are plotted operating lines for various types of feed, on the assumption that x_F, x_B, x_D, L, and D are all constant. The corresponding feed lines are shown. If the feed is a cold liquid, the feed line slopes upward and to the right; if the feed is a saturated liquid, the line is vertical; if the feed is a mixture of liquid and vapor, the line slopes upward and to the left, and the slope is the negative of the ratio of the liquid to the vapor; if the feed is a saturated vapor, the line is horizontal; and finally, if the feed is superheated vapor, the line slopes downward and to the left.

Feed plate location

After the operating lines have been plotted, the number of ideal plates is found by the usual step-by-step construction, as shown in Fig. 3. 3. The construction can start either at the bottom of the stripping line or at the top of the rectifying line. In the following it is assumed that the construction begins at the top and that a total condenser is used. As the intersection of the operating lines is approached, it must be decided when the steps should transfer from the rectifying line to the stripping line. The change should be made in such a manner that the maximum enrichment per plate is obtained, so that the number of plates is as small as possible. Fig. 3. 3 shows that this criterion is met if the transfer is made immediately after a value of x is reached that is less than the x coordinate of the intersection of the two operating lines. The feed plate is always represented by the triangle that has one corner on the rectifying line and one corner on the stripping line. At the optimum position, the triangle representing the feed plate straddles the intersection of the operating lines.

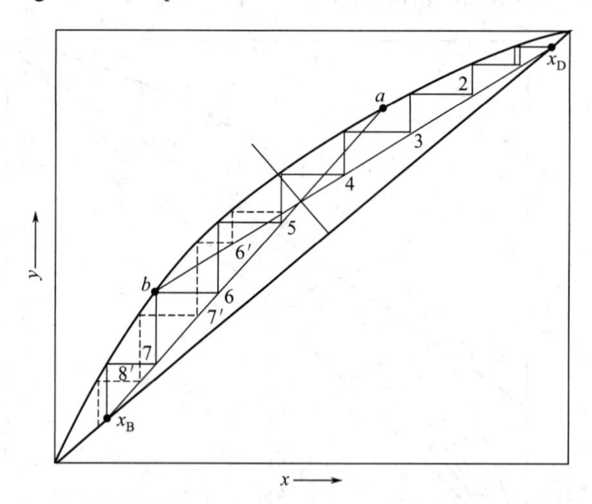

Fig. 3. 3 Optimum feed plate location: ---, with feed on plate 5 (optimum location); ——, with feed on plate 7

The transfer from one operating line to the other, and hence the feed plate location, can be made at any location between points a and b in Fig. 3. 3; but if the feed plate is placed anywhere but at the optimum point, an unnecessarily large number of plates are called for. For example, if the feed plate in Fig. 3. 3 is number 7, the smaller steps shown by the dashed lines make the number of ideal plates needed about 8 plus a reboiler, instead of 7 plus a reboiler when the feed is on plate number 5. Note that the liquid on the feed plate does not have the same composition as the feed except by coincidence, even when the feed plate location is optimum.

When we are analyzing the performance of a real column, the switch from one operating line to another must be made at a real feed plate. Because of changes in feed composition and uncertainties in plate efficiency, large columns are often operated with the feed entering a few plates above or below the optimum location. If large changes in feed composition are anticipated, alternate feed locations can be provided.

In an actual column with a fixed number of plates, feeding on the wrong plate may seriously affect the column performance. Feeding on too low a plate, for example, close to point b in Fig. 3. 3, increases the number of plates in the rectifying section; but many of them are now operating in a pinched region where the driving force is small. These plates do very little separation. The diagram, therefore, must change, lowering the quality of both the top and bottom products to reflect the poorer performance of the plates. Feeding too high in the column leads to similar consequences.

【例 3-5】 在一常压连续精馏塔内分离苯-甲苯混合物，已知进料流量为 80kmol/h，进料中苯含量为 0.40（摩尔分数，下同），泡点进料，塔顶馏出液含苯 0.90，要求塔顶苯的回收率不低于 90%。塔顶为全凝器，回流比取为 2。在操作条件下，物系的相对挥发度为 2.47，试分别用逐板计算法和图解法计算所需的理论板数。

解 (1) 根据苯的回收率计算塔顶产品的流量为：

$$D=\frac{\eta_A F x_F}{x_D}=\frac{0.9\times 80\times 0.4}{0.9}=32(\text{kmol/h})$$

由物料衡算计算塔底产品的流量及组成：

$$W=F-D=80-32=48(\text{kmol/h})$$

$$x_W=\frac{Fx_F-Dx_D}{W}=\frac{80\times 0.4-32\times 0.9}{48}=0.0667$$

已知回流比 $R=2$，所以精馏段操作线方程为：

$$y_{n+1}=\frac{R}{R+1}x_n+\frac{x_D}{R+1}=\frac{2}{2+1}x_n+\frac{0.9}{2+1}=0.667x_n+0.3 \tag{a}$$

下面求提馏段操作线方程：

提馏段上升蒸气量：

$$V'=V-(1-q)F=V=(R+1)D=(2+1)\times 32=96(\text{kmol/h})$$

下降液体量：

$$L'=L+qF=RD+qF=2\times 32+80=144(\text{kmol/h})$$

$$y_{m+1}=\frac{L'}{V'}x_m-\frac{Wx_W}{V'}=\frac{144}{96}x_m-\frac{48\times 0.0667}{96}=1.5x_m-0.033 \tag{b}$$

相平衡方程可写成：
$$x=\frac{y}{\alpha-(\alpha-1)y}=\frac{y}{2.47-1.47y} \tag{c}$$

利用操作线方程(a)、(b) 和相平衡方程(c)，可自上而下逐板计算所需要的理论板数。因塔顶为全凝器，则 $y_1=x_D=0.9$。

由式(c) 求得第一块板下降液体组成为：

$$x_1=\frac{y_1}{\alpha-(\alpha-1)y_1}=\frac{y_1}{2.47-1.47y_1}=\frac{0.9}{2.47-1.47\times 0.9}=0.785$$

利用精馏段操作线计算第二块板上升蒸气组成为：

$$y_2=0.667x_1+0.3=0.667\times 0.785+0.3=0.824$$

交替使用式(a) 和式(c) 直到 $x_n\leqslant x_F$，然后改用提馏段操作线方程，直到 $x_n\leqslant x_W$ 为止，计算结果见本题附表。

【例 3-5】附表　计算结果（各层塔板上的气液组成）

板号	1	2	3	4	5	6	7	8	9	10
y	0.9	0.824	0.737	0.652	0.587	0.515	0.419	0.306	0.194	0.101
x	0.785	0.655	0.528	0.431	$0.365<x_F$	0.301	0.226	0.151	0.089	$0.044<x_W$

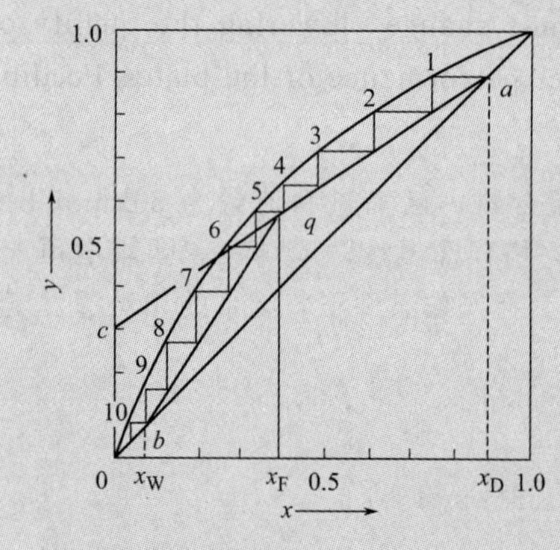

【例 3-5 附图】 图解法计算理论板数

精馏塔内理论塔板数为 10－1＝9 块，其中精馏段 4 块，第 5 块为进料板。

（2）图解法计算理论板数

在直角坐标系中绘出 x-y 相图，如本题附图所示。根据精馏段操作线方程式(a)，找到 $a(0.9,0.9)$，$c(0,0.3)$ 点，连接 ac 即得到精馏段操作线。因为泡点进料，$x_q=x_F$，由 $x=x_F$ 作垂线交精馏段操作线于 q 点，连接 $b(0.0667,0.0667)$ 点和 q 点即为提馏段操作线 bq。从 a 点开始在平衡线与操作线之间作梯级，直到 $x_n\leqslant x_W$ 为止。由附图可知，理论板数为 10 块，除去再沸器一块，塔内理论板数为 9 块，其中精馏段 4 块，第 5 块为进料板，与逐板计算法结果相一致。

3.2.6.3　几种特殊情况下 N 的计算

（1）塔顶设置分凝器

如图 3-25 所示，塔顶第一个冷凝器只是将回流液冷凝下来（即部分冷凝），叫分凝器，从分凝器出来的气相再由全凝器全部冷凝下来作为产品。

对于分凝器来说，同时离开分凝器的气液两相组成 y_0-x_0 成相平衡关系，起到一定的分离作用，相当于一块理论板。设有分凝器时，则计算理论板应从分凝器开始，精馏段操作线方程不变，只是算的理论板总数中应扣除 1 块（即为分凝器）即可。

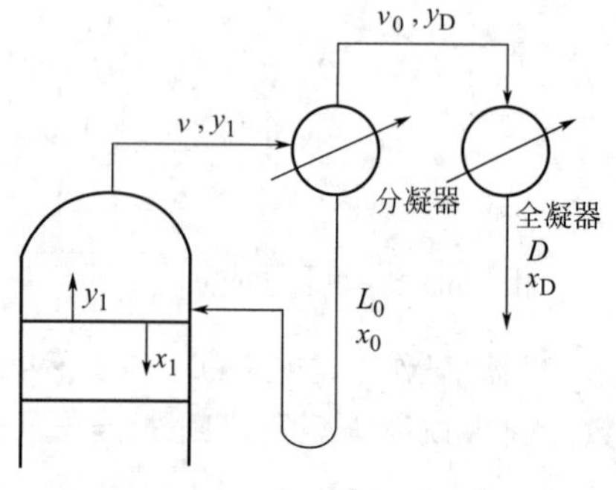

图 3-25　分凝器流程图

采用分凝器操作时，塔顶调节回流比 R 不太方便，所以生产上不常使用。

（2）塔底采用直接蒸汽加热

若对某轻组分 A 与水的混合物进行精馏分离时，由于塔底产物主要是水，可考虑在塔底直接通入水蒸气加热，以省去塔釜再沸器，其装置如图 3-26(a) 所示。

设蒸汽为饱和水蒸气，恒摩尔流假定成立，则 $V_0=V'$，此情况下，精馏段操作线、q 线与常规塔相同，但因提馏段塔底多了一股蒸汽流，所以其操作线发生变化。如图所示，由物料衡算可导出：

$$y=\frac{W}{V_0}x-\frac{W}{V_0}x_W \tag{3-55}$$

式中　V_0——直接蒸汽加热的流量，kmol/h。

当采用作图法求解时，$x=x_W$，$y=0$，即该点落在横坐标上，图解法如图 3-26(b) 所示。直接蒸汽加热与间接蒸汽加热相比，对于相同的分离要求，所需理论板数 N 稍多。这是因为直接蒸汽稀释了釜液，需增加 N 来回收 A。

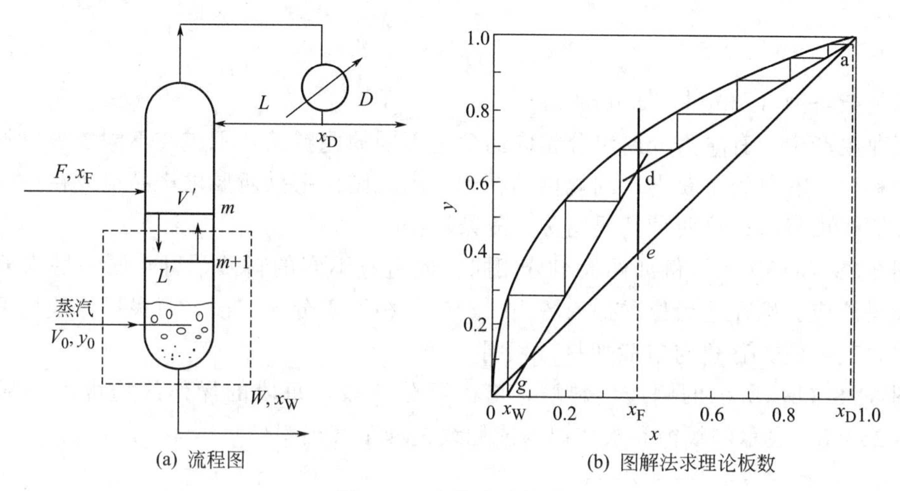

(a) 流程图　　(b) 图解法求理论板数

图 3-26　直接蒸汽加热

(3) 提馏塔

图 3-27(a) 为提馏塔装置简图。原料液从塔顶加入塔内，然后逐板下流提供塔内的液相，塔顶蒸气冷凝后全部作为馏出液产品，塔釜用间接蒸汽加热，只有提馏段而无精馏段。这种塔主要用于物系在低浓度下的相对挥发度较大，不要精馏段也可得到所希望的馏出液组成，或用于回收稀溶液中的轻组分而对馏出液组成要求不高的场合。

在设计型计算时，给定原料液流量 F、组成 x_F 及加料热状况参数 q，规定塔顶轻组分回收率 η_A 及釜残液组成 x_W，则馏出液组成 x_D 及其流量 D 由全塔物料衡算确定。此情况下的操作线方程与一般精馏塔的提馏段操作线方程相同，即：

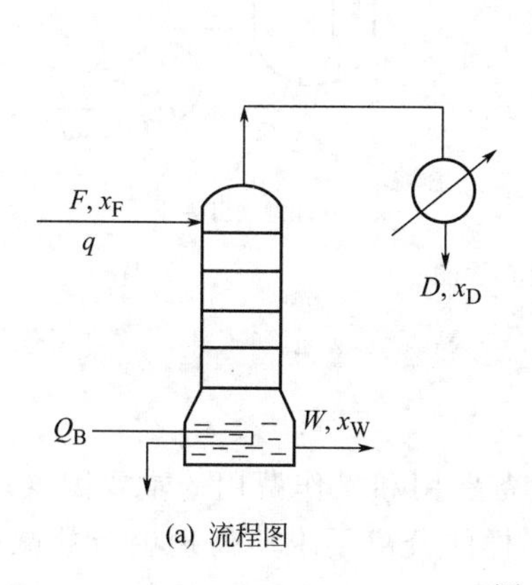

q　d　a　c　b　x_W　x_F　x_D

(a) 流程图　　(b) 图解法求理论板数

图 3-27　提馏塔

$$y_{m+1}=\frac{L'}{V'}x-\frac{W}{V'}x_W$$

$$L'=qF$$

$$V'=D+(q-1)F \quad 或 \quad V'=L'-W$$

此操作线的下端为 x-y 图的点 $b(x_W, x_W)$，上端由 q 线与 $y=x_D$ 的交点坐标 d 来确定，如图 3-27(b) 所示。然后在操作线与平衡线之间绘阶梯确定理论板数。

当泡点进料时，$L'=F$，$V'=D$，则操作线方程变为：

$$y_{m+1}=\frac{F}{D}x_m-\frac{W}{D}x_W \tag{3-56}$$

（4）复杂塔型（complex columns）

在工业生产中，有时为分离组分相同而含量不同的原料液，则应在不同塔板位置上设置相应的进料口；有时为了获得不同规格的精馏产品，则可根据所要求产品组成在塔的不同位置上开设侧线出料口，这两种情况均称为复杂塔型。

如图 3-28（a）所示，例如两股组分相同、但组成不同的料液，可在同一塔内分离。此时，两股料液应分别在适当位置加入塔内。这时将精馏塔分成三段，每段操作线均可由物料衡算推出，其 q 线方程仍与单股加料时相同。

如图 3-28（b）所示的侧线出料塔，塔内可分三段，每段的操作线均可由物料衡算导出。图 3-28（b）表示侧线饱和液体出料或侧线饱和蒸气出料。

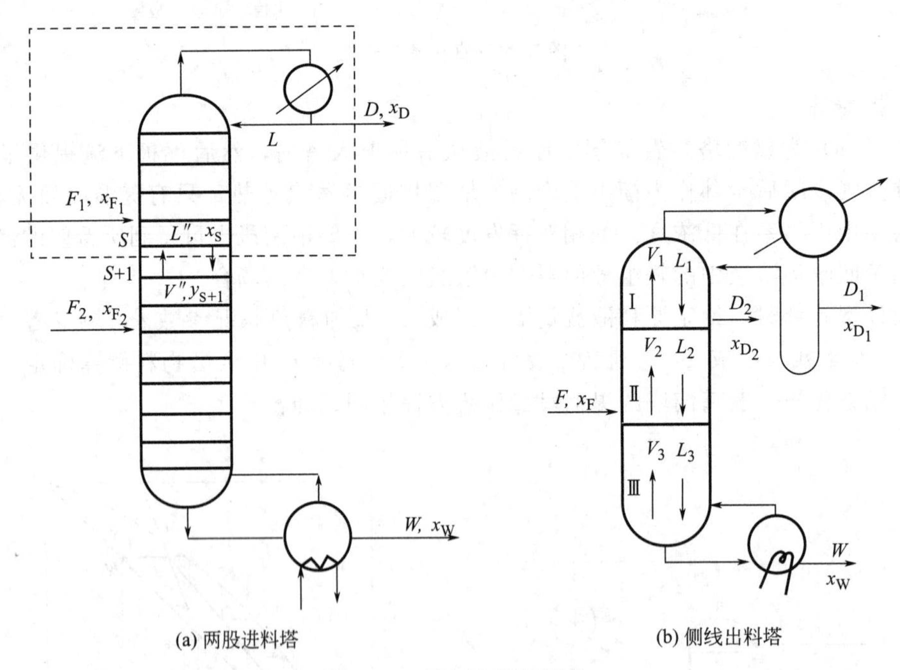

(a) 两股进料塔　　(b) 侧线出料塔

图 3-28　复杂塔型流程图

3.2.7 回流比的选择及影响

精馏的核心是回流，它的大小是影响精馏塔设备费用和操作费用的重要因素。当 R 增大时，操作线远离平衡线，梯级跨度增大，对于同样的分离要求所需理论板数减少，但同时，塔顶和塔底的热负荷增加，能耗高，即操作费用增多。因此回流比 R 是个重要的可调操作参数，要通过经济评价来讨论 R 的适宜范围。

(1) 最小回流比（minimum reflux ratio）

如图3-29所示，对于指定的分离要求，当 R 减小时，两条操作线向平衡线移动，传质推动力减小，所需 N 增多。当两条操作线的交点落在平衡线上时（图中 e 点），由于 e 点处气液两相已达到相平衡关系，即无增浓作用，所以称此区为恒浓区（挟紧区），点 e 称为挟紧点，此时，对应的回流比为最小回流比。

最小回流比的求取可用两种方法，作图法和解析法。

① 对于理想体系对应正常的平衡曲线　如图3-29所示，可由图中读得挟紧点 e 点坐标 (x_q, y_q)，则：

$$\frac{R_{min}}{R_{min}+1}=\frac{x_D-y_q}{x_D-x_q}$$

$$R_{min}=\frac{x_D-y_q}{y_q-x_q} \tag{3-57}$$

② 对于非理想体系对应的特殊平衡线　如图3-30所示，当操作线与平衡线相切，则切点 g 为挟紧点，对应的 R 为 R_{min}。d 点是精、提馏段交点，但没落在平衡线上。则由图读得 d 点 (x_q, y_q) 或点 $g(x_e, y_e)$ 的坐标值，有：

$$R_{min}=\frac{x_D-y_q}{y_q-x_q}=\frac{x_D-y_e}{y_e-x_e} \tag{3-58}$$

式中，x_q 与 y_q 不是相平衡关系，但 x_e 与 y_e 为相平衡关系。

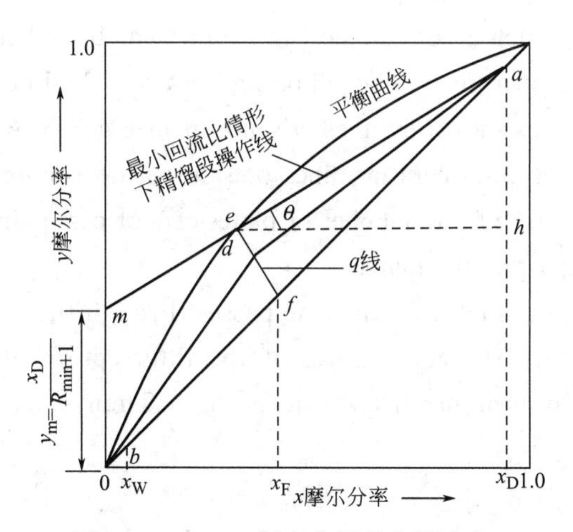

图3-29　在 x-y 图中分析最小回流比

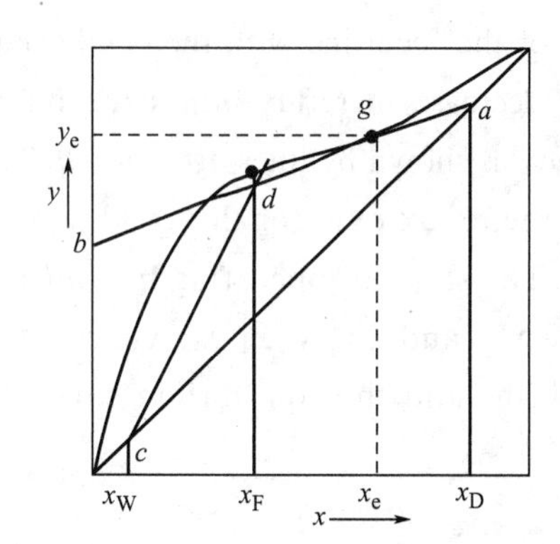

图3-30　非理想物系最小回流比确定

③ 解析法　对于理想物系，由式(3-38)和式(3-57)联立可得

$$R_{min}=\frac{1}{\alpha-1}\left[\frac{x_D}{x_q}-\frac{\alpha(1-x_D)}{1-x_q}\right] \tag{3-59}$$

应当注意，对于理想物系，若两操作线交点 d 落在相平衡线以外是理论所不允许的，但生产实际中仍可操作，这是技术上的问题。但此时，即使采用无穷多块塔板，无论怎样操作，也不可能达到指定的分离要求。

Minimum Reflux Ratio

At any reflux less than total, the number of plates needed for a given separation is larger than at total reflux and increases continuously as the reflux ratio is decreased. As the ratio becomes smaller, the number of plates becomes very large, and at a definite minimum, called the minimum reflux ratio, the number of plates becomes infinite. All actual columns producing a finite amount of desired top and bottom products must operate at a reflux ratio between the minimum, at which the number of plates is infinity, and infinity, at which the number of plates is a minimum. If L_a/D is the operating reflux ratio and $(L_a/D)_{min}$ is the minimum reflux ratio, then

$$\left(\frac{L_a}{D}\right)_{min} < \frac{L_a}{D} < \infty \qquad \text{(eq. 3-12)}$$

The minimum reflux ratio can be found by following the movement of the operating lines as the reflux is reduced. In Fig. 3. 4 both operating lines coincide with the diagonal *afb* at total reflux. For an actual operation lines *ae* and *eb* are typical operating lines. As the reflux is further reduced, the intersection of the operating lines moves along the feed line toward the equilibrium curve, the area on the diagram available for steps shrinks, and the number of steps increases. When either one or both of the operating lines touch the equilibrium curve, the number of steps necessary to cross the point of contact becomes infinite. The reflux ratio corresponding to this situation is, by definition, the minimum reflux ratio.

For the normal type of equilibrium curve, which is concave downward throughout its length, the point of contact, at minimum reflux, of the operating and equilibrium lines is at the intersection of the feed line with the equilibrium curve, as shown by lines *ad* and *db* in Fig. 3. 4. A further decrease in reflux brings the intersection of the operating lines outside of the equilibrium curve, as shown by lines *agc* and *cb*. Then even an infinite number of plates cannot pass point *g*, and the reflux ratio for this condition is less than the minimum.

The slope of operating line *ad* in Fig. 3. 4 is such that the line passes through the points (x', y') and (x_D, x_D), where x' and y' are the coordinates of the intersection of the feed line and the equilibrium curve. Let the minimum reflux ratio be R_m, Then:

$$\frac{R_m}{R_{m+1}} = \frac{x_D - y'}{x_D - x'} \qquad \text{(eq. 3-13)}$$

or

$$R_m = \frac{x_D - y'}{y' - x'}$$

Eq. (3-13) cannot be applied to all systems. Thus, if the equilibrium curve has a concavity upward, as for example, the curve for ethanol and water shown in Fig. 3. 5, it is clear that the rectifying line first touches the equilibrium curve between abscissas x_F and x_D and line *ac* corresponds to minimum reflux. Operating line *ab* is drawn for a reflux less than the minimum, even though it does intersect the feed line below point (x', y'), In such a situation the minimum reflux ratio must be computed from the slope of the operating line *ac* that is tangent to the equilibrium curve.

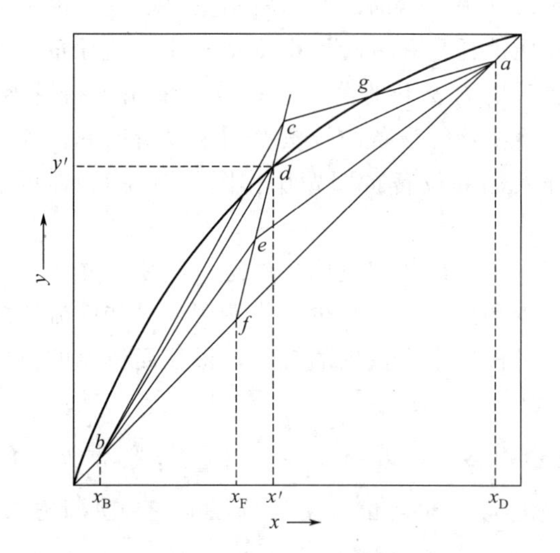

Fig. 3.4 Minimum reflux ratio

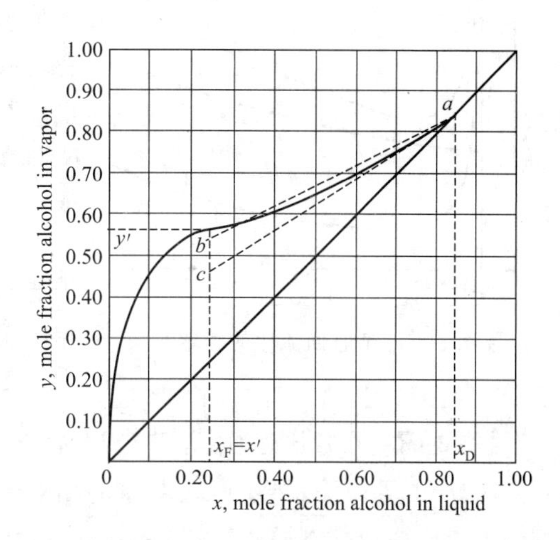

Fig. 3.5 Equilibrium diagram
(system ethanol-water)

(2) 全回流及最少理论板数（total reflux and minimum number of plates）

全回流时，反应在图上是精、提馏段操作线均与对角线重合，传质推动力最大，对于指定分离要求而言，所需理论板数最小。

全回流是回流的上限，生产能力等于零，对正常生产无意义。但开车阶段，先开全回流可尽快达到稳定操作；生产中，当精馏塔前后工序出了故障，可临时改为全回流操作；实验研究中，多采用全回流操作测定单板效率，这样可排除进料波动等不稳定因素的影响，便于不同结构塔型的比较；利用全回流还可以给出完成一定分离任务所需最少理论板数 N_{min} 的极限概念。以下介绍求解 N_{min} 的两种方法：

① 图解法 因精、提馏两条操作线均与对角线重合，所以可由点（x_D，x_D）或点（x_W，x_W）开始在平衡线与对角线之画梯级，直到跨过 x_W（或 x_D）为止，所得梯级数即为 N_{min} 的值。

② 解析法——芬斯克（Fenske）公式 依据操作线方程和理想系统的相平衡线方程可导出芬斯克公式：

$$N_{min}+1=\frac{\lg\left[\left(\frac{x_A}{x_B}\right)_D\left(\frac{x_B}{x_A}\right)_W\right]}{\lg\bar{\alpha}_m}=\frac{\lg\left[\left(\frac{x_D}{1-x_D}\right)\left(\frac{1-x_W}{x_W}\right)\right]}{\lg\bar{\alpha}_m} \tag{3-60}$$

式中，取$\bar{\alpha}_m=\sqrt{\alpha_{塔顶}\ \alpha_{塔底}}$为常数，$N_{min}$ 值不包括再沸器。

利用芬斯克公式可估计进料板位置，即：

$$N_{min,T}=\frac{\lg\left[\left(\frac{x_A}{x_B}\right)_D\left(\frac{x_B}{x_A}\right)_F\right]}{\lg\bar{\alpha}_1} \tag{3-60a}$$

式中 $N_{min,T}$——精馏段最少理论塔板数；$\bar{\alpha}_1=\sqrt{\alpha_{塔顶}\ \alpha_{进料}}$。

(3) 最佳回流比的确定（optimum reflux ratio）

最佳回流比是指总费用（即操作费用和设备费用之和）最低时的回流比，需通过经济衡

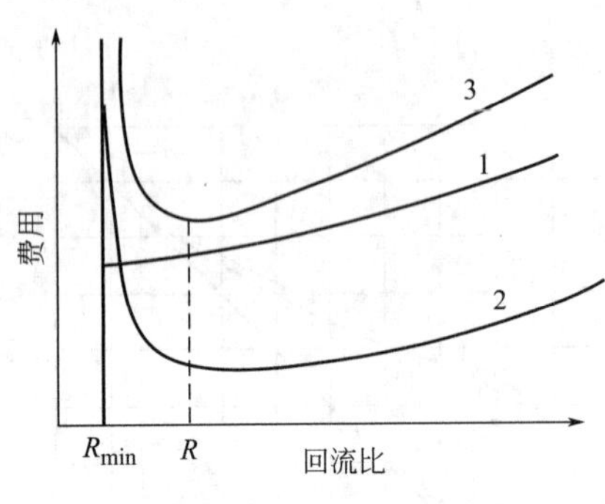

图 3-31 适宜回流比的确定

算来确定。其中，精馏的操作费用主要取决于再沸器中加热介质消耗量和冷凝器中冷却介质的消耗量，两者均取决于塔内上升的蒸气量；设备费用是指塔器、再沸器、冷凝器等设备的投资费用乘以折旧率。若将 R 对操作费用和设备折旧费用作图（图 3-31），可求最佳回流比。

长期以来，最佳回流比的确定一直令人注目。由于能源缺少和昂贵，为减少操作费用，最佳回流比应相应减少。根据生产数据的统计，最佳回流比的范围为：$R=(1.1\sim2)R_{min}$。

【例 3-6】 在常压连续精馏塔内分离某理想混合液，已知 $x_F=0.40$（摩尔分数，下同），$x_D=0.97$，$x_W=0.04$，物系的相对挥发度为 2.47。试分别计算以下三种进料方式下的最小回流比及全回流下的最少理论板数：(1) 冷液进料 $q=1.387$；(2) 泡点进料；(3) 饱和蒸气进料。

解 (1) 冷液进料 $q=1.387$ 时，则 q 线方程：

$$y=\frac{q}{q-1}x-\frac{x_F}{q-1}=\frac{1.387}{1.387-1}x-\frac{0.4}{1.387-1}=3.584x-1.034$$

相平衡方程：
$$y=\frac{\alpha x}{1+(\alpha-1)x}=\frac{2.47x}{1+1.47x}$$

联立以上两式得：
$$x_q=0.483,y_q=0.698$$

$$R_{min}=\frac{x_D-y_q}{y_q-x_q}=\frac{0.97-0.698}{0.698-0.483}=1.265$$

(2) 泡点进料，$q=1$，则 $x_q=x_F=0.4$。

$$y_q=\frac{\alpha x_q}{1+(\alpha-1)x_q}=\frac{2.47\times0.4}{1+1.47\times0.4}=0.622$$

$$R_{min}=\frac{x_D-y_q}{y_q-x_q}=\frac{0.97-0.622}{0.622-0.4}=1.568$$

(3) 饱和蒸气进料，$q=0$，则 $y_q=x_F=0.4$。

$$x_q=\frac{y_q}{\alpha-(\alpha-1)y_q}=\frac{0.4}{2.47-1.47\times0.4}=0.213$$

$$R_{min}=\frac{x_D-y_q}{y_q-x_q}=\frac{0.97-0.4}{0.4-0.213}=3.048$$

(4) 全回流下的最少理论板数：

$$N_{min}=\frac{\lg\left[\left(\frac{x_D}{1-x_D}\right)\left(\frac{1-x_W}{x_W}\right)\right]}{\lg\bar{\alpha}_m}-1=\frac{\lg\left[\left(\frac{0.97}{0.03}\right)\left(\frac{0.96}{0.04}\right)\right]}{\lg2.47}-1=6.36\text{(不包括再沸器)}$$

由以上计算可知，在分离要求一定的情况下，最小回流比与进料热状况有关，q 值越大，在满足同样分离要求条件下，最小回流比越小。

（4）简捷法求理论板数

如图3-32所示，吉利兰图对R_{min}、R、N_{min}和N四个变量进行了关联。该图方便、实用，对一定的分离任务，可大致估算理论板数；也可粗略地定量分析N与R的关系。

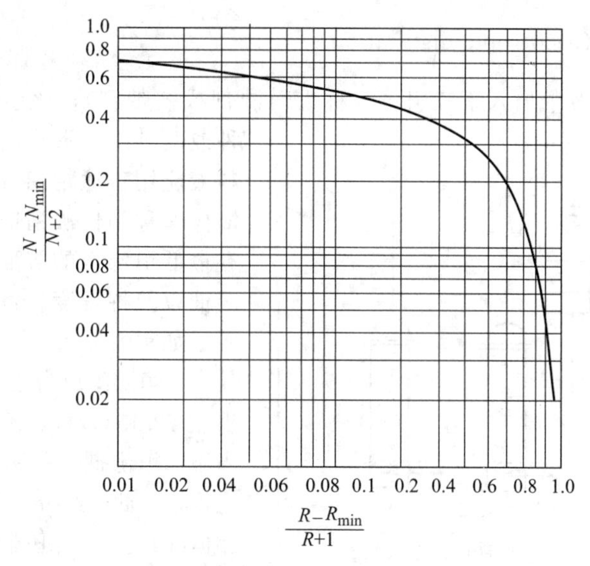

图3-32 吉利兰关联图

吉利兰关联图求理论板数的步骤为：

① 利用解析法或图解法求解R_{min}，并按上述原则选择适宜的R；

② 用芬斯克公式计算N_{min}；

③ 计算$\frac{R-R_{min}}{R+1}$，查吉利兰关联图得出纵坐标数值后可相应计算出理论板数N值。

Number of ideal plates at operating reflux

A simple empirical method due to Gilliland is much used for preliminary estimates. The correlation requires knowledge only of the minimum number of plates at total reflux and the minimum reflux ratio. The correlation is self-explanatory, The Gilliland correlation, however, is based mainly on calculations for systems with nearly constant relative volatility and may be considerably in error for nonideal systems. An alternate correlation shows that the ratio of N/N_m (for ideal binary systems) depends mainly on R/R_m, for a wide range of relative volatilities. On the other hand, for the system methanol-water, where a changes from 7.5 for dilute solutions to 2.7 for nearly pure methanol, the values of N/N_{min} are much greater and change more rapidly than with ideal systems.

3.2.8 特殊精馏（Special distillation）

在实际生产中，常有这样的情况：

① 两组分的挥发度很接近（即平衡线离对角线很近，传质推动力很小），用一般精馏方法分离，所需的理论板数N非常多。

② 液体混合物具有恒沸点，如乙醇-水的恒沸组成为乙醇0.894、水0.106（摩尔分数），所以采用一般精馏方法，无法制取无水酒精。

对于上述情况，工业上可采用特殊精馏方式进行分离，其基本原理是在二元溶液中加入第三组分，以改变原来二元物系的非理想性或提高组分间的相对挥发度。根据第三组分所起的作用，可分为恒沸精馏和萃取精馏。

3.2.8.1 恒沸精馏（azeotropic distillation）

若在两组分恒沸液中加入第三组分（称为恒沸剂或挟带剂），该组分能与原料液中的一种或两种组分形成新的恒沸液，从而使原料液能用普通精馏方法予以分离，这种精馏操作称为恒沸精馏。恒沸精馏可分离具有最低恒沸点的溶液、具有最高恒沸点的溶液以及挥发度相近的物系。

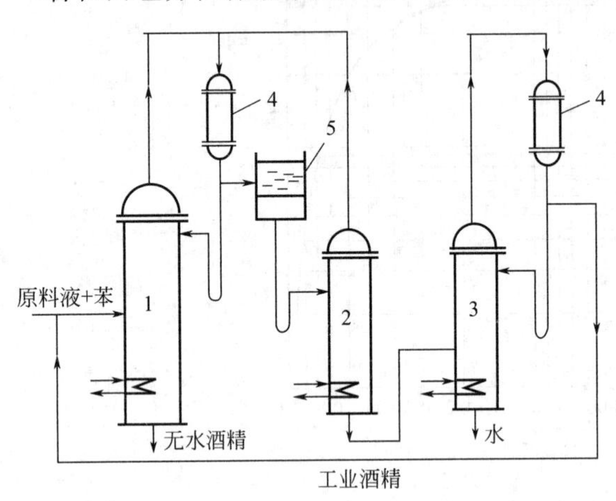

图 3-33 恒沸精馏流程示意图
1—恒沸精馏塔；2—苯回收塔；3—乙醇回收塔；4—冷凝器；5—分层器

如图 3-33 所示，以乙醇-水物系为例，其二元恒沸组成为乙醇 0.894、水 0.106，恒沸点 78.15℃，水/乙醇＝0.12（摩尔比）。当添加恒沸剂——苯后，形成苯-乙醇-水三元物系，此三组分可形成一种新的三元非均相恒沸物。三元恒沸组成为苯 0.554，乙醇 0.320，水 0.226，水/乙醇＝0.98（摩尔比），恒沸点为 64.5℃。由于新三元恒沸物沸点最低，故由恒沸精馏塔顶蒸出，塔底产品为近于纯态的乙醇。塔顶蒸气进入冷凝器 4 中冷凝后，部分液相回流到塔 1，其余的进入分层器 5，在分层器内分为轻重两层液体。轻相返回塔 1 作为补充回流。重相送入苯回收塔 2，以回收其中的苯。塔 2 的蒸气由塔顶引出也进入冷凝器 4 中，塔 2 底部的产品为稀乙醇，被送到乙醇回收塔 3 中。塔 3 中塔顶产品为乙醇-水恒沸液，送回塔 1 作为原料，塔底产品几乎为纯水。在操作中苯是循环使用的，只要有足够量的苯，就可将水全部集中于三元恒沸物中带出，从而使乙醇-水混合液得以分离。

恒沸精馏操作关键是恒沸剂的选择，它的选择原则是：

① 能与 A、B 组分中的一种或两种形成最低恒沸物，并希望与 A、B 中含量小的组分形成恒沸物从塔顶蒸出，以减少操作的热量损耗；

② 新形成的恒沸物要便于分离，以利于恒沸剂重新回收循环使用；

③ 恒沸物中，恒沸剂的相对含量要小，这样可使较少的恒沸剂夹带较多的其他组分，且使操作经济合理。

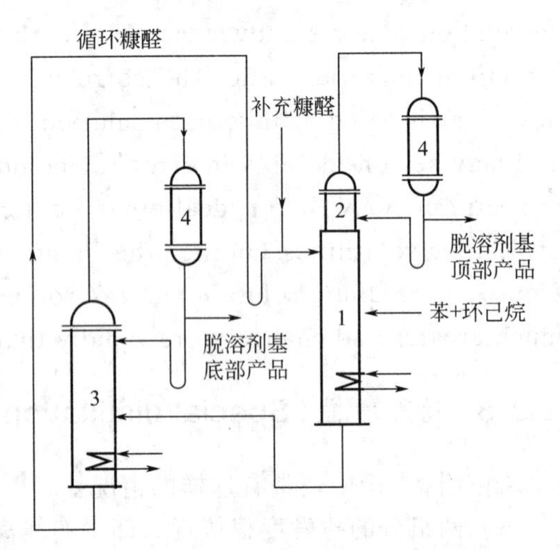

图 3-34 萃取精馏流程示意图
1—萃取精馏塔；2—萃取剂回收段；3—苯回收塔；4—冷凝器

3.2.8.2 萃取精馏（extractive distillation）

萃取精馏和恒沸精馏相似，也是向原料液中加入第三组分（称为萃取剂或溶剂），以改变原有组分间的相对挥发度而达到分离要求的特殊精馏方法。但不同的是要求萃取剂的沸点较原料液中各组分的沸点高得多，且不与组分形成恒沸液，容易回收。

以苯-环己烷为例，如图3-34所示。选择萃取剂糠醛，由于糠醛分子对苯分子的吸引力大，对环己烷分子的影响不明显。因此加入糠醛后，苯-环己烷的相对挥发度增大。例如加入摩尔分数为0.72的糠醛后，则使原苯-环己烷物系容易分离。

萃取剂的选择原则如下：

① 选择性高，即加入少量萃取剂后能使原双组分间的相对挥发度大大地增大；

② 挥发性小，即萃取剂的沸点要比原双组分A、B的沸点高得多，且不与A、B形成恒沸物；

③ 萃取剂与原溶液的A、B组分间有足够的溶解度，以避免分层。

萃取精馏中萃取剂的加入量一般较多，以保证各层塔板上足够的添加剂浓度，而且萃取精馏塔往往采用饱和蒸气加料，以使精馏段和提馏段的添加剂浓度基本相同。

Azeotropic and Extractive Distillation

Several enhanced distillation-based separation techniques have been developed for close-boiling or low-relative-volatility systems, and for systems exhibiting azeotropic behavior. All of these special techniques are ultimately based on the same differences in the vapor and liquid compositions as ordinary distillation, but, in addition, they rely on some additional mechanism to further modify the vapor-liquid behavior of the key components, such as azeotropic distillation, extractive distillation, and pressure-swing distillation, and so on. Both the azeotropic and extractive distillations are briefly introduced in this section.

Azeotropic Distillation

The term azeotropic distillation has been applied to a broad class of fractional-distillation -based separation techniques in that specific azeotropic behavior is exploited to effect a separation. The agent that causes the specific azeotropic behavior, often called the entrainer, may already be present in the feed mixture (a self-entraining mixture) or may be an added mass-separation agent. Azeotropic distillation techniques are used throughout the petrochemical and chemical processing industries for the separation of close-boiling, pinched, or azeotropic systems for which simple distillation is ether too expensive or impossible. Separation of the original mixture may be enhanced by adding a solvent that forms an azeotrope with one of the key components. This process is called azeotropic distillation. With an azeotropic feed mixture, presence of the entertainer results in the formation of a more favorable azeotropic pattern for the desired separation. For a closing-boiling or pinched feed mixture, the entrainer changes the dimensionality of the system.

The approach of azeotropic distillation, choosing an entrainer to cause azeotrope formation in combination with liquid-liquid immiscibility, relies on distillation, but also exploits another physical phenomena, liquid-liquid phase formation (phase splitting), to assist in entrainer recovery. One powerful and versatile approach exploits several physical phenomena simultaneously including enhanced vapor-liquid behavior, where possible, and liquid-liquid behavior to bypass difficult distillation separations. For example, the overall separation of close-boiling mixtures can be made easier by the addition of the entrainer that forms a heterogeneous minimum-boiling azeotrope with one (generally the lower-boiling) of the key components. Two-liquid-phase formation provides a means of breaking this azeotrope, thus simplifying the entrainer recovery and recycle process.

Extractive Distillation

The separation of components that have nearly the same boiling points is difficult by simple distillation even if the mixtures are ideal, and complete separation may be impossible because of azeotrope formation. For such systems the separation can often be improved by adding a third component to alter the relative volatility of the original components. The added component may be a miscible, higher-boiling liquid, non-volatile mass-separation agent, normally called the 'solvent', that is miscible with both of the key components but is chemically more similar to one of them. The solvent is added to an azeotropic or non-azeotropic feed mixture to alter the volatilities of the key components without the formation of any additional azeotropes. The key component that is more like the solvent will have a lower activity coefficient in the solution than the other component, so the separation is enhanced. This process is called extractive distillation and is like liquid-liquid extraction with an added vapor phase.

Extractive distillation is used throughout the petrochemical- and chemical-processing industries for the separation of close-boiling, pinched or azeotropic systems for which simplle single-feed distillation is either too expensive or impossible.

The solvent selectively alters the activity coefficients of the components being separated . To do this, a high concentration of solvent is necessary. Several features are essential:

(1) The solvent must be chosen to affect the liquid-phase behavior of the key components differently, otherwise no enhancement in separability will occur.

(2) The solvent must be higher boiling than the key components of the separation must be relatively nonvolatile in the extractive column, in order to remain largely in liquid phase.

(3) The solvent should not form additional azeotropes with the components in the mixture to be separated.

(4) The extractive column must be a double-feed column, with the solvent feed above the primary feed; the column must have an extractive section.

3.2.9 间歇精馏（Batch distillation）

间歇精馏流程如图 3-35 所示，原料液一次性加入塔釜后，将釜内的液体加热至沸腾，所生成的蒸气到达塔顶全凝器，冷凝液全部回流进塔内，待全回流操作稳定后，改为部分回流操作，并从塔顶采集产品。其主要特点是：

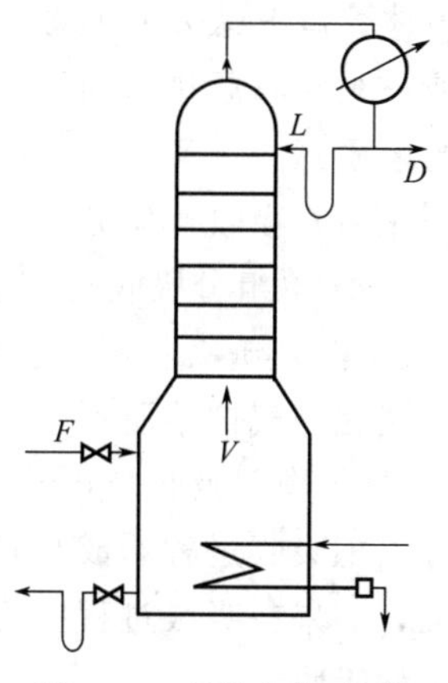

图 3-35　间歇精馏流程

① 间歇精馏是分批操作，过程为非稳态，但由于塔内气液浓度变化是连续而缓慢的，可视为“拟稳态”过程。对操作的任一瞬间，仍可用连续、稳态精馏的计算方法进行分析。

② 间歇精馏操作中，全塔均为精馏段，无提馏段，由于塔顶有液体回流，故属于精馏，而不是简单蒸馏操作。

③ 间歇精馏操作方式比较灵活机动，适用于处理物料量少、物料品种经常改变的场合，尤其是对缺乏技术资料的物系进行精馏分离开发时，可先采用间歇精馏进行小试，以获取有关的数据。但对于同样的分离要求，间歇精馏的能耗大于连续精馏。

间歇精馏有两种基本操作方式：

① 恒定回流比 R 的操作　随塔顶产物浓度 x_D 的降低，塔釜液组成 x_W 也不断降低；

② 恒定 x_D 的操作　随回流比 R 的增大，x_W 不断降低。

工业生产中，往往先恒定 R，当 x_D 降至一定值后，再增大 R，将产品分开储存。

实际生产中，按恒回流比操作，得不到高纯度产品，若按恒馏出液组成操作，需不断地变化 R，难以做到，通常将两法结合，先按前者操作一段时间，等有明显变化时，再加大 R，并使 R 恒定。如此阶跃地增加 R，可保持大致不变。

间歇精馏设备简单，通用性广，机动性大，特别适用于精细化工产品，且可方便地用于多元蒸馏。

3.3　吸收 (Absorption)

当混合气体（两组分或多组分）与某种液体相接触，气体混合物中某种或某些能溶解的组分便进入液相形成溶液，而不能溶解的组分仍然留在气相中，这种利用溶解度的差异来分离气体混合物的操作称为吸收，例如 HCl 气体溶于水生成盐酸、SO_3 溶于水生成硫酸等都是气体吸收的例子。

在过程工业中，吸收是重要的单元操作之一，主要用于分离气体混合物，就其分离目的而言，有以下两类用途：

① 一类是回收混合气体中的有用组分，制造产品。例如，硫酸吸收 SO_3 制浓硫酸，水吸收甲醛制福尔马林溶液，从焦炉气或城市煤气中分离苯，用液态烃处理裂解气以回收其中的乙烯、丙烯等。

② 另一类是净化混合气体中的有害组分，以符合工艺要求，又有利于环保。例如，用水或碱液脱除合成氨原料气中的 CO_2，硝酸尾气脱除 NO_x，磷肥生产中除去气态氟化物及皮毛消毒过程中环氧乙烷的回收等。

实际的吸收操作往往同时兼有回收与净化的双重功能，不能截然分开。

3.3.1　吸收基本原理 (Principles of absorption)

气体吸收的原理是根据混合气体中各组分在某液体溶剂中的溶解度不同而将气体混合物进行分离。吸收操作所用的液体溶剂称为吸收剂，以 S 表示；混合气体中，能够显著溶解于吸收剂的组分称为吸收物质或溶质，以 A 表示；而几乎不被溶解的组分统称为惰性组分或载体，以 B 表示；吸收操作所得到的溶液称为吸收液或溶液，它是溶质 A 在溶剂 S 中的溶液；被吸收后排出的气体称为吸收尾气，其主要成分为惰性气体 B，但仍含有少量未被吸收的溶质 A。

（1）吸收操作分类

① 单组分吸收与多组分吸收　即若混合气体中只有一种组分进入液相，其余组分不溶（或微溶）于吸收剂，这种吸收过程称为单组分吸收；反之，若在吸收过程中，混合气中进入液相的气体溶质不止一种，这样的吸收称为多组分吸收。

② 物理吸收与化学吸收　在吸收过程中，如果溶质与溶剂之间不发生显著的化学反应，可以把吸收过程看成是气体溶质单纯地溶解于液相溶剂的物理过程，则称为物理吸收；相反，如果在吸收过程中气体溶质与溶剂（或其中的活泼组分）发生显著的化学反应，则称为化学吸收。

③ 低浓度吸收与高浓度吸收　在吸收过程中，若溶质在气液两相中的摩尔分数均较低（通常不超过 0.1），这种吸收称为低浓度吸收；反之，则称为高浓度吸收。对于低浓度吸收过程，由于气相中溶质浓度较低，传递到液相中的溶质量相对于气、液相流率也较小，因此流经吸收塔的气、液相流率均可视为常数。

④ 等温吸收与非等温吸收　气体溶质溶解于液体时，常由于溶解热或化学反应热，而产生热效应，热效应使液相的温度逐渐升高，这种吸收称为非等温吸收；若吸收过程的热效应很小，或虽然热效应较大，但吸收设备的散热效果很好，能及时移出吸收过程所产生的热量，此时液相的温度变化并不显著，这种吸收称为等温吸收。

工业生产中的吸收过程以低浓度吸收为主。本节讨论单组分低浓度的等温物理吸收过程，对于其他条件下的吸收过程，可参考有关书籍。

Gas Absorption

Many chemical process materials and biological substances occur as mixtures of different components in the gas, liquid, or solid phase. In order to separate or remove one or more of the components from its original mixture, it must be contacted with another phase. The two phases are brought into more or less intimate contact with each other so that a solute or solutes can diffuse from one to the other. The two bulk phases are usually only somewhat miscible in each other. The two-phase pair can be gas-liquid, gas-solid, liquid-liquid, or liquid-solid. During the contact of the two phases the components of the original mixture redistribute themselves between the two phases. The phases are then separated by simple physical methods. By choosing the proper conditions and phases, one phase is enriched while the other is depleted in one or more components.

This section deals with the mass-transfer operations known as gas absorption and stripping, or desorption.

When the two contacting phases are a gas and a liquid, this operation is called absorption. A solute A or several solutes are absorbed from the gas phase into the liquid phase in absorption. This process involves molecular and turbulent diffusion or mass transfer of solute A through a stagnant, nondiffusing gas B into a stagnant liquid C. An example is absorption of ammonia A from air B by the liquid water C. Usually, the exit ammonia-water solution is distilled to recover relatively pure ammonia.

Another example is absorbing SO_2 from the flue gases by absorption in alkaline solutions. In the hydrogenation of edible oils in the food industry, hydrogen gas is bubbled into oil and absorbed. The hydrogen in solution then reacts with the oil in the presence of a catalyst. The reverse of absorption is called stripping or desorption, and the same theories and basic principles hold. An example is the steam stripping of nonvolatile oils, in which the steam contacts the oil and small amounts of volatile components of the oil pass out with the steam.

Principles of Absorption

Packed towers are used for continuous countercurrent contacting of gas and liquid in absorption. The tower consists of a cylindrical column containing a gas inlet and distributing space at the bottom, a liquid inlet and distributing device at the top, a gas outlet at the top, a liquid outlet at the bottom, and a packing or filling in the tower. The gas enters the distributing space below the packed section and rises upward through the openings or interstices in the packing and contacts the descending liquid flowing through the same openings. A large area of intimate contact between the liquid and gas is provided by the packing.

The diameter of a packed absorption tower depends on the quantities of gas and liquid handled, their properties, and the ratio of one stream to the other. The height of the tower, and hence the total volume of packing, depends on the magnitude of the desired concentration changes and on the rate of mass transfer per unit of packed volume. Calculations of the tower height, therefore, rest on material balances, enthalpy balances, and estimates of driving force and mass-transfer coefficients.

（2）吸收与解吸（absorption and adsorption）

应予指出，吸收过程使混合气中的溶质溶解于吸收剂中而得到一种溶液，但就溶质的存在形态而言，仍然是一种混合物，并没有得到纯度较高的气体溶质。在工业生产中，除以制取溶液产品为目的的吸收（如用水吸收 HCl 气体制取盐酸等）之外，大都要将吸收液进行解吸，以便得到纯净的溶质或使吸收剂再生后循环使用。解吸也称为脱吸，它是使溶质从吸收液中释放出来的过程，解吸通常在解吸塔中进行。图 3-36 为洗油脱除煤气中粗苯的流程简图。图中虚线左侧为吸收部分，在吸收塔中，苯系化合物蒸气溶解于洗油中，吸收了粗苯的洗油（又称富油）由吸收塔底排出，被吸收后的煤气由吸收塔顶排出。图中虚线右侧为解吸部分，在解收塔中，粗苯由液相释放出来，并被水蒸气带出，经冷凝分层后即可获得粗苯产品，解吸出粗苯的洗油（也称为贫油）经冷却后再送回吸收塔循环使用。

（3）吸收剂的选择

吸收剂选择的原则是对溶质的溶解度大、选择性好（对要求吸收的组分溶解度很大而对不希望吸收的组分溶解度很小）、溶解度对温度的变化率大（即利于吸收与解吸联合操作）、蒸气压低（吸收剂不易挥发，随气流带出塔的损耗小）、化学稳定性好、无毒、价廉、黏度小及不易起泡（利于操作）等。

大多数工业吸收操作都属于低浓度气体吸收，即指待处理混合气体中所含溶质的浓度较低（一般在5%～10%以下），因而本节重点是关于低浓度气体吸收的计算。

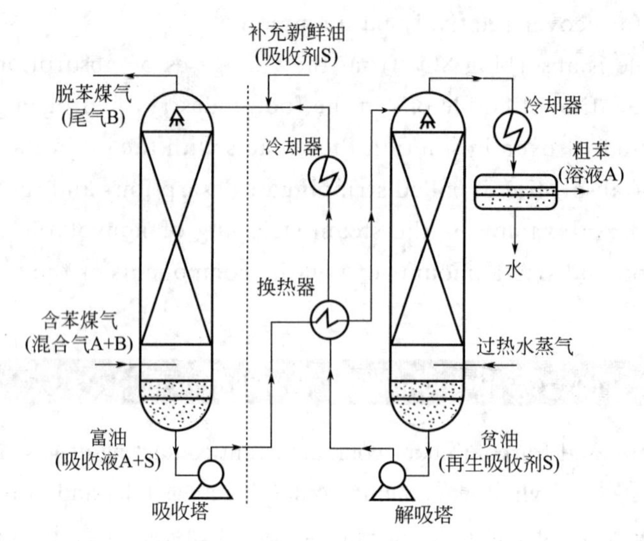

图 3-36　洗油脱除煤气中粗苯的吸收与解吸联合操作

3.3.2　气液相平衡关系（Gas-liquid equilibrium）

（1）气体在液体中的溶解度

在一定的压力及温度下，气相与液相充分接触后，两相趋于平衡状态。此时，液相组成 c_A 称为组分 A 的平衡溶解度，简称溶解度（或气相分压 p_A 的平衡浓度）；p_A 称为组分 A 的平衡分压（或 c_A 的平衡浓度），即：

$$c_A^* = f(p_A) \tag{3-61}$$

$$p_A^* = f(c_A) \tag{3-61a}$$

上两式称为吸收操作的气、液相平衡关系。对于不同物系的具体函数形式，要借助于实验研究来最后确定。

气、液两相处于平衡状态时，表示溶质在气相中分压与液相中浓度的关系曲线为溶解度曲线（平衡线），如图 3-37～图 3-39 所示。

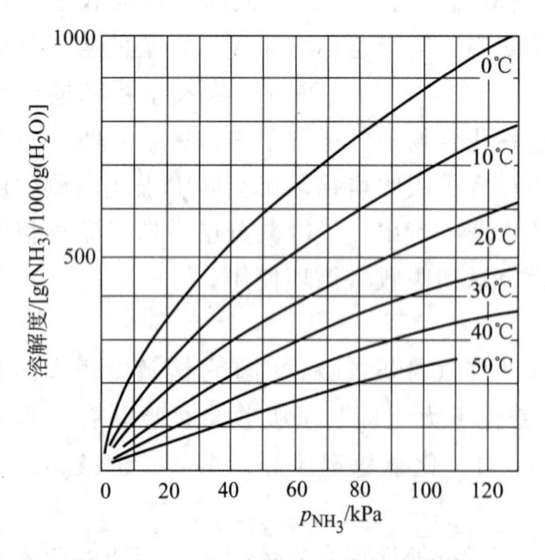

图 3-37　NH_3 在水中的溶解度曲线

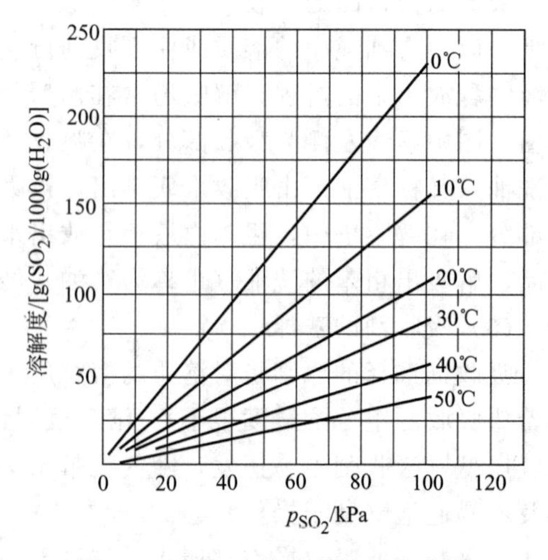

图 3-38　SO_2 在水中的溶解度曲线

由溶解度曲线图可以看出，对于同一物质在同一平衡压力下，温度越高，其对应的溶解度越低。在同一平衡压力下，不同物质，其对应的溶解度不同。溶解度大者为易溶气体，如图 3-37 中的 NH_3；溶解度小者为难溶气体，如图 3-39 中的 O_2。对于难溶气体，其相平衡关系为直线；而易溶气体及溶解度适中的气体，仅在液相为低浓度时可近似为直线。

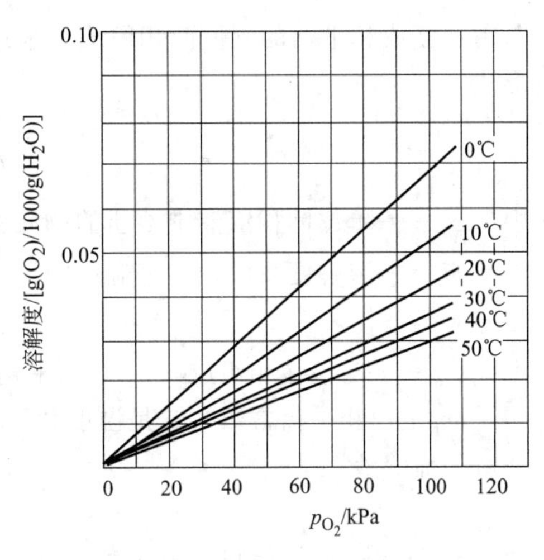

图 3-39 O_2 在水中的溶解度曲线

（2）亨利定律（Henry's law）

亨利定律是描述当系统总压不太高（$\leqslant 5\times10^5$ Pa）时，在恒定的温度下，稀溶液上方的气体溶质分压与其在液相中的浓度之间的平衡关系。

$$p_A^* = Ex_A \tag{3-62}$$

式中 p_A^*——溶质 A 在气相中的平衡分压，kPa；

x_A——溶质 A 在液相中的摩尔分数；

E——亨利系数，它是温度的函数，即 $E=f(t)$，kPa。

式(3-62) 称为亨利（Henry）定律，它表明稀溶液上方的溶质分压与该溶质在液相中的浓度成正比，其比例常数即为亨利系数。表 3-4 列出某些气体水溶液的亨利系数，可供参考。

表 3-4 某些气体水溶液的亨利系数

气体种类	温度/℃ 0	5	10	15	20	25	30	35	40	45	50	60	70	80	90	100
	$E/\times10^{-6}$ kPa															
H_2	5.87	6.16	6.44	6.70	6.92	7.16	7.39	7.52	7.61	7.70	7.75	7.75	7.71	7.65	7.61	7.55
N_2	5.35	6.05	6.77	7.48	8.15	8.76	9.36	9.98	10.5	11.0	11.4	12.2	12.7	12.8	12.8	12.8
空气	4.38	4.94	5.56	6.15	6.73	7.30	7.81	8.34	8.82	9.23	9.59	10.2	10.6	10.8	10.9	10.8
CO	3.57	4.01	4.48	4.95	5.43	5.88	6.28	6.68	7.05	7.39	7.71	8.32	8.57	8.57	8.57	8.57
O_2	2.58	2.95	3.31	3.69	4.06	4.44	4.81	5.14	5.42	5.70	5.96	6.37	6.72	6.96	7.08	7.10
CH_4	2.27	2.62	3.01	3.41	3.81	4.18	4.55	4.92	5.27	5.58	5.85	6.34	6.75	6.91	7.01	7.10
NO	1.71	1.96	2.21	2.45	2.67	2.91	3.14	3.35	3.57	3.77	3.95	4.24	4.44	4.45	4.58	4.60
C_2H_6	1.28	1.57	1.92	2.90	2.66	3.06	3.47	3.88	4.29	4.69	5.07	5.72	6.31	6.70	6.96	7.01
	$E/\times10^{-5}$ kPa															
C_2H_4	5.59	6.62	7.78	9.07	10.3	11.6	12.9	—	—	—	—	—	—	—	—	—
N_2O	—	1.19	1.43	1.68	2.01	2.28	2.62	3.06	—	—	—	—	—	—	—	—
CO_2	0.378	0.8	1.05	1.24	1.44	1.66	1.88	2.12	2.36	2.60	2.87	3.46	—	—	—	—
C_2H_2	0.73	0.85	0.97	1.09	1.23	1.35	1.48	—	—	—	—	—	—	—	—	—
Cl_2	0.272	0.334	0.399	0.461	0.537	0.604	0.669	0.74	0.80	0.86	0.90	0.97	0.99	0.97	0.96	—
H_2S	0.272	0.319	0.372	0.418	0.489	0.552	0.617	0.686	0.755	0.825	0.689	1.04	1.21	1.37	1.46	1.50
	$E/\times10^{-4}$ kPa															
SO_2	0.167	0.203	0.245	0.294	0.355	0.413	0.485	0.567	0.661	0.763	0.871	1.11	1.39	1.70	2.01	—

由于互成相平衡的气液两相组成可采用不同的表示法，因而亨利定律也有以下不同的表达形式。

$$p_A^*=\frac{c_A}{H} \tag{3-62a}$$

式中　c_A——单位体积溶液中溶质的物质的量，$kmol/m^3$；

H——溶解度系数，它是温度的函数，即 $H=f(t)$，$kmol/(m^3 \cdot kPa)$。

$$y_A^*=mx_A \tag{3-62b}$$

式中　y_A^*——与该液相组成（x_A）相平衡的气相中溶质A的摩尔分数；

m——相平衡常数，它是温度和压力的函数，即 $m=\varphi(t, P)$，无量纲。

对于稀溶液：

$$Y^* \approx mX \tag{3-63}$$

式中　Y^*，X——气相和液相的摩尔比。

$$X=\frac{x}{1-x} \tag{3-64}$$

$$Y=\frac{y}{1-y} \tag{3-64a}$$

以上各式进行换算时：

$$E=\frac{\rho_S}{M_S H} \tag{3-65}$$

$$m=\frac{E}{P} \tag{3-66}$$

式中，ρ_S 为吸收剂S的密度，kg/m^3；M_S 是吸收剂S的分子量。

【例3-7】 在总压1atm及温度为20℃的条件下，氨在水中浓度为0.582$kmol/m^3$，液面上方氨的平衡分压为6mmHg。若在此范围内符合亨利定律，试求 H、E、m 的值（因溶解浓度较低，故溶液密度可按纯水计算）。

解　(1) H 值：

$$p_A^*=\frac{c_A}{H}$$

$$p_A^*=6\times133.3=800(Pa)=\frac{6}{760}=0.0079(atm)$$

故：

$$H=\frac{0.582}{0.0079}=73.7[kmol/(m^3 \cdot atm)]$$

$$H=\frac{0.582}{800}=7.28\times10^{-4}[kmol/(m^3 \cdot Pa)]$$

(2) E 值：

$$E=\frac{\rho_S}{M_S H}=\frac{1000}{18\times7.28\times10^{-4}}=7.63\times10^4(Pa)$$

$$E=\frac{\rho_S}{M_S H}=\frac{1000}{18\times73.7}=0.753(atm)$$

(3) m 值：

$$m=\frac{E}{p}$$

$$p=101.3\times10^{3}(Pa)=1(atm)$$

故：

$$m=\frac{7.63\times10^{4}}{101.3\times10^{3}}=0.753$$

$$m=\frac{0.753}{1}=0.753$$

因为E、H仅取决于物系温度t，而$m=f(t,P)$，因此，提到m时，必须指明系统的总压强P。若m值越大，则表明该气体的溶解度越小。亨利定律中涉及的系数m、E和H随操作条件及溶质气体的性质而变化的趋势列于表3-5中。

表3-5 亨利定律中E、H及m的变化对比表

项　目	E值	H值	m值
系统温度升高	增大	减少	增大
易溶气体	较小	较大	较小
难溶气体	较大	较小	较大

由表3-5及式(3-66)可看出，加压降温对吸收有利，反之对解吸有利。

(3) 相平衡与吸收过程的关系

相平衡关系在吸收（或解吸）过程的分析与计算中是不可缺少的，其作用如下：

① 判断过程进行的方向　当气液两相接触时，可用相平衡关系确定平衡的气液组成，将平衡组成与此相的实际组成比较，可以判断出该过程是吸收还是解吸。即当以气相的组成判断时，气相中A的实际组成y高于与液相成平衡的组成y^*为吸收，反之为解吸；而以液相的组成判断时，液相中A的实际组成x低于与气相成平衡的组成x^*为吸收，反之为解吸。

② 计算吸收过程的推动力　吸收或解吸过程是相际传质过程，当互不平衡的气、液两相接触时，通常以实际组成与平衡浓度的偏离程度来表示吸收推动力。过程的传质推动力越大，其传质速率也越快，完成指定的分离要求时所需的设备尺寸越小。

③ 指明过程的极限　随着操作条件的改变，气、液两相间的传质情况会随之变化，但传质过程最终受到相平衡关系的制约，即平衡状态是过程进行的极限。例如对于逆流吸收过程，随着塔高增大，吸收剂用量增加，出口气体中溶质A的组成y_2将随之降低，但即使塔无限高，且吸收剂用量很大，y_2最终只会降低到与溶剂入口组成达成相平衡的组成y_2^*为止，不会再继续下降。同理，塔底出口的吸收液中溶质A的组成x_1也有一个最大值，即$x_{1,max}=y_1/m$。

【例3-8】 水与空气-二氧化硫的混合物中SO_2浓度为3%（摩尔分数，下同），液相中SO_2的浓度为4.13×10^{-4}。压力为1.2atm，温度10℃，试判断该过程是吸收还是解吸？过程推动力为多少？

解　查得温度10℃时，亨利系数$E=24.2$atm。

则相平衡常数：$m=\frac{E}{P}=\frac{24.2}{1.2}=20.17$

可计算出：$y^*=mx=20.17\times4.13\times10^{-4}=8.33\times10^{-3}$

由于 $y=0.03>y^*$，故该过程为吸收。

推动力：$\Delta y=y-y^*=0.03-8.33\times10^{-3}=2.17\times10^{-2}$

$$\Delta x=x^*-x=\frac{y}{m}-x=\frac{0.03}{20.17}-4.13\times10^{-4}=1.07\times10^{-3}$$

3.3.3 吸收速率

（1）双膜模型

吸收过程分三个步骤连续进行：即溶质A由气相主体传递至两相界面；A在两相界面上溶解；A由两相界面传递至液相主体。如图3-40所示，对界面两侧的气、液两相分别应用膜模型，则A由气相主体至界面，推动力为 p_G-p_i，相应阻力折合在有效气膜内；A由界面至液相主体，推动力为 C_i-C_L，相应阻力折合在有效液膜内。一般认为，两相界面上传质阻力很小，小到可以忽略，即界面上气、液两相处于平衡状态，无传质阻力存在。

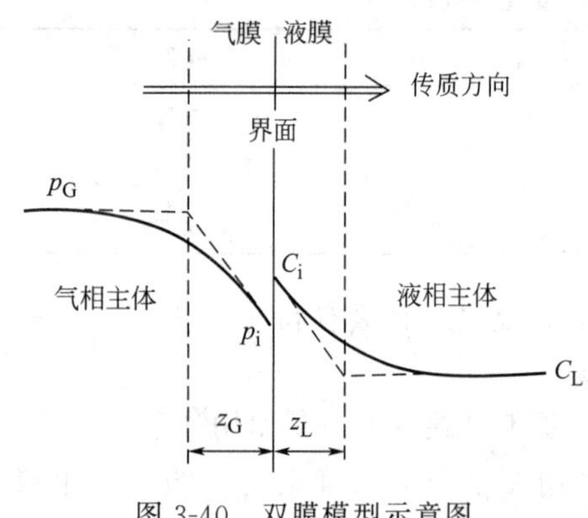

图3-40　双膜模型示意图

上述用膜模型处理界面两侧的传质及界面上无传质阻力的设想，称为双膜模型。

综上所述，双膜模型可以归纳为以下三条基本论点：

① 相互接触的气液两相流体间存在着稳定的相界面，界面两侧各有一个很薄的有效层流膜，溶质以分子扩散的方式穿过这两层膜由气相主体进入液相主体。流体的流速越快，膜越薄。

② 相界面处气液相处于平衡状态，没有传质阻力。

③ 两相主体中由于质点间充分湍动，浓度均匀，无浓度梯度，即无阻力存在。

双膜理论认定：气体吸收过程的全部阻力均集中在两层膜中，阻力的大小决定了传质速率的大小。

双膜理论适用于有固定传质界面的传质设备，如湿壁塔的吸收操作，按双膜模型推导出的传质结果与实际情况基本符合。

Two Film Theory

In many separation processes, material must diffuse from one phase into another phase, and the rates of diffusion in both phases affect the overall rate of mass transfer. In the two-film theory, proposed by Whitman in 1923, equilibrium is assumed at the interface, and the resistances to mass transfer in the two phases are added to get an overall resistance, just as is done for heat transfer. The reciprocal of the overall resistance is an overall coefficient, which is easier to use for design calculations than the individual coefficients.

What makes mass transfer between phases more complex than heat transfer is the discontinuity at the interface, which occurs because the concentration or mole fraction of diffusing solute is hardly ever the same on opposite sides of the interface. For the absorption of a very soluble gas, the mole fraction in the liquid at the interface would be greater than that in the gas (Fig. 3. 6).

In the two-film theory, the rate of transfer to the interface is set equal to the rate of the transfer from the interface:

$$N = k_x(x_A - x_{Ai}) \quad \text{(eq. 3-14)}$$

$$N = k_y(y_{Ai} - y_A) \quad \text{(eq. 3-15)}$$

The rate is also set equal to an overall coefflcient K_y times an overall driving force $y_A^* - y_A$, where y_A^*, is the composition of the vapor that would be in equilibrium with the bulk liquid of composition x_A:

$$N = K_y(y_A^* - y_A) \quad \text{(eq. 3-16)}$$

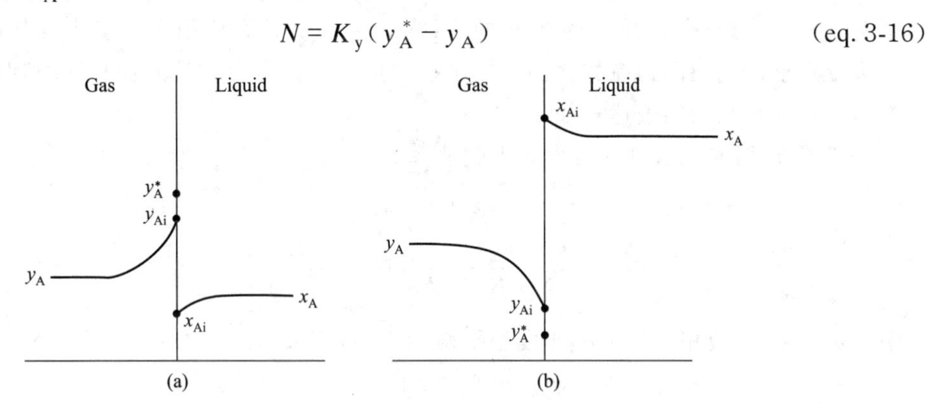

Fig. 3. 6 Concentration gradients near a gas-liquid interface: (a) distillation; (b) absorption of a very soluble gas

$$\frac{1}{K_y} = \frac{m}{k_x} + \frac{1}{k_y} \quad \text{(eq. 3-17)}$$

The term $1/K_y$, can be considered an overall resistance to mass transfer, and the terms m/k_x and $1/k_y$, are the resistances in the liquid and gas films. These "films" need not be stagnant layers of a certain thickness in order for the two-film theory to apply. Mass transfer in either film may be by diffusion through a laminar boundary layer or by unsteady-state diffusion, as in the penetration theory, and the overall coefficient is still obtained from eq. (3-17). For some problems, such as transfer through a stagnant film into a phase where the penetration theory is thought to apply, the penetration theory coefficient is slightly changed because of the varying concentration at the interface, but this effect is only of academic interest.

The essential part of the two-film theory is the method of allowing for the partition of solute between phases at the interface while combining individual coefficients to get an overall coefficient. This approach is used in the analysis of laboratory data and in equipment design for many types of mass-transfer operations, including absorption, adsorption, extraction, and distillation. The same principle of adding resistances with the appropriate distribution factors also applies to membrane separations, in which there are three resistances in series.

（2）吸收速率方程的导出

根据双膜模型，气相传质速率方程（气相主体到界面）和液相传质速率方程（界面到液相主体）分别为：

$$N_A = k_G(p_G - p_i) \tag{3-67}$$

式中 p_G——气相主体中溶质A的分压，kPa；

p_i——相界面处溶质A的分压，kPa。

$$N_A = k_L(c_i - c_L) \tag{3-68}$$

式中 c_i——相界面处溶质A的浓度，$kmol/m^3$；

c_L——液相主体中溶质A的浓度，$kmol/m^3$。

若气相主体到液相主体为稳态传质，即 N_A = 常数，则应用亨利定律或其他的相平衡关系，可分别求出气液两相在相界面上的平衡浓度，因此，由数学的加和性原理可导出吸收速率方程及总吸收系数的表达式如下：

① 以气相组成表示总推动力的吸收速率方程：

$$N_A = K_G(p_G - p_L^*) \tag{3-69}$$

$$\frac{1}{K_G} = \frac{1}{k_G} + \frac{1}{k_L H} \tag{3-70}$$

式中 p_L^*——与液相主体浓度 c 成相平衡的气相分压，kPa；

K_G——以气相浓度表示的总吸收系数，$kmol/(m^2 \cdot s \cdot kPa)$；

$1/k_G$——气相传质分阻力，相对应于推动力 $p_G - p_i$；

$1/(k_L H)$——液相传质分阻力，相对应于推动力 $p_i - p_L^*$；

$1/K_G$——总传质阻力，其相对应于总推动力 $p_G - p_L^*$。

② 以液相组成表示总推动力的吸收速率方程：

$$N_A = K_L(c_G^* - c_L) \tag{3-71}$$

$$\frac{1}{K_L} = \frac{H}{k_G} + \frac{1}{k_L} \tag{3-72}$$

式中 c_G^*——与气相主体分压 p 成相平衡的液相浓度，$kmol/m^3$；

K_L——总吸收系数，相应于总推动力 $c_G^* - c_L$；

H/k_G——气相传质分阻力，相应于推动力 $c_G^* - c_i$；

$1/k_L$——液相传质分阻力，相应于推动力 $c_i - c_L$；

$1/K_L$——总传质阻力，相应于总推动力 $c_G^* - c_L$。

（3）吸收速率方程的不同表达方式

由于传质推动力和传质阻力的表达形式各不相同，相应的吸收速率方程也有不同的表达方式，表3-6列举了各种常用的吸收速率方程式。

表3-6 吸收速率方程的各种形式

推动力表示方法	膜吸收速率方程（采用任一相主体与界面处的浓度差表示推动力，此时界面浓度难以求取）	总吸收速率方程（采用任一相主体浓度与其平衡浓度之差表示推动力，避免了难测定的界面浓度，使用更方便）

续表

吸收速率方程具体形式	$N_A=k_G(p_G-p_i)$ $=k_L(c_i-c_L)$ $=k_y(y-y_i)$ $=k_x(x_i-x)$	$N_A=K_G(p_G-p_L^*)$ $=K_L(c_G^*-c_L)$ $=K_y(Y-Y^*)$ $=K_x(X^*-X)$
吸收系数间的换算关系	$k_y=Pk_G$ $k_x=ck_L$	$\frac{1}{K_G}=\frac{1}{Hk_L}+\frac{1}{k_G}$ $\frac{1}{K_L}=\frac{H}{k_G}+\frac{1}{k_L}$ $K_y=PK_G$ $K_x=CK_L$

上述的吸收速率方程式仅适合于描述定态操作的吸收塔内任一横截面上的情况，而不能直接用来对全塔做数学描述。还应注意，在使用总吸收速率方程时，在整个吸收过程所涉及的浓度范围内，相平衡应符合亨利定律或为直线关系。

（4）气膜控制与液膜控制

依据式(3-70)，对于易溶气体，因 H 值很大，若 k_G 与 k_L 数量级相同或接近，则有 $1/(Hk_L)\ll 1/k_G$，因而可简化为 $1/K_G\approx 1/k_G$ 或 $K_G\approx k_G$。此时传质阻力的绝大部分集中于气膜中，即吸收总推动力的绝大部分用于克服气膜阻力，而液膜阻力可以忽略，这种情况称为气膜控制。例如用水吸收氨或 HCl 及用浓硫酸吸收气相中的水蒸气等过程，常被视为气膜控制的吸收过程。显然，对于气膜控制的吸收过程，尽量减少气膜阻力是提高其吸收速率的关键，这时可采取增大气相流量、设法提高气流湍动等措施，若只采取增大液相流量的方法，则收效甚微。

依据式(3-72)，对于难溶气体，因 H 值较小，若 k_G 与 k_L 数量级相同或接近，则有 $H/k_G\ll 1/k_L$，因而可简化为 $1/K_L\approx 1/k_L$ 或 $K_L\approx k_L$。此时液膜阻力控制着整个吸收过程的速率，吸收总推动力的绝大部分用于克服液膜阻力，气膜阻力可以忽略，这种情况称为液膜控制。例如用水吸收氧、CO_2 或 H_2 等气体的过程，都是液膜控制的吸收过程。对于该类吸收过程，应注意减少液膜阻力，例如增加吸收剂用量、改变液相的分散程度及增加液膜表面的更新频率，都可明显地提高其吸收速率。

一般情况下，对于具有中等溶解度的气体吸收过程，气膜阻力和液膜阻力均不可忽略。要提高吸收过程速率，必须兼顾气液两膜阻力的降低，方能得到满意的效果。

【例 3-9】 吸收塔某截面上气相中 A 组分的分压为 10.13kPa，液相主体中 A 的浓度为 2.78×10^{-3}kmol/m^3，而 $k_G=5.0\times10^{-6}$kmol/(m^2·s·kPa)，$k_L=1.5\times10^{-4}$kmol/(m^2·s·kmol/m^3)。当 $H=0.667$kmol/(m^3·kPa) 时，试问该过程是气膜控制还是液膜控制？

解 按气相总传质系数计算：

$$\frac{1}{K_G}=\frac{1}{k_G}+\frac{1}{Hk_L}=\frac{1}{5\times10^{-6}}+\frac{1}{0.667\times1.5\times10^{-4}}=20\times10^4+10^4=21\times10^4$$

$$K_G=4.76\times10^{-6}$$

即气膜阻力占总阻力的比例为：$\dfrac{\frac{1}{k_G}}{\frac{1}{K_G}}=\dfrac{20\times10^4}{21\times10^4}=0.95$

则此过程属于气膜控制。

3.3.4 填料塔简介

吸收操作既可以采用板式塔，也可以采用填料塔。本节对吸收操作的讨论主要结合填料塔进行。

如图 3-41 所示，填料塔是竖立的圆筒形设备，上下有端盖，塔体上、下端适当位置上设有气液进出口，填料塔内填充某种特定形状的固体物物——填料，以构成填料层，填料层是塔内实现气液接触的有效场所。由于填料层中有一定的空隙体积，气体可以在填料间隙所形成的曲折通道中流动，提高了湍动程度；同时，由于单位体积填料层中有一定的固体表面，使下降的液体可以分布于填料表面而形成液膜，从而提供气液接触机会。

在填料吸收塔内，气液两相流动方式原则上可为逆流也可为并流。一般情况下塔内液体作为分散相，总是靠重力作用自上而下地流动，而气体靠压强差的作用向下流经全塔，逆流时气体自塔底进入而自塔顶排出，并流时则相反。

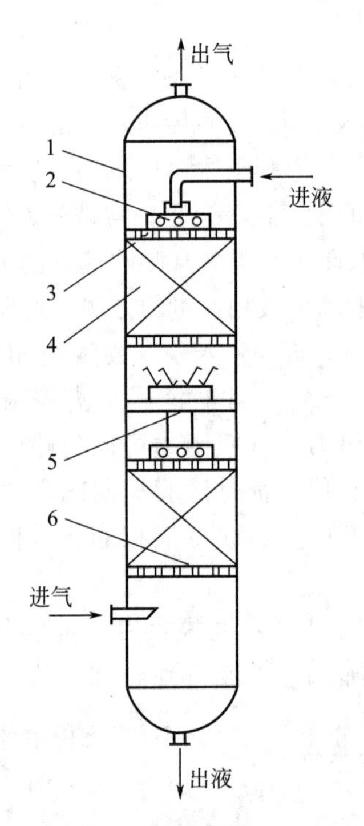

图 3-41 填料塔示意图

1—塔壳体；2—液体分布器；3—填料压板；
4—填料；5—液体再分布装置；6—填料支承板

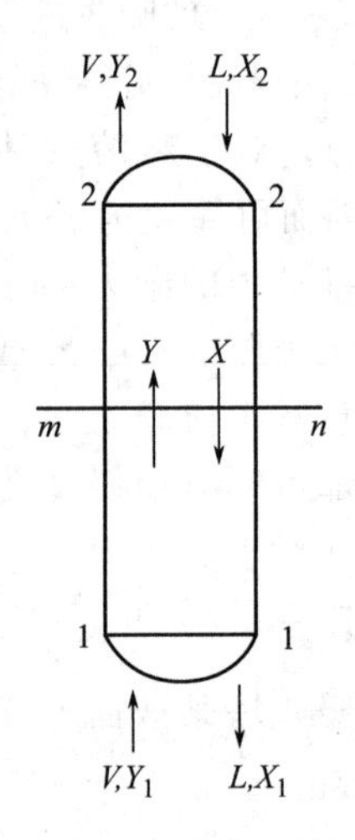

图 3-42 逆流填料吸收塔

3.3.5 吸收剂用量计算

(1) 物料衡算与操作线方程（material balances and operating lines）

① 物料衡算（material balances） 稳态、逆流操作的吸收塔如图 3-42 所示，其中以截面 2 表示塔顶，截面 1 表示塔底，图中各个符号含义如下：V 为单位时间内通过吸收塔的惰性气体摩尔流量，kmol(B)/s；L 为单位时间内通过吸收塔的吸收剂摩尔流量，kmol(S)/s；

Y_1，Y_2 分别为进塔及出塔气体中溶质 A 的摩尔比，kmol(A)/kmol(B)；X_2，X_1 分别为进塔及出塔液体中溶质 A 的摩尔比，kmol(A)/kmol(S)。

则全塔物料衡算式为：

$$VY_1+LX_2=VY_2+LX_1 \tag{3-73}$$

或：

$$V(Y_1-Y_2)=L(X_1-X_2) \tag{3-73a}$$

式中，Y_2 是设计型计算中根据分离要求指定的数值，分离要求通常有两种表达方式。当吸收目的是除去气体中的有害物质时，例如为符合环保要求，工业尾气排空前必须将其中有害气体浓度降到一定数值，即一般直接规定吸收后气体中有害溶质的残余浓度（y_2）；当吸收的目的是回收有用物质时，通常规定溶质的回收率 φ_A。它的定义为

$$\varphi_A=\frac{\text{被吸收的溶质量}}{\text{进塔混合气体中含有的溶质总量}}=\frac{V(Y_1-Y_2)}{VY_1}=\frac{Y_1-Y_2}{Y_1}$$

则：

$$Y_2=Y_1(1-\varphi_A) \tag{3-74}$$

利用式(3-74) 计算出 Y_2 代入式(3-73a) 中，则可求得吸收液出口浓度 X_1，利用物料衡算可使填料塔的塔顶、塔底两个截面上的四个进、出口浓度均为已知，在此基础上，再结合相平衡关系，决定吸收剂用量。

② 操作线方程（operating lines） 如图 3-42 所示，描述填料塔内任一截面 m—n 上气相组成 Y 和液相组成 X 之间的数学关系称为填料吸收塔的操作线方程，可由该截面与塔顶之间作物料衡算得到：

$$V(Y-Y_2)=L(X-X_2) \tag{3-75}$$

或：

$$Y=\frac{L}{V}X+\left(Y_2-\frac{L}{V}X_2\right) \tag{3-75a}$$

同理，在截面 m—n 与塔底之间做物料衡算，可得：

$$Y=\frac{L}{V}X+\left(Y_1-\frac{L}{V}X_1\right) \tag{3-75b}$$

式(3-75a) 与式(3-75b) 皆可称为逆流吸收塔的操作线方程，它表示了填料塔内任一截面上气相浓度 Y 与液相浓度 X 之间呈直线关系。如图 3-43 所示，直线的斜率为 L/V，直线的两个端点分别为 $M(X_2, Y_2)$ 及 $N(X_1, Y_1)$。因逆流吸收塔中塔底端的气液相浓度为全塔最大值，故称为"浓端"，而塔顶截面处具有最小的气液相浓度，因此称为"稀端"；操作线上任一点 $A(X, Y)$ 代表塔内任一截面上相互接触的气液相组成，该点到平衡线的垂直距离 $Y-Y^*$ 是以气相组成表示的总推动力，该点到平衡线的水平距离（X^*-X）是以液相组成表示的总推动力；对于吸收过程，操作线位于平衡线上方，如果操作线位于相平衡曲线的下方，则应进行脱吸过程。

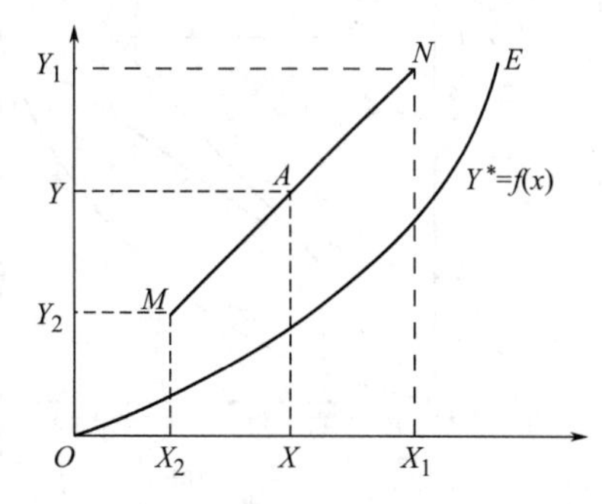

图 3-43 逆流填料吸收塔的操作线

与逆流操作类似，并流操作的吸收塔，其操作线方程也可由物料衡算导出。也就是说，操作线方程来自物料衡算，而与系统的平衡关系及设备结构型式无关。

(2) 吸收剂的用量（liquid rate）

吸收剂用量的选择，是兼顾技术可行和经济合理的综合问题。为表达方便，通常用吸收剂的

液气比表示，液气比是指处理1kmol/h惰性气体（在混合气中）所需吸收剂的用量（kmol/h）。

吸收剂的最小用量存在着技术上的限制，如图3-44（a）所示，当吸收剂用量减少到恰使操作线与相平衡线相交于N^*点时，入塔混合气与出塔吸收液达到相平衡，则塔底截面处吸收过程传质推动力为零，所以为达到相同的吸收分离目标，所需的塔高为无限大，这显然是不现实的，所以这是理论上液气比的下限，称为最小液气比，用$(L/V)_{min}$表示，相应的吸收剂用量即为最小吸收剂用量，以L_{min}表示。

最小液气比可用图解法求出，如图3-44（a）所示的一般情况，在图中找到N^*点对应的横坐标X_1^*的数值，则下式可用来计算最小液气比，即：

$$\left(\frac{L}{V}\right)_{min}=\frac{Y_1-Y_2}{X_1^*-X_2} \tag{3-76}$$

或：

$$L_{min}=V\frac{Y_1-Y_2}{X_1^*-X_2} \tag{3-76a}$$

若系统的相平衡关系符合亨利定律，即可用$Y^*=mX$表示，且进塔为纯吸收剂时（即$X_2=0$），式(3-76)可写成：

$$\left(\frac{L}{V}\right)_{min}=\frac{Y_1-Y_2}{\frac{Y_1}{m}-X_2}=m\frac{Y_1-Y_2}{Y_1}=m\varphi_A \tag{3-77}$$

若相平衡关系是如图3-44(b)中曲线所示的形状，则应过点M作平衡曲线的切线，则该切线与水平线$Y=Y_1$相交于点N'，在图中读得N'的横坐标X_1'的数值后，可代入下式计算最小液气比，即：

$$\left(\frac{L}{V}\right)_{min}=\frac{Y_1-Y_2}{X_1'-X_2} \tag{3-78}$$

因N'点未落在相平衡线上，所以式中Y_1与X_1'不成相平衡关系。

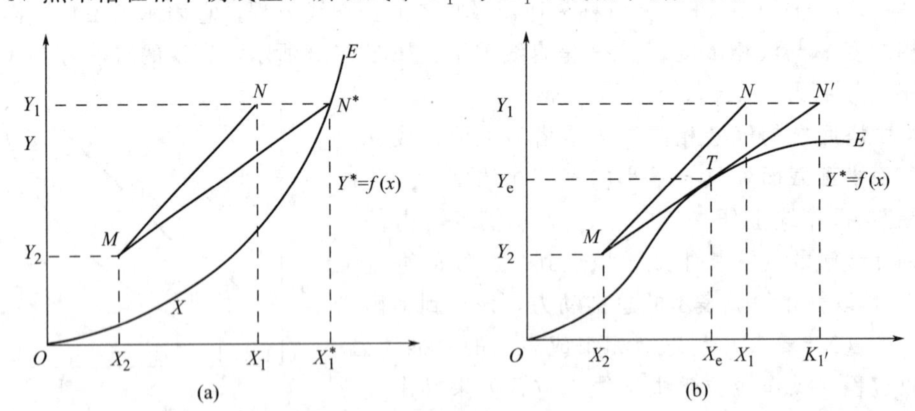

图3-44　吸收塔的最小液气比

也可由图中找到操作线与相平衡线的切点T，在图中读得T的坐标（X_e，Y_e），再用下式计算最小液气比：

$$\left(\frac{L}{V}\right)_{min}=\frac{Y_e-Y_2}{X_e-X_2} \tag{3-79}$$

因切点T落在相平衡线上，所以T点的气液相组成呈相平衡关系，即有$X_e=f(Y_e)$。

根据生产实践经验，一般情况下可按下式选择吸收剂用量，即：

$$\frac{L}{V}=(1.1\sim2.0)\left(\frac{L}{V}\right)_{\min} \tag{3-80}$$

或：

$$L=(1.1\sim2.0)L_{\min} \tag{3-80a}$$

【例 3-10】 某填料吸收塔用洗油来吸收混合气体中的苯，洗油中苯的浓度为0.0002kmol苯/kmol洗油，混合气体量为1000m^3/h，入塔气体中苯的含量为4%（体积分数），要求吸收率为80%，操作压强为101.3kPa，温度293K。溶液的平衡关系为$Y^*=30.9X$，$\left(\frac{L}{V}\right)_{操作}=1.2\left(\frac{L}{V}\right)_{\min}$。试计算吸收剂用量和溶液出口组成。

解 $Y_1=\frac{y_1}{1-y_1}=\frac{0.04}{1-0.04}=0.0417$

$Y_2=Y_1(1-\varphi_A)=0.0417\times(1-0.8)=0.00834$

$X_2=0.0002$

$$V=\frac{1000}{22.4}\times\frac{273}{293}\times(1-0.04)=39.93(\text{kmol/h})$$

$$\left(\frac{L}{V}\right)_{\min}=\frac{Y_1-Y_2}{X_1^*-X_2}=\frac{Y_1-Y_2}{Y_1/m-X_2}=\frac{0.0417-0.00834}{0.00135-0.0002}=29$$

因此：$L_{操作}=1.2\times\left(\frac{L}{V}\right)_{\min}\times V=1.2\times29\times39.93=1390$(kmol洗油/h)

$$X_1=\frac{V}{L}(Y_1-Y_2)+X_2=\frac{39.93}{1390}\times(0.0417-0.00834)+0.0002=0.00116$$

3.3.6 吸收塔高度的计算（Fundamental calculation equation of packing height）

填料塔是连续接触式的气液传质设备，因而在塔内不同位置上，气液两相的传质速率都可能有所不同，因此，需采用微元法对其传质规律进行研究。其中，吸收塔所需填料层高度的计算，是设计型问题的重点，它实质上是计算整个吸收塔的有效相际接触面积，涉及物料衡算、传质速率及相平衡关系三个方面的知识。

如图3-45所示，在填料吸收塔中任取一段高度为dZ的微元填料层，对此微元填料层作溶质A的物料衡算，又因气液相浓度变化很小，故可认为吸收速率N_A在该微元层内为定值，则有：

$$dM_A=VdY=LdX=N_AdS=N_A(a\Omega dZ) \tag{3-81}$$

式中　dM_A——单位时间内由气相转入液相的溶质A的物质的量，kmol/s；

dS——微元填料层内的有效传质面积，m^2；

a——有效比表面积，即指单位体积填料层中的有效传质面积，m^2/m^3；

Ω——塔截面积，m^2。

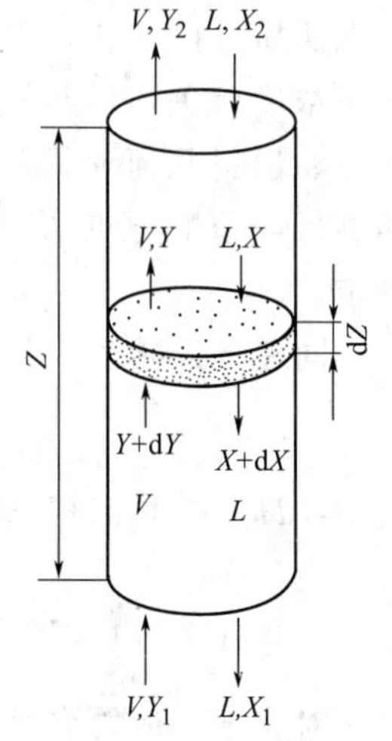

图3-45　微元填料层的物料衡算

在填料塔操作中，流体流经填料表面形成液膜（并非液体充满填料之间），液膜与气体接触传质。因为液膜很薄，可忽略液膜厚度。

但操作中，并非全部的填料表面都被润湿（如填料接触点），或被润湿而不流动处（死角），这些情况对传质无效。所以，填料层中有效传质面积小于填料表面积。$a=f$（填料的几何特性、气液物性、流动状态），难以准确知道，一般与 K_Y 或 K_X 放在一起测定。当填料、气液物性、流动状态一定时，体积总传质系数 $K_Ya\approx$常数，$K_Xa=$常数，不随塔高变化。

若将吸收速率方程式 $N_A=K_Y(Y-Y^*)=K_X(X^*-X)$ 分别代入式(3-81) 中，并在全塔范围内积分，整理得：

$$Z=\frac{V}{K_Ya\Omega}\int_{Y_2}^{Y_1}\frac{dY}{Y-Y^*} \tag{3-82}$$

$$Z=\frac{L}{K_Xa\Omega}\int_{X_2}^{X_1}\frac{dX}{X^*-X} \tag{3-83}$$

式(3-82) 和式(3-83) 称为低浓度气体吸收时计算填料层高度的基本方程式。若浓度以其他方式表示，上述的基本方程还可以写成其他形式。

(1) 传质单元（高度）与传质单元数（number of transfer units and height of a transfer unit）

现以式(3-82) 为例，令：

$$H_{OG}=\frac{V}{K_Ya\Omega} \tag{3-84}$$

$$N_{OG}=\int_{Y_2}^{Y_1}\frac{dY}{Y-Y^*} \tag{3-85}$$

$$Z=H_{OG}N_{OG} \tag{3-86}$$

式中，H_{OG} 为气相总传质单元高度，m。其物理含义为塔内取某段高度填料层，若该段填料层的气相浓度变化等于该段填料层以气相浓度表示的总推动力，则该段填料层高度为一个传质单元（高度）。

N_{OG} 是一个无量纲的数，称为气相总传质单元数。其物理含义是全塔气相浓度变化与全塔以气相浓度表示的总推动力之比。

将塔高计算式写成传质单元（高度）与传质单元数的乘积，只是变量的分离与合并，并无实质性的变化，但这样处理有以下优点：N_{OG} 中所含变量仅与相平衡关系、气相进出塔浓度有关，而与塔设备的形式无关，反映了完成分离任务的难易；H_{OG} 与操作状况、物性、填料几何形状等设备效能有关，是完成一个传质单元所需塔高，反映了设备效能的高低或填料层传质动力学性能的好坏。一般 H_{OG} 为 0.15～1.5m，具体数值由实验确定。

同理，式(3-83) 可表示成：

$$Z=H_{OL}N_{OL} \tag{3-87}$$

式中 H_{OL}——液相总传质单元高度，$H_{OL}=\dfrac{L}{K_Xa\Omega}$，m；

N_{OL}——称为液相总传质单元数，$N_{OL}=\displaystyle\int_{X_2}^{X_1}\frac{dX}{X^*-X}$，无量纲。

对应于其他的浓度表示方法，还可将传质单元高度和传质单元数写成其他形式。

(2) 传质单元数的求取

根据物系相平衡关系的不同，求取传质单元数的方法主要有以下两类。

① 相平衡方程为直线——解析法求传质单元数 若在吸收操作所涉及的浓度区间内，相平衡关系可按直线处理，即可写成 $Y^*=mX+b$ 的形式，可采用以下两种解析法求取 N_{OG}。

a. 对数平均推动力法 以 $N_{OG}=\int_{Y_2}^{Y_1}\frac{dY}{Y-Y^*}$ 的计算为例，当相平衡方程和操作线方程均为直线时，则塔内任一横截面上气相总传质推动力 $\Delta Y=Y-Y^*$ 与 Y 呈线性关系，即可写成：

$$dY=\frac{Y_1-Y_2}{\Delta Y_1-\Delta Y_2}d(\Delta Y) \tag{3-88}$$

式中 ΔY_2——塔顶气相浓度表示的总传质推动力；

ΔY_1——塔底气相浓度表示的总传质推动力。

将上式代入式(3-85) 中，则有：

$$N_{OG}=\int_{Y_2}^{Y_1}\frac{dY}{Y-Y^*}=\frac{Y_1-Y_2}{\Delta Y_m} \tag{3-89}$$

$$\Delta Y_m=\frac{\Delta Y_1-\Delta Y_2}{\ln\frac{\Delta Y_1}{\Delta Y_2}}=\frac{(Y_1-Y_1^*)-(Y_2-Y_2^*)}{\ln\frac{Y_1-Y_1^*}{Y_2-Y_2^*}} \tag{3-90}$$

式中 ΔY_m——塔顶与塔底两截面上吸收推动力对数平均值，称为对数平均推动力。

同理，可推出液相总传质单元数 N_{OL} 的相应解析式：

$$N_{OL}=\frac{X_1-X_2}{\Delta X_m} \tag{3-91}$$

$$\Delta X_m=\frac{\Delta X_1-\Delta X_2}{\ln\frac{\Delta X_1}{\Delta X_2}}=\frac{(X_1^*-X_1)-(X_2^*-X_2)}{\ln\frac{X_1^*-X_1}{X_2^*-X_2}} \tag{3-92}$$

当 $0.5<\frac{\Delta Y_1}{\Delta Y_2}<2$ 或 $0.5<\frac{\Delta X_1}{\Delta X_2}<2$ 时，相应的对数平均推动力也可用算术平均值代替，不会带来较大的误差。

b. 脱吸因数法 气相总传质单元数中的积分值还可用另一种方法求得，即将 $Y^*=mX+b$ 代入式(3-85) 中整理得：

$$N_{OG}=\frac{1}{1-S}\ln\left[(1-S)\frac{Y_1-Y_2^*}{Y_2-Y_2^*}+S\right] \tag{3-93}$$

式中 S——脱吸因数，$S=\frac{mV}{L}=\frac{m}{L/V}$。其几何意义为平衡线斜率 m 与操作线斜率 L/V 之比，S 值越大，越易于解吸。式(3-93) 也可写为

$$N_{OG}=\frac{1}{1-\frac{1}{A}}\ln\left[\left(1-\frac{1}{A}\right)\frac{Y_1-Y_2^*}{Y_2-Y_2^*}+\frac{1}{A}\right] \tag{3-93a}$$

式中 A——吸收因数，是解吸因数的倒数，$A=\frac{1}{S}$。A 值越大，越易于吸收。

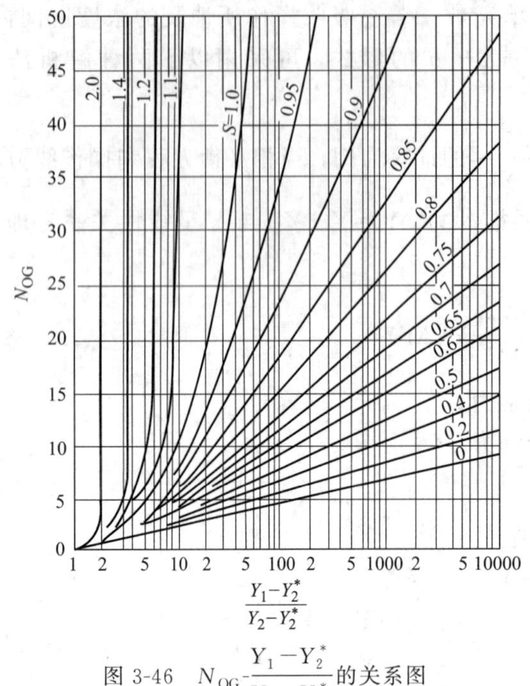

图 3-46 N_{OG}-$\dfrac{Y_1-Y_2^*}{Y_2-Y_2^*}$的关系图

由式(3-93) 可知，N_{OG} 数值的大小取决于 S 与$\dfrac{Y_1-Y_2^*}{Y_2-Y_2^*}$两个因素。为便于计算，在半对数坐标系中以 S 为参数，按式(3-93)标绘出 N_{OG}-$\dfrac{Y_1-Y_2^*}{Y_2-Y_2^*}$的函数关系，得到如图 3-46 所示的一组曲线。若已知 V、L、Y_1、Y_2、X_2 及相平衡线斜率 m 时，利用此图可方便地读出 N_{OG} 的数值；或由已知的 L、V、Y_1、X_2、N_{OG} 及 m 求出气体出口浓度 Y_2。

参数 S 则反映吸收推动力的大小。在气液进口浓度及溶质吸收率已知的条件下，若增大 S 的值，就意味着减小液气比，则使吸收液出口浓度提高而塔内吸收推动力下降，则 N_{OG} 的数值必然增大。反之亦然。通常认为 $S=0.7\sim0.8$ 是经济适宜的。

同理，可推导出液相总传质单元数 N_{OL} 的如下关系式：

$$N_{OL}=\frac{1}{1-A}\ln\left[(1-A)\frac{Y_1-Y_2^*}{Y_1-Y_1^*}+A\right] \tag{3-94}$$

式(3-93) 与式(3-94) 相比较可知，二者具有同样的函数形式，所以，图 3-46 将完全适用于表示 N_{OL}-$\dfrac{Y_1-Y_2^*}{Y_1-Y_1^*}$的关系（以 A 为参数）。

对数平均推动力法的优点是形式简明，适用于吸收塔的设计型计算；而对于已知填料层高和入塔气液流率及组成的操作型问题（或称校核型）来讲，采用脱吸因数法更为简便。所以，根据如上所述的两种方法的不同特点，可适当选择使用。

【例 3-11】 空气和氨的混合物在直径为 0.8m 的填料塔中用水吸收其中所含氨的 99.5%。混合气量为 1400kg/h。混合气体中氨与空气的摩尔比为 0.0132，所用液气比为最小液气比的 1.4 倍。操作温度为 20℃，相平衡关系为 $Y^*=0.75X$。气相体积吸收系数为 $K_Ya=0.088\text{kmol/(m}^3\cdot\text{s)}$，求每小时吸收剂用量与所需填料层高度。

解 （1）求吸收剂用量

因混合气体中氨含量很少，故用空气分子量计算出气相流率。

$$V=\frac{1400}{29}=48.3(\text{kmol/h})$$

$$Y_1=0.0132,Y_2=Y_1(1-\varphi_A)=0.0132\times(1-0.995)=0.000066$$

$$X_2=0$$

$$\left(\frac{L}{V}\right)_{\min}=\frac{Y_1-Y_2}{Y_1/m-X_2}=\frac{0.0132-0.000066}{0.0132/0.75-0}=0.746$$

则　$\frac{L}{V}=1.4\times0.746=1.04\Rightarrow L=1.04\times48.3=50.2(\text{kmol/h})$

(2) 用平均推动力法求填料层高度

$$X_1=\frac{V}{L}(Y_1-Y_2)+X_2=\frac{1}{1.04}\times(0.0132-0.000066)+0=0.0126$$

$$Y_1^*=mX_1=0.75\times0.0126=0.0095,\ Y_2^*=0$$

$$\Delta Y_1=Y_1-Y_1^*=0.0132-0.0095=0.0037$$

$$\Delta Y_2=Y_2-Y_2^*=0.000066-0=0.000066$$

$$\Delta Y_m=\frac{0.0037-0.000066}{\ln\frac{0.0037}{0.000066}}=0.000906$$

所以：$N_{OG}=\frac{Y_1-Y_2}{\Delta Y_m}=\frac{0.0132-0.000066}{0.000906}=14.5$

又有：$H_{OG}=\frac{V}{K_Y a\Omega}=\frac{48.3}{0.088\times3600\times0.785\times0.8^2}=0.307(\text{m})$

则：$Z=H_{OG}N_{OG}=14.5\times0.307=4.45(\text{m})$

(3) 用吸收因数法求填料层高度

因为：$S=\frac{mV}{L}=\frac{0.75}{1.04}=0.72$

且：$\frac{Y_1-Y_2^*}{Y_2-Y_2^*}=\frac{Y_1-mX_2}{Y_2-mX_2}=\frac{0.0132}{0.000066}=200$

由图 3-46 可查出：$N_{OG}=14.5$

则：$Z=0.307\times14.5=4.45(\text{m})$

【例 3-12】 在一逆流操作的填料塔中，用循环溶剂吸收某混合气体中的溶质。气体入塔组成为 0.025（摩尔比，下同），液气比为 1.6，操作条件下气液平衡关系为 $Y=1.2X$。若循环溶剂组成为 0.001，则出塔气体组成为 0.0025。现因脱吸不良，循环溶剂组成变为 0.01，试求此时出塔气体组成。

解　原工况：$X_1=\frac{V}{L}(Y_1-Y_2)+X_2=\frac{0.025-0.0025}{1.6}+0.001=0.0151$

$$\Delta Y_1=Y_1-Y_1^*=0.025-1.2\times0.0151=0.00688$$

$$\Delta Y_2=Y_2-Y_2^*=0.0025-1.2\times0.001=0.0013$$

$$\Delta Y_m=\frac{\Delta Y_1-\Delta Y_2}{\ln\frac{\Delta Y_1}{\Delta Y_2}}=\frac{0.00688-0.0013}{\ln\frac{0.00688}{0.0013}}=0.00335$$

$$N_{OG}=\frac{Y_1-Y_2}{\Delta Y_m}=\frac{0.025-0.0025}{0.00335}=6.72$$

新工况：$N'_{OG}=N_{OG}=6.72$

$$X'_1=\frac{V}{L}(Y_1-Y'_2)+X'_2=\frac{0.025-Y'_2}{1.6}+0.01$$

$$S=\frac{mV}{L}=\frac{1.2}{1.6}=0.75$$

$$N'_{OG}=\frac{1}{1-S}\ln\left[(1-S)\frac{Y_1-mX'_2}{Y'_2-mX'_2}+S\right]$$

$$6.72=\frac{1}{1-0.75}\ln\left[(1-0.75)\frac{0.025-1.2\times0.01}{Y'_2-1.2\times0.01}+0.75\right]$$

解得：$Y'_2=0.0127$

计算本题的关键是理解两种工况下 H_{OG} 不变，又因为填料层高度不变，故两种工况下的 N_{OG} 不变。

② 当相平衡线为曲线时——采用图解积分法　当气液相平衡线不能作为直线处理时，求传质单元数中的积分值时，通常采用图解积分或数值积分法。

采用图解积分法求解时，如图 3-47 所示，可在直角坐标系中，以 Y 为横坐标，$1/(Y-Y^*)$ 为纵坐标，将 $1/(Y-Y^*)$ 与 Y 的对应值标点描绘出曲线，所得函数曲线与 $Y=Y_2$，$Y=Y_1$ 及 $Y=1/(Y-Y^*)$ 三条直线之间所包围的面积，就是气相总传质单元数 N_{OG} 的数值，它可通过计量被积函数曲线下的面积来求得。

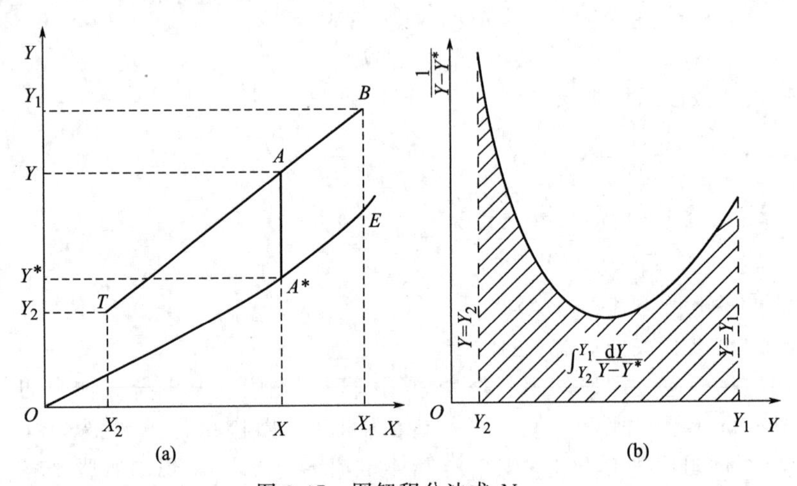

图 3-47　图解积分法求 N_{OG}

也可采用适宜的近似公式计算 N_{OG}，例如，可利用定步长的辛普森（Simpson）数值积分公式求解：

$$\int_{Y_0}^{Y_n} f(Y)\mathrm{d}Y \approx \frac{\Delta Y}{3}[f_0+f_n+4(f_1+f_3+\cdots+f_{n-1})+2(f_2+f_4+\cdots+f_{n-2})] \tag{3-95}$$

$$\Delta Y=\frac{Y_n-Y_0}{n}$$

式中　n——在 Y_0 和 Y_n 间划分的区间数目，可取任意偶数，n 值越大，则计算结果越准确；

ΔY——每个均等区间的步长；

Y_0——出塔气体组成，即 $Y_0=Y_2$；

Y_n——入塔气体组成，即 $Y_n=Y_1$；

f_i——为 $Y=Y_i$ 所对应的函数值。

若用积分法求液相总传质单元数 N_{OL} 或其他形式的传质单元数（如 N_G、N_L）时，方法与上相同。

3.3.7 其他

（1）解吸（脱吸）(adsorption or desorption)

一个完整的吸收流程通常由吸收和解吸（脱吸）联合操作。解吸的目的是使吸收剂再生后循环使用，还可以回收有价值的组分。解吸是吸收的逆过程，当液相中某一组分的平衡分压大于该组分在气相中A的分压时，可采用解吸操作。与吸收塔类似，解吸塔由塔顶（或塔底）与塔内任一截面做物料衡算，可得到解吸塔的操作线方程，不同的是，浓端在塔顶部，稀端在塔底部。

① 常用的解吸方法

a. 气提法（stripping） 该法也称为载气解吸法，其过程类似于逆流吸收，只是解吸时溶质由液相传递到气相。操作时，载气从塔底通入，与从塔顶来的吸收液逆流接触，则溶质不断地自液相扩散至气相中。一般来说，载气一般不含（或含极少）溶质的惰性气体或吸收剂蒸气。该类脱吸过程适用于溶剂的回收，不能直接得到纯净的溶质组分。

b. 提馏法 当溶质是可凝性蒸气时，而且溶质冷凝后与水不互溶，则可由塔底通入水蒸气作惰性气体进行解吸操作，水蒸气同时作为加热介质。此时，可将塔顶所得混合气体冷凝并由冷凝液中分离出水层，从而得到纯净的原溶质组分。

c. 闪蒸法 对于在加压情况下获得的吸收液，可采用一次或几次降低操作压力的方法，使溶质从吸收液中自动放出来，溶质被解吸的程度取决于解吸操作的最终压力和温度。

应予指出，在工程上很少采用单一解吸方式，往往是先升温再减压至常压，最后再采用气提法解吸。

② 减少解吸能耗的途径

a. 减少吸收剂用量 当气体流率一定时，最小吸收剂用量 L_{min} 由溶解度决定。溶解度越大，相平衡常数 m 越小，所需吸收剂用量越小，从而降低了解吸操作的能耗。

b. 减少吸收剂的温升 吸收剂的溶解度对温度变化的反应灵敏，即低温时溶解度大，但随着温度升高，溶解度迅速减小。

（2）高浓度气体吸收（absorption from rich gases）

当进口气体中溶质浓度>10%（体积分数），被吸收的溶质量较多时，称为高浓度气体吸收，此时，前面对于低浓度气体吸收的简化处理不再适用，高浓度气体吸收具有如下特点：

① 在高浓度气体吸收过程中，气体流率 V 及液相流率 L 沿塔高有明显变化，但惰性气体流率不变。若不考虑吸收剂挥发，则纯吸收剂流率也不变。

② 在高浓度气体吸收过程中，被吸收的溶质量较多，所产生的溶解热使两相温度升高，液体温度升高将对相平衡产生较大的影响。

③ 气膜及液膜吸收系数均受流动状况（包括气液流率）影响，因此在全塔不再为一个常量。

因为上述特点，使高浓度气体吸收过程的计算比低浓度气体吸收过程的计算要复杂得多。

（3）非等温吸收（non-isothermal absorption）

前面处理等温吸收时，都忽略了气液两相在吸收过程中的温度变化，即没有考虑吸收过程所伴随的热效应；而当溶质的溶解热较大尤其是伴随反应热时，使得液相温度不断升高，平衡关系不断发生变化，则不利于吸收。

非等温吸收的近似处理方法是假设所有放出的热量都被液体吸收，即忽略气相的温度变化及其他热损失。据此可以推算出液体组成与温度的对应关系，从而得到变温情况下的平衡曲线。当然，以上假设会导致对液体温升的估计偏高，因此计算出的塔高数值也偏大。

当吸收过程的热效应很大时，例如用水吸收 HCl，必须设法排除热量，以控制吸收过程的温度，通常采用以下几项措施：

① 在吸收塔内装置冷却元件，例如在板式塔上安装冷却蛇管或在板间设置冷却器；

② 引出吸收剂到外部进行冷却，例如在填料塔内不宜放置冷却元件，可将温度升高的吸收剂中途引出塔外，冷却后重新送入塔内继续进行吸收；

③ 采用边吸收边冷却的装置，例如盐酸吸收，采用管壳式换热器形式的吸收设备，使吸收过程在管内进行的同时，向壳方不断通入冷却剂以移除大量溶解热。

④ 采用大的喷淋密度，使吸收过程释放的热量以显热的形式被大量吸收剂带走。

3.4 萃取 (Extraction)

对于液体混合物的分离，除可采用蒸馏的方法外，还可采用萃取的方法，即在液体混合物（原料液）中加入一种与原料液不相混溶的液体作为溶剂，造成第二相，利用原料液中各组分在两个液相中的溶解度不同而使原料液混合物得以分离，称为液-液萃取，亦称溶剂萃取，简称萃取或抽提。选用的溶剂称为萃取剂，用 S 表示，它应对原料液中一个组分有较大溶解力，该易溶组分称为溶质，用 A 表示；对另一组分完全不溶解或部分溶解，该难溶于 S 的组分称为原溶剂（或稀释剂），用 B 表示。

如果萃取过程中，萃取剂与原料液中的有关组分不发生化学反应，则称为物理萃取，反之则称为化学萃取。

3.4.1 萃取原理 (Principles of extraction)

（1）萃取基本原理

萃取操作的基本过程如图 3-48 所示。将一定量萃取剂加入原料液中，然后加以搅拌使原料液与萃取剂充分混合，溶质通过相界面由原料液向萃取剂中扩散，所以萃取操作与精馏、吸收等过程一样，也属于两相间的传质过程。搅拌停止后，两液相因密度不同而分层：一层以溶剂 S 为主，并溶有较多的溶质，称为萃取相，以 E 表示；另一层以原溶剂（稀释剂）B 为主，且含有未被萃取完的溶质，称为萃余相，以 R 表示。若溶剂 S 和 B 为部分互溶，则萃取相中还含有少量的 B，萃余相中亦含有少量的 S。

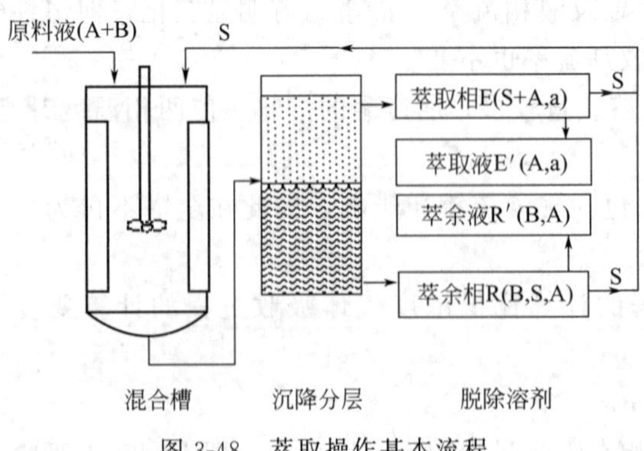

图 3-48　萃取操作基本流程

由上可知，萃取操作并未得到纯净的组分，而是新的混合液：萃取相 E 和萃余相 R。为了得到产品 A，并回收溶剂以供循环使用，尚需对这两相分别进行分离。

通常采用蒸馏或蒸发的方法，有时也可采用结晶等其他方法。脱除溶剂后的萃取相和萃余相分别称为萃取液和萃余液，以 E′和 R′表示。

对于一种液体混合物，究竟是采用蒸馏还是萃取加以分离，主要取决于技术上的可行性和经济上的合理性。一般地，在下列情况下采用萃取方法更为有利

① 原料液中各组分间的沸点非常接近，即组分间的相对挥发度接近于1，若采用蒸馏方法很不经济；

② 料液在蒸馏时形成恒沸物，用普通蒸馏方法不能达到所需的纯度；

③ 原料液中需分离的组分含量很低且为难挥发组分，若采用蒸馏方法须将大量稀释剂汽化，能耗较大；

④ 原料液中需分离的组分是热敏性物质，蒸馏时易于分解、聚合或发生其他变化。

Liquid Extraction

When separation by distillation is ineffective or very difficult, liquid extraction is one of the main alternatives to consider. Close-boiling mixtures or substances that cannot withstand the temperature of distillation, even under a vacuum, may often be separated from impurities by extraction, which utilizes chemical differences instead of vapor pressure differences. For example, penicillin is recovered from the fermentation broth by extraction with a solvent such as butyl acetate, after lowering the pH to get a favorable partition coefficient. The solvent is then treated with a buffered phosphate solution to extract the penicillin from the solvent and give a purified aqueous solution, from which penicillin is eventually produced by drying. Extraction is also used to recover acetic acid from dilute aqueous solutions; distillation would be possible in this case, but the extraction step considerably reduces the amount of water to be distilled.

One of the major uses of extraction is to separate petroleum products that have different chemical structures but about the same boiling range. Lube oil fractions (bp$>$300℃) are treated with low-boiling-point polar solvents such as phenol, furfural, or methyl pyrrolidone to extract the aromatics and leave an oil that contains mostly paraffins and naphthenes. The aromatics have poor viscosity-temperature characteristics, but they cannot be removed by distillation because of the overlapping boiling-point ranges. In a similar process, aromatics are extracted from catalytic reformate using a high-boiling-point polar solvent, and the extract is later distilled to give pure benzene, toluene, and xylenes for use as chemical intermediates. An excellent solvent for this use is the cyclic compound $C_4H_8SO_2$ (Sulfolane), which has high selectivity for aromatics and very low volatility (bp of 290℃).

When either distillation or extraction may be used, the choice is usually distillation, in spite of the fact that heating and cooling are needed. In extraction the solvent must be recovered for reuse (usually by distillation), and the combined operation is more complicated and often more expensive than ordinary distillation without extraction. However, extraction does offer greater flexibility in choice of operating conditions, since the type and amount of solvent can be varied as well as the operating temperature. In this sense, extraction is more like gas absorption than ordinary distillation. In many problems, the choice between methods should be based on a comparative study of both extraction and distillation.

Extraction may be used to separate more than two components; and mixtures of solvents, instead of a single solvent, are needed in some applications. These more complicated methods are not treated in this text.

（2）萃取剂的选择

选择合适的萃取剂是保证萃取操作能够正常进行且经济合理的关键，萃取剂的选择主要考虑以下因素。

① 萃取剂的选择性及选择性系数　萃取剂的选择性是指萃取剂 S 对原料液中两种组分溶解能力的差异。若 S 对溶质 A 的溶解能力比对原溶剂 B 的溶解能力大得多，即萃取相中 y_A 比 y_B 大得多，萃余相中 x_B 比 x_A 大得多，那么这种萃取剂的选择性就好。

萃取剂的选择性可用选择性系数 β 表示，其定义式为：

$$\beta=\frac{\dfrac{\text{萃取相中 A 的质量分数}}{\text{萃取相中 B 的质量分数}}}{\dfrac{\text{萃余相中 A 的质量分数}}{\text{萃余相中 B 的质量分数}}}=\frac{\dfrac{y_A}{y_B}}{\dfrac{x_A}{x_B}}=\frac{\dfrac{y_A}{x_A}}{\dfrac{y_B}{x_B}}=\frac{K_A}{K_B} \tag{3-96}$$

式中　β——选择性系数，无量纲；

y_A，y_B——萃取相 E 中组分 A、B 的质量分数；

x_A，x_B——萃余相 R 中组分 A、B 的质量分数；

K_A，K_B——组分 A、B 的分配系数。

由 β 的定义可知，选择性系数 β 为组分 A、B 的分配系数之比，其物理意义颇似蒸馏中的相对挥发度。若 $\beta>1$，说明组分 A 在萃取相中的相对含量比萃余相中的高，即组分 A、B 得到了一定程度的分离，显然 K_A 值越大，K_B 值越小，选择性系数 β 就越大，组分 A、B 的分离也就越容易，相应的萃取剂的选择性也就越高；若 $\beta=1$，则由上式可知，$\frac{y_A}{x_A}=\frac{y_B}{x_B}$ 或 $K_A=K_B$，即萃取相和萃余相在脱除溶剂 S 后将具有相同的组成，并且等于原料液的组成，说明 A、B 两组分不能用此萃取剂分离，换言之所选择的萃取剂是不适宜的。

萃取剂的选择性越高，则完成一定的分离任务所需的萃取剂用量也就越少，相应地用于回收溶剂操作的能耗也就越低。

由上式可知，当组分 B、S 完全不互溶时，$y_B=0$，则选择性系数趋于无穷大，显然这是最理想的情况。

② 萃取剂回收的难易与经济性　萃取后的 E 相和 R 相，通常以蒸馏的方法进行分离。萃取剂回收的难易直接影响萃取操作的费用，从而在很大程度上决定萃取过程的经济性。因此，要求萃取剂 S 与原料液中组分的相对挥发度要大，不应形成恒沸物，并且最好是组成低的组分为易挥发组分。若被萃取的溶质不挥发或挥发度很低时，则要求 S 的汽化热要小，以节省能耗。

③ 萃取剂的其他物性　为使两相在萃取器中能较快地分层，要求萃取剂与被分离混合物有较大的密度差，同时，两液相间的界面张力对萃取操作具有重要影响，且要考虑溶剂的黏度影响。此外，选择萃取剂时，还应考虑如化学稳定性和热稳定性，对设备的腐蚀性要

小，来源充分，价格较低廉，不易燃易爆等。

通常，很难找到能同时满足上述所有要求的萃取剂，这就需要根据实际情况加以权衡，以保证满足主要要求。

3.4.2 液-液相平衡（Phase equilibria of liquid-liquid）

液-液相平衡是萃取传质过程进行的极限，与气液传质相同，在讨论萃取之前，首先要了解液-液的相平衡问题。由于萃取的两相通常为三元混合物，故其组成和相平衡的图解表示法与前述气液传质不同，在此首先介绍三元混合物组成在三角形坐标图上的表示方法，然后介绍液-液平衡相图及萃取过程的基本原理。

3.4.2.1 三角形坐标图及杠杆规则

（1）三角形坐标图（rectangular coordinates diagrams）

三角形坐标图通常有等边三角形坐标图、等腰直角三角形坐标图和非等腰直角三角形坐标图，如图 3-49 所示，其中以等腰直角三角形坐标图最为常用。

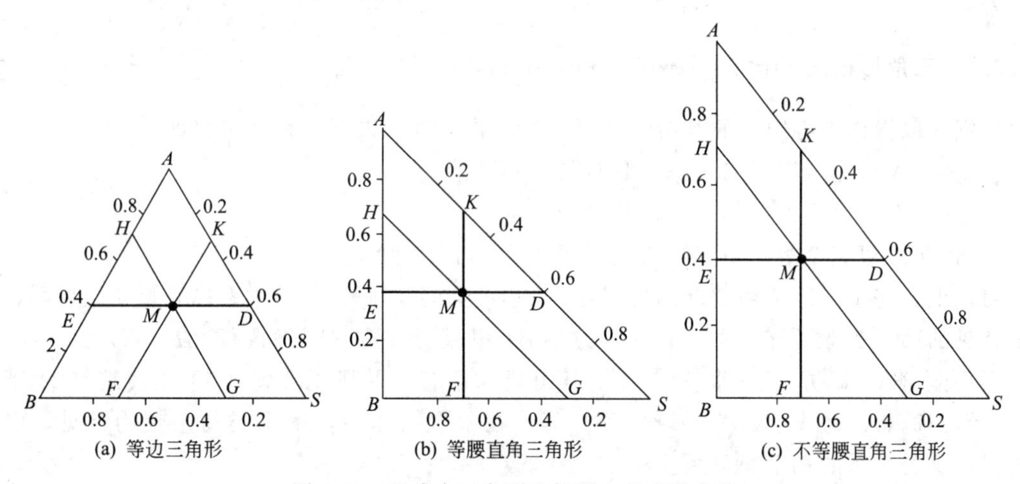

图 3-49 组成在三角形坐标图上的表示方法

一般而言，在萃取过程中很少遇到恒摩尔流的简化情况，故在三角形坐标图中混合物的组成常用质量分数表示。习惯上，在三角形坐标图中，*AB* 边以 *A* 的质量分数作为标度，*BS* 边以 *B* 的质量分数作为标度，*SA* 边以 *S* 的质量分数作为标度。

三角形坐标图的每个顶点分别代表一种纯组分，即顶点 *A* 表示纯溶质 A，顶点 *B* 表示纯原溶剂（稀释剂）B，顶点 *S* 表示纯萃取剂 S。三角形坐标图三条边上的任一点代表一个二元混合物系，第三组分的组成为零。例如 *AB* 边上的 *E* 点，表示由 A、B 组成的二元混合物系，由图可读得：A 的组成为 0.40，则 B 的组成为 1.0－0.40＝0.60，S 的组成为零。

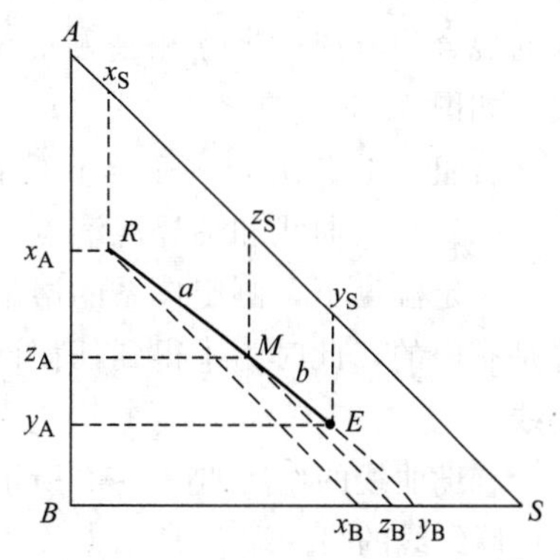

图 3-50 杠杆规则的应用

三角形坐标图内任一点代表一个三元混合物系。例如 *M* 点即表示由 A、B、S 三种组分组成的混合物系。其组成可按下法确定：过物系点 M 分别作对边的平行线 ED、HG、KF，则由点 E、G、K 可直接读得 A、B、S 的组成分别为：$x_A=0.40$、$x_B=0.30$、$x_S=0.30$。

（2）杠杆规则（lever-arm rule）

如图 3-50 所示，将质量为 $r$kg、组成为 x_A、x_B、x_S 的混合物系 R 与质量为 $e$kg、组成为 y_A、y_B、y_S 的混合物系 E 相混合，得到一个质量为 $m$kg、组成为 z_A、z_B、z_S 的新混合物系 M，其在三角形坐标图中分别以点 R、E 和 M 表示。M 点称为 R 点与 E 点的和点，R 点与 E 点称为差点。

和点 M 与差点 E、R 之间的关系可用杠杆规则描述，即：

① 几何关系　和点 M 与差点 E、R 共线，即和点在两差点的连线上，一个差点在另一差点与和点连线的延长线上。

② 数量关系　和点与差点的量 m、r、e 与线段长 a、b 之间的关系符合杠杆原理，即：以 R 为支点可得 m、e 之间的关系为 $ma=e(a+b)$；以 M 为支点可得 r、e 之间的关系为 $ra=eb$；以 E 为支点可得 r、m 之间的关系为 $r(a+b)=mb$。

根据杠杆规则，若已知两个差点，则可确定和点；若已知和点和一个差点，则可确定另一个差点。

3.4.2.2　三角形相图（triangular phase diagrams）

根据萃取操作中各组分的互溶性，可将三元物系分为以下三种情况，即：

① 溶质 A 可完全溶于 B 及 S，但 B 与 S 不互溶；

② 溶质 A 可完全溶于 B 及 S，但 B 与 S 部分互溶；

③ 溶质 A 可完全溶于 B，但 A 与 S 及 B 与 S 部分互溶。

习惯上，将①、②两种情况的物系称为第Ⅰ类物系，而将③的物系称为第Ⅱ类物系。工业上常见的第Ⅰ类物系有：丙酮（A）-水（B）-甲基异丁基酮（S）、乙酸（A）-水（B）-苯（S）及丙酮（A）-氯仿（B）-水（S）等；第Ⅱ类物系有：甲基环己烷（A）-正庚烷（B）-苯胺（S）、苯乙烯（A）-乙苯（B）-二甘醇（S）等。在萃取操作中，第Ⅰ类物系较为常见，以下主要讨论这类物系的相平衡关系。

（1）溶解度曲线及联结线

设溶质 A 可完全溶于 B 及 S，但 B 与 S 为部分互溶，其平衡相图如图 3-51 所示。此图是在一定温度下绘制的，图中曲线 $R_0R_1R_2R_iR_nKE_nE_iE_2E_1E_0$ 称为溶解度曲线，该曲线将三角形相图分为两个区域：曲线以内的区域为两相区，以外的区域为均相区。位于两相区内的混合物分成两个互相平衡的液相，称为共轭相，联结两共轭液相相点的直线称为联结线，如图 3-51 中的 R_iE_i 线（$i=0,1,2,\cdots,n$）。显然萃取操作只能在两相区内进行。

若组分 B 与组分 S 完全不互溶，则点 R_0 与 E_0 分别与三角形顶点 B 及顶点 S 相重合。

（2）辅助曲线和临界混溶点

一定温度下，测定体系的溶解度曲线时，实验测出的联结线的条数（即共轭相的对数）总是有限的，此时为了得到任何已知平衡液相的共轭相的数据，常借助辅助曲线（亦称共轭曲线）。

辅助曲线的作法如图 3-52 所示，通过已知点 R_1、R_2、…分别作 BS 边的平行线，再通过相应联结线的另一端点 E_1、E_2、…分别作 AB 边的平行线，各线分别相交于点 F、G、…，连接这些交点所得的平滑曲线即为辅助曲线。

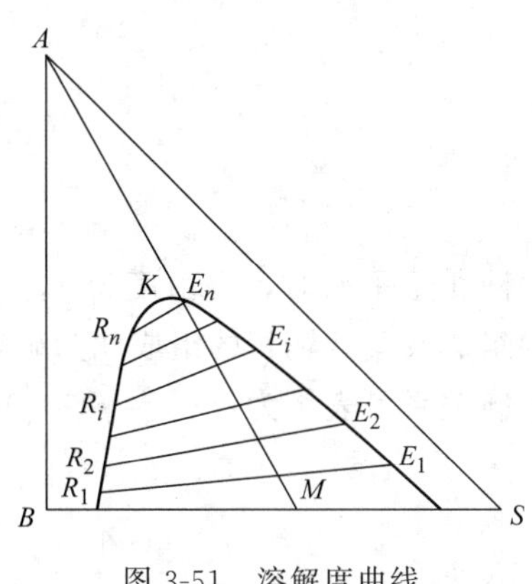

图 3-51 溶解度曲线

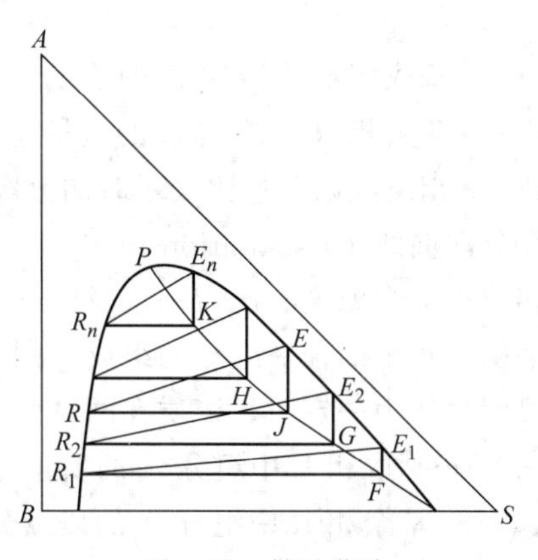

图 3-52 辅助曲线

利用辅助曲线可求任何已知平衡液相的共轭相。如图 3-52 所示，设 R 为已知平衡液相，自点 R 作 BS 边的平行线交辅助曲线于点 J，自点 J 作 AB 边的平行线，交溶解度曲线于点 E，则点 E 即为 R 的共轭相点。

辅助曲线与溶解度曲线的交点为 P，显然通过 P 点的联结线无限短，即该点所代表的平衡液相无共轭相，相当于该系统的临界状态，故称点 P 为临界混溶点。P 点将溶解度曲线分为两部分：靠原溶剂 B 一侧为萃余相部分，靠溶剂 S 一侧为萃取相部分。由于联结线通常都有一定的斜率，因而临界混溶点一般并不在溶解度曲线的顶点。临界混溶点由实验测得，但仅当已知的联结线很短即共轭相接近临界混溶点时，才可用外延辅助曲线的方法确定临界混溶点。

通常，一定温度下的三元物系溶解度曲线、联结线、辅助曲线及临界混溶点的数据均由实验测得，有时也可从手册或有关专著中查得。

3.4.2.3 分配系数和分配曲线

(1) 分配系数（distribution coefficient）

一定温度下，某组分在互相平衡的萃取相（E 相）与萃余相（R 相）中的组成之比称为该组分的分配系数，以 K 表示，即：

溶质 A：
$$K_A=\frac{y_A}{x_A} \tag{3-97}$$

原溶剂 B：
$$K_B=\frac{y_B}{x_B} \tag{3-97a}$$

式中 y_A，y_B——萃取相 E 中组分 A、B 的质量分数；

x_A，x_B——萃余相 R 中组分 A、B 的质量分数。

分配系数 K_A 表达了溶质在两个平衡液相中的分配关系。显然，K_A 值愈大，萃取分离的效果愈好。K_A 值与联结线的斜率有关。同一物系，其 K_A 值随温度和组成而变。如第Ⅰ类物系，一般 K_A 值随温度的升高或溶质组成的增大而降低。一定温度下，仅当溶质组成范围变化不大时，K_A 值才可视为常数。

对于萃取剂 S 与原溶剂 B 互不相溶的物系，溶质在两液相中的分配关系与吸收中的类似，即：

$$Y=KX \tag{3-98}$$

式中 Y——萃取相 E 中溶质 A 的质量比组成；

X——萃余相 R 中溶质 A 的质量比组成；

K——相组成以质量比表示时的分配系数。

（2）分配曲线（distribution line）

由相律可知，温度、压力一定时，三组分体系两液相呈平衡时，自由度为 1。故只要已知任一平衡液相中的任一组分的组成，则其他组分的组成及其共轭相的组成就为确定值。换言之，温度、压力一定时，溶质在两平衡液相间的平衡关系可表示为：$y_A=f(x_A)$

式中 y_A——萃取相 E 中组分 A 的质量分数；

x_A——萃余相 R 中组分 A 的质量分数。

上式即分配曲线的数学表达式。

如图 3-53 所示，若以 x_A 为横坐标，以 y_A 为纵坐标，则可在 x-y 直角坐标图上得到表示这一对共轭相组成的点 N。每一对共轭相可得一个点，将这些点连接起来即可得到曲线 ONP，称为分配曲线。曲线上的 P 点即为临界混溶点。

分配曲线表达了溶质 A 在互成平衡的 E 相与 R 相中的分配关系。若已知某液相组成，则可由分配曲线求出其共轭相的组成。若在分层区内 y 均大于 x，即分配系数 $K_A>1$，则分配曲线位于 $y=x$ 直线的上方，反之则位于 $y=x$ 直线的下方。

3.4.2.4 温度对相平衡的影响

通常物系的温度升高，溶质在溶剂中的溶解度增大，反之减小。因此，温度明显地影响溶解度曲线的形状、联结线的斜率和两相区面积，从而也影响分配曲线的形状。图 3-54 为温度对第Ⅰ类物系溶解度曲线和联结线的影响。显然，温度升高，分层区面积减小，不利于萃取分离的进行。

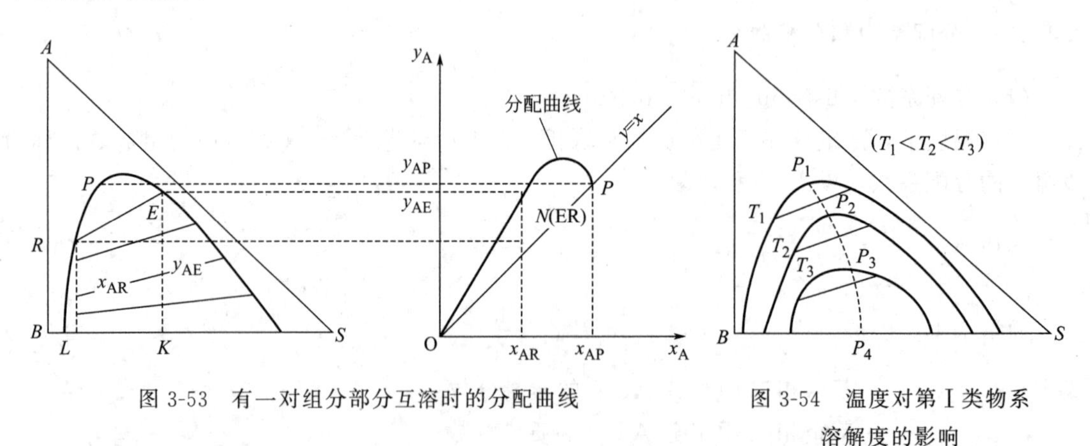

图 3-53 有一对组分部分互溶时的分配曲线

图 3-54 温度对第Ⅰ类物系溶解度的影响

3.4.3 萃取过程计算（Calculations for extraction processes）

根据两相接触方式的不同，萃取设备可分为逐级接触式和连续接触式两类。本节主要讨论逐级接触萃取过程的计算，对连续接触萃取过程的计算则仅做简要介绍。

在逐级接触萃取过程计算中，无论是单级还是多级萃取，均假设各级为理论级，即离开每一级的萃取相与萃余相互呈平衡。萃取理论级的概念类似于蒸馏中的理论板，是设备操作效率的比较基准。实际需要的级数等于理论级数除以级效率。级效率目前尚无准确的理论计算方法，一般通过实验测定。

在萃取过程计算中，通常操作条件下的平衡关系、原料液的处理量及组成均为已知。常见的计算可分为两类，一类是规定了各级的溶剂用量及组成，要求计算达到一定分离程度所需的理论级数 n；另一类是已知某多级萃取设备的理论级数 n，要求估算经该设备萃取后所能达到的分离程度。前者称为设计型计算，后者称为操作型计算。本知识点主要讨论设计型计算。

3.4.3.1 单级萃取计算（single-stage equilibrium extraction）

单级萃取是液-液萃取中最简单、最基本的操作方式，其流程如图 3-48 所示，可间歇操作也可连续操作。为方便表达，假定所有流股的组成均以溶质 A 的含量表示，故书写两相的组成时均只标注相应流股的符号，而不再标注组分的符号。

在单级萃取过程的设计型计算中，已知操作条件下的相平衡数据、原料液量 F 及组成 x_F、萃取剂的组成 y_S 和萃余相的组成 x_R，求萃取剂 S 的用量、萃取相 E 及萃余相 R 的量及萃取相组成 y_E。

（1）三角形坐标图解法

三角形坐标图解法是萃取计算的通用方法，特别是对于稀释剂 B 与萃取剂 S 部分互溶的物系，其平衡关系一般很难用简单的函数关系式表达，故其萃取计算不宜采用解析法或数值法，目前主要采用基于杠杆规则的三角形坐标图解法，其计算步骤如下：

① 由已知的相平衡数据在等腰直角三角形坐标图中绘出溶解度曲线及辅助曲线，如图 3-55 所示。

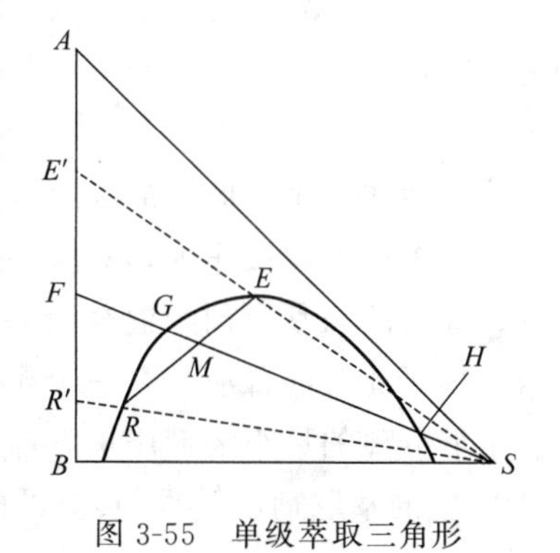

图 3-55 单级萃取三角形坐标图解

② 在三角形坐标的 AB 边上根据原料液的组成确定点 F，根据萃取剂的组成确定点 S（若为纯溶剂，则为顶点 S），连接点 F、S，则原料液与萃取剂的混合物系点 M 必落在 FS 连线上。

③ 由已知的萃余相组成 x_R，在图上确定点 R，再由点 R 利用辅助曲线求出点 E，作 R 与 E 的联结线，显然 RE 线与 FS 线的交点即为混合液的组成点 M。

④ 由质量衡算和杠杆规则求出各流股的量，即：

$$S=F\frac{\overline{MF}}{\overline{MS}} \tag{3-99}$$

$$M=F+S=R+E \tag{3-100}$$

$$E=M\frac{\overline{RM}}{\overline{RE}} \tag{3-101}$$

$$R=M-E \tag{3-102}$$

萃取相的组成可由三角形相图直接读出。

若从 E 相和 R 相中脱除全部溶剂，则得到萃取液 E′和萃余液 R′。因 E′和 R′中只含组分 A 和 B，所以它们的组成点必落于 AB 边上，且为 SE 和 SR 的延长线与 AB 的交点。可看出，E′中溶质 A 的含量比原料液 F 中的要高，而 R′中溶质 A 的含量比原料液 F 中的要低，即原料液的组分经过萃取并脱除溶剂后得到了一定程度的分离。E'和R'的数量关系可由杠杆规则来确定，即：

$$E'=F\frac{\overline{R'F}}{\overline{R'E'}} \tag{3-103}$$

$$R'=F-E' \tag{3-104}$$

以上诸式中各线段的长度可从三角形相图直接量出。上述各量亦可由质量衡算求出，组分 A 的质量衡算为：

$$Fx_F+Sy_S=Rx_R+Ey_E=Mx_M \tag{3-105}$$

联立求解上几式可得：

$$S=F\frac{x_F-x_M}{x_M-y_S} \tag{3-106}$$

$$E=M\frac{x_M-x_R}{y_E-x_R} \tag{3-107}$$

$$R=M-E \tag{3-108}$$

同理，可得萃取液和萃余液的量E'、R'，即：

$$E'=F\frac{x_F-x'_R}{y'_E-x'_R} \tag{3-109}$$

$$R'=F-E' \tag{3-110}$$

上述诸式中各股物流的组成可由三角形相图直接读出。

在单级萃取操作中，对应一定的原料液量，存在两个极限萃取剂用量，在两个极限用量下，原料液与萃取剂的混合物系点恰好落在溶解度曲线上，如图 3-55 中的点 G 和点 H 所示，由于此时混合液只有一个相，故不能起分离作用。这两个极限萃取剂用量分别表示能进行萃取分离的最小溶剂用量 S_{min}（和点 G 对应的萃取剂用量）和最大溶剂用量 S_{max}（和点 H 对应的萃取剂用量），其值可由杠杆规则计算，即：

$$S_{min}=F\left(\frac{\overline{FG}}{\overline{GS}}\right) \tag{3-111}$$

$$S_{max}=F\left(\frac{\overline{FH}}{\overline{HS}}\right) \tag{3-112}$$

显然，适宜的萃取剂用量应介于二者之间，即 $S_{min}<S<S_{max}$。

（2）解析法

对于原溶剂 B 与萃取剂 S 不互溶的物系，在萃取过程中，仅有溶质 A 发生相际转移，原溶剂 B 及溶剂 S 均只分别出现在萃余相及萃取相中，故用质量比表示两相中的组成较为方便。此时溶质在两液相间的平衡关系可以用与吸收中的气液平衡类似的方法表示，即 $Y=f(X)$。

若在操作范围内，以质量比表示相组成的分配系数 K 为常数，则平衡关系可表示为：$Y=KX$。

溶质 A 的质量衡算式为：

$$B(X_F-X_1)=S(Y_1-Y_S) \tag{3-113}$$

式中 B——原料液中原溶剂的量，kg 或 kg/h；

S——萃取剂中纯萃取剂的量，kg 或 kg/h；

X_F，Y_S——原料液和萃取剂中组分 A 的质量比组成；

X_1，Y_1——单级萃取后萃余相和萃取相中组分 A 的质量比组成。

联立求解上两式，即可求得 Y_1 与 S。

上述解法亦可在直角坐标图上表示，式(3-113) 可改写为：

$$\frac{Y_1-Y_S}{X_1-X_F}=-\frac{B}{S} \tag{3-113a}$$

式(3-113a) 即为该单级萃取的操作线方程。

由于该萃取过程中 B、S 均为常量，故操作线为过点（X_F，Y_S）、斜率为 $-B/S$ 的直线。如图 3-56 所示，当已知原料液处理量 F、组成 X_F、溶剂的组成 Y_S 和萃余相的组成 X_1 时，可由 X_1 在图中确定点（X_1，Y_1），连接点（X_1，Y_1）和点（X_F，Y_S）得操作线，计算该操作线的斜率即可求得所需的溶剂用量 S；当已知原料液处理量 F、组成 X_F、溶剂的用量 S 和组成 Y_S 时，则可在图中确定点（X_F，Y_S），过该点作斜率为 $-B/S$ 的直线（操作线）与分配曲线的交点坐标（X_1，Y_1）即为萃取相和萃余相的组成。

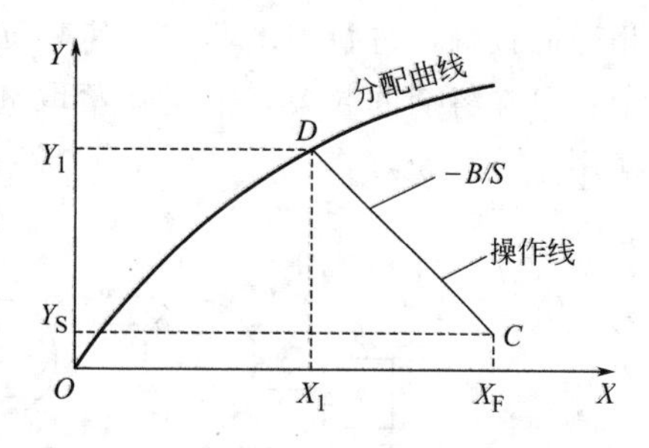

图 3-56 单级萃取的操作线方程图示

应予指出，在实际生产中，由于萃取剂都是循环使用的，故其中会含有少量的组分 A 与 B。同样，萃取液和萃余液中也会含有少量的 S。此时，图解计算的原则和方法仍然适用，但点 S 及 E'、R'的位置均在三角形坐标图的均相区内。

3.4.3.2 多级错流萃取（continuous multistage crosscurrent extraction）的计算

除了选择性系数极高的物系之外，一般单级萃取所得的萃余相中往往还含有较多的溶质，为进一步降低萃余相中溶质的含量，可采用多级错流萃取，其流程如图 3-57 所示。

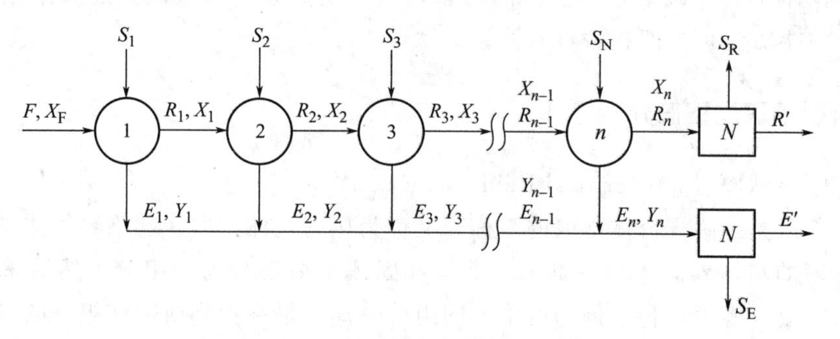

图 3-57 多级错流萃取流程示意图

在多级错流萃取操作中，每一级均加入新鲜萃取剂。原料液首先进入第一级，被萃取后，所得萃余相进入第二级作为原料液，并用新鲜萃取剂再次进行萃取，第二级萃取所得的萃余相又进入第三级作为原料液⋯，如此萃余相经多次萃取，只要级数足够多，最终可得到溶质组成低于指定值的萃余相。

多级错流萃取的总萃取剂用量为各级萃取剂用量之和，原则上，各级萃取剂用量可以相等也可以不等。但可以证明，当各级萃取剂用量相等时，达到一定的分离程度所需的总萃取剂用量最少，故在多级错流萃取操作中，一般各级萃取剂用量均相等。

在多级错流萃取过程的设计型计算中，已知操作条件下的相平衡数据、原料液量 F 及组成 x_F、萃取剂组成 y_S 和萃余相组成 x_R，求所需理论级数的具体方法包括三角形坐标图解法、直角坐标图解法及解析法。

3.4.3.3 多级逆流萃取（continuous multistage countercurrent extraction）的计算

在生产中，为了用较少的萃取剂达到较高的萃取率，常采用多级逆流萃取操作，其流程如图 3-58 所示。原料液从第 1 级进入系统，依次经过各级萃取，成为各级的萃余相，其溶质组成逐级下降，最后从第 n 级流出；萃取剂则从第 n 级进入系统，依次通过各级与萃余相逆向接触，进行多次萃取，其溶质组成逐级提高，最后从第 1 级流出。最终的萃取相与萃余相可在溶剂回收装置中脱除萃取剂从而得到萃取液与萃余液，脱除的溶剂返回系统循环使用。

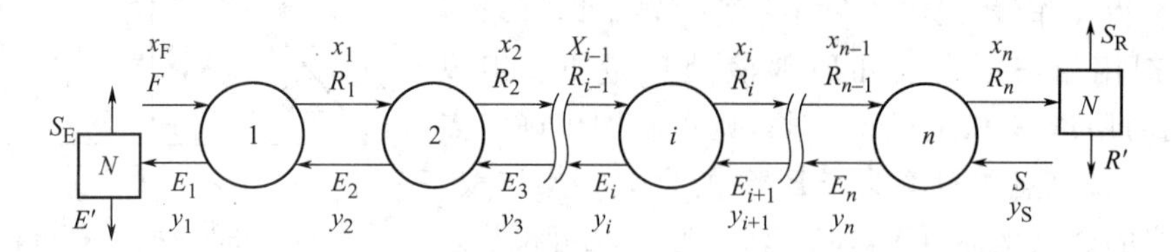

图 3-58 多级逆流萃取流程示意图

对于多级逆流萃取的设计型计算，已知原料液流量 F 和组成 x_F、萃取剂的用量 S 和组成 y_S，可求得最终萃余相中溶质组成降至一定值所需的理论级数 n。

对于原溶剂 B 与萃取剂 S 部分互溶的物系，由于其平衡关系难以用解析式表达，通常应用逐级图解法求解理论级数 n，具体方法有三角形坐标图解法和直角坐标图解法两种，它们是多级逆流萃取计算的通用方法。

对于原溶剂 B 与萃取剂 S 不互溶的物系，当平衡关系为直线时，还可采用解析法求算理论级数。以上方法可参照相关资料。

3.4.4 新型萃取过程简介

（1）超临界萃取（supercritical fluid extraction）

超临界萃取是超临界流体萃取的简称，又称为压力流体萃取、超临界气体萃取。它是以高压、高密度的超临界流体为溶剂，从液体或固体中溶解所需的组分，然后采用升温、降压、吸收（吸附）等手段将溶剂与所萃取的组分分离，最终得到所需纯组分的操作。例如从咖啡豆中脱除咖啡因，在生产链霉素时，利用超临界 CO_2 萃取去除甲醇等有机溶剂以及从单细胞蛋白游离物中提取脂类等研究均显示了超临界萃取技术的优势。

超临界流体是指超过临界温度与临界压力状态的流体。如果某种气体处于临界温度之上，则无论压力增至多高，该气体也不能被液化，称此状态的气体为超临界流体。超临界流体的密度接近于液体，黏度接近于气体，而自扩散系数介于气体和液体之间，比液体大 100 倍左右。因此，超临界流体既具有与液体相近的溶解能力，萃取时又具有远大于液态萃取剂的传质速率。二氧化碳是最常用的超临界流体。

超临界萃取过程主要包括萃取阶段和分离阶段。在萃取阶段，超临界流体将所需组分从原料中萃取出来；在分离阶段，通过改变某个参数，使萃取组分与超临界流体分离，从而得到所需的组分并可使萃取剂循环使用。根据分离方法的不同，可将超临界萃取流程分为等温变压流程、等压变温流程和等温等压吸附流程三类，如图 3-59 所示。

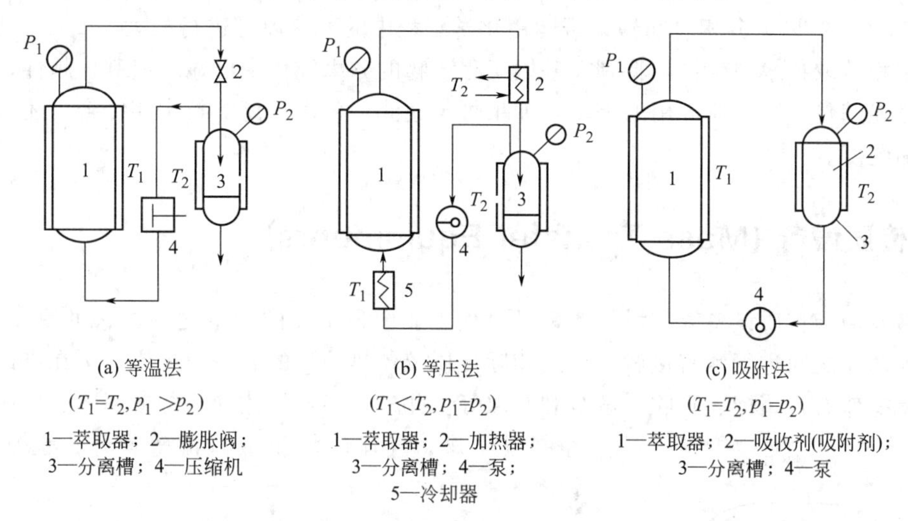

(a) 等温法 ($T_1=T_2, p_1>p_2$)
1—萃取器；2—膨胀阀；3—分离槽；4—压缩机

(b) 等压法 ($T_1<T_2, p_1=p_2$)
1—萃取器；2—加热器；3—分离槽；4—泵；5—冷却器

(c) 吸附法 ($T_1=T_2, p_1=p_2$)
1—萃取器；2—吸收剂(吸附剂)；3—分离槽；4—泵

图 3-59 超临界气体萃取的三种典型流程

(2) 反胶束萃取 (reversed micelles extraction)

传统的液液萃取技术尽管已在抗生素工业中广泛使用，却不适用于大部分基因工程的主要产品（蛋白质）的分离，这是因为难以选到一种具有良好选择性的萃取剂，需要找到一种与水不互溶，而蛋白质能溶于其中并保持活性的液相。近年来出现的反胶束萃取技术，用于萃取生物活性物质，就是采用满足上述要求的溶剂的新技术。

反胶束 (reversed micelle) 是表面活性剂在有机溶剂中自发形成的纳米尺度的一种聚集体，反胶束溶液是透明的、热力学稳定的系统。表面活性剂是由亲水的极性头和疏水的非极性尾两部分组成的两性分子。在反胶束溶液中，组成反胶束的表面活性剂的非极性尾向外伸入非极性有机溶剂主体中，而极性头则向内排列成一个极性核，此极性核具有溶解大分子（例如蛋白质）的能力。溶解了蛋白质的反胶束扩散进入有机相，从而实现了蛋白质的萃取，又由于蛋白质外表面有极性头的保护，使其避免与有机溶剂直接接触，所以蛋白质不会变性。

有机溶剂中表面活性剂浓度超过临界胶束浓度 (cmc) 时，才能形成反胶束溶液，这是体系的特性，与表面活性剂的化学结构、溶剂、温度及压力等因素有关。在非极性溶液中，cmc 值的变化范围是 $(0.1\sim1)\times10^{-3}$ mol/L。

(3) 双水相萃取 (aqueous two-phase extraction)

前面介绍的各种萃取方法，几乎都是利用溶质在水油两相的溶解度不同而达到分离目的。而该类萃取是利用两个互不相溶的水溶液相，组成双水相体系而竭力进行萃取分离。双水相体系萃取分离技术的原理是生物物质在双水相体系中选择性分配。当生物物质（例如酶、核酸、病毒等）进入双水相体系后，在上相和下相间进行选择性分配，具有一定的分配系数。在很大的浓度范围内，欲分离物质的分配系数与浓度无关，而与被分离物质本身的性质及特定的双水相体系性质有关，不同的物质在特定的体系中有着不同的分配系数。例如，在特定的双水相体系中，各种类型的细胞粒子、噬菌体等分配系数都大于 100 或小于 0.01，酶、蛋白质等生物大分子的分配系数大致在 0.1～10 之间，而小分子盐的分配系数在 1.0 左右。由此可见，双水相体系对上述物质的分配具有很大的选择性。

双水相萃取的特点：该技术可以利用不太复杂的设备，在温和条件下进行，系统的含水量多达 75%～90%，分相时间短，可运用化学工程中的萃取原理进行放大。

双水相萃取技术在生物，特别在基因工程产物的分离纯化中已显示出其优越性，所得产品纯度已能满足一般工业应用的需求，如果再与超滤、色谱分离等技术相结合，还可进一步提高产品纯度。

3.5 传质设备 (Mass Transfer Equipments)

气液及液液传质设备统称塔设备，是过程工业中最重要的设备之一，它可使气（或汽）液或液液两相之间进行紧密接触，达到相际传质及传热的目的。一般来说，可在塔设备中完成的单元操作有：蒸馏、吸收、解吸和萃取等。此外，工业气体的冷却或回收、气体的湿法净制或干燥、气体的增湿或减湿等操作，也常采用塔设备。

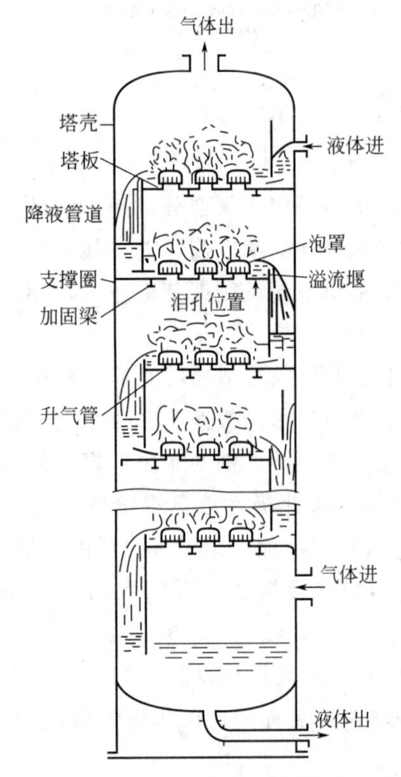

图 3-60 板式塔的典型结构

塔设备经过长期的发展，形成了各种各样的结构，以满足不同的需要。塔设备可按不同的方法分类，如按操作压力分为常压塔、减压塔和加压塔；按单元操作分为精馏塔、吸收塔、解吸塔、萃取塔、反应塔和干燥塔等。

长期以来，最常用的分类方法是按塔的内件结构分为填料塔和板式塔两大类。

3.5.1 板式塔 (Plate tower)

如图 3-60 所示，板式塔内装若干块塔板。操作时，一般是气体由塔底至塔顶，液体由塔顶至塔底，上升的气流在塔板的液层中鼓泡并带起液滴，气、液两相靠气泡表面或液滴表面传质，浓度沿塔高呈阶跃式变化，故称逐级接触（阶跃式）传质设备。

3.5.1.1 常见塔板类型

常见的塔板类型有：泡罩塔板、筛孔塔板、浮阀塔板、新型的喷射型塔板和浮动喷射塔板等。

泡罩塔板是最早在工业上大规模使用的板型，如图 3-61 所示。

泡罩塔每层塔板上开有若干个圆孔，上面覆以泡罩。操作时，液体横流过塔板时，泡罩下缘的齿缝浸没于液层之中形成液封，当上升气体通过齿缝进入液层时，被分散成许多细小的气泡型流股，为气液两相提供了大量的传质界面。

泡罩塔的优点是：不易漏液，有较大的操作弹性，易维持恒定的板效率；塔板不易堵塞，适于处理各种物料。但其缺点是结构复杂，压降大，且雾沫夹带现象较严重，限制了气

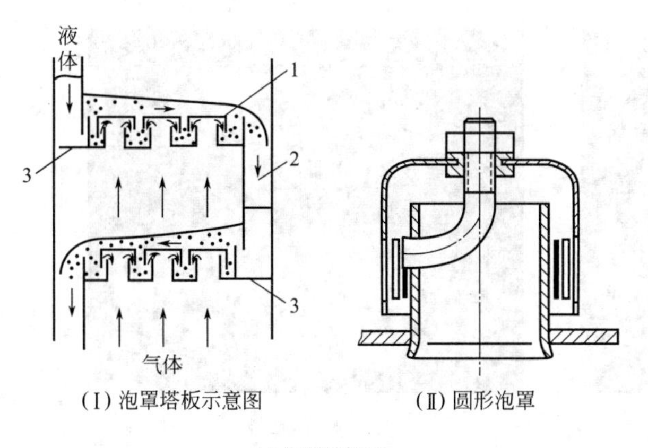

(a) 泡罩结构图

(b) 泡罩实物图

图 3-61 泡罩塔板

1—泡罩；2—降液管；3—塔板

速的提高，致使生产能力及板效率均较低，近年来泡罩塔已逐渐被筛板塔和浮阀塔所取代。

筛孔塔板结构如图 3-62 所示，塔板上开有许多均布的筛孔，孔径一般为 3～8mm，筛孔在板上呈正三角形排列。操作时，上升气流通过筛孔分散成细小的流股，在板上液层中鼓泡而出，气液间密切接触而进行传质。在通常的操作气速下，通过筛孔上升的气流，应能阻止液体经筛孔向下泄漏。

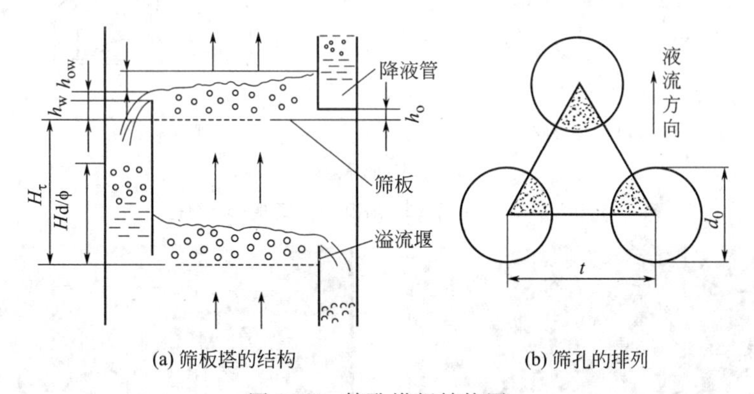

(a) 筛板塔的结构　　(b) 筛孔的排列

图 3-62 筛孔塔板结构图

筛孔塔板的优点是结构简单，造价低廉，气体压降小，板上液层落差也较小，生产能力及效率均较泡罩塔高，如图 3-63 所示；主要缺点是操作弹性小，筛孔小。筛板塔的筛孔较小时容易堵塞，近年来采用大孔径（12～25mm）筛板，可避免堵塞，而且由于气速的提高，生产能力增大，筛孔板已被广泛采用。

浮阀塔既有泡罩塔板的稳定性，又有筛孔塔板的大负荷，塔板上开大孔（标准孔径为 39mm），孔上盖有能上下移动的阀片，阀片上有三个脚，阀片周围又冲出三块略向下弯的定距片，阀片的这种特殊结构可使其按气流量大小自动上下调节，所以操作弹性很大，生产能力大，塔板效率高，气体压降及液面落差较小，且造价低，所以已被广泛采用。目前最常用的浮间型式为 F1 型和 V-4 型，F1 型浮阀（国外称为 V-1 型），如图 3-64(a)、(b) 所示，而图 3-64(c) 给出了浮阀在工业塔板上的布置情况。

上述泡罩、筛板及浮阀塔板都属于气体为分散相的塔板，塔板上在鼓泡或泡沫状态下进

(a) 筛孔布置图

(b) 设备图

图 3-63　筛孔塔板

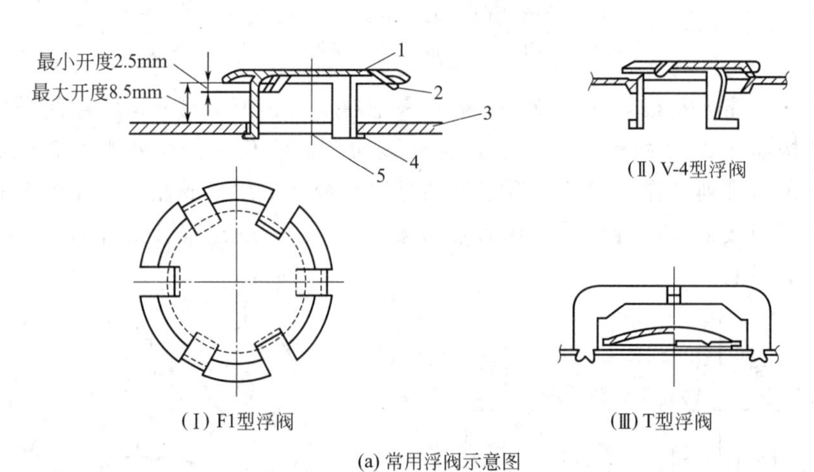

(a) 常用浮阀示意图

1—阀片；2—定距片；3—塔板；4—底脚；5—阀孔

(b) 常用浮阀实物图

(c) 浮阀塔布置图

图 3-64　浮阀塔

行气液接触。为防止严重的雾沫夹带，操作气速不可太高，故生产能力受到限制。近年发展起来的喷射型塔板克服了这个弱点。

在喷射型塔板上，由于气体喷出的方向与液体流动的方向一致，可充分利用气体的动能来减薄或打碎较深的液层，从而促进两相的接触。因而塔板压降减低，雾沫夹带量减小，不仅提高了传质效果，而且可采用较大的气速，提高了生产能力。其中具有代表性的是舌形塔板，其结构如图3-65所示。塔板上冲出许多舌形孔，舌片与板成一定角度，向塔板的溢流出口侧张开。上升气流以较高的速度（20～30m/s）沿舌片的张角向斜上方喷出，喷出的气流强烈扰动液体而形成泡沫体，从而强化了两相间的传质，能获得较高的塔板效率。板上液面较薄，塔板压强降小。

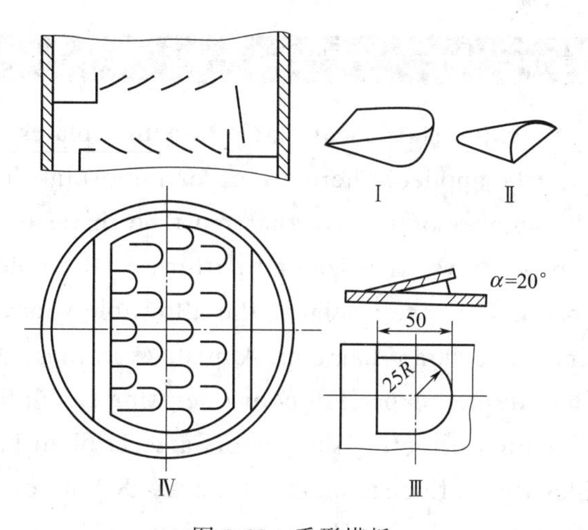

图3-65　舌形塔板

Ⅰ—三面切口舌片；Ⅱ—拱形舌片；Ⅲ—50mm×50mm定向舌片的尺寸和倾角；Ⅳ—塔板

浮动喷射塔板是兼有浮阀塔板的可变气道截面及舌形塔板的并流喷射特点的新型塔板，它允许较高的气流喷射速度，故生产能力大；由于浮动板的张开程度能随上升气体的流量而变化，使气流的喷出速度保持较高的适宜值，因而操作弹性大；此外，它还有压强降小，液面落差小等优点。其缺点是有漏液及“吹干”现象，影响传质效果，使板效率降低，且塔板结构较复杂，且例如浮舌塔板，如图3-66所示。

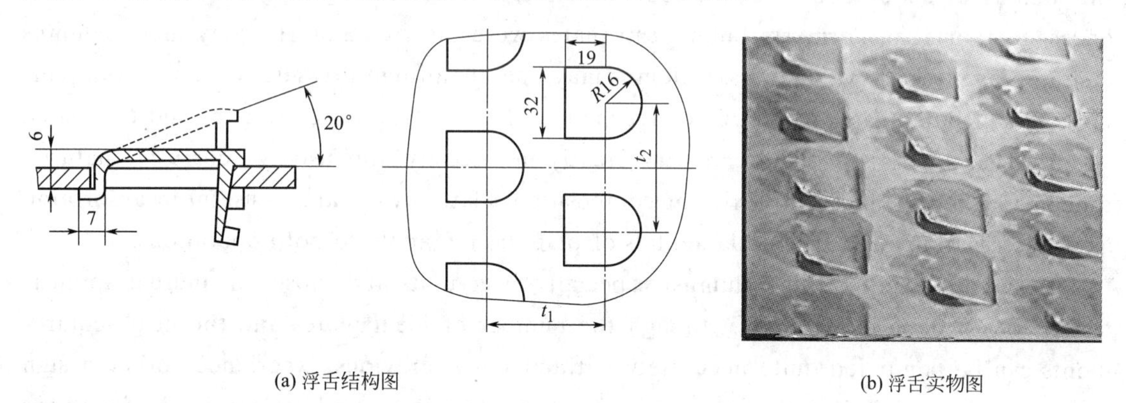

(a) 浮舌结构图　　(b) 浮舌实物图

图3-66　浮舌塔板

3.5.1.2　板式塔设备基本构件

板式塔为逐级接触式气液传质设备，它主要由圆柱形塔体（包括塔板、溢流堰、降液管及受液盘等内部部件）和以下辅助构件构成。

① 塔体　塔体是塔设备的外壳，由圆筒和上下椭圆形或圆形封头所组成。

② 塔体支座　塔体支座是塔体安放到基础上的连接部分。

③ 除沫装置　除沫装置用于捕集夹带在气流中的液滴。

④ 接管　塔设备的接管用于连接工艺管路。按接管的用途分为进液管、出液管、进气管、出气管、回流管、侧线抽出管和仪表接管等。

⑤ 人孔和手孔　人孔和手孔都是为了安装、检修和检查的需要而设置的。

⑥ 吊耳　为了起吊方便，在塔设备上焊以吊耳。

⑦ 吊柱　在塔顶设置吊柱是为了在安装和检修时，方便塔内件的运送。

Plate Tower Design

To translate ideal plates to actual plates, a correction for the efficiency of the plates must be applied. There are other important decisions, some at least as important as fixing the number of plates, that must be made before a design is complete. These include the type of trays, the size and pattern of the holes in the tray, the downcomer size, the tray spacing, the weir height, the allowable vapor and liquid rates and pressure drop per tray, and the column diameter. A mistake in these decisions results in poor fractionation, lower-than-desired capacity, poor operating flexibility, and with extreme errors, an inoperative column. Correcting such errors after a plant has been built can be costly. Since many variables that influence plate efficiency depend on the design of the individual plates, the fundamentals of plate design are discussed first.

The extent and variety of rectifying columns and their applications are enormous. The largest units are usually in the petroleum industry, but large and very complicated distillation plants are encountered in fractionating solvents, in treating liquefied air, and in general chemical processing. Tower diameters may range from 300 mm to more than 9 m and the number of plates from a few to about a hundred. Plate spacings may vary from 150 mm or less to 1 or 2 m. Formerly bubble-cap plates were most common; today most columns contain sieve trays or valve trays. Columns may operate at high pressures or low, from temperatures of liquid gases up to 900℃ reached in the rectification of sodium and potassium vapors. The materials distilled can vary greatly in viscosity, diffusivity, corrosive nature, tendency to foam, and complexity of composition. Plate towers are as useful in absorption as in rectification, and the fundamentals of plate design apply to both operations.

Designing fractionating columns, especially large units and those for unusual applications, is best done by experts. Although the number of ideal plates and the heat requirements can be computed quite accurately without much previous experience, other design factors are not precisely calculable, and a number of equally sound designs can be found for the same problem. In common with most engineering activities, sound design of fractionating columns relies on a few principles, on a number of empirical correlations (which are in a constant state of revision), and on much experience and judgement. The following discussion is limited to the usual types of column, equipped with sieve plates or valve-trays, operating at pressures not far from atmospheric, and treating mixtures having ordinary properties.

3.5.1.3 板式塔的流体力学特性(fluid mechanics of plate columns)

精馏塔的操作能否正常进行，与塔内气液两相的流体力学状况有关。板式塔的流体力学性能包括：塔板压降（vapor pressure drop）、液泛（flooding）、雾沫夹带（excessive entrainment）、漏液（weeping）及液面落差（hydraulic gradient）等。

（1）塔板上的气液流动状态

① 气液接触状态有三种：鼓泡接触状态、泡沫接触状态和喷射接触状态。

当操作气速很低时，数量较少的气泡、鼓泡穿过塔板上大量的清液层，两相接触面积为气泡表面。由于湍动程度较低，所以气液间的传质阻力较大，此时，气相为分散相而液相为连续相，称为鼓泡接触状态；随着气速增大，气泡数量急剧增加，则塔板上液体被大量气泡分隔成液膜，即两相传质表面是面积很大的液膜。由于高度湍动，所以传质阻力小。此时，气相仍为分散相而液相仍为连续相，称为泡沫接触状态。如果气速继续增大，气体射流穿过液层，将板上的液体破碎成大小不等的液滴而抛向板上空，落下后又反复被抛起，所以两相传质表面是液滴的外表面。此时，液体为分散相而气体为连续相，这是喷射接触状态与泡沫接触状态的根本区别，由泡沫状态转为喷射状态的临界点称为转相点。

工业上的操作多是控制在泡沫状态或喷射接触状态，其特征分别是有不断更新的液膜表面和液滴表面。

② 非理想流动：包括反向流动和不均匀流动。

反向流动又包括雾沫夹带和气泡夹带。所谓雾沫夹带指上升气流穿过塔板上液层时，将板上液体带入上一层板的现象；而下降液体流量过大时，将少量气泡带入下一块塔板的现象称为气泡夹带。过量的雾沫夹带或气泡夹带都会造成两相在塔板上的返混，进而导致塔板效率严重下降，因此应予以限制。例如生产中将雾沫夹带限制在一定限度以内，即控制雾沫夹带量<0.1kg(液)/kg(气)。

不均匀流动包括液膜流过塔盘时的不均匀流动和因盘上液位落差导致的气相不均匀流动。以上两种情况均可能导致气液接触不充分，所以应尽量避免。

Weeping and Hydraulic Gradient

At low vapor velocities, the pressure drop is not great enough to prevent liquid from flowing down through some of the holes. This condition is called weeping and is more likely to occur if there is a slight gradient in liquid head (hydraulic gradient) across the plate. Hydraulic gradient on plate is caused by the liquid flow resistance across the plate, including friction loss through plate surface, local resistance of parts (bubble-cap, floating valve) on plate and resistance from vapor flow. With such a gradient, vapor will tend to flow through the region where there is less liquid and therefore less resistance to flow, and liquid will flow through the section where the depth is greatest. Weeping decreases the plate efficiency, since some liquid passes to the next plate without contacting the vapor. The lower limit of operation could be extended by using smaller holes or a lower fraction open area, but these changes would increase the pressure drop and reduce the maximum flow rate. A sieve tray can usually be operated over a three- to four fold range of flow rates between the weeping and flooding points. If a greater range is desired, other types of plates such as valve trays can be used.

(2) 液泛现象及塔径的计算

若气液两相中的一相流量增大，使降液管内液体不能顺利下流，沿降液管向上逐板积累，并依次上升直至塔顶，这种现象称为液泛，亦称淹塔。此时，塔板压降急剧上升，气液的正常接触被破坏，所以操作时应避免液泛现象发生。

影响液泛速度的因素除气液流体流动特性外，塔板结构，特别是塔板间距也是重要因素，设计中采用较大的板间距，以提高液泛速度。

液泛是一种极不正常的操作，应尽量避免。设计时，先以不发生过量的雾沫夹带为原则。其设计思想为：首先计算出液泛气速 u_{max}→操作气速 u→塔截面＝气体流量/操作气速→算出降液管截面积→再用液体流量核算。

其中液泛气速：
$$u_{max}=c\frac{\sqrt{\rho_L-\rho_V}}{\sqrt{\rho_V}} \tag{3-114}$$

式中　c——气相负荷因子，m/s。c 值可由史密斯（Smith）关联图查取，如图 3-67 所示。

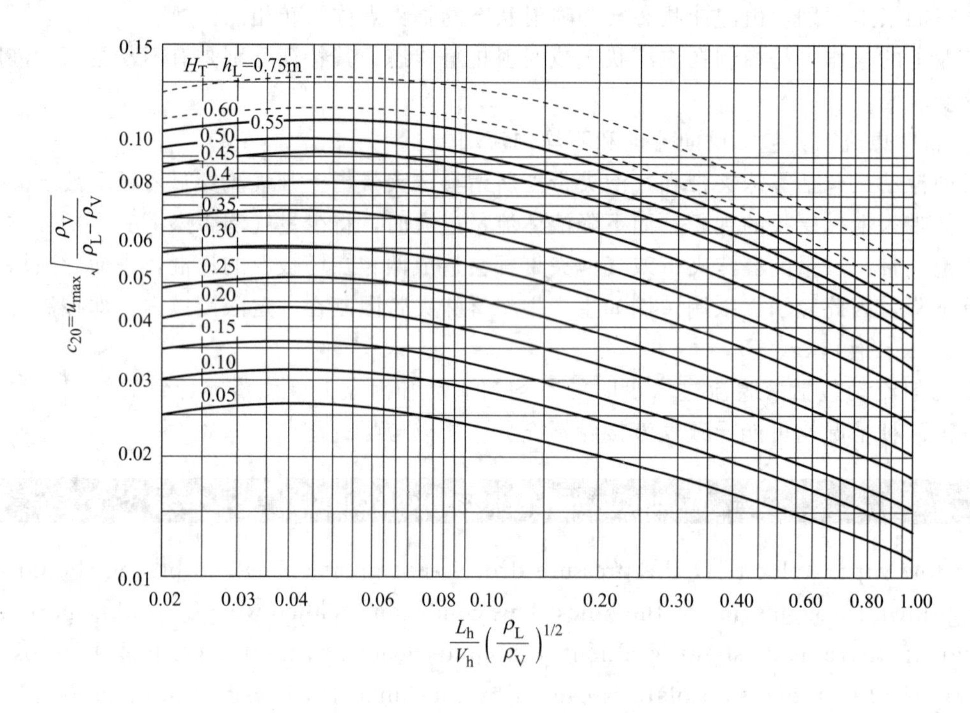

图 3-67　史密斯（Smith）关联图

再取正常的操作气速：
$$u=(0.6\sim0.8)u_{max} \tag{3-115}$$

则塔径表示为：
$$D=\sqrt{\frac{4V_s}{\pi u}} \tag{3-116}$$

式中　D——精馏塔内径，m；

u——空塔气速，m/s；

V_s——操作条件下气相体积流量，m^3/s。

若精馏操作压强较低时，气相可视为理想气体混合物，则：

$$V_s=\frac{22.4V_h}{3600}\times\frac{Tp_0}{T_0p} \tag{3-117}$$

式中　T，T_0——操作条件下的平均温度和标准状况下的热力学温度，K；

p，p_0——操作条件下的平均压强和标准状况下的压强，Pa。

由于进料状况及操作条件不同，精、提两段上升蒸气量可能不同，若计算的 $D_{精}$、$D_{提}$ 相差不大，应取大者使塔径统一，以便使塔的设计和安装简便。

Excessive Entrainment and Flooding

The upper limit of the vapor velocity in a sieve-tray column is usually determined by the velocity at which entrainment becomes excessive, causing a large drop in plate efficiency for a small increase in vapor rate. This limit is called the flooding point, or more properly, the entrainment flooding point, since the allowable vapor velocity is sometimes limited by other factors. When the pressure drop on a plate is too high and the liquid in the downcomer backs up to the plate above, the flow from that plate is inhibited. This leads to an increase in liquid level and a further increase in pressure drop. This phenomenon is called downcomer flooding; it can occur before entrainment becomes excessive. For liquids with low surface tension, the limit may be the velocity that makes the froth height equal to the plate spacing, leading to a large carryover of liquid to the plate above.

（3）适宜的气液流量操作范围——塔板负荷性能图（operating range and capacity chart）

要维持塔板正常操作，必须将塔内的气液负荷波动限制在一定范围内。通常在直角坐标系中，以气相负荷 V 及液相负荷 L 分别表示纵、横坐标，用五条曲线表示各种极限条件下的 V-L 关系，该图形称为塔板的负荷性能图，如图 3-68 所示。负荷性能图对检验塔的设计是否合理及改进塔板操作性能都具有一定的指导意义。

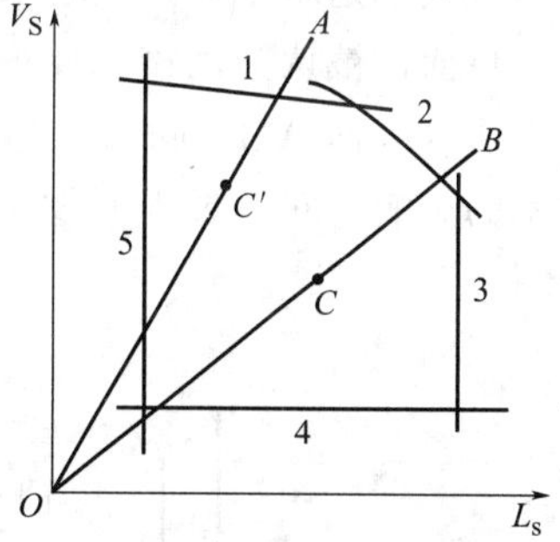

图 3-68 塔板负荷性能图

① 雾沫夹带线（又称气相负荷上限线）：当气相负荷超过此线时，雾沫夹带量将过大，甚至发生液泛现象，使板效率严重下降，塔板适宜操作区应在雾沫夹带线以下。

② 液泛线：塔板的适宜操作区应在此线以下，否则将会发生液泛现象，使塔不能正常操作。

③ 液相负荷上限线（又称降液管超负荷线）：液体流量超过此线，表明液体流量过大，液体在降液管内停留时间过短，进入降液管中的气泡来不及与液相分离而被带入下层塔板，造成气相返混，降低塔板效率。

④ 漏液线（又称气相负荷下限线）：气相负荷低于此线时，将发生严重的漏液现象，气液不能充分接触，使板效率下降。

⑤ 液相负荷下限线（又称吹干线）：液相负荷低于此线时，使塔板上液流不能均匀分布，导致板效率下降。

（4）塔板效率（plate efficiencies）

塔板效率反映了实际塔板上气液两相间传质与理论板对比的完善程度，板式塔的效率可用总板效率和单板效率表示。

① 全塔效率（又称总板效率）E_T（overall efficiency） 总板效率又称全塔效率，是指达到指定分离要求所需理论板数与实际板数的比值：

$$E_T=\frac{N_T}{N_P} \tag{3-118}$$

式中 N_T——塔内所需理论板的层数；

N_P——塔内实际板的层数。

该式将影响传质过程的动力学因素全部归结到总板效率内，它简单地反映了整个塔内的平均传质效果，而对每层塔板的传质效率没有明确表示。

② 单板效率（Murphree efficiency） 单板效率又称为默弗里（Murphree）板效率，是指气相或液相经过一层塔板前后的实际组成变化与经过该层塔板前后的理论组成变化的比值。例如，第 n 层塔板的效率有如下两种表达方式：

a. 按气相组成变化表示的单板效率为：

$$E_{MV}=\frac{y_n-y_{n+1}}{y_n^*-y_{n+1}} \tag{3-119}$$

b. 按液相组成变化表示的单板效率为：

$$E_{ML}=\frac{x_{n-1}-x_n}{x_{n-1}-x_n^*} \tag{3-120}$$

式中 y_n^*——与 x_n 成平衡的气相组成；

x_n^*——与 y_n 成平衡的液相组成。

其他符号的意义如图 3-69 所示。

单板效率可直接反映该层塔板的传质效果。一般说来，同一层塔板的 E_{MV} 与 E_{ML} 数值并不相同，各层塔板的单板效率通常也不相等。即使塔内各板效率相等，全塔效率在数值上也不等于单板效率，这是因为二者定义的基准不同，全塔效率是基于所需理论板数的概念，而单板效率是基于理论板增浓程度的概念。

目前，被认为比较能真实反映实际情况的，是美国化工学会提出的一套预测板效率的计算方法（简称 A. I. Ch. E 法），该方法不仅考虑了较多的影响因素，而且能反映塔径放大对效率的影响，对于过程开发很有意义。但是，这套计算方法程序颇为繁复，此处不做具体介绍。

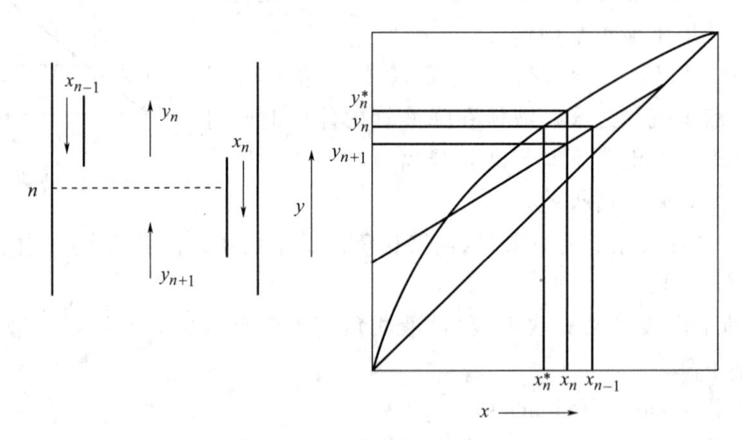

图 3-69 单板效率定义图

另一类方法是简化的经验计算法，该类方法归纳了试验数据及工业数据，得出总板效率与少数主要影响因素的关系，例如较多使用的奥康奈尔方法。

Plate efficiencies

In all the previous discussions of theoretical trays or stages in distillation, we assumed that the vapor leaving a tray was in equilibrium with the liquid leaving. However, if the time of contact and the degree of mixing on the tray are insufficient, the streams will not be in equilibrium. As a result the efficiency of the stage or tray will not be 100%. This means that we must use more actual trays for a given separation than the theoretical number of trays determined by calculation. To translate ideal plates to actual plates, the plate efficiency must be known. The following discussion applies to columns for gas absorption as well as to those for distillation.

Types of plate efficiency

Two kinds of plate efficiency are used: (1) overall efficiency, which concerns the entire column; (2) Murphree efficiency, which has to do with a single plate.

The overall efficiency η_0 is simple to use but is the least fundamental. It is defined as the ratio of the number of ideal plates needed in an entire column to the number of actual plates. For example, if six ideal plates are called for and the overall efficiency is 60 percent, the number of actual plates is $6/0.60 = 10$.

The Murphree efficiency η_M is defined by

$$\eta_M = \frac{y_n - y_{n+1}}{y_n^* - y_{n+1}} \qquad \text{(eq. 3-18)}$$

Where y_n = actual concentration of vapor leaving plate n;

y_{n+1} = actual concentration vapor entering plate n;

y_n^* = concentration of vapor in equilibrium with liquid leaving downpipe from plate n.

The Murphree efficiency is therefore the change in vapor composition from one plate to the next divided by the change that would have occurred if the vapor leaving were in equilibrium with the liquid leaving. The liquid leaving is generally not the same as the average liquid on the plate, and this distinction is important in comparing local and Murphree efficiencies.

The Murphree efficiency is defined using vapor concentrations as a matter of custom, but the measured efficiencies are rarely based on analysis of the vapor phase because of the difficulty in getting reliable samples. Instead, samples are taken of the liquid on the plates, and the vapor compositions are determined from a McCabe-Thiele diagram. A plate efficiency can be defined using liquid concentrations, but this is used only occasionally for desorption or stripping calculations.

Columns operated at high velocity will have significant entrainment, and this reduces the plate efficiency, because the drops of entrained liquid are less rich in the more volatile component than is the vapor. Although methods of allowing for entrainment have been published, most empirical correlations for the Murphree efficiency are based on liquid samples from the plates, and this includes the effect of entrainment.

3.5.2 填料塔 (Packed tower)

在填料吸收塔内，气液两相流动方式原则上可为逆流也可为并流。一般情况下塔内液体

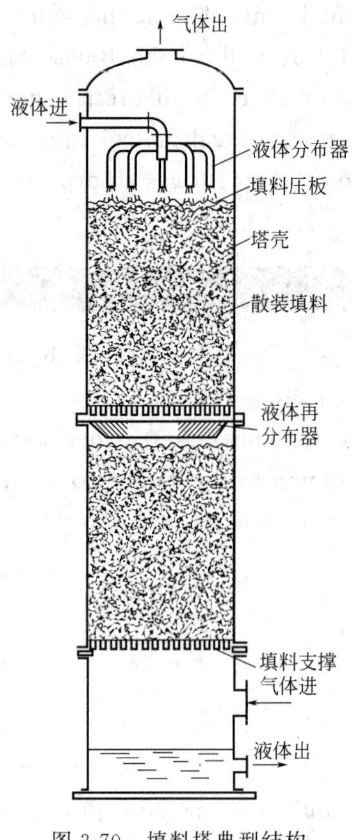

图 3-70 填料塔典型结构

作为分散相，总是靠重力作用自上而下地流动，而气体靠压强差的作用下流经全塔，逆流时气体自塔底进入而自塔顶排出，并流时则相反。在气、液相的进、出口浓度相同的条件下，逆流操作方式可获得较大的平均推动力，因而能有效地提高过程的速率。

填料塔内填充各种类型的填料，如图 3-70 所示，如拉西环、鲍尔环、波纹板等，塔底有气体或蒸气的进口及分配空间，其上为填料的支承——常用大空隙率的栅板；塔顶设有液体分布装置，使液体尽可能均匀地喷淋在填料层的顶部。操作时，气液两相一般为逆流流动，气体自下而上，液体由上向下。填料表面被下流液体润湿，湿表面与上升气流达到连续微分接触传质，故称连续接触式（微分式）传质设备。

3.5.2.1 填料的主要类型及性能

(1) 填料主要类型

填料塔的性能主要取决于填料的类型。填料（packings）有很多种形式，一般分为两大类，一类是个体填料如拉西环、鲍尔环、鞍形环等；另一类是规整填料，如栅板、θ 网环、波纹填料等。根据操作流体的不同，可分别选用陶瓷、金属、塑料、玻璃、石墨等材料的填料。图 3-71 给出了几种常见填料的形状示意图。近年来还研制出了阶梯斜壁形、套筒式、脉冲式及直通式等许多新型填料，它们不仅价格低廉，而且性能良好。

① 常见个体填料简介

a. 拉西环：拉西环填料于 1914 年由拉西（F. Rashching）发明，为外径与高度相等的圆环，如图 3-71(a) 所示。拉西环填料的气液分布较差，传质效率低，阻力大，通量小，目前工业上已较少应用。

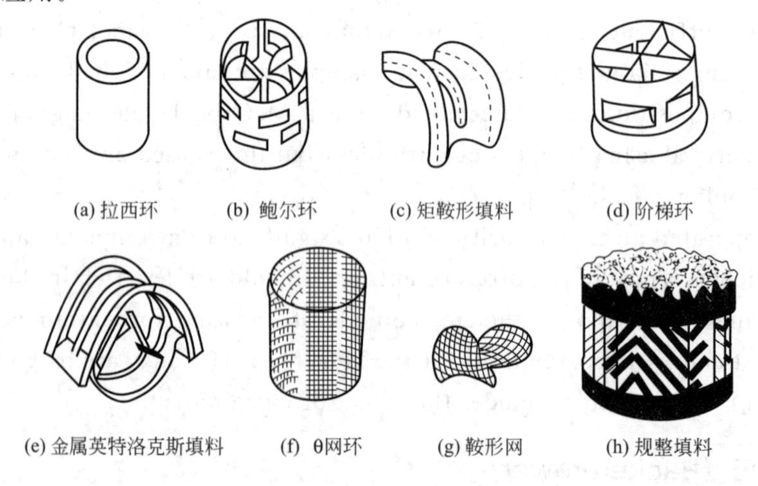

图 3-71 常见填料的形状

b. 鲍尔环：如图 3-71(b) 所示，鲍尔环是对拉西环的改进，在拉西环的侧壁上开出两排长方形的窗孔，被切开的环壁的一侧仍与壁面相连，另一侧向环内弯曲，在环中心相搭。鲍尔环由于环壁开孔，大大提高了环内空间及环内表面的利用率，气流阻力小，液体分布均匀。与拉西环相比，鲍尔环的气体通量可增加 50%以上，传质效率提高 30%左右，因此是一种应用较广的填料。

c. 矩鞍填料：将类似于马鞍形状的两端弧形面改为矩形面，且两面大小不等，即成为矩鞍填料，如图 3-71(c) 所示。矩鞍填料堆积时不会套叠，液体分布较均匀。矩鞍填料一般采用瓷质材料制成，其性能优于拉西环。目前，国内绝大多数应用瓷拉西环的场合，均已被瓷矩鞍填料所取代。

d. 阶梯环：阶梯环是对鲍尔环的改进，其高度比鲍尔环减少了一半并在一端增加了一个锥形翻边，如图 3-71(d) 所示。由于高径比减少，使得气体绕填料外壁的平均路径大为缩短，减少了气体通过填料层的阻力，同时锥形边增加了填料间的空隙，可以促进液膜的表面更新，有利于传质效率的提高。阶梯环的综合性能优于鲍尔环，成为目前所使用的环形填料中最为优良的一种。

e. 金属环矩鞍填料：环矩鞍填料（国外称为 Intalox）是兼顾环形和鞍形结构特点而设计出的一种新型填料，该填料一般以金属材质制成，故又称为金属环矩鞍填料，如图 3-71(e) 所示。环矩鞍填料将环形填料和鞍形填料的优点集于一体，其综合性能优于鲍尔环和阶梯环，在散装填料中应用较多。

② 常见规整填料简介

a. 波纹填料：该类填料是我国开发成功并于 1971 年发表的填料类型。该填料的基本件是冲压出 45°斜波纹槽的薄板，薄板的高度通常是 40～60mm。若干板片平行组合，但相邻薄板的波纹反向。当塔截面为圆形，则波形板片的组合为圆柱形。上下相邻的填料组合体，其薄板方向互呈 90°交错。该类填料为气液提供了一段带分支的直通道，气流阻力小，允许操作气速较大，即处理能力大，同时其特殊的结构可促进液膜的表面更新。

b. 金属丝网波纹填料：该类填料是 20 世纪 60 年代由瑞士苏尔寿（Sulzer）公司开发的一种规整填料，它由丝网波纹片垂直叠合组装而成，波纹倾角有 30°和 45°两种，分别为 X 型和 Y 型。丝网波纹填料在小直径塔内整盘装填，在大直径塔分块装填，相邻盘填料方向成 90°安装。常用的金属丝网填料有 BX 和 CY 两种，其中 BX 型的综合性能比 CY 型好。金属丝网波纹填料广泛应用于难分离物系，但其造价高，抗污能力差，难以清洗。

(2) 填料性能参数

填料的一般要求是比表面积大、空隙率大、对气体的流动阻力小、耐腐蚀及机械强度高。表示填料性能的参数有以下几项：

① 比表面积 σ 指单位填料层提供的填料的表面积，即：

$$\sigma=\frac{\text{填料层表面积}(\mathrm{m}^2)}{\text{填料层体积}(\mathrm{m}^3)}$$

填料的比表面积愈大，所能提供的气液传质面积愈大。同一种类的填料，尺寸愈小，则比表面积愈大。

② 空隙率 ε 单位体积填料层的空隙体积称为空隙率，即：

$$\varepsilon=\frac{\text{填料层空隙体积}(\mathrm{m}^3)}{\text{填料层体积}(\mathrm{m}^3)}$$

填料的空隙率大，气液通过能力大且气体流动阻力小。

③ 填料因子　填料因子表示填料的流体力学性能。

a. 干填料因子　指无液体喷淋时，将 σ 与 ε 组合成 σ/ε^3 的形式，称为干填料因子，单位为 1/m。

b. 湿填料因子（以后简称填料因子）φ　当填料被喷淋的液体润湿后，填料表面覆盖了一层液膜，σ 与 ε 均发生相应的变化，此时 σ/ε^3 称为湿填料因子，以 φ 表示，单位为 1/m。φ 代表实际操作时填料的流体力学特性，故进行填料塔计算时，应采用液体喷淋条件下实测的湿填料因子。φ 值小，表明流动阻力小，液泛速度可以提高。

在选择填料时，一般要求比表面积及空隙率要大，填料的润湿性能好，单位体积填料的质量轻，造价低，并有足够的机械强度。若 σ 增大，即气液两相接触面积增加时，则有利于传质。

3.5.2.2　填料塔附件

填料塔的附件主要有填料支承装置、气液体分布装置、液体再分布装置和除沫装置等。合理选择和设计填料塔的附件，对于保证塔的正常操作及良好性能十分重要。

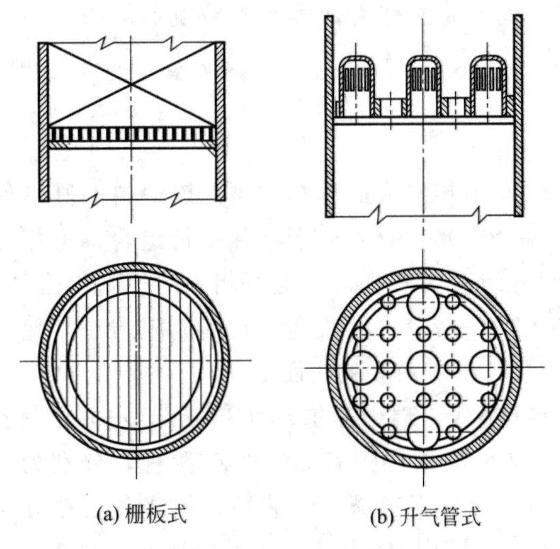

图 3-72　填料支撑装置

（1）填料支撑装置（packing supports）

填料支撑装置要有足够的机械强度，如常见的栅板式，如图 3-72（a）所示，以支撑塔内填料及其所持有的液体质量。同时，支撑装置应具有较大的自由截面积，以免此处发生液泛。常见的还有升气管式支撑装置[图 3-72(b)]。

（2）液体分布装置（liquid distributors）

液体分布装置的作用是使液体在填料塔内分布均匀，以提高分离效率。从喷淋密度考虑，应保证每 30cm^2 的塔截面上约有一个喷淋点，这样，可以防止塔内发生壁流和沟流现象。

如图 3-73 所示，常用的液体分布装置有莲蓬式、溢流管式、筛孔式、齿槽式及多孔环管式分布器等。

（3）液体再分布装置

为避免液体的偏流现象，可在填料层内每隔一定高度设置液体再分布装置。所选高度因填料种类而异，对于拉西环填料可为塔径的 2.5～3 倍，对于鲍尔环及鞍形填料可为塔径的 5～10 倍，但通常填料层高度最多不超过 6m。

对于整砌填料，因不存在偏流现象，填料不必分层安装，也无须设再分布装置，但对液体的初始分布要求较高。相比之下，乱堆填料因具有自动均匀分布液体的能力，对液体初始分布无过苛要求，却因偏流需要考虑液体再分布装置。

再分布器的形式很多。常用的为截锥式再分布器。图 3-74 即为两种截锥式再分布器。截锥式再分布器适用于直径在 0.8m 以下的塔，安排再分布器时，应注意其自由截面积不得小于填料层的自由截面积，以免当气速增大时首先在此处发生液泛。

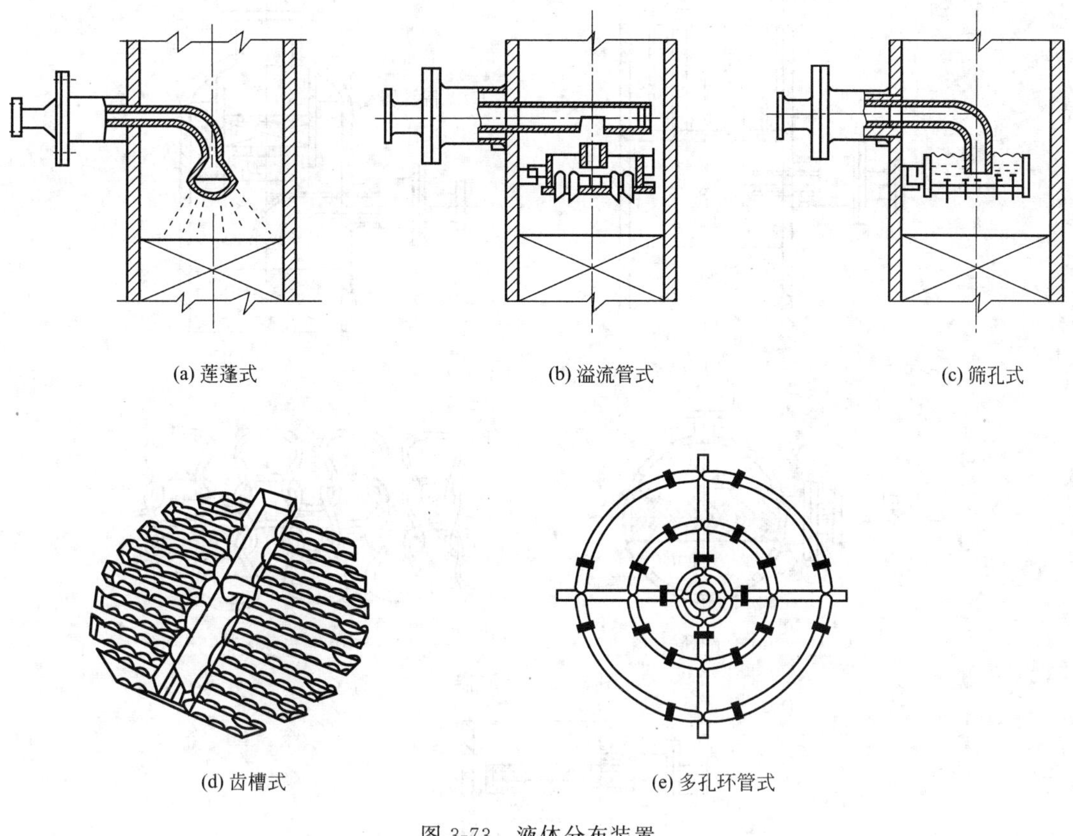

图 3-73 液体分布装置

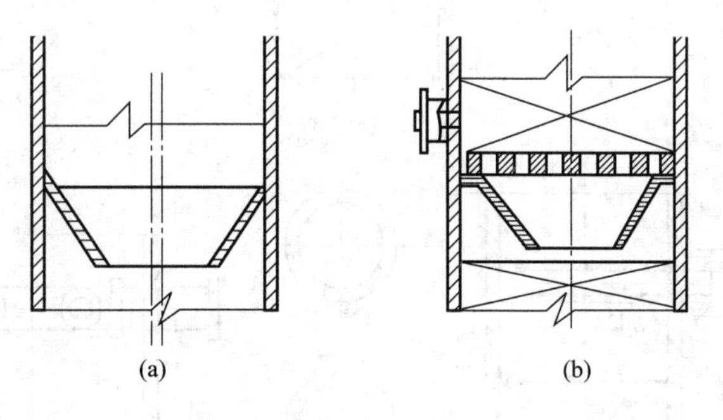

图 3-74 截锥式再分布器

（4）除沫器

除沫装置安装在液体分布器的上方，用于除去出口气流中的液滴。常用的除沫装置有折流板除沫器、丝网除沫器（图 3-75）及旋流板除沫器等。除此之外，填料层顶部常需设置填料压板或挡网，以避免操作中因气速波动而使填料被冲动及损坏。

（5）气体分布装置

填料塔的气体进口的构形，除应考虑防止液体倒灌外，更重要的是要有利于气体均匀地进入填料层，对于小塔最常见的方式是将进气管伸入塔截面中心位置，管端作成向下倾斜的切口或向下弯的喇叭口，如图 3-76 所示；对于大塔，应采取其他更有效的措施，例如管末端可做成类似图 3-76（b）、（c）的多孔直管式或多孔盘管式。

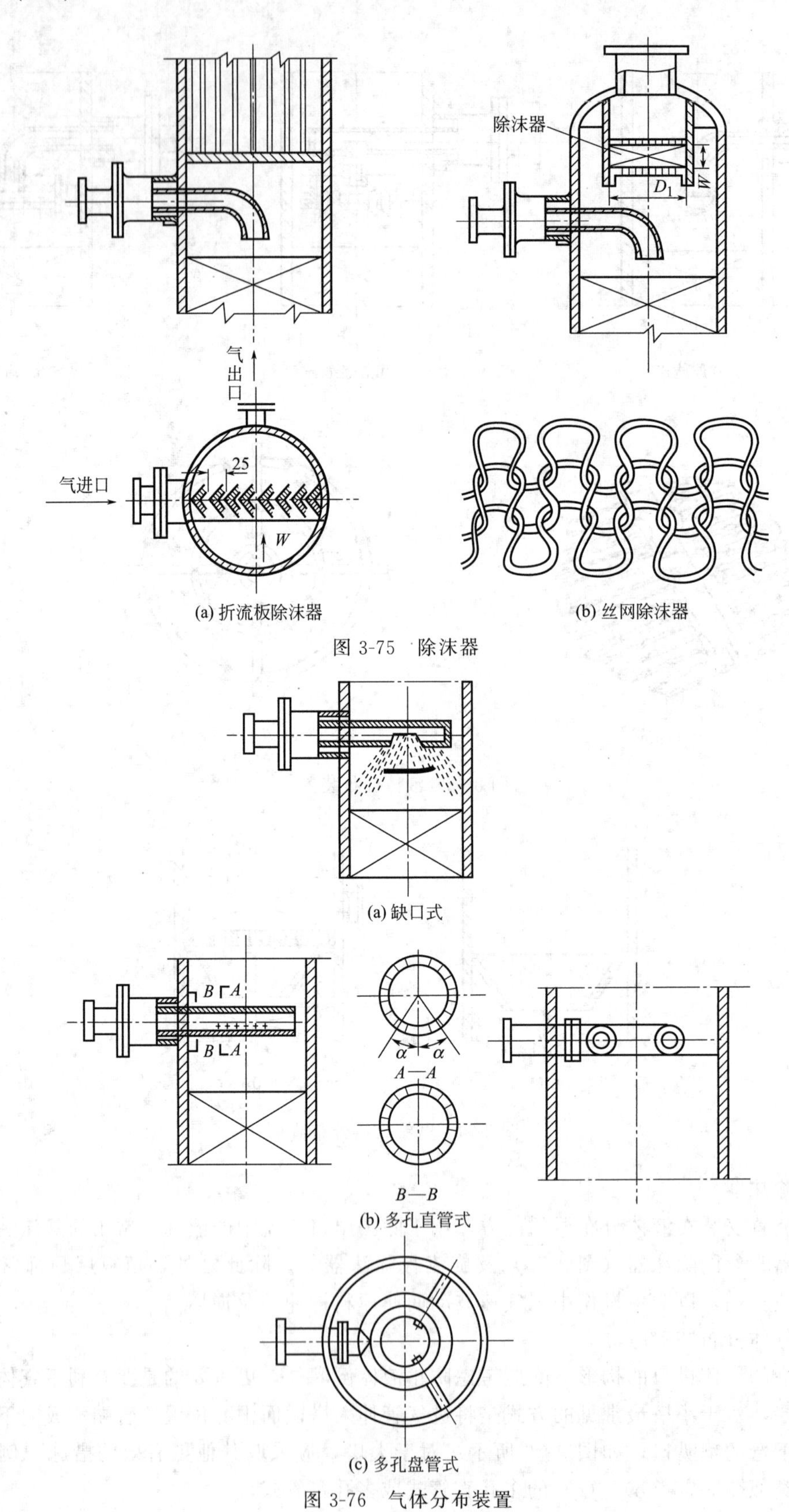

(a) 折流板除沫器

(b) 丝网除沫器

图 3-75　除沫器

(a) 缺口式

(b) 多孔直管式

(c) 多孔盘管式

图 3-76　气体分布装置

Packed Tower Design

Packed tower and packings

A common apparatus used in gas absorption and certain other operations is the packed tower. The device consists of a cylindrical column, or tower, equipped with a gas inlet and distributing space at the bottom; a liquid inlet and distributor at the top; gas and liquid outlets at the top and bottom, respectively; and a supported mass of inert solid shapes, called tower packing. The packing support is typically a screen, corrugated to give it strength, or an upper grid or riser plate, which prevents packing movement and is with a large open area so that flooding does not occur at the support. The inlet liquid, which may be pure solvent or a dilute solution of solute in the solvent and which is called the weak liquor, is distributed over the top of the packing by the distributor and, in ideal operation, uniformly wets the surfaces of the packing. In large towers, spray nozzles or distributor plates with overflow weirs are more common. For very large towers, up to 9 m in diameter, Nutter Engineering advertises a plate distributor with individual drip tubes.

The solute-containing gas, or rich gas, enters the distributing space below the packing and flows upward through the interstices in the packing countercurrent to the flow of the liquid. The packing provides a large area of contact between the liquid and gas and encourages intimate contact between the phases. The solute in the rich gas is absorbed by the fresh liquid entering the tower, and dilute, or lean, gas leaves the top. The liquid is enriched in solute as it flows down the tower, and concentrated liquid, called the strong liquor, leaves the bottom of the tower through the liquid outlet.

Tower packings are divided into three principal types: those that are dumped at random into the tower, those that must be stacked by hand, and those known as structured or ordered packings. Dumped packings consist of units 6 to 75 mm (1/4 to 3 in) in major dimension; packings smaller than 25 mm are used mainly in laboratory or pilot-plant columns. In stacked packings the units are 50 to 200 mm (2 to 8 in) in size. They are much less commonly used than dumped packings and are not discussed here.

3.5.2.3 填料塔流体力学特性（fluid mechanics of packed tower)

(1) 填料层的持液量

持液量是指在一定的操作条件下，单位体积填料层中填料表面和填料的空隙中积存的液体体积，即每立方米填料层中的液体（m^3）。持液量可分为静持液量 H_S、动持液量 H_O 和总持液量 H_t。总持液量 H_t 为静持液量与动持液量之和，即：

$$H_t = H_S + H_O \tag{3-121}$$

式中，静持液量是指当填料被充分润湿后，停止气液两相进料后，并经适当时间的排液，直至无滴液时仍存留于填料层中液体的体积。静持液量只取决于填料和流体的特性，与气液负荷无关；而动持液量是指填料塔停止气液两相进料时流出的液量，它与填料、液体特性及气液负荷有关。

填料层的持液量可由实验测出，也可由经验公式计算。一般来说，适当的持液量对填料塔的操作稳定性和传质是有益的，但持液量过大，将减少填料层的空隙，使气相的压降增大，处理能力下降。

（2）塔的压降与液泛气速

对于气液逆流接触的填料塔的操作，当液体的流量一定时，随着气速的提高，气体通过填料层的压力损失（体现为压降）也不断提高。压降是填料塔设计中的重要参数，气体通过填料层的压降的大小决定了塔的动力消耗。图 3-77 给出了每米压降（$\Delta p/z$）与空塔气速（u）及淋洒密度（L_W）的关系曲线。

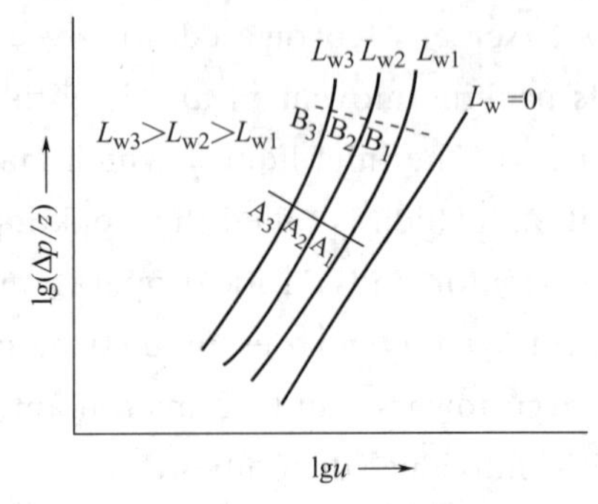

图 3-77　填料塔压降与空塔气速的关系

如图 3-77 所示，当 $L=0$ 时，即无液体喷淋（又称干填料线）时，$\Delta p/z$ 与 u 成直线关系；当有液体喷淋时，曲线都有两个转折点：第一个折点——载点；第二个折点——泛点。载点和泛点将 $\Delta p/z$ 与 u 关系曲线分成三段，即恒持液量区、载液区和液泛区。

① 恒持液量区　由于液体在填料空隙中流动，则液体占据一定的塔内空间，使气体的真实速度较通过干填料层时的真实速度高，因而压强降也较大。此区域的 $\Delta p/z$-u 线在干填料线的左侧，且两线相互平行。

② 载液区　随着气速的增大，上升气流与下降液体间的摩擦力开始阻碍液体下流，使填料层的持液量随气速的增加而增加，此现象称为拦液现象。开始发生拦液现象时的空塔气速称为载点气速。

③ 液泛区　如果气速继续增大，由于液体不能顺利下流，而使填料层内持液量不断增多，以致几乎充满了填料层中的空隙，此时压降急剧升高，$\Delta p/z$-u 线的斜率可达 10 以上。压降曲线近于垂直上升的转折点称为泛点。达到泛点时的空塔气速称为液泛气速或泛点气速。

在泛点气速下，持液量的增多使液相由分散相变为连续相，而气相则由连续相变为分散相，此时气体呈气泡形式通过液层，气流出现脉动，液体被大量带到塔顶甚至出塔，塔的操作极不稳定，甚至被破坏，所以应控制正常的操作气速在泛点气速以下。

一般认为正常操作的空塔气速 u 应在载点气速之上，在泛点气速的 0.8 倍以下，但到达载点时的症状不明显，而到达泛点气速时，塔内气液的接触状况被破坏，现象十分明显，易于辨认。由于液泛是塔操作的极限，必须避免发生，故应计算出液泛气速作为操作气速的上限，再核算出合理的正常操作气速。

工程上常用埃克特（Eckert）通用关联图来确定填料塔内的气体压降和泛点气速。如图 3-78 所示，在最上方的弦栅、整砌拉西环线下，就是乱堆填料的泛点线，与泛点线相对应的纵坐标中的空塔气速应为空塔液泛气速 u_{max}；在泛点线下面的线群则为各种乱堆填料的压降线，若已知气、液两相流量比及各相的密度，可根据规定的压降，求其相应的空塔气速，反之，根据选定的实际操作气速求压降。

埃克特通用关联图适用于各种乱堆填料，如拉西环、鲍尔环、弧鞍、矩鞍等，但需确知填料的 φ 值。

（3）液体的喷淋密度与填料的润湿性能（wettability）

填料塔中气液两相间的传质主要是在填料表面流动的液膜上进行的，因此，传质效率就与填料的润湿性能密切相关。为使填料能获得良好的润湿，应使塔内液体的喷淋量不低于最

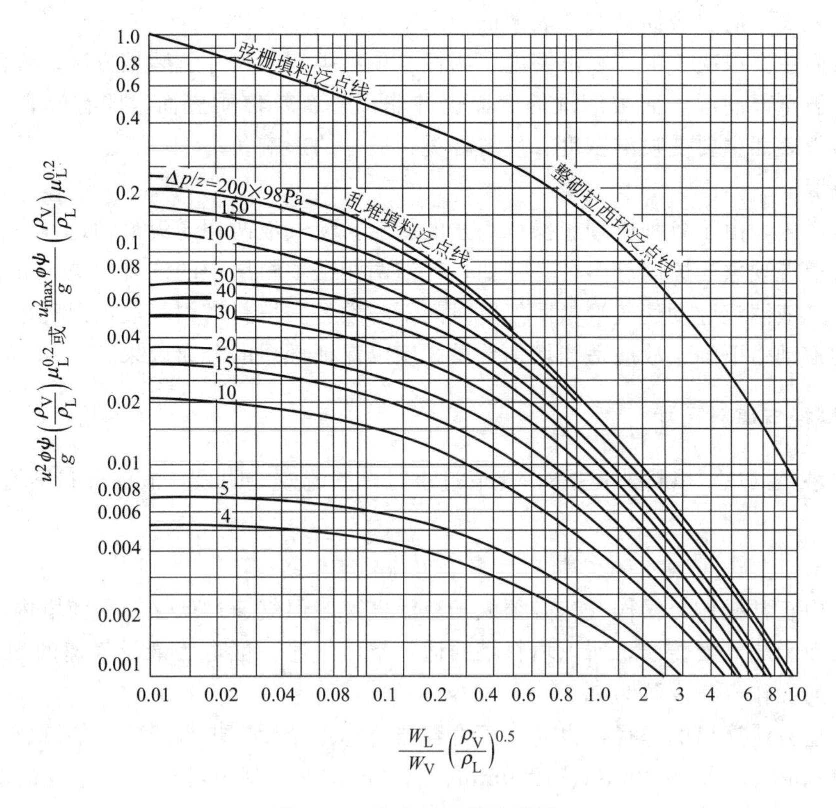

图 3-78　埃克特通用关联图

u_{max}—泛点气速，m/s；u—空塔气速，m/s；g—重力加速度，m/s^2；

ϕ—填料因子，1/m；ψ—液体密度校正系数，等于水的密度与液体密度之比，即 $\psi=\frac{\rho_水}{\rho_L}$；

ρ_L，ρ_V—液体与气体的密度，kg/m^3；μ_L—液体的黏度，mPa·s；

W_L，W_V—液相及气相的质量流量，kg/s

小喷淋密度。所谓液体的喷淋密度是指单位时间内单位塔截面上喷淋的液体体积，最小喷淋密度能维持填料的最小润湿速率，它们之间的关系为：

$$U_{min}=L_{min}\sigma \tag{3-122}$$

式中　U_{min}——最小喷淋密度，$m^3/(m^2\cdot h)$；

L_{min}——最小润湿速率 $m^3/(m\cdot h)$。

润湿速率是指在塔的横截面上，单位长度的填料周边上液体的体积流量。对于直径不超过75mm的拉西环及其他填料，可取最小润湿速率为 $0.08m^3/(m\cdot h)$；对于直径大于75mm的环形填料，应取最小润湿速率为 $0.012m^3/(m\cdot h)$。

实际操作时采用的喷淋密度应大于最小喷淋密度。若喷淋密度过小，可采用增大回流比或采用液体再循环的方法加大液体流量，以保证填料的润湿性能。也可采用减小塔径，或适当增加填料层高度的方法予以补偿。

填料的润湿性能与填料的材质有关，例如常用的陶瓷、金属及塑料三种材料中，陶瓷填料的润湿性能最好，而塑料填料的润湿性能最差。则对于金属、陶瓷等材料的填料，可采用表面处理方法，改善其表面的润湿性能。

（4）比表面、润湿表面、有效表面的含义

润湿表面主要取决于液体喷淋密度、物性及填料类型、尺寸、装填方法。液泛以前，填料表面难以被全部润湿，而被润湿的表面也并非都是有效传质表面，即有效表面<润湿表面；其中有效表面主要受喷淋密度、填料种类、尺寸的影响。

（5）返混

在填料塔内，由于各种不理想操作因素的影响，使气液两相逆流流动过程中存在着返混现象。造成返混现象的原因有多种，例如，气液两相在填料层中的沟流现象，气液的分布不均及塔内的气液湍流脉动使气液微团停留时间不一致等。填料塔内气液返混现象的发生，使得传质平均推动力下降，故应适当增加传质高度以保证理想的分离效果。

3.5.2.4 塔径与填料尺寸

一般取塔径（D）/填料外径≥8，比值过小时，壁流效应明显，液体分布不均匀。

$$D=\sqrt{\frac{4V_S}{\pi u}} \tag{3-123}$$

为了塔内正常操作，操作气速必须低于液泛气速，但气速过低又容易使塔内的液体分布不均匀，从而影响气液传质，所以填料塔的适宜空塔气速一般取为液泛气速的50%～85%，一般填料塔的操作气速大致为0.2～1.0m/s。

根据上述方法算出的塔径，也应按压力容器公称直径标准进行圆整，如圆整为400mm、500mm、600mm、…、1000mm、1200mm、1400mm等。算出塔径后，还应检验塔内的喷淋密度是否大于最小喷淋密度。

3.5.2.5 填料塔与板式塔的比较

（1）塔设备的共同要求

塔设备虽然形式繁多，但共同的要求是：

① 相际传质面积大，气液两相充分接触，以获得较高的传质效率；

② 生产能力（即气液相负荷）大，在较大的气速下不发生大量的雾沫夹带、拦液、液泛等不正常操作现象；

③ 操作稳定，操作弹性（最大负荷/最小负荷）大，传质效率高（但往往与生产能力冲突）；

④ 流体通过塔设备的压力减小（尤其是对于真空精馏等操作），且气液接触传质后，两相易于分离；

⑤ 耐腐蚀，不堵塞，易检修；

⑥ 结构简单，易于加工、安装。

（2）塔设备的选型原则

板式塔和填料塔的选择无统一标准，很大程度上取决于设计者的知识与经验。一般来说，下列的情况优先选用板式塔：

① 液相负荷较小，使用填料塔则其表面不能充分润湿，难以保证分离效率；

② 易结垢，有结晶的物料，采用板式塔不易堵塞；

③ 需要设置塔内部换热元件或多个侧线进料、出料口时，板式塔较合适；

④ 板式塔内液体滞料量大，操作弹性大，易于稳定，对进料浓度的变化不甚敏感。

下列的情况应优先选用填料塔：

① 在分离程度要求高的情况下，采用新型填料塔可降低塔高；

② 新型填料塔压降低，有利于节能；

③ 新型填料塔具有较小的持液量，适合于热敏性物料的蒸馏分离；

④ 易发泡的物料宜采用填料塔，因为在填料塔内气相主要不是以气泡形式通过液相，可减小发泡的程度；

⑤ 对于腐蚀性物料，可用耐腐蚀的填料塔。

20世纪70年代以前，板式塔的发展速度一直优先于填料塔，涌现出许多新型塔板。但是，自瑞士Sulzer公司开发出金属孔板波纹填料以来，填料塔的发展取得了突破性的进展，通过对规整填料的深入研究以及气液分布器的精心设计，目前已基本解决了大型填料塔的工业放大问题，彻底改变了填料塔不能在大直径塔上应用的传统认识，使填料塔在过程工业中被广泛应用。在旧塔改造方面，由于塔径和塔高确定，如需增加产量或提高产品质量，只能将原有的板式塔部分或全部改成高效填料塔。对新建塔器，填料塔和板式塔在不同的场合下互相竞争，设计者应依据工艺条件和具体情况，以期达到技术上可靠，经济上合理的目的。

值得注意的是，用填料改造高压下的板式蒸馏塔时要特别慎重，这是因为加压蒸馏操作时，液体流量大，容易产生不均匀流动，且液膜厚度增加，同时加压下液体的黏度也增加，这些因数都会使填料塔的分离能力下降。

另外，因为层出不穷的新型塔板结构或高效填料各具特点，所以应根据不同的工艺及生产需要来选择塔板的形式或填料的类型。

3.5.3 萃取设备（Extraction equipment）

根据两相接触方式的不同，萃取设备可分为逐级接触式和微分接触式两类。在逐级接触式设备中，每一级均进行两相的混合与分离，故两液相的组成在级间发生阶跃式变化。而在微分接触式设备中，两相逆流连续接触传质，两液相的组成则发生连续变化。

根据外界是否输入机械能，萃取设备又可分为有外加能量和无外加能量两类。若两相密度差较大，萃取时，仅依靠液体进入设备时的压力差及密度差即可使液体较好地分散和流动，此时不需外加能量即能达到较好的萃取效果；反之，若两相密度差较小，界面张力较大，液滴易聚合不易分散，此时常采用从外界输入能量的方法来改善两相的相对运动及分散状况，如搅拌、振动、离心等。

目前，工业上使用的萃取设备种类很多，在此仅介绍一些典型设备。

Extraction Equipment

In liquid-liquid extraction, as in gas absorption and distillation, two phases must be brought into good contact to permit transfer of material and then separated. In absorption and distillation, the mixing and separation are easy and rapid. In extraction, however, the two phases have comparable densities, so that the energy available for mixing and separation-if gravity flow is used-is small, much smaller than when one phase is a liquid and the other is a gas. The two phases are often hard to mix and harder to separate. The viscosities of both phases also are relatively high, and linear velocities through most extraction equipment are low. In some types of extractors, therefore, energy for mixing and separation is supplied mechanically.

Extraction equipment may be operated batchwise or continuously. A quantity of feed liquid may be mixed with a quantity of solvent in an agitated vessel, after which the layers are settled and separated. The extract is the layer of solvent plus extracted solute, and the raffinate is the layer from which solute has been removed. The extract may be lighter or heavier than the raffinate, and so the extract may be shown coming from the top of the equipment in some cases and from the bottom in others. The operation may, of course, be repeated if more than one contact is required; but when the quantities involved are large and several contacts are needed, continuous flow becomes economical. Most extraction equipment is continuous with either successive stage contacts or differential contacts. Representative types are mixer-settlers, vertical towers of various kinds that operate by gravity flow, agitated tower extractors, and centrifugal extractors. Liquid-liquid extraction can also be carried out using porous membranes. This method has promise for difficult separations.

3.5.3.1 混合澄清器（mixer-settlers）

混合澄清器是使用最早，而且目前仍广泛应用的一种萃取设备，它由混合器与澄清器组成，典型装置如图 3-79 所示。

在混合器中，大多应用机械搅拌，有时也可将压缩气体通入底部进行气流搅拌，还可以利用流动混合器或静态混合器。两相分散体系在混合器内停留一定时间后，流入澄清器，轻、重两相依靠密度差进行重力沉降（或升浮），并在界面张力的作用下凝聚分层，形成萃取相和萃余相。

混合澄清器可以单级使用，也可以组成多级逆流或错流串联流程，图 3-80 为水平排列的三级逆流混合澄清萃取装置示意图，也可以将几个级上下重叠。

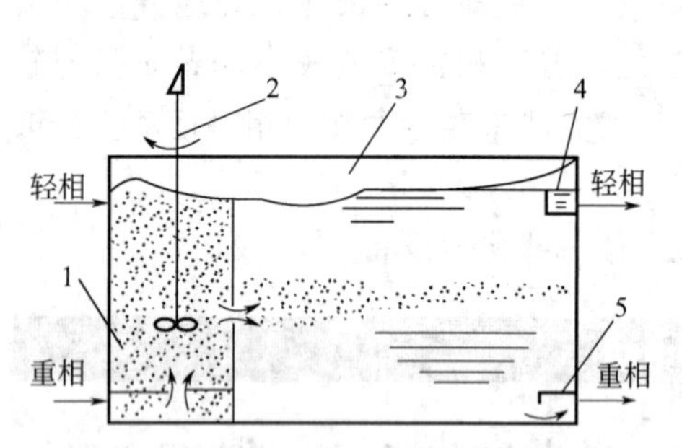

图 3-79　混合器与澄清器组合装置

1—混合器；2—搅拌器；3—澄清器；
4—轻相液出口；5—重相液出口

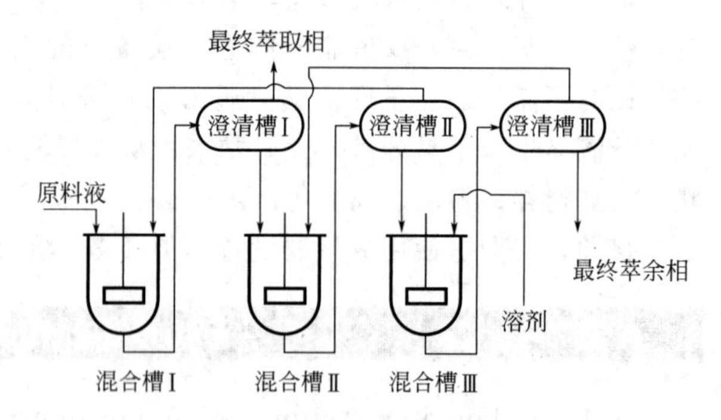

图 3-80　三级逆流混合澄清萃取装置

混合澄清器具有处理量大、传质效率高、两液相流量比范围大、设备结构简单、操作方便，易实现多级连续操作等优点；但缺点是水平排列的设备占地面积大，每级内都设有搅拌装置，液体在级间流动需输送泵，设备费和操作费都较高。

Mixer-settlers

For batchwise extraction the mixer and settler may be the same unit. A tank containing a turbine or propeller agitator is most common. At the end of the mixing cycle the agitator is shut off, the layers are allowed to separate by gravity, and extract and raffinate are drawn off to separate receivers through a bottom drain line carrying a sight glass. The mixing and settling times required for a given extraction can be determined only by experiment; 5min for mixing and 10min for settling are typical, but both shorter and much longer times are common.

For continuous flow the mixer and settler are usually separate pieces of equipment. The mixer may be a small agitated tank provided with inlets and a drawoff line and baffles to prevent short-circuiting, or it may be a motionless mixer or other flow mixer. The settler is often a simple continuous gravity decanter. With liquids that emulsify easily and have nearly the same density it may be necessary to pass the mixer discharge through a screen or pad of glass fiber to coalesce the droplets of the dispersed phase before gravity settling is feasible. For even more difficult separations, tubular or disk-type centrifuges are employed.

3.5.3.2　萃取塔（extraction towers）

通常将高径比较大的萃取装置统称为塔式萃取设备，简称萃取塔。根据两相混合和分散所采用的措施不同，萃取塔的结构形式也多种多样。下面简介几类工业上常用的萃取塔。

（1）重力流动型萃取塔

重力流动型萃取塔是两液相靠重力作逆流流动而不断输入机械能的萃取塔，结构简单，适用于界面张力不大、要求的理论级数不多（如不超过3～4级）的场合，主要有以下一些类型。

① 喷洒塔　喷洒塔又称喷淋塔，是最简单的萃取塔，如图3-81所示，轻、重两液分别从塔底和塔顶进入。若以重液为分散相，则重液经塔顶的分布装置分散为液滴后进入轻液，与轻液逆流接触传质，重液液滴降至塔底分离段处聚合形成重液液层排出，而轻液上升至塔顶并与重液分离后排出［图3-81(a)］；若以轻液为分散相，则轻液经塔底的分布装置分散为液滴后进入连续的重液，与重液进行逆流接触传质，轻液升至塔顶分离段处聚合形成轻液层排出，而重液流至塔底与轻相分离后排出［图3-81(b)］。

喷洒塔结构简单，塔体内除进出各流股物料的接管和分散装置外，无其他内部构件。缺点是轴向返混严重，传质效率较低，因而适用于仅需一、二个理论级的场合，如水洗、中和或处理含有固体的物系。

② 填料萃取塔（packed extraction towers）　填料萃取塔的结构与精馏和吸收填料塔基本相同，如图3-82所示，塔内装有适宜的填料，轻、重两液分别由塔底和塔顶进入，由塔顶和塔底排出。萃取时，连续相充满整个填料塔，分散相由分布器分散成液滴进入填料层中的连续相，在与连续相逆流接触中进行传质。

填料的作用是使液滴不断发生凝聚与再分散，以促进液滴的表面更新，填料也能起到减少轴向返混的作用。

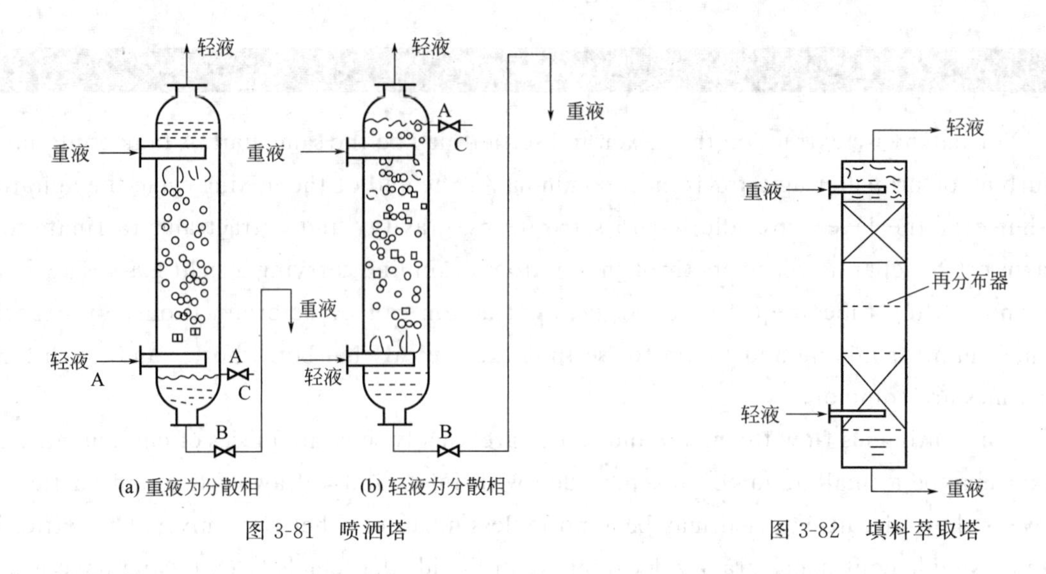

图 3-81　喷洒塔

图 3-82　填料萃取塔

填料萃取塔的优点是结构简单、操作方便，适合于处理腐蚀性料液；缺点是传质效率低，一般用于所需理论级数较少（如 3 个萃取理论级）的场合。

Packed Extraction Towers

Tower extractors give differential contacts, not stage contacts, and mixing and settling proceed continuously and simultaneously. Extraction can be carried out in an open tower, with drops of heavy liquid falling through the rising light liquid or vice versa; however, such towers, called spray towers, are rarely used because of pronounced axial mixing in the continuous phase. Instead, the tower is filled with packing such as rings or saddles, which causes the drops to coalesce and reform, and tends to limit axial dispersion.

In an extraction tower there is continuous transfer of material between phases, and the composition of each phase changes as it flows through the tower. At any given level, of course, equilibrium is not reached; indeed, it is the departure from equilibrium that provides the driving force for mass transfer. The design procedure for extraction towers is similar to that for packed absorption towers, but the height of a transfer unit is generally greater than for a typical absorber.

③ 筛板萃取塔（perforated-plate extraction towers）　筛板萃取塔如图 3-83(a) 所示，塔内装有若干层筛板，筛板的孔径一般为 3～9mm。

筛板萃取塔是逐级接触式萃取设备，两相依靠密度差，在重力的作用下，进行分散和逆向流动。若以轻液为分散相，则其通过塔板上的筛孔而被分散成细小的液滴，与塔板上的连续相充分接触进行传质；若以重液为分散相，则重液穿过板上的筛孔，分散成液滴落入连续的轻液中进行传质，轻液则连续地从筛板下侧横向流过，从升液管进入上层塔板，如图 3-83(b) 所示。

筛板萃取塔由于塔板的限制，减小了轴向返混，同时由于分散相的多次分散和聚集，液滴表面不断更新，使筛板萃取塔的效率比填料塔有所提高，加之筛板塔结构简单，造价低廉，可处理腐蚀性料液，因而应用较广。

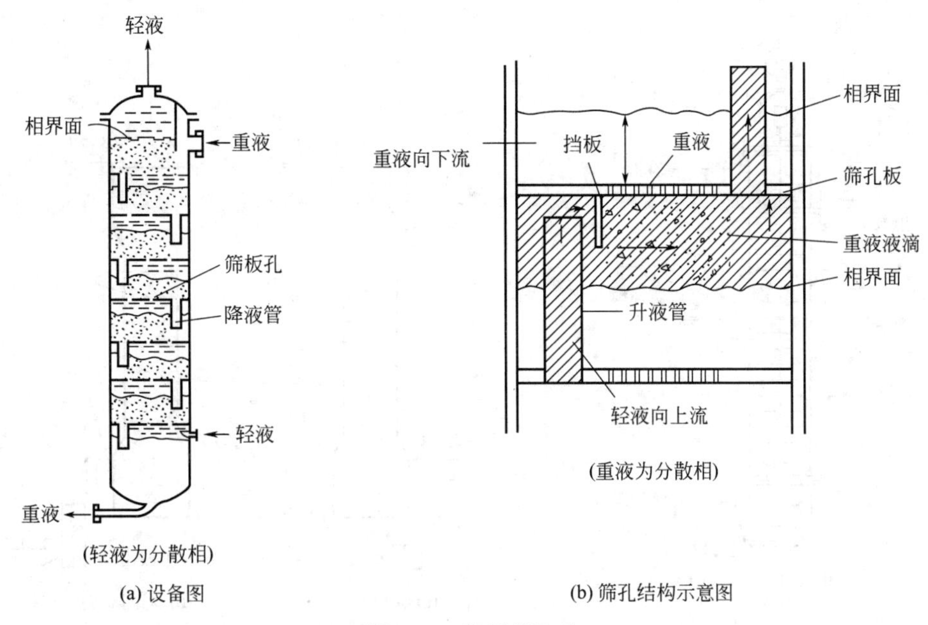

图 3-83 筛板萃取塔

（2）输入机械能量型萃取塔

① 往复筛板萃取塔（reciprocating perforated-plate extraction towers） 往复筛板萃取塔的结构如图 3-84 所示，将若干层筛板按一定间距固定在中心轴上，由塔顶的传动机构驱动而做上下往复运动。往复筛板的孔径要比脉动筛板的大些，一般为 7～16mm。为防止液体沿筛板与塔壁间的缝隙短路，每隔若干块筛板，在塔内壁应设置一块环形挡板。

往复筛板萃取塔可较大幅度地增加相际接触面积和提高液体的湍动程度，传质效率高，流体阻力小，操作方便，生产能力大，在石油化工、食品、制药和湿法冶金工业中应用日益广泛。

② 脉冲筛板塔（pulse -plate extraction columns） 脉冲筛板塔是指在外力作用下，液体在塔内产生脉冲运动的筛板塔，其结构与气液传质过程中无降液管的筛板塔类似。在塔的下澄清段装有脉冲管，萃取操作时，由脉冲发生器提供的脉冲使塔内液体做上下往复运动，迫使液体经过筛板上的小孔，使分散相破碎成较小的液滴分散在连续相中，并形成强烈的湍动，从而促进传质过程的进行。图 3-85 为两种常见类型，图 3-85（a）是直接将发生脉冲的往复泵连接在轻液入口管中，图 3-85（b）则是使往复泵发生的脉冲通过隔膜输入塔底。

脉冲萃取塔的优点是结构简单，传质效率高，适用于有腐蚀性或含有悬浮固体的液体；但其生产能力一般有所下降，在化工生产中的应用受到一定限制。

③ 转盘萃取塔（RDC 塔） 转盘萃取塔的基本结构如图 3-86 所示，在塔体内壁面上按一定间距装有若干个环形挡板，称为固定环，两固定环之间均装一转盘，转盘固定在中心轴上，转轴由塔顶的电机驱动。

萃取操作时，转盘随中心轴高速旋转，其在液体中产生的剪应力将分散相破裂成许多细小的液滴，在液相中产生强烈的涡旋运动，从而增大了相际接触面积和传质系数。同时固定环的存在在一定程度上抑制了轴向返混，因而转盘萃取塔的传质效率较高。

转盘萃取塔结构简单，传质效率高，生产能力大，因而在石油化工中应用比较广泛。

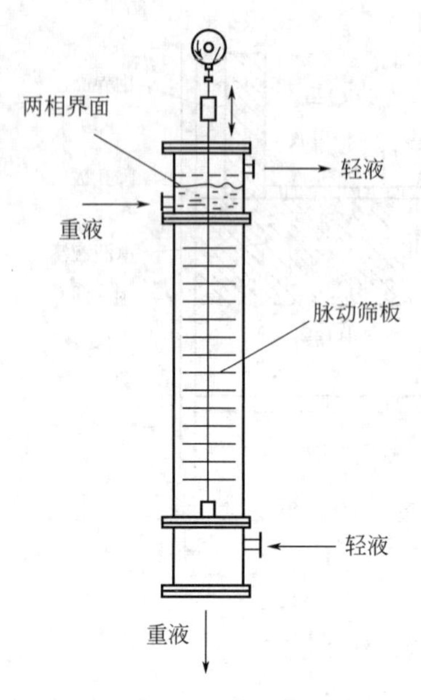

图 3-84　往复筛板萃取塔

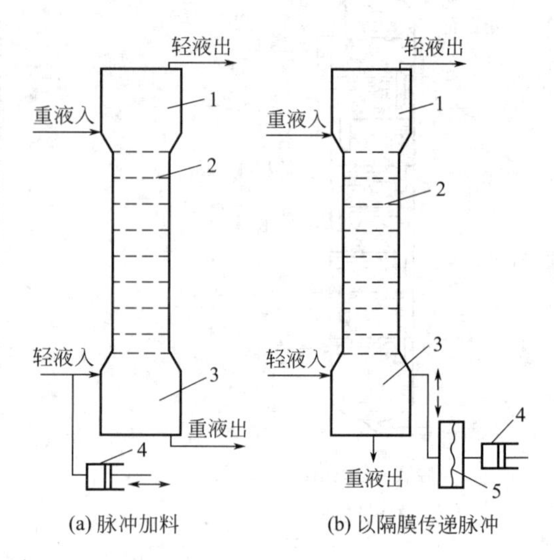

图 3-85　脉冲筛板塔

1—塔顶分层段；2—无溢流筛板；

3—塔底分层段；4—脉冲发生器；5—隔膜

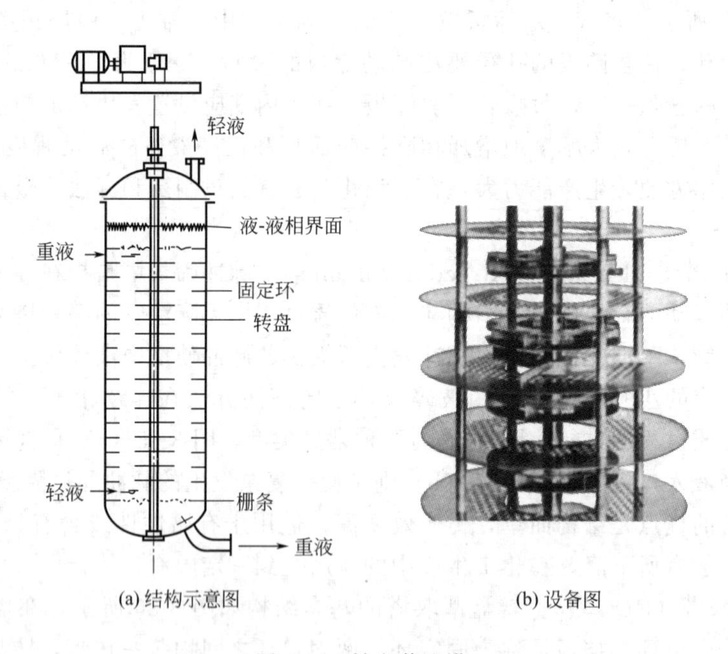

图 3-86　转盘萃取塔

Pulse Columns

Agitation may also be provided by external means, as in a pulse column. A reciprocating pump "pulses" the entire contents of the column at frequent intervals, so that a rapid reciprocating motion of relatively small amplitude is superimposed on the usual flow of the liquid phases. The tower may contain ordinary packing or special sieve plates. In a packed tower the pulsation disperses the liquids and eliminates channeling, and the contact between the phases is greatly improved. In sieve-plate pulse towers the holes are smaller than in non pulsing towers, ranging from 1.5 to 3 mm in diameter, with a total open area in each plate of 6 to 23 percent of the cross-sectional area of the tower. Such towers are used almost entirely for processing highly corrosive radioactive liquids. No downcomers are used. Ideally the pulsation causes light liquid to be dispersed into the heavy phase on the upward stroke and the heavy phase to jet into the light phase on the downward stroke. Under these conditions the stage efficiency may reach 70 percent. This is possible, however, only when the volumes of the two phases are nearly the same and when there is almost no volume change during extraction. In the more usual case, the successive dispersions are less effective, and there is backmixing of one phase in one direction. The plate efficiency then drops to about 30 percent. Nevertheless, in both packed and sieve-plate pulse columns, the height required for a given number of theoretical contacts is often less than one-third that required in an unpulsed column.

(3) 离心萃取器（centrifugal extractors）

离心萃取器是利用离心力的作用使两相快速混合、分离的萃取装置。离心萃取器的类型较多，按两相接触方式可分为逐级接触式和微分接触式两类。在逐级接触式萃取器中，两相的作用过程与混合澄清器类似；而在微分接触式萃取器中，两相接触方式则与连续逆流萃取塔类似。

① 转筒式离心萃取器（drum centrifugal extractors） 它是单级接触式离心萃取器，其结构如图 3-87 所示。重液和轻液由底部的三通进入混合室，在搅拌桨的剧烈搅拌下，两相充分混合进行传质，然后共同进入高速旋转的转筒。在转筒中，混合液在离心力的作用下，重相被甩向转鼓外缘，而轻相则被挤向转鼓的中心。两相分别经轻、重相堰流至相应的收集室，并经各自的排出口排出。

转筒式离心萃取器结构简单，效率高，易于控制，运行可靠。

② 芦威式离心萃取器（Luwesta centrifugal extractors） 芦威式离心萃取器简称 LUWE 离心萃取器，它是立式逐级接触式离心萃取器的一种，图 3-88 为三级离心萃取器，其主体是固定在壳体上并随之做高速旋转的环形盘。壳体中央有固定不动的垂直空心轴，轴上也装有圆形盘，盘上开有若干个喷出孔。

萃取操作时，原料液与萃取剂均由空心轴的顶部加入，重液沿空心轴的通道向下流至萃取器的底部而进入第三级的外壳内，轻液由空心轴的通道流入第一级。在空心轴内，轻液与来自下一级的重液混合，再经空心轴上的喷嘴沿转盘与上方固定盘之间的通道被甩至外壳的四周。重液由外部沿转盘与下方固定盘之间的通道而进入轴的中心，并由顶部排出，其流向为由第三级经第二级再到第一级，然后进入空心轴的排出通道，如图中实线所示；轻液则由第一级经第二级再到第三级，然后进入空心轴的排出通道，如图中虚线所示。两相均由萃取器顶部排出。

该类萃取器主要用于制药工业，其处理能力为 7～49m^3/h，在一定条件下，级效率可接近 100％。

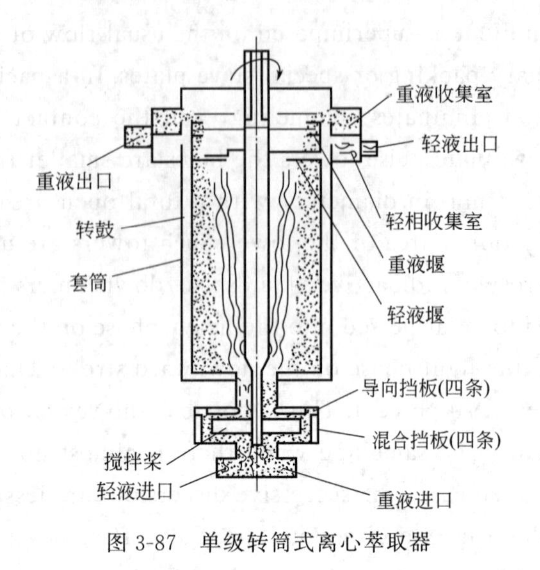

图 3-87 单级转筒式离心萃取器

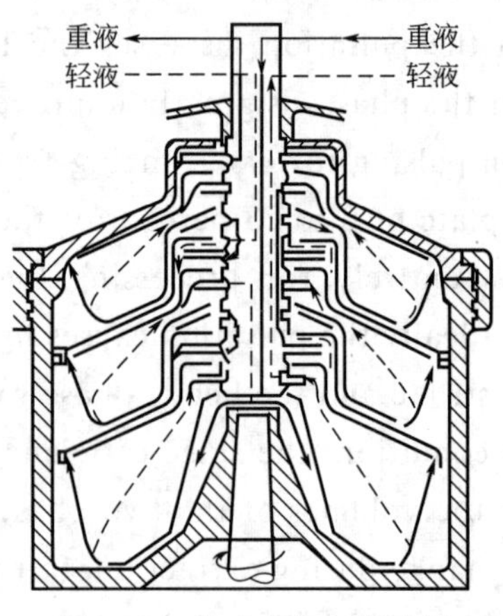

图 3-88 芦威式离心萃取器

Centrifugal extractors

The dispersion and separation of the phases may be greatly accelerated by centrifugal force, and several commercial extractors make use of this. In the Podbielniak extractor a perforated spiral ribbon inside a heavy metal casing is wound about a hollow horizontal shaft through which the liquids enter and leave. Light liquid is pumped to the outside of the spiral at a pressure between 3 and 12 atm to overcome the centrifugal force; heavy liquid is fed to the center. The liquids flow countercurrently through the passage formed by the ribbons and the casing walls. Heavy liquid moves outward along the outer face of the spiral; light liquid is forced by displacement to flow inward along the inner face. The high shear at the liquid-liquid interface results in rapid mass transfer. In addition, some liquid sprays through the perforations in the ribbon and increases the turbulence. Up to 20 theoretical contacts may be obtained in a single machine, although 3 to 10 contacts are more common. Centrifugal extractors are expensive and find relatively limited use. They have the advantages of providing many theoretical contacts in a small space and of very short holdup times-about 4s. Thus they are valuable in the extraction of sensitive products such as vitamins and antibiotics.

3.5.3.3 萃取设备的选择

萃取设备的类型较多，特点各异，物系性质对操作的影响错综复杂。对于具体的萃取过程，选择萃取设备的原则是：在满足工艺条件和要求的前提下，使设备费和操作费之和趋于最低，通常选择萃取设备时应考虑以下因素：

（1）系统特性

对密度差较大、界面张力较小的物系，可选用无外加能量的设备；对密度差较小、界面张力较大的物系，宜选用有外加能量的设备；对密度差甚小、界面张力小、易乳化的物系，应选用离心萃取器。

对有较强腐蚀性的物系，宜选用结构简单的填料塔或脉冲填料塔。对于放射性元素的提取，脉冲塔和混合澄清器用得较多。

物系中有固体悬浮物或在操作过程中产生沉淀物时，需定期清洗，此时一般选用混合澄清器或转盘塔。另外，往复筛板塔和脉冲筛板塔本身具有一定的自清洗能力，在某些场合也可考虑使用。

(2) 处理量

处理量较小时，可选用填料塔、脉冲塔；处理量较大时，可选用混合澄清器、筛板塔及转盘塔，离心萃取器的处理能力也相当大。

(3) 理论级数

当需要的理论级数不超过2～3级时，各种萃取设备均可满足要求；当需要的理论级数较多（如超过4～5级）时，可选用筛板塔；当需要的理论级数更多（如10～20级）时，可选用有外加能量的设备，如混合澄清器、脉冲塔、往复筛板塔、转盘塔等。

(4) 物系的稳定性和液体在设备内的停留时间

对生产中要考虑物料的稳定性、要求在设备内停留时间短的物系（如抗生素）的生产，宜选用离心萃取器；反之，若萃取物系中伴有缓慢的化学反应，要求有足够长的反应时间，则宜选用混合澄清器。

在选用萃取设备时，还应考虑其他一些因素，如能源供应情况，在电力紧张地区应尽可能选用依靠重力流动的设备；当厂房面积受到限制时，宜选用塔式设备，而当厂房高度受到限制时，则宜选用混合澄清器。

Auxiliary Equipment

The dispersed phase in an extraction tower is allowed to coalesce at some point into a continuous layer from which one product stream is withdrawn. The interface between this layer and the predominant continuous phase is set in an open section at the top or bottom of a packed tower; in a sieve-plate tower it is set in an open section near the top of the tower when the light phase is dispersed. If the heavy phase is dispersed, the interface is kept near the bottom of the tower. The interface level may be automatically controlled by a vented overflow leg for the heavy phase, as in a continuous gravity decanter. In large columns the interface is often held at the desired point by a level controller actuating a valve in the heavy-liquid discharge line.

In liquid-liquid extraction the solvent must nearly always be removed from the extract or raffinate or both. Thus auxiliary stills, evaporators, heaters, and condensers form an essential part of most extraction systems and often cost much more than the extraction device itself. As mentioned at the beginning of this section, if a given separation can be done by either extraction or distillation, economic considerations usually favor distillation. Extraction provides a solution to problems that cannot be solved by distillation alone but does not usually eliminate the need for distillation or evaporation in some part of the separation system.

工程案例分析

吸收塔的改造

目前，全球每年排放的 SO_2 大约为 3 亿吨，主要来源于化工、电力和冶炼等行业。SO_2 的排放严重污染了大气，影响人体的健康，产生酸雨，危及农作物。因此，治理 SO_2 排放问题十分重要。根据环境保护部规定二氧化硫排放总量要逐渐降低，所以部分单位对二氧化硫排放超标的设备进行改造。

某金属冶炼厂冶炼炉排放的含有 1% SO_2（摩尔分数，下同）的混合气体用清水在装有陶瓷拉西环的填料塔逆流吸收，经过一段时间分析吸收塔尾气 SO_2 超标，工厂组织技术人员分析原因，并采取方便而又有效的措施进行改造（原来要求尾气排放 SO_2 不超过 0.1%，当时的排放组成是 0.5%，原设计液气比为 10，操作条件下平衡关系为 $Y^*=8.0X$）。

首先分析 SO_2 超标的原因，从两方面入手。一方面可能是操作条件不当，如因管路等原因引起气量和液量的变化，使得吸收操作采用的液气比变小，也可能是矿石组成变化使含硫量增加导致进塔组成提高，吸收剂温度升高，吸收压力降低；另一方面是设备出了问题，传质系数下降，传质阻力增大。针对可能的原因进行检测确定，对进气量和清水量检测发现波动很小，分析矿石组成变化也不大，按设计时的富裕程度出口 SO_2 不可能超标。当时正值冬季，不可能是清水温度变化所致，吸收压力为常压没有变化。唯一可能是填料使用时间长，有破损，液体分布不均，填料性能下降，传质阻力增加。

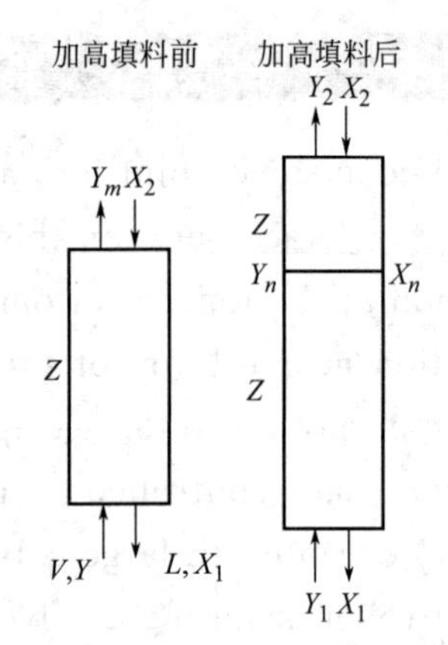

附图 1　填料层高度计算示意

采取什么措施控制 SO_2 超标，有人首先提出增加清水流量。这一措施看起来应该有效又方便，但也有人反对，理由是对于清水吸收 SO_2 的体系，溶解度适中，气、液两相的阻力都不能忽略，这样双膜控制的吸收过程，提高液体流量，传质系数能提高，但只有提高很大，总传质系数才能显著增加，何况在填料已经破碎的情况下，若采用较大的喷淋量很可能引起液泛，该措施不宜采取。

也有人提出加高一段填料，并采用新型填料，小试实验测得新型填料传质单元高度为 0.8m。那么增加填料层的高度为多少米才能使得 SO_2 达到排放标准呢？

通过附图 1 设计计算如下：

原塔：$$Z=H_{OG}N_{OG}$$

对于加高填料层高度后的原塔部分，其气量和传质系数没变化，故传质单元高度 H_{OG} 不变，塔高未变，所以传质单元数 N_{OG} 没变化。吸收温度可以认为近似不变，所以，解吸因数 S 不变。即：

原塔段：

$$N_{OG}=\frac{1}{1-S}\ln\left[(1-S)\frac{Y_1}{Y_m}+S\right] \tag{1}$$

加高后的原塔段：

$$N'_{OG}=\frac{1}{1-S}\ln\left[(1-S)\frac{Y_1-mX_n}{Y_n-mX_n}+S\right] \tag{2}$$

$$S=\frac{m}{\frac{L}{V}}=\frac{8}{10}=0.8$$

由式(1)、式(2) 得到：
$$\frac{Y_1}{Y_m}=\frac{Y_1-mX_n}{Y_n-mX_n} \tag{3}$$

添加的塔高段：
$$N''_{OG}=\frac{1}{1-S}\ln\left[(1-S)\frac{Y_n}{Y_2}+S\right] \tag{4}$$

对添加的塔高段做物料衡算： $$(Y_n-Y_2)V=LX_n \tag{5}$$

$$\frac{(Y_n-Y_2)V}{L}=X_n \tag{6}$$

将式(6) 代入式(3)，整理得

$$Y_n=\frac{Y_1+SY_2-\frac{SY_1Y_2}{Y_m}}{\frac{Y_1}{Y_m}(1-S)+S}=\frac{0.01+0.8\times0.001-\frac{0.8\times0.01\times0.001}{0.005}}{\frac{0.01}{0.005}\times(1-0.8)+0.8}$$
$$=0.0077 \tag{7}$$

将式(7) 代入式(4)，整理得：

$$N''_{OG}=\frac{1}{1-0.8}\ln\left[(1-0.8)\times\frac{0.0077}{0.001}+0.8\right]=4.24$$
$$Z'=N''_{OG}H'_{OG}=4.24\times0.8=3.4(\text{m})$$

即在原塔的基础上增加 3.4m 的塔段，但这一方案有人提出质疑，增加塔高会带来清水离心泵扬程不够的问题，要解决此问题还要在原离心泵的管路上串联一个离心泵或换一台扬程大的离心泵，若这样，不如在原吸收流程中增加一个小吸收塔，其塔径与原塔的塔径相同，同时填料还用新型填料，增加一台离心泵，采用清水。流程如附图 2 所示，具体塔高设计如下：

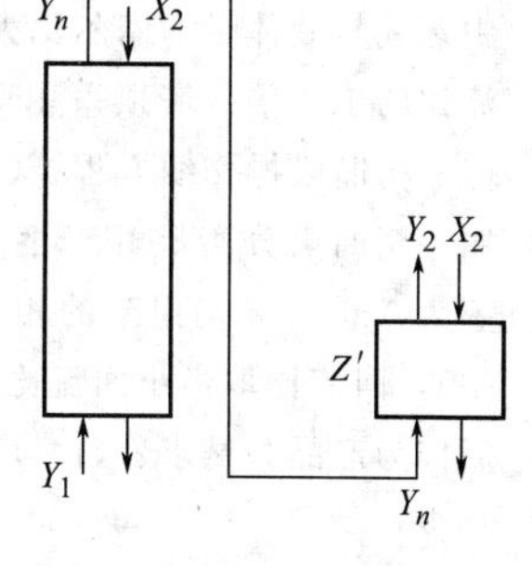

附图 2 流程

$$N_{OG}=\frac{1}{1-S}\ln\left[(1-S)\frac{Y_n}{Y_2}+S\right]$$

$$N_{OG}=\frac{1}{1-0.8}\ln\left[(1-0.8)\times\frac{0.005}{0.001}+0.8\right]=2.9$$

串联塔的塔高 $Z'=0.8\times2.9=2.3(\text{m})$。对于当地低廉的水价来说，该方案较为经济，若水价高，要进行具体的经济核算。

最后技术改造方案确定串联一段 2.3m 装有新型填料的吸收塔，经过实际运行，对吸收塔出口混合气体进行测试，发现 SO_2 排放浓度达到了原工艺要求。

习 题

1. 计算甲醇在 30℃水中的扩散系数。
2. 正庚烷（A）和正辛烷（B）所组成的混合液，在 388K 时沸腾，外界压力为 101.3kPa，根据实验测定，在该温度条件下的 $p_A^0=160\text{kPa}$，$p_B^0=74.8\text{kPa}$，试求相平衡时气、液相中正庚烷的组成。

3. 苯酚（A）和对甲苯酚（B）的饱和蒸气压数据如下：

温度 t/℃	苯酚蒸气压 p_A^0/kPa	对甲苯酚蒸气压 p_B^0/kPa	温度 t/℃	苯酚蒸气压 p_A^0/kPa	对甲苯酚蒸气压 p_B^0/kPa
113.7	10.0	7.70	117.8	11.99	9.06
114.6	10.4	7.94	118.6	12.43	9.39
115.4	10.8	8.20	119.4	12.85	9.70
116.3	11.19	8.50	120.0	13.26	10.0
117.0	11.58	8.76			

试按总压 $p=75$mmHg（绝压）计算该物系的 t-x-y 相平衡数据并画出相图，该物系为理想物系。

4. 将含苯摩尔分数为 0.5，甲苯摩尔分数为 0.5 的溶液加以汽化，汽化率为 1/3，已知物系的相对挥发度为 2.47，试计算：

（1）简单蒸馏时，气相与液相产物的组成；

（2）平衡蒸馏时，气相与液相产物的组成。

5. 每小时将 15000kg 含苯 40％和甲苯 60％的溶液，在连续精馏塔中进行分离。操作压力为 101.3kPa，要求馏出液能回收原料中的 97.1％的苯，釜液中含苯不高于 2％（以上均为质量分数）。求馏出液和釜液的摩尔流率及摩尔组成。

6. 在一连续精馏塔中分离某二元理想混合液。原料液流量为 100kmol/h，浓度为 0.4（摩尔分数，下同）。要求塔顶产品浓度为 0.9，塔釜浓度为 0.1。试求：（1）馏出液与釜残液的流量；（2）若每小时从塔顶采出 50kmol 馏出液，工艺要求应作何改变？

7. 某液体混合物易挥发性组分含量为 0.6，在泡点状态下连续送入精馏塔，加料量为 100kmol/h，易挥发性组分的回收率为 99％，釜液易挥发性组分含量为 0.05，回流比为 3，以上均为摩尔分数。试求：（1）塔顶产品与塔底产品的摩尔流率；（2）精馏段和提馏段内上升蒸气及下降液体的摩尔流率；（3）精馏段和提馏段的操作线方程。

8. 采用常压连续精馏塔分离苯-甲苯混合物。原料中含苯 0.44（摩尔分数，下同），进料为气液混合物，其中蒸气与液体量的摩尔比为 1∶2。已知操作条件下物系的平均相对挥发度为 2.5，操作回流比为最小回流比的 1.5 倍。塔顶采用全凝器冷凝，泡点回流，塔顶产品液中含苯 0.96。试求：（1）操作回流比；（2）精馏段操作线方程；（3）塔顶第二层理论板的气液相组成。

9. 用一精馏塔分离苯-甲苯溶液，进料为气液混合物，气相占 50％（摩尔分数），进料混合物中苯的摩尔分率为 0.60，苯与甲苯的相对挥发度为 2.5，现要求塔顶、塔底产品组成分别为 0.95 和 0.05（摩尔分率），回流比取最小回流比的 1.5 倍。塔顶分凝器所得的冷凝液全部回流，未冷凝的蒸气经过冷凝冷却后作为产品。试求：（1）塔顶塔底产品分别为进料量的多少倍？（2）塔顶第一理论板上升的蒸气组成为多少？

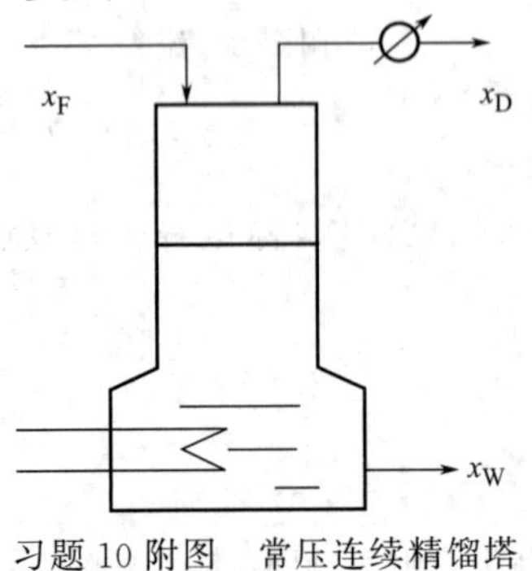

习题 10 附图　常压连续精馏塔组成示意图

10. 如附图所示，常压连续精馏塔具有一层实际塔板及一台蒸馏釜，原料预热到泡点由塔顶加入，进料组成 $x_F=0.20$（易挥发性组分的摩尔分率，下同）。塔顶上升蒸气经全凝器全部冷凝后作为产品，已知塔顶馏出液的组成为 0.28，塔顶易挥发性组分的回收率为 80％。系统的相对挥发度为 2.5。试求釜液组成及塔板的默弗里效率。

11. 将氨气通入水中，平衡后测得氨在 1kg 水中的溶解量为 10g，当绝对压强为 100kPa、温度为 20℃时，液相上方氨的分压为 780Pa，当绝对压强为 100kPa、温度为 40℃时，液相上方氨的分压为 980Pa。试求两种条件下氨的溶解度系数 H、亨利系数 E、相平衡常数 m。

12. 理想气体混合物中溶质A的含量为0.06（体积分数），与溶质A含量为0.012（摩尔比）的水溶液相接触，此系统的平衡关系为$Y^*=2.52X$。试判断传质进行的方向并计算过程的传质推动力。

13. 用填料塔进行吸收操作，在操作条件下$k_G=1.2\times10^{-2}$kmol/(m^2·h·kPa)，$k_L=1.2$m/h，已知液相吸收系数$k_L \propto L^{0.67}$。若操作中气体流量不变，认为k_G也不变，而液体流量增加1倍。求当溶解度系数$H=1.00\times10^{-3}$kmol/(m^3·Pa)及$H=2.00\times10^{-3}$kmol/(m^3·Pa)时，K_G、K_L分别增加的百分数。

14. 用清水在填料吸收塔内吸收混合于空气中的甲醇，操作温度为27℃，压强为101.3kPa。稳态操作下，塔内某截面上的气相甲醇分压为5kPa，液相中甲醇的浓度为2.11kmol/m^3。已知吸收系数$k_G=1.22\times10^{-5}$kmol/(m^2·s·kPa)，溶解度系数$H=1.955$kmol/(m^3·kPa)。试计算该截面上的吸收速率。

15. 用清水在填料吸收塔内吸收混合气体中的溶质A，已知操作条件下体系的相平衡常数$m=2.5$，进塔气体浓度为0.05（摩尔比），当操作液气比为5时，试分别计算逆流操作与并流操作情况下气体出口与液体出口极限浓度。如果平衡常数为5，而操作的液气比为2.5，结果又会如何？

16. 在填料吸收塔内用清水吸收混合气体中的组分A，气体流量为5000N m^3/h，已知混合气体中A的摩尔比为0.12，系统平衡关系为$Y^*=2.5X$，清水的用量是最小用量的1.6倍，逆流操作，组分A的吸收率为96%。试计算：(1) 清水用量；(2) 吸收液出塔浓度；(3) 分别采用平均推动力法和脱吸因数法计算气相总体积传质系数为0.15kmol/(m^3·s)，塔截面积为0.5m^{2}时，所需填料层高度。

17. 用填料塔从一混合气体中吸收所含的苯。混合气体中含苯5%（体积分数），其余为空气，要求苯的回收率为90%，吸收塔为常压操作，温度为25℃，入塔混合气体为每小时940m^3（标准状况），入塔吸收剂为纯煤油，煤油的耗用量为最小耗用量的1.5倍，已知该系统的平衡关系$Y=0.14X$（其中Y、X为摩尔比），气相总体积吸收系数$K_Ya=0.035$kmol/(m^3·s)，纯煤油的平均分子量为$M_s=170$，塔径$D=0.6$m。求：(1) 吸收剂使用量（单位为kg/h）；(2) 吸收液出塔浓度；(3) 填料层高度。

18. 在单级萃取器中以异丙醚为萃取剂，从乙酸组成为0.50（质量分数）的乙酸-水溶液中萃取乙酸。乙酸-水溶液量为500kg，异丙醚量为600kg，试作如下各项：(1) 在直角三角形相图上绘出溶解度曲线和辅助曲线；(2) 确定原料液与萃取剂混合后，混合液的坐标位置；(3) 求萃取相与萃余相间溶质（乙酸）的分配系数及溶剂的选择性系数。参考数据列于附表中。

习题18附表　20℃时乙酸(A)-水(B)-异丙醚(S)的相平衡数据(质量分数)

水相			有机相		
A	B	S	A	B	S
0.69	98.1	1.2	0.18	0.5	99.3
1.41	97.1	1.5	0.37	0.7	98.9
2.89	95.5	1.6	0.79	0.8	98.4
6.42	91.7	1.9	1.9	1.0	97.1
13.34	84.4	2.3	4.8	1.9	93.3
25.50	71.7	3.4	11.40	3.9	84.7
36.7	58.9	4.4	21.60	6.9	71.5
44.3	45.1	10.6	31.10	10.8	58.1
46.40	37.1	16.5	36.20	15.1	48.7

19. 某二元混合物含A 40kg，B 60kg，现在加入溶剂S进行单级萃取，萃取相中$y_A/y_B=3$（质量比），$k_A=6$，$k_B=0.5$，则脱除溶剂后可得萃取液量为多少？

20. 以水为溶剂从丙酮-乙酸乙酯中萃取丙酮，通过单级萃取，使丙酮含量由原料液中的0.3降至萃余液中的0.15。平衡数据如附表所示，若原料液量为100kg，求：(1) 溶剂水的使用量；(2) 所得到的萃取相的量及组成；(3) 为得到含丙酮浓度最大的萃取液所需的溶剂用量。

习题20附表　丙酮(A)-乙酸乙酯(B)-水(S)在30℃下的相平衡数据(质量分数)

乙酸乙酯相			水　相		
A	B	S	A	B	S
0	96.5	3.50	0	7.40	92.6
4.80	91.0	4.20	3.20	8.30	88.5
9.40	85.6	5.00	6.00	8.00	86.0
13.50	80.5	6.00	9.50	8.30	82.2
16.6	77.2	6.20	12.8	9.20	78.0
20.0	73.0	7.00	14.8	9.80	75.4
22.4	70.0	7.60	17.5	10.2	72.3
26.0	65.0	9.00	19.8	12.2	68.0
27.8	62.0	10.2	21.2	11.8	67.0
32.6	51.0	13.4	26.4	15.0	58.6

思　考　题

1. 压力对气液相平衡关系有何影响？精馏塔的操作压力增大，其他条件不变，塔顶温度、塔底温度和浓度如何变化？
2. 精馏塔的进料量对塔板层数有无影响？为什么？
3. 比较精馏塔的不同塔顶冷凝方式(全凝器冷凝和分凝器冷凝）及塔底加热方式(直接蒸汽加热和间接蒸汽加热）各有何特点？
4. 影响精馏操作的主要因素有哪些？它们遵循哪些基本关系？
5. 比较温度、压力对亨利系数、溶解度系数及相平衡常数的影响。
6. 吸收剂进入吸收塔前先被冷却与直接进入吸收塔两种情况，吸收效果有何不同？
7. 什么是气膜控制及液膜控制，各有何特点？用水吸收混合气体中的CO_2属于什么控制过程？提高其吸收速率的有效措施是什么？
8. 确定适宜液气比的理论依据是什么？
9. 对于均相液体混合物的分离，根据哪些因素来决定是采用蒸馏方法还是萃取方法进行分离？
10. 温度对于萃取分离效果有何影响？
11. 塔板上有哪些异常操作现象？它们对传质性能有何影响？
12. 填料塔的流体力学性能包括哪些？对塔的传质性能有何影响？

名　词　术　语

英文/中文

a battery of tank/一组槽
absorber/吸收设备
absorption factor/吸收因子
activity gradient/活度梯度
adsorption/吸收
agitated tower extractor/搅拌塔萃取器

英文/中文

association parameter for solvent/溶剂的缔合参数
average molecular velocity/平均分子速度
average plate efficiency/平均塔效率
axial dispersion/轴向扩散
azeotrope /恒沸物
azeotropic composition/恒沸组成

azeotropic distillation/恒沸精馏
benzene-toluene/苯-甲苯
boiling-point diagram/沸点图
Bollman extractor/立式吊篮（波耳曼）萃取器
bottom product or bottoms/塔底产品
boundary layer/边界层
bubble-point line/泡点线
buffered phosphate solution/磷酸盐缓冲溶液
butyl acetate/乙酸丁酯
cascades/多级操作
centrifugal extractor/离心萃取器
chlorinated hydrocarbon/氯化烃类化合物
close-boiling liquid/沸点相近的液体
close-boiling mixture/沸点相近的混合物
cocurrent flow/并流
collision integral/碰撞积分
complete miscibility/完全互溶
concentration gradient/浓度梯度
condensate/冷凝液
condenser/冷凝器
constant molar overflow/恒摩尔流
continuous countercurrent leaching/连续逆流浸提
continuous distillation/连续蒸馏
continuous gravity decanter/连续重力分层器
convective bulk flow/对流总体流动
convective flow/对流流动
countercurrent multistage/多级逆流
countercurrent/逆流
degree of freedom/自由度
desorption/解吸
dew-point line/露点线
differential-contact method/微分接触法
differential distillation/微分蒸馏
diffusion battery/扩散组
diffusivity/扩散系数
direct steam injection/直接蒸汽加热
dispersed phase/分散相
distillation/蒸馏
distribution coefficient/分配系数
dome-shaped curve/圆顶曲线
down-comer/降液管
downhill/向下
eddy diffusion/涡流扩散
eddy diffusivity/涡流扩散系数
effective collision diameter/有效碰撞直径
effective film thickness/有效膜厚
emulsify/乳化
enthalpy balance/焓衡算
equilibrium relate/平衡关系
equilibrium relations/相平衡方程
equilibrium-stage cascade/多级平衡操作
equilibrium stage/平衡级
equimolar counter diffusion/等摩尔反向扩散
equimolar diffusion/等分子扩散
extraction battery/萃取组
extractive distillation/萃取精馏
extract layer/萃取层
extract phase/萃取相
feed line/进料线
feed plate/进料板
feed plate location/进料板位置
fermentation broth/发酵液
film coefficient/膜系数
flash distillation/闪蒸
flooding point/泛点
flow-weighted average concentration/流动权重平均浓度
forced diffusion/强制扩散
gas absorption/气体吸收
gas film resistance/气膜阻力
gas-liquid interface/气液界面
geometric series/几何级数
graphical method/图解法
graphical step-by-step/逐级图解
heated kettle/加热釜
heavy liquid/重液体（重相）
height of a transfer unit（HTU)/传质单元高度
height of the packed section/填料层高度
Henry's law/亨利定律
hexane-heptane /己烷-庚烷
high boiler/高沸物
high-boiling component/高沸点组分
high-boiling-point polar solvent/高沸点极性溶剂
high selectivity/高选择性
holdup/持液
horizontal tie line/水平连接线
hydrocarbon/碳氯化合物
ideal liquid/理想液体
ideal plate/理论板
ideal-solution law/理想溶液定律
ideal stage/理论级
individual coefficient/分系数
inert gas/惰性气体
instantaneous flux/瞬时通量
kinematic viscosity/动力黏度
leaching/浸提

less volatile component/难挥发组分
lever-arm rule/杠杆定律
light liquid/轻液体（轻相）
liquid film resistance/液膜阻力
liquid-liquid extraction/液-液萃取
liquid side-draw/液相侧线抽料
local efficiency/点效率
logarithmic mean/对数平均
low boiler/低沸物
low-boiling-point polar solvent/低沸点极性溶剂
lower boiling constituent/低沸物
mass fraction/质量分数
mass-transfer coefficients/传质系数
mass transfer/传质
maximum boiling azeotrope/最高恒沸点恒沸物
mean free path/平均自由程
methyl alcohol-ethyl alcohol/甲醇-乙醇
minimum boiling azeotrope/最低恒沸点恒沸物
minimum reflux ratio/最小回流比
mixer-settler/混合澄清槽
mixing zone/混合区
molar density/摩尔密度
molar flux/摩尔通量
molar heat capacity/摩尔比热容
molar volume/摩尔体积
molecular diffusion/分子扩散
mole fraction of solute/溶质的摩尔分数
mole fraction/摩尔分数
mole ratio/比摩尔
more volatile component/易挥发组分
moving-bed continuous leaching/移动床式连续浸取
multicomponent diffusion/多组分扩散
multicomponent distillation/多组分蒸馏
multiple feed column/多点进料
multiple-stage extraction/多级萃取
multistage crosscurrent extraction/多级错流萃取
Murphree efficiency/默弗里效率
nonstagewise contacting device/非逐级接触设备
nonvolatile solvent/不挥发性溶剂
number of transfer units（NTU）/传质单元数
one-dimensional diffusion/一维扩散
one-way diffusion/单向扩散
one-way mass transfer/单向传质
operating costs/操作费用
operating line/操作线
optimum gas-liquid ratio/最适宜气液比
optimum reflux ratio/最佳回流比
ordinary distillation/普通蒸馏
overall coefficient/总系数
overall efficiency/总板效率
overflow/溢流
overhead product /塔顶产品
overlapping boiling-point range/相互交叠的沸点范围
packed column/填料塔
packed extraction tower/填料萃取塔
paddle conveyor/桨式输送机
partial condensation/部分冷凝
partial condenser/分凝器
partial miscibility/部分互溶
partial pressure driving force/分压推动力
partial pressure/分压
partial vaporization/部分汽化
penetration theory/渗透理论
perfect plate/理论板
perorate false bottom/多孔假底
petroleum product/石油产品
pinch point/夹点
plait point /临界混溶点（结点）
plate columns/板式塔
plate efficiency/板效率
plate-to-plate calculation/逐板计算法
pressure drop/压降
pressure gradient/压力梯度
pseudo-binary mixtures/拟二元混合物
pulse column 脉冲塔
raffinate layer/萃余层
raffinate phase/萃余相
raffinate/萃余液
random motion/随机运动
Raoult's law/拉乌尔定律
rate of absorption/吸收速率
Rayleigh equation/雷利方程
reboiler/再沸器
rectangular coordinate/直角坐标
rectangular extraction-phase-diagram/直角形萃取相图
rectifying line/精馏段操作线
rectifying section/精馏段
reference plane/参考面
reflux ratio/回流比
reflux splitter/回流分配器
reflux/回流
regeneration/再生
relative volatility/相对挥发度
reverse osmosis/反渗透

Rotocel extractor/洛特赛萃取机
saddle packing/鞍形填料
Schmidt number/施密特数
selectivity coefficient/选择系数
separator/分离器
settling zone/沉降区
short-circuiting/短路
sidestreams/侧线
sieve-plate distillation tower/筛板蒸馏塔
sieve-plate pulse column/筛板脉冲塔
simple batch distillation/简单间歇蒸馏
simple steam distillation/简单水蒸气蒸馏
solid-bowl helical-conveyor centrifuge/固体转筒螺旋式输送离心器
solid extraction/固体萃取
soluble material/易溶物质
solute-free solid/无溶质固体
solute/溶质
solvent/溶剂
stage efficiency or plate efficiency/板效率
stagnant film/滞流膜
stagnant gas layer/气体滞流层
stationary solid-bed leaching/静态床浸提
steady-state diffusion/稳态扩散
still/釜
stripping/脱吸
stripping-column distillation/提馏塔
stripping factor/解吸因子
stripping line/提馏段操作线
stripping or enriching section/ 提馏段
stripping section/提馏线
subcooling/过冷
supercritical fluid extraction/超临界流体萃取
superheated vapor/过热蒸气
temperature gradient/温度梯度
thermal condition of the feed/进料状态
thermal diffusion/热扩散
thermal diffusivity/热扩散系数
tie line/联络线
total condenser/全凝器
total pressure/总压
total resistance/总阻力
transient diffusion/瞬时扩散
triangular diagram/三角相图
two-film theory/双膜理论
two-phase region/两相区
unagitated bed of solid/无搅拌的固体床
underflow/下溢
unsteady-state diffusion/非稳态扩散
vacuum distillation/真空蒸馏
vented overflow leg/排放溢流腿
vertical tower/垂直塔
volumetric diffusivity/体积扩散系数
volumetric film mass-transfer coefficient/膜体积传质系数
volumetric overall mass-transfer coefficient/总体积传质系数
washing of filtered solid/过滤后固体的洗涤
weir/堰

第4章

固体颗粒流体力学基础与机械分离

Chapter 4 Solid Particles' Fluid Mechanics Basis and Mechanical Separations

学习目标

通过本章学习，学生应具备下列知识和能力：

1. 掌握固体颗粒流体力学基础原理，能够基于该基础原理，建立复杂工程问题中关于固体颗粒流体力学涉及的数学模型。

2. 掌握非均相物系相关单元操作沉降、过滤、固体流态化等分离的基本原理，熟悉复杂工程问题中关于非均相物系分离单元操作设备的结构特点、工作原理、操作方法。

3. 掌握上述操作设备的设计计算及设计选型；能够针对复杂化学工程中的相关问题，使用恰当的技术、资源、现代工程工具和信息技术工具，通过对复杂工程问建立数学模型，设计合适的分离设备，并能够针对相关化学工程所涉及相关的问题进行分析，提出合理的解决措施。

4. 熟悉相关领域的英文表达，增强国际交流能力，拓宽国际视野。

学习要求

1. 能够基于固体颗粒流体力学基础原理，建立复杂工程问题中关于固体颗粒流体力学涉及的数学模型。

2. 能够基于沉降、过滤、固体流态化等单元操作的基本原理，能够针对复杂化学工程中的相关问题进行分析，选择合适的单元操作方法，并掌握沉降、过滤、固体流态化等非均相物系分离单元设备的操作操作方法。

3. 能够熟悉相关领域的英文表达，具有国际交流能力。

化工生产中，经常需要将混合物加以分离。为了实现不同的分离目的，必须根据混合物性质的不同而采用不同的方法。一般将混合物分为两大类，即均相混合物（homogeneous mixture）和非均相混合物（heterogeneous mixture）。其中，非均相混合物由两相或两相以上构成，相界面两侧物质的性质截然不同。这种混合物的分离就是将不同的相分开，所以通常可采用能耗较低的机械方法加以分离。例如，要对锅炉等装置进行尾气除尘，由于空气与固体粉尘的密度差别很大，所以可采用重力及离心力场中的沉降操作。

非均相混合物由具有不同物理性质（例如密度差别）的分散物质和连续介质组成。其中处于分散状态的物质，如分散于流体中的固体颗粒、液滴或气泡，称为分散相；而包围分散物质且处于连续状态的物质称为分散介质或连续相。

在过程工业中，对非均相混合物进行分离的目的是：

① 净化分散介质　如原料气在进入催化反应器前必须除去其中的尘粒和有害杂质，以保证催化剂的活性。

② 回收分散物质　如从气流干燥器出来的气体或从结晶器出来的晶浆中，常含有有用的固体颗粒状产品，必须回收。

③ 环境保护　生产中的废气、废液在排放前，必须把其中所含的有害物质分离出来，使其达到规定的排放标准，以保护环境。

本章只讨论分离非均相混合物所采用的机械分离方法。

Heterogeneous Mixture Separation

In chemical production, it is often necessary to separate the mixture. In order to achieve different separation purposes, different methods must be used according to the properties of the mixture. Generally, mixtures are divided into two categories, namely, homogeneous mixture and heterogeneous mixture. The heterogeneous mixture consists of two or more phases, and the properties of substances on both sides of the phase interface are quite different. The separation of the mixture is the separation of the different phases, so the separation can usually be done mechanically with less energy consumption. For example, to the boiler and other devices exhaust dust, because the density of air and solid dust is very different, so gravity and centrifugal force field can be used in the settlement operation.

Heterogeneous mixtures consist of dispersed substances and continuum media with different physical properties, such as differences in density. A substance in a dispersed state, such as a solid particle, droplet or bubble dispersed in a fluid, is called a dispersed phase. The substance surrounded by the dispersed substance and in a continuous state is called the disperse medium or continuous phase.

In the process industry, the purpose of separating heterogeneous mixtures is to:

(1) Purification of disperse media. For example, dust particles and harmful impurities must be removed before the raw gas enters the catalytic reactor to ensure the activity of the catalyst.

(2) Recovery of dispersed substances. For example, the gas from the air dryer or the crystal slurry from the crystallizer often contains useful solid granular products that must be recovered.

(3) Environmental protection. Before discharge, the waste gas and waste liquid in production must be separated from the harmful substances contained therein so as to meet

the prescribed discharge standards，so as to protect the environment.

This chapter only discusses the mechanical separation methods used to separate heterogeneous mixtures.

4.1 固体颗粒特性（Characteristics of Solid Particles）

表述固体颗粒特性的主要参数为颗粒的形状、大小（体积）和表面积。

4.1.1 单一颗粒特性（Characteristics of single particle）

（1）球形颗粒（spherical particle）

球形颗粒通常用直径（粒径）表示其大小。球形颗粒的各有关特性均可用单一的参数，即直径 d 来全面表示，如：

$$V=\frac{\pi}{6}d^3 \tag{4-1}$$

$$S=\pi d^2 \tag{4-2}$$

$$a=\frac{6}{d} \tag{4-3}$$

式中 d——颗粒直径，m；

V——球形颗粒的体积，m^3；

S——球形颗粒的表面积，m^2；

a——比表面积（单位体积颗粒具有的表面积），m^2/m^3。

（2）非球形颗粒（non-spherical particle）

工业上遇到的固体颗粒大多是非球形的，非球形颗粒可用当量直径及形状系数来表示其特性。

当量直径是根据实际颗粒与球体的某种等效性而确定的。根据测量方法及在不同方面的等效性，当量直径有不同的表示方法。工程上，体积当量直径应用比较多。令实际颗粒的体积等于当量球形颗粒的体积，则体积当量直径定义为：

$$d_e=\sqrt[3]{\frac{6V_p}{\pi}} \tag{4-4}$$

式中 d_e——体积当量直径，m；

V_p——非球形颗粒的实际体积，m^3。

形状系数又称球形度，它表征颗粒的形状与球形的差异程度，定义为：

$$\varphi_s=\frac{S}{S_p} \tag{4-5}$$

式中 φ_s——颗粒的形状系数或球形度；

S_p——颗粒的表面积，m^2；

S——与该颗粒体积相等的圆球的表面积，m^2。

因体积相同时球形颗粒的表面积最小，所以任何非球形颗粒的形状系数均小于 1。对于球形颗粒，$\varphi_s=1$。颗粒形状与球形差别越大，φ_s 值越小。

对于非球形颗粒，必须有两个参数才能确定其特性。通常选用体积当量直径和形状系数来表征颗粒的体积、表面积和比表面积，即：

$$V_p = \frac{\pi}{6} d_e^3 \tag{4-6}$$

$$S_p = \frac{\pi d_e^2}{\phi_s} \tag{4-7}$$

$$a_p = \frac{6}{d_e \phi_s} \tag{4-8}$$

4.1.2　颗粒群的特性

工业中遇到的颗粒大多是由大小不同的粒子组成的集合体，称为非均一性粒子或多分散性粒子。与此相对应，将具有同一粒径的颗粒称为单一性粒子或单分散性粒子。颗粒群的特性可用粒度分布和平均直径来表示。

（1）粒度分布（particle size distribution）

不同粒径范围内所含粒子的个数或质量，称为粒度分布。可采用多种方法测量多分散性粒子的粒度分布。对于粒径大于 40μm 的颗粒，通常采用一套标准筛进行测量，这种方法称为筛分分析。颗粒的尺寸，通常用标准筛的目数来表征。所谓目数，是指物料的粒度或粗细度，一般指在 1in×1in 的面积内有多少个网孔数，即筛网的网孔数。物料能通过该网孔即定义为多少目数。如 200 目，就是该物料能通过 1in×1in 内有 200 个网孔的筛网。以此类推，目数越大，说明物料粒度越小；目数越小，说明物料粒度越大。各国标准筛的规格不尽相同，常用的泰勒标准筛的目数与对应的孔径如表 4-1 所示。

表 4-1　泰勒标准筛数据

目数	孔径		目数	孔径	
	/in	/μm		/in	/μm
3	0.263	6680	48	0.0116	295
4	0.185	4699	65	0.0082	208
6	0.131	3327	100	0.0058	147
8	0.093	2362	150	0.0041	104
10	0.065	1651	200	0.0029	74
14	0.046	1168	270	0.0021	53
20	0.0328	833	400	0.0015	38
35	0.0164	417			

当使用某一号筛子时，通过筛孔的颗粒量称为筛过量，截留于筛面上的颗粒量则称为筛余量。称取各号筛面上的颗粒筛余量即得筛分分析的基本数据。目前各种筛制正向国际标准组织 ISO 筛系进行统一。

（2）颗粒的平均直径（average particle diameter）

颗粒平均直径的计算方法很多，其中最常用的是平均比表面积直径。设有一批大小不等的球形颗粒，其总质量为 G，经筛分分析得到相邻两号筛直径的颗粒质量为 G_i，筛分直径（两筛号筛孔的算术平均值）为 d_i。根据比表面积相等原则，颗粒群的平均比表面积直径可写为：

$$\frac{1}{d_a} = \sum \frac{1}{d_i} \times \frac{G_i}{G} = \sum \frac{x_i}{d_i}$$

或

$$d_{a}=\frac{1}{\sum \frac{x_{i}}{d_{i}}} \tag{4-9}$$

式中 d_a——平均比表面积直径，m；

d_i——筛分直径，m；

x_i——d_i 粒径段内颗粒的质量分数。

4.1.3 粒径测量（Particle size measurement）

粒径是颗粒占据空间大小的线性尺度，测定方法种类繁多。因原理不同，所测粒径范围及参数各异，应根据使用目的及方法的适用性做出选择。测量及表达粒径的方法可分为长度、质量、横截面、表面积及体积五类，常用方法见表 4-2 和表 4-3。粒径测量的结果应指明所采用的方法和表示法。

表 4-2 常用粒径测量法

测量方法	粒径范围/μm	参数类别	粒径表达	分布基准	介质	测量依据
标准筛 微目筛	＞38 5～40	长度	筛分	质量	干、湿	筛孔
光学显微镜 电子显微镜 全息照相	0.25～250 0.001～5 2～500	—	投影面积	面积或个数	干	通常是颗粒投影像的某种尺寸或某种相当尺寸
空气中沉降 液体中沉降 离心沉降 喷射冲击器 空气中抛射 淘析	3～250 2～150 0.01～10 0.3～50 ＞100 1～100	质量	同沉降速度的球直径(层流区)	质量	干 湿 干、湿 — 干 干、湿	沉降效应、沉积量、悬浮液浓度、密度或吸光度等随时间或位置的变化
光散射 X射线小角度散射 比浊计	0.3～50 0.008～0.2 0.05～100	横截面	等效球直径	质量或个数	湿 干 湿	颗粒对光的散射或吸光(散射和吸收)，颗粒对X射线的散射
吸附法 透过法(层流) 透过法(分子流) 扩散法	0.002～20 1～100 0.001～1 0.003～0.3	表面积	比表面积直径	—	干、湿	气体分子在颗粒表面的吸附，床层中颗粒表面对气流的阻力
Coulter 计数器 声学法	0.2～800 50～200	体积	常为等效球直径	体积或个数	湿	颗粒在小孔电阻传感区引起的电阻变化

表 4-3 超细粉尘的粒径分布测量

测量方法	测量仪器	粒径范围/μm
电子显微镜	透射式电子显微镜 扫描式电子显微镜	＞0.001 ＞0.006
X射线小角度散射	X射线小角测角仪	0.005～0.2
扩散法	筛网式扩散分级仪	0.005～0.2
电迁移率法	静电气溶胶测定仪 EAA 微分电迁移率式粒径分布测定仪 DMPS	0.01～1.0
惯性沉降法	低压级联冲击器	＞0.05

4.2　固体颗粒在流体中运动时的阻力（The Drag of Solid Particles Moving in the Fluid）

如图 4-1 所示，当流体以一定速度绕过静止的固体颗粒流动时，黏性流体会对颗粒施加一定的作用力。反之，当固体颗粒在静止流体中移动时，流体同样会对颗粒施加作用力。这两种情况的作用力性质相同，通常称为曳力或阻力（drag）。

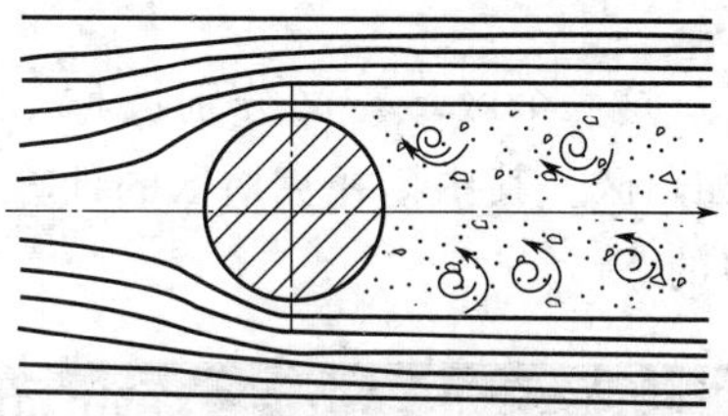

图 4-1　流体与颗粒间的相对运动

颗粒在静止流体中作沉降运动，或运动着的颗粒与流动着的流体之间的相对运动，均会产生这种阻力。对于一定的颗粒和流体，不论是哪一种相对运动，只要相对运动速度相同，流体对颗粒的阻力就一样。对于密度为 ρ、黏度为 μ 的流体，如果直径为 d_p 的颗粒在运动方向上的投影面积为 A，且颗粒与流体间的相对运动速度为 u，则颗粒所受到的阻力 F_d 可用下式来计算：

$$F_d=\zeta A\frac{\rho u^2}{2} \tag{4-10}$$

式中，ζ 为无量纲阻力系数，是流体相对于颗粒运动时的雷诺数（$Re=d_p u\rho/\mu$）的函数，即：

$$\zeta=f(Re)=f\left(\frac{d_p u\rho}{\mu}\right) \tag{4-11}$$

此函数关系需由实验测定。不同球形度颗粒 ζ 的实测数据，示于图 4-2 中。图中曲线大致可分为三个区域。对于球形颗粒（$\varphi_s=1$），各区域的曲线可分别用不同的计算式表示。

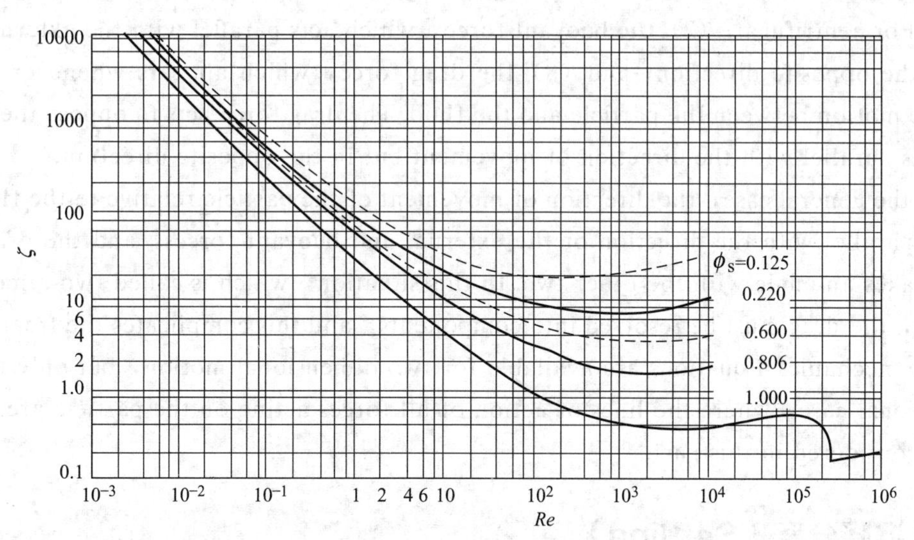

图 4-2　ζ-Re 关系曲线

① 层流区（$10^{-4}<Re<1$）：

$$\zeta=\frac{24}{Re} \tag{4-12}$$

② 过渡区（$1<Re<10^3$）：

$$\zeta=\frac{18.5}{Re^{0.6}} \tag{4-13}$$

③ 湍流区（$Re>10^3$）：

$$\zeta=0.44 \tag{4-14}$$

这三个区域，又分别被称为斯托克斯（Stokes）区、阿伦（Allen）区和牛顿（Newton）区。其中斯托克斯区的计算公式［式（4-12）］是准确的，其他两个区域的计算式是近似的。

Motion of Particles through Fluids

Many processing steps, especially mechanical separations, involve the movement of solid particles or liquid drops through a fluid. The fluid may be gas or liquid, and it may be flowing or at rest. Examples are the elimination of dust and fumes from air or flue gas, the removal of solids from liquid wastes, and the recovery of acid mists from the waste gas of an acid plant.

Mechanics of particle motion

The movement of a particle through a fluid requires an external force acting on the particle. This force may come from a density difference between the particle and the fluid, or it may be the result of electric or magnetic fields. In this section only gravitational or centrifugal forces, which arise from density differences, are considered.

Three forces act on a particle moving through a fluid: (1) the external force, gravitational or centrifugal; (2) the buoyant force, which acts parallel with the external force but in the opposite direction; and (3) the drag force, which appears whenever there is relative motion between the particle and the fluid. The drag force acts to oppose the motion and acts parallel with the direction of movement but in the opposite direction.

In the general case, the direction of movement of the particle relative to the fluid may not be parallel with the direction of the external and buoyant forces, and the drag force then makes an angle with the other two. In this situation, which is called two-dimensional motion, the drag must be resolved into components, and this complicates the treatment of particle mechanics. Equations are available for two dimensional motion, but only the one-dimensional case, where the lines of action of all forces acting on the particle are collinear, is considered in this part.

4.3 沉降分离（Settling）

沉降是指在某种力场中利用分散相和连续相间密度之差，使之发生相对运动而实现分离的操作过程。实现沉降操作的作用力可以是重力，也可以是惯性离心力。因此，沉降过程分为重力沉降和离心沉降两种方式。

4.3.1　重力沉降（Gravitational settling）

在重力场中，利用连续相与分散相的密度差异而产生相对运动，使两相得以分离的过程称为重力沉降。

（1）光滑球形颗粒在静止流体中的自由沉降

如果颗粒的沉降未受到其他颗粒沉降及器壁的影响，则称为自由沉降，反之称为干扰沉降。在静止流体中，颗粒将沿重力的方向作沉降运动。颗粒在沉降过程中受到三个力的作用，即重力、浮力及阻力，如图4-3所示。假设颗粒初速度等于零，此时颗粒仅受到重力和浮力的作用。此时，由于颗粒的密度大于流体的密度，则重力大于浮力，颗粒必产生加速度，称为沉降的加速阶段。由于颗粒和流体的密度都是一定的，所以颗粒受到的重力和浮力的大小是不变的，但颗粒所受阻力却随颗粒沉降速度的增大而增大。当三力的合力恰等于零时，颗粒开始作匀速沉降运动，称为沉降的等速阶段。一般来说，加速段很短，工程上可忽略不计，故沉降速度专指等速阶段中颗粒相对于流体的运动速度，以 u_t 表示。下面是 u_t 计算式的推导过程。

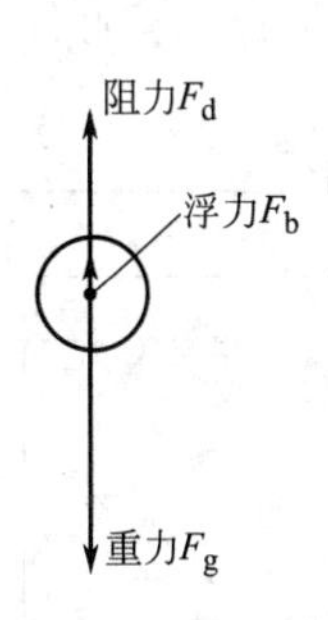

图4-3　沉降颗粒的受力

图4-3中的三个力可分别表示为：

重力（gravitational force）：　$F_g=\frac{\pi}{6}d^3\rho_s g$

浮力（buoyant force）：　$F_b=\frac{\pi}{6}d^3\rho g$

阻力（drag force）：　$F_d=\zeta\frac{\pi}{4}d^2\frac{\rho u_t^2}{2}$

令三力之和为零，则：

$$u_t=\sqrt{\frac{4gd(\rho_s-\rho)}{3\rho\zeta}} \tag{4-15}$$

式中　u_t——颗粒的自由沉降速度，m/s；

d——颗粒的直径，m；

ρ_s——颗粒的密度，kg/m^3；

ρ——流体的密度，kg/m^3。

由式（4-15）可以看出，颗粒直径越大，沉降速度越大，说明大直径颗粒较小直径颗粒更容易沉降。所以，在沉降小直径颗粒前，通常预先将大直径颗粒沉降下来。还可以看出，固体与流体的密度差越大，沉降越快，这正是非均相混合物机械分离的基本依据。

影响沉降速度的其他因素还有壁效应和干扰沉降。当颗粒在靠近器壁的位置沉降时，由于受器壁的影响，其沉降速度比自由沉降速度小，这种影响称为壁效应。若颗粒在流体中的体积分数较高，使得颗粒沉降过程中相互影响较大，或因沉降过快而引起涡流，增大阻力，从而影响沉降的现象均称为干扰沉降。干扰沉降的速度比自由沉降的速度小得多。

当分散相是液滴或气泡时，其沉降运动与固体颗粒的运动有所不同。其主要差别是液滴或气泡在曳力作用下产生形变，使阻力增大；或者液滴或气泡内部的流体产生循环运动，降低了相界面上的相对速度，使阻力减少。

上述颗粒沉降速度的计算方法，适用于多种情况下颗粒与流体在重力作用下的相对运动计算。例如既可适用于 $\rho_s>\rho$ 的情况，即沉降操作，也可适用于 $\rho_s<\rho$ 的颗粒浮升运动；既可适用于静止流体中颗粒的沉降，也可适用于流体相对于静止颗粒的运动；既可适用于颗粒与流体的逆向运动情况，也可适用于颗粒与流体同向运动但具有不同速度的相对运动速度的计算。

式（4-15）中的阻力系数 ζ 根据沉降时的雷诺数 $Re_t=du_t\rho/\mu$ 来计算［式（4-12）～式（4-14）］，此时各区域内的曲线分别用相应的函数式表达，如表 4-4 所示。

表 4-4　球形颗粒沉降速度在各区内的表达式

区域	定　律	Re_t 范围，阻力系数 ζ	计算公式
层流区	Stokes(斯托克斯)式	$10^{-4}<Re_t<1, \zeta=\dfrac{24}{Re_t}$	$u_t=\dfrac{d^2(\rho_s-\rho)g}{18\mu}$
过渡状态区	Allen(阿伦)式	$1<Re_t<10^3, \zeta=\dfrac{18.5}{Re_t^{0.6}}$	$u_t=0.27\sqrt{\dfrac{d(\rho_s-\rho)gRe_t^{0.6}}{\rho}}$
湍流区	Newton(牛顿)式	$Re>10^3, \zeta=0.44$	$u_t=1.74\sqrt{\dfrac{d(\rho_s-\rho)g}{\rho}}$

根据表 4-4 计算沉降速度 u_t 时，需要预先知道沉降雷诺数 Re_t 值才能选用相应的计算式。但是，u_t 待求，Re_t 值也就未知。所以，沉降速度 u_t 的计算需要采用试差法。即先假设沉降属于某一流型（如层流区），则可直接选用与该流型相应的沉降速度公式计算 u_t，然后再按 u_t 检验 Re_t 值是否在原假设的流型范围内。如果与原假设一致，则求得的 u_t 有效；否则，依照算出的 Re_t 值另选流型，并改用另外相应的公式求 u_t，直到算出的 Re_t 与所选用公式的 Re_t 值范围相符为止。由表 4-4 可以看出，该试差过程最多需要三次假设即可完成。

【例 4-1】 求直径为 10μm、密度为 2000kg/m³ 的颗粒在 20℃空气中的沉降速度。已知 20℃空气的密度为 1.205kg/m³，黏度为 0.0181mPa·s。

解 先假定颗粒沉降时处于层流区，则可采用斯托克斯公式求沉降速度：

$$u_t=\frac{d^2(\rho_s-\rho)g}{18\mu}=\frac{(10\times10^{-6})^2\times(2000-1.205)\times9.81}{18\times0.0181\times10^{-3}}=0.0060(\mathrm{m/s})$$

校核流型：

$$Re_t=\frac{du_t\rho}{\mu}=\frac{10\times10^{-6}\times0.0060\times1.205}{0.0181\times10^{-3}}=0.00399<1$$

所以假设成立，沉降速度计算正确。

Settling Separation of the Solids from Gases

Separations are extremely important, much processing equipment is devoted to separating one phase or one material from another.

Mechanical separations are applicable to heterogeneous mixtures not to homogeneous solutions. The techniques are based on physical differences between the particles such as size, shape, or density. They are applicable to separating solids from gases, liquid drops from gases, and solids from liquids.

Many mechanical separations are based on the sedimentation of solid particles or liquid drops through a fluid, impelled by the force of gravity or by centrifugal force. The fluid may be a gas or liquid; it may be flowing or at rest. In some situations the subjective of the process is to remove particles from a stream in order to eliminate contaminants from the fluid or to recover the particles, as in the elimination of dust and fumes from air or flue gas or the removal of solids from liquid wastes.

Gravitational Settling Processes

If a particle starts at rest with respect to the fluid in which it is immersed and is then moved through the fluid by a gravitational force, its motion can be divided into two stages. The first stage is a short period of acceleration, during which the velocity increases from zero to the terminal velocity, the second stage is the period during which the particle is at its terminal velocity.

Since the period of initial acceleration is short, usually on the order of tenths of a second or less, initial acceleration affects are short-range. Terminal velocities, on the other hand, can be maintained as long as the particle is under treatment in the equipment.

Particles heavier than the suspending fluid may be remove from a gas or liquid in a large settling box or settling tank, in which the fluid velocity is low and particles have ample time to settle out.

Dust particles may be removed from gases by a variety of methods. For coarse solid particles, larger than about 325-mesh (43μm diameter), which may be removed by a gravity settling.

(2) 重力沉降设备（gravitational settling equipment）

① 降尘室（settling chamber）　降尘室是分离气固混合物的一种十分常见的重力分离设备，如图 4-4 所示。

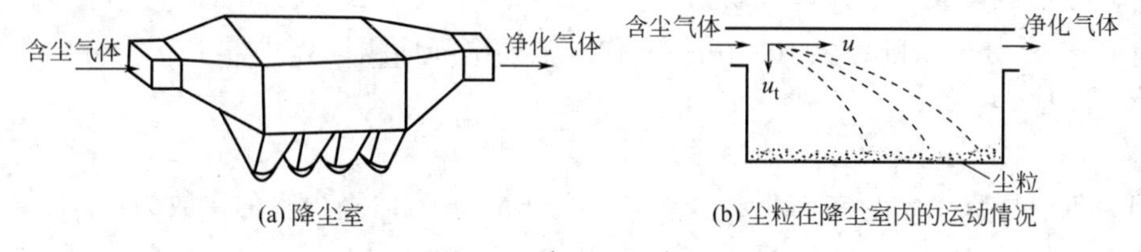

(a) 降尘室　　(b) 尘粒在降尘室内的运动情况

图 4-4　降尘室示意图

含尘气体进入降尘室后，因流道截面积扩大而速度减慢。气体中的固体颗粒一方面随气流作水平运动，另一方面在重力作用下作垂直运动，其运动轨迹如图 4-4(b) 所示。在每一瞬间，单个颗粒的真实速度可分解为两个速度：a. 水平方向上的分速度（等于气流进口速度 u）；b. 在垂直方向上的分速度（等于颗粒沉降速度 u_t）。此时，降尘室内颗粒的最长停留时间（或气流的停留时间）为 L/u，颗粒的最长沉降时间为 H/u_t，则颗粒可从气流中分离出来的条件是：

$$\frac{L}{u} \geqslant \frac{H}{u_t} \tag{4-16}$$

式中　H——降尘室的高度，m；

L——降尘室的长度，m；

u——气流速度，m/s。

降尘室生产能力用单位时间内通过降尘室的含尘气体的体积流量来表示，即：

$$V_s = \text{气体流通截面} \times \text{气速} = BHu \tag{4-17}$$

式中，B 为降尘室的宽度，m。

对于式（4-16），在停留时间等于沉降时间时，有 $Lu_t = Hu$。代入式（4-17）中，则：

$$V_s = BLu_t = Au_t \tag{4-18}$$

式中，A 为降尘面积，m^2。

式（4-18）说明，降尘室的生产能力与颗粒沉降速度和降尘室的降尘面积有关，而与降尘室的高度无关。基于这一原理，工业上的降尘室应设计成扁平状或在室内设置多层水平隔板。一般来说，降尘室可分离粒径为 50～75μm 以上的颗粒。

【例 4-2】 拟用降沉室除去矿石焙烧炉炉气中的氧化铁粉尘（$\rho = 4500 kg/m^3$），要求净化后的气体中不含粒径大于 40μm 的尘粒。操作条件下的气体体积流量为 $10800 m^3/h$，气体密度为 $1.6 kg/m^3$，黏度为 0.03mPa·s。求：(1) 所需的降尘面积至少为多少（m^2）？若降尘室底面宽 1.5m，长 4m，则降尘室需几层？(2) 假设气流中颗粒均匀分布，则直径为 20μm 的尘粒能除去的百分率是多少？

解 （1）由式（4-18）可知，降尘室所需的沉降面积为：

$$A = V_s / u_t$$

为求上式中的 u_t，首先假设该颗粒的沉降速度符合斯托克斯公式（层流区），即

$$u_t = \frac{d^2(\rho_s - \rho)g}{18\mu} = \frac{(40 \times 10^{-6})^2 \times (4500 - 1.6) \times 9.81}{18 \times 0.03 \times 10^{-3}} = 0.13(m/s)$$

检验：

$$Re_t = \frac{du_t\rho}{\mu} = \frac{40 \times 10^{-6} \times 0.13 \times 1.6}{0.03 \times 10^{-3}} = 0.277 < 1$$

所以上述假设成立，计算有效。

将 $u_t = 0.13 m/s$ 和 $V_s = 10800 m^3/h$ 代入上述沉降面积计算式得：

$$A = V_s / u_t = \frac{10800}{0.13 \times 3600} = 23.1(m^2)$$

因为每一层降尘室的底面积 $A_0 = 1.5 \times 4 = 6 m^2$，所以降尘室的层数为：

$$A/A_0 = 23.1/6 = 3.8\text{层} \approx 4\text{层}$$

（2）依据降尘室除尘原理，则直径为 20μm 的尘粒中能被除去的部分，一定满足“停留时间≥沉降时间”的条件。即当停留时间等于沉降时间时，颗粒则是刚好被除去的。

设直径为 20μm 的尘粒中，刚好能被除去的颗粒在降尘室入口处的高度为 h，则那些高度超过 h 的颗粒将无法除去。因此，则直径为 20μm 的尘粒被除去的比例为 h/H。又因为对于高度 h 处的颗粒，停留时间＝沉降时间，则：

$$\frac{h}{u_t'} = \frac{L}{u}$$

又对于 40μm 的尘粒有：

$$\frac{H}{u_t} = \frac{L}{u}$$

两式对比可得：

$$\frac{h}{H}=\frac{u'_{t}}{u_{t}}$$

由第（1）问知，$d=40\mu m$ 颗粒的沉降速度符合斯托克斯区，所以 $d=20\mu m$ 颗粒的沉降也必在该区。最终，根据斯托克斯公式可知，直径为 $20\mu m$ 的尘粒能除去的百分率是：

$$\frac{h}{H}=\frac{u'_{t}}{u_{t}}=\frac{(d')^{2}}{d^{2}}=\left(\frac{20}{40}\right)^{2}=25\%$$

② 沉降槽（settling tank）　沉降槽是用来提高悬浮液浓度并同时得到澄清液体的重力沉降设备。沉降槽又称增浓器或澄清器。沉降槽可间歇或连续操作。

连续沉降槽是底部略成锥状的大直径浅槽，如图 4-5 所示。料浆经中央进料口送到液面以下 0.3～1.0m 处，在尽可能减小扰动的条件下，迅速分散到整个横截面上。液体向上流动，清液经由槽顶端四周的溢流堰连续流出，称为溢流。固体颗粒下沉至底部，槽底有徐徐旋转的耙将沉渣缓慢地聚拢到底部中央的排渣口连续排出，排出的稠浆称为底流。

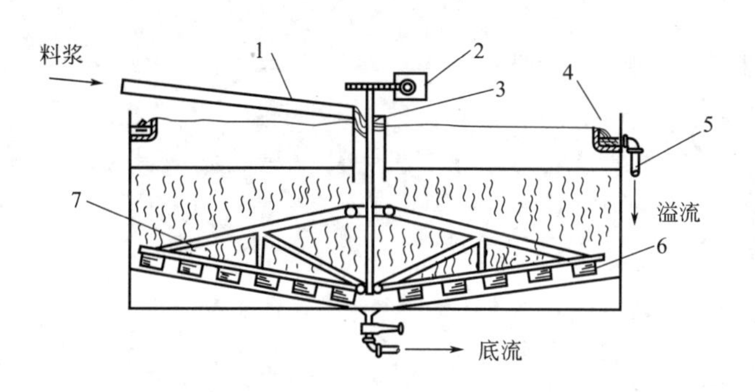

(a) 示意图

1—进料槽道；2—转动机构；3—料井；4—溢流槽；5—溢流管；6—叶片；7—转耙

(b) 设备图

图 4-5　沉降槽

连续沉降槽适用于处理量大而浓度不高且颗粒不甚细微的悬浮料浆，常见的污水处理就是一例。经过这种设备处理后的沉渣中还含有约 50%的液体。

③ 分级器（classifier）　利用重力沉降可将悬浮液中不同粒度的颗粒进行粗略的分离，或将两种不同密度的颗粒进行分类，这样的过程统称为分级。实现分级操作的设备称为分级器。

图 4-6 为一个双锥分级器，利用它可将密度不同或尺寸不同的粒子混合物分开。混合粒子由上部加入，水经可调锥与外壁的环形间隙向上流过。沉降速度大于水在环隙处上升流速的颗粒进入底流，而沉降速度小于该流速的颗粒则被溢流带出。

4.3.2　离心沉降（Centrifugal settling）

在离心力场，依靠惯性离心力的作用而实现沉降的过程称为离心沉降。

（1）离心沉降速度（centrifugal settling velocity）

在离心力场，当流体携带着颗粒旋转时，由于颗粒密度大于流体的密度，则惯性离心力将使颗粒在径向上与流体发生相对运动而飞离中心，如图 4-7 所示。颗粒在离心力场中的运

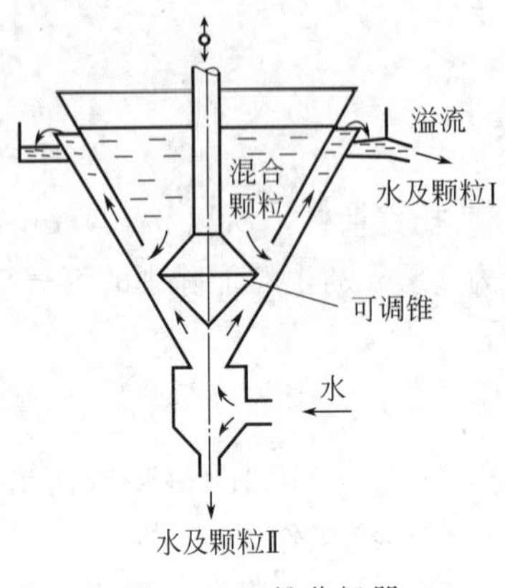

图 4-6　双锥分级器

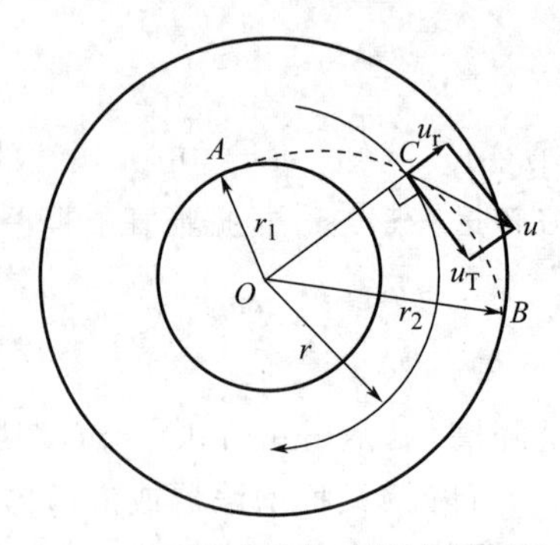

图 4-7　颗粒在旋转流体中的运动

动速度 u 可分解为径向速度 u_r 和切向速度 u_T，其中颗粒的径向速度 u_r 称为离心沉降速度。显然，u_r 不是颗粒运动的真实速度，而是颗粒沿着半径逐渐扩大的螺旋形轨道运动时的一种分离效果的表示方法。

在径向上对颗粒做受力分析，颗粒受到的三个力是：

离心力（方向是沿着半径向外）：$\frac{\pi}{6}d^3\rho_s\frac{u_T^2}{r}$

向心力（方向是沿着半径向里）：$\frac{\pi}{6}d^3\rho\frac{u_T^2}{r}$

阻力（方向是沿着半径向里）：$\zeta A\frac{\rho u_r^2}{2}=\zeta\frac{\pi}{4}d^2\frac{\rho u_r^2}{2}$

当三力平衡时，可导出离心沉降速度的表达式为：

$$u_r=\sqrt{\frac{4d(\rho_s-\rho)}{3\zeta\rho}\times\frac{u_T^2}{r}} \tag{4-19}$$

式中，u_T 指气体的切线速度，一般可用气体进口速度近似计算。

离心沉降时，沉降速度中的阻力系数 ζ 同样也是根据流型的不同区域来确定。如果颗粒与流体的相对运动属于层流，则有 $\zeta=24/Re_t$，所以沉降速度变为：

$$u_r=\frac{d^2(\rho_s-\rho)}{18\mu}\times\frac{u_T^2}{r} \tag{4-20}$$

对比该式与表 4-4 中的斯托克斯式可以看出，颗粒的离心沉降速度与重力沉降速度形式上很相似，只是用离心加速度代替了重力加速度。通常将两个加速度的比值定义为离心分离因数 K_c（centrifugal separation factor）：

$$K_c=\frac{u_T^2}{rg}=\frac{r\omega^2}{g}=\frac{\text{离心加速度}}{\text{重力加速度}}$$

K_c 值的大小是反映离心分离设备性能的重要指标。一般用于气固分离的旋风分离器和用于液固分离的旋液分离器的 K_c 值在 5～2500 之间，而用于液固分离的离心机的 K_c 值可达到几万甚至十几万。

应该注意的是，重力加速度是恒定值，而离心加速度可人为调节，且离心加速度值远高于 9.81m/s^2。在离心力场中，颗粒所受到的离心力随着旋转半径 r 和角速度 ω 的增大而增大。以离心机为例，增大转鼓直径和转速均有利于提高离心分离效率，但从设备的机械强度考虑，离心机采取的措施是尽可能地增加转速而减小转鼓的直径。

离心分离消耗能量较大，一般对于两相的密度差较小且颗粒粒度较细的非均相物系，可采用离心沉降加快沉降过程，提高分离效率。

(2) 离心沉降设备（centrifugal settling equipment）

① 旋风分离器（cyclone separator）　旋风分离器是利用离心沉降原理从气流中分离固体颗粒的设备。其结构形式很多，标准的旋风分离器结构如图 4-8 所示。

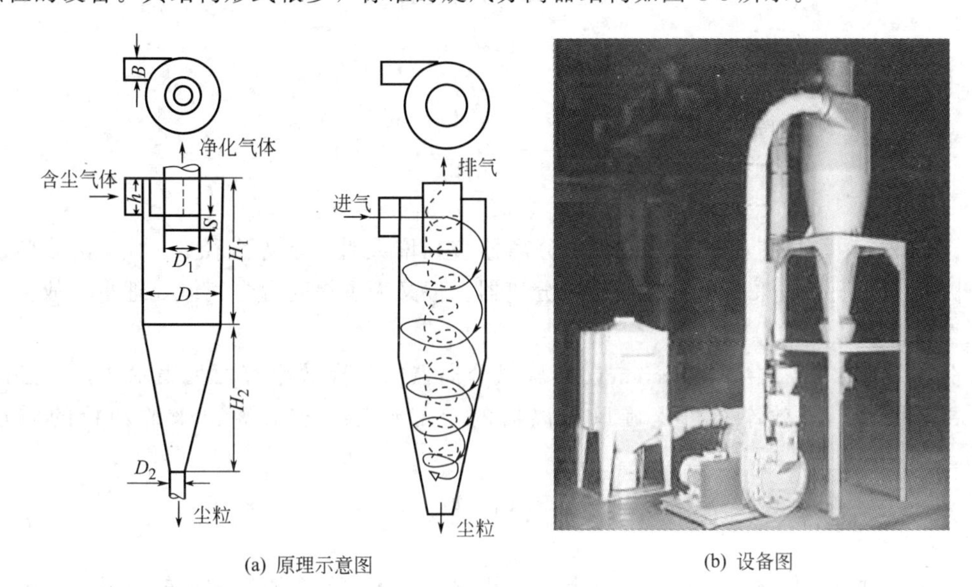

(a) 原理示意图　　(b) 设备图

图 4-8　标准旋风分离器

图 4-8 中，$h=D/2$，$B=D/4$，$D_1=D/2$，$H_1=2D$，$H_2=2D$，$S=D/8$，$D_2=D/4$。

旋风分离器的上半部为圆筒形，下半部为圆锥形。含尘气体从圆筒上侧的长方形进气管切向进入，在分离器内作旋转运动。分离出灰尘颗粒后的气流由圆筒形上半部的排气管排出，而灰尘则落入锥底的灰斗中。旋风分离器要求的气体流速为 10～25 m/s，所产生的离心力可以分离出 5μm 的颗粒。旋风分离器内上行的螺旋形气流（内圈）称为内旋流（又称气芯），下行的螺旋形气流（外圈）称为外旋流。内外旋流的旋转方向相同，但外旋流的上部是主要的除尘区。

旋风分离器是化工生产中使用很广的设备，常用于厂房的通风除尘系统。它的缺点是气流的阻力较大，处理硬质颗粒时容易被磨损。所以，当处理 $d>200\mu m$ 的颗粒时，应先用重力沉降，再用旋风分离器。表示旋风分离器分离效能的主要参数包括临界直径、分离效率和阻力损失：

a. 临界直径（critical diameter）　颗粒在旋风分离器中能被分离下来的条件为 $\tau_{停} \geqslant \tau_{沉}$，当取分离条件 $\tau_{停}=\tau_{沉}$ 时，旋风分离器能除去的最小颗粒直径称为临界直径。

颗粒在旋风分离器中的沉降速度 u_r 指颗粒沿径向穿过气流主体而到达器壁的运动速度。假设颗粒与气体的相对运动为层流，则有：

$$u_r=\frac{d^2(\rho_s-\rho)u_T^2}{18\mu r_m}\approx\frac{d^2\rho_s u_i^2}{18\mu r_m} \tag{4-21}$$

式中　r_m——气流旋转平均半径，$r_m=(D-B)/2$，m；

u_i——气体进口气速，m/s。

气体在旋风分离器内的停留时间为：

$$\tau_{停}=\frac{2\pi r_m N_e}{u_i}$$

式中，N_e 指气体在器内有效旋转圈数，对于标准的旋风分离器，$N_e=5$。

根据式（4-21），颗粒在旋风分离器内的沉降时间为：

$$\tau_{沉}=\frac{B}{u_r}=\frac{18\mu r_m B}{d^2\rho_s u_i^2}$$

式中，B 指进气口宽度。

令 $\tau_{停}=\tau_{沉}$，则可得到临界直径：

$$d_c=\sqrt{\frac{9B\mu}{\pi N_e \rho_s u_i}} \tag{4-22}$$

式（4-22）中，B 正比于 D，说明当分离器直径增大时，临界直径 d_c 增大，分离效果降低，因此工业上一般采用小直径的旋风分离器。但又为了提高含尘气体处理量，故多采用小直径多台并联以形成分离器组。

b. 分离效率（separation efficiency） 旋风分离器的分离效率有两种表示法，一是总效率，二是分效率。总效率指进入旋风分离器的全部颗粒中被分离出来的颗粒的质量分率，即：

$$\eta_0=\frac{c_1-c_2}{c_1} \tag{4-23}$$

式中 c_1——进口气体含尘浓度，是指单位体积含尘气体中所含固体颗粒的质量，g/m³；

c_2——出口气体含尘浓度，g/m³。

总效率是工业上最常用的，也是最易于测定的分离效率，但它不能表明旋风分离器对各种尺寸颗粒的不同分离效果。由于含尘气体中的颗粒通常是大小不均的，所以经过分离器后各种颗粒的分离效率也各有不同。按各种粒度分别表明其被分离下来的质量分数，称为粒级效率（又称分效率），即：

$$\eta_{pi}=\frac{c_{i,1}-c_{i,2}}{c_{i,1}} \tag{4-24}$$

式中 $c_{i,1}$——进口气体中粒径在第 i 小段范围内的颗粒的浓度，g/m³；

$c_{i,2}$——出口气体中粒径在第 i 小段范围内的颗粒的浓度，g/m³。

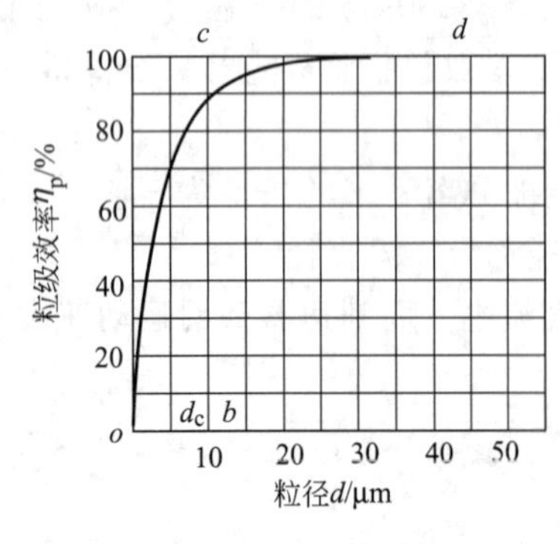

图 4-9 粒级效率曲线

为形象地表明各种尺寸颗粒被分离出的质量分数，可用粒级效率曲线描述，如图 4-9 所示。从图 4-9 中可以看出，在理论上 $d_i \geqslant d_c$ 的应能被全部分离的颗粒，在实际中只能部分分离；在理论上 $d_i < d_c$ 的不应被分离的颗粒也能被分离出一些。这是因为，在 $d_i < d_c$ 的颗粒中，有些在入口处已很靠近壁面，在停留时间内能到达壁面上，或在器内聚结成了大颗粒，因而具有较大的沉降速度；而在 $d_i \geqslant d_c$ 的颗粒中，有部分受气体涡流的影响未能达到壁面，或沉降后又被气流重新卷起而带走，因而不能被全

部分离下来。

总效率与粒级效率有关，还与气体中粉尘的浓度分布有关，即：

$$\eta_0 = \sum_{i=1}^{n} x_i \eta_{pi} \tag{4-25}$$

式中 x_i——在第 i 小段粒径范围内的颗粒占全部颗粒的质量分数；

η_{pi}——第 i 小段粒径范围内颗粒的粒级效率。

有时也用分割粒径 d_{50} 来表示分离效率，d_{50} 是粒级效率恰为50%的颗粒直径。把旋风分离器的粒级效率 η_{pi} 标绘成粒径 d/d_{50} 的函数曲线，该曲线对于同一形式且尺寸比例相同的旋风分离器都适用，这给旋风分离器效率的估算带来很大方便。

c. 阻力损失 气体通过旋风分离器时受器壁的摩擦阻力、流动时局部阻力以及气体旋转运动所产生的动能损失影响，造成气体的压强降低，该压降可表示为：

$$\Delta p = \zeta \frac{\rho u_i^2}{2} \tag{4-26}$$

式中，ζ 为阻力系数，可由实验测定，标准旋风分离器的 $\zeta=8.0$。旋风分离器的压降一般为500～2000Pa。

② 旋液分离器（cyclone hydraulic separators） 旋液分离器是一种利用离心力从液流中分离出固体颗粒的分离设备，其工作原理、结构和操作特性与旋风分离器（旋风除尘器）十分相似。

旋液分离器的主体由圆筒和圆锥两部分构成，如图4-10所示。悬浮液经入口管切向进入圆筒，形成螺旋状向下运动的旋流。固体颗粒受惯性离心力作用被甩向器壁，并随旋流降至锥底的出口，由底部排出（增浓液称为底流）。清液或含有细微颗粒的液体则为上升的内旋流，从顶部的中心管排出（顶流）。

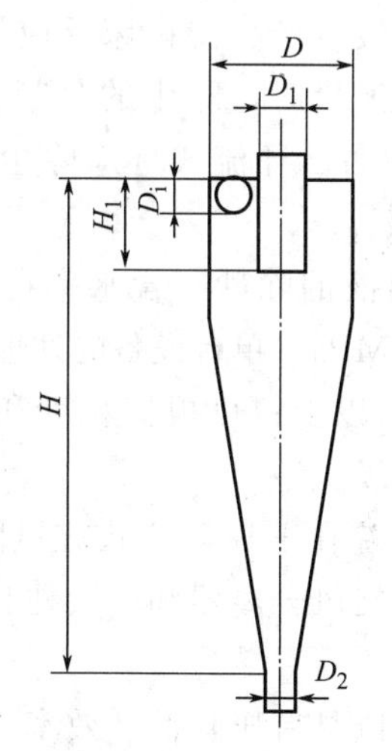

图4-10 旋液分离器

与旋风分离器相比，旋液分离器的结构特点是圆筒直径小，而圆锥部分长。这是由于液固密度差比气固密度差小得多，在一定的切线进口速度下，较小的旋转半径可使固体颗粒受到较大的离心力，从而提高离心沉降速度。另外，适当地增加圆锥部分的长度，可延长悬浮液在器内的停留时间，有利于液固分离。

旋液分离器结构简单，设备费用低，占地面积小，处理能力大，可用于悬浮液的增浓、分级操作，也可用于不互溶液体的分离、气液分离、传热、传质和雾化等操作中，在化工、石油、冶金、环保、制药等工业部门被广泛采用。其缺点是进料泵的动能消耗大，内壁磨损大，进料流量和浓度的变化容易影响分离性能。所以，旋液分离器一般采用耐磨材料制造，或采用耐磨材料作内衬。与旋风分离器相同，在给定处理量时，为了提高分离效率，应选用若干小直径的旋液分离器并联运行。超小型的旋液分离器的圆筒直径小于 15mm，可分离的固体颗粒直径小到 2～5μm。

图 4-10 中，$D_i=D/4$，$D_1=D/3$，$H=5D$，$H_1=(0.3\sim0.4)D$，锥形段倾斜角一般为 10°～20°。

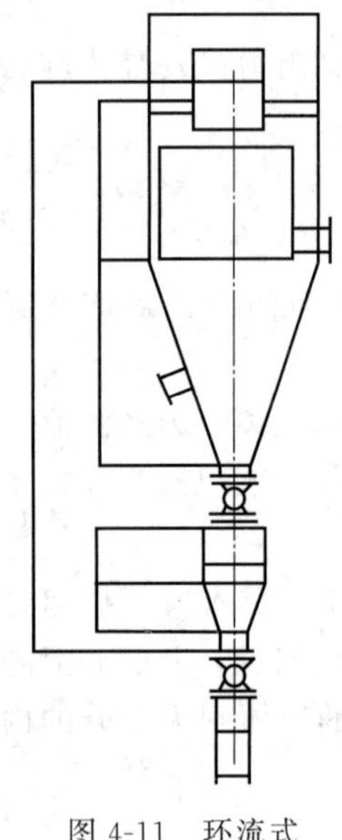

图 4-11　环流式旋风除尘器

③ 环流式旋风除尘器与液固分离器（circulation type cyclone dust collector and liquid-solid separator）　环流式旋风除尘器与液固分离器是青岛科技大学化工学院开发的新一代高效、节能型气固或液固分离设备，如图 4-11 所示。该类设备的外型与旋风除尘器或旋液分离器相似，但器内增设了强化分离效率的内件。启用时，流体介质从直筒段下部以切向方式进入器内，在直筒段进行一次分离，达到要求的流体介质直接从顶部溢流口排出。部分流体连同固体颗粒由顶部特设旁路引入锥体，在锥体内得到二次分离。分离后的流体在锥部沿轴向返回一次分离区，固体颗粒在锥体底部富集并从底流口排入砂包或排向器外。此新型分离器压降低、放大效应小，且由于特殊的流路设计，防止了流体的短路及锥体内颗粒的卷扬，使分离效率大幅度提高，具有操作弹性大、操作稳定性好的特点。

环流式旋风除尘器的工业应用结果表明，当用于合成氨造气系统时，除尘效率比国家定型设计的除尘器效率提高 17.8%。此外，它还被成功地用于复合肥干燥、硫酸生产、流化床反应器等系统的除尘，处理量可达 30000m^3/h。低处理量时，除尘的分割粒径可达 2μm 以下；高处理量时，可达 3μm 左右。环流式旋风除尘器亦曾成功地将 5μm 以上和以下硅胶颗粒、1μm 以上和以下石墨分级。

旋液环流式液固分离器用于高含水、原油、采出液的除砂（含水率在 85%左右，沙粒直径为 117μm），除砂效率可达 92%，压降仅为 0.04MPa，单台设备的处理量可达 3000m^3/d 以上；分离水中 20μm 以上砂粒的分离效率可达 90%以上；可根据粒度和密度的差别进行选煤、选矿及液体中晶体物质及固体颗粒的去除或分级。

为了进一步提高除尘效率，青岛科技大学化工学院在上述环流式旋风除尘器的基础上，又开发成功环流循环除尘系统。该系统由两个环流式旋风分离器和一段柱状旋风分离段组合而成，其组合流程见图 4-12。

该除尘系统因采用了二级环流式旋风分离器，所以具有压降低、效率高、放大效应小及操作弹性大的优点，系统排出的气体中不含 1μm 以上的粉尘，分割粒径 d_{50} 达到了 0.5～0.7μm。环流式旋风分离器结构简单、无转动部件，系统只增加一台小的风机，故可连续高效运行。

该环流式旋风除尘器已在工业上推广应用，工业应用的单台最大处理风量达到了 35000m^3/h，直径为 2000mm。将该除尘系统用于分子筛生产时，与其他除尘方法相比价格低得多。

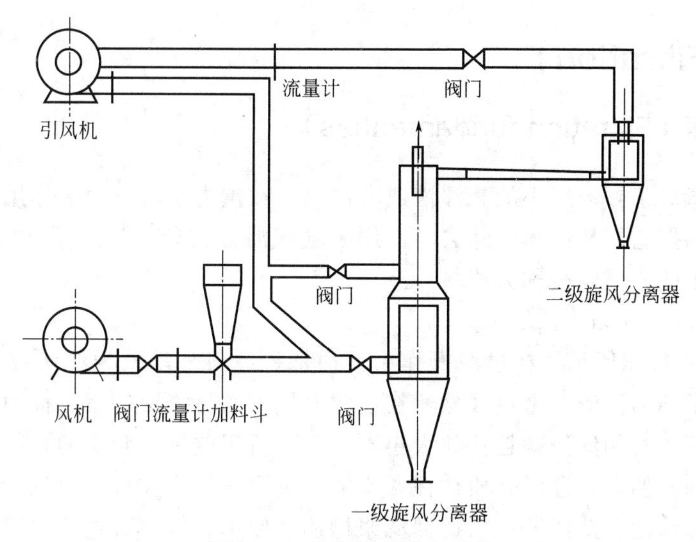

图 4-12　环流循环除尘系统

④ 离心沉降机（centrifugal settler）

离心沉降机是分离悬浮液和乳浊液的有效设备。它是在转动机械的带动下，使悬浮液产生高速旋转运动。在强大的离心力场中，液体中极细的颗粒，或颗粒密度与液体密度相差很小的悬浮液、乳浊液都可以在离心沉降机中得到有效的分离。离心沉降机的种类很多，有连续操作的，也有间歇操作的。

图 4-13 为一转鼓式离心沉降机。这种离心沉降机有一中空转鼓，转鼓壁上是不开孔的。当悬浮液随转鼓一起转动时，物料受离心力的作用，按密度大小不同分层沉淀。密度大、颗粒粗的物料直接附于鼓壁上，而密度小、颗粒细的物料则在靠旋转中心的内层沉降，清液则从转鼓上部溢流。

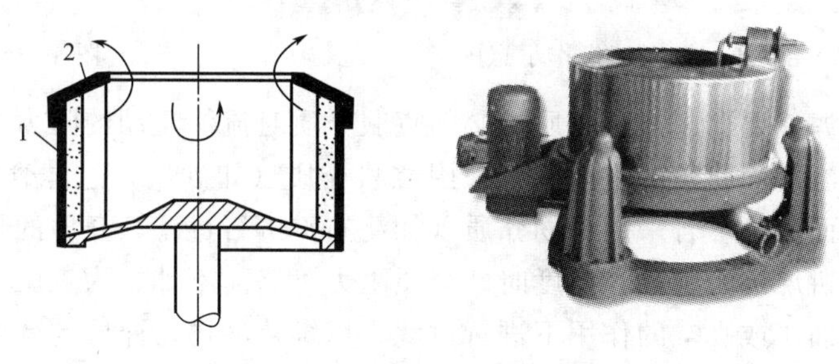

图 4-13　离心沉降机

1—固体；2—液体

离心沉降机中物料的沉降过程可分为两个阶段：

a. 沉降　悬浮液中的固体颗粒由于受离心力的作用而向转鼓壁沉降，从而积聚在外层。而液体留在内层，形成一个中空的垂直圆筒状液柱。

b. 渣层压紧　沉降在鼓壁上的颗粒层，在离心力的作用下被逐渐压紧。当悬浮液中含固体量较多时，沉降的颗粒大量积集，渣层很快堆厚，因此必须考虑连续排渣的问题。

当悬浮液中含固体量不多，渣层堆积很慢，此时没有渣层的压紧阶段，因此可以采用间歇排渣法，这种情况下主要起到离心澄清作用。

4.4 过滤（Filtration）

4.4.1 过滤原理（Filtration fundamentals）

一般过滤是指对于颗粒细小的悬浮液或乳浊液，在重力、离心力的作用下，采用多孔介质构成的障碍物场把它们从流体中除去。而用于气固混合物分离的过滤则常称为袋滤除尘。被作为障碍物的多孔介质则被称为滤材。

（1）过滤过程（filtration process）

过滤操作如图 4-14 所示。在过滤操作中，通常称原料悬浮液为滤浆或料浆；滤浆中过滤出来的固体粒子称为滤渣；透过被截留在过滤介质上的滤渣层的液体称为滤液。

由于滤浆中所含有的滤渣颗粒往往大小不一，而所用的过滤介质的孔径较一部分微粒的直径大，故在过滤开始时，过滤介质往往并不能完全阻止细小颗粒的通过。因此，开始阶段所得的滤液常是浑浊的，此滤液可以送回滤浆槽循环使用。但随着过滤的继续进行，细小的颗粒便可能在孔道上及孔道中发生“架桥”现象，拦截住后来的颗粒，使其不能通过，如图 4-15 所示。此时滤饼开始形成，由于滤饼中的通道（孔道）通常比介质的孔道细小，便能起到截留粒子的作用。所以，一般过滤进行一段时间后，便可得到澄清的滤液。

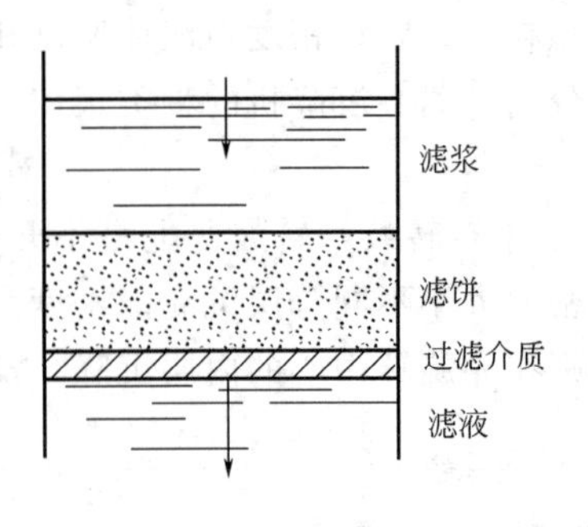

图 4-14　过滤操作示意图

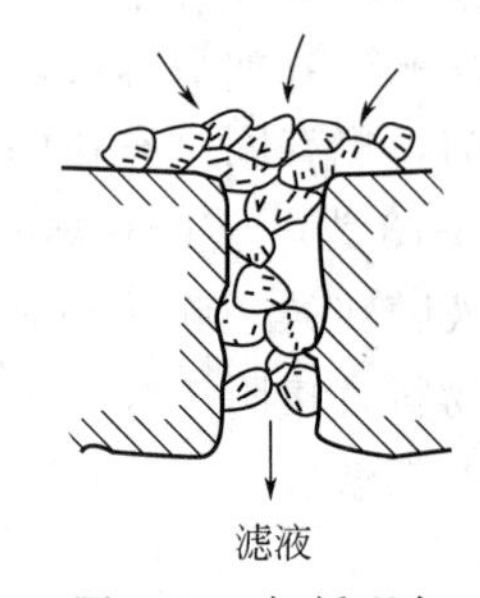

图 4-15　架桥现象

过滤开始时，滤液要通过过滤介质，必须克服介质对流体流动的阻力。当介质上形成滤饼后，还必须克服滤饼和介质的阻力。若采用粒状介质（如砂层）过滤含滤渣很少的料浆，则滤饼的阻力可略而不计。若采用织物介质（如滤布）进行过滤，介质的阻力仅在过滤开始时较为显著，至滤饼层沉积到相当厚度时，介质阻力可略而不计。为克服过滤阻力，悬浮液可以在重力作用、加压或真空的作用下进行过滤，以维持滤饼与介质的两侧间存在一定的压力差，作为过滤过程的推动力。为了提高重力的位能、形成离心力场或压力场，都需要对流体做功，因此机械能是分离所必需的能量。

过滤操作进行了一定时间后，滤饼的厚度不断增加，过滤的阻力也不断增加，过滤速度变得很低，以至于过程不能进行。如果再进行下去则动力消耗太大，不够经济。所以过滤到一定滤饼厚度后，需将滤饼移走，重新开始过滤。在移去滤饼以前，有时须将滤饼进行洗涤，洗涤所得的溶液称为洗涤液（或洗液）。洗涤完毕后，有时还要将滤饼中所含的液体除去，称为去湿。最后将滤饼从滤布上卸下来，卸料要尽可能不留残渣，这样不但可以最大限度地得到滤饼，而且也是为了清净滤布以减少下次过滤的阻力。采用压缩空气从过滤介质后面反吹是卸除滤饼的好方法，可以同时达到上述两个目的。当滤布使用一定时间后，其小孔被细小颗粒所堵塞，而使阻力大大增加，此时应取下滤布进行清洗。

由上可见，过滤操作包括过滤、洗涤、去湿及卸料四个阶段，如此周而复始循环进行。

（2）过滤操作分类

常见的过滤操作分为饼层过滤（cake filtration）和深床过滤（deep bed filtration）两种方式，如图4-16所示。其中，饼层过滤是指过滤中，固体物质沉积于介质表面而形成滤饼层，滤液穿过滤饼层时即被过滤，所以滤饼层是有效过滤介质，现在工业上一般多采用此方法。该法要求悬浮液中固体颗粒体积含量大于1%。而在深床过滤中，固体颗粒并不形成滤饼层，而是沉积于较厚的粒状过滤介质床层内部。这种过滤适用于生产能力大而悬浮液中颗粒小且含量甚微（固相分率小于0.1%）的场合。

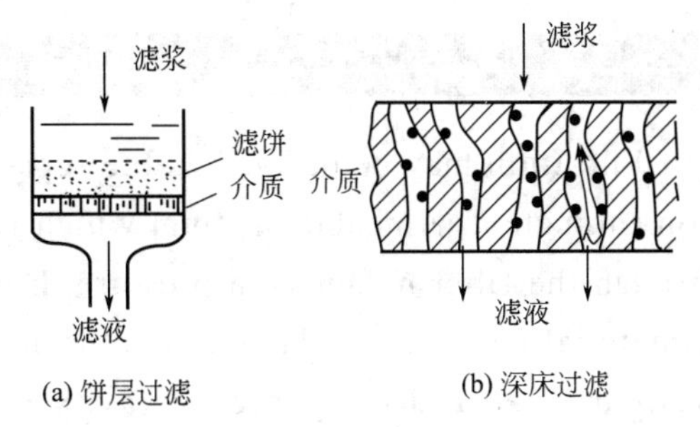

图4-16　过滤操作分类

（3）过滤介质

过滤介质是滤饼的支撑物，它应具有足够的机械强度和尽可能小的流动阻力。同时，它还应有相应的耐腐蚀性和耐热性。工业上常用的过滤介质有织物、堆积的粒状介质和多孔固体介质。滤饼的压缩性和助滤剂也是应考虑的问题，因为滤饼是悬浮液中固体颗粒在过滤介质上堆积而成的，所以随着过滤的进行而逐渐变厚。若采用正压法过滤，则滤饼可分为可压缩滤饼及不可压缩滤饼两种。可压缩滤饼有可能将通道堵塞，不利于过滤，而且流动阻力增大。为减少压缩滤饼的阻力，将某种质地坚硬又能形成疏松滤饼层的另一种固体颗粒混入悬浮液或涂于介质上，使滤液畅流，这种粒状物称为助滤剂，例如硅藻土。表4-5示出了常用的过滤介质及其能截留的最小颗粒直径。

表4-5　过滤介质的种类及能阻挡的最小颗粒直径

型式	种类	能阻挡的最小颗粒直径/μm
固定组合滤件	扁平楔形-金属丝网	100
	金属丝绕管	100
	层叠环	5
硬质多孔介质	多孔陶瓷	1
	烧结金属	3
	多孔塑料	3
金属片材	打孔板	100
	编织金属丝网	5
塑料片材	聚合模	0.1
	单丝织造网	
织物	天然及合成纤维织物	10
滤芯	片材制品	3
	绕线管	2
	粘叠层	2
非织造介质	纤维板	0.1
	毡及针刺毡	10
	纤维素滤纸	5
	玻璃滤纸	2
	黏结的介质	10
松散介质	纤维（如石棉、纤维素）	超微颗粒（<1）
	粉粒（如硅藻土、膨胀珍珠岩、炭、砂、吸附剂等）	超微颗粒（<1）

Filtration Fundamentals

Filtration may be defined as the separation of solids from liquids by passing a suspension through a permeable medium which retains the particles. In order to obtain fluid flow through the filter medium, a pressure drop Δp has to be applied across the medium. It is immaterial from the fundamental point of view how this pressure drop is achieved but there are four types of driving force: gravity, vacuum, pressure and centrifugal.

There are basically two types of filtration used in practice: the so-called surface filters are used for cake filtration in which the solids are deposited in the form of a cake on the up-stream side of a relatively thin filter medium, while depth filter are used for deep bed filtration in which particle deposition takes place inside the medium and cake deposition on the surface is undesirable.

Surface filter

In a surface filter, the filter medium has a relatively low initial pressure drop and particles of the same size as or larger than the openings wedge into the openings and create smaller passages which remove even smaller particles from the fluid. A filter cake is thus formed, which in turn functions as a medium for the filtration of subsequent input suspension. In order to prevent blinding of the medium, filter aids are used.

Surface filters are usually used for suspensions with higher concentrations of solids, say above 1% by volume, because of the blinding of the medium that occurs in the filtration of dilute suspensions. This can be, however, sometimes be avoided by an artificial increase of the input concentration, in particular by adding a filter aid; as filter aids are very porous, their presence in the cake improves permeability and often makes cake filtration of dilute and generally difficult slurries possible.

Deep bed filtration

The principles of deep bed filtration are quite well-known. It is a clarification process using a deep bed of granular media, usually sand. As a sewage tertiary treatment process it can frequently produce filtrates containing only 5mg/L or less of suspend matter.

In a depth filter, particles are smaller than the medium openings and hence they proceed through relatively long and tortuous pores where they are collected by a number of mechanisms (gravity, diffusion and inertia) and attached to the medium by molecular and electrostatic forces.

The initial pressure drop across the depth filter is generally small than that across a surface filter of comparable efficiency but the build-up of pressure drop as particles are collected is more gradual for a depth filter. Depth filters are commonly used for clarification, i. e. for the separation of fine particles from very dilute suspension, say less than 0. 1% by volume.

4.4.2　过滤基本方程式

（1）过滤基本方程式表述

液体通过饼层（包括滤饼和过滤介质）空隙的流动与普通管内流动相仿。由于过滤操作所涉及的颗粒尺寸一般很小，故形成的通道呈不规则网状结构。由于孔道很细小，流动类型可认为在层流范围。

仿照圆管中层流流动时计算压降的范宁公式：

$$\Delta p_f = \lambda \frac{l}{d} \times \frac{\rho u^2}{2} = \frac{64\mu}{du\rho} \times \frac{l}{d} \times \frac{\rho u^2}{2} = \frac{32\mu l u}{d^2} \tag{4-27}$$

在过滤操作中，Δp_f 就是液体通过饼层克服流动阻力的压强差 Δp。由于过滤通道曲折多变，可将滤液通过饼层的流动看作液体以速度 u 通过许多平均直径为 d_0、长度等于饼层厚度（$L+L_e$）的小管内的流动（L 为滤饼厚度，L_e 为过滤介质的当量滤饼厚度）。液体通过饼层的瞬间平均速度为：

$$u = \frac{1}{A_0} \frac{dV}{dt} \tag{4-28}$$

$$A_0 = \varepsilon A \tag{4-29}$$

式中　A_0——饼层空隙的平均截面积，m^2；

A——过滤面积，m^2；

ε——饼层空隙率，对不可压缩滤饼为定值；

t——过滤时间，s；

V——滤液量，m^3；

dV/dt——单位时间获得的滤液体积，m^3/s。

将式（4-28）和式（4-29）代入式（4-27）得

$$\Delta p = \frac{32\mu (L+L_e) \frac{dV}{dt}}{d_0^2 \varepsilon A} \tag{4-30}$$

式中　μ——滤液的黏度，Pa·s。

整理式（4-30）后得到：

$$\frac{dV}{A dt} = \frac{\varepsilon d_0^2 \Delta p}{32\mu (L+L_e)} \tag{4-31}$$

令 $r = \frac{32}{\varepsilon d_0^2}$，则：

$$\frac{dV}{A dt} = \frac{\Delta p}{r\mu (L+L_e)} \tag{4-32}$$

式中　r——滤饼比阻，反映滤饼结构特征的参数，$1/m^2$。

将滤饼体积 AL 与滤液体积 V 的比值用 ν 表示，意义为每获得 $1m^3$ 滤液所形成滤饼的体积，则式（4-32）变为：

$$\frac{dV}{dt} = \frac{A^2 \Delta p}{r\mu\nu (V+V_e)} \tag{4-33}$$

式中　V_e——过滤介质的当量滤液体积，m^3。

式（4-33）称为过滤基本方程式，表示过滤过程中任意瞬间的过滤速度与有关因素间的关系，是过滤计算及强化过滤操作的基本依据。该式适用于不可压缩滤饼。对于大多数可压缩滤饼，式中 $r=r'\Delta p'$，r'为单位压强差下的滤饼比阻。

过滤操作有两种典型方式，即恒压过滤、恒速过滤。恒压过滤时维持操作压强差不变，但过滤速率将逐渐下降；恒速过滤则逐渐加大压强差，保持过滤速率不变。对于可压缩滤饼，随着过滤时间的延长，压强差会增加许多，因此恒速过滤无法进行到底。有时，为了避免过滤初期压强差过高而引起滤液浑浊，可采用先恒速后恒压的操作方式，即开始时以较低的恒定速率操作，当表压升至给定值后，转入恒压操作。由于工业中大多数过滤属恒压过滤，因此以下讨论恒压过滤的基本计算。

（2）恒压过滤基本方程式

在恒压过滤中，压强差 Δp 为定值。对于一定的悬浮液和过滤介质，r、μ、ν、V_e 也可视为定值，故可对式（4-33）进行积分：

$$\int_0^V (V+V_e)\mathrm{d}V=\frac{A^2\Delta p}{r\mu\nu}\int_0^t \mathrm{d}t \tag{4-34}$$

令 $K=\dfrac{2\Delta p}{r\mu\nu}$，$q=\dfrac{V}{A}$，$q_e=\dfrac{V_e}{A}$，则式（4-34）变为：

$$q^2+2q_eq=Kt \tag{4-35}$$

式（4-35）即为恒压过滤方程，该式表达了过滤时间 t 与获得滤液体积 V 或单位过滤面积上获得的滤液体积的关系。式中的 K、q_e 为一定条件下的过滤参数：K 与物料特性及压强差有关，单位为 m^2/s；q_e 与过滤介质阻力大小有关，单位为 m^3/m^2。二者均可由实验测定。

4.4.3 过滤设备（Filtering equipment）

工业上应用的过滤设备称为过滤机。过滤机的类型很多，按操作方法可分为间歇式和连续式；按过滤推动力可分为加压过滤机和真空过滤机。工业上应用最广泛的板框压滤机和加压叶滤机为间歇式过滤机，转筒真空过滤机则为连续式过滤机。

（1）板框压滤机（plate-and-frame filter presses）

板框压滤机是最早为工业所使用的压滤机，至今仍沿用不衰。板框压滤机主要由机架、滤板、滤框和压紧装置等组成，其结构如图 4-17 所示。

板框压滤机的板和框多做成正方形，滤板、框上左右角均开有圆孔，组装并压紧后即构成供悬浮液或洗涤液流通的孔道。框的右上角的圆孔内侧面还开有一个暗孔，与框内空间相通，供悬浮液进入。框的两侧覆以滤布，使空框与两侧的滤布围成一个空间，称为过滤室。滤板分为过滤板和洗涤板两种。滤板的表面上制成各种凹凸的沟槽，凸者起支撑滤布的作用，凹者形成滤液或洗涤液的流道。

过滤时，悬浮液在一定的压差作用下，经悬浮液通道由滤框右上角圆孔侧面的暗孔进入框内进行过滤。而滤液则分别通过两侧滤布，再沿相邻滤板的板面凹槽下流汇合至滤液出口排出，滤渣则被截留于框内。待滤渣全部充满框后，即停止过滤。若滤渣需要洗涤，则将洗涤液压入洗涤液通道，并经洗涤板左上角圆孔侧面的暗孔进入板与滤布之间。洗涤结束后，将压紧装置松开，卸下滤渣，清洗滤布，整理板、框，重新组装，即可进行下一个操作循环。

板框压滤机结构简单，制造方便，占地面积较小，且过滤面积较大。该设备的操作压强

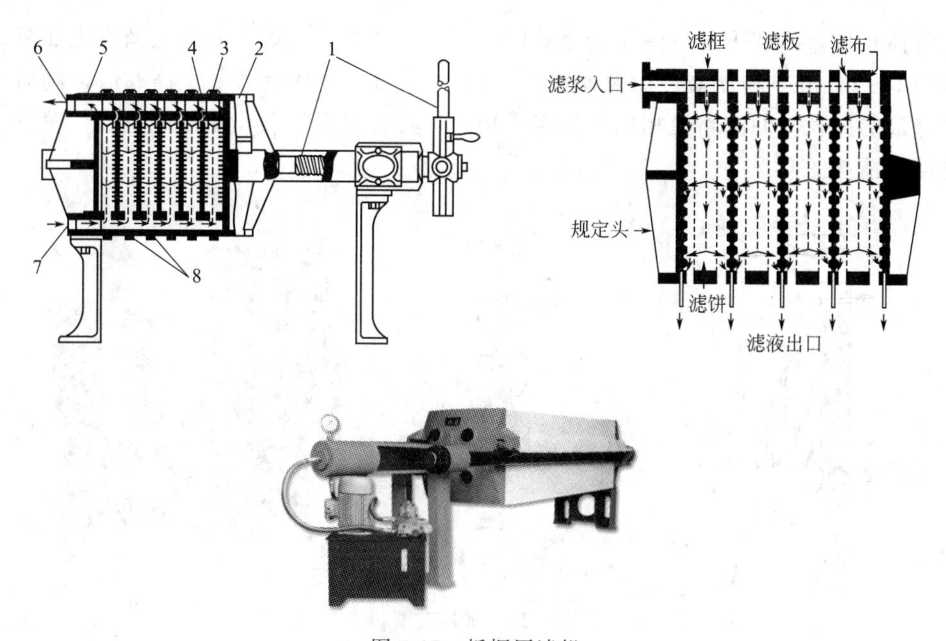

图 4-17　板框压滤机

1—压紧装置；2—可动头；3—滤框；4—滤板；5—规定头；6—滤液出口；7—滤浆出口；8—滤布

高，适应能力强，故应用颇为广泛。它的主要缺点是间歇操作，生产效率低，劳动强度大，滤布损耗也较快。各种自动操作板框压滤机的不断出现，使上述缺点在一定程度上得到改善。

Discontinuous pressure filters

Pressure filters can apply a large pressure differential across the septum to give economically rapid filtration with viscous liquid or dine solids. The most common types of pressure filters are presses filter and shell-and-leaf filters.

Filter press. The structure of plate-and-frame press filter is shown above. A filter press contains a set of plate designed to provide a series of chambers or compartments in which solids may collect. The plates are covered with a filter medium such as canvas. Slurry is admitted to each compartment under pressure; liquor passes through the canvas and out a discharge pipe, leaving a wet cake of solids behind. After assembly of the press, slurry is admitted to one end of the assembly of plates and frames from a pump or blow case under a pressure of 3 to 10 atm. Auxiliary channels carry slurry from the main inlet channel into each frame. Here the solids are deposited on the cloth-covered faces of the plates. Liquor passes through the cloth, down grooves or corrugations in the plate faces, and out of the press-filtration is continued until liquor no longer flows out the discharge or the filtration pressure suddenly raises. These occur when the frames are full of solid and no more slurry can enter.

（2）加压叶滤机（pressure leaf filter）

加压叶滤机（叶滤机）主要由一个垂直或水平放置的密封圆柱形滤槽和许多滤叶组成，如图 4-18 所示。滤叶是叶滤机的过滤元件，它是一个金属筛网框架或带沟槽的滤板（作用与板框压滤机的滤板相类似），外覆一层滤布。在滤叶的一端装有短管，供滤液流出，同时供安装时悬挂滤叶之用。过滤时，将许多滤叶安装在密闭机壳内，并浸没在悬浮液中。悬浮液中的液体在压差作用下穿过滤布沿金属网流至出口短管，而滤渣则被截留在滤布外，其厚度通常为5～35mm（视滤渣性质及操作情况而定）。过滤完毕后若要洗涤滤渣，则通入洗涤液，洗涤液的路径与滤液路径相同。洗涤完毕后，打开机壳下盖，用压缩空气、蒸汽或清水卸除滤渣并清洗滤叶。

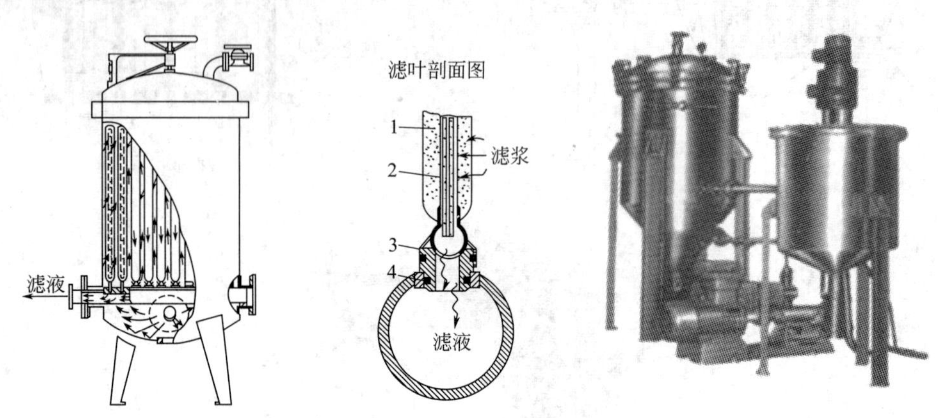

图4-18　加压叶滤机

1—滤饼；2—滤布；3—拔出装置；4—橡胶圈

叶滤机也是间歇操作的过滤机，其优点是过滤推动力大，单位体积的生产能力和过滤面积都较大。另外，叶滤机的洗涤效果也较好，劳动强度也比板框压滤机小。又由于它是在密闭条件下操作，故劳动环境也得到改善。其缺点是构造较复杂，造价较高，更换滤布麻烦。而且当过滤粒径大小不同的微粒时，会使它们聚集于滤叶的不同高度处。这样在洗涤时，大部分洗液会由粗颗粒中通过，造成洗涤不均匀。加压式叶滤机的过滤面积一般为20～$100m^2$，主要用于含固体量少（约为1%）的悬浮液和需要液体而不是固体的场合，如各种酒类、食品饮料以及植物油等的过滤。

（3）转筒真空过滤机（rotary drum filter）

前面介绍的板框过滤机和叶滤机都是间歇式过滤机。转筒真空过滤机则是连续式过滤机。此机主体是一个转动的圆筒，在其回转一周的过程中，即可完成过滤、洗涤、卸饼等各项工作。这几项工作虽是同时进行的，但却是在转筒的不同部位完成的。

转筒真空过滤机的主要部件为转筒，水平安装在中空转轴上，其长径比为1/2～2，如图4-19所示。

转筒的侧壁上覆盖有金属网，滤布支撑在网上，浸没在滤浆中的过滤面积约占全部面积

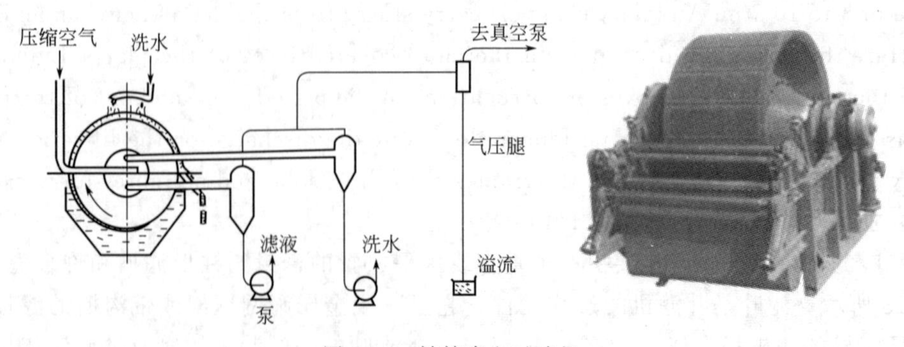

图4-19　转筒真空过滤机

的30%～40%，转速为0.1～3r/min。筒壁按周边平分为若干段，各段均有管通至轴心处，但各段的筒内并不相通。圆筒的一端有分配头装于轴心处，与从筒壁各段引来的连通管相接。通过分配头，圆筒旋转时其壁面的每一段可以依次与过滤装置中的滤液罐、洗水罐（以上两者处于真空之下）、鼓风机稳定罐（正压下）相通。因而，在回转一周的过程中，每个扇形格表面均可顺序进行过滤、洗涤、吸干、吹松、卸饼等操作。

转筒过滤机的优点是：操作连续、自动，能大规模处理固体物含量很大的悬浮液。其缺点是：转筒体积大，过滤面积小；用真空过滤，过滤的推动力不大；悬浮液温度不能高，否则真空会失去效应；滤饼含湿量大，洗涤不彻底。

（4）袋式过滤器（bag filter）

袋式过滤器是工业过滤除尘设备中使用最广的一类。它的捕集效率高，一般达到99%以上，而且可以捕集不同性质的粉尘，适用性广，处理气体量可由每小时几百立方米到数十万立方米，使用灵活，结构简单，性能稳定，维修也较方便。但其应用范围主要受滤材的耐温、耐腐蚀性的限制，一般用于300℃以下，也不适用于黏性很强及吸湿性强的粉尘，设备尺寸及占地面积也很大。

袋式过滤器的结构形式很多，按滤袋形状可分为圆袋及扁袋两种。前者结构简单，清灰容易，应用最广。后者可大大提高单位体积内的过滤面积。袋式过滤器的清灰方式有：机械清灰、逆气流清灰、脉冲喷吹清灰及逆气流振动联合清灰等形式。

在袋式过滤器中，过滤过程分成两个阶段。首先是含尘气体通过清洁滤材。由于惯性碰撞、拦截、扩散、沉降等各种机理的联合作用，把气体中的粉尘颗粒捕集在滤材上。当这些捕集的粉尘不断增加时，一部分粉尘嵌入或附着在滤材上形成粉尘层，此时的过滤主要是依靠粉尘层的筛滤效应，捕集效率显著提高，但压降也随之增大。由此可见，工业袋式过滤器的除尘性能受滤材上粉尘层的影响很大，所以根据粉尘的性质而合理地选用滤材是保证过滤效率的关键。一般当滤材孔径与粉尘直径之比小于10时，粉尘就易在滤材孔上架桥堆积而形成粉尘层。

通常滤材上沉积的粉尘负荷量达到0.1～0.3kg/m^3、压降达到1000～2000Pa时，便需进行清灰。应尽量缩短清灰的时间，延长两次清灰的间隔时间，这是当今过滤问题研究中的关键问题之一。

袋式过滤器常见的形式有：机械振动袋式过滤器、逆气流清灰袋式过滤器（图4-20）、脉冲喷吹清灰袋式过滤器（图4-21）、扁袋脉冲喷吹清灰式过滤器及静电袋式过滤器等。

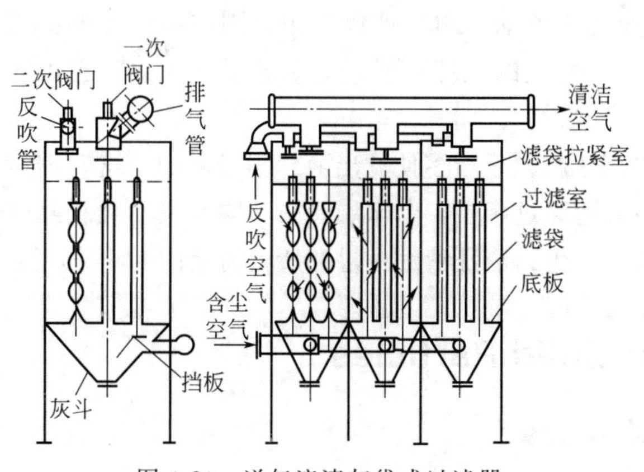

图4-20　逆气流清灰袋式过滤器

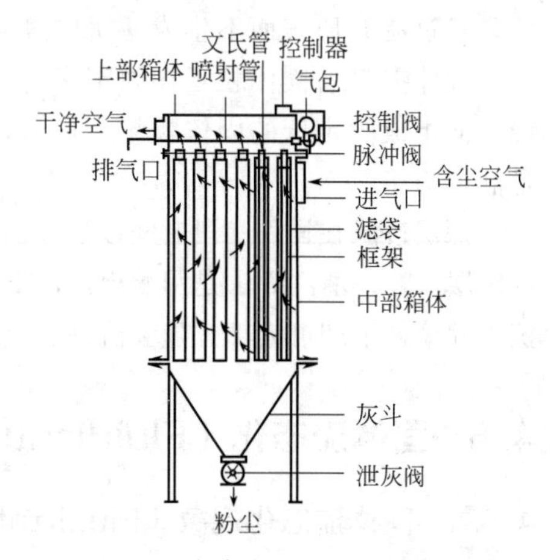

图4-21　脉冲喷吹清灰袋式过滤器

4.4.4 过滤操作的改进（Improvement of filtering operations）

（1）使用助滤剂

生产上在遇到滤饼压缩性大的情况时，因滤饼空隙率小，阻力大，过滤速率低，且难以靠加大压差改善操作，这时多使用助滤剂改进过滤性能。常用的助滤剂有硅藻土（单细胞水生植物的沉积化石经干燥或煅烧而得的颗粒，含 SiO_2 在 85%以上）、珍珠岩（玻璃状熔融火山岩倾入水中形成的中空颗粒）或石棉粉、炭粉等刚性好的颗粒。助滤剂有两种使用方法，一种是把助滤剂与水混合成的悬浮液在滤布上先进行预滤，使滤布上形成 1～3mm 助滤剂层然后正式过滤，此法可防止滤布堵塞；另一种是将助滤剂混在料浆中一道过滤，此法只能用于滤饼不予回收的情况。

（2）动态过滤（dynamic filter）

传统的滤饼过滤随着过滤的进行，滤饼不断积厚，过滤阻力不断增加。在恒定压差推动下过滤，则取得滤液的速率必不断降低。为了在过滤过程中限制滤饼的增厚，蒂勒（Tiller）于 1977 年提出了一种新的过滤方式，即料浆沿过滤介质平面的平行方向高速流动，使滤饼在剪切力作用下被大部分铲除，以维持较高的过滤能力。因滤液与料浆流动方向呈错流，故称错流式过滤，又因滤饼被基本铲除，称为无滤饼过滤，但较多的称法为动态过滤。

欲使料浆高速流过过滤介质表面，一般采用设置旋转圆盘的方法。令圆盘与过滤介质表面平行，在圆盘面向过滤介质方向设有凸起的筋以带动圆盘与过滤介质间的料浆高速旋转。可令筋的端面与过滤介质表面间距离在 20mm 范围内适当调小，圆盘以小于 1000r/min 的转速旋转。

动态过滤需多耗机械能，但对许多难过滤的悬浮液能明显改善过滤性能，故有较好的推广价值。

（3）深层过滤（deep filter）

当悬浮液中固体的体积分数在 0.1%以下，且固体粒子的粒度很小时，用滤布或金属丝网作过滤介质难以形成有效截留固体粒子的滤饼层，且滤布或丝网易堵。这时若采用粒径为 1mm 左右的石英砂固定床层作为过滤介质，则效果较好。这种以固体颗粒固定床层作为过滤介质，将悬浮液中的固体粒子（为区别于过滤介质颗粒，故称为“粒子”）截留在床层内部，且过滤介质表面不生成滤饼的过滤称为深层过滤。

液体夹带着固体粒子在固定床内的弯曲通道中流动，由于惯性作用，固体粒子会偏离流线趋向组成固定床的固体颗粒，称为颗粒捕捉。捕捉的机理一般认为是分子作用力，也可能是静电力。

深层过滤在操作一段时间以后，因床层内积存的粒子增多使滤出液含固量增加，这时需清洗床层颗粒。清洗方法是用滤出液由下而上高流速穿过床层，令床层膨胀、颗粒翻动且相互摩擦，固体粒子即可大部分随溢流液流走，使床层再生。清洗液约占过滤所得清液的 3%～5%。

4.5 固体流态化（Fluidization of Solid Particles）

4.5.1 固体流态化现象（Fluidization of solids）

在一个容器内装一块分布板（筛板），板上铺一层细颗粒物料，当流体自分布板下面通过颗粒床层时，根据流体流速的不同，表现出以下几种现象：

（1）固定床阶段（fixed bed）

如图 4-22(a) 所示，当流体的流速较小时，流体从固体颗粒之间的空隙穿过，颗粒在原处不动，床层高度不变，这时的床层称为固定床。

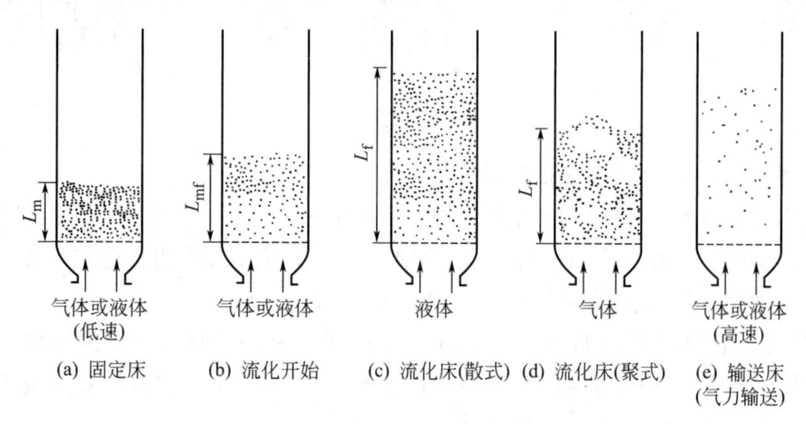

图 4-22　固体流态化现象

（2）流化床（fluidized bed）阶段

当流体的流速增大到一定值时，床层开始松动和膨胀，但颗粒仍不能自由运动，这时的床层称为初始流化床，如图 4-22(b) 所示。若继续增大流体流速，固体颗粒被流体浮起，并上下翻滚，随机运动，好像沸腾的液体一样，床层明显增高，如图 4-22(c)、(d) 所示。这时床层具有类似于流体的性质，故称为流化床。流化床现象可在一定的流体空速范围内出现。在该流速范围内，随着流速的增加，流化床高度增大，床层空隙率增大。流化床有散式流化与聚式流化两种流化形式。

① 散式流化（disperse fluidization）　若流化床中固体颗粒均匀地分散于流体，床层中各处空隙率大致相等，床层有稳定的上界面，这种流化形式称为散式流化。固体与流体密度差别较小的体系流化时可发生散式流化，液固系统的流化基本上属于散式流化，情况如图 4-22(c) 所示。

② 聚式流化（aggregative fluidization）　一般气固系统在流化操作时，因固体与气体密度差别很大，气体对颗粒的浮力很小，气体对颗粒的支托主要靠曳力，这时床层会产生不均匀现象，在床层内形成若干空穴。空穴内固体含量很少，是气体排开固体颗粒后占据的空间，称为气泡相。气体通过床层时优先通过各空穴，但空穴并不是稳定不变的，气体支撑的空穴上方的颗粒会落下，使空穴位置上升，最后在上界面处破裂。当床层产生空穴时，非空穴部位的颗粒床层仍维持在刚发生流化时的状态，通过的气流量较少，这部分称为乳化相。在发生聚式流化时，细颗粒被气体带到上方，形成稀相区，而较大颗粒留在下部，形成浓相区，两个区之间有分界面。一般的流化床层主要指浓相区，床层高度 L 指浓相区高度。聚式流化如图 4-22(d) 所示。

（3）输送床（pneumatic transport）阶段

如图 4-22(e) 所示，当流体流速继续增大到某一数值后，流化床上界面消失，固体颗粒被流体带走，这时床层称为输送床。

如图 4-23 所示，在流化床阶段，整个气固系统或液固系统的很多方面都呈现出类似流体的性质：

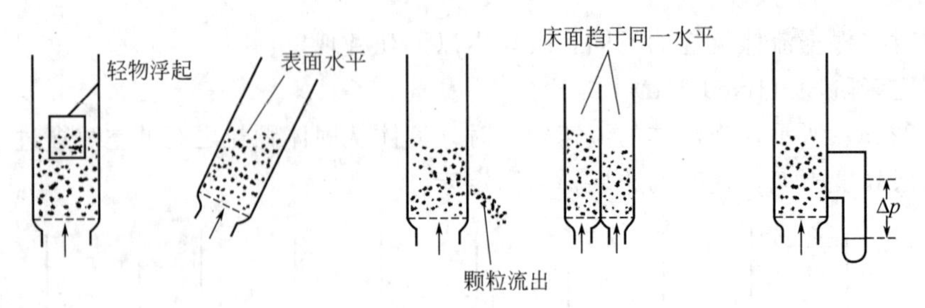

图 4-23　流化床的流动性

① 当容器倾斜时，床层上表面保持水平。

② 当两床层连通时，它们的床面能自行调至同一水平面。床层中某两点之间的压力差大致服从流体静力学的关系式 $\Delta p=\rho gL$，其中 ρ、L 分别为床层的密度与高度。

③ 具有流动性，颗粒能像液体那样从器壁小孔流出。

因流化床呈现流体的某些性质，所以在一定的状态下，它具有一定的密度、热导率、比热容、黏度等。而且，利用其类似于流体的流动性，可以实现固体颗粒在设备内或设备间的流动，易于实现生产过程的连续化和自动化。

流化床中的不正常现象如下。

（1）腾涌现象

腾涌现象主要发生于气固流化床中。当床层直径较小且气速过高时，气泡容易相互聚并而成为大气泡。若气泡直径长大到等于或接近床层直径时，气泡层与颗粒层相互隔开，颗粒层被气泡层像推动活塞一样推至向上移动。其中，部分颗粒在气泡周围落下或者推到一定高度后气泡突然破裂，颗粒在整个截面上洒落，这种现象称为腾涌或节涌，如图 4-24 所示。

（2）沟流

流体通过床层时形成短路，部分流体与固体颗粒接触不良而经沟道上升的现象称为沟流。引起沟流现象主要是颗粒粒度过细，易黏结，以及流体初始分布不均匀等原因所造成的。

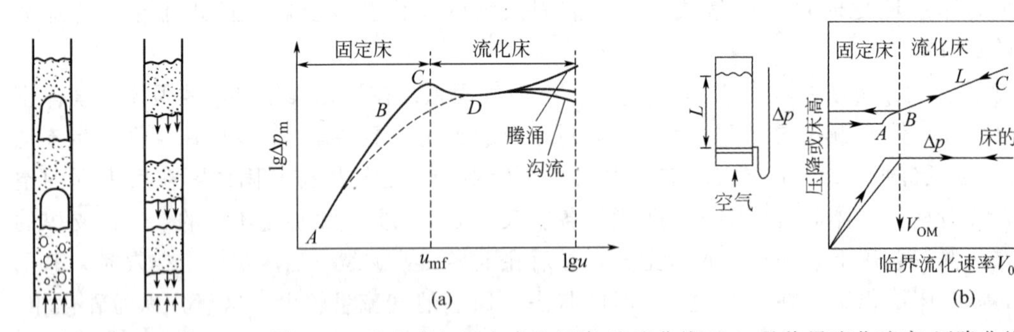

图 4-24　腾涌现象　　图 4-25　流化床与固定床的压降-流速曲线（a）及临界流化速率-压降曲线（b）

4.5.2　固体流态化流体力学特性

（1）压降与流速的关系

固体颗粒床层随流体空速 u 的增大，先后出现的固定床与流化床的压降（Δp_m）对 u 的实验曲线示于图 4-25。图中 AB 段颗粒静止不动，为固定床阶段。BC 段床层膨胀，颗粒松动，由原来堆积状况调整成疏松堆积状况。C 点表示颗粒群保持接触的最松堆置，这时流

体空速为 u_{mf}，称为"临界流化速度"。从 C 点开始，随着空速增大，床层进入流化阶段。

在 C 点时，颗粒虽相互接触，但颗粒质量正好为流体的曳力与浮力支托，颗粒间没有重力的向下传递。自 C 点以后的整个流化阶段中，颗粒质量都靠流体的曳力与浮力支撑。CD 阶段是床层颗粒自上而下逐粒浮起的过程。由于颗粒间的摩擦及部分叠置，床层压降比纯支撑颗粒质量时稍高。

若流化阶段是散式流化，则流化阶段床层修正压降等于单位截面积床层固体颗粒的净重，即：

$$\Delta p_m=\frac{m}{A\rho_s}(\rho_s-\rho)g=L(1-\varepsilon)(\rho_s-\rho)g \tag{4-36}$$

式中 Δp_m——流化床的修正压降，Pa；

m——整个床层内颗粒的质量，kg；

A——床层横截面积，m^2；

ρ_s——颗粒密度，kg/m^3；

ρ——流体密度，kg/m^3。

式（4-36）表明，散式流化过程中，床层压降不随流体空速的变化而改变，这一点已被实验基本证实。实际上，由于颗粒与器壁的摩擦，随空速的增大，流化床层的压降略微升高。

对于聚式流化，由于气穴的形成与破裂，流化床层的压降会有起伏。此外，还可能发生两种不正常的操作状况，即腾涌与沟流，使其压降曲线形状与散式流化的压降曲线形状有一定差别。发生腾涌时，气、固接触不良，而且由于固体颗粒的抛起与落下，易损坏设备。腾涌的流化压降高于散式流化压降。发生沟流时，同样气、固接触不良，其流化压降比散式流化压降低。

（2）流化床的流体空速范围

流化床的操作范围一般应在临界流化速度（临界速度）之上，而又不能大于带出速度。对于某一流化床操作，确定临界速度和带出速度是很重要的。

① 临界流化速度 u_{mf} 可以通过实测和计算两种方法确定临界流化速度，下面介绍实测法。

实测法不受计算公式精确程度和使用条件的限制，是得到临界流化速度的既准确又可靠的一种方法。该方法中，测取固体颗粒床层从固定状态到流化状态的一系列压降与气体流速的对应数值，将这些数据标在对数坐标上，得到如图4-25(a)所示的 $ABCD$ 曲线。若在床层达到流化状态后，再继续降低气速，则床层高度下降至 C 点所对应的床层高度时，固体颗粒互相接触而成为静止的固定床。气速再降低则流速与压降曲线沿 AC 变化。与 C 点对应的流速即为所测的临界流化速度。

测定时常用空气作为流化介质，最后要根据实际生产中的不同条件将测得的值加以校正。若 u'_{mf} 代表以空气为流化介质时测出的临界流化速度，则实际生产中的 u_{mf} 可按下式推算：

$$u_{mf}=u'_{mf}\frac{(\rho_s-\rho)}{(\rho_s-\rho_a)}\frac{\mu_a}{\mu} \tag{4-37}$$

式中 ρ——实际流化介质的密度，kg/m^3；

ρ_a——空气的密度，kg/m^3；

μ——实际流化介质的黏度，Pa·s；

μ_a——空气的黏度，Pa・s。

② 带出速度　当床层的表观速度达到颗粒的沉降速度时，大量颗粒将被流体带出器外，故流化床中颗粒的带出速度为单个颗粒的沉降速度 u_t，可用式（4-15）来计算。

但应注意，计算 u_{mf} 时要用实际存在于床层中不同粒度颗粒的平均直径，而计算 u_t 时必须用最小的颗粒直径。

（3）流化床的操作范围

流化床的操作范围，为空塔速度的上下极限，用比值 u_t/u_{mf} 的大小来衡量。u_t/u_{mf} 称为流化数。对于细颗粒，$u_t/u_{mf}=91.7$；对于粗颗粒，$u_t/u_{mf}=8.62$。

研究表明，上面两个 u_t/u_{mf} 的上下限值与实验数据基本相符，u_t/u_{mf} 值常在 10～90 之间。细颗粒流化床较粗颗粒流化床有更宽的流速操作范围。

实际上，对于不同工业生产过程中的流化床来说，u_t/u_{mf} 的差别很大。有些流化床的流化数高达数百，远远超过上述 u_t/u_{mf} 的上限值。在操作气速几乎超过床层的所有颗粒带出速度的条件下，虽有夹带现象，但未必严重。这种情况之所以可能发生，是因为气流的大部分作为几乎不含固相的大气泡通过床层，而床层中的大部分颗粒则是悬浮在气速依然很低的乳化相中。此外，在许多流化床中都配有内部或外部旋风分离器以捕集被夹带的颗粒，并使之返回床层，因此可以采用较高的气速以提高生产能力。

Fluidization

When a liquid or a gas is passed at very low velocity up through a bed of solid particles, the particles do not move. If the fluid velocity is steadily increased, the pressure drop and the drag on individual particles increase, and eventually the particles start to move and become suspended in the fluid. The terms fluidization and fluidized bed are used to describe the condition of fully suspended particles, since the suspension behaves as a dense fluid. If the bed is tilted, the top surface remains horizontal and large objects will either float or sink in the bed depending on their density relative to the suspension. The fluidized solids can be drained from the bed through pipes and valves just as a liquid can, and this fluidity is one of the main advantages of the use of fluidization for handling solids.

Conditions for fluidization

Consider a vertical tube partly filed with a fine granular material such as catalytic cracking catalyst, as shown schematically in Fig. 4-25(b). The tube is open at the top and has a porous plate at the bottom to support the bed of catalyst and to distribute the flow uniformly over the entire cross section. Air is admitted below the distributor plate at a low flow rate and passes upward through the bed without causing any particle motion. If the particles are quite small, flow in the channels between the particles will be laminar and the pressure drop across the bed will be proportional to the superficial velocity u. As the velocity is gradually increased, the pressure drop increases, but the particles do not move and the bed height remains the same. At a certain velocity, the pressure drop across the bed counterbalances the force of gravity on the particles or the weight of the bed, and any further increase in velocity causes the particles to move.

4.5.3 分布板对流化质量的影响

流化质量是指流化床均匀的程度，即气固接触的均匀程度。一般来说，流化床内形成的气泡愈小，气固接触的情况愈好。此外，床底部分布板的形式对流化质量也有影响。

一般流化设备是由圆柱体和它底部的锥体组成。在锥体与圆柱体间设置多孔的分布板，当气体从锥体部分进入后，通过分布板上的筛孔上升，使颗粒流化。

(1) 分布板的作用

在流化床中，分布板不仅支撑了固体颗粒，防止漏料和使气体均匀分布，还有分散气流、使气流在分布板上方产生较小气泡的作用。然而，分布板对气体分布的影响是有限的，通常只能局限在分布板上方不超过0.5m的区域内。

设计良好的分布板，应对通过它的气流有足够大的阻力，以保证气流均匀分布在整个床层截面上。也只有当分布板的阻力足够大时，才能克服聚式流化的不稳定性，抑制床层中出现沟流现象的趋势。实验证明，当采用某种致密的多孔介质或低开孔率的分布板时，可使气固接触良好。但气体通过这种分布板的阻力必然要大，这会大大增加鼓风机的动力消耗。因此，通过分布板的压降应有个适宜值。适宜分布板的稳定压降应等于或大于床层压降的10%，且满足绝对值不低于3.5kPa的条件。床层压降可取为单位截面上的床层重力。

(2) 分布板的形式

工业生产用的分布板的形式很多，常见的有直流式、侧流式和填充式。

① 直流式分布板　单层直流式分布板如图4-26(a)所示。这种分布板结构简单，便于设计和制造。其缺点是：气流方向与床层相垂直，易使床层形成沟流；小孔易于堵塞，停车时易漏料。图4-26(b)所示的多层孔板能避免漏料，但结构稍复杂。图4-26(c)为凹形多孔分布板，它能承受固体颗粒的重荷和热应力，还有助于抑制鼓泡和沟流现象的发生。

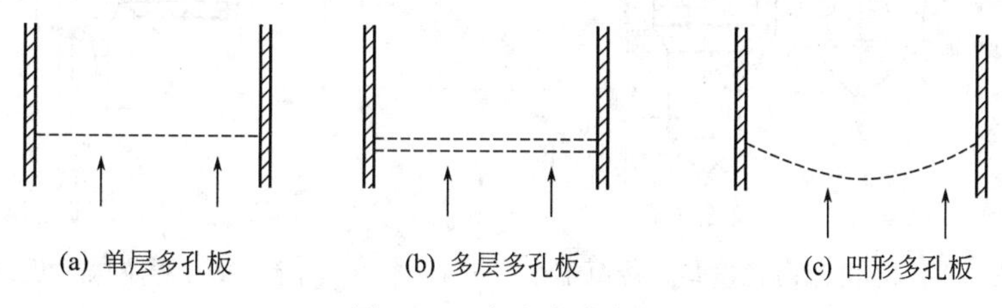

图4-26　直流式分布板

② 侧流式分布板　如图4-27所示，在分布板的孔上装有锥形风帽，气流从锥形风帽底部的侧缝或锥帽四周的侧孔流出。目前，这种带锥帽的分布板应用最广，效果也最好。其中侧缝式锥帽应用最多，它具有下列优点：

a. 固体颗粒不会在锥帽顶部堆成死床。每三个锥帽之间形成一个小锥形床，由此形成许多小锥形床，改善了床层的流化质量。

b. 气体紧贴分布板面从侧缝吹出而进入床层，在板面上形成一层“气垫”，使颗粒不能停留在板面上，这就减少了在板上形成死床和发生烧结现象的可能性。

③ 填充式分布板　如图4-28所示，填充式分布板是在直孔筛板或栅板和金属丝网层间铺上卵石、石英砂、卵石。这种分布板结构简单，能达到均匀分布气体的要求。

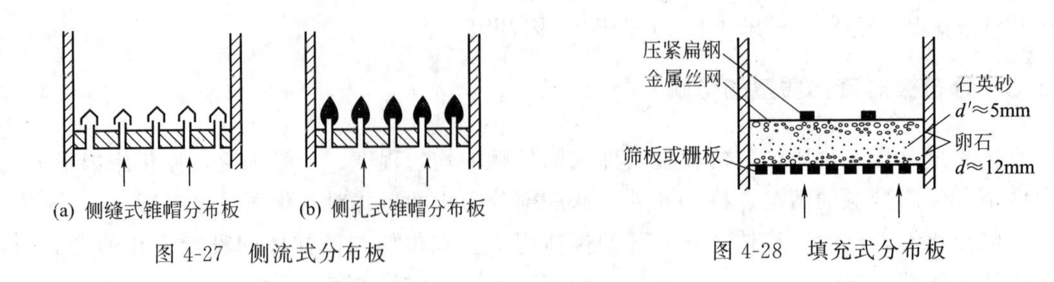

图 4-27　侧流式分布板　　　　图 4-28　填充式分布板

4.5.4　固体流态化技术的应用（Application of solid fluidization technology）

固体流态化技术可应用于传热、颗粒干燥、颗粒混合、颗粒分级、气体吸收等物理过程。例如，加热和冷却是逆向过程，具有相似性，生产过程中采用流化床进行气固直接接触，可对固体颗粒实行快速加热或冷却，图 4-29 为苯酐流态化冷却、冷凝器。

固体流态化技术也可用于催化或非催化反应过程。例如，用流化床气化法制造合成氨所需的原料气（半水煤气），是流态化技术在反应过程中的重要应用，如图 4-30 所示。水蒸气和富氧空气的混合气以 2～3m/s 的速度从流化床底部通入，床层内的煤粒处于流化状态，氧与碳反应放出大量的热，同时水蒸气也与碳反应生成 CO 和 H_2（半水煤气的主要成分）。

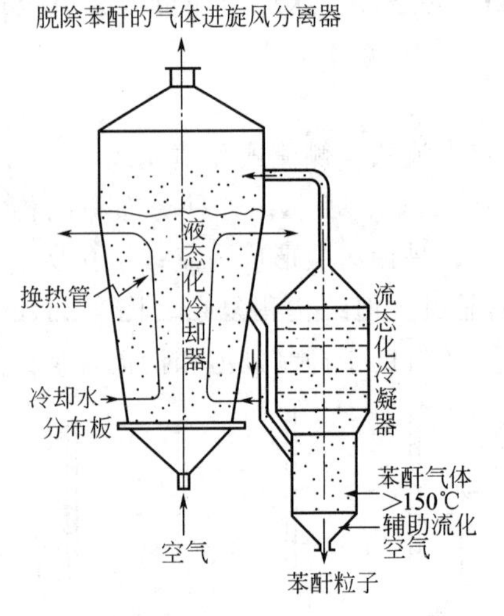

图 4-29　苯酐流态化冷却、冷凝器

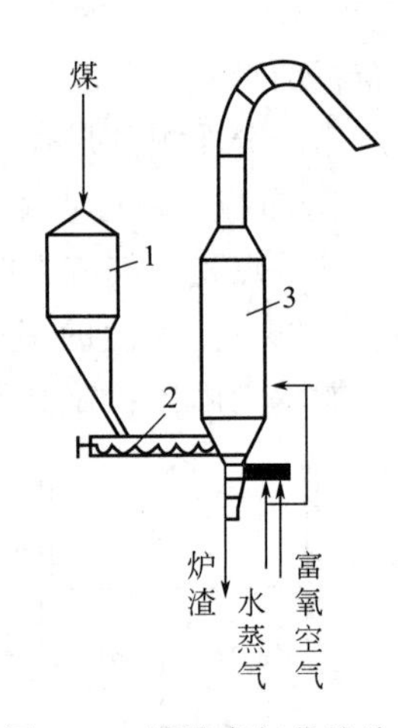

图 4-30　沸腾床气化炉造气

1—煤斗；2—螺旋输送机；3—气化炉

近一个世纪来，流态化技术的发展大致经历了两个阶段。第一阶段是以气泡现象为主要特征的鼓泡流态化及液固流态化，第二阶段是以颗粒团聚为主要特征的快速流态化及气固流态化的散式化。近年来，气、液、固三相流态化和外力场下的流态化得到新的发展，成为引人注目的前沿研究领域，如图 4-31 所示。

固体颗粒的输送是固体流态化技术的又一重要应用。固体物料的流体输送方式很多，按输送介质可分为水力输送和气力输送；按输送位置可分为水平输送和垂直输送；按流动状态可分为浓相输送和稀相输送；按操作方式可分为正压输送和负压输送；按供料进料方式可分为连续给料和脉冲注料。工业上，固体颗粒物料的流体输送系统由动力机械（风机或水泵）、

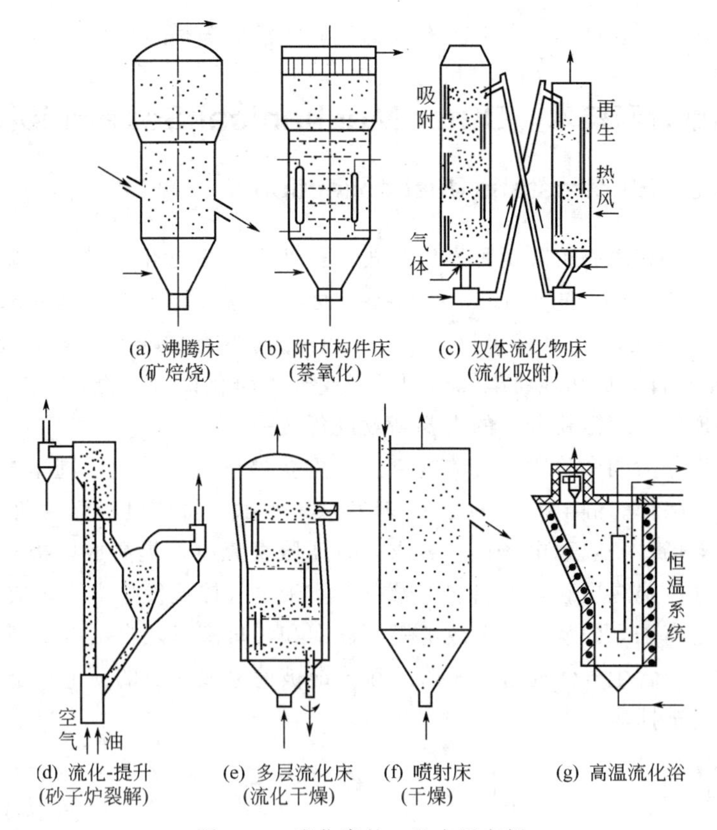

图 4-31　流化床的一些应用实例

供料机、管路和分离回收装置构成，通常应根据物料的特性、输送距离、能量消耗及现场其他条件选择合适的输送装置和设计合理的输送系统。图 4-32 为吸引式气力输送的典型装置。这种装置往往在物料吸入口处设有带吸嘴的挠性管，以便将分散于各处的或在低处、深处的散装物料收集至储仓。这种输送方式适用于需要避免粉尘飞扬的场合。

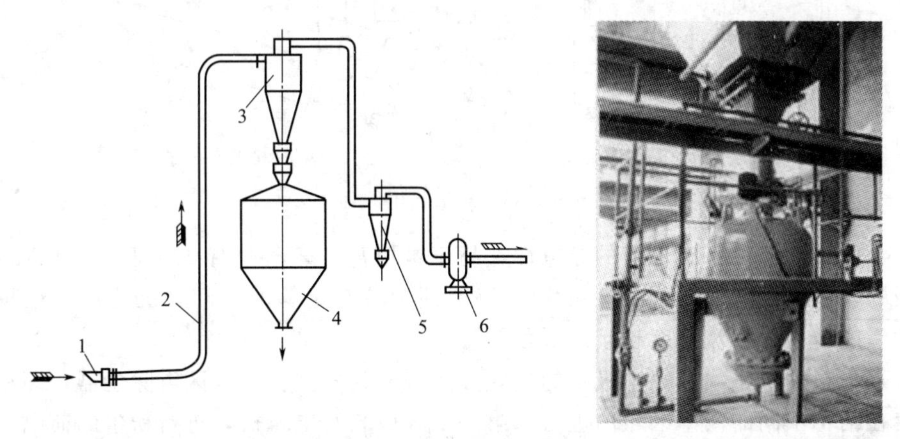

图 4-32　吸引式气力输送典型装置

1—吸嘴；2—输送管；3—一次旋风分离器；4—料仓；5—二次旋风分离器；6—抽风机

利用流体流动来输送固体颗粒的操作主要具有以下优点：①系统密闭操作，可防止环境污染，安全可靠；②在输送过程中可同时对物料进行加热、冷却、混合、干燥等操作；③设备简单，操作方便，而且容易实现自动化控制。其缺点是动力消耗较大，大颗粒或黏结性的颗粒物料难以输送，而且必须配备完善有效的气固或液固分离装置。

4.6 其他机械分离技术（Other Mechanical Separations）

4.6.1 静电除尘（Electrostatic dust collection）

前述的重力沉降和离心沉降两种操作方式，虽然能用于含尘气体或含颗粒溶液的分离，但是前者能够分离的粒子不能小于 50～70μm，而后者也不能小于 1～3μm。对于更小的颗粒，其常用分离方法之一就是静电除尘，即在电力场中将微小粒子集中起来再除去。自 Cottrell（1907 年）首先成功地将电除尘用于工业气体净化以来，经过一个世纪的发展，静电除尘器已成为现代处理微粉分离的主要高效设备之一。

静电除尘过程分为四个阶段：气体电离、粉尘获得离子而荷电、荷电粉尘向电极移动、将电极上的粉尘清除掉。静电除尘的原理如图 4-33 所示。将放电极作为负极、平板集尘极作为正极而构成电场，一般对电场施加 60kV 的高压直流电，提高放电极附近的电场强度，可将电极周围的气体绝缘层破坏，引起电晕放电。于是气体便发生电离，成为负离子和正离子及自由电子。正离子立即就被吸至放电极而被中和，负离子及自由电子则向集尘极移动并形成负离子屏障。当含尘气体通过这里时，粒子即被荷电成为负的荷电粒子，在库仑力的作用下移向集尘极而被捕集。

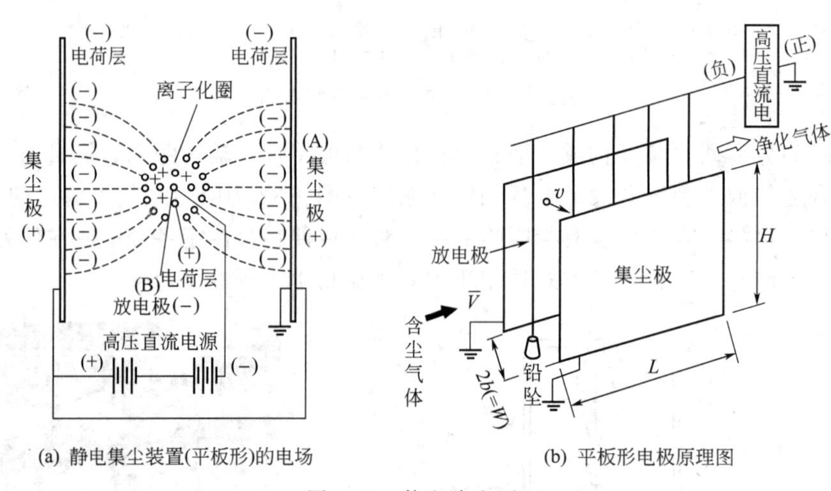

(a) 静电集尘装置(平板形)的电场　　(b) 平板形电极原理图

图 4-33 静电除尘原理

大多数的工业气体都有足够的导电性，易于被电离。若气体导电率低，可以加水蒸气。流过电极的气体速度宜低（0.3～2m/s），以保证尘粒有足够的时间来沉降。颗粒越细，要求分离的程度越高，气流速度越接近低限。

静电除尘根据电晕线和收尘极板的配置方法可分为两类：①双区电除尘器，即粉尘的荷电和收集在结构不同的两个区域内进行。第一个区域装有电晕极，进行粉尘的荷电；第二个区域装有收尘极，把粉尘收集起来。这种除尘器一般用于空气净化。②单区电除尘器，即电晕极和收尘极都在同一区域内。粉尘的荷电和收集都在同一区域内进行，工业生产中大多采用这种除尘器。其中按收尘极的分类，又可分成管式电除尘器（图 4-34）和板式电除尘器（图 4-35）两种。管式电除尘器的收尘极为 ϕ200mm～300mm 的圆管或蜂窝管，其特点是电场强度比较均匀，有较高的电场强度。但其粉尘的清理比较困难，一般不宜用于干式除尘，而通常用于湿式除尘。而板式电除尘器具有各种形式的收尘极板，极间距离一般为250～400mm。电晕极安放在板的中间悬挂在框架上，电除尘器的长度根据对除尘效率的要求确定。它是工业中最广泛采用的形式。

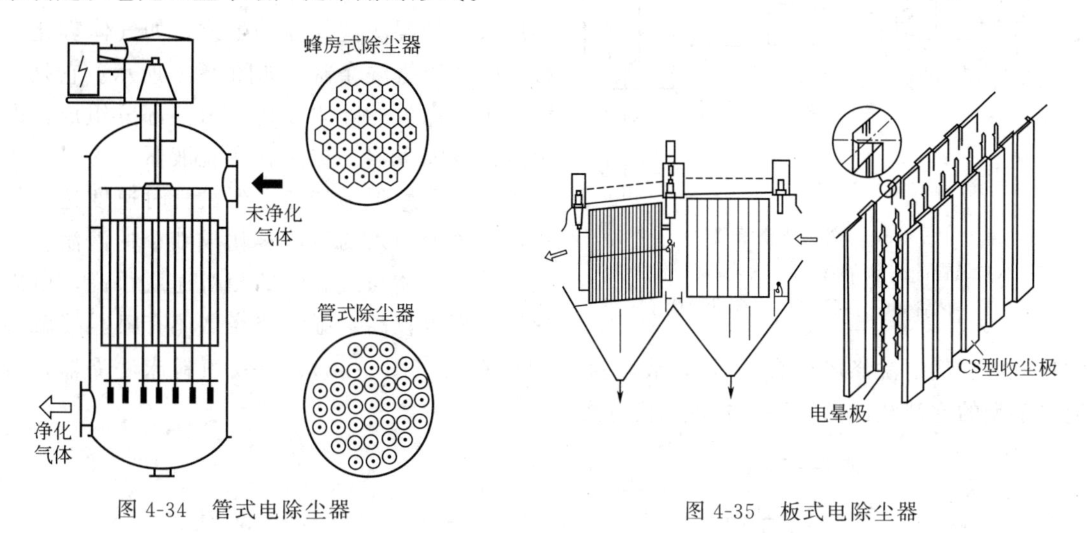

图4-34　管式电除尘器　　图4-35　板式电除尘器

在化学工业中，电除尘器常用于硫酸、氯化铵、炭黑、焦油沥青及石油油水分离等生产过程，用于除去粉尘或烟雾。其中使用最多的是硫酸中的干、湿法静电除尘器。电除尘器的设备复杂，价格昂贵，但因能够除去极细小的颗粒，除尘效率很高，所以在工业生产中已得到应用。

4.6.2　湿法捕集

湿法捕集是利用液体作为捕集体，将气体内所含颗粒捕集下来的一类方法，所用设备统称湿法洗涤器。它与干法捕集相比，具有如下特点。

① 在捕集气体内悬浮物的同时，只要液体选用适当，还可吸收而除去气体内的有害组分，所以用途更为广泛。

② 在消耗同样能量的情况下，湿法捕集一般可比干法捕集（如旋风分离器）的除尘效率高，而且能处理黏性大的固体颗粒，但不适于处理憎水性和水硬性颗粒。

③ 洗涤器设备本身较简单，费用也不高，但却要有一套液体的供给及回收系统，所以总的造价及操作维修等费用也不低。

湿法洗涤器的捕集机理是一种流体动力捕集机理，其捕集体可分为三种形式，即液滴、液膜及液层。其中，液滴的产生基本上有两种方法：一种是使液体通过喷嘴而雾化；一种是用高速气流将液体雾化。液体呈分散相，含有固体颗粒的气体则呈连续相，两相间存在着相对速度，依靠颗粒对于液滴的惯性碰撞、拦截、扩散、重力、静电吸引等效应而把颗粒捕集下来。液膜是将液体淋洒在填料上并在填料表面形成很薄的液体网络。此时，液体和气体都是连续相，气体在通过这些液体网络时也会产生上述各种捕集效应。而液层的作用是使气体通过液层时生成气泡，气体变为分散相，液体则为连续相。颗粒在气泡中依靠惯性、扩散和重力等机理而产生沉降，被液体带走。

湿法洗涤器设计的关键问题是要使含尘气与液体接触。气液接触的方法有很多，可以是

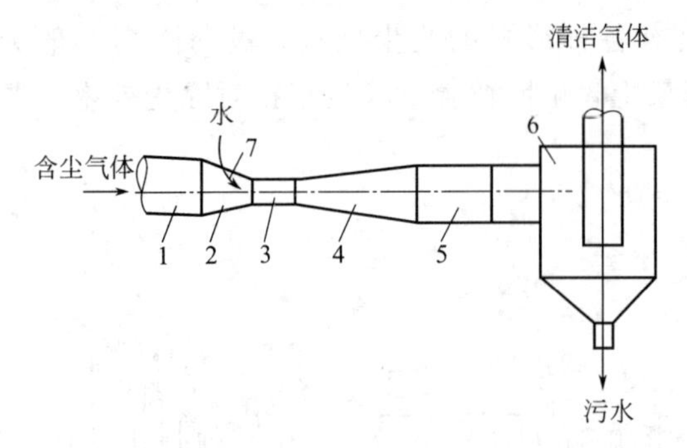

图 4-36　文氏管洗涤器

1—入口风管；2—渐缩管；3—喉管；4—渐扩管；5—风管；6—脱液器；7—雾化喷嘴

将液体雾化成细小液滴；也可以是使气体鼓泡进入液体内；还可以是气体与很薄的液膜接触；更可以是这几种方法的综合应用。所以可形成很多种湿法洗涤器的形式，如气体雾化接触型的文氏管洗涤器，如图 4-36 所示。它是一种简单而高效的除尘设备，由三部分组成，即引液器或喷雾器、文氏管及脱液器。含尘气体进入文氏管后逐渐加速，到喉管时速度达到最高，将该处引入的液体雾化成细小的液滴。在喉管处，气体与液滴间的相对速度很高，所以捕集效率也较高。湿法洗涤器还有喷雾接触型的喷淋塔、喷射洗涤塔（图 4-37），液膜接触型填料塔（图 4-38）、浮动填料床洗涤器，鼓泡接触型的泡沫洗涤器、冲击式泡沫洗涤器等。

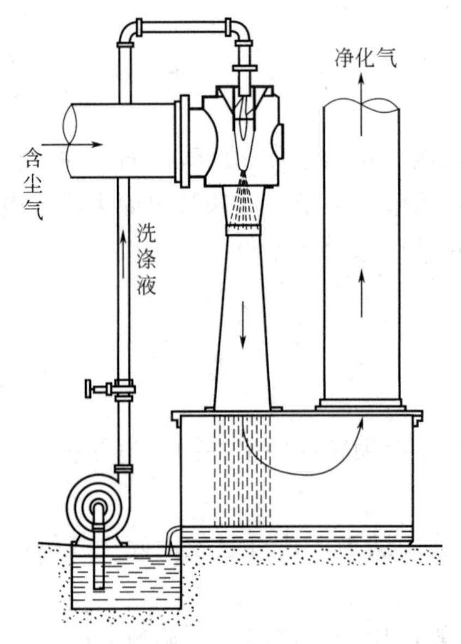

图 4-37　喷射洗涤塔

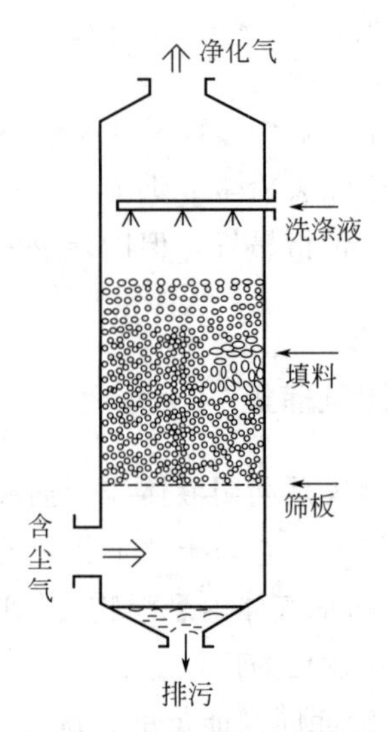

图 4-38　液膜接触型填料塔

一般的板式塔和填料塔都可用作洗涤器来除去气体中的尘粒及有害成分，如泡沫洗涤器，如图 4-39 所示，它的典型结构是筛板塔。若液体与气体都是通过筛孔而逆流接触，则称为无溢流式，又称淋降塔板，如图 4-39(a) 所示。若液体横流过筛板而从一侧降液管中流下，气体通过筛孔穿入液层内，则称为溢流式，如图 4-39(b) 所示。作为除尘器，一般以淋降塔板为主。将填料塔作为洗涤除尘设备时，入口含尘浓度不能过高，否则易产生堵塞现象。若将静止的填充床改变为流化床，就可解决此问题，这就是湍球塔，其典型结构如图 4-40 所示。在两块开孔率很高的孔板间，放置有直径为 ϕ15mm～76mm 的轻质空心球，含尘气体从下部进入，以较大速度进入床层将空心球吹起形成流化床。洗涤液从上面喷淋下来。被流化状态下的小球激烈扰动，小球在湍动旋转及相互碰撞中，又使液膜表面不断更新，从而强化了气液两相的接触，可大大提高除尘效率。另外，还有冲击式洗涤器、强化型洗涤器等。

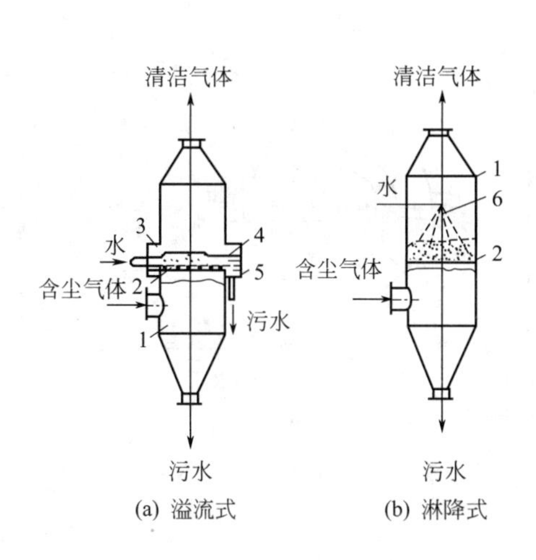

图 4-39 泡沫洗涤器

1—塔体；2—筛板；3—液雾区；4—泡沫区；5—降液管；6—喷淋头

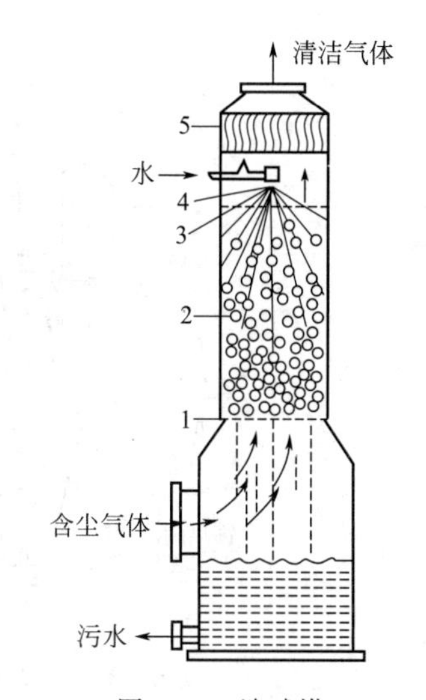

图 4-40 湍球塔

1—支撑筛板；2—填料塔；3—挡板筛板；4—喷嘴；5—除沫板

工程案例分析

应用双出风口旋风分离器改造旋风选粉机

某水泥厂1号$\phi 2.4m\times 13m$水泥磨，1995年改为闭路磨，配$\phi 2.8mm$旋风选粉机。实施水泥新标准后该厂为保强度，只好将混合材料掺量降到5%，实际上等于增加了生产成本。

1.技改方案和措施

2001年8月，该厂对1号水泥磨系统测试考察，拟订了改进选粉机分级能力、提高旋风分离器捕集细粉能力、强化磨机的粉磨能力、改善磨内料流和通风能力为主线的改造方案。

改造时将回转小风叶拆除，改成笼形转子。分离力场从平面圆环变成圆柱体，径向变化范围较小，分离力场均匀，分级临界粒径范围较窄，因而混料窜料较少，分级效果较好。为使分离临界粒径可调，将转子电动机改为调频电动机。

原选粉机外围是6个$\phi 1.3m$的单出风口旋风分离器，它一般只能收集$20\mu m$以上的颗粒。研究和实践表明：现行单出风口旋风分离器核心强制涡的无用压力损失占旋风分离器总压力损失的65%～85%；出风口下端附近有较大的径向速度，呈短路流现象；锥体下部有较大的偏心流，造成排尘口粉尘返混，而使分离效率下降。为解决上述问题，选用导流口可调式双出风口旋风分离器（双出风口旋风分离器）。

2.双出风口旋风分离器结构

双出风口旋风分离器结构如附图所示。含尘气体从进风口切向进入筒体，在筒体与出风管、导流管之间的环形空间旋转向下并逐渐进入锥体，气体中的粉尘受旋转力场中的离心力作用而与气流分离并碰撞筒壁，失去动能，沿筒壁进入锁风阀后及时排出。气固分离后的净化气流，经反射屏强制改向，沿导流管外壁旋转向上并经导流口从上、下出风口排出。

与传统旋风分离器相比，该旋风分离器具有如下优点：

① 出口风速降低近半，压力损失显著降低；

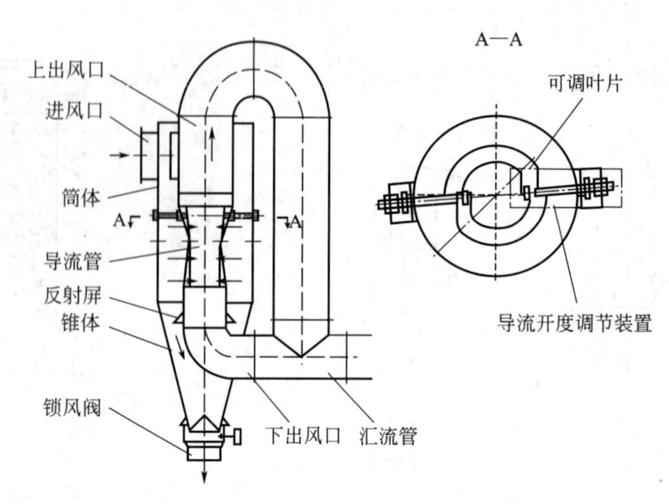

附图　导流口可调式双出风口旋风分离器结构

② 筒身纵向开设多个导流口，可基本消除核心强制涡；

③ 导流筒上口与上出风口下端连接，可消除短路流；

④ 导流筒下口与下出风口上端连接，并设置反射屏，可显著降低粉尘返混现象，分离效率可进一步提高。

3. 应用及效果

2002 年 2 月 10 日，在对选粉机 7 个旋风筒完成上述改造后，粉磨系统未作其他调整即投料生产 42.5 级普通硅酸盐水泥，磨机产量从改造前 18t/h（细度≤3%）提高到 25t/h（细度≤2.1%），说明捕集细粉能力明显增强。

2002 年 2 月 27 日，对粉磨系统稍作调整，将风机风量增大 10%，隔仓板风孔率增大 5%，磨球级配稍加调整，改善系统漏风情况。改后粉磨 42.5 级普通硅酸盐水泥产量提高到 27t/h（细度≤2.6%）。水泥比表面积从原来的 $325m^2/kg$ 提高到 $335m^2/kg$。

本技术改造投资 12 万元，水泥磨产量按 42.5 级和 32.5 级普通硅酸盐水泥的加权平均提高 4t/h 计算，全年可增产 28000t。每吨水泥的增效部分按 30 元计算，全年效益 84 万元，扣除 12 万元成本，全年收效 72 万元。

习　　题

1. 颗粒的直径为 30μm，密度为 $2000kg/m^3$，求它在 20℃水中的沉降速度。
2. 直径分别为 100μm 与 180μm 球形颗粒 A、B 在 20℃的水中做重力沉降，已知颗粒的密度为 $3000kg/m^3$，试计算两者的沉降速度。
3. 一种测量液体黏度的方法是测量金属球在其中沉降一定距离所用时间。现测得密度为 $8000kg/m^3$、直径为 5.6mm 的钢球在某种密度为 $920kg/m^3$ 的油品中重力沉降 300mm 所用的时间为 12.1s，问此种油品的黏度是多少？
4. 用高 2m、宽 2.5m、长 5m 的重力降尘室分离空气中的粉尘。在操作条件下，空气的密度为 $0.779kg/m^3$，黏度为 $2.53\times10^{-5}Pa\cdot s$，流量为 $5.0\times10^4m^3/h$，粉尘的密度为 $2000kg/m^3$。试求所能分离下来的最小颗粒直径。
5. 多层降尘室宽 2 m，长 4 m，总高 4 m，每层高 0.2 m。用它来处理含尘气体，要求把直径大于 20μm 的尘粒全部除去。试问此多层降尘室的最大处理气量为多少？已知气体的密度为 $1kg/m^3$，黏度为 $3\times10^{-5}Pa\cdot s$，尘粒的密度为 $3000kg/m^3$。

6. 用降尘室净化烟气，已知烟气流量为3600m^3/h，密度为0.6kg/m^3，黏度为0.03mPa·s，尘粒的密度为4300kg/m^3，要求除去6μm以上的尘粒。试求：(1) 所需的沉降面积；(2) 若降尘室底面积为15m^2，则需多少层？
7. 某除尘室高2m、宽2m、长6m，用于炉气除尘。球形矿尘颗粒的密度为4500kg/m^3，操作条件下气体流量为25000m^3/h，气体密度为0.6kg/m^3、黏度为0.03mPa·s。试求理论上能完全除去的最小矿粒直径。现由于生产工艺改进，需要完全除去粒径大于20μm的颗粒，如果采用将降尘室改为多层结构的方法，试计算多层降尘室的层数。
8. 有一降尘室，长6m，宽3m，共20层，每层高100mm，用以除去炉气中的矿尘。矿尘密度为3000kg/m^3，炉气密度为0.5kg/m^3，黏度为0.035mPa·s，现要除去炉气中的10μm以上的颗粒。试求：(1) 为完成上述任务，可允许的最大气流速度；(2) 每小时最多可送入的炉气量；(3) 若取消隔板，为完成任务该降尘室的最大处理量为多少？
9. 用降尘室实现某气体的除尘，气流流量为3000m^3/h，其黏度为0.03mPa·s，密度为0.5kg/m^3。气流中的粉尘颗粒均匀分布，密度为4000kg/m^3。试求：(1) 设计一个能够保证30μm以上颗粒完全除去的降尘室，计算该降尘室的面积。(2) 采用该降尘室时，直径为15μm的颗粒被除去的百分数是多少？(3) 如果气量减少20%，该降尘室所能全部分离的最小颗粒的直径为多少？(4) 如果将降尘室改为两层，对颗粒的处理要求不变，则所能处理的气流流量变为多少？
10. 方铅矿石和硅石的混合物，其粒度分布为0.075～0.65mm，方铅矿石和硅石的密度分别为7500kg/m^{3}与2650kg/m^3，用温度为20℃的向上的水流进行分离。试问：(1) 要得到纯净的方铅矿石产品，水的流速应为多大？(2) 纯净产品的粒度分布如何？
11. 速溶咖啡粉（密度为1050kg/m^3）的直径为60μm，用250℃的热空气送入标准旋风分离器中。进入时的切线速度为20 m/s，在器内的旋转半径为0.5m。试求：(1) 咖啡粉粒的径向沉降速度；(2) 若此咖啡在同温度的静止空气中沉降，其沉降速度为多少？
12. 已知含尘气体中尘粒的密度为2300kg/m^3，气体的流量为1000m^3/h，密度为0.874kg/m^3，黏度为3.6×10^{-5}Pa·s。拟采用标准旋风分离器进行除尘，分离器直径为400mm。

试求：(1) 计算其临界粒径及压降；

(2) 若含尘气体原始含尘量为150g/m^3，尘粒的粒度分布列于附表中。

习题12附表 尘粒的粒度分布

粒径 d/μm	<3	3～5	5～10	10～20	20～30	30～40	40～50	50～60	60～70
质量分数/%	3	11	17	27	12	9.5	7.5	6.4	6.6

求出口气体的含尘浓度及除尘效率。

(3) 若临界直径不变，经过旋风分离器的压降可以增大到1.1kPa，求旋风分离器的直径及并联个数。

13. 拟采用标准旋风分离器收集流化床出口处的碳酸钾粉尘。旋风分离器入口处的空气温度为200℃，分离器的处理能力为4400m^3/h，粉尘密度为2290kg/m^3，旋风分离器的直径为700mm。试求：(1) 此分离器所收集粉尘的临界粒径；(2) 如分别以2台直径为500mm及4台直径为400mm的同类旋风分离器并联操作以代替原来的分离器，则临界粒径又将为多少？
14. 某一旋风分离器，在某一进口气速下进入分离器的总粉尘量为300.0g，收集到的粉尘量为262.5g。投料粉尘和收集粉尘经分级和称重，其结果列于附表中。

习题14附表 投料粉尘和收集粉尘分级和称重

粒径范围/μm	0～5	5～10	10～20	20～40	40～60	>60
进入 C_{1i}/g	30.3	40.5	42.3	45.6	51.0	90.3
收集 C_{2i}/g	12.7	32.8	36.8	41.5	48.5	90.2

试计算该分离器的总效率和粒级效率。

15. 用一板框式过滤机过滤某固体悬浮液，在 101.3kPa（表压）下过滤 20min，单位过滤面积上得到滤液 0.197m^3，继续过滤 20min，单位过滤面积上又得到滤液 0.09m^3。如果过滤 1h，单位过滤面积上所能得滤液为多少？

思考题

1. 直径为 100μm 的球形石英颗粒在 20℃的水中自由沉降，试计算该颗粒由静止状态开始加速到沉降速度的 99%时所需经历的时间和沉降的距离。
2. 什么是离心分离因数？分离因数的大小说明什么问题？为提高分离因数，可采取什么措施？
3. 为什么高分离因数的离心机都是采用高转速和小直径的转鼓呢？
4. 试分析旋风分离器的直径及入口气速的大小对临界颗粒直径的影响。
5. 同一流化床，用空气和水作流化介质时，床层压降是否相同？为什么？

名词术语

英文/中文

air current/空气对流
Archimede's principle/阿基米德原理
backwashing/反洗
boundary layer separation/边界层分离
boundary layer thickness/边界层厚度
Brownian movement/布朗运动
cake filtration/滤饼过滤
cake resistance/滤饼阻力
canvas/帆布
catalytic reactor/催化反应器
constant-pressure filtration/恒压过滤
constant-rate filtration/恒速过滤
contaminant/污染物
continuous pressure filter/连续压滤机
correction factor/校正因子
corrugation/波纹
creeping flow/爬流
criterion/判断标准
critical diameter/临界直径
cross-sectional area/横截面积
cyclone separator/旋风分离器
deep bed filtration/深床过滤
density difference/密度差
depth filter/深床过滤机
drag coefficient/阻力系数
drag force/曳力
eddy or vortices/旋涡
equivalent channel diameter/孔道的当量直径
filter-area/过滤面积

英文/中文

filter-cake/滤饼
filter-cloth/过滤机滤布
filter medium resistance/过滤介质阻力
filter medium/过滤介质
filtrate/滤液
filtration/过滤
filtration constant/过滤常数
filtration equation/过滤方程
fluidization/流态化
fluidized bed/流化床
form drag/型体阻力
free settling/自由沉降
granular media/颗粒介质
gravitational settling/重力沉降
heterogeneous mixture/非均相混合物
hindered settling/干扰沉降
horizontal knife/水平刮刀
incompressible cake/不可压缩滤饼
ion-exchange/离子交换
low Reynolds number/低雷诺数
mechanical separation/机械分离
minimum fluidization velocity/最小流化速度
molecular and electrostatic force/分子力和静电力
motion from gravitational force/重力场作用下的运动
motion in a centrifugal field/离心力场作用下的运动
nominal size/公称尺寸
nonspherical particle/非球形颗粒
open gravity filter/敞口式重力过滤器
packed tower/填料塔

particulate fluidization/颗粒流化
permeable medium/可渗透介质
plate-and-frame press filter/板框压滤机
pressure drop/压降
pressure filter or presses filter/压滤机
projected area/投影面积
residence time/停留时间
rotary drum/转鼓
sedimentation/沉降
separation factor/分离因子
settling chamber/沉降室
settling tank/沉降槽
settling time/沉降时间
shell-and-leaf filter/叶滤机
slurry/滤浆
smooth sphere/光滑球体
specific cake resistance/滤饼比阻
sphericity/球形度
stagnation point/驻点
superficial velocity/表观速度
surface filter/表面过滤机
surface-volume ratio/比表面
suspensions/悬浮物
tangential velocity/切向速度
terminal velocity/沉降速度
tortuosity factor/曲率因子
tortuous channel/弯曲流道
tortuous pore/弯曲孔道
total void fraction/总空隙率
transition/过渡
two-dimensional motion/二维运动
two-phase countercurrent flow/两相逆流流动
up-flow filter/上流式过滤器
up-stream side/上游一侧
void volume/空隙
volume fraction/体积分数
wake formation/尾流的形成
wall drag/壁面阻力
washing/洗涤

第5章

其他单元操作

Chapter 5 Others Unit Operation

学习目标

通过本章学习，学生应具备下列知识和能力：

1. 掌握蒸发与干燥等其他单元操作的基本原理；能够基于相关的基本原理，清晰复杂工程问题中相关问题所涉及的数学模型，分析、解决化工生产中相关的复杂化学工程问题。

2. 了解相关单元操作设备的结构特点、工作原理、操作方法；能够针对复杂化学工程问题中所涉及的单元操作问题，提出合理的解决措施。

3. 了解相关单元操作设备的设计计算及选型；能够针对复杂化学工程问题，使用恰当的技术、资源、现代工程工具和信息技术工具，通过对相关复杂工程问题进行数学建模，计算并设计合适的单元操作设备。

4. 熟悉相关单元操作的英文表达，增强国际交流能力，拓宽国际视野。

学习要求

通过本章学习：

1. 掌握相关单元操作的特点、相关单元操作设备的类型，能够基于相关基本原理，进行相关过程的计算。

2. 能够根据生产工艺要求和物料特性，合理选择单元操作设备类型，并确定适宜操作流程和操作条件。

3. 能够对实际工程中的单元操作问题进行分析，解决实际工程问题。

4. 能够对相关单元操作进行英文表达，具有国际交流能力。

除前几章所涉及的单元操作之外，蒸发、干燥、结晶、吸附、离子交换以及膜分离等单元操作都是化工生产中常用的单元操作。其中蒸发、干燥涉及物料中液体的去除，都是传热、传质耦合的单元操作过程。和传热相耦合的溶液的浓缩常采用蒸发操作，和传热相耦合的固体中液相的去除常常采用干燥操作。

在实际化工生产中，为节省成本，经常采用多种单元操作耦合的方式。例如高浓盐水的处理，经常先采用蒸发浓缩至饱和然后结晶，再通过过滤或离心分离得到固体盐，为便于固体物料的储运和加工，往往根据要求将固体中的湿分（水分或其他溶剂）除去，这类操作称为除湿操作。常用的除湿方法有机械除湿法（如压榨、过滤、压滤等）、热物理法（如加热或冷冻除湿）、吸附法（如用干燥剂除湿）等。

其中利用热能使湿物料中湿分汽化并及时排出生成蒸汽，以获得湿分含量达到规定的成品过程，称为干燥操作。因为完全采用干燥操作除去湿分的方法能耗很大，所以湿物料都应先采用机械分离方法除湿，然后再干燥。例如，硫酸铵结晶经离心分离后，约含水分3%，再经干燥后得到含水量约为0.1%的产品。

5.1 蒸发（Evaporation）

当溶液中溶质的挥发性甚小，而溶剂又具有明显的挥发性时，工业上常用加热的方法，使溶剂汽化达到溶液浓缩或挥发性物质回收的目的，这样的操作称为蒸发。蒸发常常将稀溶液加以浓缩，以便得到工艺要求的产品。它是过程工业中应用最广泛的浓缩方法之一，常用于烧碱、抗生素、糖、盐以及淡水等生产中。蒸发操作可以除去各种溶剂，其中以浓缩含有不挥发性物质的水溶液最为普遍，所以本章仅介绍水溶液的蒸发。

水溶液在低于沸点温度下也会有水分蒸发，但这种蒸发速度很慢，效率不高，所以工程上采用在溶液沸腾状态下的蒸发过程。蒸发过程中的热源常采用新鲜的饱和水蒸气，又称生蒸汽；而将从溶液中蒸出的水蒸气称为二次蒸汽，以区别作为热源的生蒸汽。若将二次蒸汽直接冷凝，而不利用其冷凝热，这样的操作称为单效蒸发；若将二次蒸汽引到下一级蒸发器作为加热蒸汽，这样的串联蒸发操作称为多效蒸发。

蒸发操作可以在加压、常压及减压下进行。常压蒸发时，二次蒸汽直接排入大气中；而加压或减压蒸发时，为了保持蒸发器内的操作压力，必须将二次蒸汽送入冷凝器中用冷却水冷凝的方法排出。

Evaporation

Evaporation is one of the oldest unit operations, it is also an area in which much has changed in the last quarter century. Heat transfer to a boiling liquid has been discussed generally in Chap. 2. Especial case occurs so often that it is considered an evaporation and is the subject of this chapter.

The objective of evaporation is to concentrate a solution consisting of a nonvolatile solute and a volatile solvent. In the overwhelming majority of evaporations the solvent is water. Evaporation is conducted by vaporizing a portion of the solvent to produce a concentrated solution of thick liquor. Evaporation differs from drying in that the residue is a liquid-sometimes a highly viscous one—rather than a solid; it differs from distillation in that the vapor usually is a single component, and even when the vapor is a mixture, no attempt is made in the evaporation step to separate the vapor into fractions; it differs from crystallization in that emphasis is placed on concentrating a solution rather than forming and building crystals. In certain situations, for example, in the evaporation of brine to produce common salt, the line between evaporation and crystallization is far from sharp. Evaporation sometimes produces a slurry of crystals in a saturated mother liquor.

Normally, in evaporation the thick liquor is the valuable product, and the vapor is condensed and discarded. In one specific situation, however, the reverse is true. Mineral-bearing water often is evaporated to give a solid-free product for boiler feed, for special process requirements, or for human consumption. This technique is often called water distillation, but technically it is evaporation. Large-scale evaporation processes have been developed and used for recovering potable water from seawater. Here the condensed water is the desired product. Only a fraction of the total water in the feed is recovered, and the remainder is returned to the sea.

5.1.1 蒸发设备（Evaporation equipments）

蒸发器主要由加热室和分离室两部分组成，如图 5-1 所示。加热室的作用是利用水蒸气热源来加热被浓缩的料液，分离室的作用是将二次蒸汽中夹带的物沫分离出来。按加热室的结构和操作时溶液的流动状况，可将工业中常用的加热蒸发器分为循环型和单程型两大类。

图 5-1 中各符号的含义如下：

F——原料液流量，kg/h；

W——水的蒸发量，kg/h；

D——生蒸汽消耗量，kg/h；

x_0，x_1——原料液和浓缩液中溶质的质量分数；

t_0，t_1——原料液和浓缩液的温度，℃；

h_0，h_1——原料液和浓缩液的焓，kJ/kg；

T，T'——生蒸汽和二次蒸汽的温度，℃；

H，h_w，H'——生蒸汽、冷凝水和二次蒸汽的焓，kJ/kg。

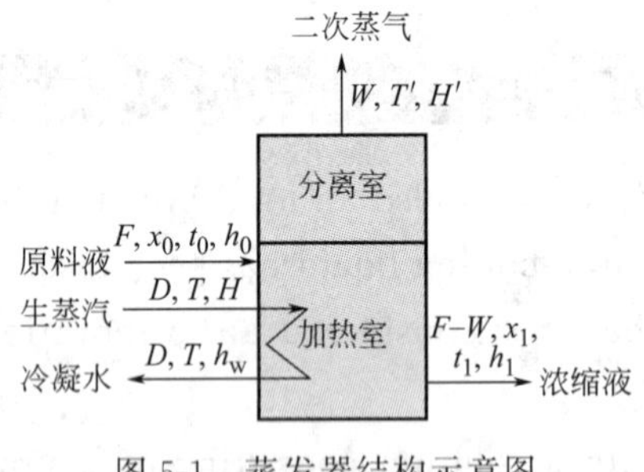

图 5-1 蒸发器结构示意图

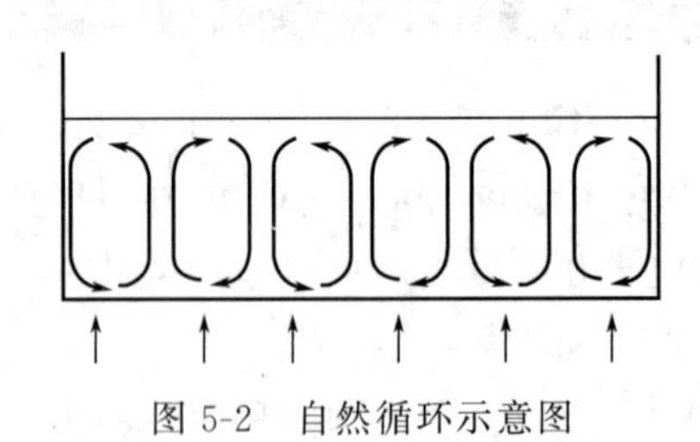

图 5-2 自然循环示意图

5.1.1.1 循环型蒸发器

循环型蒸发器的特点是溶液在蒸发器内作连续的循环运动，以提高传热效果，具体可分为自然循环和强制循环两种类型。

（1）自然循环蒸发器（natural circulation evaporator）

在自然循环蒸发器中，溶液因被加热的情况不同而产生密度差，形成自然循环（图 5-2）。

加热室有横卧式和竖式两种，其中以竖式应用最为广泛。

① 中央循环管式（标准式）蒸发器（vertical type evaporator） 如图5-3所示，标准式蒸发器由上、下两部分构成，下部为加热室，上部为分离室。加热室由直立的沸腾管束组成，管长1～2m，管长与管径之比为20～40。加热室的中央有一根很粗的管子，它的截面积大约等于沸腾管束总面积的40%～100%，称为中央循环管。加热蒸汽在管间冷凝放热，而待分离的溶液由中央管下降，沿细管上升，形成连续的自然循环运动。

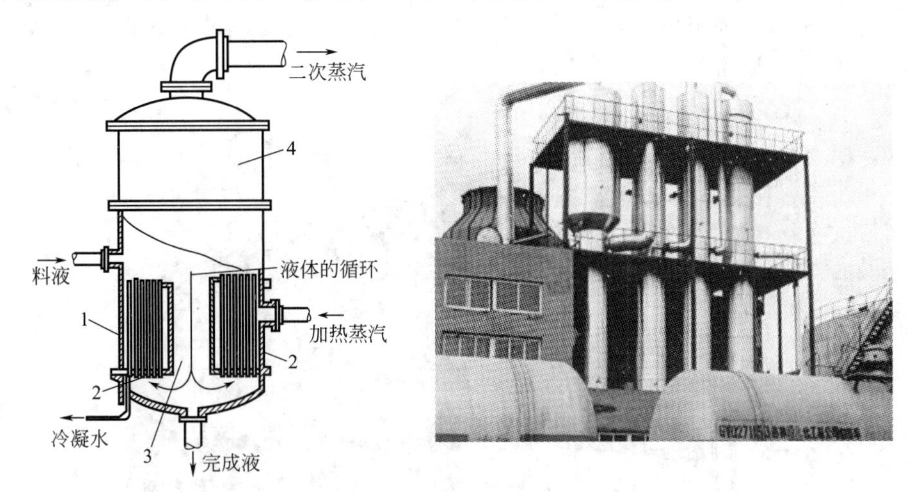

图5-3 中央循环管式蒸发器

1—外壳；2—加热室；3—中央循环管；4—分离室

标准式蒸发器结构简单，制造方便，操作可靠；但溶液循环速度低（0.3～1.0m/s），沸腾液一侧的对流传热系数小，不便清洗和更换管子。

② 外热式蒸发器 加热室在外的蒸发器是现代蒸发器发展的一个特点，图5-4即属于此类。这类蒸发器由加热室、分离室和循环管三部分组成。一方面，加热器和分离器分开，可调节料液的循环速度，使加热管仅用于加热，而沸腾恰在高出加热管顶端处进行，这样可防止管子内表面析出结晶堵塞管道，又可改变分离雾沫的条件（如采用离心分离的形式）；

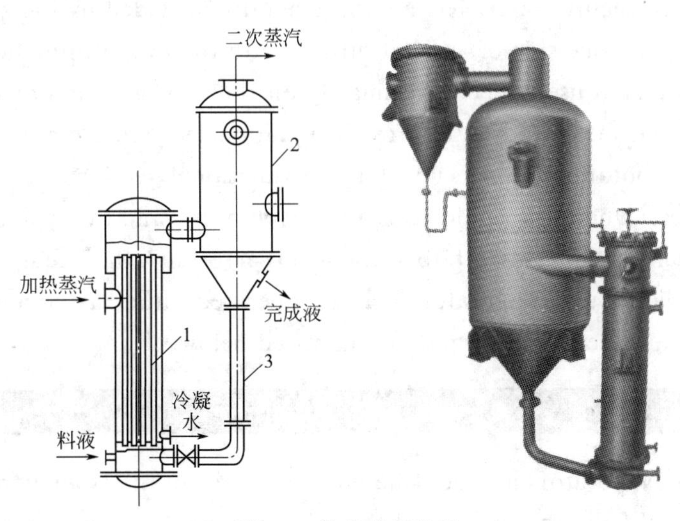

图5-4 外热式蒸发器

1—加热室；2—分离室；3—循环管

另一方面，加热管与循环管分开，由于循环管内的溶液未受蒸汽加热，其密度比加热管内的大，因此沿循环管下降而沿加热管上升形成的溶液自然循环速度较大（可达到 1.5m/s）。这种蒸发器相对于标准式而言，循环条件和分离条件都得到了改善，检修、清洗也比较方便。

（2）强制循环蒸发器（forced circulation evaporator）

自然循环蒸发器的缺点是溶液循环速度较低，传热效果差。当处理黏度大、易结垢或易结晶的溶液时，可采用图 5-5 所示的强制循环蒸发器。这种蒸发器内的溶液利用外加动力进行循环，图中表示用泵强迫溶液沿一个方向以 2～5m/s 的速度通过加热管。这种蒸发器的缺点是动力消耗大，通常为 0.4～0.8kW/m^2。

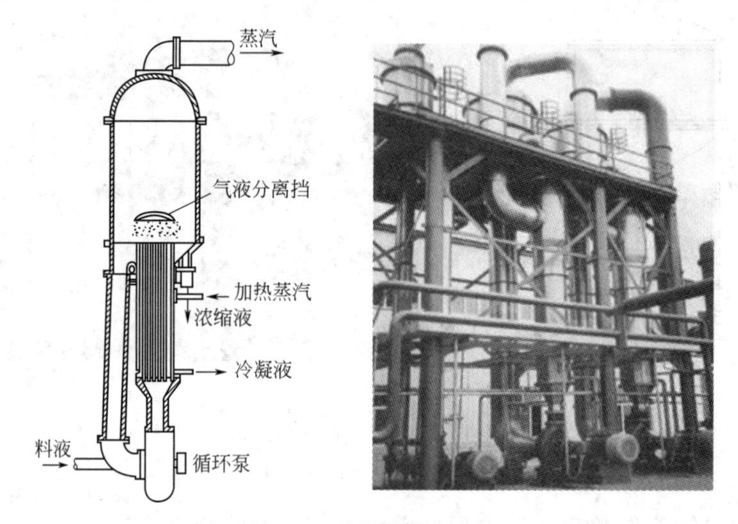

图 5-5　强制循环蒸发器

Evaporation Equipment

The evaporator has two principal functions, to exchange heat and to separate the vapor that is formed from the liquid. In evaporation, heat is added to a solution to vaporize the solvent which is usually water. The heat is generally provided by the condensation of a vapor such as steam on one side of a metal surface with the evaporating liquid on the other side. The type equipment used depends primarily on the configuration of the heat-transfer surface and on the means employed to provide agitation or circulation of the liquid.

The typical evaporator is made up of three functional sections: the heat exchanger evaporating section, where the liquid boils and evaporates, and the separator in which vapor leaves the liquid and passes off to the condenser or to other equipment. Considered piece of process plant, the evaporators may be operated either as circulation units or as once-through units. These general types are discussed below.

Circulation Evaporators

In circulation evaporators a pool of liquid is held within the equipment. Incoming feed mixes with the liquid from the pool, and the mixture passes through the tubes. Unevaporated liquid discharged from the tubes returns to the pool, so that only part of the total evaporation occurs in one pass. All forced-circulation evaporators are operated in this way; climbing-film evaporators are usually circulation units.

The thick liquor from a circulation evaporator is withdrawn from the pool. All the liquor in the pool must therefore be at the maximum concentration. Since the liquid entering tubes may contain several parts of thick liquor for each part of feed, its viscosity is high the heat-transfer coefficient tends to be low.

Circulation evaporators are not well suited to concentrating heat-sensitive liquids. With a reasonably good vacuum the temperature of the bulk of the liquid may be nondestructive, but the liquid is repeatedly exposed to contact with hot tubes. Some of the liquid, therefore, may be heated to an excessively high temperature. Although the average residence time of the liquid in the heating zone may be short, part of the liquid is retained in the evaporator for a considerable time. Prolonged heating of even a small part of a heat-sensitive material such a food can ruin the entire product.

Circulation evaporators, however, can operate over a wide range of concentration between feed and thick liquor in a single unit, and are well adapted to single effect evaporation. They may operate either with natural circulation, with the flow through the tubes induced by density differences, or with forced circulation, with flow provided by a pump.

5.1.1.2 单程型蒸发器（once-through evaporator）

上述各种蒸发器的主要缺点是加热室内物料的停留时间长，当处理物料为热敏性物质时容易变质。而在单程型蒸发器内，溶液只通过加热室一次即可达到需要的浓度，蒸发速度快（仅为数秒或十余秒），这种好的传热效果来源于溶液在加热管壁上呈薄膜状流动。

（1）升膜式蒸发器（climbing film evaporator）

升膜式蒸发器结构如图5-6所示，加热室由单根或多根垂直管组成，加热管长径之比为100～150，管径在25～50mm之间。原料液经预热必须达到沸点或接近沸点后，由加热室

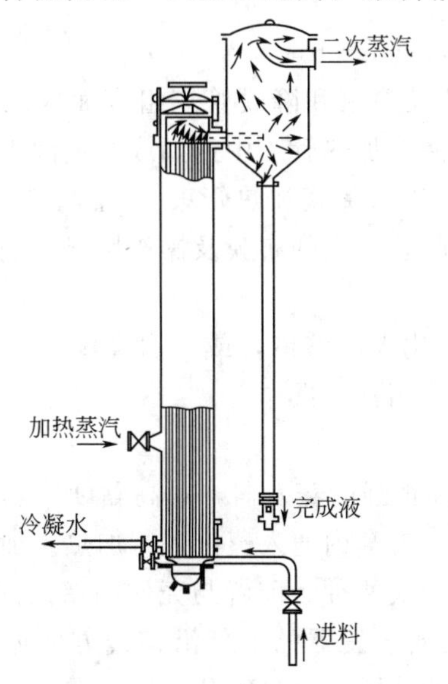

图5-6　升膜式蒸发器

底部引入管内，然后用高速上升的二次蒸汽带动沿壁面边呈膜状流动，在加热室顶部浓缩液可达到所需的浓度，气、液两相则在分离室中分开。二次蒸汽在加热管内的速度通常大于10m/s，一般为20～50m/s，减压下可高达100～160m/s或更高。

这种蒸发器适用于处理蒸发量较大的稀溶液及热敏性或易发泡的溶液，不适用于处理高黏度、有晶体析出或易结垢的溶液。

（2）降膜式蒸发器（falling-film evaporator）

蒸发浓度或黏度较大的溶液，可采用如图5-7所示的降膜式蒸发器，它的加热室与升膜式蒸发器类似。原料液由加热室的顶部加入，均匀分布后，在溶液本身重力的作用下，液体沿管内壁呈膜状流下，同时进行蒸发操作。为了使液体能在壁上均匀布膜，且防止二次蒸汽由加热管顶端直接窜出，加热管顶部必须设置性能良好的液体分布器。

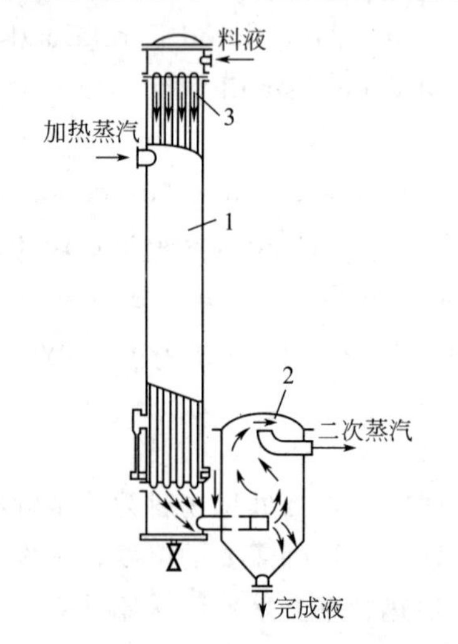

图5-7 降膜式蒸发器

降膜式蒸发器适用于处理热敏性溶液，不适用于处理高黏度、有晶体析出或易结垢的溶液。

（3）升-降膜式蒸发器（climbing-falling-film evaporator）

升-降膜式蒸发器的结构如图5-8所示，由升膜管束和降膜管束组合而成。蒸发器的底部封头内有一隔板，将加热管束均分为二。原料液在加热管1中达到或接近沸点后，引入升膜加热管束2的底部。气液混合物经管束由顶部流入降膜加热管束3，然后转入分离器4，浓缩液由分离器底部排出。溶液在升膜管束和降膜管束内的布膜及操作情况分别与前述的升膜及降膜蒸发器内的情况完全相同。

升-降膜式蒸发器一般用于浓缩过程中黏度变化大的溶液，或厂房高度有一定限制的场合。若蒸发过程中溶液的黏度变化较大，则推荐采用常压操作。

（4）刮板式蒸发器（scraper evaporator）

刮板式蒸发器如图5-9所示，它是一种适应性很强的蒸发器，对高黏度、热敏性和易结晶、易结垢的物料都适用。蒸发器外壳带有夹套，夹套内通入蒸汽进行加热。加热段的壳体内装有可旋转的叶片及刮板，可分为固定式和转子式两种。前者与壳体内壁的间隙为0.5～1.5mm，后者的间隙随转子的转速而变。原料液从蒸发器的上部沿切线方向加入，在重力和旋转刮板的刮带下，沿壳体内壁形成旋转下降的液膜，经过水分蒸发，在蒸发器底部得到完成液。结构复杂，动力消耗大是刮板式蒸发器的主要缺点。

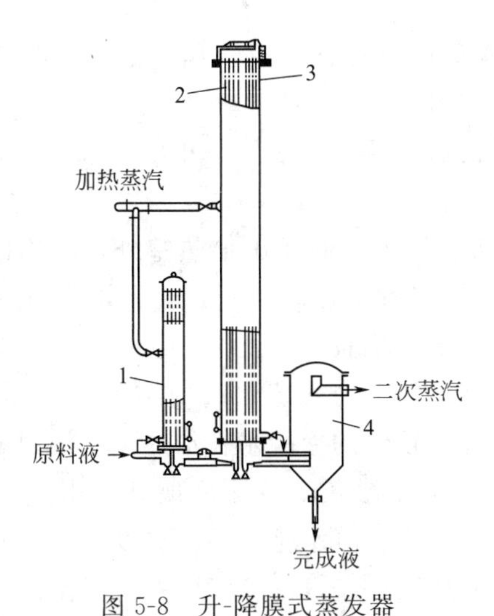

图 5-8 升-降膜式蒸发器

1—加热管；2—升膜加热管束；3—降膜加热管束；4—分离器

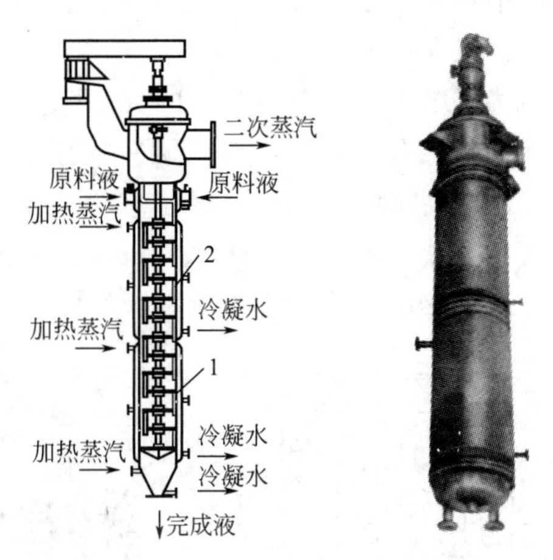

图 5-9 刮板式蒸发器

1—夹套；2—刮板

5.1.1.3 蒸发附属设备及技术新进展（new development of evaporation equipment and technology）

（1）蒸发附属装置（evaporation accessory）

① 除沫器（demister） 离开加热室的蒸汽夹带有大量的液体，夹带液体与一次蒸汽的分离主要是在蒸发室内进行。同时，还在蒸汽出口处装设了除沫器以进一步捕集蒸汽中的液体。否则，会造成产品损失，污染冷凝液或堵塞管道。除沫器的形式很多，如图 5-10 所示。它们的原理都是利用液沫的惯性以实现气液的分离。

② 冷凝器及真空装置（condenser and vacuum unit） 产生的二次蒸汽如果不加以利用，则应将其冷凝。由于二次蒸汽多为水蒸气，故一般是采用直接接触的混合式冷凝器进行冷凝。图 5-11 是逆流高位混合式冷凝器。二次蒸汽与从顶部喷淋下来的冷却水直接接触冷凝，

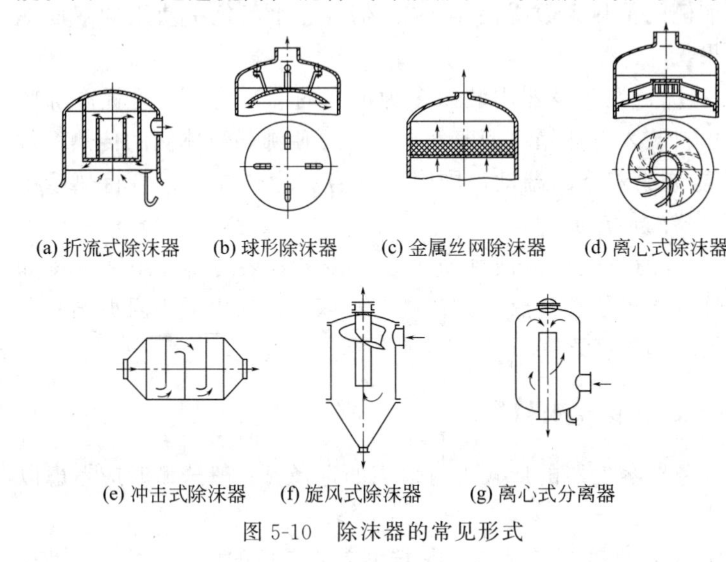

图 5-10 除沫器的常见形式

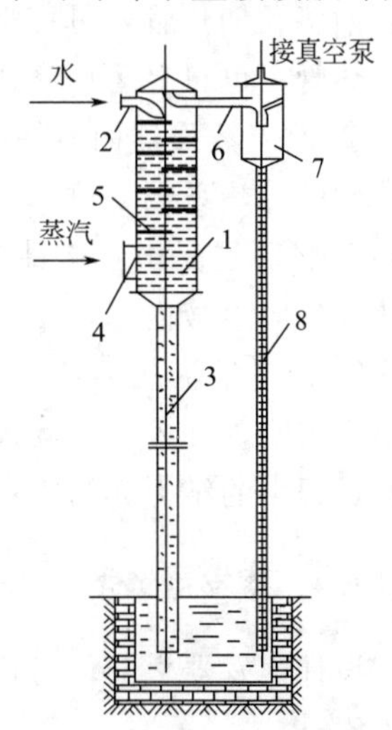

图 5-11 逆流高位混合式冷凝器

1—外壳；2—进水口；3,8—气压管；4—蒸汽进口；5—淋水板；6—不凝性气体管；7—分离器

冷凝液和水一起沿气压管流入地沟。由于冷凝器在负压下操作，故气压管必须有足够的高度，一般在10m以上，以便使液体借助自身的位能由低压排向大气。

为了维持蒸发器所需要的真空度，一般在冷凝器后设置真空装置以排出不凝性气体，常用的真空装置有水环式真空泵、喷射泵及往复式真空泵。

③ 疏水阀（steam trap） 为了防止加热蒸汽和冷凝水一起排出加热室外，在冷凝水出口管路上装有疏水阀。疏水阀的形式很多，常用的有三种：热动力式、钟形浮子式和脉冲式。其中热动力式结构简单，操作性能好，因而生产上使用较为广泛。

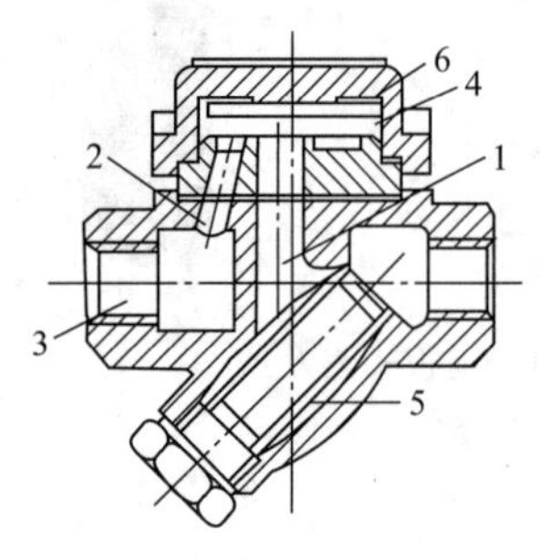

图5-12 热动力式疏水阀
1—冷凝水入口；2—冷凝水出口；3—排出管；
4—被压室；5—滤网；6—阀片

热动力式疏水阀的结构示于图5-12。温度较低的冷凝水在加热蒸汽压强的推动下流入通道，将阀片顶开，由排水孔排出。当冷凝水排尽后，温度较高的蒸汽将通过通道流入阀片背面的背压室。由于气体的黏度小、流速高，容易使阀片与阀座间形成负压，因而使阀片上面的压力高于阀片下面的压力，加上阀片自身的质量，使阀片落在阀座上，切断通道。经一定时间后，当疏水阀中积存了一定的冷凝水后，阀片又重新开启，从而实现周期性排水。

（2）蒸发技术进展（development of evaporation technology）

近年来，国内外对于蒸发器的研究十分活跃，归结起来主要有以下几个方面：

① 开发新型蒸发器 在这方面主要是通过改进加热管的表面形状来提高传热效果，例如新近发展起来的板式蒸发器，不但具有体积小、传热效率高、溶液滞留时间短等优点，而且其加热面积可根据需要而增减，拆卸和清洗方便。又如，在石油化工、天然气液化中使用的表面多孔加热管，可使沸腾溶液侧的传热系数提高10～20倍。海水淡化中使用的双面纵槽加热管，也可显著提高传热效果。

② 改善蒸发器内液体的流动状况 在蒸发器内装入多种形式的湍流构件，可提高沸腾液体侧的传热系数。例如将铜质填料装入自然循环型蒸发器后，可使沸腾液体侧的传热系数提高50%。这是由于构件或填料能造成液体的湍动，同时其本身亦为热导体，可将热量由加热管传向溶液内部，增加了蒸发器的传热面积。

③ 改进溶液的性质 近年来亦有通过改进溶液性质来改善传热效果的研究报道。例如有研究表明，加入适当的表面活性剂，可使总传热系数提高1倍以上。加入适当阻垢剂减少蒸发过程中的结垢亦为提高传热效率的途径之一。

5.1.1.4 蒸发器选型（choice of the evaporators）

设计蒸发器之前，必须根据任务对蒸发器的形式进行恰当的选择。一般选型时应考虑以下因素：

① 溶液的黏度 蒸发过程中溶液黏度变化的范围，是选型首要考虑的因素。

② 溶液的热稳定性 对长时间受热易分解、易聚合以及易结垢的溶液蒸发时，应采用滞料量少、停留时间短的蒸发器。

③ 有晶体析出的溶液　对蒸发时有晶体析出的溶液应采用外热式蒸发器或强制循环式蒸发器。

④ 易发泡的溶液　易发泡的溶液在蒸发时会生成大量层层重叠不易破碎的泡沫，充满了整个分离室后即随二次蒸汽排出，不但损失物料，而且污染冷凝器。蒸发这种溶液宜采用外热式蒸发器、强制循环式蒸发器或升膜式蒸发器。若将中央循环管式蒸发器和悬筐式蒸发器的分离室设计大一些，也可用于这种溶液的蒸发。

⑤ 有腐蚀性的溶液　蒸发腐蚀性溶液时，加热管应采用特殊材质制成，或内壁衬以耐腐蚀材料。若溶液不怕污染，也可采用直接加热蒸发器。

⑥ 易结垢的溶液　无论蒸发何种溶液，蒸发器长久使用后，传热表面上总会有污垢生成。垢层的热导率小，因此对易结垢的溶液，应考虑选择便于清洗和溶液循环速度大的蒸发器。

⑦ 溶液的处理量　溶液的处理量也是选型应考虑的因素。要求传热表面大于 $10m^2$ 时，不宜选用刮板搅拌薄膜式蒸发器；要求传热表面在 $20m^2$ 以上时，宜采用多效蒸发操作。

总之，应视具体情况选型，选用时首先保证产品质量和生产任务，然后考虑上述诸因素选用适宜的蒸发器，表 5-1 列出常见蒸发器的一些性能，以供参考。

表 5-1　常见蒸发器的主要性能列表

蒸发器形式	造价	总传热系数		溶液在管内流速/(m/s)	停留时间	完成溶液能否恒定	浓缩比	处理量	对溶液性质的适应性					
		稀溶液	高黏度						稀溶液	高黏度	易生泡沫	易结垢	热敏性	有结晶析出
标准型	最低	良好	低	0.1～0.5	长	能	良好	一般	适	适	适	尚适	尚适	稍适
外热式（自然循环）	低	高	良好	0.4～1.5	较长	能	良好	较大	适	尚适	较好	尚适	尚适	稍适
强制循环式	高	高	高	2.0～3.5	—	能	较高	大	适	好	好	适	尚适	适
升膜式	低	良好	良好	0.4～1.0	短	较难	高	大	适	尚适	好	尚适	良好	不适
降膜式	低	高	高	0.4～1.0	短	尚能	高	大	较适	好	适	不适	良好	不适
刮板式	最高	高	高	—	短	尚能	高	较小	较适	好	较好	不适	良好	不适
浸没燃烧	低	高	高	—	短	较难	良好	较大	适	适	适	适	不适	适

5.1.2 单效蒸发（Single-effect evaporation）

5.1.2.1 单效蒸发物料衡算与热量衡算

针对图 5-1 所示的蒸发器，根据物料衡算、热量衡算式和传热速率方程可求得单效蒸发操作中的水蒸发量、加热蒸汽消耗量和蒸发器传热面积。

（1）水的蒸发量

做溶质的物料衡算：

$$Fx_0=(F-W)x_1 \tag{5-1}$$

由上式得到水的蒸发量：

$$W=F\left(1-\frac{x_0}{x_1}\right) \tag{5-2}$$

（2）加热蒸汽消耗量

蒸发操作中，加热蒸汽的热量一般包括将溶液加热至沸腾，将水分蒸发为蒸汽以及向周围散失的热量。对图 5-1 所示的蒸发器进行焓衡算，得到：

$$D=\frac{WH'+(F-W)h_1-Fh_0+Q_L}{H-h_w} \tag{5-3}$$

式中　Q_L——热损失，kJ/h。

若加热蒸汽的冷凝液在蒸汽的饱和温度下排出，则有 $H-h_w=r$，式（5-3）变为：

$$D=\frac{WH'+(F-W)h_1-Fh_0+Q_L}{r} \tag{5-4}$$

式中　r——加热蒸汽的汽化潜热，kJ/kg。

假设 $H'-h_1\approx r'$，则热量衡算式可简化为：

$$D=\frac{Wr'+FC_{p_0}(t_1-t_0)+Q_L}{r} \tag{5-5}$$

式中　r'——二次蒸汽的汽化潜热，kJ/kg；

C_{p0}——原料液的比热容，kJ/(kg·℃)。

式（5-5）说明了加热蒸汽的热量包括将溶液加热至沸腾、将水分蒸发为蒸汽以及向周围散失的热量。若原料液预热至沸点再进入蒸发器，且忽略热损失，该式可简化为：

$$D=\frac{Wr'}{r} \tag{5-6}$$

或：

$$e=\frac{D}{W}=\frac{r'}{r} \tag{5-6a}$$

式中　e——蒸发 1kg 水分时，加热蒸汽的消耗量，称为单位蒸汽消耗量，kg/kg。

由于蒸汽的汽化潜热随压强变化不大，即 $r\approx r'$，故单效蒸发操作中 $e\approx 1$，即每蒸发 1kg 的水分约消耗 1kg 的加热蒸汽。但实际蒸发操作中因有热损失等的影响，e 值约为 1.1 或更大些，所以 e 值是衡量蒸发装置经济程度的指标。

5.1.2.2　蒸发器的传热面积

蒸发器的传热过程可认为是恒温传热，可用恒温传热方程式求传热面积 S_o，即：

$$S_o=\frac{Q}{K_o(T-t)} \tag{5-7}$$

式中　S_o——蒸发器的传热外表面积，m^2；

K_o——基于外表面的总传热系数，W/(m^2·℃)；

Q——蒸发器的热负荷，即蒸发器的传热速率，W；

t——操作条件下溶液的沸点，℃。

Evaporator Capacity

The principal measures of the performance of a steam-heated tubular evaporator are the capacity and the economy. Capacity is defined as the number of kilograms of water vaporized per hour. Economy is the number of kilograms vaporized per kilogram of steam fed to the unit. In a single-effect evaporator the economy is nearly always less than 1, but in multiple-effect equipment it may be considerably greater. The steam consumption, in kilograms per hour, is also important. It equals the capacity divided by the economy.

The rate of heat transfer is the product of three factors: the area of the heat-transfer surface, the overall heat-transfer coefficient and the overall temperature drop.

If the feed to the evaporator is at the boiling temperature corresponding to the absolute pressure in the vapor space, almost all the heat transferred through the heating surface is available for evaporation, and the capacity is nearly proportional to the rate of heat transfer.

As evaporation proceeds the boiling point rises and some sensible heat is required, but the gain in sensible heat is usually small in comparison with the heat of vaporization. If the feed is cold, the heat required to heat it to its boiling point may be quite large and the capacity for a given value of the rate of heat transfer is reduced accordingly, as heat used to heat the feed is not available for evaporation. Conversely, if the feed is at a temperature above the boiling point in the vapor space, a portion of the feed evaporates spontaneously by adiabatic equilibration with the vapor-space pressure and the capacity is greater than that corresponding to the rate of heat transfer. This process is called flash evaporation.

The actual temperature drop across the heating surface depends on the solution being evaporated, the difference in pressure between the steam chest and the vapor space above the boiling liquid, and the depth of liquid over the heating surface. In some evaporators the velocity of the liquid in the tubes also influences the temperature drop because the frictional loss in the tubes increases the effective pressure of the liquid. When the solution has the characteristics of pure water, its boiling point can be read from steam tables if the pressure is known, as can the temperature of the condensing steam. In actual evaporators, however, the boiling point of a solution is affected by two factors, boiling point elevation and liquid head.

5.1.3 多效蒸发（Multiple-effect evaporators）

蒸发操作要蒸发溶液中的大量水分，就要消耗大量的加热蒸汽。为了节约蒸汽，可采用多效蒸发。即二次蒸汽的压力和温度虽比原来所用加热蒸汽的压力和温度低，但还可以用作另一蒸发器的加热剂。这时后一蒸发器的加热室就相当于前一蒸发器的冷凝器。按此原则顺次连接起来的一组蒸发器就称为多效蒸发器。每一个蒸发器称为一效，通入加热蒸汽的蒸发器称为第一效，用第一效的二次蒸汽作为加热剂的蒸发器称为第二效，用第二效的二次蒸汽作为加热剂的蒸发器称为第三效，依此类推。

各效的操作压力是自动分配的，为了得到必要的传热温度差，一般多效蒸发器的末效或最后几效总是在真空下操作的。显然，各效的压力和沸点是逐渐降低的。由于各效（末效除外）的二次蒸汽都作为下一效蒸发器的加热蒸汽，故提高了生蒸汽的利用效率，即提高了经济效益。

5.1.3.1 多效蒸发的流程

根据原料液的加入方法不同，多效蒸发操作的流程可分为顺流、逆流和平流三种。

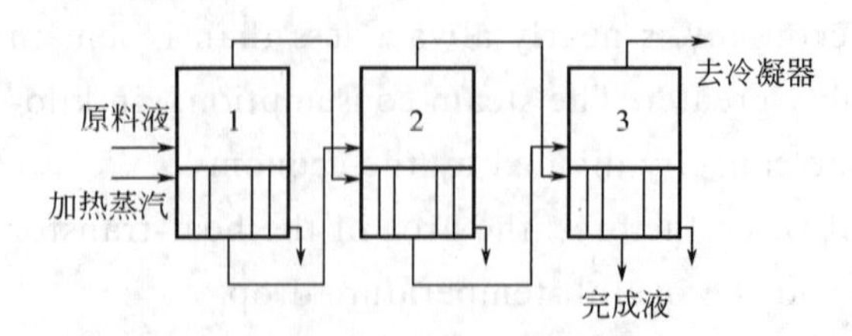

图 5-13　顺流加料的三效蒸发装置流程示意图

（1）顺流加料（也称并流）(forward feed multiple-effect evaporator)

顺流加料的流程如图 5-13 所示，是工业上常用的进料方法。原料液和蒸汽都加入第一效，然后溶液顺次流过第一效、第二效和第三效，由第三效取出浓缩液；而加热蒸汽在第一效加热室中冷凝后，经冷凝水排除器排出；由第一效溶液中蒸发出来的二次蒸汽送入第二效加热室供加热用；而第二效的二次蒸汽送入第三效加热室，第三效的二次蒸汽送入冷凝器中冷凝后排出。

顺流加料的优点是：因各效的压力依次降低，溶液可以自动地由前一效流入后一效，无需用泵输送；因各效溶液的沸点依次降低，前一效的溶液流入后一效时，将发生自蒸发而蒸发出更多的二次蒸汽。

顺流加料的缺点是：随着溶液的逐效蒸浓，温度逐效降低，溶液的黏度则逐渐提高，致使传热系数逐渐减小。因此，在处理黏度随浓度的增加而迅速加大的溶液时，不宜采用顺流加料的操作方法。

Forward-feed multiple-effect evaporator

A simplified diagram of a forward-feed multiple-effect evaporation system is shown in Fig. 5. 13. The usual method of feeding a multiple-effect evaporator is to pump the thin liquid into the first effect and send it in turn through the other effects, the concentration of the liquid increases from the first effect to the last. This pattern of liquid flow is the simplest. It requires a pump for feeding dilute solution to the first effect, since this effect is often at about atmospheric pressure, and a pump to remove thick liquor from the last effect. The transfer from effect to effect, however, can be done without pumps, since the flow is in the direction of decreasing pressure, and control valves in the transfer line are all that is required.

（2）逆流加料

图 5-14 是三效逆流加料流程。原料液由末效进入，用泵依次输送至前效，最后浓缩液由第一效的底部取出，而加热蒸汽的流向仍是由第一效顺序至末效。因蒸汽和溶液的流动方向相反，故称为逆流进料法。

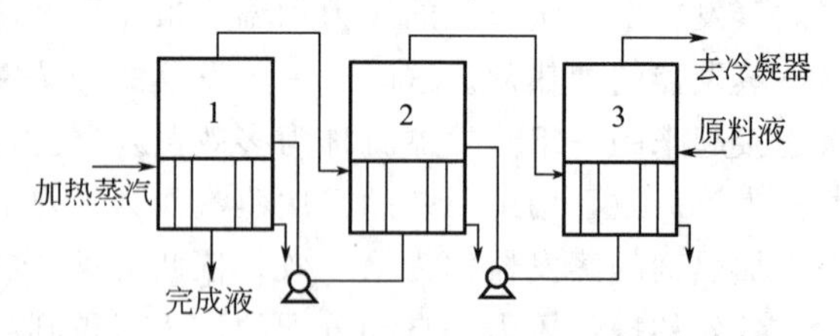

图 5-14　逆流加料的三效蒸发装置流程示意图

逆流加料法蒸发流程的主要优点是：溶液的浓度沿着流动方向不断提高，同时温度也逐渐上升，因此各效溶液的黏度较为接近，使各效的传热系数也大致相同。其缺点是：效间的溶液需用泵输送，能量消耗较大，且因各效的加料温度均低于沸点，与顺流加料法相比，产生的二次蒸汽量也较少。

一般来说，逆流加料法适宜处理黏度随温度和浓度变化较大的溶液，而不宜于处理热敏性的料液。

Backward-feed multiple-effect evaporators

Another common method is backward-feed, in which dilute liquid is fed to the last effect and then pumped through the successive effects to the first. This method requires a pump between each pair of effects in addition to the thick-liquor pump, since the flow is from low pressure to high pressure. Backward feed often gives a higher capacity than forward feed when the thick liquor is viscous, but it may give a lower economy than forward feed when the feed liquor is cold.

(3) 平流加料（parallel-feed multiple-effect evaporators）

平流加料操作流程如图5-15所示。在每一效中都送入原料液，放出浓缩液，而蒸汽的流向仍是由第一效流至末效。这种进料方法主要处理在蒸发过程中有晶体析出的情况。

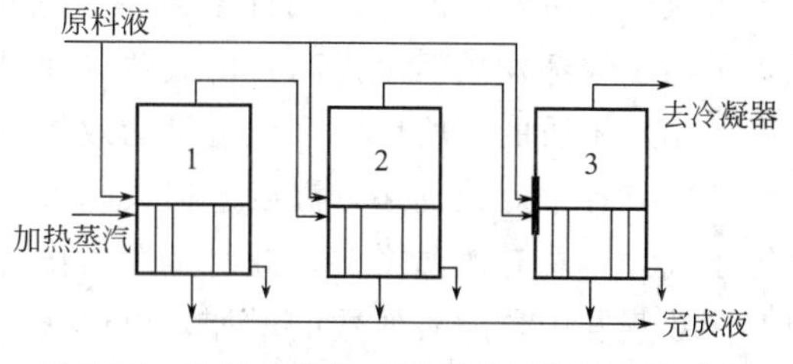

图5-15 平流加料的三效蒸发装置流程示意图

Multiple-Effect Calculations

In designing a multiple-effect evaporator the results usually desired are the amount of steam consumed, the area of the heating surface required, the approximate temperatures in the various effects, and the amount of vapor leaving the last effect. As in a single-effect evaporator, these quantities are found from material balances, enthalpy balances, and the capacity equation. In a multiple-effect evaporator however, a trial-and-error method is used in place of a direct algebraic solution.

Another method of calculation is as follows:

(1) Assume values for the boiling temperatures in the first and second effects.

(2) From enthalpy balances find the rates of steam flow and of liquid from effect to effect.

(3) Calculate the heating surface needed in each effect from the capacity equations.

(4) If the heating areas so found are not nearly equal, estimate new values for the boiling temperatures and repeat items 2 and 3 until the heating surfaces are equal.

In practice these calculation are done by computer.

5.1.3.2 多效蒸发的适宜效数

若多效蒸发和单效蒸发的操作条件相同，即第一效（或单效）的加热蒸汽压强和冷凝器的操作压强各自相同，则多效蒸发的温度差因经过多次的损失，使总温度差损失较单效蒸发时大。多效蒸发提高了加热蒸汽的利用效率，即经济效益。对于蒸发等量的水分而言，采用多效蒸发时所需的加热蒸汽较单效蒸发时为少。在工业生产中，若需蒸发大量的水分，宜采用多效蒸发。

多效蒸发的经济性，需从生产能力和生产强度两个方面来均衡考虑。

蒸发器的生产能力是指单位时间内蒸发的水分量，即蒸发量。通常可认为蒸发量是与蒸发器的传热速率成正比。由传热速率方程式知：

单效：$Q=KS\Delta t$

三效：$Q_1=K_1S_1\Delta t_1$，$Q_2=K_2S_2\Delta t_2$，$Q_3=K_3S_3\Delta t_3$

若各效的总传热系数取平均值 K，且各效的传热面积相等，则三效的总传热速率为：

$$Q=Q_1+Q_2+Q_3\approx KS(\Delta t_1+\Delta t_2+\Delta t_3)=KS\sum\Delta t$$

当蒸发操作中没有温度差损失时，由上式可知，三效蒸发和单效蒸发的传热速率基本上相同，因此生产能力也大致相同。但是，两者的生产强度（单位传热面积的蒸发量）是不相同的，三效蒸发时的生产强度约为单效蒸发时的三分之一。实际上，由于多效蒸发时的温度差损失较单效蒸发时的大，因此多效蒸发时的生产能力和生产强度均较单效蒸发时小。可见，采用多效蒸发虽然可提高经济效益（即提高加热蒸汽的利用效率），但降低了生产强度，两者是相互矛盾的。所以，多效蒸发的效数应权衡决定。

蒸发装置中效数越多，温度差损失越大，而且某些浓溶液的蒸发还可能发生总温度差损失等于或大于总有效温度差，此时蒸发操作就无法进行，所以多效蒸发的效数应有一定的限制。多效蒸发中，随着效数的增加，单位蒸汽的消耗量减少，使操作费用降低；而且效数越多，装置的投资费用也越大。由表 5-2 给出了效数对最小单位蒸汽消耗量 $(D/W)_{\min}$ 的影响。可以看出，随着效数的增加，虽然 $(D/W)_{\min}$ 不断减小，但减小的速度越来越慢。例如，由单效增至双效，可节省的生蒸汽量约为 50%，而由四效增至五效，可节省的生蒸汽量约为 10%。综合上面分析可知，最佳效数要通过经济权衡决定，单位生产能力的总费用为最低时的效数即为最佳效数。

表 5-2　单位蒸汽消耗量

效数	单效	双效	三效	四效	五效
$(D/W)_{\min}$	1.1	0.57	0.4	0.3	0.27

通常，工业中的多效蒸发操作的效数并不是很多。例如，NaOH、NH_4NO_3 等电解质溶液，由于其沸点升高（即温度差损失）较大，故取 2～3 效；对于非电解质溶液，如有机溶液等，其沸点升高较小，所用效数可取 4～6 效；海水淡化的温度差损失为零，故蒸发装置可达 20～30 效之多。

蒸发器的设计任务中往往只给出溶液性质、要求达到的浓缩液浓度及可提供的加热蒸汽压强等。设计者首先应根据溶液性质选定蒸发器形式、冷凝器压强、进料方式及最佳效数（最佳效数由设备投资费、折旧费及经常操作费间的经济衡算确定），再根据经验数据选出或算出总传热系数后，按前述方法算出传热面积，最后再选定或算出蒸发器的主要工艺尺寸（包括加热管尺寸及管数、循环管尺寸、加热室外壳直径、分离室尺寸及附属设备的计算或选用）。

Optimum Number of Effects

The cost of each effect of an evaporator per square meter or square foot of surface is a function of its total area and decreases with area, approaching an asymptote for very large installations. Thus the investment required for an N-effect evaporator is about N times that for a single-effect evaporator of the same capacity. The optimum number of effects must be found from an economic balance between the savings in steam obtained by multiple-effect operation and the added investment required.

5.1.3.3 蒸发过程的生产能力和生产强度（production capacity and production intensity of evaporation process）

（1）生产能力（production capacity）

蒸发器的生产能力是用单位时间内蒸发的水分量，即蒸发量来表示，其单位为 kg/h。蒸发器生产能力的大小取决于通过蒸发器传热面的传热速率 Q，因此也可以用蒸发器的传热速率来衡量其生产能力。

根据传热速率方程，单效蒸发时的传热速率为：

$$Q=KS\Delta t=KS(T-t_1) \tag{5-8}$$

若蒸发器的热损失可忽略不计，且原料液在沸点下进入蒸发器，则由蒸发器的热量衡算可知，通过传热面所传递的热量全部用于蒸发水分，这时蒸发器的生产能力随传热速率的增大而增大。蒸发器的生产能力，还与原料液的入口温度有关。若原料液在低于沸点下进料，则需要消耗部分热量将冷溶液加热至沸点，因而降低了蒸发器的生产能力；若原料液在高于沸点下进入蒸发器，则由于部分原料液的自动蒸发，使得蒸发器的生产能力有所增加。

（2）生产强度（production intensity）

蒸发器的生产强度是评价蒸发器性能的重要指标。蒸发器的生产强度 U 是指单位传热面积上单位时间内所蒸发的水量，其单位为 $kg/(m^2 \cdot h)$，即：

$$U=\frac{W}{S} \tag{5-9}$$

若为沸点进料，且忽略蒸发器的热损失，将式（5-8）和式（5-9）代入上式得：

$$U=\frac{Q}{Sr'}=\frac{K}{\Delta tr'} \tag{5-10}$$

由式（5-10）可以看出，欲提高蒸发器的生产强度，必须设法提高蒸发器的总传热系数和传热温差。

（3）提高生产强度的途径（ways to increase production intensity）

传热温差 Δt 主要取决于加热蒸汽的压力和冷凝器的真空度。加热蒸汽压力越高，其饱和温度也越高，但是加热蒸汽压力常受具体的供气条件限制，其压力范围一般为 300～500kPa（绝压），压力高时为 600～800kPa（绝压）。若提高冷凝器的真空度，使溶液的沸点降低，则也可以增大温差。但是这样做的结果，不仅增加真空泵的功率消耗，而且还会因溶液的沸点降低，使其黏度增大，导致沸腾传热系数下降。因此一般冷凝器中的压力不低于 10～20kPa。此外，为了控制沸腾操作处于泡核沸腾区，也不宜采用过高的传热温差。由以上分析可知，传热温差的提高是有一定限度的。

因此，增大总传热系数是提高蒸发器生产强度的主要途径。总传热系数值取决于传热面两侧的对流传热系数和污垢热阻。现分析如下：

① 蒸汽冷凝传热系数 α_o 通常比溶液沸腾传热系数 α_i 大，即总传热热阻中，蒸汽冷凝侧的热阻较小。不过在蒸发器的设计和操作中，必须考虑蒸汽中不凝性气体的及时排除；否则，其热阻将大大地增加，导致总传热系数下降。

② 管内溶液侧的污垢热阻往往是影响总传热系数的重要因素。尤其在处理结垢和有结晶析出的溶液时，在传热面上很快形成垢层，使 K 值急剧下降。为了减小垢层热阻，蒸发

器必须定期清洗。减小垢层热阻的措施还有：选用对溶液扰动程度较大的强制循环蒸发器等；或是在溶液中加入晶种或微量阻垢剂，以阻止在传热面上形成垢层。

③ 管内溶液沸腾传热系数 α_i 是影响总传热系数的主要因素。影响沸腾传热系数的因素很多，如溶液的性质、蒸发操作条件及蒸发器的类型等。故必须根据蒸发任务的具体情况，选定适宜的操作条件和蒸发器的形式，才能提高蒸发的生产强度。

5.1.4 蒸发操作的其他节能措施（Other energy saving measures for evaporation operations）

蒸发过程是一个消耗热能较多的单元操作，因而有必要介绍它的常用节能措施。除采用多效蒸发可提高热能的利用效率外，工业上还常采用以下方法。

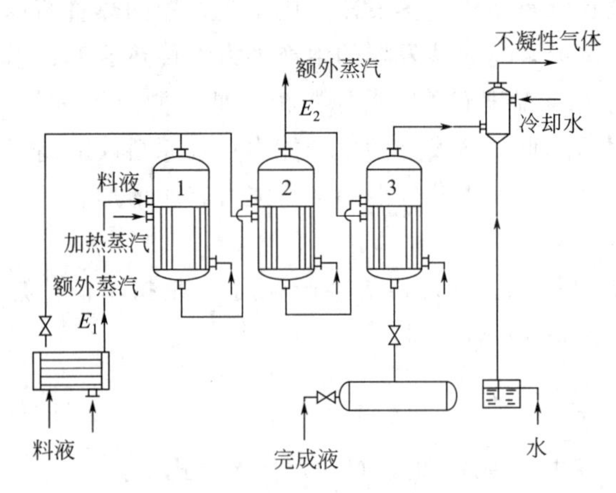

图 5-16 引出额外蒸汽的蒸发流程

（1）抽取额外蒸汽

在有些场合中，将多效蒸发器中的某一效的二次蒸汽引出一部分，作为其他换热器的加热剂，这部分引出的蒸汽称为额外蒸汽。能否引出额外蒸汽，关键是看二次蒸汽的温度（即能位）。多效蒸发的末效大多处于负压，而且绝对压力较低，故末效二次蒸汽难以再利用，往往可在前几效引出额外蒸汽。它的流程如图 5-16 所示。目前国内制糖厂中已有所应用。

（2）二次蒸汽的再压缩

在单效蒸发中，如将二次蒸汽压缩，则其温度升高，与沸腾的料液间形成足够的传热温差，送回加热器冷凝后可放出大量潜热。如此用少量的外加压缩功可回收二次蒸汽的潜热，流程如图 5-17 所示。在连续操作过程中，开始时需供给加热蒸汽，当产生二次蒸汽使压缩机运行后，几乎无需补充蒸汽。

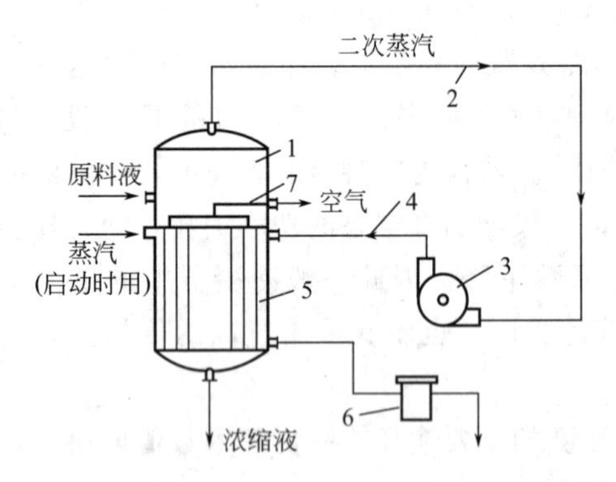

图 5-17 热泵蒸发操作简图

1—蒸发器；2,4—二次蒸汽管；3—压缩机；5—加热室；6—疏水阀；7—不凝性气体放空管

Vapor Recompression

The energy in the vapor evolved from a boiling solution can be used to vaporize more water, provided there is a temperature drop for heat transfer in the desired direction. In a multiple-effect evaporator this temperature drop is created by progressively lowering the pressures. The desired driving force can also be obtained by increasing the pressure (and, therefore, the condensing temperature) of the evolved vapor by mechanical or thermal recompression. The compressed vapor is then condensed in the steam chest of the evaporator from which it came.

Thermal Recompression

In a thermal recompression system the vapor is compressed by acting on it with high-pressure steam in a jet ejector. This results in more steam than is needed for boiling the solution, so that excess steam must be vented or condensed. The ratio of motive steam to the vapor from the solution depends on the evaporation pressure; for many low-temperature operations, with steam at 8 to 10 atm pressure, the ratio of steam required to the mass of water evaporated is about 0.5.

Since steam jets can handle large volumes of low-density vapor, thermal recompression is better suited than mechanical recompression to vacuum evaporation. Jets are cheaper and easier to maintain than blowers and compressors. The chief disadvantages of thermal recompression are the low efficiency of the jets and lack of flexibility in the system toward changed operating conditions.

(3) 冷凝水的利用

蒸发装置消耗的蒸汽是可观的，因此会产生大量冷凝水。冷凝水排出加热室后，除可用于预热料液外，还可使其减压进行自蒸发。自蒸发产生的蒸汽与二次蒸汽混合后一同进入下一效蒸发器的加热室，使得冷凝水的显热得到部分回收利用，流程如图 5-18 所示。

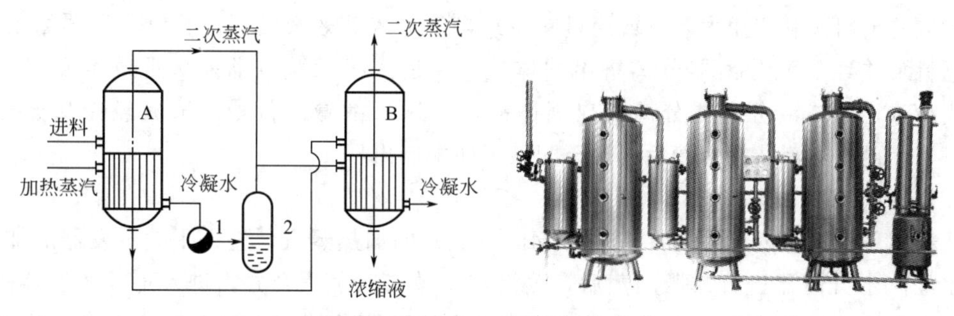

图 5-18 冷凝水自蒸发的应用

工业上还常将冷的料液与热的浓缩液或蒸汽冷凝液进行热交换，以提高料液进入蒸发器的温度，因此可减少蒸发器的传热面积和蒸汽消耗量。

5.1.5 蒸发应用

5.1.5.1 烧碱增浓

在烧碱（NaOH）的生产过程中，由隔膜电解槽阴极室流出的电解液中含氢氧化钠10%左右，而氯化钠却高达 15%～16%。要得到符合商品规格的烧碱（NaOH≥30%，

NaCl≤4.7%），必须进行浓缩。在浓缩过程中，同时将析出的盐进行分离和回收。因此，电解液的蒸发是烧碱生产系统的一个重要环节，它的主要任务有如下几方面：

① 浓缩　将电解液中的 NaOH 含量从 10%浓缩到 30%或 45%。

② 分盐　将浓缩过程中析出的结晶盐分离。

③ 回收盐　将分离碱液后的固体盐溶解成接近饱和的盐水，送化盐工序重新利用。

电解液蒸发是一个耗能较多的过程，其能量消耗约占烧碱生产综合能耗的 30%左右。因此电解液蒸发的运行情况和生产技术，直接影响整个氯碱系统的能耗水平和经济效益。

图 5-19 为常用的三效顺流碱液蒸发系统。

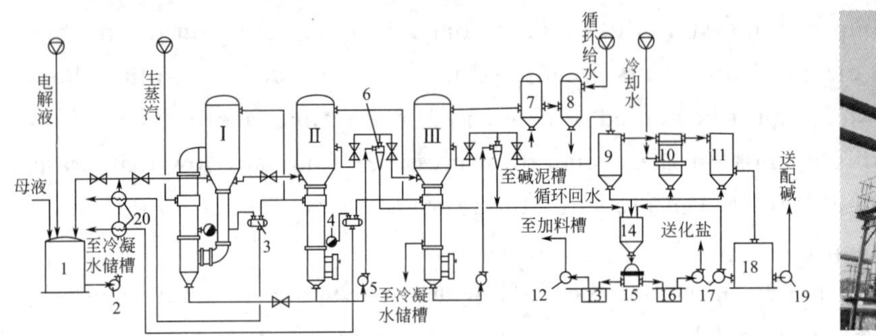

图 5-19　三效顺流部分强制循环蒸发工艺流程

1—电解液储槽；2—加料泵；3—汽水分离器；4—强制循环泵；5—过料泵；6—旋液分离器；7—捕沫器；8—大气冷凝器；9—浓碱高位槽；10—碱液冷却器；11—中间槽；12—母液泵；13—母液槽；14—碱泥槽；15—离心机；16—盐水回收槽；17—回收盐水泵；18—澄清桶；19—打碱泵；20—预热器；Ⅰ、Ⅱ、Ⅲ—蒸发器

（1）碱系统

加料泵将电解液储槽内的电解液送入预热器，预热至 100℃以上后进入Ⅰ效蒸发器。Ⅰ效蒸发器的出料液利用压力差（或用过料泵）自动进入Ⅱ效蒸发器。Ⅱ效蒸发器的出料液，利用过料泵并经旋液分离器分离析出固体盐后，送入Ⅲ效蒸发器。从Ⅲ效蒸发器出来的 30%成品碱，经旋液分离器分离析出固体盐后，送入浓碱高位槽。浓碱经冷却器冷却至 45℃以下，再经过澄清桶澄清后，送至配碱工序配制出厂。

（2）蒸汽系统

从蒸汽总管道来的压力为 0.6～0.8MPa（表）的加热蒸汽，进入Ⅰ效蒸发器的加热室。冷凝水经汽水分离器进行减压闪蒸后，蒸汽与Ⅰ效二次蒸汽合并，进入Ⅰ效蒸发器的加热室，作为Ⅱ效的加热蒸汽。冷凝水则流经两段电解液预热器，预热电解液后送往冷凝水储槽。Ⅱ效蒸发器的冷凝水，经汽水分离器及减压闪蒸后，蒸汽部分Ⅱ效的二次蒸汽合并后作为Ⅲ效蒸发器的加热蒸汽。未汽化的冷凝水流经一段电解液预热器后，送往冷凝水储槽。Ⅰ效蒸发器内的冷凝水直接流至冷凝水储槽。Ⅰ效的二次蒸汽经过捕沫器分离出夹带的碱沫后，由大气冷凝器冷凝后排入下水池，并借此使Ⅰ效蒸发室获得负压。

（3）盐泥系统

从Ⅰ效、Ⅱ效旋液分离器分离出来的盐泥，送至碱泥高位槽。在浓碱高位槽内经沉清、冷却后的盐泥也放至碱泥高位槽。经离心机分离后，母液送回电解液储槽，固体盐经洗涤后用蒸汽冷凝水溶解成接近饱和的含碱盐水，送化盐工段。

三效顺流工艺不但适用于生产浓度为30%的烧碱，也适用于生产浓度为42%的烧碱。为改变蒸发器的传热状况，大部分工厂在Ⅱ、Ⅲ效安装了强制循环泵，这样蒸发器的生产能力就有较大的提高。在三效顺流蒸发工艺中，两次利用了二次蒸汽，只有Ⅲ效的二次蒸汽被冷凝排放（它仅占总蒸发水量的1/3左右）。故该工艺的热量利用率高，蒸汽消耗低。在生产30%碱时，每吨100%烧碱的蒸发汽耗仅2.8～3.0t左右，在生产42%碱时的汽耗也只有3.5～3.7t。三效顺流工艺操作容易制，对设备、材料也无特殊要求，故而应用较为广泛。

5.1.5.2　废水处理

废水的蒸发法处理是指加热废水，使水分子大量汽化逸出，废水中的溶质被浓缩以便进一步回收利用，水蒸气冷凝后可获得纯水的一种物理化学过程。废水进行蒸发处理时，既有传热过程，又有传质过程。根据蒸发前后的物料和热量衡算原理，可以推算出蒸发操作的基本关系式。

图5-20为浸没燃烧蒸发器的构造示意图。它是热气与废水直接接触式蒸发器，以高温烟气为热源。燃料（煤气或油）在燃料室中燃烧产生的高温烟气（约1200℃）从浸于废水中的喷嘴喷出，加热和搅拌废水，二次蒸汽和燃烧空气由蒸发器顶出口排出，浓缩液由蒸发器底用空气喷射泵抽出。浸没燃烧蒸发器结构简单，传热效率高，适用于蒸发强腐蚀性和易结垢的废液，但不适于热敏性物料和易被烟气污染的物料蒸发。

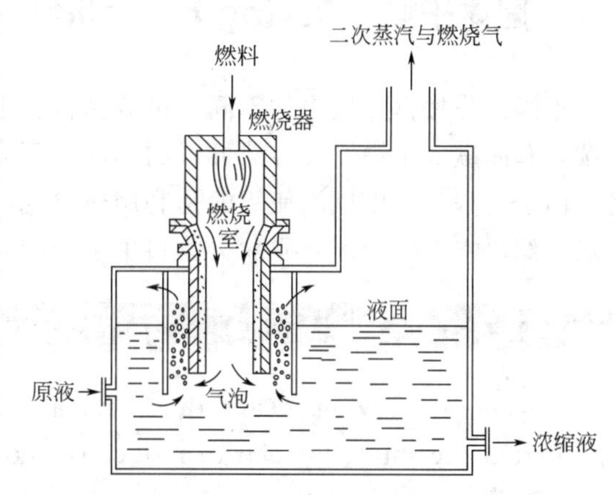

图5-20　浸没燃烧蒸发器结构示意图

蒸发法在废水处理中的应用主要包括如下几个方面。

（1）浓缩高浓度有机废水

高浓度有机废水，如酒精废液、造纸黑液、酿酒业蒸馏残液等可用蒸发法浓缩，然后将浓缩液加以综合利用或焚化处理。例如，在酸法纸浆厂，将亚硫酸盐纤维素废液蒸发浓缩后，可用作道路黏结剂、砂模减水剂、鞣剂和生产杀虫剂的原料等。

（2）浓缩放射性废水

废水中绝大多数放射性污染物质是不挥发的，可用蒸发法浓缩，然后将浓缩液密封存放，让其自然衰变。一般经二次蒸发，废水体积可减小为原来的1/500～1/200，这样大大减少了昂贵的储罐容积，从而降低了处理费用。

（3）浓缩废酸、废碱

酸洗废液可用浸没燃烧法进行浓缩和回收。例如，某钢铁厂的废酸液中含H_2SO_4 100～200g/L、$FeSO_4$ 220～250g/L，经浸没燃烧蒸发浓缩后，母液含H_2SO_4增至600g/L，而$FeSO_4$减至60g/L。

5.1.5.3　生物溶液的蒸发

（1）生物溶液的特性

生物溶液具有如下特性：

① 大多数生物溶液（如果汁及中药浸出液）为热敏性物料，且其黏度随着溶液中溶质含量的增加而显著（或急剧）加大。溶液中的溶质具有粘连到传热壁面上的趋向，造成局部过热，从而导致溶液中有效成分的破坏，甚至焦化。

② 在蒸发过程中，有些物料中细菌生长很快，而且附着在设备壁面上。因此，要求设备应便于清洗。

③ 溶液的沸点升量小，常可忽略。

（2）蒸发设备和操作条件的选择

物料在蒸发中受损的程度取决于操作温度和受热时间的长短。为了降低操作温度，宜采用真空蒸发；为了缩短受热时间，设备必须提供很高的传热速率。例如，果汁的蒸发大都选用单程型（如降膜式、搅拌薄膜型等）、强制循环型（如垂直长管强制循环、搅拌强制循环）以及热泵循环蒸发器，以实现传热表面上物料的高速循环。

5.2 固体干燥（Drying of Solid）

干燥操作按操作压力的不同，可分为常压干燥和真空干燥；按操作方式的不同，可分为连续式和间歇式干燥；按传热方式的不同，可分为对流式、传导式、辐射式及介电加热式干燥。由于过程工业中广泛使用的是利用热空气与湿物料做相对运动的对流式干燥，所以本章重点介绍以空气为干燥介质除去物料中水分的对流式干燥操作。

Drying

In general, drying means the removal of relatively small amounts of water or other liquid from the solid material to reduce the content of residual liquid to an acceptably low value. Drying is usually the final step in a series of operations, and the product from a dryer is often ready for final packaging.

The discussions of drying in this chapter are concerned with the removal of water from process materials and other substances. The term “drying” is also used to refer to removal of other organic liquid, such as benzene or organic solvents, from solids. Many of the types of equipment and calculation methods discussed for removal of water can also be used for removal of organic liquids.

Water or other liquids may be removed from solids mechanically by presses or centrifuges or thermally by vaporization. This chapter is restricted to drying by thermal vaporization. It is generally cheaper to remove liquid mechanically than thermally, and thus it is advisable to reduce the liquid content as much as practicable before feeding the material to a heated dryer.

Drying generally means removal of relatively small amounts of water from material. Evaporation refers to removal of relatively large amounts of water from material. In evaporation the water is removed as vapor at its boiling point. In drying the water is usually removed as a vapor by air.

In some cases water may be removed mechanically from solid materials by means of presses, centrifuging, and other methods. This is cheaper than drying by thermal means for removal of water, which will be discussed here. The moisture content of the final dried product varies depending upon the type of product. Dried salt contains about 0.5% water, coal about 4%, and many food products about 5%. Drying is usually the final processing step before packaging and makes many materials, such as soap powders and dyestuffs, more suitable for handling.

The liquid content of a dried substance varies from product to product. Occasionally the product contains no liquid and is called bone-dry. More commonly, the product does contain some liquid. Dried table salt, for example, contains about 0.5 percent water, dried coal about 4 percent, and dried casein about 8 percent. Dryness is a relative term, and drying means merely reducing the moisture content from an initial value to some acceptable final value.

Drying or dehydration of biological materials, especially foods, is used as a preservation technique. Microorganisms that cause food spoilage and decay cannot grow and multiply in the absence of water. Also, many enzymes that cause chemical changes in food and other biological materials cannot function without water. When the water content is reduced below about 10 wt%, the microorganisms are not active. However, it is usually necessary to lower the moisture content below 5wt% in foods to preserve flavor and nutrition. Dried foods can be stored for extended periods of time.

5.2.1 干燥基本概念（Basic conception of drying）

图5-21所示的气流干燥器是常见的典型干燥设备。

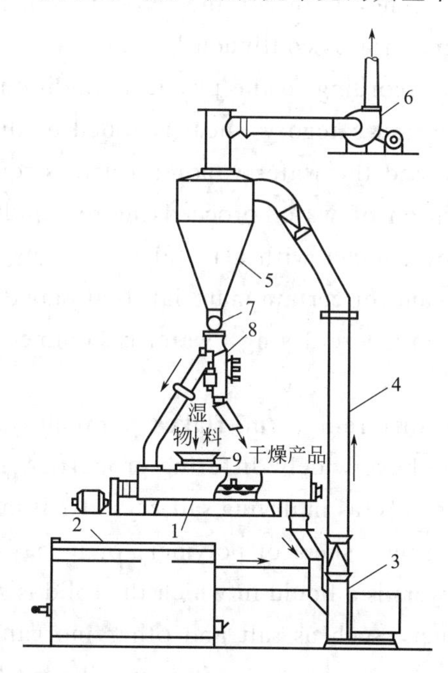

图5-21 装有粉碎机的气流干燥装置的流程图

1—螺旋输送混合器；2—燃烧炉；3—粉碎机；4—气流干燥器；5—旋风分离器；6—风机；7—星式加料器；8—流动固体物料的分配器；9—加料斗

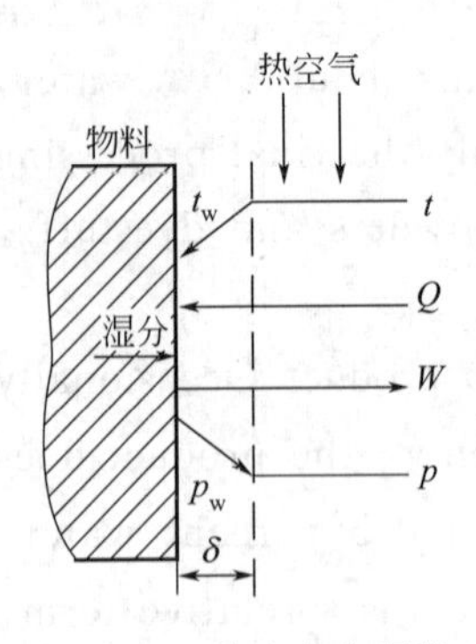

图 5-22　热空气与湿物料间的传热和传质过程

t—空气主体温度；
t_w—物料表面温度；
p—空气中的水汽分压；
p_w—物料表面的水汽分压；
Q—由气体传给物料的热流量；
W—由物料汽化的水分质量流量

它的主体是直立圆管，湿物料由加料斗 9 加入螺旋输送混合器 1 内，与一定量的干燥物料混合后进入粉碎机 3。从燃烧炉 2 来的烟道气（也可以是热空气）也同时进入粉碎机，将颗粒状的固体吹入气流干燥器中。由于热空气作高速运动，使物料颗粒分散并悬浮于气流中。热空气与物料间进行传质和传热，物料得以干燥，并随气流进入旋风分离器 5 中，经分离后由底部排出，再借分配器 8 的作用，定时地排出作为产品或送入螺旋输送混合器 1 中供循环使用。这种气流干燥方法，适合于除去能在气体中自由流动的颗粒物料中的水分。

图 5-22 为上述气流干燥器 4 中，热空气与湿物料间的传热和传质过程。当作为干燥介质的空气的温度高于湿物料的温度，而物料表面水汽分压 p_w 又大于空气中的水汽分压 p 时，热量由热空气传递到物料表面，然后再传递到物料内部；而水分则由物料内部传递到物料表面，然后再传递到空气流主体，且传质和传热过程同时发生。显然，在干燥过程中，干燥介质既是载热体，又是载湿体。

General methods of drying

Drying methods and processes can be classified in several different ways. Drying processes can be classified as batch, where the material is inserted into the drying equipment and drying proceeds for a given period of time, or as continuous, where the material is continuously added to the dryer and dried material is continuously removed.

Drying processes can also be categorized according to the physical conditions used to add heat and remove water vapor: (1) in the first category, heat is added by direct contact with heated air at atmospheric pressure, and the water vapor formed is removed by the air; (2) in vacuum drying, the evaporation of water proceeds more rapidly at low pressures, and the heat is added indirectly by contact with a metal wall or by radiation (low temperatures can also be used under vacuum for certain materials that may discolor or decompose at higher temperatures); and (3) in freeze-drying, water is sublimed from the frozen material.

The solids to be dried may be in many different forms-flakes, granules, crystals, powders, slabs, or continuous sheets-and may have widely differing properties. The liquid to be vaporized may be on the surface of the solid, as in drying salt crystals; it may be entirely inside the solid, as in solvent removal from a sheet of polymer; or it may be partly outside and partly inside. The feed to some dryers is a liquid in which the solid is suspended as particles or is in solution. The dried product, such as salt and other inorganic solids, may be able to stand rough handling and high temperatures, or it may, like food or pharmaceuticals, require gentle treatment at low or moderate temperatures. Consequently, a multitude of types of dryers are on the market for commercial drying. They differ chiefly in the way the solids are moved through the drying zone and in the way heat is transferred.

5.2.2　湿空气性质（Properties of moist air）

含有湿分的空气称为湿空气，湿空气除去水分的能力与它的性质有关。从干燥操作的角度考虑，湿空气包含绝干空气和水汽两部分。

（1）湿空气中水汽含量的表示法

① 湿度 H（humidity）　湿空气中水汽的质量与湿空气中绝干空气的质量之比，称为湿度 H（又称湿含量、绝对湿度）。对于理想气体，湿度可表示为：

$$H=\frac{m_v}{m_g}=\frac{M_v}{M_g}\times\frac{p_v}{P-p_v}=0.622\frac{p_v}{P-p_v} \tag{5-11}$$

式中　m_v，m_g——水汽和绝干空气的质量，kg；

M_v，M_g——水汽和绝干空气的摩尔质量，kg/kmol；

p_v——水汽分压，Pa；

P——湿空气总压，Pa；

H——湿空气的湿度，kg水/kg绝干空气。

当空气达到饱和状态时，$p_v=p_s$，式（5-11a）变为：

$$H_s=0.622\frac{p_s}{P-p_s} \tag{5-11a}$$

式中　p_s——空气温度下水汽的饱和蒸气压，Pa；

H_s——湿空气的饱和湿度，kg水/kg绝干空气。

② 相对湿度 φ　湿空气中水汽分压与同温度下水的饱和蒸气压之比称为湿空气的相对湿度，即：

$$\varphi=\frac{p_v}{p_s} \tag{5-12}$$

φ 值表示了湿空气偏离饱和空气的程度，φ 值越小，说明其吸湿能力越强，更适合作为干燥介质，所以相对湿度的概念在各个领域中得到了广泛应用。其中当 $\varphi=0$ 时，表示湿空气中不含水汽，称为绝干空气。而 $\varphi=1$ 时，表示湿空气被水汽所饱和，不能作为干燥介质。由于 p_v 和 p_s 均随温度升高而增大，故当 p_v 一定时，φ 值随温度升高而减小。

将式（5-12）代入式（5-11a），可得 φ 与 H 之间的关系式：

$$H=\frac{0.622\varphi p_s}{P-\varphi p_s} \tag{5-13}$$

（2）湿空气的比热容、比容和焓

① 湿比热容 C_H（humid heat of an air-water vapor mixture）　以1kg绝干空气为基准，对应1kg绝干空气和 H kg水汽温度升高（或降低）1℃所需吸收（或放出）的总热量，称为湿比热容，其单位为kJ/(kg绝干空气·℃)。在0～200℃的温度范围内，可近似把绝干空气的比热容 C_g 和水汽的比热容 C_v 看作常数，其值分别为1.01kJ/(kg绝干空气·℃）和1.88 kJ/(kg水·℃)

$$C_H=1.01+1.88H \tag{5-14}$$

上式表明，湿空气的比热容只是湿度的函数。

② 湿比容 v_H（humid volume of an air-water vapor mixture） 以1kg绝干空气为基准，对应1kg绝干空气和 H kg水汽所占的总体积，称为湿比容，其单位为 m^3/kg绝干空气。若按理想气体处理，湿空气的比容可表示为：

$$v_H = 22.4\left(\frac{1}{M_g}+\frac{H}{M_v}\right)\times\frac{101.3}{P}\times\frac{273+t}{273}$$

$$=(0.772+1.244H)\times\frac{101.3}{P}\times\frac{273+t}{273} \tag{5-15}$$

式中 P——湿空气总压，kPa；

t——湿空气温度，℃。

③ 湿空气的焓 I（total enthalpy of an air-water vapor mixture） 以0℃时气体焓为基准，对应1kg绝干空气的焓 I_g 和其中 H kg水汽的焓 I_v 之和，称为湿空气的焓，其单位为kJ/kg绝干空气。

$$I = I_g + I_v H = C_g t + (C_v t + r_0)H = (C_g + C_v H)t + r_0 H$$

$$=(1.01+1.88H)t+2490H \tag{5-16}$$

式中 r_0——0℃时水的汽化潜热，$r_0 \approx 2490$ kJ/kg。

(3) 湿空气的几种温度表示法

① 干球温度与湿球温度（dry-bulb temperature and wet-bulb temperature） 图5-23所示的两支温度计，一支温度计的感温球暴露在空气中，称为干球温度计，其显示的是干球温度 t。干球温度指湿空气的真实温度，可用普通温度计测量。另一支温度计的感温球用纱布包裹，纱布下部浸于水中使之保持湿润，这就是湿球温度计。当空气至湿纱布的传热速率与水分汽化传向空气的传质速率恰好相等时，其显示的温度称为空气的湿球温度 t_w。t_w 不代表空气的真实温度，是表示空气状态或性质的一种参数。

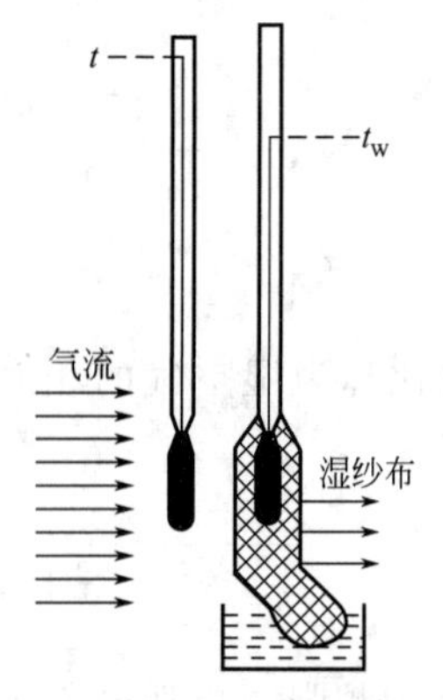

图5-23 湿球温度的测量

根据热量和质量传递的关系可以推出，干球温度与湿球温度的关系式：

$$t_w = t - \frac{k_H r_{t_w}}{\alpha}(H_{s,t_w} - H) \tag{5-17}$$

式中 α——空气向湿纱布的对流传热系数，W/(m^2·℃)；

t——空气的干球温度，℃；

t_w——空气的湿球温度，℃；

k_H——以湿度差为推动力的传质系数，kg/(m^2·s·ΔH)；

r_{t_w}——湿球温度下水汽的汽化潜热，kJ/kg；

H_{s,t_w}——湿球温度下空气的饱和湿度，kg/(kg绝干空气)。

实验表明，k_H 与 α 都与空气流速的0.8次方成正比。一般在气速为3.8～10.2 m/s的范围内，比值 α/k_H 近似为一常数。对水蒸气与空气的系统，$\alpha/k_H \approx 1.09$。另外，H_{s,t_w}、r_{t_w} 只决定于 t_w，于是当 α/k_H 为常数时，t_w 是 t 和 H 的函数。当 t 和 H 一定时，则 t_w 必为定值。反之，当测得了湿空气的 t 和 t_w 后，即可确定空气的 H。

Dry-bulb Temperature and Wet-bulb Temperature

Dry-bulb temperature

Commonly, an uncovered thermometer is used along with the wet bulb to measure T, the actual gas temperature, and the gas temperature is usually called the dry-bulb temperature.

Wet-bulb temperature

The properties discussed above are static or equilibrium quantities. Equally important are the rates at which mass and heat are transferred between the gas and liquid phases that are not in equilibrium. The driving forces for mass and heat transfer are concentration and temperature differences, which can be predicted by using a quantity called the wet-bulb temperature.

The wet-bulb temperature is the steady-state, nonequilibrium temperature reached by a small mass of liquid exposed under adiabatic conditions to a continuous stream of gas. Because the gas flow is continuous, the properties of the gas are constant and are usually evaluated at inlet conditions. If the gas is not saturated, some liquid evaporates, cooling the remaining liquid until the rate of heat transfer to the liquid just balances the heat needed for evaporation. The liquid temperature when steady state is reached is the wet-bulb temperature.

The method of measuring the wet-bulb temperature is shown in Fig. 5. 1(a). A thermometer or other temperature-measuring device, such as a thermocouple, is covered by a wick, which is saturated with pure liquid and immersed in a stream of gas having a definite temperature T and humidity H. Assume that initially the temperature of the liquid is about that of the gas. Since the gas is not saturated, liquid evaporates; and because the process is adiabatic, the latent heat is supplied at first by cooling the liquid. As the temperature of the liquid decreases below that of the gas, sensible heat is transferred to the liquid. Ultimately a steady state is reached at such a liquid temperature that the heat needed to evaporate the liquid and heat the vapor to gas temperature is exactly balanced by the sensible heat flowing from the gas to the liquid. It is this steady-state temperature, denoted by

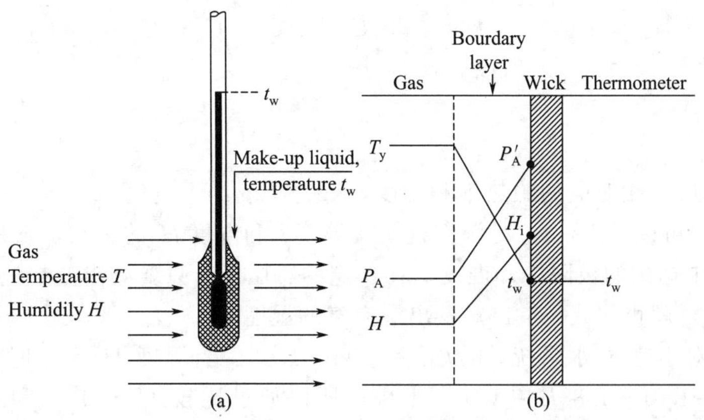

Fig. 5. 1 (a) Wet-bulb thermometer; (b) Gradients in the gas boundary layer

T_w，that is called the wet-bulb temperature. It is a function of both T and H. The temperature and concentration gradients at steady state are shown in Fig. 5.1（b）.

Measurement of wet-bulb temperature

To measure the wet-bulb temperature with precision，three precautions are necessary：（1）The wick must be completely wet，so no dry areas of the wick are in contact with the gas；（2）the velocity of the gas should be large enough（at least 5 m/s）to ensure that the rate of heat flow by radiation from warmer surroundings to the bulb is negligible in comparison with the rate of sensible heat flow by conduction and convection from the gas to the bulb；（3）if make-up liquid is supplied to the bulb，it should be at the wet-bulb temperature. When these precautions are taken，the wet-bulb temperature is independent of gas velocity over a wide range of flow rates.

The wet-bulb temperature superficially resembles the adiabatic saturation temperature T_s. Indeed，for air-water mixtures the two temperatures are nearly equal.

This is fortuitous，however，and is not true of mixtures other than air and water. The wet-bulb temperature differs fundamentally from the adiabatic saturation temperature. The temperature and humidity of the gas vary during adiabatic saturation，and the endpoint is a true equilibrium rather than a dynamic steady state.

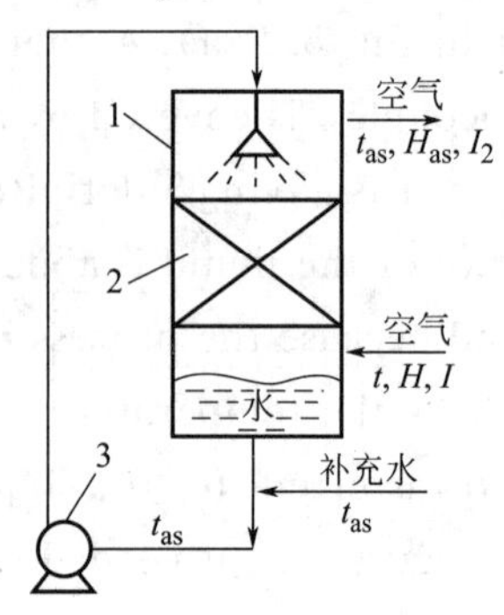

图 5-24　绝热饱和冷却塔示意图

1—塔身；2—填料；3—循环泵

② 绝热饱和温度 t_{as}（adiabatic saturation temperatures）

图 5-24 为绝热饱和冷却塔，含有水汽的不饱和空气（温度为 t，湿度为 H）连续地通过塔内填料与大量喷洒的水接触，水用泵循环。假定水温完全均匀，饱和冷却塔处于绝热状态，故水汽化所需要的潜热只能取自空气中的显热，使空气绝热增湿而降温，直至空气被水所饱和，则空气的温度不再下降而等于循环水的温度，此温度即为空气的绝热饱和温度 t_{as}，对应的饱和湿度为 H_{as}。因为该过程中空气经历等焓过程，所以：

$$(C_g+HC_v)t+Hr_0=(C_g+H_{as}C_v)t_{as}+H_{as}r_0$$

一般 H 及 H_{as} 值均很小，故可认为：

$$C_g+HC_v\approx C_g+H_{as}C_v=C_H$$

所以

$$t_{as}=t-\frac{r_0}{C_H}(H_{as}-H) \tag{5-18}$$

式中　H_{as}——与 t_{as} 相对应的绝热饱和湿度，kg 水/kg 绝干空气。

由式（5-18）可以看出，t_{as} 是湿空气初始温度 t 和湿度 H 的函数，它是湿空气在绝热、冷却、增湿过程中达到的极限冷却温度。在一定的总压下，只要测出湿空气的初始温度 t 和绝热饱和温度 t_{as}，就可用式（5-11）算出湿空气的湿度 H。

前已指出，对于空气-水系统，$\alpha/k_H=0.96\sim1.05$。当湿度 H 不大时（一般干燥过程 $H<0.01$），$C_H=1.01+1.88H=1.01\sim1.03$。所以通过比较式（5-17）和式（5-18）可知，$t_w\approx t_{as}$。但对其他物系，$\alpha/k_H=1.5\sim2$，与 C_H 相差很大，则湿球温度高于绝热饱和温度。

在绝热条件下，用湿空气干燥湿物料的过程中，气体温度的变化是趋向于绝热饱和温度

t_{as} 的。如果湿物料足够湿润，则其表面温度也就是湿空气的绝热饱和温度 t_{as}，即湿球温度 t_w，因此这两个温度在干燥器的计算中有着极其重要的实用意义。因为湿球温度容易测定，这就给干燥过程的计算和控制带来了较大的方便。

③ 露点温度 t_d（dew point of an air-water vapor mixture） 不饱和湿空气在总压 p 和湿度 H 保持不变的情况下，使其冷却达到饱和状态时的温度称为露点温度 t_d。露点温度仅取决于湿度。

湿空气的四个温度参数（干球温度 t、湿球温度 t_w、绝热饱和温度 t_{as} 和露点温度 t_d）都可用来确定空气状态。而对于一定状态的不饱和湿空气，它们之间的关系是：$t>t_w\approx t_{as}>t_d$；而当空气被水所饱和时：$t=t_w=t_{as}=t_d$。

（4）湿空气的 H-I 图及其应用（humidity chart and its application）

为简捷清晰地描述湿空气性质的各项参数（p,t,φ,H,I,t_w 等），可用图的形式表示各性质间的关系，这里采用的是湿焓图（H-I 图），如图 5-25 所示。

在常压下，以湿空气的焓为纵坐标，湿度为横坐标，两轴采用斜角坐标系，其间夹角 135°。

H-I 图上共有五类线或线群：

① 等湿（H）线群 等湿线为一系列平行于纵轴的直线。

② 等焓（I）线群 等焓线为一系列平行于横轴的直线。

③ 等温（t）线群 将式（5-16）改写为：

$$I=1.01t+(1.88t+2490)H$$

可见，若 t 为定值，则 I 与 H 成直线关系。由于直线的斜率 $1.88t+2490$ 随 t 而变，故一系列的等温线并不平行。

④ 等相对湿度（φ）线群 由式（5-13）可知，当总压 P 一定时，对某一固定的 φ 值，由任一温度 t 可查到一个相对的饱和水蒸气压力 p_s，进而可算出对应的 H 值。将许多（t，H）点连接起来，即成为一条等相对湿度线。图 5-25 中标绘了由 $\varphi=5\%$ 至 $\varphi=100\%$ 的一系列等相对湿度线。

$\varphi=100\%$ 的等相对湿度线称为饱和空气线，此时空气完全被水汽所饱和。饱和线以上（$\varphi<100\%$）为不饱和空气区域。显然，只有位于不饱和区域的湿空气才能作为干燥介质。由图 5-25 可知，当湿空气的 H 一定时，温度愈高，其相对湿度 φ 值愈小，即用作干燥介质时去湿能力愈强。所以，湿空气在进入干燥器之前必须先经预热以提高温度。预热空气除了可提高湿空气的焓值使其作为载热体外，同时也可为了降低其相对湿度而作为载湿体。

⑤ 蒸汽分压线 式（5-11）可改写成

$$p_v=\frac{PH}{0.622+H}$$

可见，当总压 P 一定时，水汽分压 p_v 仅随湿度而变。因 $H\ll 0.622$，故 p_v 与 H 近似成直线关系。此直线关系标绘在饱和空气线的下方。

在使用 H-I 图时，图上任何一点都代表一定的空气状态，只要规定其中两个互相独立的参数，湿空气的性质就被唯一地确定下来了。反之，利用表示湿空气性质的任意两个相对独立变量（即两个在图上有交点的参数），就可以在图上定出一个点，如图 5-26 所示。该点即表示湿空气所处的状态点，由此点可查出其他各项参数。通常给出以下条件来确定湿空气的状态点：(t,t_w)、(t,t_d)、(t,φ)、(t,p)、(t,H) 和 (t_w,p) 等。

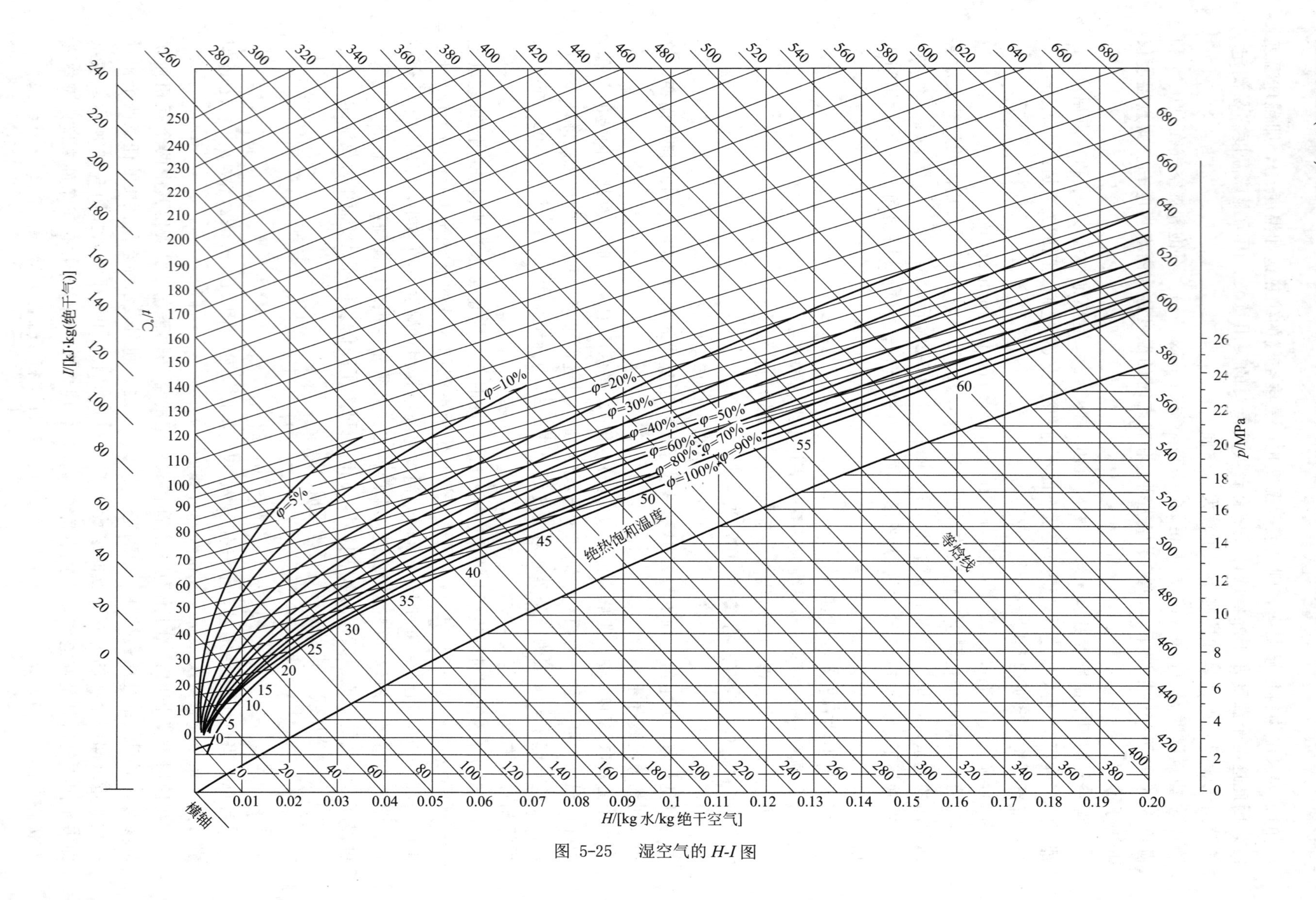

图 5-25　湿空气的 *H-I* 图

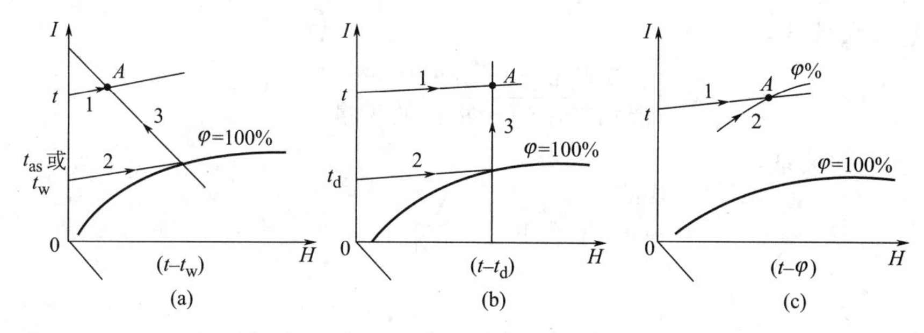

图 5-26 在 H-I 图中确定湿空气的状态点

5.2.3 干燥工艺计算（Calculation of drying technology）

典型的用热空气作为干燥介质的一般对流干燥流程图如图 5-27。该流程主要由空气预热器对空气进行加热，提高其热焓值，并降低它的相对湿度，以便使空气更适宜作为干燥介质，而干燥器则是对物料除湿的主要场所。

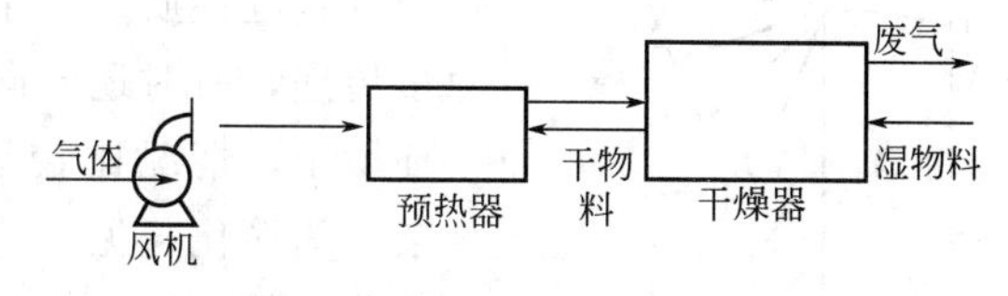

图 5-27 对流干燥流程示意图

对干燥流程的设计中，通过物料衡算可计算出物料汽化的水分量 W（或称为空气带走的水分量）、和空气的消耗量（包括绝干空气消耗量 L 和新鲜空气消耗量 L_0），而通过热量衡算计算干燥流程的热能耗用量及各项热量分配量（即预热器换热量 Q_P，干燥器供热量 Q_D 及干燥器热损失 Q_L）。图 5-28 为连续干燥过程的物料和热量衡算示意图。

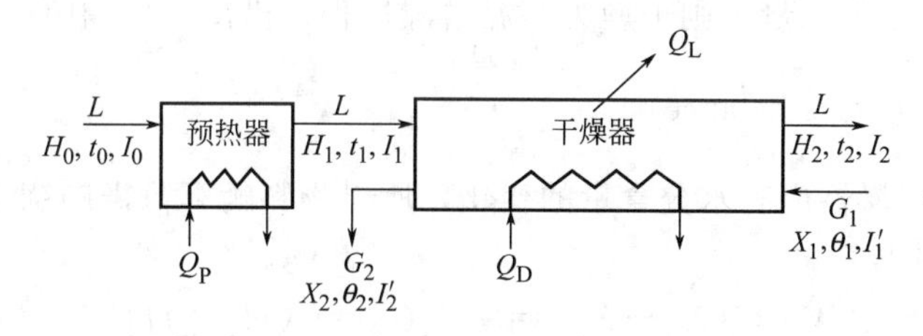

图 5-28 连续干燥过程的物料和热量衡算示意图

H_0，H_1，H_2—新鲜湿空气进入预热器、离开预热器（进入干燥器）和离开干燥器时的湿度，kg 水/kg 绝干空气；I_0，I_1，I_2—新鲜湿空气进入预热器、离开预热器（进入干燥器）和离开干燥器时的焓，kJ/kg 绝干空气；t_0，t_1，t_2—新鲜湿空气进入预热器、离开预热器（进入干燥器）和离开干燥器时的温度,℃；L—绝干空气的流量，kg 绝干空气/s；Q_P—单位时间内预热器消耗的热量，kW；Q_D—单位时间内向干燥器补充的热量，kW；Q_L—干燥器的热损失速率，kW；G_1，G_2—湿物料进入和离开干燥器时的流量，kg 湿物料/s；X_1，X_2—湿物料进入和离开干燥器时的干基含水量，kg 水/kg 绝干物料；θ_1，θ_2—湿物料进入和离开干燥器时的温度,℃；I'_1，I'_2—湿物料进入和离开干燥器时的焓，kJ/kg 绝干物料

湿物料含水量是物料衡算和热量衡算中的重要变量。湿物料分成水分和绝干物料两部分，其中水含量有湿基含水量和干基含水量两种表示法。

① 湿基含水量 w：

$$w=\frac{\text{湿物料中水分的质量}}{\text{湿物料的总质量}}\times 100\% \tag{5-19}$$

② 干基含水量 X：

$$X=\frac{\text{湿物料中水分的质量}}{\text{湿物料中绝干物料的质量}}\times 100\% \tag{5-20}$$

③ w 与 X 的关系：

$$w=\frac{X}{1+X} \tag{5-21a}$$

或：

$$X=\frac{w}{1-w} \tag{5-21b}$$

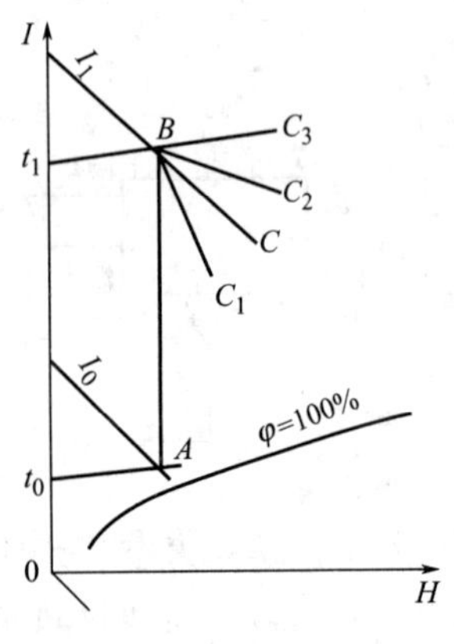

图 5-29　干燥过程中湿空气的状态变化图

此外，对干燥系统进行物料和热量衡算时，必须知道空气离开干燥器时的状态参数，确定这些参数涉及空气在干燥器内所经历的过程性质。在干燥器内，空气与物料间既有质量传递也有热量传递，有时还要向干燥器补充热量，而且又有热量损失于周围环境中，情况比较复杂，故确定干燥器出口处空气状态较为复杂。为简化起见，一般根据空气在干燥器内焓的变化，将干燥过程分为等焓过程与非等焓过程两大类。

图 5-28 中，若 $I_2=I_1$，则空气在干燥器内经历等焓干燥过程，又称理想干燥过程，如图 5-29 所示。过程沿等焓线 BC 进行，整个干燥过程由状态 $A \rightarrow B \rightarrow C$ 点。图 5-28 中，若 $I_2 \neq I_1$，则干燥为非等焓过程，过程沿图 5-29 中的 BC_1 或 BC_2 线变化。此过程中，若外界向干燥器补充的热量恰好等于水分蒸发所需热量，则干燥为等温增湿过程，沿图 5-29 中的等温线 BC_3 变化。

5.2.3.1　物料衡算（mass balance）

图 5-28 中，仅干燥器内有水分含量的变化，所以物料衡算只需围绕干燥器进行。

(1) 水分蒸发量 W

$$W=G(X_1-X_2)=G_1-G_2=L(H_2-H_1) \tag{5-22}$$

其中：

$$G=G_1(1-w_1)=G_2(1-w_2) \tag{5-23}$$

(2) 空气消耗量 L

由式 (5-22) 得：

$$L=\frac{G(X_1-X_2)}{H_2-H_1}=\frac{W}{H_2-H_1} \tag{5-24}$$

$$l=\frac{L}{W}=\frac{1}{H_2-H_1} \tag{5-25}$$

其含义为每蒸发 1kg 水所消耗的绝干空气量，称为单位空气消耗量，单位为 kg 绝干空气/kg 水。

新鲜空气用量：

$$L_0=L(1+H_0) \tag{5-26}$$

利用风机向预热器入口输送新鲜空气时，则风机入口风量就是新鲜空气的体积流量 V_0（单位为 m^3 新鲜空气/s），可根据式 (5-15) 得到：

$$V_0=L\nu_H=L(0.772+1.244H)\frac{101.3}{P}\times\frac{t+273}{273} \tag{5-27}$$

（3）干燥产品流量 G_2

$$G_1(1-w_1)=G_2(1-w_2)$$

所以

$$G_2=G_1\frac{1-w_1}{1-w_2} \tag{5-28}$$

5.2.3.2　热量衡算（thermal balance）

若忽略预热器的热损失，则对图 5-28 中的预热器进行热量衡算，得：

$$Q_P=L(I_1-I_0) \tag{5-29}$$

再对图 5-28 中的干燥器进行热量衡算，得：

$$Q_D=L(I_2-I_1)+G(I'_2-I'_1)+Q_L \tag{5-30}$$

所以，干燥过程所需总热量为：

$$Q=Q_P+Q_D=L(I_2-I_0)+G(I'_2-I'_1)+Q_L \tag{5-31}$$

假设新鲜空气中水汽的焓等于离开干燥器废气中水汽的焓，则：

$$Q\approx L\times1.01(t_2-t_0)+W(2490+1.88t_2)+GC_m(\theta_2-\theta_1)+Q_L \tag{5-32}$$

即：

Q＝加热空气所需热量＋物料中水分蒸发所需热量＋加热湿物料所需热量＋热损失

式中　C_m——湿物料的平均比热容，kJ/(kg 绝干物料·℃)。

干燥系统的热效率 η 定义为蒸发水分所需热量与向干燥系统输入的总热量之比，即：

$$\eta=\frac{W(2490+1.88t_2)}{Q}\times100\% \tag{5-33}$$

当空气出干燥器时，若温度 t_2 降低而湿度 H_2 增大，则 η 会提高。但 t_2 过低而 H_2 过高，会使物料返潮，所以应综合考虑。

5.2.4　干燥动力学（Dynamics of drying）

5.2.4.1　物料中水分分类

物料中水分可以按两种形式分类：

（1）平衡水分与自由水分（free and equilibrium moisture of a substance）

当物料与一定状态的湿空气充分接触后，物料中不能除去的水分称为平衡水分（平衡含水量），用 X^* 表示；而物料中超过平衡水分的那部分水分称作自由水分，这种水分可用干燥操作除去。由于物料达到平衡水分时，干燥过程达到此操作条件下的平衡状态，所以 X^* 与物料性质、空气状态及两者接触状态有关。平衡水分 X^* 可用平衡曲线描述，如图 5-30 所示。

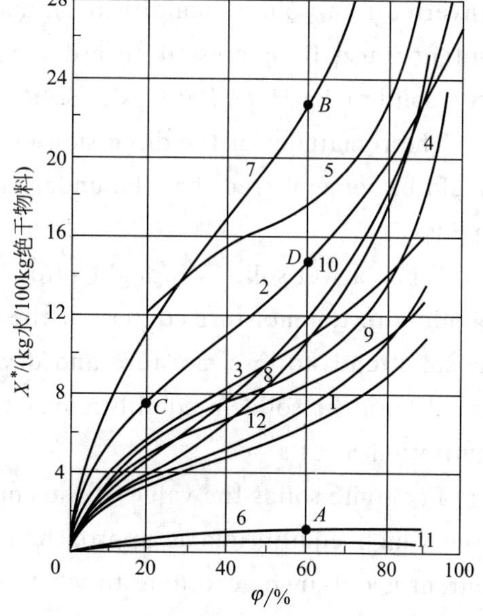

图 5-30　25℃时某些物料的平衡含水量 X^* 与空气相对湿度 φ 的关系

由图 5-30 可知：当空气状态恒定时，不同物料的平衡水分相差很大；同一物料的平衡水分随空气状态而变化；当 $\varphi=0$ 时，X^* 均为 0，说明湿物料只有与绝干空气相接触，才有可能得到绝干物料。

物料中平衡含水量随空气温度升高而略有减少，可由实验测得。可由于缺乏各种温度下平衡含水量的实验数据，因此只要温度变化范围不太大，一般可近似认为物料的平衡含水量与空气温度无关。

Phase Equilibria

As in other transfer processes, such as mass transfer, the process of drying of materials must be approached from the viewpoint of the equilibrium relationships together with the rate relationships. In most of the drying apparatus discussed here, material is dried in contact with an air-water vapor mixture. The equilibrium relationships between the air-water vapor and the solid material will be discussed in this section.

An important variable in the drying of materials is the humidity of the air in contact with a solid of given moisture content. Suppose that a wet solid containing moisture is brought into contact with a stream of air having a constant humidity H and temperature. A large excess of air is used, so its conditions remain constant. Eventually, after exposure of the solid sufficiently long for equilibrium to be reached, the solid will attain a definite moisture content. This is known as the equilibrium moisture content of the material under the specified humidity and temperature of the air.

Equilibrium data for moist solids are commonly given as relationships between the relative humidity of the gas and the liquid content of the solid, in mass of liquid per unit mass of bone-dry solid. Examples of equilibrium relationships are shown in Fig. 5-30. Curves of this type are nearly independent of temperature. The abscissas of such curves are readily converted to absolute humidities, in mass of vapor per unit mass of dry gas. The moisture content is usually expressed on a dry basis as kg of water per kg of moisture-free (bone-dry) solid or kg H_2O/100 kg dry solid.

The remainder of the discussion in this section is based on the air-water system, but it should be remembered that the underlying principles apply equally well to other gases and liquids.

When a wet solid is brought into contact with air of lower humidity than that corresponding to the moisture content of the solid, as shown by the humidity equilibrium curve, the solid tends to lose moisture and dry to equilibrium with the air. When the air is more humid than the solid in equilibrium with it, the solid absorbs moisture from the air until equilibrium is attained.

For some solids the value of the equilibrium moisture content depends on the direction from which equilibrium is approached. The different value for the equilibrium moisture content is obtained according to whether a wet sample is allowed to dry by desorption or a dry sample adsorbs moisture by adsorption. For drying calculations it is the desorption equilibrium that is the larger value and is of particular interest.

Porous solids such as catalysts or adsorbents often have an appreciable equilibrium moisture content at moderate relative humidity. Liquid water in fine capillaries exerts an abnormally low vapor pressure because of the highly concave surface of the meniscus.

Adsorbents such as silica or alumina have monolayers of water strongly adsorbed on the surface, and this water has a much lower vapor pressure than liquid water. Beds of nonporous particles such as sand have negligible equilibrium moisture content in humid air unless the particles are so small that the fillets of liquid where particles touch have a very small radius of curvature.

In fluid phases diffusion is governed by concentration differences expressed in mole fractions. In a wet solid, however, the term mole fraction may have little meaning, and for ease in drying calculations the moisture content is nearly always expressed in mass of water per unit mass of bone-dry solid. This practice is followed throughout this chapter.

(2) 结合水与非结合水（bound and unbound water in solids）

结合水是指与物料以化学力和物理化学力等强结合力相结合的水分，它们大多存在于如细胞壁、微孔中，不易除去；而结合力较弱、机械附着于固体表面的水分称为非结合水，这种水分容易由干燥操作除去。结合水与非结合水的关系示于图5-31中。由于直接测定物料的结合水与非结合水很困难，所以可利用平衡曲线外延至与$\varphi=100\%$相交而得到的X为结合水量，高于它的部分为非结合水量。非结合水极易除去，而结合水中高于平衡水分的部分也能被除去，但很困难。

Bound and unbound water in solids

If an equilibrium curve like those in Fig. 5-31 is continued to its intersection with the axis for 100 percent humidity, the moisture content so defined is the minimum moisture this material can carry and still exert a vapor pressure as great as that exerted by liquid water at the same temperature. If such a material contains more water than that indicated by this intersection, it can still exert only the vapor pressure of water at the solids temperature. This makes possible a distinction between two types of water held by a given material. The water corresponding to concentrations lower than that indicated by the intersection of the curves in Fig. 5-31 with the line for 100 percent humidity is called bound water, because it exerts a vapor pressure less than that of liquid water at the same temperature. Substances containing bound water are often called hygroscopic substances.

Bound water may exist in several conditions. Liquid water in fine capillaries exerts an abnormally low vapor pressure because of the highly concave curvature of the surface; moisture in cell or fiber walls may suffer a vapor pressure lowering because of solids dissolved in it; water in natural organic substances is in physical and chemical combination, the nature and strength of which vary with the nature and moisture content of the solid.

Water corresponding to concentrations greater than that indicated by the intersections is called unbound water. Unbound water, on the other hand, exerts its full vapor pressure and is largely held in the voids of the solid. Large nonporous particles, such as coarse sand, contain only unbound water.

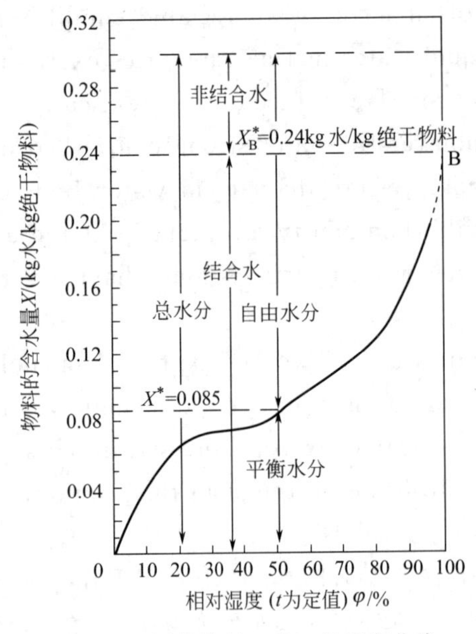

图 5-31　固体物料（丝）的平衡曲线

（3）平衡曲线的应用

① 判断过程进行的方向　当干基含水量为 X 的湿物料与一定温度及相对湿度为 φ 的湿空气相接触时，可在干燥平衡曲线（图 5-30、图 5-31）上找到与该湿空气相应的平衡水分 X^*，比较湿物料的含水量 X 与平衡含水量 X^* 的大小，可判断过程进行的方向。

若物料含水量 X 高于平衡含水量 X^*，则物料脱水而被干燥；若物料的含水量 X 低于平衡含水量 X^*，则物料将吸水而增湿。

② 确定过程进行的极限　平衡水分是物料在一定空气条件下被干燥的极限，利用平衡曲线，可确定一定含水量的物料与指定状态空气相接触时平衡水分与自由水分的大小。

图 5-31 为一定温度下某种固体物料（丝）的平衡曲线。当将干基含水量为 $X=0.30$kg 水/kg 绝干物料的物料与相对湿度为 50% 的空气相接触时，由平衡曲线可查得平衡水分为 $X^*=0.084$kg 水/kg 绝干物料，相应自由水分为 $X-X^*=0.216$kg 水/kg 绝干物料。

③ 判断水分去除的难易程度　利用平衡曲线可确定结合水分与非结合水分的大小。图 5-31 中，平衡曲线与 $\varphi=100\%$相交于 S 点，查得结合水分为 0.24kg 水/kg 绝干物料，此部分水较难去除。相应非结合水分为 0.06kg 水/kg 绝干物料，此部分水较易去除。

应予指出，平衡水分与自由水分是依据物料在一定干燥条件下，其水分能否用干燥方法除去而划分，既与物料的种类有关，也与空气的状态有关。而结合水分与非结合水分是依据物料与水分的结合方式（或物料中所含水分去除的难易）而划分，仅与物料的性质有关，而与空气的状态无关。

5.2.4.2　恒定干燥条件下的干燥速率

恒定干燥条件是指干燥介质的温度、湿度、流速及与物料的接触方式等，在整个干燥过程中均保持恒定。这是一种对问题的简化处理方式，可适用于大量空气干燥少量湿物料的情况，空气的定性温度取进、出口温度平均值。

（1）干燥速率（rate-of-drying）

干燥速率的定义为单位时间单位干燥面积上汽化的水分量，即：

$$U=\frac{\mathrm{d}W'}{S\,\mathrm{d}\tau} \tag{5-34}$$

式中　U——干燥速率，又称干燥通量，kg/(m^2·s)；

S——干燥面积，m^2；

W'——一批操作中汽化的水分量，kg；

τ——干燥时间，s。

又因为：

$$\mathrm{d}W'=-G'\mathrm{d}X$$

式中　G'——一批操作中绝干物料的质量，kg。

所以：

$$U=\frac{-G\mathrm{d}X}{S\mathrm{d}\tau} \tag{5-35}$$

上式中的负号表示 X 随干燥时间的增加而减小。

(2) 干燥速率曲线（rate-of-drying curves）

干燥过程的计算主要包括确定干燥的操作条件，计算干燥时间和干燥器尺寸，这就必须求得干燥速率。干燥速率通常通过实验测得，即将实验数据计算处理后描点绘图，得到干燥速率曲线以供参考。

图 5-32 为恒定干燥条件下一种典型的干燥速率曲线，AB 是预热段，很快进入恒速干燥阶段（BC 段），这是除去非结合水的阶段，所以速率大且不随 X 而变化；CD 和 DE 段是降速阶段，是除去结合力很强的结合水的过程，所以干燥速率随 X 减小而迅速下降，情况比较复杂；最后当 $U=0$ 时，达到该操作条件下的平衡含水量 。其中 C 点是临界点，X_c 称为临界含水量（kg 水/kg 绝干物料），U_c 称为恒速段干燥速率 [kg 水/(m^2 · s)]。若 X_c 增大，则恒速段变短，不利于干燥操作。

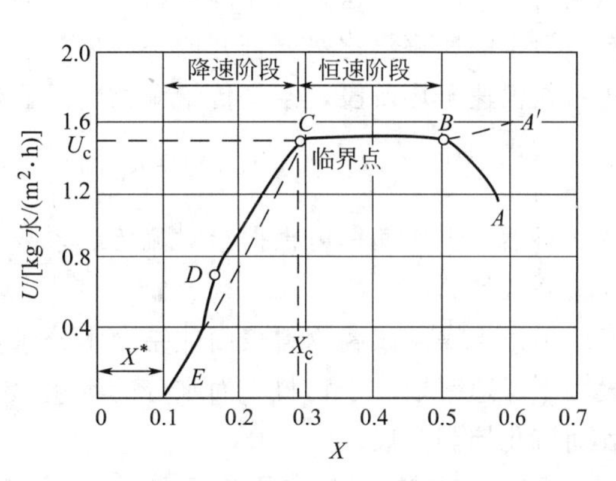

图 5-32 恒定干燥条件下干燥速率曲线

临界含水量随物料的性质、厚度及干燥速率不同而异。对同一种物料，如干燥速率增大，则其临界含水量值亦增大；对同一干燥速率，物料层愈厚，X_c 值也愈大。物料的临界含水量通常由实验测定，在缺乏实验数据的条件下，可按表 5-3 所列的 X_c 值估计。

表 5-3 不同物料的临界含水量范围

有机物料		无机物料		临界含水量（干基）/%
特征	实例	特征	实例	
很粗的纤维	未染过的羊毛	粗粒无孔的物料，大于 50 目	石英	3～5
		晶体的、粒状的、孔隙较少的物料，颗粒大小为 50～325 目	食盐、海砂、矿石	5～15
晶体的、粒状的、孔隙较小的物料	麸酸结晶	细晶体有孔物料	硝石、细砂、黏土料、细泥	15～25
粗纤维细粉	粗毛线、醋酸纤维、印刷纸、碳素颜料	细沉淀物、无定形和胶体状态的物料、无机颜料	碳酸钙、细陶土、普鲁士蓝	25～50
细纤维、无定形和均匀状态的压紧物料	淀粉、亚硫酸、纸浆、厚皮革	浆状，有机物的无机盐	碳酸钙、碳酸镁、二氧化钛、硬脂酸钙	50～100
分散的压紧物料、胶体状态和凝胶状态的物料	鞣制皮革、糊墙纸、动物胶	有机物的无机盐、催化剂、吸附剂	硬脂酸锌、四氯化锡、硅胶、氢氧化铝	100～3000

经实验测定和理论分析可知，空气平行流过物料表面时，U_c 与 $(L')^{0.8}$ 成正比；空气垂直流过物料表面时，U_c 与 $(L')^{0.37}$ 成正比。式中 L' 指湿气质量流速，单位为 kg/(m^2 · h)，$L'=u\rho$。

(3) 干燥速率的影响因素

在恒速阶段与降速阶段内，物料干燥的机理不同，从而影响因素也不同，分别讨论如下。

① 恒速干燥阶段（drying in the constant-rate period） 恒速干燥阶段中，当干燥条件恒定时，物料表面与空气之间的传热和传质情况与测定湿球温度时相同。

在恒定干燥条件下，空气的温度、湿度、速度及气固两相的接触方式均应保持不变，故 α 和 k_H 亦应为定值。这阶段中，由于物料表面保持完全润湿，若不考虑热辐射对物料温度的影响，则湿物料表面达到的稳定温度即为空气的湿球温度 t_w，与 t_w 对应的 H_{s,t_w} 值也应恒定不变。所以，这种情况湿物料和空气间的传热速率和传质速率均保持不变。这样，湿物料以恒定的速率汽化水分，并向空气中扩散。

在恒速干燥阶段，空气传给湿物料的显热等于水分汽化所需的潜热，由此可推出：

$$U=k_H(H_{s,t_w}-H)=\frac{\alpha}{r_w}(t-t_w) \tag{5-36}$$

显然，干燥速率可根据给对流传热系数 α 确定。对于静止的物料层，α 可根据相应的经验关联式计算。

恒速干燥阶段属物料表面非结合水分汽化过程，与自由液面汽化水分情况相同。这个阶段的干燥速率取决于物料表面水分的汽化速率，亦即取决于物料外部的干燥条件，故又称为表面汽化控制阶段。

由式（5-36）可知，影响恒速干燥速率的因素有 α、k_H、$t-t_w$、$H_{s,t_w}-H$。提高空气流速能增大 α 和 k_H，而升高空气温度、降低空气湿度可增大传热和传质的推动力 $t-t_w$ 和 $H_{s,t_w}-H$。此外，水分从物料表面汽化的速率与空气同物料接触方式有关。图 5-33 所示的 3 种接触方式中，以图 5-33(c) 接触效果最佳，不仅 α 和 k_H 最大，而且单位质量物料的干燥面积也最大；图 5-33(b) 次之；图 5-33(a) 最差。

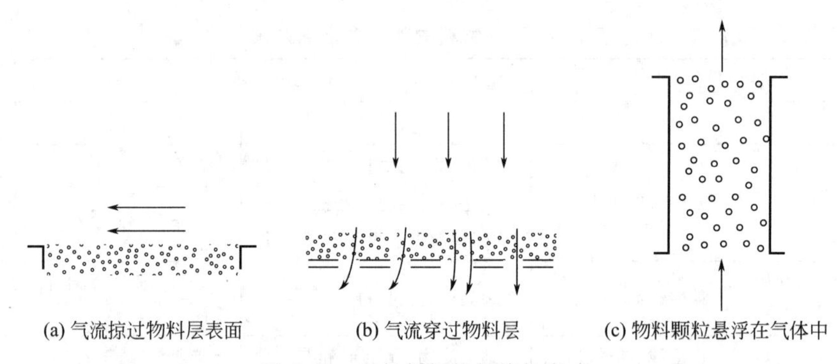

图 5-33 空气与物料的接触方式

应注意的是干燥操作不仅要求有较大的汽化速率，而且还要考虑气流的阻力、物料的粉碎情况、粉尘的回收、物料耐温程度以及物料在高温、低湿度气流中的变形或收缩等问题。所以，对于具体干燥物系，应根据物料特性及经济核算等来确定适宜的气流速度、温度和湿度等。

Drying in the constant-rate period

Drying of different solids under different constant conditions of drying will often give curves of different shapes in the falling-rate period, but in general the two major portions of the drying-rate curve, constant-rate period and falling-rate period, are present.

In the constant-rate drying period, the surface of the solid is initially very wet and a continuous film of water exists on the drying surface. This water is entirely unbound water and it acts as if the solid were not present. The rate of evaporation under the given air conditions is independent of the solid and essentially the same as the rate from a free liquid surface. Increased roughness of the solid surface, however, may lead to higher rates than from a flat surface.

If the solid is porous, most of the water evaporated in the constant-rate period is supplied from the interior of the solid. This period continues only as long as the water is supplied to the surface as fast as it is evaporated. Evaporation during this period is similar to that in determining the wet bulb temperature, and in the absence of heat transfer by radiation or conduction, the surface temperature is approximately the same as the wet bulb temperature.

② 降速干燥阶段（drying in the falling-rate period）　如图5-32所示，当物料的含水量降至临界含水量 X_c 以下时，物料的干燥速率随其含水量的减小而降低。此时，因水分自物料内部向表面迁移的速率低于物料表面水分的汽化速率，所以湿物料表面逐渐变干，汽化表面逐渐向内部移动，表面温度逐渐上升。随着物料内部水分含量的不断减少，物料内部水分迁移速率不断降低，直至物料的含水量降至平衡含水量 X^* 时，物料的干燥过程便停止。

在降速干燥阶段中，干燥速率的大小主要取决于物料本身结构、形状和尺寸，与外界干燥条件关系不大，故降速干燥阶段又称为物料内部迁移控制阶段。

综上所述，恒速干燥阶段和降速干燥阶段速率的影响因素不同。因此，在强化干燥过程时，首先要确定在某一定干燥条件下物料的临界含水量 X_c，再区分干燥过程属于哪个阶段，然后采取相应措施以强化干燥操作。

（4）干燥时间的计算

恒定干燥条件下，干燥时间等于恒速阶段干燥时间 τ_1 与降速阶段干燥时间 τ_2 之和，即：

$$\tau_{总}=\tau_1+\tau_2 \tag{5-37}$$

① τ_1 的计算：

因为：

$$U_c=\frac{-G'\mathrm{d}X}{S\mathrm{d}\tau}\Rightarrow\int_0^{\tau_1}\mathrm{d}\tau=-\frac{G'}{U_cS}\int_{X_1}^{X_c}\mathrm{d}X$$

所以：

$$\tau_1=\frac{G'}{U_cS}(X_1-X_c) \tag{5-38}$$

式中　X_1——物料的初始含水量，kg水/kg绝干物料；

G'/S——单位干燥面积上的绝干物料质量，kg绝干物料/m^2。

又因为：

$$U_c=\frac{\alpha(t-t_w)}{r_{t_w}}$$

所以：

$$\tau_1=\frac{Gr_{t_w}}{S\alpha}\times\frac{X_1-X_2}{t-t_w} \tag{5-39}$$

② τ_2 的计算：

$$\tau_2=\int_0^{\tau_2}\mathrm{d}\tau=-\frac{G'}{S}\int_{X_c}^{X_2}\frac{\mathrm{d}X}{U} \tag{5-40}$$

式中　X_2——降速阶段终了时物料的含水量，kg 水/kg 绝干物料；

　　U——降速阶段的瞬时干燥速度，kg/(m^2·s)。

若 U 与 X 呈线性关系，即：　$U=K_x(X-X^*)$

则：
$$\tau_2=\frac{G'}{S}\int_{X_2}^{X_c}\frac{\mathrm{d}X}{K_x(X-X^*)}=\frac{G'}{SK_x}\ln\frac{X_c-X^*}{X_2-X^*} \tag{5-41}$$

又因为：

$$U_c=K_x(X_c-X^*)\Rightarrow K_x=\frac{U_c}{X_c-X^*}$$

代入式（5-41）中得到：

$$\tau_2=\frac{G'(X_c-X^*)}{SU_c}\ln\frac{X_c-X^*}{X_2-X^*} \tag{5-42}$$

式中　K_x——降速阶段干燥速率线的斜率，kg 绝干物料/(m^2·s)。

若 U 与 X 呈非线性关系，则可采用图解积分法求解。

5.2.5　干燥设备（Drying equipment）

5.2.5.1　干燥器的类型（types of dryers）

工业上常用干燥器类型多种多样，除气流干燥器外，还有以下几种主要类型：

（1）厢式干燥器（tray dryers）

厢式干燥器又称盘架式干燥器，如图 5-34 所示。其外形像一个箱子，外壁为绝热层，物料装在浅盘中，置于支架上，层叠放置。新鲜空气由风机引入，经预热器加热后沿挡板均匀进入各层挡板之间，吹过处于静止状态的物料而起干燥作用。部分废气经排除管排出，余下的废气循环使用以提高热效率。这种干燥器采用常压间歇式操作，可以干燥多种不同形态的物料，一般在下列情况下使用才合理：①小规模生产；②物料停留时间长时不影响产品质量；③同时干燥几种产品。

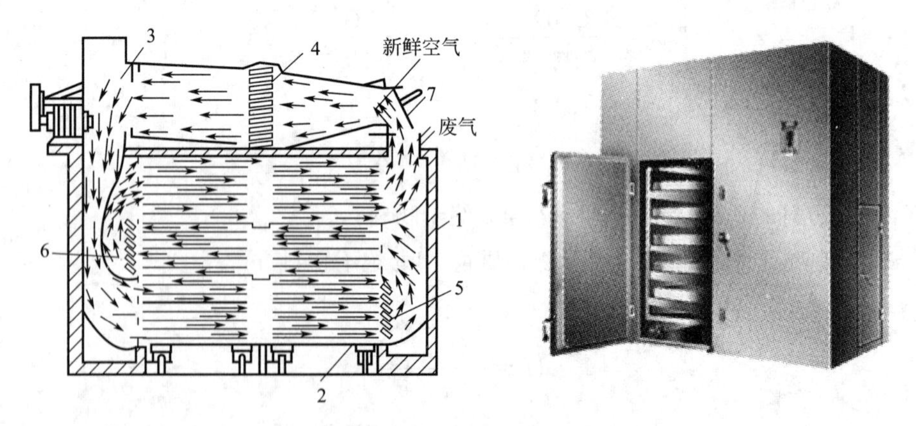

图 5-34　厢式干燥器

1—干燥室；2—小板车；3—送风机；4～6—空气预热器；7—调节门

（2）洞道式干燥器（screen-conveyor dryers）

洞道式干燥器是一种连续操作的干燥设备，如图5-35所示。外形为狭长的隧道，两端有门，底部有铁轨。待干燥的物料置于轨道的小车上，每隔一定时间用推车机推动小车前进，小车上的物料与热空气接触被干燥。这种干燥器容积大，常用来干燥陶瓷、木材、耐火制品等，缺点是干燥品种单纯，造价高。

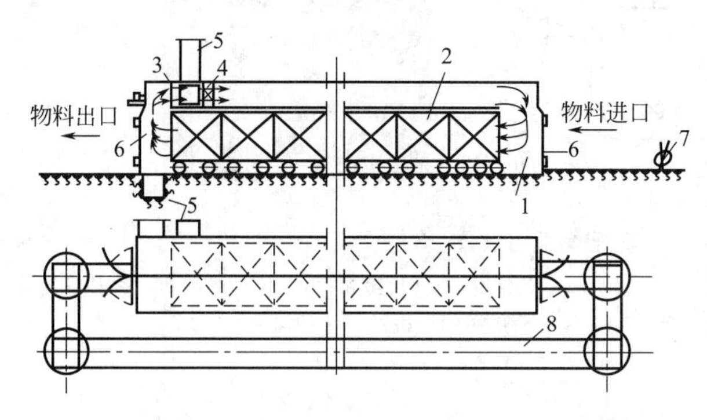

图5-35　洞道式干燥器

1—洞道；2—运输车；3—送风机；4—空气预热器；5—废气出口；6—封闭门；7—推送运输车的绞车；8—铁轨

（3）转筒干燥器（rotary dryers）

转筒干燥器的主体是与水平线稍成倾斜的可转动的圆筒，如图5-36所示。湿物料自转筒高的一端加入，自低的一端排出。转筒内壁安装有翻动物料的各式抄板，可使物料均匀分散，同时也使物料向低处流动。干燥介质常用热空气、烟道气等，被干燥的物料多为颗粒状或块状，操作方式可采用逆流或并流。这种干燥器对物料适应性强，生产能力大，操作控制方便，产品质量均匀；但设备复杂庞大，一次性投资大，占地面积大。

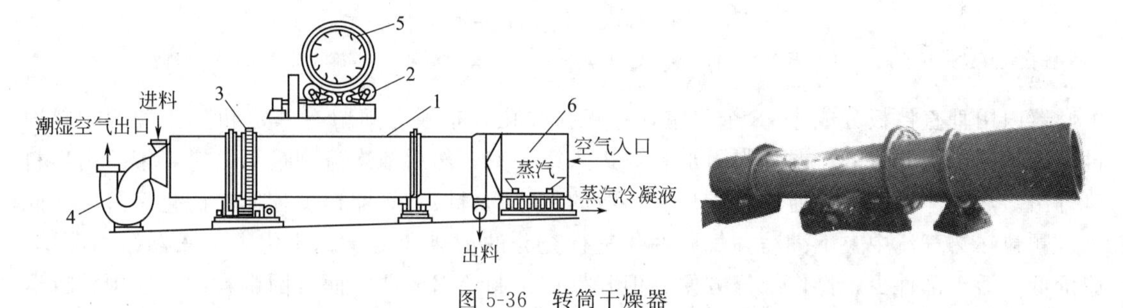

图5-36　转筒干燥器

1—圆筒；2—支架；3—驱动齿轮；4—风机；5—抄板；6—蒸汽加热器

（4）喷雾干燥器（spray dryers）

喷雾干燥器是连续式常压干燥器的一种，用于溶液、悬浮液或泥浆状物料的干燥，如图5-37所示。料液用泵送到喷雾器，在圆筒形的干燥室中喷成雾滴而分散于热气流中。物料与热气流以并流、逆流或混流的方式相互接触，使水分迅速汽化达到干燥的目的。干燥后可获得30～50μm粒径的干燥产品。产品经器壁落到器底，由风机吸至旋风分离器中被回收，废气经风机排出。这种干燥器干燥时间短，产品质量高，便于自动化控制；但对流传热系数小，热利用低，能量消耗大。

（5）流化床干燥器（fluid-bed dryers）

流化床干燥器（沸腾床干燥器）是流态化技术在干燥操作中的应用，如图5-38所示。

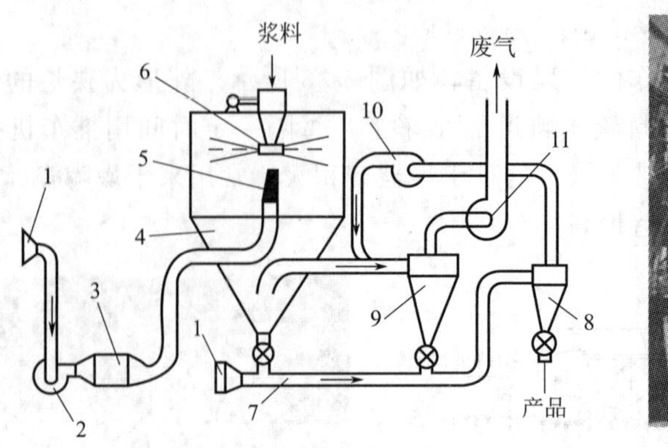

图 5-37　喷雾干燥器流程图

1—空气过滤器；2—送风机；3—预热器；4—干燥室；5—热空气分散器；6—雾化器；7—产品输送及冷却管道；8—1号分离器；9—2号分离器；10—气流输送用的风机；11—抽风机

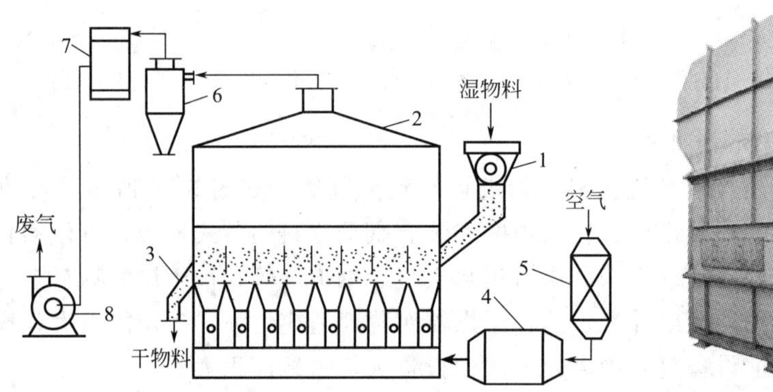

图 5-38　卧式多室沸腾床干燥器

1—摇摆式颗粒进料器；2—干燥器；3—卸料器；4—加热器；5—空气过滤器；6—旋风分离器；7—袋滤器；8—风机

干燥器内用垂直挡板分成4～8个隔室，挡板与多孔分布板之间留有一定间隙让粒状物料通过，以达到干燥的目的。湿物料由加料口进入第一室，然后依次流到最后一室，最后由出料口排出。热气体自下而上通过分布板和松散的粒状物料层，气流速度控制在流化床阶段。此时，颗粒在热气流中上下翻动，气固两相进行充分的传热和传质。流化床干燥器结构简单，造价低，活动部件少，操作维修方便；但其操作控制要求较严，而且因颗粒在床层中随机运动，可能引起物料的返混或短路。

5.2.5.2　干燥器的选型（selection of drying equipment）

干燥操作是一种比较复杂的过程，很多问题还不能从理论上解决，需要借助于经验。干燥器的类型和种类也很多，主要由物料的性质决定其所适用的干燥器。间歇操作的干燥器生产能力低，设备笨重，物料层是静止的，不适合现代化大生产的要求，只适合干燥小批量或多品种的产品。间歇操作的干燥器已逐渐被连续操作的干燥器所代替。连续操作的干燥器可以缩短干燥时间，提高产品质量，操作稳定，容易控制。

选择干燥器时，首先根据被干燥物料的性质和工艺要求选用几种可用的干燥器，然后通过对所选的干燥器的基建费和操作费进行经济核算，比较后最终选定一种最适用的干燥器。表5-4可作为干燥器选型的参考。

表 5-4 干燥器选型参考

项目		物料							
		溶液	泥浆	膏糊状	粒径 100 目以下	粒径 100 目以上	特殊形状	薄膜状	片状
加热方式	干燥器	无机盐、牛奶、萃取液、橡胶乳液等	颜料、纯碱、洗涤剂、石灰、高岭土、黏土等	滤饼、沉淀物、淀粉、染料等	离心机滤饼、颜料、黏土、水泥等	合成纤维、结晶、矿砂、合成橡胶等	陶瓷、砖瓦、木材、填料等	塑料薄膜、玻璃纸、纸张、布匹等	薄板、泡沫塑料、照相材料、印刷材料、皮革、三夹板
对流加热	气流	5	3	3	4	1	5	5	5
	流化床	5	3	3	4	1	5	5	5
	喷雾	1	1	4	5	5	5	5	5
	转筒	5	5	3	1	1	5	5	5
	盘架	5	4	1	1	1	1	5	1
传导加热	耙式真空	4	1	1	1	1	5	5	5
	滚筒	1	1	4	4	5	5	多滚筒	5
	冷冻	2	2	2	2	2	5	5	5
辐射加热	红外线	2	2	2	2	2	1	1	1
介电加热	微波	2	2	2	2	2	1	2	2

注：1 为适合；2 为经费许可时才适合；3 为特定条件下适合；4 为适当条件时可应用；5 为不适合。

干燥器的设计计算采用物料衡算、热量衡算、速率关系和平衡关系四类基本方程，但由于干燥过程的机理比较复杂，因此干燥器的设计仍借助经验或半经验的方法进行。各种干燥器的设计方法差别很大，但设计的基本原则是物料在干燥器内的停留时间必须等于或稍大于所需的干燥时间。

Selection of Drying Equipments

The first consideration in selecting a dryer is its operability; above all else, the equipment must produce the desired product in the desired form at the desired rate. Despite the variety of commercial dryers on the market, the various types are largely complementary, not competitive, and the nature of the drying problem dictates the type of dryer that must be used, or at least limits the choice to perhaps two or three possibilities. The final choice is then made on the basis of capital and operating costs. Attention must be paid, however, to the costs of the entire isolation system, not just the drying unit alone.

General considerations

There are some general guidelines for selecting a dryer, but it should be recognized that the rules are far from rigid and exceptions not uncommon. Batch dryers, for example, are most often used when the production rate of dried solid is less than 150 to 200kg/h; continuous dryers are nearly always chosen for production rates greater than 1 or 2 t/h. At intermediate production rates other factors must be considered. Thermally sensitive materials must be dried at low temperature under vacuum, with a low-temperature heating medium, or very rapidly as in a flash or spray dryer. Fragile crystals must be handled gently as in a tray dryer, a screen-conveyor dryer, or tower dryer.

The dryer must also operate reliably, safely, and economically. Operating and maintenance costs must not be excessive; pollution must be controlled; energy consumption must be minimized. As with other equipment, these considerations may conflict with one another, and a compromise must be reached in finding the optimum dryer for a given service.

As far as the drying operation itself is concerned, adiabatic dryers are generally less expensive than non-adiabatic dryers, in spite of the lower thermal efficiency of adiabatic units. Unfortunately there is usually a lot of dust carry over from adiabatic dryers, and these entrained particles must be removed almost quantitatively from the drying gas. Elaborate particle removal equipment may be needed, equipment that may cost as much as the dryer itself. This often makes adiabatic dryers less economical than a 'buttoned-up' non-adiabatic system in which little or no gas is used. Rotary dryers are an example; they were once the most common type of continuous dryer, but because of the inevitable entrainment, other types of dryers which avoid the problem of dust carryover would now, if possible, be selected in their place. Non-adiabatic dryers are always chosen for very fine particles or for solids that are too chemically reactive to be exposed to a stream of gas. They are also widely used for solvent removal and recovery.

5.3 膜分离技术（Membrane Separation Technology）

膜分离（membrane separation）利用天然或人工合成的、具有选择透过能力的薄膜，以外界能量或化学位差为推动力，对双组分或多组分体系进行分离、分级、提纯或富集。分离膜可以是固体或液体。反应膜（reaction membrane）除起到反应体系中物质分离作用外，还作为催化剂或催化剂的固载体，改变反应进程，提高反应效率。膜分离技术由于具有常温下操作、无相态变化、高效节能、在生产过程中不产生污染等特点，因此在饮用水净化、工业用水处理，食品、饮料用水净化、除菌，生物活性物质回收、精制等方面得到广泛应用，并迅速推广到纺织、化工、电力、食品、冶金、石油、机械、生物、制药、发酵等各个领域。当利用常规分离方法不能经济、合理地进行分离时，膜分离过程（membrane separation process）作为一种分离技术就有可能特别适用。由于膜分离技术特别适用于热敏性物质的处理，所以在食品加工、医药、生化技术等领域有其独特的适用性。它也可以和常规的分离单元结合起来作为一个单元操作来运用。例如，膜渗透单元操作可用于蒸馏塔加料前破坏恒沸点混合物。

5.3.1 膜分离技术发展简介

在人类的生活和生产实践中，人们早已不自觉地接触和应用了膜分离技术。在我国汉代的《淮南子》中已有制豆腐的记叙。后来，人们又知道了制豆腐皮、粉皮的方法。这可以说是人类利用天然物制得食用“人工薄膜”的最早记载。对膜分离技术的利用，最早的记述也可以追溯到2000多年以前。我国古代的先民们在酿造、烹饪、炼丹和制药的实践中，就利用了天然生物膜的分离特性。但随后的漫长历史进程中，我国的膜技术没有得到应有的发展。在国外，200多年前，Nollet在1748年就注意到了水能自发地扩散穿过猪膀胱而进入到酒精中的渗透现象。但由于受到人们认识能力和当时科技条件的限制，过了100多年后，1864年Traube才成功研制成人类历史上第一片人造膜——亚铁氰化铜膜。随后，研究工作一直徘徊不前。虽有Gibbs的渗透压理论及别的热力学理论作基础，但由于没有可靠的膜可供采用，研究工作曾一度被迫停顿。

到了20世纪，由于物理化学、聚合物化学、生物学、医学和生理学等学科的深入发展，新型膜材料及制膜技术的不断开拓，各种膜分离技术才相继出现和发展，并渗入研究和工业生产的各个领域，反渗透、超滤、微滤、电渗析和气体膜分离等技术开始在水的脱盐和纯化、石油化工、轻工、纺织、食品、生物技术、医药、环境保护等领域得到应用。1925年世界上第一个滤膜公司（Sartorius）在德国哥廷根成立。1930年，Treorell Meyer、Sievers等对膜电动势的研究，为电渗析和膜电极的发明打下了一定的基础。1950年W. Juda等试制成功第一张具有实用价值的离子交换膜，电渗析过程得到迅速发展。

膜分离技术的快速发展和工业应用是在20世纪60年代以后，当时在大规模生产高通量、无缺陷的膜和紧凑的、高比表面积的膜分离器上取得突破，开发了水中脱盐的反渗透过程，获得巨大的经济效益和社会效益。l960年洛布和索里拉简首次研制成世界上具有历史意义的非对称反渗透膜，这在膜分离技术发展中是一个重要的突破，使膜分离技术进入了大规模工业化应用的时代。我国膜科学技术的发展是从1958年研究离子交换膜开始的；20世纪60年代进入开创阶段；1965年着手反渗透的探索，1967年开始的全国海水淡化会战，大大促进了我国膜科技的发展；20世纪70年代进入开发阶段，这时期，微滤、电渗析、反渗透和超滤等各种膜和组器件都相继研究开发出来；20世纪80年代跨入了推广应用阶段。80年代又是气体分离和其他新型膜分离技术的开发阶段。

20世纪膜分离技术发展的大致历史如图5-39所示。此外，以膜为基础的其他新型分离过程，以及膜分离与其他分离过程结合的集成过程也日益得到重视和发展。40多年来，作为一门新型的高分离、浓缩、提纯及净化技术，新的膜技术不断地得到开发研究，如渗透汽化、膜蒸馏、支撑液膜、膜萃取、膜生物反应器、控制释放膜、仿生膜及生物膜等过程的研究工作不断深入。这些合成膜技术主要应用在四大方面：①分离（微滤、超滤、反渗透、电渗析、气体分离、渗透汽化、渗析）；②控制释放（治疗装置、药物释放装置、农药持续释放、人工器官）；③膜反应器（酶和催化剂反应器、生物反应传感装置、移植的免疫隔离）；④能量转换（电池隔膜、燃料电池隔膜、电解池隔膜、同体聚电解质）。这些膜过程，有的已经在生产上应用，有的即将进入实用阶段。同时，由于各种膜分离过程在使用中的最大问题是膜的污染与劣化，所以膜的形成机理、合成材料和条件以及如何控制其结构、料液的预处理、组件的流体力学条件的优化、膜的清洗等，就成为膜科学技术领域中的重要内容。

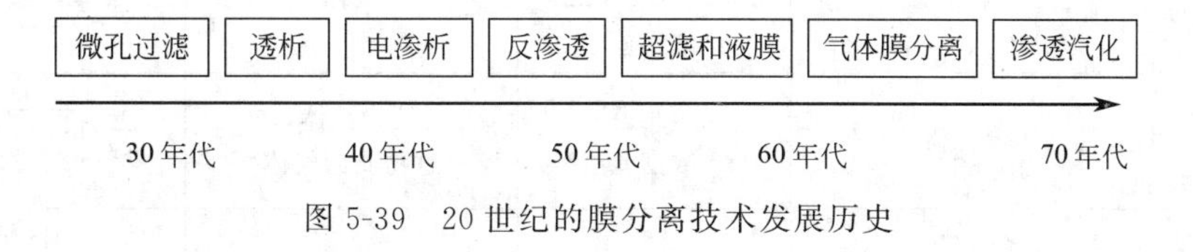

图5-39 20世纪的膜分离技术发展历史

膜科学目前的主要发展方向为：①集成膜过程；②杂化过程；③水的电渗离解；④细胞培养的免疫隔离；⑤膜反应器；⑥催化膜；⑦手征膜（chirale）。1994年世界膜和膜组件的销售总值为35亿美元（我国为0.24亿美元），且每年以14%～30%的速度继续增长，最大的市场为生物医学市场。更严的环保法规、更高的能源和原材料价格进一步刺激了膜分离技术市场的发展。可以说，膜分离技术已成为解决当代能源、资源和环境污染问题的重要高新技术及可持续发展技术的基础。

5.3.2 各种固体膜分离过程简介（Introduction of various solid membrane separation process）

物质通过膜的分离过程是复杂的，膜的传递模型可分为两类。第一类以假定的传递机理为基础，其中包含了被分离物的物化性质和传递特性。这类模型又可分为两种不同情况：一是通过多孔型膜的流动；另一是通过非多孔型膜的渗透。前者有孔模型、微孔扩散模型和优先吸附-毛细孔流动模型等，后者有溶解-扩散模型和不完全的溶解-扩散模型等。第二类以不可逆热力学为基础，称为不可逆热力学模型。它从不可逆热力学唯象理论出发，统一关联了压力差、浓度差、电位差等对渗透流率的关系。

工业化应用膜分离技术的分类及其基本特征性见表 5-5。

表 5-5 工业化应用膜分离技术的分类及其基本特性

过程	分离目的	推动力	进料和透过物质的相态	结构简图
微滤 MF	溶液脱粒子气体脱粒子	压力差 100kPa	液体或气体	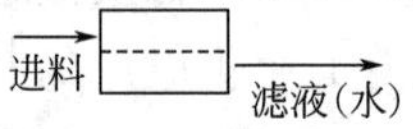 进料 滤液(水)
超滤 UF	溶液脱大分子、大分子溶液脱小分子、大分子分级	压力差 100～1000kPa	液体	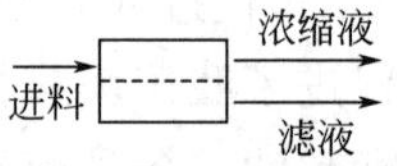 浓缩液 进料 滤液
纳滤 NF	溶剂脱有机组分、脱高价粒子、软化、脱色、浓缩、分离	压力差 500～1500kPa	液体	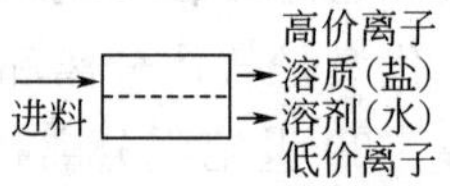 高价离子 溶质(盐) 进料 溶剂(水) 低价离子
反渗透 RO	溶剂脱溶质、含小分子溶质的溶液浓缩	压力差 1000～10000kPa	液体	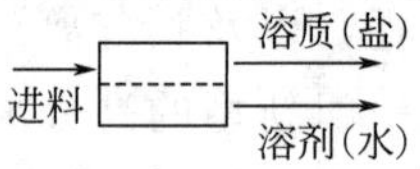 溶质(盐) 进料 溶剂(水)
渗析 D	大分子溶质溶液脱小分子、小分子溶质溶液脱大分子	浓度差	液体	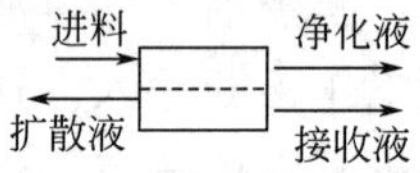 进料 净化液 扩散液 接收液
电渗析 ED	溶液脱小离子、小离子溶质的浓缩、小离子的分级	电化学势电渗透	液体	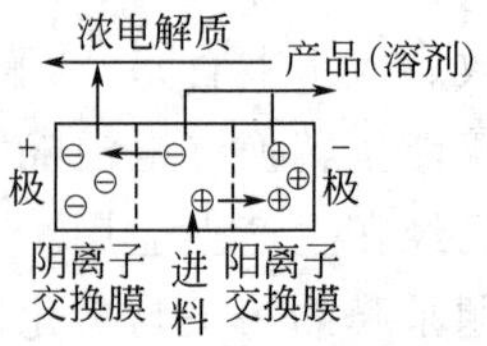 浓电解质 产品(溶剂) +极 −极 阴离子交换膜 进料 阳离子交换膜
气体分离 GS	气体混合物分离、富集或特殊组分脱除	压力差 1000～10000kPa 浓度差	气体	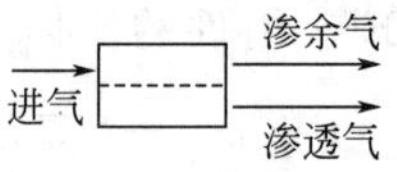 渗余气 进气 渗透气
渗透汽化 PVAP	挥发性液体混合物分离	分压差 浓度差	料液为液体，透过物质为气体	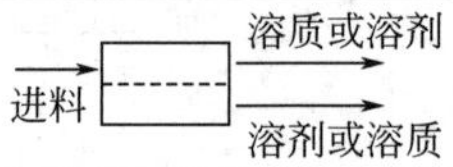 溶质或溶剂 进料 溶剂或溶质
乳化液膜 ELM(ET)	液体混合物或气体混合物分离、富集、特殊组分脱除	浓度差 pH 差	通常都为液体，也可分为气体	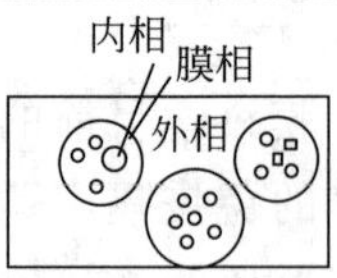 内相 膜相 外相

5.3.2.1　超滤

超滤（ultrafiltration，UF）属于以压力差为推动力的膜过程。一般地说超滤膜表面孔径范围大体在5nm到几百毫米之间，主要用于水溶液中大分子、胶体、蛋白质等的分离。超滤膜对溶质的截留作用是由于：①在膜表面及微孔内的吸附（一次吸附）；②在孔中的停留（阻塞）；③在膜表面的机械截留（筛分）。其中，一次吸附或阻塞会发生到什么程度，与膜及溶质之间的相互作用和溶质浓度、操作压力、过滤总量等因素有关。超滤最早出现于20世纪初，最初的超滤膜是纤维素衍生物膜，60年代以后开始有其他的高聚物商品超滤膜，近10年来无机膜成为受人重视的超滤膜新系列。超滤组件主要形式有管式、板框式、中空纤维式和螺旋卷式四种。中空纤维是目前最广泛应用的一种，近年来具有错流结构的螺旋卷式组件受到重视，管式和板框式水力学流动状态较好，适用于黏度大、高浓度或不是很干净的废料液。

超滤特别适用于热敏性和生物活性物质的分离和浓缩，其主要应用有以下几个方面：①食品加工，如乳制品工业是国外食品工业中应用膜技术最多的部门。1988年世界上用于乳品加工的超滤膜已超过150000m^2，每年以30%的速度增长。还可用于饮料和酒类的加工和食醋、酱油的精制除菌。②生物工程上的酶的精制、生物活性物质的浓缩分离。③在环境工程中用于处理工业废水，已在纺织、造纸、金属加工、食品及电泳材料等部门推广应用。④医疗卫生及超纯水制备等。

5.3.2.2　微孔过滤

微孔过滤（微滤）(microfiltration，MF)、超滤和反渗透都是以压力差为推动力的液相膜分离过程，三者之间并无严格的界限，它们组成了一个可分离固态颗粒到离子的三级膜分离过程。一般认为，微孔过滤的有效过滤范围为直径0.1～10μm的颗粒，操作静压差为0.01～0.2MPa。微孔过滤的分离机理主要是筛分效应。因膜的结构不同，截留作用大体分为：机械截留、吸附截留、架桥作用及网络型膜的网络内部截留作用。微孔过滤是开发最早、应用最广泛的滤膜技术。1925年在德国哥廷根成立了世界上第一个滤膜公司。目前世界上产品销售额达到15亿美元以上，年增长率约为15%。微孔过滤膜的主要特点为孔径的均一性、高空隙率和滤材薄。长期以来，微孔过滤膜的材料是高分子材料，随着膜分离技术在工业领域应用的迅速扩展，近十几年来无机微滤膜引起重视，并取得了很大进步。

微孔过滤膜组件也由框式、圆管式、螺旋卷式和中空纤维式等四类组成。工业上应用的微孔过滤装置主要为板框式，它们大多仿照普通过滤器的形式设计。在微孔过滤中有死端过滤和错流过滤两种操作，错流过滤的原料液流动方向与过滤液的流动方向呈直角交叉状态，可使膜表面沉降物不断被原料液的横向液流冲跑，因而不易将膜表面覆盖，避免过滤速度下降。对于固含量高于0.5%、易产生堵塞的原料应采用错流过滤。

我国的微滤技术起步较晚，直到20世纪70年代末还只有个别单位形成小批量生产能力，但近十年来我国微孔过滤膜及过滤器件的研究和开发取得了很大的进步。目前，我国已形成正式商品生产的微孔过滤膜，仍然以纤维素体系滤膜为主。聚酰胺、聚偏氟乙烯、聚砜、聚丙烯等微孔过滤膜中的少数品种已有商品出售。

微滤技术在我国纯水制造上已广泛应用，1985年前，我国使用的微孔膜滤芯主要靠进口，近年来这一状况已基本改变。目前正处在以国产滤芯替代进口滤芯的阶段；用于高产值生物制品分离的微孔过滤膜，主要仍从国外进口，在制造无菌液体和用于饮料、医药制品分离方面基本上还处于小规模试验阶段，少部分已用于生产线上。微孔过滤技术在我国国民经济中发挥更大作用，有赖于国产微孔过滤膜和滤芯品种、规模、性能的进一步提高。

Ultrafiltration

Ultrafiltration retains particles of submicron size by ultramicroporous membranes. Typically, ultrafiltration retains solutes in the 300-500,000 molecular weight range, including biomolecules, polymers, sugars, and colloidal particles.

Microfiltration

Microfiltration is a pressure-driven, microporous membrane process used to retain matter as low as 0.02 micron in size, but more commonly of 0.1-10 microns. The matter may include large colloids, small and solid particles, blood cells, yeast, bacteria and other microbial cells, and very large and soluble macromolecules. Membrane structures for microfiltration include screen filters that collect retained matter on the surface and depth filters that trap particles at constrictions within the membrane.

5.3.2.3 反渗透

反渗透（reverse osmosis，RO）是用足够的压力使溶液中的溶剂（一般常指水）通过反渗透膜（一种半透膜）而分离出来，方向与渗透方向相反，可使用大于渗透压的反渗透法进行分离、提纯和浓缩溶液。利用反渗透技术可以有效地去除水中的溶解盐、胶体、细菌、病毒、细菌内毒素和大部分有机物等杂质。反渗透膜的主要分离对象是溶液中的离子，不需要化学品即可有效脱除水中盐分，系统除盐率一般为98%以上。所以反渗透是最先进的也是最节能、环保的一种脱盐方式，也已成为主流的预脱盐工艺。

1953年美国佛罗里达大学的Reid等人最早提出反渗透海水淡化，1960年美国加利福尼亚大学的Loeb和Sourirajan研制出第一张可实用的反渗透膜。从此以后，反渗透膜开发有了重大突破，膜材料从初期单一的醋酸纤维素非对称膜发展到用表面聚合技术制成的交联芳香族聚酰胺复合膜。操作压力也扩展到高压（海水淡化）膜、中压（醋酸纤维素）膜、低压（复合）膜和超低压（复合）膜。20世纪80年代以来，又开发出多种材质的纳滤膜，目前已广泛运用于科研、医药、食品、饮料、海水淡化等领域。

Reverse Osmosis

Osmosis, from the Greek word for 'push', refers to the passage of a solvent, such as water, through a membrane that is much more permeable to the solvent (A) than to the solute (s)(B)(e.g., inorganic ions). The first recorded account of osmosis was given in 1748 by Nollet, whose experiments were conducted with water, an alcohol, and an animal-bladder membrane. Reverse osmosis can be used to partially remove a solvent from a solute-solvent mixture. An important factor in developing a reverse osmosis separation process is the osmotic pressure.

5.3.2.4 渗透汽化

渗透汽化（pervaporation，PVAP）是指液体混合物在膜两侧组分的蒸气分压差作用下，其中组分以不同速度透过膜并蒸发除去，从而达到分离目的的一种膜分离方法。正因为这一过程是由“渗透”和“蒸发”两个步骤组成，又称为渗透蒸发。渗透汽化技术是膜分离技术的一个新的分支，也是热驱动的蒸馏法与膜法相结合的分离过程。不过，它不同于反渗透等膜分离技术，因为它在渗透过程中将产生由液相到气相的相变。渗透汽化的原理是在膜的上游连续输入经过加热的液体，而在膜的下游则以真空泵抽吸造成负压，从而使特定的液体组分不断透过分离膜变成蒸气，然后，再将此蒸气冷凝成液体而得以从原料液中分离除去。渗透汽化与反渗透、超滤及气体分离等膜分离方法的最大区别在于，前者透过膜时，物料将产生相变。因此，在操作过程中，必须不断加入至少相当于透过物潜热的热量，才能维持一定的操作温度。

渗透汽化过程有如下特点：①渗透汽化的单级选择性好，这是它的最大特点。从理论上讲，渗透汽化的分离程度无极限，适合分离沸点相近的物质，尤其适用于恒沸物的分离。对于回收含量少的溶剂也不失为一种好的方法。②由于渗透汽化过程有相变发生，所以能耗较高。③渗透汽化过程操作简单，易于掌握。④在操作过程中，进料侧原则上不需加压，所以不会导致膜的压密，因而其透射率不会随时间的增加而减小。而且，在操作过程中将形成活性层及所谓的“干区”，膜可自动转化为非对称膜，此特点对膜的透射率及寿命有益。⑤渗透汽化的通量较小。基于上述特点，在一般情况下，渗透汽化技术尚难与常规的分离技术相比，但由于它所特有的高选择性，在某些特定的场合，例如在混合液中分离少量难以分离的组分（例如对一些结构相似和沸点接近的有机溶剂——苯）和在常规分离手段无法解决或虽能解决但能耗太大的情况下，采用该技术则十分合适。

Pervaporation

Pervaporation is a separation process in which one or more components of a liquid mixture diffuse through a selective membrane, evaporate under low pressure on the downstream side, and are removed by a vacuum pump or a chilled condenser. Composite membranes are used with the dense layer in contact with the liquid and the porous supporting layer exposed to the vapor. The phase change occurs in the membrane, and the heat of vaporization is supplied by the sensible heat of the liquid conducted through the thin dense layer. The decrease in temperature of the liquid as it passes through the separator lowers the rate of permeation, and this usually limits the application of pervaporation to removal is needed, several stages are used in series with intermediate heaters. Commercial units generally use flat-sheet membranes stacked in a filter-press arrangement, with spacers acting as product channels, although spiral-wound membranes could also be used. Hollow-fiber membranes are not as suitable because of the pressure drop from the permeate flow through the small-bore fibers.

5.3.2.5 气体膜分离

气体膜分离（gas membrane separation，GS）技术大规模工业化只是近15年的事。通常的气体分离膜可分成多孔质及非多孔质两种，它们分别由无机物膜材质和有机高分子膜材质组成。选择性高分子膜材质的性能对气体渗透的影响是十分明显的。气体分离用聚合物的选择通常是同时兼顾其渗透性与选择性。

以膜法分离气体的基本原理，主要是根据混合原料气中各组分在压力作用下，通过半透膜的相对传递速率不同而得以分离。与反渗透、超滤等液相膜分离过程类似，气体分离膜在实际应用中也是以压差作为推动力的。一般需要把分离膜组装到一定形式的组件中操作，气体膜组件由板式、圆管式、螺旋卷式和中空纤维式等四种类组成，不过，工业上用得比较多的是后面两种，特别是中空纤维式膜组件。

气体膜分离技术在以下方面得到应用，如在从合成氨弛放气中回收氢、合成气的比例调节、工业用锅炉及玻璃窑的富氧空气燃烧、油田气中分离回收 CO_2、从工业废气中回收有机蒸气、空气等气体的脱除等方面都做出了贡献。

5.3.3 液膜分离技术（Technology of liquid membrane separation）

液膜（liquid membrane）是液体膜的简称。和固体膜分离法相比，液膜分离法具有选择性高、相传质速率大的特点，因此被各界日益重视。

5.3.3.1 液膜分类（classification of liquid membrane）

液膜是由悬浮在液体中一层很薄的乳液微粒构成的。乳液通常由溶剂（水或有机溶剂）、表面活性剂（乳化剂）和添加剂组成。其中溶剂构成膜的基体；表面活性剂含有亲水基和疏水基，它可以定向排列以固定油水分界面，使膜的形状得以稳定。通常膜的内相（分散相）试剂与液膜是不互溶的，而膜的内相与膜的外相（连续相）是互溶的。将乳液分散在第三相（连续相），就形成了液膜。液膜从形态上可分为乳化液膜和支撑液膜。

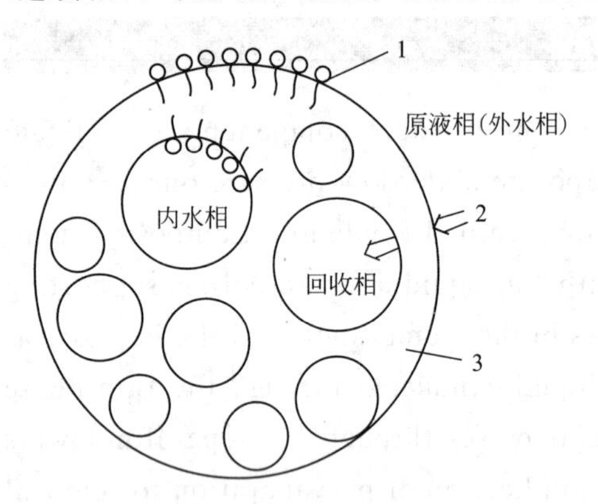

图 5-40 乳化液膜的形成

1—表面活性剂；2—溶质；3—液膜相（载体、表面活性剂、膜强化剂、有机溶剂）

① 乳化液膜（emulsion liquid membrane, ELM） 乳化液膜的形成如图 5-40 所示。首先将回收液（内相）同液膜溶液充分乳化制成 W/O（油包水）型乳液，然后令其分散于原液（外相）中形成 W/O/W（水包油包水）型多相乳液。介于被包封的内相与连续的外相之间即是液膜相。由于液膜对各种物质的选择渗透能力不同，因此它能将溶液中的某种物质捕集到内相或外相，而达到分离的目的。通常，内相的微滴直径为数微米，而 W/O 型乳液的滴径约为 0.1～1mm，膜的有效厚度为 1～10μm。因而，单位体积中膜的总面积非常大，溶液组分的透过速度相当快，其传质速度比一般的聚合物膜高数十至数百倍。

表面活性膜剂在乳化液膜中是不可缺少的组分。它是由亲水基和疏水基两部分组成，两种活性基团构成了不同的亲油平衡值。一般，以 HLB（hydrophile lipophile balance）表示。若想形成油包水型液膜，可选用 HLB 值为 3～6 的表面活性剂；若想形成水包油型液膜，则选用 HLB 值为 8～18 的表面活性剂。

为了重新使用已用过的乳液，必须将已形成的并经过分离操作的乳液进行破坏，称为“破乳（demulsification）”，从中分出膜相和内相，以分别进行处理。破乳的成功与否关系到乳化液膜的分离成败，它是整个分离操作的关键。破乳的方法通常有：化学法、静电法、离心法与加热法。从迄今使用的效果来看，其中以静电法较好。图 5-41 为破乳装置示意图。

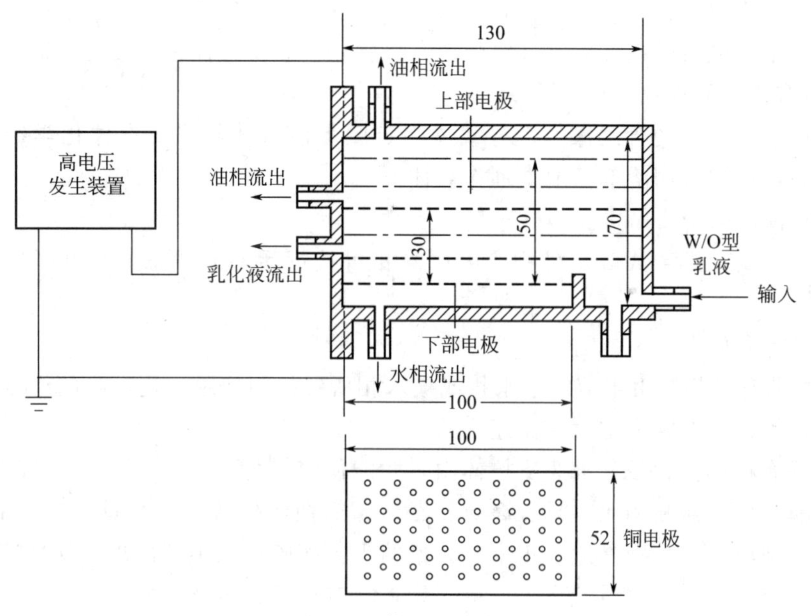

图 5-41　静电破乳装置

静电破乳的原理是让 W/O 型乳液通过两块平行的裸电极板之间，在两极板间外加的脉冲式直流高电压的作用下，达到破乳目的。这种装置的特点是两极板间的距离可在 1～5cm 内改变，被破乳后的油相能通过上部极板的孔，而水相则由下部极板的孔分别由装置中不断排出，未被破乳的乳液根据需要也可被导出。

② 支撑液膜（supported liquid membrane，SLM）　用于水溶液处理的支撑液膜如图 5-42 所示。液膜支撑体主要采用疏水性多孔膜，液膜溶液借助微孔的毛细管力含浸于其内。

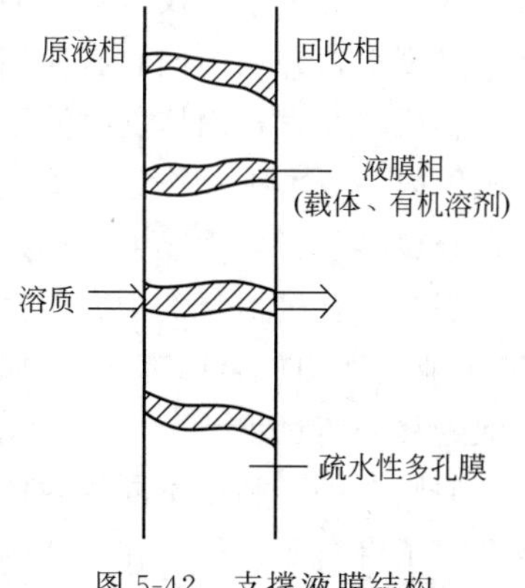

图 5-42　支撑液膜结构

支撑液膜的主要问题是性能衰减。其产生的主要原因有两个：一是被微量污物污染；二是微孔中的液膜溶液不断地向水中流失，从而导致液膜性能一再劣化。

目前，为使支撑液膜实用化而采用的膜组件主要有中空丝式和螺卷式。为使液膜达到大规模应用的水平，尚须进一步探索更新的液膜材料，开发最佳成膜工艺，彻底解决液膜的稳定性在高效破乳技术等方面的问题。

③ 乳化液膜与支撑液膜的比较　综上所述，乳化液膜与支撑液膜有各自的优缺点，归纳如下。

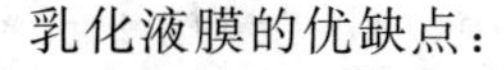

乳化液膜的优缺点：

a. 要使 W/O 型乳液分成有机膜相和内水相，需设置乳化器与破乳器。

b. 其分离过程相当复杂。

c. 膜面积非常大，一般为 1000～3000m^2/m^3（液体体积），因此透过速度十分高。

d. 若想使原液相接近活塞流状态，必需进行多级化分离。

e. 在溶剂萃取中，由于油水两相基本上已达平衡状态，因而采取逆流操作的方法，且分离各级必需设置沉降器。相反，对乳化液膜法来说，因内水相的反萃取剂捕集透过物的容量比较大，可采取顺流操作方法，各萃取槽不必设沉降器。

支撑液膜的优缺点：

a. 不必为相分离而设置沉降器（但当使有机膜相向膜支撑体中含浸时，需要小型油水分离器），因而有机相的损耗较少。

b. 膜的调制是单一的含浸操作，与载体固定膜（使载体与膜支撑体化学结合的膜）相比工序简单，可广泛采用溶剂萃取试剂作载体。

c. 多孔支撑体的价格昂贵。

d. 在中空丝式、螺卷式等组件中，因原液相可接近活塞流状态，所以可达到高脱除率效果。

e. 容易设计和放大。

f. 细孔内的有机相，由于溶解于水相或被水相所置换而使膜劣化，膜的厚度难以达到数十微米以下，因此膜的阻力增大，流速小。

g. 必须令有机相再含浸于液膜支撑体内，才能使液膜再生。

支撑液膜与乳状液膜相比，具有高效、经济、简便的优势，因此具有广泛的应用价值。自支撑液膜出现以来，特别是近十年来，大量的支撑液膜体系不断出现，支撑液膜体系已被用于分离有毒金属离子、放射性离子、稀土元素、有机酸、生物活性物质、药物、气体和手性物质。支撑液膜也成功地应用于分析过程中的样品预处理。

5.3.3.2 液膜应用（application of liquid membrane）

（1）金属离子处理（metal ion processing）

废水的处理，尤其是对含有金属离子的工业废水的处理，在环保事业中占有较大的比重。因为这类废水不仅量大，而且对生态环境污染十分严重。因此，采用较为有效的方法处理这类废水，并从中回收有使用价值的金属是当务之急。

相比较而言，固体支撑液膜处理这类工业废水有其独特优势。透过支撑液膜的受促迁移已被国内外专家推荐作为从溶液中选择性分离、浓缩和回收金属的一种新技术。在这类迁移中，金属离子可以“爬坡”透过液膜，即逆浓度梯度进行迁移。将可以流动的载体溶于同水不相混溶的有机稀释剂中并吸附于微孔聚丙烯薄膜上，该载体可以同水溶液中的金属离子（或工业废液中的金属离子），如 Zn^{2+}、Cd^{2+} 等形成膜的可溶性金属络合物，从而实现膜的受促迁移分离过程。

原则上讲，利用不同的载体可实现浓缩周期表中所有元素的目的。这又为微量元素的提取开辟了崭新的途径。在一些稀土元素的分离中，由于它们具有相似的性质，所以很难用一般的方法进行分离，支撑液膜技术则提供了有效的分离方法。

（2）有机酸处理（treatment of organic acid）

用支撑液膜分离有机酸与分离金属离子具有相似的机理。Aroca. G 等采用了三辛胺（TCA）作载体制成的支撑液膜体系，采用 Na_2CO_3 作解析试剂，对废水中的有机酸进行迁移并建立了定量的迁移模型。Molinari Raffacle 利用支撑液膜体系提取氨基酸，对应用条件进行了广泛研究，所建立的体系使用寿命较长，温度范围较宽，效果良好。Bryjak Marek 建立了聚乙烯多孔膜作支撑体的支撑液膜，该体系对不同立体结构的氨基酸进行分离，效果良好。

(3) 手性物质处理（treatment of chiral material）

由于手性化合物的性质极为相似，很难进行外消旋混合物的分离，但在制药工业上分离和提纯这些物质至关重要。在许多情况下，只有一种异构体是有效的，其他的是无效的或是有副作用的。例如，镇静剂的一种异构体是有效的，但另一种却是毒性很强的物质。因此，完全分离这样的异构体非常重要。这用一般的分离方法很难进行，但是用支撑液膜却可以得到很好的分离效果。

(4) 其他物质（other substances）

一些气体，如 NH_3、CO、NO、CO_2、H_2S、O_2、烯烃、炔烃等，都可以成功地用支撑液膜进行分离。据报道，用 $AgNO_3$ 作载体可以从含 33%乙烷、34%乙烯和 33%丙烯的混合物中分离得到 99%～99.9%的乙烯。

支撑液膜法现在逐渐开始应用于分析化学中，主要是用在分析试样前期处理，即分析成分的浓缩过程。

支撑液膜技术也开始应用于无机酸溶液的处理过程，为其应用拓宽了前景。

Liquid Membranes

For more than 30 years now, liquid membranes have been in focus of research. Since diffusivities in liquids in comparison to solids are higher by several orders of magnitude, enhanced permeabilities of liquid in comparison to solid membranes can be expected.

Generally, liquid membranes with and without supports can be differentiated. For those not employing supports, the so-called bulk liquid membranes (BLM) and emulsion liquid membranes (ELM) are found. The liquid membranes employing a support can be subdivided into immobilized liquid membranes (ILM), supported liquid membranes (SLM) and contained liquid membranes (CLM).

The simplest form of liquid membrane without support is given by the BLM consisting of a U-tube and three non-miscible liquids. BLM are mainly used to study mass transfer from the donor phase through the membrane phase into the acceptor phase but do not have any relevance for large-scale separation processes due to their large thickness.

In principle, ELM represent a double emulsion consisting of an acceptor phase being dispersed in a membrane phase and this emulsion again being dispersed in a donor phase. A species from the donor phase is absorbed into the membrane phase, diffuses towards the acceptor phase and finally is desorbed into the latter. To obtain the permeate, the double emulsion is disintegrated and the species is extracted from the acceptor phase. For liquid membranes employing supports, the most compact form of a LM is given by an ILM where a liquid is held inside the pores of a porous support (e. g. a porous solid membrane) by means of capillary forces. The support has to be wettable by the liquid for this configuration. If the support is not wetted by the liquid, a SLM can be prepare where a liquid is located on top of the porous support. The CLM represents a SLM with two porous supports on both sides.

This configuration offers the possibility of replenishing or regenerating the membrane phase during operation. Thus a breakdown of the membrane function caused by evaporation of the membrane liquid can be avoided by means of continuous liquid replenishment.

Liquid membranes work according to a solution-diffusion mass transfer mechanism as do dense solid membranes. Including the mass transfer steps in the respective feed and permeate phases, a gas molecule is transported across the membrane in seven steps:

(1) Convective transport of the molecule towards the membranes;

(2) Diffusion of the molecule through the boundary layer at the feed-membrane interface;

(3) Absorption into the membrane phase;

(4) Diffusion through the liquid membranes;

(5) Desorption into the permeate phase;

(6) Diffusion of the molecule through the boundary layer at the permeate-membrane interface;

(7) Convective transport of the molecule into the permeate phase.

The actual solution-diffusion mechanism is given by the steps (3)～(5) only.

5.3.4 膜蒸馏

初期的膜蒸馏（membrane distillation，MD）研究对象均为稀盐水溶液，1985 年后开始用于化学物质浓缩和回收及处理发酵液的新方向，以有机物水溶液及恒沸混合物为对象的膜蒸馏研究工作已有报道。膜蒸馏的分离模型如图 5-43(a) 所示。

在疏水性多孔膜的一侧与高温原料水溶液相接（即暖侧），而在膜的另一侧则与低温冷壁相邻（即冷侧）。正是借助这种相当于暖侧与冷侧之间温度差的蒸气压差，促使暖侧产生的水蒸气通过膜的细孔，再经扩散到冷侧的冷壁表面被凝缩下来，而液相水溶液由于多孔膜的疏水作用无法透过膜被留在暖侧，从而达到与气相水分离的目的。需要指出的是，这里的冷侧既可如图 5-43(b) 所示设一与膜保持一定距离（Z）的冷壁（间接接触法），也可不设冷壁而直接与冷却水相接触（直接接触法）。膜蒸馏过程的特征：膜为微孔膜，膜不能被所处理的溶液所润湿，在膜孔内没有毛细管冷凝现象发生，只有蒸气能通过膜孔传质，所用膜

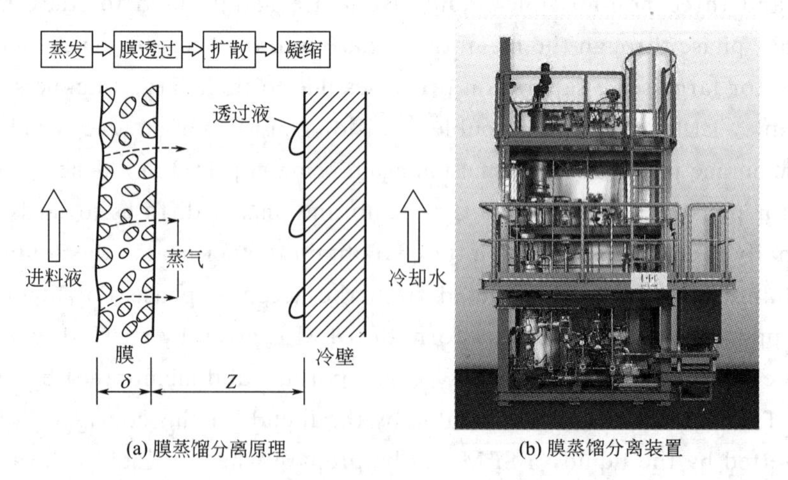

(a) 膜蒸馏分离原理　　(b) 膜蒸馏分离装置

图 5-43　膜蒸馏法的分离原理与实验装置

不能改变所处理液体中所有组分的气液平衡，膜至少有一面与所处理的液体接触，对于任何组分该膜分离过程的推动力是该组分在气相中的分压差。

膜蒸馏的优点为过程在常压和较低温度（40℃左右）下进行，设备简单、操作方便，有可能利用太阳能、地热、温泉、工厂的余热和温热的工业废水等廉价能源，因为只有水蒸气能透过膜孔，所以蒸馏液十分纯净，可望成为大规模、低成本制备超纯水的有效手段，是目前唯一能从溶液中直接分离出结晶产物的膜分离过程，膜蒸馏组件很容易设计成潜热回收形式，并具有以高效的小型组件构成大规模生产体系的灵活性。膜蒸馏的缺点是过程中有相变，热能的利用效率较低，考虑到潜热的回收，膜蒸馏在有廉价能源可利用的情况下才更有实际意义，膜蒸馏与制备纯水的其他膜分离过程相比通量较小，所以目前尚未实现在工业生产中应用。

Membrane Distillation

Membrane distillation (MD) is an emerging nonisothermal separation technique that uses microporous hydrophobic membrane in contact with an aqueous heated solution on the hand (feed or retentate) and a condensing phase (permeate or distillate) on the other. This technique belongs to the class of membrane contactors in which a nonwetting membrane does not act as a conventional barrier or filter, but promotes mass and energy exchange between two opposite interfaces according to principles of phase equilibrium.

In MD, the hydrophobic nature of the membrane prevents the mass transfer in liquid phase and creates a vapor-liquid interface at the entrance of each pore. Here, volatile compounds (most commonly water) evaporate, diffuse and/or convect across the membrane, and are condensed and/or removed on the opposite side of the system.

5.3.5 膜的性能参数

膜的性能包括膜的理化稳定性（physical and chemical stability）和分离透过特性（separation permeability）两方面。膜的理化稳定性是指膜对压力、温度、pH以及有机溶剂和各种化学药品的耐受性，它是决定膜的使用寿命的主要因素。膜的分离透过特性包括分离效率、渗透通量和通量衰减系数三个方面。

(1) 分离效率（separation efficiency）

分离效率是指膜对特定分离体系的分离能力。对不同的膜分离过程和分离对象，可以用不同的方法来表示膜的分离效率。

① 截留率 R　其定义为：

$$R=\frac{c_0-c_p}{c_0}\times 100\% \tag{5-43}$$

式中　c_0——原料液主体中被分离物质的摩尔浓度，$kmol/m^3$；

c_p——透过液中被分离物质的摩尔浓度，$kmol/m^3$。

② 截留分子量（molecular weight cut off，MWCO）　以截留率对截留物的分子量作图，可获得膜的截留分子量曲线，该曲线越陡，则被截留物质的分子量越窄，膜的孔径分布也越集中；反之，膜的孔径分布越宽。膜的截留分子量一般指截留率为90%时的物质的分子量。

③ 分离因子 α 或分离系数 β　其定义分别为：

$$\alpha=\frac{y_A/(1-y_A)}{x_A/(1-x_A)} \tag{5-44}$$

$$\beta=\frac{y_A}{x_A} \tag{5-45}$$

式中　x_A——原料液主体中组分 A 的摩尔分数；

y_A——透过液中组分 A 的摩尔分数。

由式(5-44) 和式(5-45) 可以看出，α 与 β 分别类似于蒸馏中的相对挥发度和相平衡常数的概念。所以，这两个物理量也反映了待分离物质的可分离性能。

(2) 渗透通量（permeate flux）

渗透通量是指单位时间内通过单位膜面积的透过物量，常用单位为 kmol/(m^2·s) 或 kg/(m^2·s)。膜的渗透通量大小直接决定了分离设备的大小。

(3) 通量衰减系数（flux attenuation coefficient）

因为分离过程的浓差极化、膜的压密以及膜孔的堵塞等原因，膜的渗透通量将随时间而衰减，可以用下式表示：

$$J=J_0\tau^m \tag{5-46}$$

式中　J——操作时间为 τ 时的渗透通量，kmol/(m^2·s)或 kg/(m^2·s)；

J_0——操作初始时的渗透通量，kmol/(m^2·s)或 kg/(m^2·s)

τ——操作时间，s；

m——通量衰减系数。

对于任何一种膜分离过程，总希望膜的分离效率高，渗透通量大，但实际上这两者往往存在矛盾。分离效率高的膜，其渗透通量小；渗透通量大的膜，分离效率低。所以，在选择膜时常需在两者之间进行权衡。

5.3.6　典型膜分离设备简介

各种膜分离设备主要包括膜分离器、泵、过滤器、阀、仪表及管路等。所谓膜分离器是以某种形式组装在一个基本单元设备内，在外界驱动力作用下能实现对混合物中指定各组分分离的设备，这类单元设备称为膜分离器或膜组件（module)。在膜分离的工业装置中，根据生产需要，通常可设置数个或是数百个膜组件。

目前，工业上常用的膜组件形式主要有：板框式、圆管式、螺旋卷式和中空纤维式等四种类型。一种性能良好的膜组件应具备以下条件：①对膜能提供足够的机械支撑并可使高压原料液和低压透过液严格分开；②在能耗最小的条件下，使原料液在膜表面上的流动状态均匀合理，以减少浓差极化；③具有尽可能高的装填密度（即单位体积的膜组件中填充较多的有效膜面积）并使膜的安装和更换方便；④装置牢固、安全可靠、价格低廉和容易维护。

5.3.6.1　板框式膜组件（panel membrane module）

板框式膜组件的最大特点是构造比较简单而且可以单独更换膜片。这不仅有利于降低设备投资和运行成本，而且还可作为试验机将各种膜样品同时安装在一起进行性能检验。此外，由于原料液流道的截面积可以适当增大，因此板框式膜组件压降较小，线速度可高达 1～5m/s，而且不易被纤维等异物堵塞。它具体包括系紧螺栓式（图 5-44）和耐压容器式（图 5-45）

两类。对于要求处理量更大的板框式微滤装置，可采用多层板框式过滤机。

系紧螺栓式如图5-44所示，是由圆形承压板、多孔支撑板和膜，经黏结密封构成脱盐板，再将一定数量的这种脱盐板多层堆积起来并放入O形密封圈，最后用上、下头盖（法兰）以系紧螺栓固定组成的。原水由上头盖进口流经脱盐板的分配孔，在诸多脱盐板的膜面上逐层流动，最后从下头盖的出口流出。与此同时，透过膜的淡水在流经多孔支撑板后，分别于承压板的侧面管口处流出。

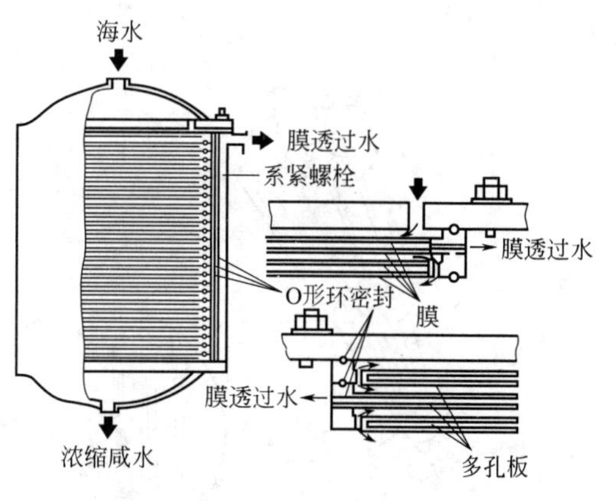

图5-44 系紧螺栓式板框式膜组件构造

耐压容器式如图5-45所示，是把多层脱盐板堆积组装后，放入一个耐压容器中而成。原水从容器的一端进入，分离后的浓水和淡水则由容器的另一端排出。容器内的大量脱盐板是根据设计要求串、并联相结合构成，其板数从进口到出口依次递减，目的是保持原水的线速度变化不大以减轻浓差极化影响。

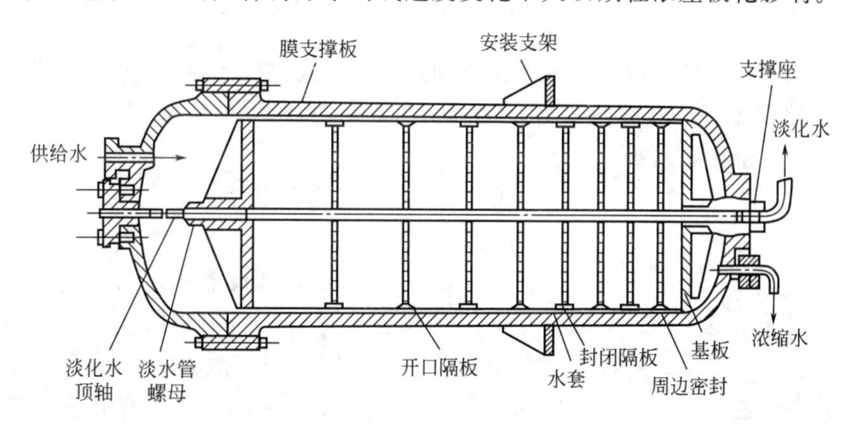

图5-45 耐压容器式板框式膜组件构造

板框式膜分离装置用途十分广泛，除用于盐水脱盐外，还有其他方面应用。如针头过滤器，即装在注射针筒和针头之间的一种微型过滤器，以微孔滤膜为过滤介质，用来除去微粒和细菌，常作静脉注射液的无菌处理。操作时，以注射针筒推注进行过滤，不需外加推动力。此外板框式膜分离装置还在电渗析、气体分离及渗透蒸发过程中有广泛应用。

5.3.6.2 圆管式膜组件（tubular membrane module）

所谓圆管式膜组件（管式膜组件），是指在圆筒状支撑体的内侧或外侧刮制半透膜而得到的圆管形分离膜组件。其支撑体的构造或半透膜的刮制方法随处理原料液的输入法及透过液的取出法而异。管式膜组件的形式较多，按其联接方式一般可分为内压单管式和管束式；按其作用方式又可分为内压式和外压式。即管式膜组件包括内压型内压式、内压型管束式（图5-46）和外压型管壳式(图5-47)。

管式组件中的接头和密封是一个关键的问题。单管式组件用U形管连接，采用喇叭口形，再用O形环进行密封。管式组件的优点是：流动状态好，流速易于控制，安装、拆卸、换膜和维修均较方便，而且能够处理含有悬浮固体的溶液；同时，机械清除杂质也较容易；

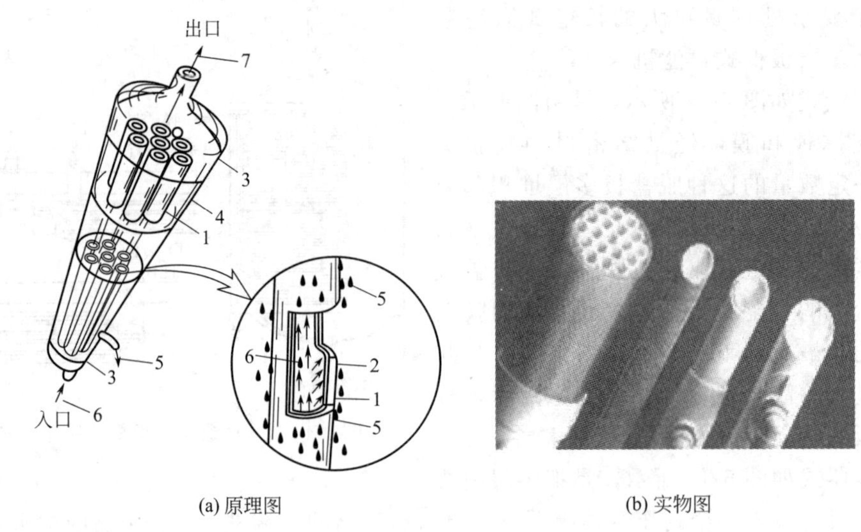

图 5-46　内压型管束式反渗透膜组件

1—玻璃纤维管；2—反渗透膜；3—末端配件；4—PVC 淡化水收集外套；5—淡化水；6—供给水；7—浓缩水

此外，合适的流动状态还可以防止浓差极化和污染。管式组件的缺点是：与平板膜相比，管式膜的制备条件较难控制；采用普通的管径（1.27cm）时，单位体积内有效膜面积较小；此外，管口的密封也比较困难。

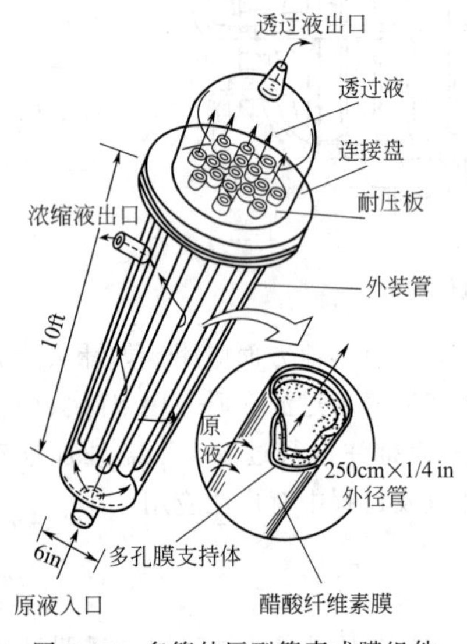

图 5-47　多管外压型管壳式膜组件

5.3.6.3　螺旋卷式膜组件（spiral-wound membrane module）

如图 5-48 所示，螺旋卷式（简称卷式）膜组件的中间为多孔支撑材料，两侧是膜。它的三边被密封成信封状膜袋，其另一个开放边与一根多孔的中心产品水收集管（集水管）密封连接，在膜袋外部的原水侧再垫一层网眼型间隔材料（隔网），把膜袋-原水侧隔网依次叠合，绕中心集水管紧密地卷起来，形成一个膜卷（或称膜元件），再装进圆柱形压力容器内，就构成一个螺旋卷式膜组件。

在实际应用中，通常是把几个膜组件的中心管密封串联起来，再安装到压力容器中，组成一个单元。原料液沿着与中心管平行的方向在隔网中流动，浓缩后由压力容器的另一侧引出。透过液（产品水）则沿着螺旋方向在两层间（膜袋内）的多孔支撑体中流动，最后汇集到中心集水管中而被导出。为了增加膜的面积，可以增加膜袋的长度，但膜袋长度的增加，透过液流向中心集水管的路程就要加长，从而阻力就要增大。为了避免该问题，在一个膜组件内可以安装几叶的膜袋，这样既能增加膜的面积，又不增大透过液的流动阻力。

目前，螺旋卷式膜组件在国外已实现机械化生产。即采用一种 0.91m 的滚压机，连续喷胶使膜与支撑材料黏结密封在一起并卷成筒，牢固后不必打开即可使用。这种制作方法避免了人工制作时的许多缺点，大大提高了卷筒质量。

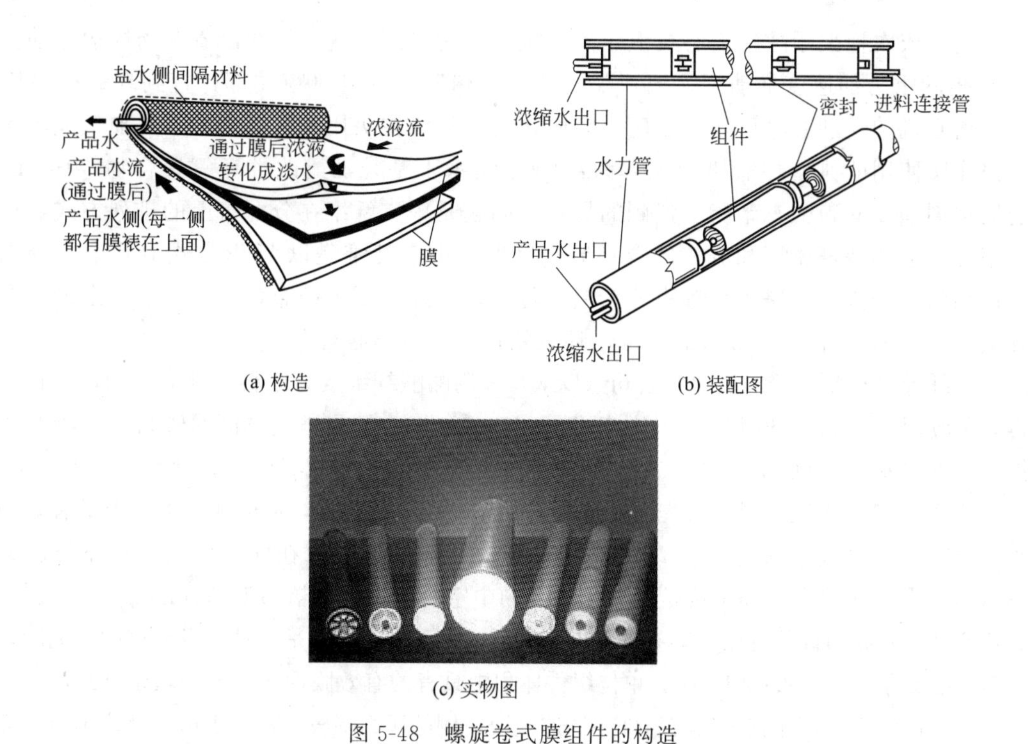

(a) 构造　　(b) 装配图

(c) 实物图

图 5-48　螺旋卷式膜组件的构造

5.3.6.4　中空纤维式膜组件（hollow-fiber membrane module）

中空纤维式膜组件是一种极细的空心膜管，无须支撑材料，本身即可耐很高的压力。如图 5-49 所示，纤维的外径有的细如人发，约为 50～200μm，内径为 25～42μm，其特点是具有在高压下不产生形变的强度。中空纤维膜组件的组装是把大量（多达几十万根或更多根）的中空纤维膜、纤维束的开口端用环氧树脂铸成管板。纤维束的中心轴处安装一根原料液分布管，使原料液径向均匀流过纤维束。纤维束外部包以网布使纤维束固定并促进原液的湍流流动。淡水透过纤维膜的管壁后，沿纤维的中空内腔经管板放出，浓缩原水则在容器的另一端排出。高压原料液在中空纤维的外侧流动的好处是：原料液在纤维的外侧流动时，如果纤维强度不够，只能被压扁，将中空内腔堵死，但不会破裂，可避免透过液被原料液污染。若把原料液引入这样细的纤维内腔，则很难避免膜面污染以至流道被堵塞。

在组装中空纤维束的过程中，遇到的一个问题是如何把非常细的中空纤维放入环氧管板里，而且不使纤维破损泄漏或干燥皱缩。也就是要解决中空纤维的保护和环氧管板的浇注问题。环氧树脂的浇注是采用离心力的原理才得以解决的。将中空纤维束（插在保护套里）放

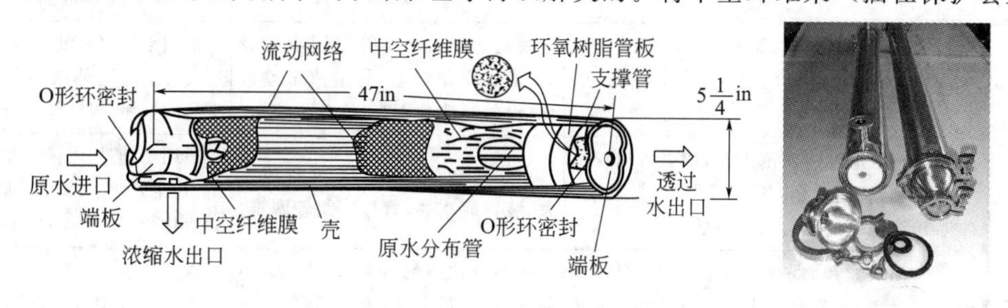

图 5-49　中空纤维式膜组件结构及外观

在架子上，依次把架子放在离心机上。离心机转动时，浇注的纤维端沿着圆周周边运动，在50～60倍的重力加速度推动下，把新配制的环氧树脂加入纤维束端部，直到环氧树脂固化才停止离心机的转动。最后，在车床上将固化的环氧浇注头用非常锋利的刀具加工成环氧管板。浇注所使用的典型环氧树脂配方为：100g缩水丁基甘油醚改性环氧聚合物，16g改性脂肪固化剂和20g磷酸三苯脂。遇到的另一个问题是中空纤维在分配管上的排列方式，这涉及中空纤维束的装填密度和流体的合理分布问题。中空纤维是以U形方式沿着中心分配管径方向均匀紧密排列，整个纤维束分十层，每一层最外边用无纺布包一层，纤维束最外层包有导流网。目前，纤维U形弯曲端也用环氧粘接，使流体合理分布。

中空纤维最早是由美国道化学公司（Dow）采用醋酸纤维素为原料研制成功的，并在工业上得到了应用。20世纪50年代末，杜邦公司（Dupont）也开展了这方面的研制工作，于1967年提出了以尼龙-66为膜原料的B-5Permasep渗透器。更具重要意义的是，杜邦公司于1970年12月以芳香聚酰胺为膜原料，首先研制成功B-9Permasep渗透器，从而找到了一个最有效的苦咸水淡化方法。该成果获得了1971年Kirkpatrick最高化工奖。在此基础上，杜邦公司又于1973年9月公布了可用于高浓度盐水淡化的新的中空纤维反渗透器B-10Permasep。

中空纤维式膜组件的特点：①不需要支撑材料，由于中空纤维是一种自身支撑的分离膜，所以不需考虑支撑问题。②结构紧凑，单位组件体积中具有的有效膜面积（即装填密度）高，一般可达16～30m^2/m^3，高于其他所有组件。③透过液一侧的压力降较大，由于透过液流出时，需通过极细的纤维内腔，因而流动阻力较大，压力降有时达到数个大气压。④再生清洗困难，同管式组件相比，因无法进行机械清洗（如以清洗球），所以一旦膜被污染，膜表面的污垢排除十分困难，因此要求原料液有严格的前处理。中空纤维式膜常用于氮氧分离、盐水淡化等。

5.3.6.5 各种膜组件形式的优缺点对比

一般来说，它们各有所长，当然从检测单位体积的产液量来看，螺旋卷式和中空纤维式是所有类型中最高的，因此，工业上大型实用装置大多数都是采用这两种形式。然而应当指出的是，从装置的膜表面清洗角度来说，圆管式膜组件有它独特的特点。另外如板框式膜组件，尽管它是最古老的一种形式，本身有不足之处，但由于它仍具有一定的特色而被广泛采用。综上所述，究竟采用哪种组件形式，尚须根据原料液情况和产品要求等实际条件具体分析、全面权衡、优化选定。

各类膜组件的比较见表5-6。

表5-6 各种膜组件优缺点和应用范围比较

类型	优点	缺点	应用范围
板框式	组装简单，结构紧凑、牢固，能承受高压，易实现模块化；性能稳定，工艺成熟，换膜方便	膜比表面积小，单程回收率较低，液流状态较差，易造成浓差极化，设备制造费用高，能耗大	适用于产水百吨/天以下的水厂及含悬浮固体或高黏度液体产品的浓缩、提纯等，已商业化
管式	料液流速可调范围大，浓差极化较易控制，流道通畅，压力损失小，易安装，易清洗，易拆换，工艺成熟	单位体积膜面积小，设备体积大，装置成本高，管口密封较困难	适用于建造中小型水厂及医药、化工产品的浓缩、提纯，已商业化
螺旋式	结构紧凑，膜比表面积很大，组件产水量大，工艺较成熟，设备费用低，可使用强度好的平板膜	浓差极化不易控制，易堵塞，不易清洗，换膜困难，密封困难，不宜在高压下操作	适用于大型水厂，已商业化

续表

类型	优点	缺点	应用范围
中空纤维式	膜比表面积最大，不需外加支撑材料，设备结构紧凑，制造费用低	易堵塞，不易清洗，原料液的预处理要求高，换膜费用高	适用于大型水厂，已商业化
槽条式	单位体积膜面积较大，设备费用低，易装配，易换膜，放大容易	运行经验较少	已商业化

5.4 结晶（Crystallization）

结晶是固体物质以晶体状态从蒸气、溶液或熔融物中析出的过程。与其他化工分离（chemical separation engineering）单元操作相比，结晶过程具有如下特点：①能从杂质含量相对多的溶液或多组分的熔融混合物中产生纯净的晶体，对于许多使用其他方法难以分离的混合物系，例如同分异构体（isomer）混合物、共沸物系（azeotropic system）、热敏性物系（thermosensitive system）等，采用结晶分离往往更为有效；②能量消耗少，操作温度低，对设备材质要求不高，一般很少有“三废”排放，有利于环境保护；③结晶产品包装、运输、存储或使用都很方便。

5.4.1 结晶的基本概念（Basic concepts of crystallization）

5.4.1.1 结晶过程分类（classification of crystallization process）

结晶过程可分为溶液结晶（solution crystallization）、熔融结晶（melt crystallization）、升华结晶（sublimation crystallization）和沉淀结晶（precipitation crystallization）四大类，其中溶液结晶和熔融结晶是化学工业中最常采用的结晶方法。

溶液结晶操作中，溶液在结晶器中结晶出来的晶体和剩余的溶液所构成的混悬物称为晶浆（magma），去除悬浮于其中的晶体后剩下的溶液称为母液（mother liquor）。工业上，通常对晶浆进行固液分离以后，再用适当的溶剂对固体进行洗涤，以尽量除去由于黏附和包藏母液所带来的杂质。在工业上的结晶过程中，不仅晶粒的产率（yield）和纯度（purity）是重要的，晶粒的形状和大小也是很重要的。通常，希望晶粒大小均匀一致，使包装中的结块（caking）现象减至最低限度，且便于倒出、洗涤（washing）和过滤（filtration）。同时，在使用时其性能也能均匀一致。

熔融结晶是根据待分离物质之间的凝固点（freezing point）不同而实现物质结晶分离的过程。与溶液结晶过程比较，熔融结晶过程的特点见表5-7。

表5-7 熔融结晶过程与溶液结晶过程的比较

项目	溶液结晶过程	熔融结晶过程
原理	冷却或除去部分溶剂，使溶质从溶液中结晶出来	利用待分离组分凝固点的不同，使之结晶分离
操作温度	取决于物系的溶解度特性	在结晶组分的熔点附近
推动力	过饱和度，过冷度	过冷度
过程的主要控制因素	传质及结晶速率	传热、传质及结晶速率

续表

项目	溶液结晶过程	熔融结晶过程
目的	分离、纯化，产品晶粒化	分离、纯化
产品形式	呈一定分布的晶体颗粒	液体或固体
结晶器形式	釜式为主	釜式或塔式

熔融结晶过程主要应用于有机物的分离提纯（purification），例如将萘与杂质（甲基萘等）分离可制得纯度（质量分数）达99.9%的精萘，而专门用于冶金材料精制或高分子材料加工的区域熔炼过程也属于熔融结晶。熔融结晶的产物外形，往往是液体或整体固相，而非颗粒。

Crystallization

Crystallization is a solid-fluid separation operation in which crystalline particles are formed from a homogeneous fluid phase. Ideally, the crystals are a pure chemical, obtained in a high yield with a desirable shape and a reasonably uniform and desirable size. Crystallization is one of the oldest known separation operations, with the recovery of sodium chloride as salt crystals from water by evaporation dating back to antiquity. Even today, the most common applications are the crystallization from aqueous solution of various inorganic salts. All these cases are referred to as solution crystallization because the inorganic salt is clearly the solute, which is crystallized, and water is the solvent, which remains in the liquid phase.

When both components of a homogeneous, binary solution have melting (freezing) points not far removed from each other, the solution is referred to as a melt if the phase diagram for the melt exhibits a eutectic point, it is possible to obtain, in one step called melt crystallization, pure crystals of one component or the other, depending on whether the composition of the melt is to the left or right of the eutectic composition.

Crystallization of a compound from a dilute, aqueous solution is often preceded by evaporation in one or more vessels, called reflects, to concentrate the solution, and followed by partial separation and washing of the crystals from the resulting slurry, called the magma, by centrifugation or filtration. The process is completed by drying the crystals to a specified moisture content.

5.4.1.2 晶体的几何形态（geometrical morphology of crystal）

晶体可以定义为按一定的规则重复排列的原子、离子或分子所组成的固体，它是一种有严格组织的物质形式。晶体是具有几个面和角的多面体。相同物质的不同晶体，其各个面和棱的相对尺寸可能有很大差别。然而，同一种物质的所有晶体，其对应面之间的夹角相同。晶体正是根据这些面的夹角来分类的。这种晶体点阵或这些空间晶格的结构，在各个方向上是重复延续的。晶体按其与夹角有关的晶轴（crystal axis）的排列情况分为七种类型（图 5-50）：①立方晶系，三条相等的晶轴相互垂直；②四方晶系，三条晶轴相互垂直，其中一条晶轴比另外两条长；③正交晶系，三条晶轴相互垂直，长度各不相等；④六方晶系，三条相等的晶轴在一个平面上彼此成60°夹角，第四条晶轴与该平面垂直，晶轴的长度与其他三条不一定相等；⑤单斜晶系，三条晶轴各不相等，其中两条在一个平面内垂直相交，第三条与该平面成某个角度；⑥三斜晶系，三条互不相等的晶轴彼此夹角互不相等，而且夹角不是30°、60°或90°；⑦三方晶系，有三条相等并且等倾斜度的晶轴。

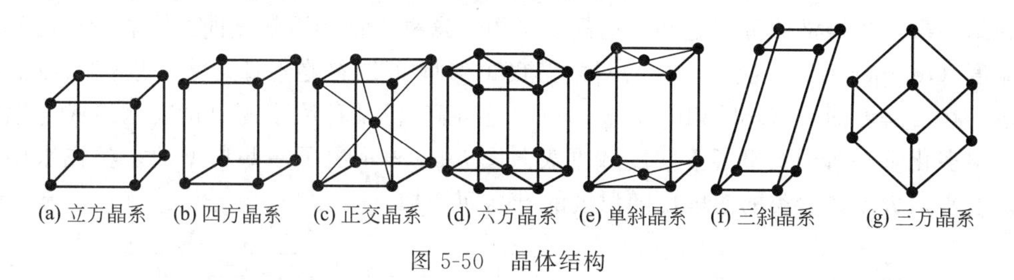

图 5-50　晶体结构

对于给定溶质的结晶过程，晶体不同形式的面的成长可能是不同的。NaCl从水溶液中结晶仅具有立方晶面。如果NaCl从含有少量某种杂质的水溶液中结晶，晶体将具有八个面。这两种晶形都属于立方晶系。但晶体结构并不相同。一般来说，结成片状晶还是针状晶，通常取决于晶体生长的工艺条件，而与晶体结构和晶系无关。

5.4.1.3　结晶水合物与结晶水

由于水合作用（hydratation），从溶液中析出的晶体含有一定数量的溶剂（水）分子，称为结晶水（crystal water），含有结晶水的晶体称为结晶水合物（crystalline hydrate）。晶体中结晶水的含量不同，会影响晶体的性质和外形。例如，虽然Na_2SO_4、$Na_2SO_4 \cdot 7H_2O$和$Na_2SO_4 \cdot 10H_2O$都是无色的，但无水硫酸钠晶体是斜方晶系，七水硫酸钠晶体是正方或斜方晶系，而十水硫酸钠晶体则是单斜晶系。又如，硫酸铜水溶液在240℃以上结晶时，得到的是白色三棱形针状的无水硫酸铜（$CuSO_4$）晶体，但若在常温下结晶，得到的是亮蓝色的三斜晶系晶体，该晶体中含有5个分子的结晶水（$CuSO_4 \cdot 5H_2O$）。

结晶水合物具有一定的蒸气压，此类晶体在空气中储存时，若其蒸气压高于周围空气中的水蒸气压，则易失去结晶水而风化（efflorescence）；若其蒸气压低于周围空气中的水蒸气压，则易吸取水分而分解。

Crystal geometry

In a solid, the motion of molecules, atoms, or ions is restricted largely to oscillations about fixed positions. If the solid is amorphous, these positions are not arranged in a regular or lattice pattern; if the solid is crystalline, they are. Amorphous solids are isotropic, such that physical properties are independent of the direction of measurement; crystalline solids are anisotropic, unless the crystals are cubic in structure.

Crystals consist of many units, each shaped like the larger crystal. This led to the concept of a space lattice as a regular arrangement of points (molecules, atoms, or ions) such that if a line is drawn between any two points and then extended in both directions, the line will pass through other lattice points with an identical spacing.

5.4.2 相平衡与溶解度（Phase equilibrium and solubility）

固体与其溶液相达到固、液相平衡时，单位质量的溶剂（solvent）所能溶解的固体的质量，称为固体在溶剂中的溶解度（solubility）。工业上通常采用以每 100 份质量的总溶剂（大多数情况为水）中，所含的无水溶质的分数来表示溶解度的大小。

在结晶中，当溶液或母液饱和时达到了平衡。这种平衡用溶解度曲线来表示，溶解度主要由温度（temperature）决定，压力对溶解度的影响可以忽略不计。溶解度数据以曲线形式表示。在溶解度曲线（solubility curve）中，溶解度以某种浓度的单位表示组成与温度的关系。很多化学手册中给出了溶解度数据表。图 5-51 表示出了一些典型的盐类在水中的溶解度。通常，大多数盐类的溶解度随温度的升高而增加。

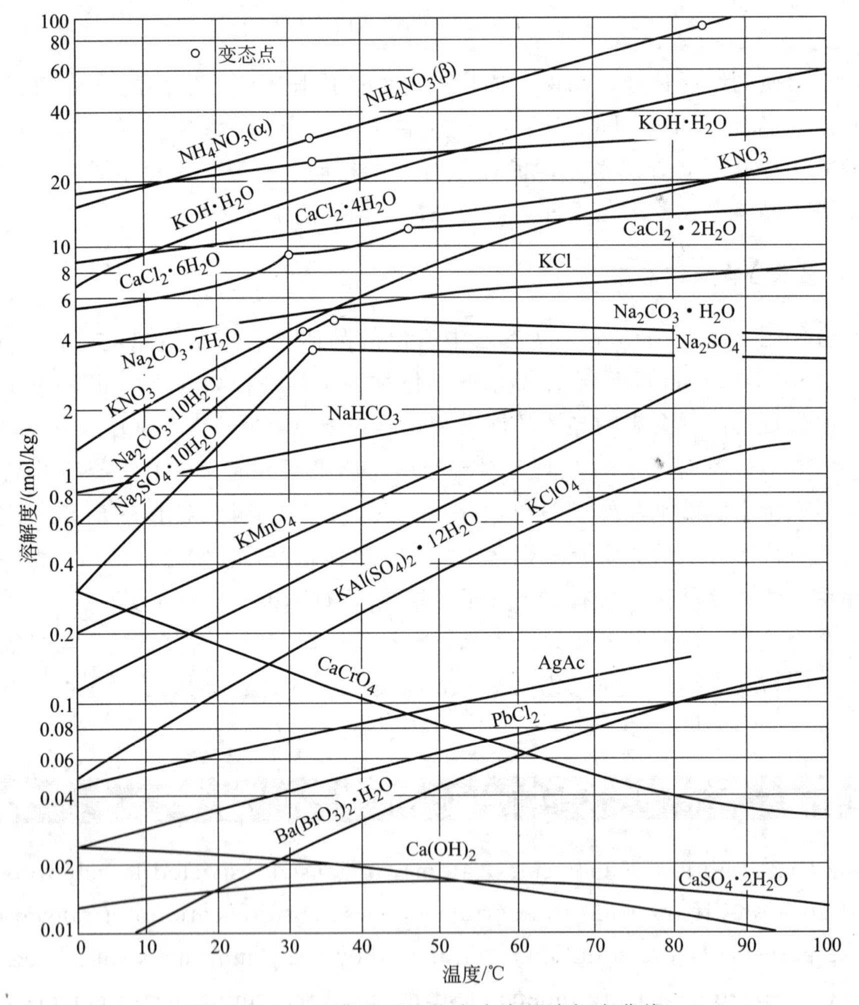

图 5-51 一些典型无机物在水中的溶解度曲线

根据溶解度随温度的变化特征，可将物质分为不同的类型。有些物质的溶解度随温度升高显著地增加，如 KNO_3、$NaNO_3$ 等；有些物质的溶解度随温度升高以中等速度增加，如 KCl、$(NH_4)_2SO_4$ 等；还有一类物质，如 NaCl 等，其溶解度的特点是随温度的变化很小。上述物质在溶解过程中需要吸收热量（heat），即具有正溶解度特性（positive solubility property）。另外有一些物质，如 Na_2SO_4 等，其溶解度随温度升高反而下降，它们在溶解过程中放出热量，即具有逆溶解度特性（inverse solubility property）。此外，从图中还可以看出，还有一些形成水合物的物质，在其溶解度曲线上有转折点，物质在转折点两侧含有的水分子数不等，故转折点又称为变态点（transformation point）。例如低于32.4℃时，从 Na_2SO_4 水溶液结晶出来的固体是 $Na_2SO_4 \cdot 10H_2O$，而在这个温度以上结晶出来的固体是 Na_2SO_4。

物质的溶解度特性对于结晶方法的选择起决定性的作用。对于溶解度随温度变化敏感的物质，适合用变温结晶方法分离；对于溶解度随温度变化缓慢的物质，适合用蒸发结晶法分离。另外，根据在不同温度下的溶解度数据还可计算出结晶过程的理论产量（theoretical yield）。

溶液的过饱和度（supersaturation）是工业结晶的主要推动力。溶液浓度恰好等于溶质的溶解度，即达到液固相平衡（liquid-solid equilibrium）状态时，称为饱和溶液（saturated solution）。溶液含有超过饱和量的溶质，则称为过饱和溶液（supersaturated solution）。Wilhelem Ostwald 第一个观察到过饱和现象。将一个完全纯净的溶液在不受任何扰动（无搅拌、无震荡）及无任何刺激（无超声波等作用）条件下，徐徐冷却，就可以得到过饱和溶液。但超过一定限度后，澄清的过饱和溶液就会开始析出晶核（crystal nucleus），Ostwald 称这种不稳定状态区段为“不稳区”（unstable region）。标志溶液过饱和而欲自发地产生晶核的极限浓度曲线称为超溶解度曲线（super solubility curve）。溶解度平衡曲线与超溶解度曲线之间的区域为结晶的介稳区。在介稳区（metastable region）内溶液不会自发成核。在图5-52中划出了这几个区域。图中的 AB 线段是溶解度平衡曲线，超溶解度曲线应是一簇曲线 $C'D'$，其位置在 CD 线之下，而与 CD 的趋势大体一致。图中的 E 点代表一个欲结晶物系，可分别使用冷却法（cooling methods）、真空绝热（vacuum insulation）冷却法或蒸发法（evaporation methods）进行结晶，所经途径相应为 EFH、$EF''G''$ 及 $E'FG'$。

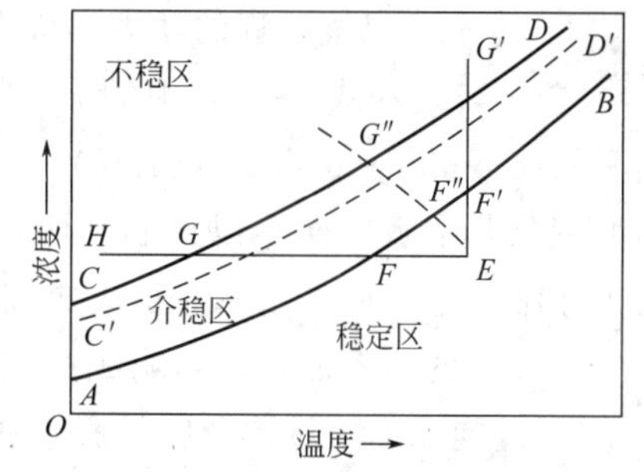

图5-52　溶液的超溶解度曲线

工业结晶过程要避免自发成核，才能保证得到平均粒度大的结晶产品。只有尽量控制在介稳区内结晶才能达到这个目的。所以只有按工业结晶过程条件来测定出超溶解度曲线，并给出介稳区才有实用价值。

溶质从溶液中结晶出来，要经历两个步骤：①首先将产生的被称为晶核的微小晶粒作为结晶的核心，这种过程称为成核（nucleation）；②然后晶核长大，成为宏观的晶体，这个过程称为晶体成长（crystal growth）。无论是成核过程还是晶体成长过程，都必须以溶液浓度与平衡溶解度（equilibrium solubility）之差即溶液的过饱和度（degree of supersaturation）作为推动力。溶液过饱和度的大小直接影响以上两个过程的快慢，而这两过程的速度又影响着晶体产品的粒度分布（particle size distribution）。因此，过饱和度是结晶过程中一个极其重要的参数。

5.4.3　结晶动力学简介

结晶动力学（crystallization kinetics）主要讨论结晶的速率问题。由于结晶过程有成核

（nucleation）和晶体成长（crystal growth）两个阶段，所以结晶速率包括成核速率和晶体成长速率。

（1）成核速率及其影响因素

成核速率（nucleation rate）是指单位时间、单位体积溶液内所产生的晶核数，即：

$$r_N=\frac{dN}{d\tau} \tag{5-47}$$

式中　r_N——成核速率，$1/(m^3 \cdot s)$；

N——单位体积晶浆中的晶核数，$1/m^3$；

τ——时间，s。

由于影响成核速率的因素很多，而对其机理的了解也有限，所以一般都使用比较简单的经验关系式来描述。由于溶液的过饱和度是结晶过程的推动力，通常将成核速率表示成过饱和度的幂函数，即：

$$r_N=K_N\Delta c^n \tag{5-48}$$

式中　K_N——成核的速率常数；

n——成核过程的动力学级数。

式(5-47）和式(5-48）中的 r_N 为初级成核速率（primary nucleation rate）与二次成核速率（secondary nucleation rate）之和。初级成核速率比二次成核速率大很多，而且对过饱和度变化非常敏感而难以控制。因此，除了超细粒子（ultrafine particle）的制备外，一般的工业结晶过程应尽量避免初级成核。

对一定的物系而言，影响成核速率的因素很多，主要的影响因素有如下几种：

① 过饱和度　过饱和度是影响成核速率的关键因素，通常成核速率特别是初级成核速率随过饱和度的增加而增大。对不同的物系，过饱和度对成核速率的影响也不相同。在结晶操作中，应根据具体需要将过饱和度控制在适当的范围内。

② 杂质（impurity）　溶液中杂质的存在会对成核速率产生很大的影响，但这种影响不能事先预测。它的存在对溶质的溶解度起促进或抑制的作用，可使成核速率增加或减小。目前的研究结果还没有总结出具有普遍性的规律。

③ 搅拌作用（stirring action）　搅拌等机械作用对提高成核速率有利。但对其作用的认识目前还仅处于定性阶段。对均相成核，在过饱和溶液中的任何轻微震动均可使成核速率增加。对二次成核（second nucleation），搅拌的作用也非常明显，搅拌时碰撞的次数及冲击能量的增加能使成核速率加大。

（2）晶体成长速率及其影响因素

晶体的成长速率（crystal growth rate）是指单位时间内结晶出来的溶质量，即：

$$r_m=\frac{dm}{d\tau} \tag{5-49}$$

式中　r_m——晶体的成长速率，kg/s；

m——晶体的质量，kg；

τ——时间，s。

目前，对晶体的成长机理的研究尚不成熟，通常也将晶体的成长速率表示成过饱和度的幂函数，采用一般的简单速率表达式来进行描述，即：

$$r_{m}=K_{m}\Delta c^{m} \tag{5-50}$$

式中 K_{m}——晶体的成长速率常数；

m——晶体的成长过程的动力学级数。

对一定的物系，晶体成长速率的大小与溶液的过饱和度、结晶温度、溶液黏度和密度、杂质、结晶位置、晶体粒度和搅拌强度（stirring Intensity）等许多因素有关。

① 过饱和度　过饱和度对晶体的成长速率影响很大，对不同的物系，过饱和度对晶体的成长速率影响不同。过饱和度还影响晶体的形状、粒度及粒度分布。例如，在低过饱和度下，β-石英晶体呈短而粗的外形，均匀性较好；但在高过饱和度下，晶体的外形呈细长状，且均匀性（uniformity）也较差。

② 溶液黏度和密度　溶液的黏度大则流动性（liquidity）差，使得溶质向晶面的传质速率（mass transfer rate）下降，这时晶体顶角和棱边较易获得溶质而生长较快，导致结晶的形成。晶体的析出和放出结晶热使溶液产生局部的密度差，从而造成溶液的涡流，涡流在晶体周围造成浓度的不均，从而导致不规则晶体的生成。

③ 搅拌作用　搅拌是影响结晶粒度分布的重要因素。大搅拌强度使晶体粒度变小，温和而又均匀的搅拌，则易获得粗颗粒结晶。

④ 晶体粒度　大晶体的表面能较小，其成长速率比小晶体大。另外，对控制步骤为扩散过程（diffusion process）的结晶，由于大晶体在溶液中的沉降速度（setting velocity）较大，可以减小晶体表面静止液膜的厚度，从而降低溶质的扩散阻力（diffusion resistance），使其更易获得溶质而继续长大。

Crystallization Kinetics

Crystallization is a complex phenomenon that involves three steps: nucleation, mass transfer of the solute to the crystal surface, and incorporation of the solute into the crystal lattice structure. Collectively, these phenomena are referred to as crystallization kinetics.

Nucleation

To determine the volume or residence time of the magma in a crystallizer, the rate of nucleation (birth) of crystals and their rate of growth must be known or estimated. The relative rates of nucleation and growth are very important because they determine crystal size and size distribution. Nucleation may be primary or secondary depending on whether the supersaturated solution is free of crystalline surfaces or contains crystals, respectively. Primary nucleation requires a high degree of super-saturation and is the principal mechanism occurring in precipitation. The theory of primary nucleation is well developed and applies as well to the condensation of liquid droplets from a supersaturated vapor and the formation of droplets of a second liquid phase from an initial liquid phase. However, secondary nucleation is the principal mechanism in commercial crystallizers, where crystalline surfaces are present and large crystals are desirable.

5.4.4 工业结晶方法与设备（Industrial crystallization methods and equipments）

5.4.4.1 工业结晶方法

工业上的结晶操作可分为冷却法（cooling method）、蒸发法（evaporation method）、真空冷却法（vacuum cooling method）及加压法（pressure method）四类。

最简单的冷却结晶过程是将热的结晶溶液置于无搅拌的敞口的结晶釜（crystal kettle）中，靠自然冷却作用降温结晶（decrease temperature crystalline）。所得产品纯度（purity）较低，粒度分布不均，容易发生结块现象。设备所占空间大，容积产生能力较低。但由于该种设备造价低，安装使用条件不高，所以至今仍在使用。

蒸发结晶是去除一部分溶剂的结晶过程，主要是使溶液在常压（normal pressures）或减压（decompression）下蒸发浓缩而变成过饱和。此法适用于溶解度随温度降低而变化不大或具有逆溶解度特性的物系。利用太阳能晒盐就是最古老而简单的蒸发结晶操作。蒸发结晶器与一般的溶液浓缩蒸发器在原理、设备结构及操作上并无不同。但一般的蒸发器用于蒸发结晶操作时，对晶体的粒度不能有效加以控制。

真空冷却结晶是使溶剂在真空下闪蒸而使溶液绝热冷却的结晶法。此法适用于具有正溶解度特性而溶解度随温度的变化率中等变化的物系。真空冷却结晶器的操作原理是：把热浓溶液送入绝热保温的密闭结晶器中，器内维持较高的真空度（degree of vaccum），由于对应的溶液沸点低于原料液（feed solution）温度，溶液势必闪蒸而绝热冷却到与器内压强相对应的平衡温度（equilibrium temperature）。实质上，溶液通过蒸发浓缩及冷却两种效应来产生过饱和度。真空冷却结晶过程的特点是主体设备结构相对简单，无换热面，操作比较稳定，不存在内表面严重结垢现象。

加压结晶是靠加大压力改变相平衡曲线进行结晶的方法。该方法已受工业界重视，装置见图 5-53。

5.4.4.2 工业结晶设备

几种主要的通用结晶器如下。

（1）强迫外循环型结晶器

图 5-54 是一台连续操作的强迫外循环型结晶器。部分晶浆由结晶室的锥形底排出后，

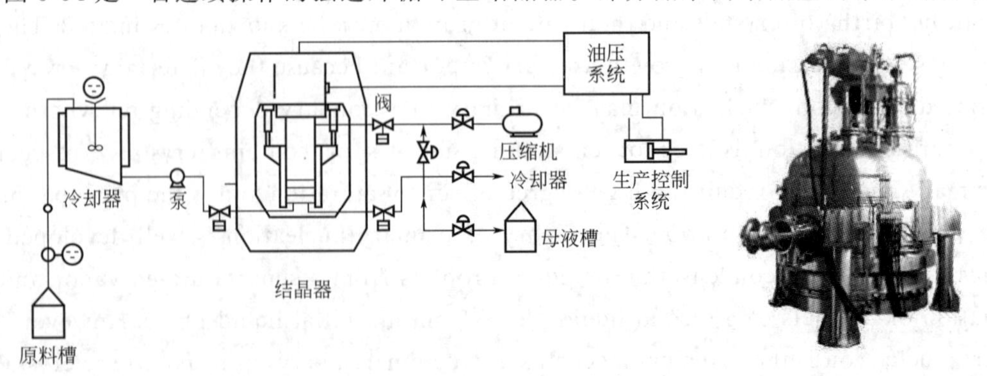

图 5-53 加压结晶装置

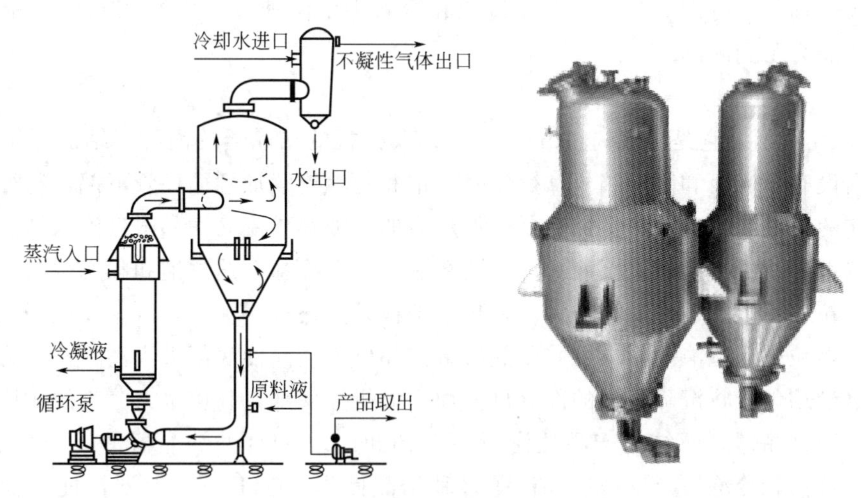

图 5-54　强迫外循环型结晶器

经循环管与原料液（material solution）一起通过换热器（heat exchanger）加热，沿切线方向重新返回结晶室。这种结晶器可用于间接冷却法、蒸发法及真空冷却法结晶过程。它的特点是生产能力很大，但由于外循环管路较长，所需的压头较高，循环量较低，结晶室内的晶浆混合不太均匀，存在局部过浓（overrich）现象，因此，所得产品平均粒度较小，分布不均。

（2）流化床型结晶器

图 5-55 是流化床型蒸发结晶器及冷却结晶器的示意图。因结晶室的上部比底部截面积大，所以流体向上的流速逐渐降低，其中悬浮晶体的粒度越往上越小，因此结晶室成为粒度分级的流化床（fluidized bed）。在结晶室的顶部，基本上已不含有晶粒，作为澄清的母液进入循环管路，与热浓料液混合后，在换热器中加热并送入汽化室（vaporizing chamber）蒸发浓缩（对蒸发结晶器），或在冷却器中冷却（对冷却结晶器）而产生过饱和度。过饱和的溶液通过中央降液管流至结晶室底部，与富集于底层的粒度较大的晶体接触，晶体长得更大。溶液在向上穿过晶体流化床时，逐步解除其过饱和度。

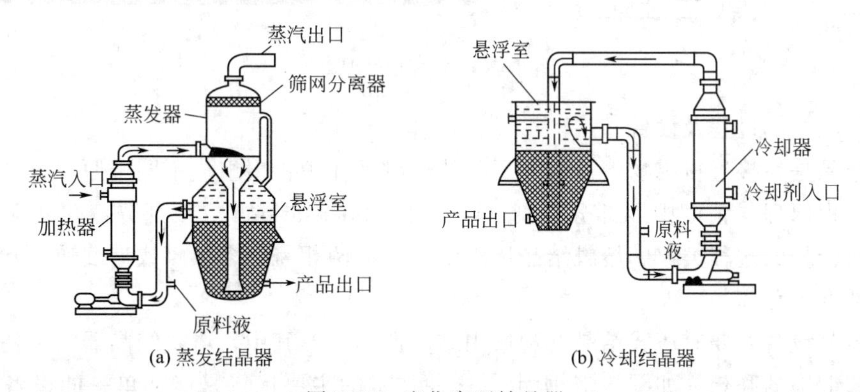

图 5-55　流化床型结晶器

流化床型结晶器的特点是将过饱和度产生的区域与晶体成长区分别设置在结晶器的两处，由于采用母液循环式，循环液中基本上不含有晶粒，从而避免发生叶轮与晶体间的接触成核现象，再加上结晶室的粒度分级（particle size classification）作用，使这种结晶器所产生的晶体大而均匀，特别适合于生产在过饱和溶液中沉降速度大于 0.02m/s 的晶粒。其缺点在于生产能力受到限制。

(3) DTB 型结晶器

如图 5-56 所示，它是导流筒（draft tube）及挡板的结晶器的简称。结晶器下部接有淘析柱，器内设有导流筒和筒形挡板，操作时热饱和料液连续加到循环管下部，与循环管内夹带有小晶体的母液混合后泵送至加热器。加热后的溶液在导流筒底部附近流入结晶器，并由缓慢转动的螺旋桨沿导流筒送至液面。溶液在液面蒸发冷却，达过饱和状态，其中部分溶质在悬浮的颗粒表面沉积，使晶体长大。在环形挡板外围还有一个沉降区。在沉降区内大颗粒沉降，而小颗粒则随母液入循环管并受热溶解。晶体于结晶器底部进入淘析柱。为使结晶产品的粒度尽量均匀，将沉降区来的部分母液加到淘析柱底部，利用水力分级的作用，使小颗粒随液流返回结晶器，而结晶产品从淘析柱下部卸出。DTB 型结晶器可用于真空冷却法、蒸发法、直接接触冷冻法及反应结晶法等多种结晶操作。DTB 型结晶器性能优良，生产强度高，能产生粒度达 600～1200μm 的大粒结晶产品，结晶器内不易结晶疤，已成为连续结晶器的最主要形式之一。

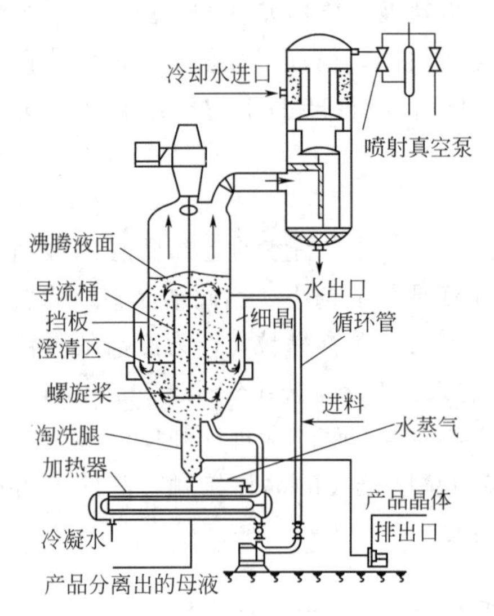

图 5-56　DTB 型结晶器

(4) 熔融结晶过程及设备

熔融结晶是在接近析出物熔点温度下，从熔融液体中析出组成不同于原混合物的晶体的操作。其过程原理与精馏中因部分冷凝（partial condensation）（或部分汽化）而形成组成不同于原混合物的过程类似。熔融结晶过程中，固液两相需经多级（或连续逆流）接触后才能获得高纯度的分离。

熔融结晶设备主要有塔式结晶器和通用结晶器两类。其中，塔式结晶器的结构和操作原理源于对精馏塔的联想，如图 5-57 所示。这种技术的主要优点是能在单一的设备内达到相当于若干个分离级的分离效果，有较高的生产速率。如图 5-56 所示，料液由塔的中部加入，晶粒在塔底被加热熔化，部分作为高熔点产物流出，部分作为液相回流向上流动。部分液相在塔顶作为低熔点产物采出，部分被冷却析出结晶向下运动。这种液固两相连续接触传质的方式又称分步结晶（fractional crystallization）。

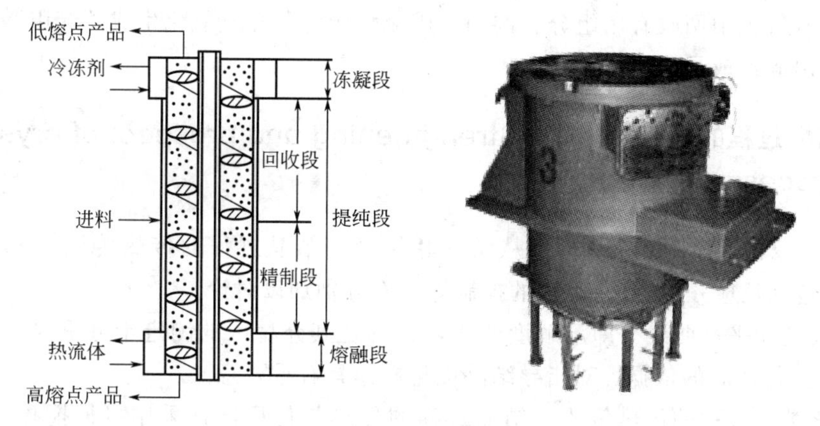

图 5-57　塔式结晶器

而通用结晶器有苏尔寿 MWB 结晶器（图 5-58）、布朗迪提纯器（图 5-59）等。此外，其他结晶方法还包括：反应沉淀、盐析（salting out）及升华结晶等。

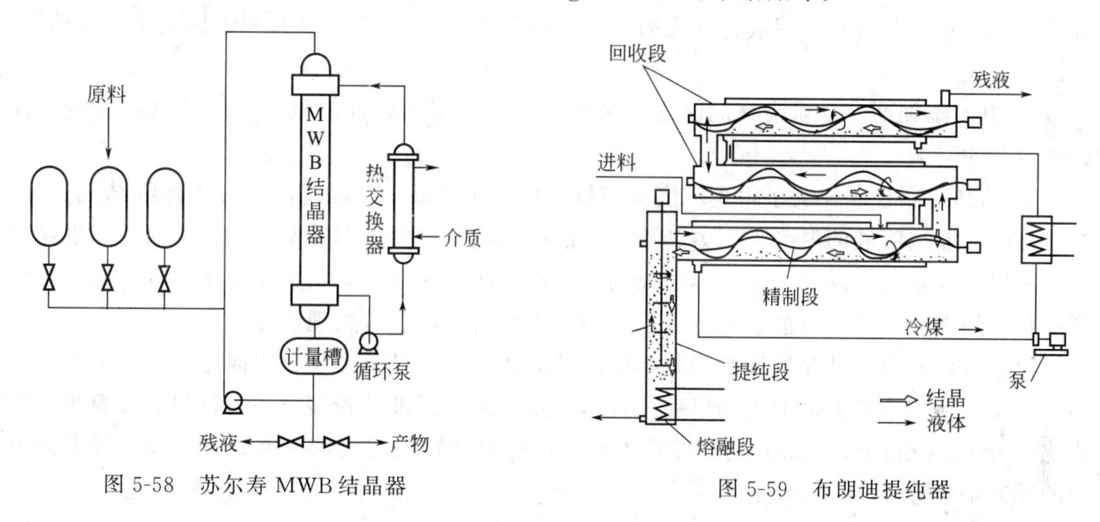

图 5-58　苏尔寿 MWB 结晶器

图 5-59　布朗迪提纯器

反应沉淀是液相中发生化学反应生成的产物以结晶或无定形物析出的过程。例如，硫酸吸收焦炉气中的氨生成硫酸铵并以结晶析出，经进一步固液分离、干燥后获得产品。沉淀过程首先是反应形成过饱和度，然后成核、晶体成长。与此同时，还往往包含了微小晶粒的成簇及熟化（curing）现象。显然，沉淀必须以反应产物在液相中的浓度超过溶解度为条件，此时的过饱和度取决于反应速率。因此，反应条件（包括反应物浓度、温度、pH 值及混合方式等）对最终产物晶粒的粒度和晶形有很大影响。

盐析是一种在混合液中加入盐类或其他物质以降低溶质的溶解度，从而析出溶质的方法。例如，向氯化铵母液中加盐（氯化钠），母液中的氯化铵因溶解度降低而结晶析出。盐析剂（salting-out agent）也可以是液体。例如，向有机混合液中加水，使其中不溶于水的有机溶质析出，这种盐析方法又称为水析（elutriation）。盐析的优点是直接改变固液相平衡，降低溶解度，从而提高溶质的回收率。此外，还可以避免加热浓缩对热敏物（heat sensitive material）的破坏。

升华结晶通常指蒸气经骤冷而直接凝结成固态晶体，如含水的湿空气骤冷形成雪。升华结晶常用来从气体中回收有用组分。例如，用流化床将萘蒸气氧化生成邻苯甲酸酐，混合气经冷却后析出固体成品。

5.4.5 结晶过程的强化与展望（Strengthening and prospect of crystallization process）

结晶过程及其强化的研究可以从结晶相平衡、结晶过程的传热传质（mass and heat transfer）（包括反应）、设备及过程的控制等方面分别加以讨论。

① 溶液的相平衡曲线　即溶解度曲线，尤其是其介稳区的测定十分重要，因为它是实现工业结晶获得产品的依据，对指导结晶优化操作具有重要意义。

② 强化结晶过程的传热传质　结晶过程的传热与传质通常采用机械搅拌、气流喷射、外循环加热等方法来实现。但是应该注意控制速率，否则晶粒易被破碎，过大的速率也不利于晶体成长。

③ 改良结晶器结构　在结晶器内采用导流筒或挡筒是改良结晶器最常用的也是十分有效的方法，它们既有利于溶液在导流筒中的传热传质（及反应），又有利于导流筒（或挡筒）外晶体的成长。

④ 引入添加剂、杂质或其他能量　最近，有文献报道，外加磁场、声场对结晶过程能产生显著的影响。

⑤ 结晶过程控制　为了得到粒度分布特性好、纯度高的结晶产品，对于连续结晶过程，控制好结晶器内溶液的温度、压力、液面、进料及晶浆出料速率等十分重要。对于间歇结晶过程来讲，计量加入晶种并采用程序控制以及控制冷却速率等，均是获得高纯度产品、控制产品粒度的重要手段。目前，工业上已应用计算机对结晶过程实现监控。

由上可以看出，结晶过程的强化不仅涉及流体力学、粒子力学、表面化学、热力学、结晶动力学、形态学等方面的机理研究和技术支持，同时还涉及新型设备与材料、计算机过程优化（process optimization）与测控技术等方面的综合知识与技术。因此进一步开展上述方面的研究是十分重要和必要的。

5.5 吸附（Adsorption）

5.5.1 吸附现象及其工业应用

吸附现象早已被人们发现。两千多年前中国人民已经采用木炭来吸湿和除臭，如湖南省长沙市附近出土的汉代古墓中就放有木炭，显然墓主当时是用木炭吸收潮气等作为防腐措施的。之后，在18世纪已有人注意到热的木炭冷却下来会捕集几倍于自身体积的气体。稍后，又认识到不同的木炭对不同的气体所捕集的体积是不一样的，并指出木炭捕集气体的效率依赖于暴露的表面积，进而强调了木炭中孔的作用。现在，人们认识到吸附现象中的两个重要因素（即表面积和孔隙率）不仅存在于木炭中，也存在于其他多孔固体颗粒中。所以，可以从气体或蒸气的吸附测量，来获得有关固体表面积和孔结构的信息。

“吸附”这个词最早由Kayser于1881于提出，用于描述气体在自由表面的凝聚（coagulation）。它与“吸收”（Absorb）不一样，吸附只发生于表面，吸收则指气体进入固体或液体本体中。1909年，McBain提议用词“吸着”来指表面吸附、吸收和孔中毛细凝聚的总和。国际上对上述三个词已有严格的定义。一般而言，吸附包括表面的吸附和孔中的毛细凝聚两部分。

5.5.1.1 概述

吸附过程是指多孔固体吸附剂与流体相（液体或气体）相接触，流体中的单一或多种溶剂向多孔固体颗粒表面选择性传递（selective transmission），积累于多孔固体吸附剂微孔表面的过程。类似的逆向操作过程称为解吸过程（desorption process），它可以使已吸附于多孔固体吸附剂表面的各类溶质有选择性地脱出。通过吸附和解吸可以达到分离、精制的目的。吸附中，具有一定吸附能力的固体材料称为吸附剂（adsorbent），被吸附的物质称为吸附质（adsorbate）。

吸附分离操作，多数情况下都是间歇式进行的。混合气体通过填充着吸附剂的固定床层，首先是易被吸收组分的大部分在床层入口附近就已被吸附，随着气体进入床层深处，该组分的其余部分也会被吸附掉。吸附剂的吸附容量（adsorption capacity）（指的是滤料或离子交换剂吸附某种物质或离子的能力）被饱和后，必须要将吸附质从吸附剂中除去。这时对吸附质是回收还是废弃，要根据它的浓度或纯度及价值来确定。一般是采用置换吸附质或使之脱吸的方法来再生吸附剂，若当吸附质被牢固地吸附于吸附剂上，还可采用燃烧法使吸附剂再生。一般情况下，随温度上升或压力下降，吸附剂的吸附容量是减少的，所以再生吸附剂时，就可采用升温或减压的操作方法，而且利用水蒸气的升温操作是工业上最为便利的方法。

提高吸附过程的处理量需要反复进行吸附和解吸操作，增加循环操作的次数。通常采用的吸附及解吸再生循环操作的方法有：①变温吸附，提高温度使吸附剂的吸附容量减少而解吸，利用温度的变化完成循环操作。小型的吸附设备常直接通入水蒸气加热床层，取其传热系数高，加热升温迅速，又可以清扫床层的优点。②变压吸附（pressure swing adsorption），降低压力或抽真空使吸附剂解吸（desorb），升高压力使之吸附，利用压力的变化完成循环操作。③变浓度吸附，待分离溶质为热敏性物质时可利用溶剂冲洗或用萃取剂抽提来完成解吸再生。④色谱分离，可分为迎头分离操作、冲洗分离操作和置换分离操作等几种形式。混合气通入吸附柱后，不同组分按吸附能力的强弱顺序流出，称为迎头分离操作。在连续通入惰性溶剂的同时，脉冲（pulse）送入混合溶液，各组分由于吸附能力大小不同，得到有一定间隔的谱峰（spectral peak），称为冲洗分离操作。用吸附能力最强的溶质组分通入达到吸附饱和的床层，依次将吸附能力强弱不同的各组分置换下来，吸附能力最强的置换溶质组分则可采用加热或其他方法解吸，称为置换分离操作。工业上采用何种操作方式可根据处理量大小、产品和溶剂的价格等因素综合确定。

Adsorption

In an adsorption process, molecules or atoms or ions, in a gas or liquid diffuse to the surface of a solid, where they bond with the solid surface or are held there by weak intermolecular forces. The adsorbed solutes are referred to as adsorbate, whereas the solid material is the adsorbent. To achieve a very large surface area for adsorption per unit volume, highly porous solid particles with small-diameter interconnected pores are used, with the bulk of the adsorption occurring within the pores.

Adsorption processes may be classified as purification or bulk separation, depending on the concentration in the feed fluid of the components to be adsorbed. In a bulk material, all the bonding requirements of the constituent atoms of the material are filled. But atoms on the surface experience a bond deficiency, because they are not wholly surrounded by other atoms. Thus it is energetically favorable for them to bong with whatever happens to be available. The exact nature of the bonding depends on the details of the species involved, but the adsorbed material is generally classified as exhibiting physic-sorption or chemi-sorption.

Physi-sorption or physical adsorption is a type of adsorption in which the adsorbate adheres to the surface only through Van der Waals interactions, which are also responsible for the non-ideal behavior of real gases. Chemi-sorption is a type of adsorption whereby a molecule adheres to a surface through the formation of a chemical bond, as opposed to the Van der Waals forces which cause physic-sorption. Adsorption is usually described through isotherms, that is, functions which connect the amount of adsorbate on the adsorbent, with its pressure (if gas) or concentration (if liquid).

Any potential application of adsorption has to be considered along with alternatives such as distillation, absorption and liquid extraction. Each separation process exploits a difference between properties of the components to be separated. In distillation, it is volatility. In absorption, it is solubility. In extraction, it is a distribution coefficient. Separation by adsorption depends on one component being more readily adsorbed than another. The selection of a suitable process may also depend on the ease with which the separated components can be recovered. Separating *n*-paraffins and *iso*-paraffins by distillation requires a large number of stages because of the low relative volatility of the components. It may be economic, however, to use a selective adsorbent which separates on the basis of slight differences in mean molecular diameters, where for example, *n*-pentane and *iso*-pentane have diameters of 0.489nm and 0.558nm respectively. When an adsorbent with pore size of 0.5nm is exposed to a mixture of the gases, the smaller molecules diffuse to the adsorbent surface and are retained whilst the larger molecules are excluded. In another stage of the process, the retained molecules are desorbed by reducing the total pressure or increasing the temperature.

5.5.1.2 吸附剂（adsorbent）

吸附过程设计中，吸附剂的选择是十分重要的。吸附剂的种类很多，可分成无机吸附剂和有机吸附剂，合成吸附剂和天然吸附剂。吸附剂可以根据需要加以改性修饰，使之对分离体系具有更高的选择性，以满足对结构类似或浓度很低的组分的分离回收的要求。吸附剂的吸附容量有限，在1%～40%（质量分数）之间。

常见的吸附剂种类有：

① 天然吸附剂　天然矿产如活性白土、蒙脱土、漂白土、黏土和硅藻土等，经适当加工处理.就可直接作为吸附剂使用。天然吸附剂虽价廉易得，但活性较低，一般使用一次失效后不再回收。天然高分子物质，如纤维素、木质素、甲壳素和淀粉等，经过反应交联或引

进官能团，也可制成吸附树脂（adsorbent resins）。

② 活性炭（activated carbon） 活性炭是一种多孔结构具有吸附性能的炭基物质的总称。将煤、椰子壳、果壳、木材等进行炭化，再经过活化（excitation）处理，可制成各种不同性能的活性炭，其比表面积可达 $1500m^2/g$。活性炭的结构除石墨化晶态炭外，还有大量的过渡态炭。过渡态炭有三种基本结构单元，即乱层石墨、无定形和高度有规则的结构（图 5-60）。活性炭性能稳定，耐酸、耐碱、耐腐蚀，可用于回收混合气体中的溶剂蒸气、各种油品和用于糖液的脱色、水的净化、气体的脱臭和作为催化剂载体等方面。

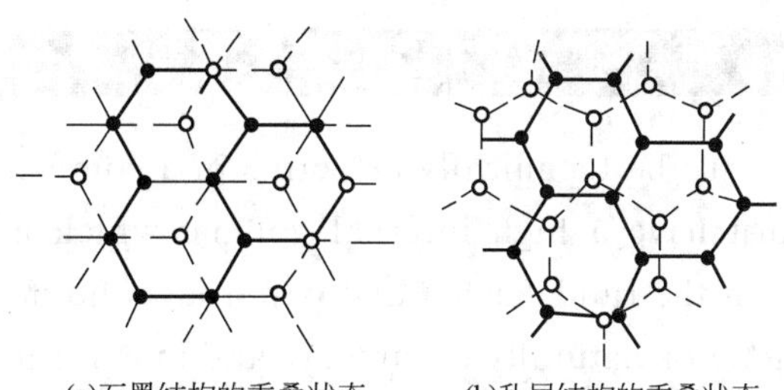
(a)石墨结构的重叠状态 (b)乱层结构的重叠状态

图 5-60 活性炭结构

③ 活性炭纤维（activated carbon fibre） 活性炭纤维可以编织成各种织物，使装置更为紧凑并减少流动阻力。它的吸附能力比一般的活性炭高出 1～10 倍，对有机物的吸附量较高，例如对恶臭物如丁硫醇等的吸附量比活性炭高出 40 倍以上。活性炭纤维适用于脱除气体中的恶臭物和废水中的污染物以及制作防护用具和服装等。在脱附阶段，当温度较高时，活性炭纤维的脱附速度比颗粒活性炭要快得多，且无拖尾现象。

④ 分子筛 分子筛是具有许多孔径大小均一微孔的物质，能够选择性吸附直径小于其孔径的分子，起到筛选分子的作用。这种固体吸附剂很多，如炭分子筛、沸石分子筛、微孔玻璃分子筛等。分子筛广泛用于气体和液体的干燥、脱水、净化、分离和回收等，应用后可以再生。此外，分子筛也可用作催化剂，如用于石油裂化（oil cracking）等。

⑤ 硅胶（silica gel） 硅胶是一种坚硬、无定形链状和网状结构的硅酸聚合物，其分子式是 $SiO_2 \cdot nH_2O$。硅胶是一种亲水性的球形吸附剂，其比表面积可达 $600m^2/g$，易于吸附极性物质（如水、甲醇等）。它吸附气体中的水分可达其本身质量的 50%，即使在相对湿度为 60%的空气流中，微孔硅胶的吸湿量也可达 24%（质量分数）。因此，硅胶常用于高湿含量气体的干燥。吸附脱水（adsorption dewatering）时，放出大量的吸附热，常易使其破碎。

⑥ 活性氧化铝（activated aluminium oxide） 氧化铝水合物经不同温度的热处理，可得 8 种亚稳态的氧化铝，其中以 $\gamma\text{-}Al_2O_3$ 和 $\eta\text{-}Al_2O_3$ 的化学活性最高，习惯上称为活性氧化铝。活性氧化铝是一种极性吸附剂，它一般不是纯粹的 Al_2O_3，而是部分水合无定形的多孔结构物质，其中不仅有无定形的凝胶，还有氢氧化物的晶体。活性氧化铝的孔径分布范围较宽，约为 10～10000Å（1Å=0.1nm），宜在 177～316℃下再生。活性氧化铝用作脱水和吸湿（moisture absorption）的干燥剂或作为催化剂的载体，它对水分的吸附容量大，常用于高湿度气体的脱湿（dehumidification）和干燥。

工业吸附对吸附剂具有如下要求：

① 吸附容量要大 吸附容量主要决定于内表面，内表面越大则比表面积越大，即吸附容量（adsorption capacity）越大。

② 选择性要好 吸附剂对不同的吸附质具有选择性吸附作用，影响选择性的因素主要是吸附剂的种类、结构和吸附机理等。

③ 应具有一定的机械强度和物理特性要求（如颗粒大小等）。

④ 再生容易　具有良好的化学稳定性和热稳定性、价廉易得等。

Adsorbents

To be technically effective in a commercial separation process, an adsorbent material must have a high internal volume which is accessible to the components being removed from the fluid. Such a highly porous solid may be carbonaceous or inorganic in nature, synthetic or naturally occurring, and in certain circumstances may have true molecular sieving properties. The adsorbent must also have good mechanical properties such as strength and resistance to attrition and it must have good kinetic properties, that is, it must be capable of transferring adsorbing molecules rapidly to the adsorption sites. In most applications the adsorbent must be regenerated after use and therefore it is desirable that regeneration can be carried out efficiently and without damage to mechanical and adsorptive properties.

5.5.1.3　吸附剂的再生（regeneration of adsorbents）

工业上能否实现吸附分离，除了取决于所选用的吸附剂是否具有良好的吸附性能外，吸附剂的再生也非常关键。吸附剂的再生程度不但决定了再生后吸附处理产品的纯度，而且也关系到处理量。因此，确定适宜的再生方法和工艺条件是吸附过程在工业上应用的一个很重要的问题。

吸附剂的再生应根据吸附工艺要求及操作条件而采取不同的方法。一般吸附操作可以分为吸附净化（adsorption purification）和吸附分离（adsorption separation）两类。

① 吸附净化　处理流体中的吸附质体积分数小于3%，如脱水（dehydration）、脱色（decoloration）、脱臭（deodorization）等，吸附后的吸附质可以抛弃不用。这类吸附-脱附再生的操作循环周期较长，再生时应尽可能除去吸附剂上所有的吸附质。

② 吸附分离　处理流体中的吸附质含量较大，一般体积分数为3%～50%，被吸附的吸附质需要回收为高纯度产品。如从合成氨弛放气中回收氢，混合二甲苯中分离对二甲苯和间二甲苯等。这类吸附-脱附操作都采用小的吸附量差和快速循环操作，主要考虑吸附-脱附处于最高效率点范围的情况。

吸附量差是指吸附后吸附剂上的平均吸附量和再生后留在吸附剂上的平均吸附量之间的差，它取决于吸附与脱附操作状况及操作循环时间。

吸附剂的脱附再生的基本方法主要有以下四种：

① 升温脱附　因为吸附剂对吸附质的吸附容量随温度升高而下降，用升高温度的方法可使被吸附的吸附质从吸附剂上脱附，从而使吸附剂再生。由于固体吸附剂的热容量较大，传热系数较小，升温、降温速度慢，故循环周期长。

② 降压脱附（step-down stripping）　因为吸附剂对吸附质的吸附容量随吸附质的分压下降而下降，用降低系统压力或抽真空的方法可使被吸附的吸附质从吸附剂上脱附，从而使吸附剂再生。

③ 置换脱附　这种方法用其他吸附质（称为脱附剂）把原吸附质从吸附剂上置换下来，特别适用于热敏性物质。当然，采用置换脱附时，还需将脱附剂进行脱附，才能使吸附剂再生。

④ 冲洗脱附　用吸附剂不吸附的惰性气体冲洗吸附剂床层，使吸附剂再生。

在工业上常根据情况将上述各方法综合使用，特别是经常把降压、升温和通气冲洗联合使用以达到吸附剂再生的目的。

5.5.2 吸附的工业应用

多孔介质固体颗粒的吸附剂具有极大的比表面积，如硅胶吸附剂颗粒的比表面积（specific surface area）高达800m^2/g。由于吸附剂本身化学结构的极性、化学键能等物理化学性质，而形成了物理吸附或化学吸附性能。吸附剂对某些组分有很强的选择吸附性能，并有极强的脱除痕量物质的能力，这对气体或液体混合物中组分的分离提纯、深度加工精制和废气废液的污染防治都有重要的意义。吸附分离过程的应用范围大致为如下几方面：

① 气体或溶液的脱水和深度干燥　水分常是一些催化剂的毒物，例如在中压下乙烯催化合成中压聚乙烯时，乙烯气体中痕量的水分可使催化剂的活性严重下降，以致影响聚合物产品的收率和性能。对于液体或溶液，如冷冻机或家用冰箱用的冷冻剂，亦需严格脱水干燥。微量的水常在管道中结冰堵塞管道，导致管道的流体阻力、冷冻剂输送的动力消耗增加，并影响节流阀的正常运转。同时，微量的水也可能被冷冻剂分解，生成氯化氢之类的酸性物质而腐蚀管道和设备。

② 气体或溶液的除臭、脱色和溶剂蒸气的回收　吸附分离常用于油品或糖液的脱色、除臭以及从排放气体中回收少量的溶剂蒸气。在喷漆工业中，常有大量的有机溶剂如苯、丙酮等挥发逸出，用活性炭处理排放气体，不仅可以减少周围环境的污染，同时还可回收此部分有价值的溶剂。

③ 气体的预处理（pretreatment）和气体中痕量物质的吸附分离精制　在气体工业中，气体未进入压缩机之前需预处理，脱除气体中的CO_2、水分、炔烃等杂质，以保证后续过程的顺利进行。

④ 气体本体组分分离　例如，从空气中分离制取富氧、富氮和纯氧、纯氮，从油田气或天然气中分离甲烷，从高炉气中回收一氧化碳和二氧化碳，从裂解气或合成氨弛放气中回收氢，从其他各种原料气或排放气体中分离回收低碳烷烃等各种气体组分。

⑤ 烷烃、烯烃和芳烃馏分的分离　石油化工、轻工和医药等精细化工都需要大量的直链烷烃或烯烃作为合成材料、洗涤剂、医药和染料的原料。例如，轻纺工业的聚酯纤维的基础原料是对二甲苯。从重整（reforming）、热裂解（thermal cracking）或炼焦煤油等所得的混合二甲苯为乙苯，间位、邻位和对位二甲苯各种异构体的混合物。其中除乙苯是塑料聚苯乙烯的原料外，其他三种异构体均是染料、医药、油漆等工业的原料。由于邻二甲苯与其他三种异构体沸点（boiling point）相差较大，所以可用一般精馏塔分离。其他三种，特别是间二甲苯与对二甲苯的沸点极为接近（在101.32kPa下，二者相差仅为0.75℃），不能用一般的精馏（rectification）过程分离。采用冷冻结晶法，其设备材料要求较高，投资较大，能量的消耗也很多。当模拟移动床吸附分离法工业化并广泛采用后，在世界上基本上已取代了结晶法。

⑥ 食品工业的分离精制　在食品工业和发酵产品中有各种异构体和性质相类似的产物。例如，果糖和葡萄糖等左旋和右旋的糖类化合物（carbohydrate），其性质类似，热敏性高，在不太高的温度下受热都易于分离变色。用色谱分离柱吸附分离果糖、葡萄糖浆，可取得果糖浓度含量在90%以上的第三代果糖糖浆，其生产能力已达年产果糖浆万吨以上。在其他食品工业中，产品的精制加工也常采用色谱分离柱吸附分离法。

⑦ 环境保护和水处理　加强副产物的综合利用回收和三废（three wastes）的处理，不

仅仅涉及环境保护、生态平衡和增进人民的身体健康，还直接关系到资源利用、降低能耗、增产节约和提高经济效益的问题。例如，从高炉废气中回收一氧化碳和二氧化碳；从煤燃烧后废气中回收二氧化硫，再氧化制成硫酸；从合成氨厂废气中脱除 NO_x；从炼厂废水中脱除大量含氧（酚）含氮（吡啶）等化合物和有害组分，可使大气和河流水源免遭污染，并具有较高的社会效益和经济效益。

⑧ 海水工业和湿法冶金等工业中的应用　由于吸附剂有很强的富集能力，从海水中回收富集某些金属离子，如钾、铀等，对国民经济都有很高的效益。众所周知，我国化肥中氮和钾的比例不当，钾元素的用量过低，因此在我国如何从海水中提取钾肥是重要的课题。我国的贵金属（黄金）和稀土金属资源丰富，采用活性炭吸附回收常是有效的方法。其他如能源利用吸附剂的特性，在太阳能的收集制冷、稀土金属的储氢材料方面，作为能量转换等都为人们所注意。

如上所述，吸附分离技术过去只作为脱色、除臭和干燥脱水等的辅助过程。由于新型吸附剂如合成沸石分子筛等的开发，并又经过了各种改性，所以提高了吸附剂对各种性质近似的物质和组分的选择性系数。随着适宜的连续吸附分离工艺的开发，相继建立了各种大型的生产装置，满足了工业生产的需要。吸附分离工艺得到迅速发展，日益成为重要的单元操作。对于液相吸附，我国已建立了多套年产万吨以上对二甲苯的生产装置。对于气相吸附分离，大型的炼厂气或合成氨弛放气变压吸附分离氢气的装置，大型空气变压吸附分离制氧和氮的装置均已工业化并普通推广。

5.5.3 工业吸附方法与设备

5.5.3.1 固定床吸附

间歇式固定床吸附分离设备，其设备简单，容易操作，是中等处理量以下最常用的设备，广泛应用于液体或气体混合物的分离与纯化。

图 5-61 表示用固定床吸附操作回收工业废气中的苯蒸气，吸附剂选用活性炭。含苯的混合气进入固定床 1，苯被活性炭吸附，废气则放空。操作一段时间后，由于活性炭上所吸附的苯量增多，废气排放浓度达到限定数值，此时切换使用固定床 2。与此同时，水蒸气送入固定床 1 进行脱附操作。从活性炭上脱附下的苯随水蒸气一起进入冷凝器 3，冷凝液排放

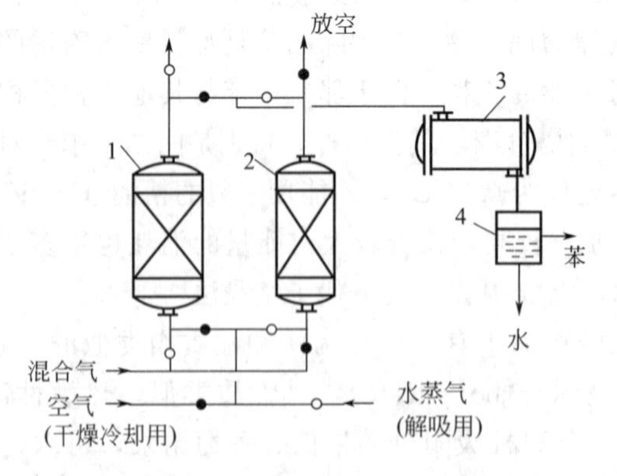

图 5-61　固定床吸附流程

1，2—装有活性炭的固定床；3—冷凝器；4—静止分离器

（图中○表示开着的阀门；●表示关着的阀门）

到静止分离器4使水与苯分离。脱附操作完成以后，还要将空气通入固定床1使活性炭干燥、冷却以备再用。

5.5.3.2 流化床吸附

流态化除了大量用于化学反应方面之外，还可用于流化吸附分离单元操作（unit operations）。用于吸附操作的流化床一般为双体流化床，即该系统由吸附单元与脱附单元组成（图5-62）。

含有吸附质的流体由吸附塔底进入，由下而上流动，使向下流动的吸附剂流态化。净化后的流体由吸附塔顶部排出，吸附了吸附质的吸附剂由吸附塔底部排出，进入脱附单元顶部。在脱附单元，用加热吸附剂或用其他方法使吸附质解吸，从而使脱附后的吸附剂再返回吸附单元顶部，继续进行吸附操作。

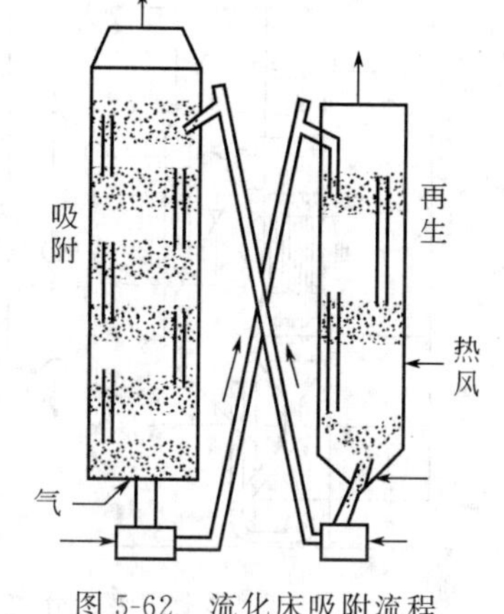

图5-62 流化床吸附流程

为了使颗粒处于流态化状态，流体的速度必须大于使吸附剂颗粒呈流态化所需的最低速度。所以它适用于处理大量流体的场合。

流化床吸附分离的优点是床内流体的流速高，传质系数大，床层浅，因而压力损失小。其缺点是吸附剂磨损比较大，操作弹性很窄，设备比较复杂，费用高。

5.5.3.3 移动床和模拟移动床吸附

移动床连续吸附分离又称超吸附（hypersorption），如图5-63所示。移动床连续吸附是充分利用吸附剂的选择性能高的优点，同时考虑到固体颗粒难于连续化和吸附容量低的缺点而设计的。固体吸附剂在重力作用下，自上而下移动，通过吸附、精馏、脱附、冷却等单元过程，使吸附剂连续循环操作。移动床早期用于含烃类原料气（如焦炉气）中提取烯烃等组分，目前在液相（如糖液脱色或润滑油精制）吸附精制中仍有采用。移动床吸附分离一般在接近常压和室温下操作。对比深冷精馏法，此法对设备和钢材的要求不高，投资费用也较低，但对吸附剂的要求较高，除要求吸附性能良好外，还要强度高、耐磨性能好，才能实现其优点。

液相模拟移动床吸附操作是大型吸附分离装置，如图5-64所示。它可分离各种异构体，如分离C_8芳烃中的对位二甲苯PX、间位二甲苯MX、邻位二甲苯OX及乙基苯EB等过程已实现工业化。模拟移动床综合各操作的优势，用脱附剂冲洗置换代替移动床使用的升温脱附，定期启闭切换各塔节进液料的阀门和脱附剂的进出口。在塔节足够多时，各液流进出口位置不断改变，相当于在吸附剂颗粒微孔内的液流和循环泵输送的循环液不断逆流接触。在各进出口未切换的时间内，各塔节是固定床；但对整条吸附塔在进出口不断切换时，却是连续操作的移动床。

5.5.3.4 变压吸附

变压吸附（pressure swing adsorption，PSA）分离过程是以压力为热力学参量，不断循环变换的过程，广泛用于气体混合物的分离精制。该过程在常温下进行，故又称为无热源吸

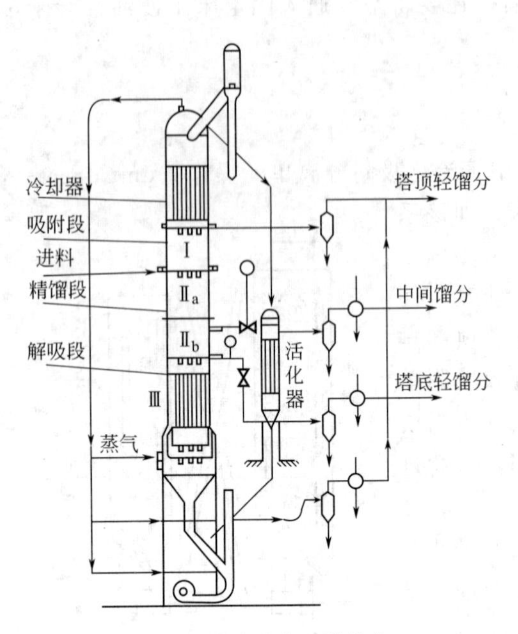

图 5-63　移动床吸附分离

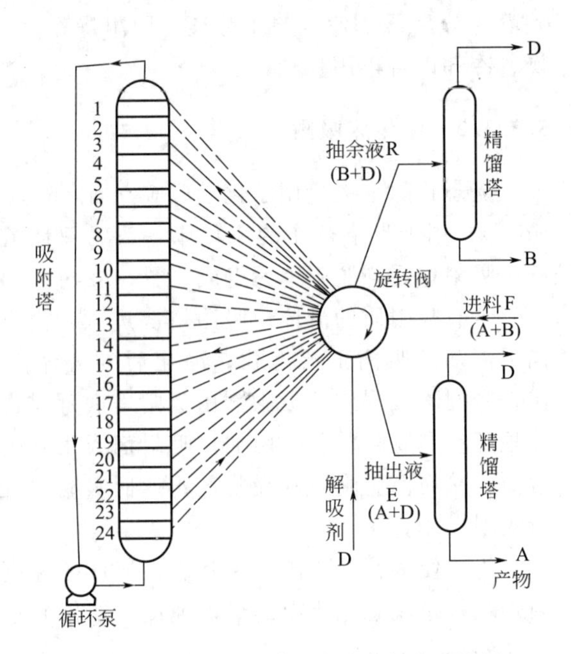

图 5-64　模拟移动床吸附分离操作流程图

附分离过程，可用于气体混合物的本体分离，或脱除不纯组分（如脱水干燥）精制气体，还可以将去除杂质（水分和 CO_2）作为预处理与本体吸附分离同时进行。

变压吸附是以压力变化为推动力的热力学参量泵吸附分离过程，在取得产品的纯度一定，回收率较高的情况下，要尽量降低能耗，以降低操作费用。为了回收床层和管道等死空间内气体的压力能量，除增加储罐外，增加了均压、升压等步骤。变压吸附工艺对氢气的回收和精制是特别成功的，这是由于氢气和其他组分如 CH_4、CO、CO_2 等在分子筛和活性炭吸附剂上的选择性系数相差很大。例如，原来的脱氢装置，多数是四床层系统，基本阶段是吸附、均压、并流清洗、下吹、清洗和升压，过程在室温下进行，典型压力为 $10\sim40\mathrm{kgf/cm^2}$，可获得 99.999%～99.9999%的高纯度氢，回收率在 85%～90%之间。而目前，吸附剂的利用效率（即每单位质量吸附剂回收氢）从 10%增至 20%，单一装置的容量有的已达 $2.8\times10^7\mathrm{m^3}$ 氢气的规模，床层发展到 7 个以至 10 个。

图 5-65 为变压吸附分离空气制氧的流程图。空气经过压缩、冷却和分离出液滴后进入以沸石为吸附剂的已经升压好的吸附器（以吸附器 A 为例），在吸附过程（以 ADS 表示）中，空气中的氮气被截留，离开吸附器的为氧产品。完成吸附步骤的吸附器 A 经过与吸附器 B 之间的顺向均压过程（顺向指与原料气流向相同，均压即压力平衡，以↑EQ1 表示）、顺向降压过程（以↑CD 表示），与吸附器 C 之间的顺向均压过程（以↑EQ2 表示）、逆向降压过程（以↓CD 表示）、冲洗过程（以↓PUR 表示）迫使停留在吸附剂表面的氮气脱附出来。然后经过与吸附器 D 之间的逆向均压过程（以↓EQ2 表示），与吸附器 B 之间的逆向均压过程（以↓EQ1 表示）后，对吸附塔 A 进行逆向升压（以↓R 表示），为下一次吸附做好准备。四台吸附器通过自动控制的阀门开关自动切换，轮流进行吸附-再生过程，只是在时间上相互错开，以保证分离过程连续进行。图 5-65 流程中各吸附器的操作步骤如表 5-8 所示。

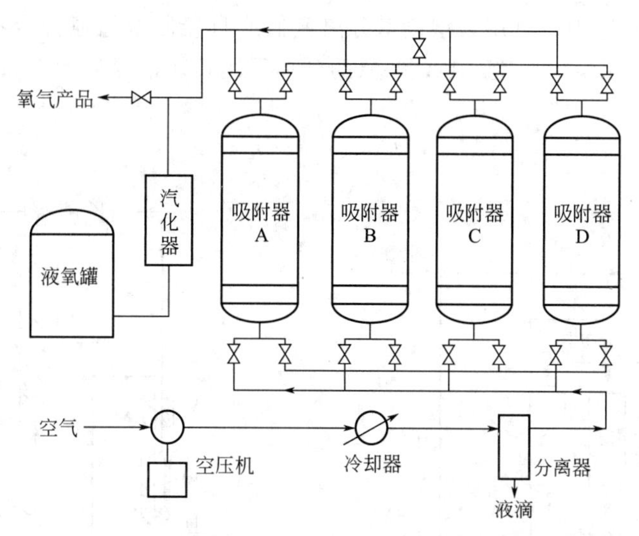

图 5-65 变压吸附分离空气制氧的流程图

表 5-8 四塔变压吸附分离空气制氧工艺中各吸附器的操作步骤

<table>
<tr><th>吸附器</th><th colspan="12">操作步骤</th></tr>
<tr><td>A</td><td colspan="3">ADS</td><td>↑EQ1</td><td>↑CD</td><td>↑EQ2</td><td>↓CD</td><td>↓PUR</td><td>↓EQ2</td><td>↓EQ1</td><td colspan="2">↓R</td></tr>
<tr><td>B</td><td>↓CD</td><td>↓PUR</td><td>↓EQ2</td><td>↓EQ1</td><td colspan="2">↓R</td><td colspan="3">ADS</td><td>↑EQ1</td><td>↑CD</td><td>↑EQ2</td></tr>
<tr><td>C</td><td>↑EQ1</td><td>↑CD</td><td>↑EQ2</td><td>↓CD</td><td>↓PUR</td><td>↓EQ2</td><td>↓EQ1</td><td colspan="2">↓R</td><td colspan="3">ADS</td></tr>
<tr><td>D</td><td>↓EQ1</td><td colspan="2">↓R</td><td colspan="3">ADS</td><td>↑EQ1</td><td>↑CD</td><td>↑EQ2</td><td>↓CD</td><td>↓PUR</td><td>↓EQ2</td></tr>
</table>

5.5.3.5 色谱吸附分离

扩大气体色谱处理量，用在实验室规模的分离上，称为制备色谱。但把它扩大到工厂规模，经过无数次尝试后，20 世纪 80 年代初期才获得成功。大型色谱分离法是在色谱分析的基础上发展起来的，即将一种气体混合物加入载气中，通过色谱柱的色谱分离（chromatographic separation），可得到若干个二元气体混合物（载气和某一组分），然后再将这些二元混合物逐一分离而得到产品。由于产品易被载气所污染，所以选择载气特别重要。一般选择不易冷凝或不吸附的气体，如氦气和氢气作载气，它们很容易从产品中分离出来，可循环使用。

色谱法比吸附法分离效率高，是由于气体和吸附剂之间的不断接触和平衡，每次接触相当于一个理论板（theoretical plate），通常在短柱中可能达到几百到几千的理论板数。因此，对于精馏来说，相对挥发度特别低，而对一般吸附来说，分离系数特别小的混合物系统采用色谱是最有效的。

放大到生产规模色谱分离的主要困难是色谱柱的均匀装填。装填不均匀会造成类似沟流（channeling）的非理想流动，或导致短路，从而使分离效率降低。此外，对于直径大的色谱柱希望径向分散高，轴向分散低。为了提高分离效率，可在柱内加挡板。

Elf-SRTI 法目前已商业化，其分离规模已扩大到 100t/a 的香料原料，同时已有厂家声称能分离 10×10^4 t/a 的正构和异构烷烃。此方法见图 5-66。

Asahi 公司也声称改良的大沸石柱的装填方法可用于工厂规模的液相二甲苯和乙苯的色谱分离。

在稀有气体的制造中，可用大型色谱分离氪氙或直接制取氪氙混合物。流程见图 5-67，以纯氢为载气（carrier gas），氢-氪或氢-氙的分离采用负压液氮温度下冻结的方法。

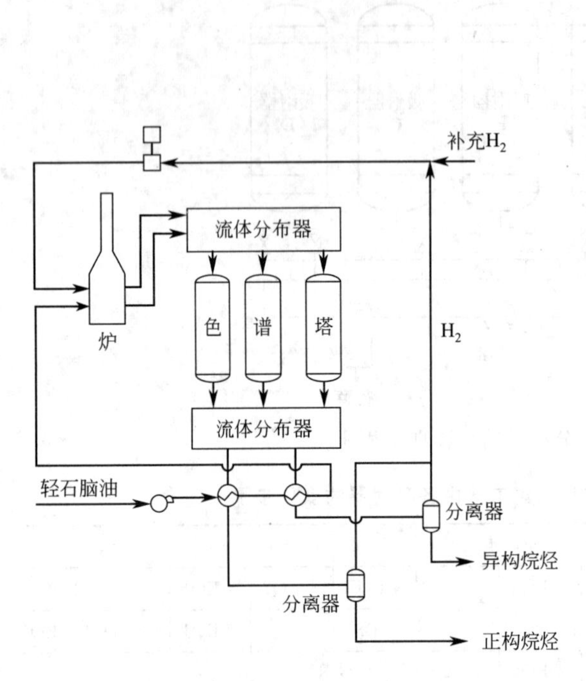

图 5-66　以轻石脑油为原料分离正构和异构烷烃的色谱分离器

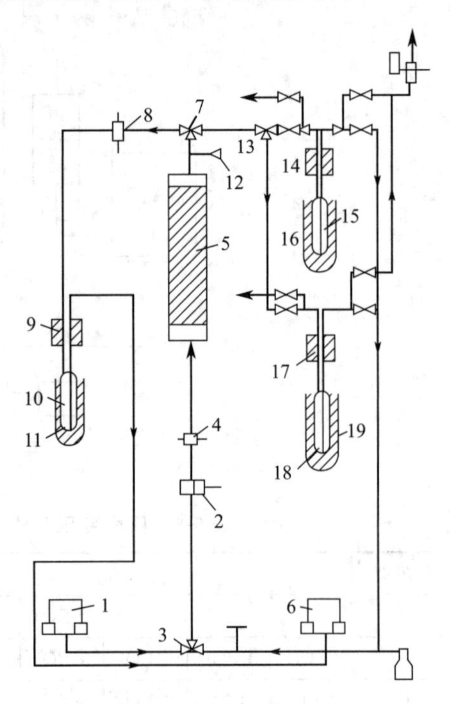

图 5-67　大型色谱法分离氪氙混合物流程

1,6—气柜；2—压缩机；3,7,13—阀；4,8—流量计；5—塔；9,14,17—换热器；10,15,18—吸附器；11—液氮瓶；12—气体分析仪；16,19—杜瓦瓶

单个的吸附塔都是间歇式操作的，但如果把两个或两个以上的塔连接起来，使吸附再生交换进行，这样一来整个吸附系统就能够连续操作了。气体吸附和液体吸附都是这样的流程。例如，在工业上废水处理过程中，先用生物的手段除去废水中部分有机物，再通过絮凝沉降（flocculating setting）、砂滤（sand filtration）等方法，除去水中悬浮物之后，这时就可用活性炭吸附来处理那些残留于水中的有机物。

另外，液体吸附和气体吸附在操作上没有本质的区别，所以有关的分离原理可参考前面的内容。其中有较大差别的是对吸附剂的再生方法。如果是气体吸附，常使用蒸气加热或减压等方法，比较简单就可实现再生。而对于液体吸附，由于吸附组分被牢固地吸附于吸附剂上，所以方法要复杂一些，主要是加热法、燃烧再生法和药物处理法（也称为提取置换法）、湿式氧化分解法、微生物分解法。

5.6　离子交换（Ion Exchange）

5.6.1　离子交换原理

离子交换现象早已为人们所知道，如海水经砂层过滤后会变得淡些，并以此来制备饮用水。英国农业化学家 H. S. Thompson 首先发现了离子交换现象：用硫酸铵溶液处理土壤时，大部分铵被土壤吸附，且不能用水洗下，同时有硫酸钙析出。1848 年，他将研究成果报告了皇家农学会顾问，化学家 J. T. Way 对此进行了全面审查并向皇家农学会作了总结性的报告。之后，法国化学家 Gans 首先合成了 $Na_2Al_2Si_3O_{10}$ 型无机离子交换剂，其中 Na^+ 能进行交换反应，它可代替天然沸石应用于软化水和糖类加工工业。他的工作对早期离子交换工艺的发展是很有意义的。后来，人们发现某些天然有机物，如泥煤、褐煤、无烟煤等，同土壤一样也具有某些离子交换能力，而且在酸、碱性介质中比较稳定。经过人工处理的磺化煤，其性能更好，一经出现立即在软化水等工业中被广泛采用。由于其价格便宜，直至今天仍有使用。Adams 和 Holmes 不仅仔细地研究了磺化煤的吸附性能，还于 1935 年首先合成了有机离子交换树脂（ion exchange resin）。这是离子交换材料制备和应用的理论和实践发展的一个新起点。如今，离子交换在离子交换剂、生产设备及工艺操作等方面的理论研究与实践上已取得很大的进展，已成为一个具有固-液非均相（heterogeneous）扩散传质规律而不同于传统分离过程的新型化工单元操作。

离子交换技术作为一种液相组分独特的分离技术，具有优异的分离选择性与很高的浓缩倍数，操作方便，效果突出。因此，在各种回收、富集与纯化作业中得到广泛应用。特别是第二次世界大战后，强碱性阴离子交换树脂在核燃料提炼过程中取得的历史性成功，以及作为换代技术，离子交换树脂继沸石、磺化煤之后，在热电站给水工程中的大规模应用，使离子交换技术很快推广到许多现代工业的分离工程中，发挥了卓越的技术功能。例如，在化工、医药、食品、水处理、冶金、环境保护以及核燃料后处理等方面，离子交换技术作为一种新型提取、浓缩、精制手段，得到广泛应用和迅速发展。20 世纪 50 年代，固定床离子交换设备的大量应用，曾被誉为离子交换技术发展的“黄金时代”。70 年代以后，离子交换分离技术与流态化技术相结合，出现了塔型多级流化床连续逆流离子交换设备，则被认为是离子交换技术发展史上的丰碑。

在许多水溶液组分的分离过程中，如稀贵金属的回收、高附加值生物制品与产品的提取、高纯水的制备以及污染控制等工艺中，所处理的工艺物料一般来说数量大，成分复杂，杂质含量很高，甚至含有大量悬浮固体，待分离、待提取的有用组分含量往往很低，而分离过程中又常常要求很高的回收率和选择性。面对生产发展中这种苛刻的要求，常规分离方法往往难以奏效。因此，具有独特操作行为和分离性能的离子交换技术，便应运而兴，成为一种颇引人注目的分离手段。今天，在应用的广度与研究的深度上，离子交换技术的发展潜力，日益引起人们的关注，不断更新人们的观念。

Ion Exchange

Ion exchange consists of the interchange of ions between two phases. With ion-exchange resins, the resin-a cross-linked polymer network-is the insoluble phase to which an ion is electro-statically bound; when contacted with a solution containing ions of the same charge, an exchange can occur, the extent to which depends on the concentration of the ions in solution and the affinity of the ions for the insoluble phase relative to the solution phase. Both cation-and anion-exchange resins have been prepared. The different ion-exchange resins consist of different polymers and different ligands covalently bound to the polymers.

Ion-exchange resins are useful because of the insolubility of the resin phase. After contacting with the ion-contacting solution, the resin can be separated by filtration. They are also adaptable to continuous processes involving columns and chromatographic separations Their insolubility renders them to be used for many years. Ion-exchange resins have been used in water softening, removal of toxic metals from water in the environment, wastewater treatment, hydrometallurgy, sensors, chromatography, and bio-molecular separations. They have also been used as catalysts, both in place of homogeneous catalysts such as sulfuric acid and to immobilize metallic catalysts.

离子交换过程是被分离组分（即被提取、被纯化的离子）在水溶液与固体交换剂之间发生的一种化学计量分配过程。离子交换反应发生在固相的离子交换树脂和水相的离子之间。离子交换时，有些离子负载到树脂上，与溶液中其他离子分离，然后再从树脂上洗脱下来回收；同时使树脂再生，反复使用，如同萃取（extraction）和反萃（recovering）一样。不过在交换和洗脱两个操作之间，要进行洗涤，以除去树脂上残留的溶液。

离子交换树脂为带有极性基团的高分子化合物，它不溶于水、酸、碱及任何溶剂。其结构可分为三部分：①不溶性高分子骨架；②骨架上有极性基团；③极性基团可以电离的离子。连于母体上的离子叫固定离子（fixed ion），与固定离子荷电相反的离子叫反离子（counter-ion）（也叫可交换离子或可扩散离子）。

反离子能与溶液中的同性离子发生交换。例如，用 HR 代表阳离子交换树脂，它在水中能电离出 H^+（即反离子），可与溶液中的正离子（如 Na^+）发生如下交换反应：

$$\overline{H^+R^-}+Na^+ \rightleftharpoons \overline{Na^+R^-}+H^+$$

其中，带上划线的代表树脂相。

树脂虽为固体，但当与水接触而水分子进入树脂内部后，其极性基闭上的离子水化而溶胀。溶胀的树脂为凝胶状，含有大量水分子，就像浸在水中一样，故而可视为电解质。这类电解质的许多性质与溶液中的电解质很相似，主要不同点是树脂的一种离子（固定离子）太大，不能进入溶液。所以，树脂虽为固体，但与溶液中离子的交换反应可作为液相中的反应来处理。

由于电中性的缘故，交换到树脂上 Na^+ 的物质的量与树脂上释放出的 H^+ 的物质的量完全相等，即交换反应是严格按照定量关系进行的。而且，离子交换反应是可逆的。例如，H 型阳离子交换树脂（HR）与 Na^+ 发生交换时，H^+ 又能置换出 Na^+。树脂的再生正是利用了交换反应的可返性。

离子交换过程表现出的反应速率不如溶液中离子互换反应速率那样快。这是因为树脂与溶液接触进行的离子交换反应，不仅发生在树脂颗粒表面，更主要的是在颗粒内部进行。当溶液中的欲交换离子扩散到树脂表面后，还需经过 5 个步骤，才能完成一个交换过程：

① 溶液中欲交换离子穿过树脂颗粒表面的液膜（liquid membrane）（称为膜扩散）；

② 继续在树脂相（即树脂颗粒内）扩散（称为粒扩散），达到交换位置；

③ 离子交换；

④ 交换下来的离子在树脂相扩散，扩散至颗粒表面；

⑤ 穿过颗粒表面的液膜进入溶液。

图 5-68 给出了离子交换反应的上述步骤。

以上 5 个步骤都可能成为离子交换反应的控制步骤，但是究竟哪一步最慢，取决于许多条件。

滞留层（stagnant layer）（液膜）中的扩散系数与一般溶液中无太大差异，离子通过滞留层的速度取决于层的厚度。如果滞留层中的扩散是控制步骤（control step），就称为颗粒外扩散控制，或液膜控制。提高外面溶液的流速，可以在一定程度上减小滞留层厚度，提高颗粒外扩散的速度。滞留层厚度大概在 0.1～1mm 之间。反应离子的浓度很低时（如小于 0.001mol/L），扩散的浓差推动力十分小，容易造成液膜控制。颗粒外扩散控制的一个特点是交换速度与树脂颗粒直径成反比。

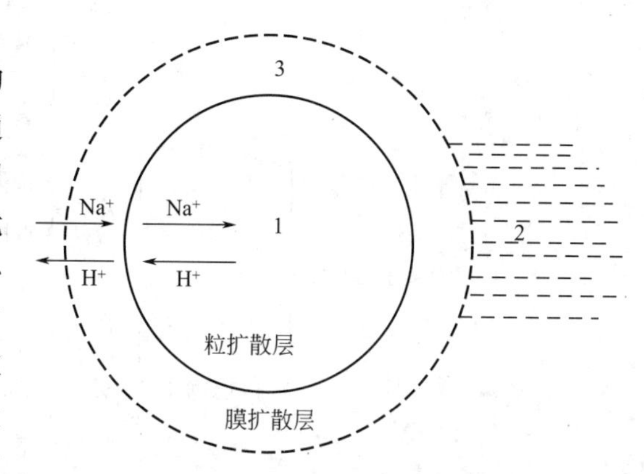

图 5-68　离子交换过程示意图

1—树脂相；2—溶液相；3—液膜

由于孔隙狭窄而且曲折，分布着许多高分子链，对离子扩散造成非常大的阻力，扩散系数比一般溶液中的低几个数量级，往往是最慢的一步。颗粒内扩散除了受制于浓度差（concentration gradient），还受孔内电场（internal electric field）的影响，所以高价离子颗粒内扩散速度更慢。颗粒内扩散为控制步骤时，离子交换速度与颗粒直径的平方成反比。另外，交换和被交换离子的选择系数差别越大，交换的推动力越大，速度越快。

如果化学反应速率非常慢，成为控制步骤，则称为反应控制。此时具有一般化学反应动力学的特点，比如温度对交换速度的影响比扩散控制显著。不过，这几乎仅仅发生在反应速率很慢的情况下，比如螯合树脂与金属离子的交换。反应控制的重要特征是交换速度不受树脂颗粒大小的影响。

5.6.2　离子交换剂的种类

离子交换技术的早期应用，是以沸石类天然矿物净化水质开始的。离子交换剂的发展是离子交换技术进步的标志。离子交换剂是一种带有可交换离子（阳离子或阴离子）的不溶性固体（insoluble solid）。它具有一定的空间网络结构，在与水溶液接触时，就与溶液中的离子进行交换，即其中的可交换离子由溶液中的同符号离子取代。不溶性固体骨架在这一交换过程中不发生任何化学变化。带有阳性可交换离子的交换剂，称为阳离子交换剂（cation-exchanger）；带有阴性可交换离子的交换剂，称为阴离子交换剂（anion exchanger）。

固体离子交换剂的分类见图 5-69。其中，SA、WA 分别代表强酸性（strong acidity）与弱酸性（weak acidity），SB、WB 分别代表强碱性（strong alkaline）与弱碱性（weak alkaline）。

（1）天然无机离子交换剂

天然无机离子交换剂主要是一些具有一定结晶构造的硅铝酸盐，最具代表性的是沸石类，如方沸石 $Na(Si_2AlO_6)_2 \cdot H_2O$、菱沸石 $(Na,Ca)(Si_2AlO_6)_2 \cdot 6H_2O$、丝光沸石 $(Na,Ca)(Si_5AlO_{12})_3 \cdot 3H_2O$、交沸石 $(K,Ba)(Si_5Al_2O_{14}) \cdot 5H_2O$、钠沸石 $Na_2(Si_3Al_2O_{10}) \cdot 2H_2O$。沸石的晶格由 SiO_2、AlO_2 的四面体构成。由于 Al 是 3 价，因此晶格中带有负电荷，此负电荷可由晶格骨架中的碱金属、碱土金属离子来平衡。这些碱金属、碱土金属离子虽然不占据固定位置，却可在晶格骨架通道中自由运动。因此，这些碱金属、碱土金属离子

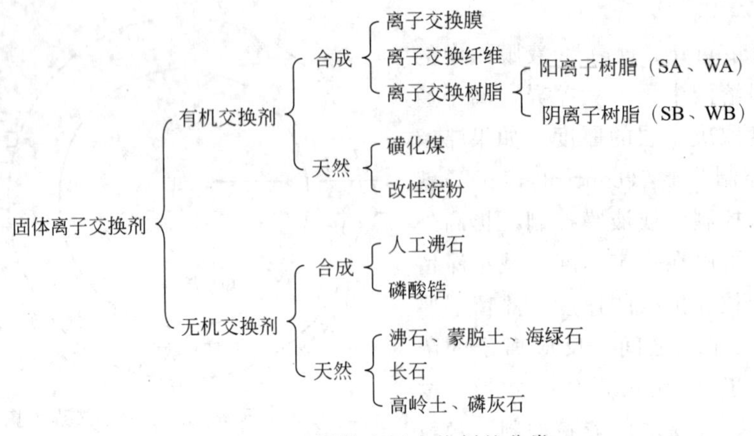

图 5-69　固体离子交换剂的分类

也就是可被交换的反离子。它们能与其他外部阳离子进行交换，即沸石类天然矿物具有阳离子交换剂特性。

具有阳离子交换性质的天然硅铝酸盐还有蒙脱土与绿砂。前者具有层状结构，组成为 $Al_2[Si_4O_{10}(OH)_2]_n H_2O$；后者称为海绿石，有可供交换的钾离子，因结晶构造较密实，故阳离子交换行为只能发生于结晶表面处。

另一种天然硅铝酸盐是长石类矿物，如正长石($K_2O \cdot Al_2O_3 \cdot 6SiO_2$)、钠长石($Na_2O \cdot Al_2O_3 \cdot 6SiO_2$)、灰长石($NaAlSi_3O_5 + CaAl_2Si_2O_3$)、钙长石($CaO \cdot Al_2O_3 \cdot 2SiO_2$)。它们也具有一定的阳离子交换作用，可作为阳离子交换剂使用。

某些蒙脱土、高岭土，特别是磷灰石$[Ca_5(PO_4)_3]F$、羟基磷灰石$[Ca_5(PO_4)_3](OH)$，都具有阴离子交换特性。

(2) 合成无机离子交换剂

① 合成沸石（synthetic zeolite）　将钠、钾、长石、高岭土等的混合物熔融，可制得具有天然沸石行为的人工沸石，即熔融型沸石。碱与硫酸铝、硅酸钠的酸性溶液反应析出沉淀，沉淀物经过适当干燥，可制得另一种类似天然沸石的凝胶型沸石。这两种合成沸石都是无定形结构。

② 分子筛（molecular sieve）　将含有铝、硅的碱溶液在较高温度下进行结晶，可制得具有规则结晶构造的分子筛。这样制备的分子筛，具有严格确定的微孔结构与孔尺寸，主要用作高选择性吸附剂。

③ 氢氧化物凝胶（hydroxide gel）　许多两性金属氢氧化物凝胶（Fe_2O_3、Al_2O_3、Cr_2O_3、Bi_2O_3、TiO_2、ZrO_2、ThO_2、SnO_2、MoO_3、WO_3）的水化物，在高于其等电点pH值的条件下，具有阳离子交换性质。TiO_2 的水化物凝胶称为钛胶，曾用于从海水中提取钠。在海水提铀的研究中，还应用各种复合型的金属氢氧化物凝胶。

④ 磷酸锆类无机离子交换剂　将氯氧化锆（$ZrOCl_2$）水溶液用磷酸或碱性磷酸盐沉淀，可制得不同锆磷比例（ZrO_2/P_2O_5）的磷酸锆，以 ZrP 表示，代号为 Phozir。这是一种阳离子交换剂。

上述合成产品多是非化学计量化合物，其化学组成取决于制备条件，通常完全不溶于水。作为离子交换剂，其特点是交换容量较高，离子交换速度也比较快，而且热稳定性与辐射稳定性皆优于合成树脂。

(3) 离子交换树脂

离子交换树脂是一种具有活性交换基团的不溶性高分子共聚物（copolymer），由惰性骨架（母体）、固定基团与活动离子（又称交换离子或反离子）三部分构成，其结构模型如图5-70所示。其交换基团使用失效后，经过再生可以恢复交换能力，重复使用。树脂的母体构架称为骨架，由高分子碳链构成，是一种三维多孔性海绵状不规则网状结构，不溶于酸溶液、碱溶液及有机溶剂。

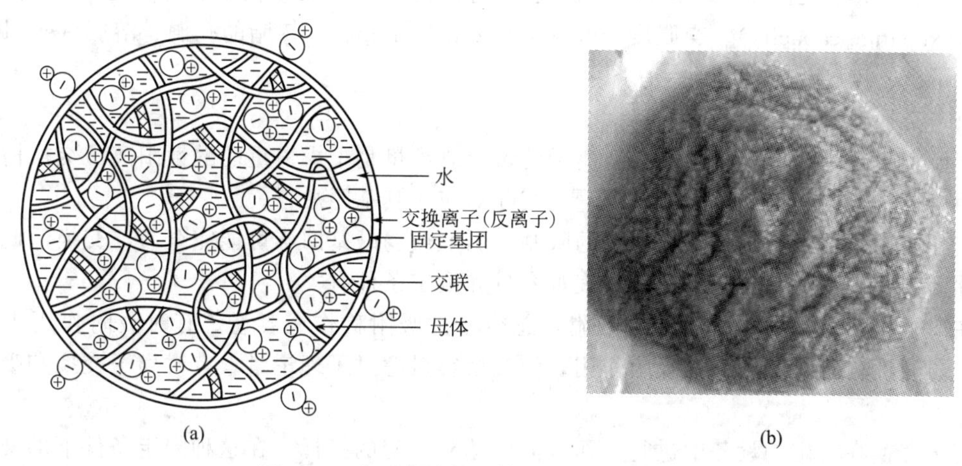

图5-70 离子交换树脂结构（a）及外观（b）

市售工业树脂是粒度为0.3～1.2mm（14～48目）的均匀球形颗粒。

离子交换树脂的颜色及颜色的深浅与其种类、结构有关。例如，凝胶型树脂具有松香样光泽，大孔树脂则无光泽。

商品离子交换树脂有各种类型，按交换基团分，有阳离子树脂（含有$—SO_3H$、$—COOH$、$—PO_3H_2$等基团）和阴离子交换树脂（含有$—N^+R_3$、$—NR_2$等基团）；按骨架材料分为聚苯乙烯型、酚醛型、丙烯酸型、乙烯吡啶型等；按结构形式分，有凝胶型、大孔型、载体型，蛇笼型、布丁型等；按功能分，有常规树脂、螯合树脂、两性树脂、热再生树脂、磁性树脂、氧化还原树脂等。

Type of Ion Exchangers

Ion exchange materials are insoluble substances containing loosely held ions which are able to be exchanged with other ions in solutions which come in contact with them. These exchanges take place without any physical alteration to the ion exchange material. Ion exchangers are insoluble acids or bases which have salts which are also insoluble, and this enables them to exchange either positively charged ions (cation exchangers) or negatively charged ones (anion exchangers). Many natural substances such as proteins, cellulose, living cells and soil particles exhibit ion exchange properties which play an important role in the way the function in nature.

Depending on the type of the functional group, ion exchangers can be divided into several types: strong acidic, strong basic, weak acidic and weak basic. Ion exchangers containing sulpho- and phospho-acidic groups and those containing tetra-ammonium basic groups are strong acidic and strong basic exchangers, respectively, whereas those containing phenolic and primary amino groups are weak acidic and weak basic exchangers, respectively.

5.6.3 离子交换树脂的基本性质

在上述的离子交换剂中，最主要的是离子交换树脂。离子交换树脂的基本性质可分为交换容量（exchange capacity）、交联度（degree of crosslinking）、树脂的孔洞、溶胀（swelling）和含水量、密度、稳定性等。

（1）交换容量

交换容量是指一定量树脂可交换离子的量。总容量是干燥恒重的酸性阳离子交换树脂或氯型阴离子交换树脂可交换的离子总量，实际对应于其基团总量。

实际使用条件下，由于解离常数的限制，弱酸或弱碱基团不可能完全参与交换，此时的容量称为表观容量或有效容量。显然表观容量和测定条件有关。

在离子交换过程中常同时伴有吸附，总容量加吸附称作全容量。有时吸附作用十分显著，如弱碱性树脂对苯酚的强吸附作用，使交换容量超过了理论量。有时离子交换和吸附同时存在。

离子交换在一定的设备中进行，而且需要达到一定的指标，在这种限定条件下的交换容量为工作容量或使用容量

（2）交联度

树脂骨架由线形高分子互相交联而成，合成的高分子之间的交联度不同，导致树脂的密度、强度、孔隙率和溶胀性不同。交联度并不能直接测定，所以只能以合成时加入的交联剂的量表达。如合成聚苯乙烯时以二乙烯苯为交联剂，常以二乙烯苯在总量中的百分含量表示交联度。

交联度越大，树脂强度越大，树脂结构越紧密，功能基越难进行反应，交换速度越小。

（3）树脂的孔洞

描述树脂孔洞的参数包括孔隙率、孔径和比表面积。孔隙率是孔的总体积和树脂体积的比值。另外，描述树脂孔洞的参数也可以用孔度表示，即单位质量或体积的树脂所具有的孔体积，常用单位是 mL/g 或 mL/mL。

凝胶树脂的孔径平均约为 2～4nm，大孔树脂的孔径达到 20～500nm，外观呈半透明或不透明状。

比表面积易于测定，是衡量孔隙率的重要指标。凝胶树脂的比表面积在 $1m^2/g$ 左右，而大孔树脂达到每克几至几十立方米。

（4）溶胀和含水量

尽管合成树脂的骨架是碳氢链，是憎水的。但是内外表面均布满功能基，功能基都具有很强的极性，从而使树脂整体上具有亲水性。干树脂浸泡于水中，功能基与水分子相互作用，活性基团（active group）充分水化，水渗透进入孔道之中，使孔道中充满水。而水化的功能基几乎完全处于水溶液的环境之中，从而能够与溶液中的离子进行交换。功能基吸引水渗入孔道，使骨架高分子的链被挤开、伸长，孔道扩大，从而使整个树脂体积膨胀。

另外，当功能团的离子由水化半径小的离子交换为水化半径大的离子时，也将导致体积膨胀。显然，随着交换的离子水化半径增加，树脂体积也随之增加。这种体积增量称为溶

胀率。

膨胀的树脂内部产生很大的压力，称为溶胀压。树脂在应用中，功能基上的离子反复改变，不断胀大和收缩，溶胀压也随之反复变化，常导致树脂老化而破碎。

(5) 密度

每个树脂球内部有孔隙，球之间又存在空隙，如图5-71所示，加上树脂有干湿之分，树脂的密度存在不同表达方式。

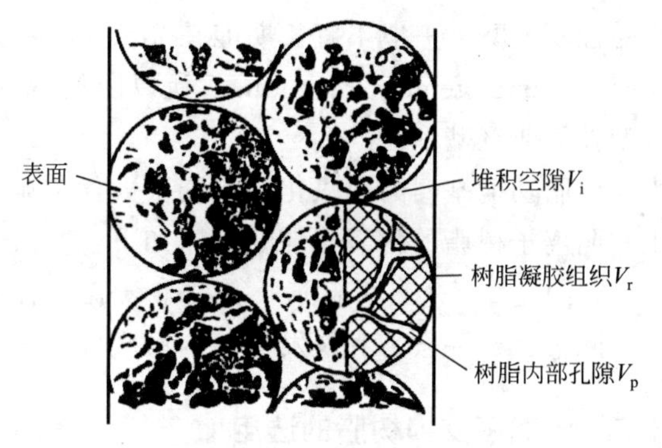

图5-71 填装树脂的剖面

显然，树脂体积等于树脂材料自身组织的体积加上树脂空隙体积，即：

$$V_s = V_r + V_p \tag{5-51}$$

树脂在容器中的体积为树脂体积加上其间的空隙，从而有：

$$V_b = V_s + V_i \tag{5-52}$$

干树脂材料的真密度为：

$$\rho_r = \frac{W_r}{V_r} \tag{5-53}$$

式中 W_r——树脂干质量，kg。

一般阳离子树脂的真密度为1.2～1.4g/mL，强酸性阳离子树脂的真密度为1.3g/mL；一般阴离子树脂的真密度为1.1～1.3g/mL，强碱性的阴离子树脂的真密度为1.1g/mL。

湿树脂的质量除以其体积，即为湿真密度：

$$\rho_s = \frac{W_s}{V_{r(w)}} \tag{5-54}$$

树脂的表观密度又称视密度（apparent density），是一定量的树脂按规定方式加入特定容器，自然产生的堆积状态而形成的密度，故又称堆积密度或松装密度，可表示为：

$$\rho_a = \frac{W_s}{V_b} \tag{5-55}$$

式中 V_b——树脂在容器中的体积，m^3。

一般树脂的表观密度为0.6～0.8g/mL。阴离子树脂较轻，偏于下限；阳离子树脂偏于上限。

有时为了提高树脂的密度，在合成时加入高密度的物料如氧化锆等，或者在骨架上引入一个卤素原子，称为高密树脂或加重树脂。高密树脂的密度达到1.3g/L以上。

有的树脂在合成时加入具有磁性的γ-Fe_2O_3，使树脂能在磁场作用下聚集，沉降速度可以提高10倍，称为磁性树脂，特别适用于浆状介质中的离子交换。

(6) 稳定性

① 物理稳定性 树脂在设备中，特别是流动中不断发生碰撞、摩擦以及受到流体的冲击。合成树脂需要具有一定的机械强度，才能够在这些条件下经久操作。

树脂都有相当的耐热性，阳离子树脂一般可以耐受100℃以上的温度，但是阴离子树脂的实际操作温度不能高于60℃。一般来说，盐型树脂比酸型或碱型树脂的热稳定性好。

② 化学稳定性 合成树脂对非氧化性酸碱都有较强的稳定性。不过阳离子树脂对碱的

耐受性不及酸，一般不宜长期泡在2mol/L以上的碱溶液中。由于有机胺在碱中易发生重排反应（rearrangement reaction），所以阴离子树脂对碱的耐受性也不好，保存时应该转化为氯型，而非羟基型。

树脂功能基团的耐氧化性差别很大，浓硝酸、次氯酸、铬酸、高锗酸都能使一些树脂氧化。阴离子树脂的耐氧化性的顺序如下：

叔胺＞氯型季铵盐＞伯胺、仲胺＞羟基型季铵盐

除伯胺、仲胺易和醛发生缩合反应外，其他树脂都有很强的耐还原能力。

5.6.4 离子交换树脂的选用

选用树脂与分离组分的性质（离子种类与形式）、体系特点（浓度、pH值等介质条件）以及分离要求等因素有关。一般说来，应选容量大、选择性好、交换速度快、强度高、易再生、价廉易得的树脂。由于这些因素与条件可能相互制约，因此需根据实验结果综合考虑，权衡决定。

（1）基本要求

① 交换容量　树脂容量越大越好，因为容量大则用量少、投资省，且设备紧凑、体积小。由于单位体积树脂处理的料液量大（料液浓度相同时），相对来说洗脱、再生时的试剂消耗量也就低。

② 选择性与交换速度　作为分离、提取的一种手段，树脂的选择性与交换速度越高越好。选择性高则分离效果好，设备效率高，设备级数少和高度低。考虑树脂的选择性，也应同时考虑再生效果与交换速度的制约。

③ 强度与稳定性　考虑到操作与成本，离子交换树脂应能长期、重复使用，要求所选树脂耐冷热、干湿、胀缩的变化，不破碎、不流失、耐酸碱、抗氧化、抗污染。

④ 洗脱与再生　吸附操作本身并非目的，只是分离过程的一个环节，因此吸附后还要将分离目的物从树脂上有效地洗脱下来。一般说来，离子越易被树脂吸附，则洗涤也就越困难。考虑吸附时，也应同时考虑洗脱与再生，有时为了兼顾洗脱操作，不得不放弃容量或选择性，也就是全面考虑后有时宁可选用吸附性能略差一点的树脂。

（2）树脂种类的选择

被分离组分是阳离子，如无机阳离子、有机碱阳离子、络合阳离子时，可用阳离子树脂分离；被分离组分是阴离子，如无机阴离子、有机酸阴离子、络合阴离子时，可用阴离子树脂处理。但是这并非绝对的。为了提高分离效果或进行有效分离，可调节介质的条件，使分离组分改变存在形态。例如，硫酸溶液中的六价铀以UO_2^{2+}形式存在，由于有大量其他阳离子共存，很难用阳离子树脂进行有效的分离。调整溶液中的SO_4^{2-}浓度与pH值，使铀以$UO_2(SO_4)_2^{2-}$、$UO_2(SO_4)_3^{4-}$形式存在，用阴离子树脂则可进行有效的分离。同样，强碱性阴离子树脂也可从碳酸盐溶液中成功地分离$UO_2(CO_3)_3^{4-}$。

对于两性氨基酸的提取，可根据其等电点调整溶液pH值，以所要求的离子形式进行分离。例如，谷氨酸在大于等电点pH值（3.22）时，以阴离子形式存在，可用阴离子树脂提取；在pH值低于3.22时，谷氨酸以阳离子形式存在，可用阳离子树脂处理。

对于大分子化合物，例如蛋白质、色素等物质，可用大孔型树脂分离。因常规树脂骨架疏水且电荷密度较大，易使蛋白质类不稳定性生物高分子产生不可逆变性。

(3) 树脂功能基团的选择

选定阳离子树脂或阴离子树脂后，还要选择合适的功能基团与使用的离子形式。

功能基团的选择以与交换离子亲和力强弱、选择性大小为依据，这是分离的基础。例如，金的提取与分离可用硫脲基树脂，对Cu、Hg的提取可用巯基树脂或羧氨基树脂，对Ni的分离可用肟基树脂等。

在中性盐溶液中要完全除去其中的阳离子与阴离子，可用强酸性阳离子树脂与强碱性阴离子树脂。例如，要求完全除去Ca^{2+}、Mg^{2+}时，可用强酸性阳离子树脂；若只要求部分除去其中的阳离子与阴离子，则可用弱酸、弱碱性树脂，或强酸、弱碱性树脂，或弱酸、强碱性树脂。

按功能基团酸碱强度（以pK_a表示）的大小选择树脂，固然是分离的基础，但pK_a值的大小并非选择功能基团的唯一标准，因为还要兼顾洗脱、再生行为，这在实际生产中往往很重要。这种情形在离子交换工艺中不乏其例。例如链霉素的提取，因其分子结构中含有两个强碱性胍基和一个弱碱性葡糖氨基，用强酸性树脂时，吸附效果虽好但洗脱困难，故生产中采用选择性与含量都很高的羧酸型弱酸性树脂（或弱碱性树脂）。有机碱、生物碱与强酸性树脂作用太强，故通常用弱酸性树脂分离。多价金属阳离子在中性或碱性介质中用弱酸性树脂处理，既易吸附又易洗脱。因一般弱酸性树脂只能在高pH值条件下操作，pH值越高其容量越大。在弱酸性介质中的阴离子可用弱碱性树脂处理。一般极性低的阳离子多用弱酸性树脂处理。乳清的净化用丙烯酸类弱酸性树脂处理时，既易吸附又易洗脱。糖液净化时，用弱酸性树脂可避免糖的转化。含铬废水中铬以$Cr_2O_7^{2-}$的形式存在，可选用大孔型阴离子树脂处理，这种树脂容量大、再生性好且抗氧化。含CN^-废水可用丙烯酰胺类弱碱性阴离子树脂处理，这种树脂易洗脱、抗污染。

树脂离子形式的选择与被分离离子的性质有关。例如，葡萄糖与果糖的分离可用钙型阳离子树脂，因为果糖与钙离子之间通过羟基可形成配位络合物，从而果糖与水进行配位交换，实现葡萄糖与果糖的分离。

树脂离子形式的选样更与交换体系的性质有关。例如，若离子交换前后要求溶液的pH值不发生变化，则应该用盐型树脂。从等电点结晶母液中回收谷氨酸时，因母液中含有大量NH_4^+，常规离子交换法用H型树脂无能为力，可用NH_4^+型树脂进行有效提取。在链霉素的提取过程中，链霉素在发酵液中呈中性的硫酸盐形式存在，用Na型或NH_4^+型弱酸性树脂处理，可将溶液的pH值维持在5～7，以利于交换。

树脂粒度与密度的选择，应依据操作系统的工艺条件而决定。对于传质速度较慢的交换体系，特别是颗粒扩散控制时，可选用细粒度树脂处理以增加两相接触表面积，缩短固相扩散路程，提高交换速度。这时，需考虑床层阻力增加的制约与树脂流失问题。通常，流化床操作中可选用大粒度树脂和高密度树脂。在混合床操作中，选用一种大密度树脂可增加两种树脂间的差异，有利于两种树脂的分离。

5.6.5 离子交换工业流程及设备

5.6.5.1 离子交换的操作程序

(1) 交换树脂使用前的处理

① 溶胀 树脂使用前需要做许多准备工作，首先是以水浸泡，使其充分溶胀，然后装

入设备，再用水淋洗，直至洗出的水清澈为止。如果树脂已经完全干燥，则不能直接以水浸泡，而应该用浓氯化钠溶液浸泡，逐渐稀释，以减缓溶胀速度。最后用水清洗，防止树脂胀裂。以上所用水都应该是纯水，至少不含钙、镁等高价离子。

② 清洗　新树脂使用前需要经过清洗。通常先在容器中从底部进水，向上冲去悬浮物。然后依次用4%氢氧化钠溶液洗，再用水洗，而后用4%的盐酸溶液洗，再经水洗。酸、碱液的用量为树脂体积的3～5倍。对于有机物敏感的工艺，最后需用95%的乙醇浸泡，溶出合成时的残余有机物。这样的清洗过程需循环反复几次，视使用的工艺要求而定。

③ 转型　清洗过的树脂要根据工艺要求经过转型才能使用。氯型强碱性阴离子树脂转化为氢氧根型比较困难，需要用树脂体积6～8倍的1mol/L的NaOH溶液处理，而后以清水洗至流出水无碱性（酚酞指示剂不变色）。所用的碱中不应含碳酸根，否则交换于树脂上，使其容量下降。氢氧根型强碱性阴离子树脂转化为氯型则十分容易，可用氯化钠溶液处理，当流出液pH值升至8～8.5时即已完成。

清洗过的强酸性阳离子树脂转化为H型，常用1mol/L的HCl处理，而后水洗至无酸性。从H型转化为其他离子型可用相应的盐溶液处理，最常用的是钠型。在冶金工艺流程中，有时为了避免将其他离子引入溶液，需要特别离子形式的树脂，也同样可以用相应盐溶液进行转型。

弱碱性阴离子树脂易于转化为氢氧根型，除了氢氧化钠，也可采用氨水进行转换。如氢氧根型需转变为其他形式，可进一步与含相应阴离子的酸反应。弱酸性阳离子树脂易于转化为H型，再转变为其他离子形式时，除了加相应的盐，同时要加入氢氧化钠，中和产生的酸，有利于反应进行完全。

（2）交换吸附

将前述经过清洗、转型的树脂与料液接触，被交换的离子和树脂的反离子进行交换反应，使溶液相中需要除去离子的浓度下降到期望值，同时树脂达到相当的负载容量。以钠型强酸性阳离子树脂从料液中除钙为例，总的反应可表示如下：

$$2\overline{RNa}+CaCl_2 \longrightarrow \overline{R_2Ca}+2NaCl$$

显然，目的是既要将钙离子尽量除去，又要使树脂得到充分利用。

在实际应用中，被交换负载于树脂上的离子，既可以是打算从溶液中除去的杂质，如前述示例中的钙，也可以是要回收的有价金属。

（3）洗脱和再生（elution and regeneration）

对于上述反应的离子交换，交换结束后，树脂负载了钙离子，成为钙型树脂。首先要把钙从树脂上脱除下来；如果负载的离子是打算回收的金属，洗脱之后的溶液就是中间产品，应该满足一定的工艺要求。在实际条件下，树脂上负载的往往不只是一种金属离子，洗脱就要分段进行，分别回收为不同产品。其次是要把树脂恢复为钠型，以便返回使用。最好这两个任务能同时完成，不然就要先脱钙，而后再转为钠型。

对于强酸性阳离子树脂，只要用浓度较高的氯化钠溶液就能使上述反应反向进行，即同时达到脱吸和树脂再生两个目的。但是有时并不能一步完成，如氯型弱碱性的阴离子树脂交换为硫酸根型后，不论用HCl或NaCl溶液洗脱，再生效率都十分低，因为硫酸根的交换势远高于氯离子，所以先要用NaOH洗脱硫酸根，使树脂转变为氢氧根型，再以HCl使之转为氯型，如下列方程所示：

$$\overline{[RN(CH_3)_3]_2SO_4}+2NaOH \longrightarrow 2\overline{RN(CH_3)_3OH}+Na_2SO_4$$

$$\overline{RN(CH_3)_3OH}+HCl = \overline{RN(CH_3)_3Cl}+H_2O$$

后一步包含了中和反应，因此十分有利于反应向右进行。

（4）洗涤

在进行了交换和洗脱再生之后，首先要把残留在树脂床层中的料液、洗脱液等残液冲出床层之外。这些残液应该返回合并到相应的原始溶液中去。另外，反应之后有一些固体颗粒和余液残留在树脂球的间隙、表面甚至孔内，需要用水经过反复清洗，才能进行下一步工序。

5.6.5.2　树脂的中毒问题（poisoning of resin）

树脂在使用过程中由于交换、吸附了不能用通常方法洗脱的物质或者被污染，导致交换能力明显下降，甚至不能继续使用，就称为“中毒”。中毒的原因可以归结为以下几点：

① 溶液中的大分子有机物污染　矿石浸取液和其他料液中有时含有天然有机化合物，如植物腐烂产生的腐殖酸等，还有前面工序加入的选矿药剂、絮凝剂等。它们具有多个活性基团，又有较大的碳氢链或稠环（fused ring），因此有的有较强的交换能力，有的易于吸附于骨架之上，是常见的导致中毒的原因。中毒树脂往往颜色加深，甚至呈现棕色，乃至黑色。

② 负载的离子交换势极高　当负载的离子交换势极高时，通常的洗脱剂难以洗脱，逐渐积累，占据了树脂的大部分功能团，使其不能再与其他离子交换而失效。对于阳离子交换树脂，Fe^{3+}交换势很高，而且用 NaCl、HCl 溶液等一般洗脱剂难以将其从树脂上置换下来。料液中含Fe^{3+}就可能积累而引起中毒。

③ 负载离子发生变化　仍以前述负载了Fe^{3+}的阳离子交换树脂为例。当以 pH 值较高的溶液洗涤树脂时，Fe^{3+}发生水解，产生氢氧化铁凝胶，沉淀于树脂孔隙之中，造成孔道堵塞，产生中毒现象。同样，阴离子树脂负载正钨酸根，遇酸性溶液，正钨酸聚合为多钨酸，分子体积庞大，也会造成相似的中毒。

④ 溶液中的固体颗粒　料液中的固体颗粒、胶体附着在树脂表面，甚至进入孔道，降低了树脂的有效表面积（effective surface area）。

当料液中存在引起中毒的因素时，为减少树脂中毒，一方面应该尽量净化料液，如去除固体颗粒；另一方面应该选择适当的树脂。许多大孔径的树脂比较耐受污染物，甚至可以去除溶液中的有机物，而且可以洗脱，反复使用。有的树脂厂开发了抗污染的专门树脂，孔径特别大，达到几千纳米，孔容量约 1mL/g，不易被有机物毒化。有时选用这些树脂放置在前面，吸附有机物，可以保护后面的树脂。另外，应该避免出现上述的附着离子发生反应，产生胶体或固体，堵塞树脂孔道。

对于已经中毒的树脂，通常需要选择适当的清洗剂，对症下药去除污染物，使树脂恢复正常性能，主要的解毒方法如下。

① 对于因固体沉淀而中毒的树脂，使用能使沉淀溶解的溶液洗涤。如氢氧化铁胶体可以用较浓的盐酸溶液浸泡，使之溶解。可使用 10%～20%的盐酸溶液浸泡，但浸泡时间不宜太长，更不要加热，以免损伤树脂。而且不能在酸洗后，紧接着就用清水淋洗。因为树脂内外的浓度差过大，会产生很大的渗透压，导致树脂破裂。正确的操作方法是浓酸洗之后，用稀酸溶液淋洗，逐渐降低浓度，最终过渡到纯水。硅酸胶体也常出现在湿法冶金过程的料液之中，吸附在树脂之上，可以用氢氧化钠溶液浸泡除去。

② 吸附的有机物很难洗脱，可以使用氧化剂将其氧化，分解为小分子化合物溶于水或生成二氧化碳而除去。次氯酸钠和过氧化氢是常用的氧化剂。次氯酸钠的浓度可在0.2%～1%之间，过氧化氢浓度可略高。有些负载的离子难以用其他离子置换的方法洗脱，采用化学方法分解，才能洗脱。如硫代钼酸根配阴离子交换势特别大，负载于强碱性树脂就不能用通常的方法洗脱，可用1%的次氯酸钠溶液将其中的硫根氧化为硫酸根，使硫代钼酸根分解而洗脱。

5.6.5.3 离子交换设备

离子交换树脂是实现离子交换分离过程的基础。如何合理有效地使用树脂，充分发挥每一粒树脂在操作中的交换作用是离子交换设备设计的关键。这就要求离子交换设备必须具备相宜的结构，以提供有效的固液相接触方式与条件，以完成所要求的分离任务。

目前，已投产应用的工业规模离子交换设备与尚处于研究开发阶段的实验室离子交换设备种类很多，设计各异。按结构类型分，有罐式、塔式与槽式；按操作方式分，有间歇式、周期式与连续式；按两相接触方式分，有固定床、移动床与流化床。流化床又分为液流流化床、气流流化床与搅拌流化床。固定床分为单床、多床、复床与混合床，正流操作型与反流操作型，以及重力流动型与加压流动型。

（1）固定床离子交换柱设备

固定床离子交换柱是一种应用广泛的工业设备，由于树脂床在柱内不移动，所以称为固定床。离子树脂交换柱通常是高径比在2～5之间的圆柱体，底部和顶部多为球形或椭球形。不但顶部和底部有进出液管，而且柱的中间也有支管，用于进液或出液，如图5-72所示。

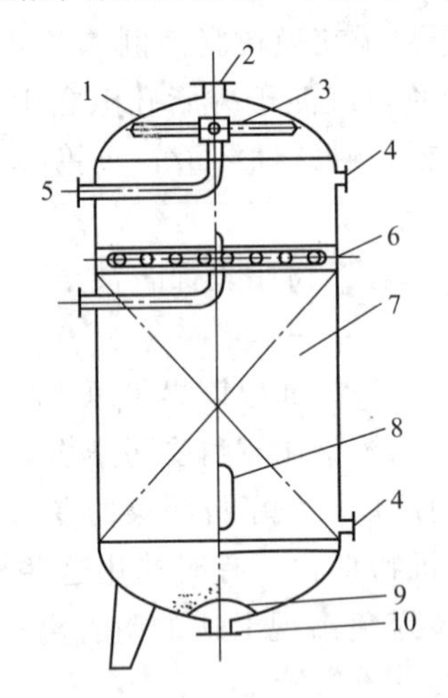

图5-72 固定床离子交换柱设备的结构与外观

1—柱体；2—排气管；3—上布水器；4—树脂装卸口；5—进液口；
6—中排液管；7—树脂床；8—窥视孔；9—下布水器；10—出液口

小型设备可采用聚氯乙烯、玻璃钢等制作柱体和管道。直径大于1m的设备多采用碳钢，内衬氯丁橡胶、聚氯乙烯、环氧树脂等防腐材料。主体高度大致为树脂层高度的1.8～2倍，以留给树脂足够的膨胀空间。交换时，阳离子树脂的层高可能膨胀75%，而阴离子树脂则可能膨胀1倍。

树脂层的高度取决于树脂和交换体系的性质，一般在1～3m之间。有的高达几米。顶部留有一段空间，操作时充入空气或溶液。底部是底板，用于支承树脂和透过溶液，多采用具有排水帽的孔板，多数为平板型，有的采用下凹球面型。有的柱底采用铺垫石英砂作为树脂承载层，底板用多孔板。石英层的高度随柱径增大而提高，总高在0.5～1m，分为4、5层铺垫，粒度由下而上，逐层变细。粗砂粒径约为20mm，细砂粒径为1～2mm。

为了使溶液在柱内沿径向均匀分布，使树脂都能接触到溶液，交换柱的顶部、中部和底部都安装了布液装置。顶部的布液器包括喷头、喇叭斗、多孔管和带排水帽的多孔板等。

交换柱在逆流操作时从中间排液管出液，为了达到均匀排出贫再生液的目的，中间排液管具有多个出口孔，呈水平状均匀分布在交换柱的中上部。一般采用平行垂直排布于一根母管的多个支管或者环形管结构。下布液管多采用孔板。

固定床的优点是结构比较简单，易于操作，而且易于设计放大，另外，树脂之间没有很强的摩擦，也有利于减少机械损失。但是，固定床填装的树脂量大，增大了投资；部分树脂在交换过程中，不能参与反应，树脂利用率低。固定床还不适用于处理固体含量高的料液。

(2) 移动树脂床设备

固定床在工作过程中，树脂在床层间不发生移动，只有液体流过床层。与此不同，移动床在交换过程中，液体和树脂都发生移动。移动床的基本思想是：树脂在设备中移动，能够在一个设备的不同部位，分别完成交换、洗脱、再生等过程，使过程趋于连续化，从而使树脂的利用效率大为提高，显著减少树脂的用量。而且设备中树脂和液体都在运动，既强化传质，又消除树脂床的死角。

图5-73为一种环形移动床，由交换段、清洗段、洗脱再生段和另一个清洗段组成，各段之间有阀门分隔，操作时分为工作周期和树脂转移周期。在工作周期，阀门A、E、F、G、I、K打开，各段之间的阀门B、C、D和脉冲进口阀H全部关闭，各段分别同时进行交换、清洗、洗脱再生等作业。交换段采用降流，交换完成后的尾液从本段的底部排走。环形的下半部分是洗脱再生段和一个脱除洗脱剂的清洗段。右侧上部为负载树脂储存和清洗段，清洗采用反上柱，以清除夹带的固体杂质。工作周期结束后，旋即开始树脂转移周期。关闭储存室和脉冲空间的阀门A及E、F、G、I、K各进出口的阀门，打开各室间阀门B、C、D

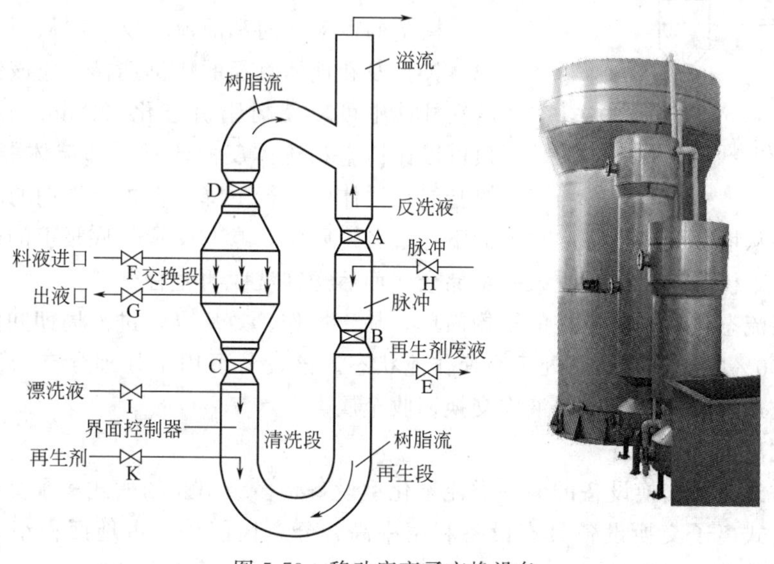

图5-73 移动床离子交换设备

和脉冲进口阀 H。开启脉冲，可以使树脂按顺时针方向移动，从一段进入下一段仅需要约半分钟。这样，清洗完毕的树脂流入下面的脉冲室，交换室的负载树脂进入清洗储存段，储存于脉冲段的负载树脂进入下面的洗脱段。而经过清洗的洗脱再生树脂则向上进入交换段。

移动床离子交换的工作和转移周期连续交替进行。由于转移周期非常短，所以工作周期几乎是在连续运行，不断地产出成品液。对于同样生产能力的设备，环形移动床使用的树脂量仅为固定床的 15%～30%，使树脂利用率大为提高。设备的效率也远高于固定床，操作线速度达 180m/h，比一般设备高出 5～10 倍。它特别适合于处理稀溶液。此种设备洗涤效率非常高，产生的废液比较少。但是，树脂的频繁迁移，阀门反复开关，设备结构比较复杂，而且易导致树脂破碎。

（3）流态化离子交换设备

借助流态化方法使树脂悬浮流动，不断更新。此时，树脂和料液逆流运动，既可以连续作业，还可以使交换反应的效率大为提高，显著减少树脂的使用量，从而也节约了洗脱剂和洗涤用水。与移动床不同，一个流态化设备只完成一个工序，即交换、洗脱分别在不同的交换柱中进行。由于液固两相流态化技术的日臻成熟，各种流态化离子交换柱的设计、制作呈现出多样化，其中不少已经广泛应用于各种工业。

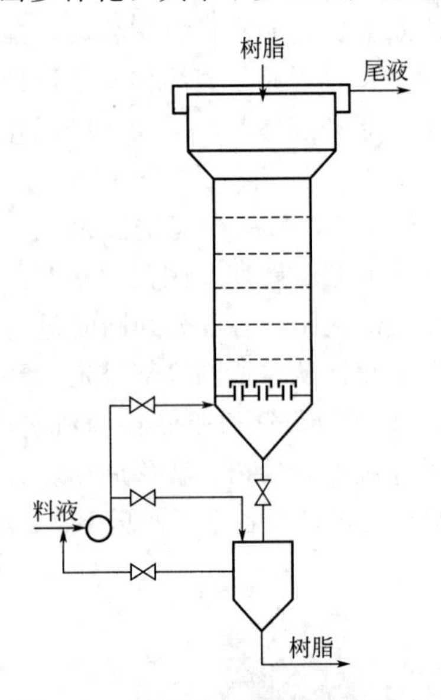

图 5-74　多层流态化离子交换设备

图 5-74 为一种应用于铀和黄金等的湿法冶金过程的多层流态化离子交换设备。它的特点是柱体中设置了若干水平多孔隔板，分为许多室，每个室都装填树脂，成为多级流态化床。柱顶有一个扩大段，减缓液流线速度，防止树脂随液体从顶部溢出。底部是带泡罩的液流分布板。

设备运行时，料液周期性地进入底部，通过分流板向上流动，自下而上地通过各个室进行交换。液流速度达到可以使各个室中的树脂发生流态化。水平隔板减少了液体和树脂的返混，提高了交换柱的效率。完成交换后的液流从扩大段上部溢流流出。

树脂的运动方向和液流相反，即从顶部加入交换柱中。扩大段和柱体直径的比例选择，应该保证流体线速度在柱体中可以使树脂流态化（fluidization），而在柱顶可以让树脂和流体反向流动。即流体能从顶部溢流，树脂能在柱体中自行下降。在进液周期的间隙，液流停止上行，树脂从塔板孔隙下降，自上而下穿过各个隔室，参与交换。底板上的泡罩可以防止树脂全部排尽。树脂排出柱后进入一个储仓，而后送往洗脱再生柱。

这种多层流态化离子交换柱的结构简单，易于操作，效率高，进液周期和停止期之比约为 9∶1，即 90%的操作时间是处于交换作业状态。该设备可用于处理含有少量固体悬浮物的料液，因此适合于从矿石堆浸液中交换回收金属。

（4）槽式离子交换设备

不同于塔式离子交换设备的另一类流态化型设备是多级串联的槽式离子交换设备。近代多级流化床塔式离子交换设备具有设备利用率高、分离性能好、占地面积小、投资省等优势，但是如何将其应用于悬浮固体含量高的料液系统中，仍是一个难题。由于槽式设备可以直接处理浸取矿浆，省去湿法冶金中操作烦冗、费用昂贵的固液分离操作，可节约巨额投资，因此在铀、金等湿法冶金方面，各国都在大力开发这种适用于矿浆离子交换等过程的槽式串联连续离子交换设备。

槽式设备中，由于搅拌作用（机械的、液流的或气流的），大大减少了因塔式设备中树脂与矿浆（矿粒）间的密度差而产生的一系列问题，如树脂与矿浆的混合、沉降问题等。只要提供足够的搅拌强度以维持所需的树脂悬浮状态及充分的接触时间，此类设备就可在很大范围内操作。

这类槽式离子交换设备的主要基特点是：设备简单，操作方便，适应性强，处理能力大。

图 5-75 为一槽式离子交换设备。它利用升流矿浆所具有的曳力将树脂悬浮在设备的中上部，即树脂处于松散的悬浮状态与流过的矿浆进行接触。矿浆出口处设有筛网，以阻挡树脂被矿浆带走。为了使矿浆均匀分布于设备中，在矿浆入口分布管以上有一层石英填充层。有时为了松动树脂，槽中还通入少量搅拌空气。运行时，一般是 3～4 个槽串联操作。当首槽达饱和时，由系统中切下，进行洗脱。因此，不需要进行槽间的树脂转移与输送。为了提高设备的生产能力，可以使用增重树脂与大粒度树脂。

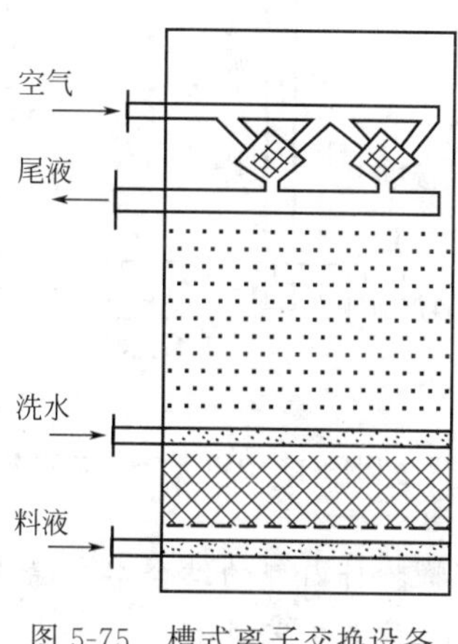

图 5-75 槽式离子交换设备

这种设备的优点是结构简单，适用性强，树脂磨损少，缺点是吸附线速度较低。因为树脂-矿浆的分离通过筛网进行，因此矿浆通过筛网的能力决定了设备的生产能力。

工程案例分析

一、气流干燥器与流化干燥器的联用

在化工生产中，由于被干燥物料的形状和性质各不相同，用单一形式的干燥器来干燥物料，常常不能达到对物料湿分的要求。有时即使能满足物料湿分的要求，单一设备体积也过大，或消耗过多的能量。此时，可将两种或多种形式的干燥器组合起来构成组合式干燥器，各发挥其长处，从而达到节省能量、减小干燥器尺寸或满足干燥产品质量要求的目的。在聚氯乙烯（PVC）的生产工艺中，PVC 的干燥过程即采用气流干燥器与流化床干燥器联用的组合式干燥设备。

PVC 树脂是一种热敏性、黏性小且多孔性的粉末状物料，其干燥过程包括非结合水分与结合水分的干燥，即经历表面汽化及内部扩散的不同控制阶段。为此，在干燥过程中采用两级装置。第一级主要用于表面水分的汽化，采用气流干燥器，利用其快速干燥的特点，使物料在很短的停留时间内，除去大部分表面水分。此时，干燥强度取决于引入的热量，通过加大风量和温度，使较高的湿含量能迅速地降至临界湿含量附近。第二级主要用于内部水分扩散，以降低风速和延长时间为宜，故采用流化床干燥器，使湿含量达到最终干燥的要求。

工业上 PVC 干燥的气流-流化组合操作中，第二级多采用卧式流化干燥器。某工厂经技术改造，用旋风流化干燥器替代卧式流化干燥器，获得了较好的效果，其干燥系统工艺流程如附图 1 所示。含水量约为 15%的 PVC 树脂湿料，经螺旋加料器送至第一级气流干燥器中干燥，离开气流干燥器的物料含水量为 3%。然后，物料再进入下一级旋风流化干燥器进一

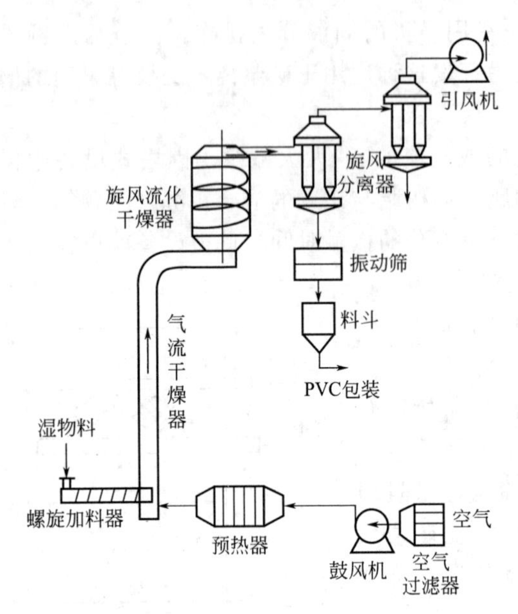

附图1 PVC气流-旋风流化干燥系统

步干燥，离开物料的含水量降至0.3%以下。干燥后的物料颗粒经旋风分离器分离下来，经振动筛过筛，进行成品包装。少量细料再经过下一级旋风分离器分离下来，湿空气则由引风机出口排出。

在气流干燥器中，物料以粉粒状分散于气流中，呈悬浮状态，被气流输送而向上运动。在此输送过程中，二者之间发生传热及传质过程，使物料干燥。由于气速很高，物料在气流干燥器中的停留时间极短（一般在2～10s），除去的是物料表面的非结合水分。

在旋风干燥器中，气流夹带物料颗粒沿切线方向进入，在干燥器中旋流上升。与气流干燥器相比，物料在旋风干燥器中的停留（干燥）时间较长。同时颗粒在热空气中处于流化状态，气、固接触面积大，故干燥强度很大，可将物料内部的结合水分除去，使干燥产品含水量更低、质量更均匀。

该新工艺具有如下特点：

① 旋风干燥器结构简单，操作容易，平稳，简化了干燥流程和操作控制；

② 降低了蒸汽消耗，一般节能50%左右；

③ 卧式流化干燥器结构复杂，易积存物料，导致PVC树脂黑黄点较高，而旋风干燥器无死角，不积存物料，使树脂合格率提高，提高了产品质量；

④ 气流-卧式流化干燥工艺中，树脂出口温度高达80℃，故干燥后的树脂需增加冷风输送工艺使其冷却；而在气流-旋风流化干燥工艺中，旋风干燥器内的空气温度降至50℃左右，树脂出口温度为45℃左右，不需要再进行冷却，也不存在树脂的热降解问题，既提高了产品质量，又降低了动力消耗。

二、300m^3/h变压吸附制氮装置的改造

氮气是目前我国化工行业广泛使用的置换介质，主要来源于压缩空气。1998年某厂为1500t/a聚丙烯装置安装了一套300m^3/h的变压吸附制氮装置，以保证聚合釜进料前的安全置换，达到平稳生产的目的。因此氮气质量是否合格，将直接影响聚丙烯装置的正常操作和生产。

变压吸附制氮原理为：压缩空气干燥后进入吸附塔，由于吸附剂（炭分子筛）的微孔直径稍小于氮分子的直径，故氧分子以较快的速度向微孔扩散，优先被吸附剂所吸附，从而实现氧、氮的分离，此时的氮气是粗氮。粗氮气中少量氧通过钯催化剂，在一定的温度下与氢气反应生成水，然后再干燥即为产品精氮（含氮99.99%以上）。由此可见，粗氮合格与否不仅影响氢气的耗量，而且直接影响精氮的质量。

（1）存在问题及分析

制氮装置刚刚运行时比较平稳，基本能达到设计要求。运行一段时间后，发现在室外排空处有吸附剂粉尘排出，并且量在不断增加。粗氮含氧严重超标，为保证生产，只有加大氢气量，严重时不得不停车添加吸附剂。同时由于粉尘的排出，使仪表控制阀磨损严重，经常

失灵，从而引起设备故障频繁发生，甚至直接影响聚丙烯装置的正常运行。

在每次拆开吸附塔进行检修时，都发现塔内吸附剂缺量，并且吸附剂粒度变小，粉尘增多。对此分析认为，本装置所设计的压实结构不尽完善（附图2）。当吸附剂装满后，尽管也压实了，但当设备运行时，在0.7MPa风压的吸附、再生、逆放过程中，吸附剂要进一步充实，因此产生一定的小空间，而固定的压实装置对吸附剂不能再产生压实作用，使之运行时处于剧烈运动状态。这样吸附剂之间必然要相互摩擦，而且国内所生产的吸附剂强度较低，这样长时间就会使吸附剂颗粒变小，粉尘增加，粉尘又随逆放过程排至室外，缺吸附剂现象会越来越严重，粗氮就会不合格。

(2) 改造方案

经过上述分析，技术人员提出了一个具体改造方案（附图3）。

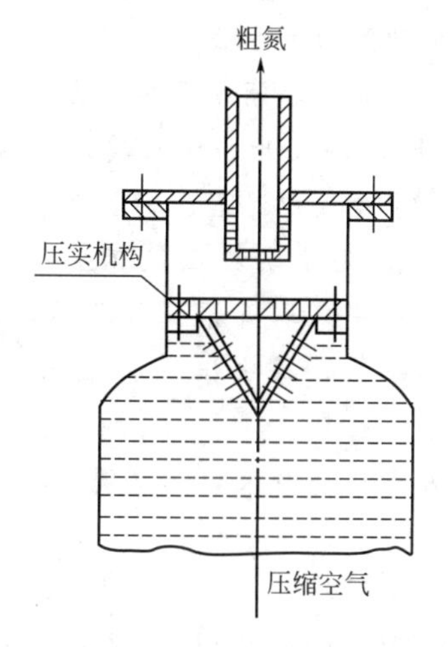

附图2 改造前的压实结构（1）

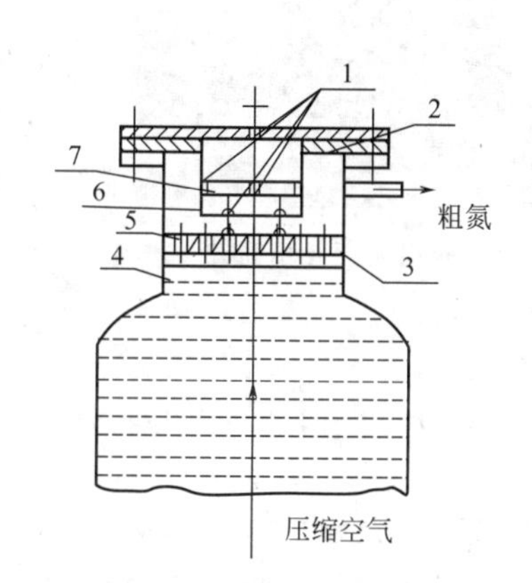

附图3 改造前的压实结构（2）

1—密封；2—单向阀；3—胶圈；4—挡板；5—压实板；6—汽缸；7—活塞

改造方案主要是将塔内固定压实结构改造，即拆除原有的压实机构，重新安装由压实板、活塞、汽缸、固定杆等组成的新型自动压实机构，是利用塔内的压缩空气自动完成对吸附剂的再次压实过程。装完吸附剂时，压实板即压在吸附剂上面，从而保证了汽缸内压力平衡。当吸附剂缺少时，压实板与吸附剂就会出现空间，这时汽缸内的压缩空气就压着活塞带着实板向下运动，对吸附剂进行了再次压实。如此循环运动，保证吸附剂始终处于静止状态。当活塞向下移动量达到设定值（150mm）时，自动报警，需停机重新填充吸附剂。

(3) 改造后效果

① 改造后，设备及仪表控制阀的故障率大大减小，从而降低了工人的劳动强度，保证了产品的产量和质量。氢气及吸附剂的消耗明显降低，每年可为工厂节约原料费约40万元。

② 改造后，由于吸附剂处于相对静止状态（stationary state），颗粒之间没有了相互摩擦，在逆放时没有粉尘，室内的环境大为改善。

通过对300m^3/h变压吸附压实装置存在的问题进行分析，提出具体改造方案，经过长时间的生产实践证明，效果良好，消除了改造前所存在的问题，收到了很好的经济效益和社会效益。

三、沙盐三井套管环空盐结晶事故的处理

中国凿井采盐的方法早在战国时期就已出现，李冰在兴建都江堰水利工程时发现了盐卤，开凿了中国第一口大口径盐井——广都盐井，开始了最早的盐业生产。但这种盐井采盐量有限。直到公元1041～1048年，也就是北宋庆历年间，勤劳智慧的四川大英县卓筒井镇人发明了开凿小口径盐井的方法（附图4）。至此，地下盐卤开采进入了一个崭新的时期。

附图4　四川大英县卓筒井

沙隆达盐矿位于湖北荆州市沙市东郊，盐层埋深2400～2850m，属陆相碎屑岩地质构造，盐层底板机械强度差，极易垮塌。沙盐三井井深2850m，采取单井对流两管油垫正循环的采卤工艺进行生产。

（1）结晶事故的形成

因盐矿采卤生产能力大于公司电化厂6万吨烧碱的生产需要，故沙盐三井采取间断开车的方式生产，停井时间一般为1～2周，每次开井生产均发现卤水中带有大量的结晶盐。由于盐层中泥岩、钙芒硝、硬石膏等不溶物含量超过35%，无法实施反循环生产。为防止套管盐结晶堵管，生产中采取开泵正循环后，每4h定期向套管环空压注水一次，每次压注水$35m^3$，直到生产参数稳定为止。

1999年2月11日停井，1999年2月23日上午8：30开泵采卤。发现卤水中带有大量的浮盐，经过11：00、16：00两次向套管反压注清水。晚上20：00再次发现卤水中带有结晶盐，再次反压注清水时套管环空因结晶盐形成盐桥将套管堵塞，并进一步结晶将堵塞的程度加剧。事故后通过水泥压裂车向套管环空注水，注水压力为17MPa，不能进水。套管堵塞后，中心管并未因此受到影响，中心管与溶腔连通较好。通过中心管向溶腔注水，中心管压力为10MPa，进水量为$30m^3/h$，同时打开套管出卤阀门，套管出水量为50～80L/h，且呈逐步减少趋势。因套管结晶堵塞。循环通道破坏，致使生产中断。

（2）处理措施

盐矿认真分析了结晶堵塞的原因。由于钙芒硝、硫酸根、泥沙、盐等物质均可引起在套管内结晶堵管，但不同的结晶其特征各不相同。通过分析沙盐三井近一段时间的钙芒硝、硫酸根的化验分析数据，其浓度值均很低，且含量很稳定，排除了钙芒硝、硫酸根结晶的可能。泥沙不溶物搭桥堵管的可能性也不存在。因沙盐三井结晶堵管是在开泵运行12h后，且循环方式为正循环（positive cycle），套管卤水流速低，携砂量极低。所以，盐矿认为此次堵管主要是大颗粒结晶盐在套管环空内逐步形成的盐桥（salt bridge），致使循环中断，并伴随进一步结晶而加剧套管堵塞。

根据以上堵管事故，分析制定了相应的处理措施。

首先，始终保持中心管的畅通。因为本地区盐层品位低，水不溶物多，加上盐井停井十多天后中心管压力已达到12MPa，中心管一旦泄压将很容易形成中心管内因沉砂搭桥、盐结晶而形成堵管事故。其次，由于溶腔已采盐达12万吨，通过中心管向溶腔内强行注水4h，共注水$100m^3$，注水压力达到14MPa。提高溶腔的压力后，观察记录套管出水量的大小，流量为80～100L/h。又由于套管出卤量很小，加上结晶的井段比较长，造成套管内阻力损失较大，故通过套管向溶腔里面压水作业，这样会促使流量过大，容易造成压力迅速上

升，无法连续作业，因而不能将结晶盐解堵。通过分析，盐矿认为：如果能找到一种压力足够克服结晶后套管内新的阻力损失值，且注水量在100L/h左右的高压泵，则可以逐步解决套管的结晶堵管问题。最后选用了四川简阳生产的DSY-35型电动试压泵，该泵基本具备处理事故所需的工艺技术条件。

1999年3月12日，在事故发生后的第17天，将试压泵接到沙盐三井井口高压管口上，关闭中心管阀门，利用试压泵向套管环空注水，注水约40L后压力由10MPa上升到16MPa，停泵。待压力降为10MPa后，再次开泵注水40L，压力再次上升到16MPa，停泵。待压力降为10MPa，第三次开泵，注水20L，压力上升到16MPa。随后压力逐步下降为10MPa，试压泵在10MPa的压力下稳定工作1h，共注水200L后，试压泵压力下降到了8MPa，已接近此时的正常套管压力值。这时，技术人员认为套管已和溶腔连通，所以立即停下试压泵，启动采卤泵连续大排量通过套管向溶腔内注水，泵压为11.5MPa。20min后向套管内注水13m^3，至此沙盐三井套管的盐结晶堵管得到了基本解堵。

(3) 事故的后期处理

沙盐三井从1999年2月23日，因结晶堵管到3月12日共停井17天。套管内聚积了大量的结晶盐，排出的卤水中固体结晶盐约占20%，最大的结晶盐块达30mm×50mm，加上溶腔经过长达29天的关井静溶，卤水浓度和溶腔温度均很高，随时都有因重结晶造成套管堵塞的可能。盐矿采取在正循环生产的同时，每4h向环空压注清水1次。经过4天的处理，沙盐三井卤水中NaCl含量为310g/L，卤水中不再携带浮盐。井口中心管压力为10MPa，注水量为35m^3/h，生产稳定。

至此，沙盐三井套管结晶事故得到了彻底处理，全部修井处理费用仅1万元，达到了安全、经济修井的目的。此次事故的处理，虽然在施工措施上比较简单，但却很实用，避免了一次复杂的修井作业。这种处理方法，对于井矿盐开采中因套管环空盐结晶堵塞，关闭阀门后套管压力能够上升，打开套管出口阀门有少量出卤，但又不能建立正常生产循环的盐井事故，具有一定的参考借鉴价值。

四、嵊山500t/a反渗透海水淡化示范工程

海天佛国，群岛之城，舟山是我国唯一以群岛设立的地级市。全市共有大小岛屿1390个，其中住人岛屿103个，素有“东海鱼仓”“祖国渔都”的美誉。舟山地处海岛，山低源短，无过境客水，水资源全靠降水补给。“如何解决缺水，保障全市经济社会可持续发展”一直是舟山上下孜孜以求的课题。

(1) 跨海输水

“如何把水从多的地方引到少的地方”，这在过去是一件想得到却做不到的事情。如今，舟山大陆引水工程，从宁波姚江引水，是解决舟山缺水的一项应急工程。舟山大陆饮水一期工程投资3亿元，引水规模为1m^3/s，年平均引水量2160万立方米。工程管路全长67公里，管径1m，其中跨海段长36公里，是迄今国内最长、最大的跨海输水工程。工程自1999年8月16日开工建设，2001年8月完成海上段钢管敷设，2003年8月建成通水，日引水6万～8万立方米，至2008年累计引水7450万立方米，有效提升了舟山市水资源的保障能力。目前投资14.6亿元，引水总规模2.8个流量，年均引水能力0.6亿立方米的大陆引水二期工程正在加紧建设之中。同时，舟山积极构建本岛和附近重要岛屿之间跨区域引水网络，已建成4处岛际引水工程，有效解决了重要岛屿水资源的供需矛盾，实现了水资源在更大空间上的开发利用和优化配置。舟山大陆引水工程建设现场见附图5。

附图 5　舟山大陆引水工程建设现场

（2）海水淡化

舟山四面环海，海水资源丰富，开发海水代用和海水淡化，是解决水资源供需矛盾的现实选择和战略举措。近 20 年来，舟山市各地积极采用海水制冰厂、电厂等用海水作冷却水，水产企业中大量使用海水作清洁洗鱼用水等途径，把海水代用逐步推广到各个领域。同时为进一步确保全市水资源供给安全，舟山充分利用丰富的海水资源积极实施海水淡化工程，缓解水资源的供需矛盾。1997 年 10 月，在嵊山建起全国第一座 500t 级反渗透海水淡化示范工程，2003 年大旱时为嵊山岛居民提供了 60％的饮用水。至 2008 年底，全市已投资 2.5 亿元，在重要岛屿建成海水淡化工程 12 处，海水淡化总能力 2.84 万吨/日。目前，投资 2 亿元的 3 处海水淡化工程正在抓紧建设中，至今年底可新增海水淡化能力 2.50 万吨/日，使全市海水淡化能力达到 5.34 万吨/日。海水淡化已成为舟山市边远小岛解决居民生活、生产用水的重要水源，为解决海岛缺水问题开辟了一条创新之路。嵊山海水淡化厂见附图 6。

附图 6　嵊山海水淡化厂

（3）嵊山海水淡化工程

嵊山又称“尽山”，意为海水于此而尽。嵊山岛地处东海边缘，东临太平洋，北濒黄海。该海水淡化工程采用海滩打双柱式沉井，以多级离心潜水泵取水，高架管引水的设计方案。项目采用反渗透海水淡化技术，工艺流程见图附图 7。

反渗透膜元件选用美国陶氏化学公司 Filmtec 卷式膜元件 SW630HR- 8040，它是反渗透海水淡化的核心部件。该元件具有大于 99％ 的高脱盐率，产水量 $15m^3/d$，有较好的耐压性和抗氧化耐污染性能。反渗透装置采用 6 个膜元件串联，6 个压力母管并联的一级一段组合结构，并配置低压自动冲洗排放和淡化水低压自动冲洗置换，不合格产水自动切换排放。

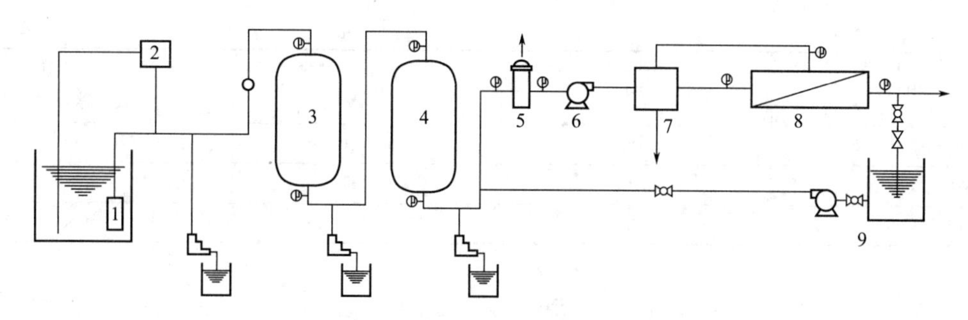

附图7 500t/d反渗透海水淡化工艺流程简图

1—深井潜水泵；2—次氯酸钠发生器；3—多介质滤器；4—活性炭滤器；5—保安滤器；6—高压泵；7—水力透平机；8—反渗透装置；9—清洗置换装置

在该系统中还设置高、低压保护和高压泵、清洗泵的联锁。

整个反渗透海水淡化控制系统设计采用先进的计算机程序控制，由工控机操作站和可编程控制器PLC组成一个分散采样控制、集中监视操作的控制系统。该控制系统的主要功能有：

① 按工艺参数设置高低压保护开关、自动切换装置，电导、流量和压力出现异常时，能实现自动切换、自动联锁报警、停机，以保护高压泵和反渗透膜元件。

② 变频控制高压泵的启动和关停，实现高压泵的软操作，节省能耗，防止由于水锤或反压造成高压泵和膜元件损坏。

③ 在反渗透装置开机和停机前后，能实现低压自动冲洗。特别在停运时，浓缩海水的亚稳定状态会转化出现沉淀，污染膜面，低压淡化水自动冲洗能置换出浓缩海水，保护膜面不受污染，延长膜的使用寿命。

④ 对系统的温度、流量、水质、产量等相关参数能实现显示、储存、统计、制表和打印功能。监视操作中的动态工艺流程画面清晰直观，系统控制简化人工操作，确保系统能自动、安全、可靠地运行。

$500m^3/d$反渗透海水淡化示范工程经过5个月精心设计和施工，于1997年10月16日竣工，整套系统投入试运行。从1998年7月5日开始500 h的系统性能考核运行，测试记录近4000个运行参数数据，并对原海水、反渗透产水和浓水水质进行了分析测定。通过考核运行和测试结果表明，该系统运行参数稳定，设备运行正常，自控系统满足工艺要求，性能指标达到设计要求，产品水符合国家生活用水标准（附表）。

附表 海水淡化产水水质与生活用水水质标准对照 (mg/L)

项目	海水	淡化产水	生活饮用水卫生指标(GB 5749—2006)
总硬度(以$CaCO_3$计)	4939.0	12.41	450
TDS	33434.9	211.95	1000
硫酸盐	2158.0	2.18	250
氯化物	15624.36	122.10	250
铁	0.08	<0.01	0.3
锰	0.04	<0.01	0.1
COD	2.43	0.89	—

续表

项目	海水	淡化产水	生活饮用水卫生指标(GB 5749—2006)
亚硝酸盐(以 N 计)	0.003	0.001	<0.002
硝酸盐(以 N 计)	0.43	0.004	20
pH 值	0.22	5.88～6.31	6.5～8.5

思 考 题

1. 某单效蒸发操作，因真空泵损坏而使冷凝器压强由某真空度升至常压，此时有效传热温差有何变化？若真空泵损坏后料液流量及状态不变，但仍要求保证完成液浓度不变，可采取什么办法？
2. 一单效蒸发器，原来在加热蒸汽压力为 $2kgf/m^2$、蒸发室压力为 $0.2kgf/m^2$ 的条件下连续操作。现发现完成液的浓度变稀，当即检查加料情况，得知加料流量下组成和温度均未变。试问可能是哪些原因引起完成液浓度变稀？这些原因所产生的结果使蒸发器内溶液沸点是升高还是降低？为什么？
3. 并流加料的蒸发装置中，一般各效的总传热系数逐效减小，而蒸发量却逐效略有增加，试分析原因。
4. 欲设计多效蒸发装置将 NaOH 水溶液自 10%浓缩到 60%，宜采用何种加料方式？已知料液温度为 30℃。
5. 溶液的哪些性质对确定多效蒸发效数有影响？进行简单分析。
6. 烧碱（NaOH）溶液的蒸发浓缩过程中，会析出一定的 NaCl 晶体，采用何种设备将其分离出来？
7. 指出下列基本概念之间的联系和区别：

（1）绝对湿度与相对湿度；

（2）露点温度与沸点温度；

（3）干球温度与湿球温度；

（4）绝热饱和温度与湿球温度。

8. 在 H-I 图上分析湿空气的 t、t_d 及 t_w（或 t_{as}）之间的大小顺序。在何种条件下，三者相等？
9. 试说明湿空气 H-I 图上等干球温度线、等相对湿度线、蒸汽分压线是如何标绘出来的？
10. 当湿空气总压变化时，湿空气 H-I 图上的各种曲线将如何变化？如保持 t、H 不变而将总压提高，这对干燥操作是否有利？为什么？
11. 如何区分结合水分与非结合水分？说明理由。
12. 对一定的水分蒸发量及空气的出口湿度，试问应按夏季还是按冬季的大气条件来选择干燥系统的风机？
13. 根据干燥器热效率 η 的定义式讨论提高 η 的途径。
14. 什么是膜分离？膜分离过程有哪几种类型？膜分离过程有什么特点？
15. 常见的膜分离技术有哪些？各适用于哪些分离场合？
16. 性能良好的膜组件应具备哪些条件？
17. 简要说明各类膜组件的结构和特点。
18. 说明反渗透分离的基本原理和流程。
19. 超滤和微滤在工业生产中主要用于何种产品的生产？
20. 电渗析器的组装方式有哪些？
21. 什么是浓差极化？
22. 什么是溶液的渗透压？
23. 气体膜分离主要用于哪些气体组分的分离？
24. 家用制氧机的工作原理是什么？
25. 普通精馏很难分离具有恒沸点的液体混合物，利用膜分离技术可以分离这些混合物吗？
26. 按照造成膜两侧蒸气压差方法的不同，可以将渗透汽化过程分为哪几类？
27. 与乳化液膜相比，支撑液膜具有哪些优点？它主要应用于哪些物质的分离？

28. 膜蒸馏与精馏分离的区别是什么？
29. 简要说明膜反应器的工作原理。
30. 试比较反渗透、超滤和微滤的差异和共同点。
31. 试说明反渗透、电渗和渗透气化的基本原理。
32. 常规的膜组件有几种？其特性和应用范围是什么？
33. 与其他化工分离单元操作相比，结晶过程具有哪些特点？
34. 结晶操作分为哪几类？
35. 简要分析结晶水是如何影响晶体性能的。
36. 溶解度如何随温度而变化？
37. 结晶成长速度与哪些因素有关？
38. 举例给出工业生产中常用的几类结晶器形式。
39. 试分析吸附与吸收的相似之处。
40. 常见的吸附剂有哪几类？
41. 工业吸附对吸附剂有哪些具体要求？
42. 吸附单元操作主要的工业用途有哪些？
43. 说明吸附过程的具体步骤。
44. 说明变压吸附的原理和工业用途。
45. 简要叙述离子交换技术的发展历史。
46. 离子交换的原理是什么？
47. 离子交换剂分为哪几类？主要的离子交换剂是哪种？
48. 如何选用离子交换树脂？
49. 说明离子交换的具体操作程序。

名 词 术 语

英文/中文

absolute humidity/绝对湿度
adiabatic/绝热
adiabatic saturation temperature/绝热饱和温度
adsorbent/吸附剂
adsorption isotherms/吸附等温线
agitated-film evaporator/搅拌薄膜蒸发器
agitator/搅拌器
alumina/氧化铝
atomizer/喷雾器
backward-feed multiple-effect evaporator/逆流加料法多效蒸发器
basket type evaporator/ 悬筐蒸发器
batch dryer/间歇干燥器
batchwise/间歇式
boiling point/沸点
boiling-point elevation/溶液沸点升高
bone-dry/绝干
boundary layer/边界层
bound water/结合水

英文/中文

bulk/主体
capillary/毛细管
capillary force/毛细作用力
catalyst/催化剂
circulation evaporator/循环蒸发器
climbing-film evaporator/升膜蒸发器
colloidal gels /凝胶体
compartment dryer/厢式干燥器
concave/凹面的
concentration difference/浓度差
constant-drying condition/恒定干燥条件
constant-rate period/恒速阶段
countercurrent/逆流
counterflow/逆流
critical moisture content/临界湿含量
cross-circulation drying/交叉循环干燥
crystal/结晶体
crystalline/结晶
cyclone/旋风分离器

dehumidification/去湿
dehydration/脱水
detergent/洗涤剂
dew point/露点
dew-point temperature/露点温度
diffusivity/扩散系数
dry bulb temperature/干球温度
drying/干燥
Duhring's rule/杜林规则
energy balance/能量衡算
enthalpy/焓
enzyme/酶
equilibrium curve/平衡曲线
equilibrium moisture content/平衡湿含量
equilibrium water/平衡水分
evaporation/蒸发
exhaust fan/排气扇
Fahrenheit/华氏温度
falling-film evaporator/降膜蒸发器
falling-rate period/降速阶段
feedstock/进料
fermentation/发酵
fine chemicals/精细化学品
flake/薄片
flash dryer/气流干燥器
flash evaporation/ 闪蒸
fluid-bed/流化床
fluidize/流（态）化
forced-circulation evaporator/强制循环蒸发器
forward-feed multiple-effect evaporator/顺流加料法多效蒸发器
free moisture content/自由湿含量
free water/自由水分
freeze-drying/冷冻干燥
geometry factor/几何因子
granular/颗粒状的
granule/颗粒
hardening/硬化
heating medium/加热介质
heat of dilution/溶液稀释热
heat of vaporization/汽化热
heat-sensitive/热敏性的
humid air/湿空气
humid heat/湿比热容
humidification/加湿
humidity/湿度
humidity chart/湿度图
humid volume/湿比容
hygroscopic/吸湿性的
inorganic/无机的
interfacial tension/界面张力
interior channel/内部通道
latent heat/潜热
liquid head/液柱静压强
logarithm mean temperature difference/对数平均温度差
lyophilization/冻干法
material balance/物料衡算
mechanical separator/机械分离器
mechanism of drying/干燥机理
meniscus/半月形的
microorganism/微生物
mixed-feed multiple-effect evaporator/混合加料法多效蒸发器
moist air/湿空气
moisture content/湿含量
moisture equilibrium/湿度平衡
moisture-free (bone-dry) solid/绝干固体
monolayer/单层
multilayers/多层
multiple-effect evaporation/多效蒸发
nonadiabatic/非绝热
nonporous/无孔的
numerical integration/数值积分
operating line/操作线
parallel-feed multiple-effect evaporator/平流加料法多效蒸发器
percentage humidity/湿度百分数
polymeric material/聚合材料
porous/多孔的
powders/粉体
pulverized/研磨成粉状的
rate-of-drying curve/干燥速率曲线
refrigerant/制冷剂
relative humidity/相对湿度
required packed height/所需的填料层高度
rotary dryer/旋转干燥器
rum dryer/转筒干燥器
saturated gas/饱和空气

saturation humidity/饱和湿度
screen/筛
screen-conveyor dryer/带式干燥器
screw-conveyor/螺旋杆
semiquantitative/半定量的
sensible heat/显热
settles/沉降
shell/壳
shrinkage/收缩
single-effect evaporation/单效蒸发
slabs/厚片
slurry/浆体
spray disk/喷雾圆盘
spray dryer/喷雾干燥机
spray nozzle/喷头
steady-state operation/稳态操作
stock/物料
sublimation/升华
surface tension/表面张力
temperature drop/温度差损失
thermal efficiency/热效率
thermometer/温度计
the saturation vapor pressure/饱和蒸气压
the specific heat/比热容
the total moisture content/总湿含量
thin-film dryer/薄膜干燥器
through-circulation drying/循环干燥
tray drying/盘式干燥
trial-and-error/试差法
turbo dryer/涡轮干燥器
unbound water/非结合水
undersaturated air/不饱和空气
vacuum drying/真空干燥
vacuum evaporation/ 真空蒸发
vapor-free gas /不含蒸汽的空气，绝干空气
vapor pressure of water/水的蒸气压
vertical type evaporator/中央循环管式蒸发器
void fraction/空隙率
volumetric heat-transfer coefficient for gas/气体体积传热系数
warp/弯曲
wet-bulb temperature/湿球温度

附录

1. 常用法定计量单位

（1）常用单位

基本单位			具有专门名称的导出单位				允许并用的其他单位			
物理量	基本单位	单位符号	物理量	单位名称	单位符号	与基本单位关系式	物理量	单位名称	单位符号	与基本单位关系式
长度	米	m	力	牛[顿]	N	$1N=1kg\cdot m/s^2$	时间	分	min	$1min=60s$
质量	千克(公斤)	kg	压强、应力	帕[斯卡]	Pa	$1Pa=1N/m^2$		时	h	$1h=3600s$
时间	秒	s	能、功、热量	焦[耳]	J	$1J=1N\cdot m$		日	d	$1d=86400s$
热力学温度	开[尔文]	K	功率	瓦[特]	W	$1W=1J/s$	体积	升	L	$1L=10^{-3}m^3$
物质的量	摩[尔]	mol					质量	吨	t	$1t=10^3kg$

（2）常用十进倍数单位及分数单位的词头

词头符号	M	k	d	c	m	μ
词头名称	兆	千	分	厘	毫	微
表示因数	10^6	10^3	10^{-3}	10^{-2}	10^{-3}	10^{-6}

2. 单位换算表

说明：下列表格中，各单位名称上的数字标志代表所属的单位制度：①cgs 制；②SI；③工程制。没有标志的是制外单位。有＊号的是英制单位。

（1）长度

①cm 厘米	②m 米	＊ ft 英尺	＊ in 英寸
1	10^{-2}	0.03281	0.3937
100	1	3.281	39.37
30.48	0.3048	1	12
2.54	0.0254	0.08333	1

（2）面积

①cm^2 厘米^2	②m^2 米^2	＊ ft^2 英尺^2	＊ in^2 英寸^2
1	10^{-2}	0.001076	0.1550
10^4	1	10.76	1550
929.0	0.0929	1	144.0
6.452	0.0006452	0.006944	1

(3) 体积

①cm^3 厘米^3	②m^3 米^3	1 公升	* ft^3 英尺^3	* Imperial gal 英加仑	* U. S. gal 美加仑
1	10^{-6}	10^{-3}	3.531×10^{-5}	0.0002200	0.0002642
10^6	1	10^3	35.31	220.0	264.2
10^3	10^{-3}	1	0.03531	0.2200	0.2642
28320	0.02832	28.32	1	6.228	7.481
4546	0.004546	4.546	0.1605	1	1.201
3785	0.003785	3.785	0.1337	0.8327	1

(4) 质量

①g 克	②kg 千克	③$kgf\cdot s^2/m$ 千克(力)·秒^2/米	ton 吨	* lb 磅
1	10^{-3}	1.020×10^{-4}	10^{-6}	0.002205
1000	1	0.1020	10^{-3}	2.205
9807	9.807	1		
453.6	0.4536		4.536×10^{-4}	1

(5) 重量或力

①dyn 达因	②N 牛顿	③kgf 千克(力)	* lbf 磅(力)
1	10^{-5}	1.020×10^{-6}	2.248×10^{-6}
10^5	1	0.1020	0.2248
9.807×10^5	9.807	1	2.205
4.448×10^5	4.448	0.4536	1

(6)密度

①g/cm^3 克/厘米^3	②kg/m^3 千克/米^3	③$kgf\cdot s^2/m^4$ 千克(力)·秒^2/米^4	* lb/ft^3 磅/英尺^3
1	1000	102.0	62.43
10^{-3}	1	0.1020	0.06243
0.009807	9.807	1	
0.01602	16.02		1

(7) 压力

①bar= $10^6 dyn/cm^2$ 巴	②$Pa=N/m^2$ 帕斯卡=牛顿/米^2	③kgf/m^2=mm H_2O 千克力/米^2	atm 物理大气压	kgf/cm^2 工程大气压	mmHg(0℃) 毫米汞柱	* lbf/in^2 磅力/英寸^2
1	10^5	10200	0.9869	1.020	750.0	14.5
10^{-5}	1	0.1020	9.869×10^{-6}	1.020×10^{-5}	0.007500	1.45×10^{-4}
9.807×10^{-5}	9.807	1	9.678×10^{-5}	10^{-4}	0.07355	0.001422
1.013	1.013×10^5	10330	1	1.033	760.0	14.70
0.9807	9.807×10^4	10000	0.9678	1	735.5	14.22
0.001333	133.3	13.60	0.001316	0.00136	1	0.0193
0.06895	6895	703.1	0.06804	0.07031	51.72	1

(8) 能量、功、热

①erg=dyn·cm 尔格	②J=N·m 焦耳	③kgf·m 千克力·米	③kcal= 1000cal 千卡	kW·h 千瓦·时	* ft·lbf 英尺·磅力	* B. t. u. 英热单位
1	10^{-7}					
10^7	1	0.1020	2.39×10^{-4}	2.778×10^{-7}	0.7376	9.486×10^{-4}
	9.807	1	2.344×10^{-3}	2.724×10^{-6}	7.233	0.009296
	4187	426.8	1	1.162×10^{-3}	3088	3.968
	3.6×10^6	3.671×10^5	860.0	1	2.655×10^6	3413
	1.356	0.1383	3.239×10^{-4}	3.766×10^{-7}	1	0.001285
	1055	107.6	0.2520	2.928×10^{-4}	778.1	1

（9）功率、传热速率

①erg/s 尔格/秒	②kW=1000J/s 千瓦	③kgf·m/s 千克力·米/秒	③kcal/s=1000cal/s 千卡/秒	* ft·lbf/s 英尺·磅力/秒	* B.t.u./s 英热单位/秒
1	10^{-10}				
10^{10}	1	102	0.2389	737.6	0.9486
	0.009807	1	0.002344	7.233	0.009296
	4.187	426.8	1	3088	3.963
	0.001356	0.1383	3.293×10^{-4}	1	0.001285
	1.055	107.6	0.2520	778.1	1

（10）黏度

①P=dyn·s/cm² =g/cm·s 泊	②N·s/m²=Pa·s 牛·秒/米²	③kgf·s/m² 千克力·秒/米²	cP 厘泊	*lb/ft·s 磅/英尺·秒
1	0.1	0.01020	100	0.06719
10	1	0.1020	1000	0.6719
98.07	9.807	1	9807	6.589
10^{-2}	10^{-3}	1.020×10^{-4}	1	6.719×10^{-4}
14.88	1.488	0.1517	1488	1

（11）运动黏度、扩散系数

①cm²/s 厘米²/秒	②③ m²/s 米²/秒	m²/h 米²/时	* ft²/h 英尺²/时
1	10^{-4}	0.36	3.875
10^{4}	1	3600	38750
2.778	2.778×10^{-4}	1	10.76
0.2581	2.581×10^{-5}	0.0929	1

（12）表面张力

①dyn/cm 达因/厘米	②N/m 牛顿/米	③kgf/m 千克力/米	* lbf/ft 磅力/英尺
1	0.001	1.020×10^{-4}	6.852×10^{-5}
1000	1	0.1020	0.06852
9807	9.807	1	0.672
14590	14.59	1.488	1

（13）热导率

①cal/(cm·s·℃) 卡/(厘米·秒·℃)	②W/(m·K) 瓦/(米·开)	③kcal/(m·s·℃) 千卡/(米·秒·℃)	kcal/(m·h·℃) 千卡/(米·时·℃)	* B.t.u./(ft·h·℉) 英热单位/(英尺·时·℉)
1	418.7	0.1	360	241.9
2.388×10^{-2}	1	2.388×10^{-4}	0.8598	0.5788
10	4187	1	3600	2419
2.778×10^{-3}	1.163	2.778×10^{-4}	1	0.6720
4.134×10^{-3}	1.731	4.139×10^{-4}	1.488	1

（14）焓、潜热

①cal/g 卡/克	②J/kg 焦耳/千克	③kcal/kgf 千卡/千克力	* B.t.u./lb 英热单位/磅
1	4187	(1)	1.8
2.389×10^{-4}	1	(2.389×10^{-4})	4.299×10^{-4}
0.5556	2326	(0.5556)	1

(15) 比热容、熵

① cal/(g·℃) 卡/(克·℃)	②J/(kg·K) 焦耳/(千克·开)	③kcal/(kgf·℃) 千卡/(千克力·℃)	* B. t. u. /(lb·℉) 英热单位/(磅·℉)
1	4187	(1)	1
2.389×10^{-4}	1	(2.389×10^{-4})	2.389×10^{-4}

(16) 传热系数

①cal/(cm²·s·℃) 卡/(厘米²·秒·℃)	②W/(m²·K) 瓦/(米²·开)	③kcal/(m²·s·℃) 千卡/(米²·秒·℃)	kcal/(m²·h·℃) 千卡/(米²·时·℃)	* B. t. u. /(ft²·h·℉) 英热单位/(英尺²·时·℉)
1	4.187×10^{4}	10	3.6×10^{4}	7376
2.388×10^{-5}	1	2.388×10^{-4}	8598	1761
0.1	4187	1	3600	737.6
2.778×10^{-5}	1.163	2.778×10^{-4}	1	2049
1.356×10^{-4}	5.678	1.356×10^{-3}	4.882	1

(17) 标准重力加速度

$$g=980.7\text{cm/s}^2 \text{①}$$
$$=9.807\text{m/s}^2 \text{②③}$$
$$=32.17\text{ft/s}^2 *$$

(18) 通用气体常数

$$R=1.987\text{cal/(mol·K)} \text{①}$$
$$=8.314\text{kJ/(kmol·K)} \text{②}$$
$$=848\text{kgf·m/(kmol·K)} \text{③}$$
$$=82.06\text{atm·cm}^3\text{/(mol·K)}$$
$$=0.08206\text{atm·m}^3\text{/(kmol·K)}$$
$$=0.08206\text{atm·L/(mol·K)}$$
$$=1.987\text{kcal/(kmol·K)}$$
$$=1.987\text{B. t. u. /(lbmol·℉)} *$$
$$=1544\text{lbf·ft/(lbmol·℉)} *$$

(19) 斯蒂芬-波尔兹曼常数

$$\sigma_0=5.71\times10^{-5}\text{erg/(s·cm}^2\text{·K}^4\text{)①}$$
$$=5.67\times10^{-8}\text{W/(m}^2\text{·K}^4\text{)②}$$
$$=4.88\times10^{-8}\text{kcal/(h·m}^2\text{·K)③}$$
$$=1.73\times10^{-9}\text{B. t. u. /(h·ft}^2\text{·℉)} *$$

(20) 温度

$$℃=(℉-32)\times\frac{5}{9},℉=℃\times\frac{9}{5}+32,K=273.3+℃$$

3. 某些气体的重要物理性质（0℃，101.3kPa）

序号	名称	分子式	分子量/(kg/kmol)	密度/(kg/m³)	定压比热容		$K=\frac{C_p}{C_V}$	黏度/10^{-3} cP 或 μPa·s	沸点(101.3kPa)/℃	汽化潜热(760mmHg)		临界点		热导率	
					/[kcal/(kg·℃)]	/[kJ/(kg·℃)]				/(kJ/kg)	/(kcal/kgf)	温度/℃	压力/atm	/[W/(m·K)]	/[kcal/(m·h·℃)]
1	空气	—	28.95	1.293	0.241	1.009	1.40	17.3	−195	197	47	−140.7	37.20	0.0244	0.021
2	氧气	O_2	32	1.429	0.218	0.653	1.40	20.3	−132.98	213	50.92	−118.82	49.72	0.0240	0.0206
3	氮气	N_2	28.02	1.251	0.250	0.745	1.40	17.0	−195.78	199.2	47.58	−147.13	33.49	0.0228	0.0196
4	氢气	H_2	2.016	0.0899	3.408	10.130	1.407	8.42	−252.75	454.2	108.5	−239.9	12.80	0.163	0.140
5	氦气	He	4.00	0.1785	1.260	3.180	1.66	18.8	−268.95	19.5	4.66	−267.96	2.26	0.144	0.124
6	氩气	Ar	39.94	1.782	0.127	0.322	1.66	20.9	−185.87	163	38.9	−122.44	48.00	0.0173	0.0149
7	氯气	Cl_2	70.91	3.217	0.115	0.355	1.36	12.9(16℃)	−33.8	305	72.95	144.0	76.10	0.0072	0.0062
8	氨气	NH_3	17.03	0.771	0.530	0.670	1.29	9.18	−33.4	1373	328	132.4	111.50	0.0215	0.0185
9	一氧化碳	CO	28.01	1.250	0.250	0.754	1.40	16.6	−191.48	211	50.5	−140.2	34.53	0.0226	0.0194
10	二氧化碳	CO_2	44.01	1.976	0.200	0.653	1.30	13.7	−78.2	574	137	31.1	72.90	0.0137	0.0118
11	二氧化硫	SO_2	64.07	2.927	0.151	0.502	1.25	11.7	−10.8	394	94	157.5	77.78	0.0077	0.0066
12	二氧化氮	NO_2	46.01	—	0.192	0.615	1.31	—	21.2	712	170	158.2	100.00	0.0400	0.0344
13	硫化氢	H_2S	34.08	1.539	0.253	0.804	1.30	11.66	−60.2	548	131	100.4	188.90	0.0131	0.0113
14	甲烷	CH_4	16.04	0.717	0.531	1.700	1.31	10.3	−161.58	511	122	−82.15	45.60	0.0300	0.0258
15	乙烷	C_2H_6	30.07	1.357	0.413	1.440	1.20	8.5	−88.50	486	116	32.1	48.85	0.0180	0.0155
16	丙烷	C_3H_8	44.1	2.020	0.445	1.650	1.13	7.95(18℃)	−42.1	427	102	95.6	43.00	0.0148	0.0127
17	丁烷(正)	C_4H_{10}	58.12	2.673	0.458	1.730	1.108	8.1	−0.5	386	92.3	152	37.50	0.0135	0.0116
18	戊烷(正)	C_5H_{12}	72.15	—	0.410	1.570	1.09	8.74	−36.08	151	36	197.1	33.00	0.0128	0.0110
19	乙烯	C_2H_4	28.05	1.261	0.365	1.222	1.25	9.85	−103.7	481	115	9.7	50.70	0.0164	0.0141
20	丙烯	C_3H_6	42.08	1.914	0.390	1.436	1.17	8.35(20℃)	−47.7	440	105	91.4	45.40	—	—
21	乙炔	C_2H_2	26.04	1.171	0.402	1.352	1.24	9.35	−83.66(升华)	829	198	35.7	61.60	0.0184	0.0158
22	一氯甲烷	CH_3Cl	50.49	2.308	0.177	0.582	1.28	9.89	−24.1	406	96.9	143	66.00	0.0085	0.0073
23	苯	C_6H_6	78.11	—	0.299	1.139	1.10	7.2	80.2	394	94	288.5	47.70	0.0088	0.0076

4. 某些液体的重要物理性质（20℃，101.3kPa）

名称	分子式	密度 ρ /(kg/m^3)	沸点 T_b /℃	汽化焓 ΔH_v /(kJ/kg)	比热容 C_p /[kJ/(kg·℃)]	黏度 μ /mPa·s	热导率 λ /[W/(m·℃)]	体胀系数 $\beta/\times10^{-4}$℃$^{-1}$	表面张力 $\sigma/(\times10^{-3}$N/m)
水	H_2O	998	100	2258	4.183	1.005	0.599	1.82	72.8
氯化钠盐水(25%)	—	1186(25℃)	107	—	3.39	2.3	0.57(30℃)	4.4	—
氯化钙盐水(25%)	—	1228	107	—	3.39	2.5	0.57	(3.4)	—
硫酸	H_2SO_4	1831	340(分解)	—	1.47(98%)	23	0.38	5.7	—
硝酸	HNO_3	1513	86	481.1	—	1.17(10℃)	—	—	—
盐酸(30%)	HCl	1149	—	—	2.55	2(31.5%)	0.42	—	—
二硫化碳	CS_2	1262	46.3	352	1.005	0.38	0.16	12.1	32
戊烷	C_5H_{12}	626	36.07	357.4	2.24(15.6℃)	0.229	0.113	15.9	16.2
己烷	C_6H_{14}	659	68.74	335.1	2.31(15.6℃)	0.313	0.119	—	18.2
庚烷	C_7H_{16}	684	98.43	316.5	2.21(15.6℃)	0.411	0.123	—	20.1
辛烷	C_8H_{18}	703	125.67	306.4	2.19(15.6℃)	0.540	0.131	—	21.8
三氯甲烷	$CHCl_3$	1489	61.2	253.7	0.992	0.58	0.138(30℃)	12.6	28.5(10℃)
四氯化碳	CCl_4	1594	76.8	195	0.850	1.0	0.12	—	26.8
1,2-二氯乙烷	$C_2H_4Cl_2$	1253	83.6	324	1.260	0.83	0.14(50℃)	—	30.8
苯	C_6H_6	879	80.10	393.9	1.704	0.737	0.148	12.4	28.6
甲苯	C_7H_8	867	110.63	363	1.70	0.675	0.138	10.9	27.9
邻二甲苯	C_8H_{10}	880	144.42	347	1.74	0.811	0.142	—	30.2
间二甲苯	C_8H_{10}	864	139.10	343	1.70	0.611	0.167	0.1	29.0
对二甲苯	C_8H_{10}	861	138.35	340	1.704	0.643	0.129	—	28.0
苯乙烯	C_8H_9	911(15.6℃)	145.2	(352)	1.733	0.72	—	—	—
氯苯	C_6H_5Cl	1106	131.8	325	1.298	0.85	0.14(30℃)	—	32
硝基苯	$C_6H_5NO_2$	1203	210.9	396	1.47	2.1	0.15	—	41
苯胺	$C_6H_5NH_2$	1022	184.4	448	2.07	4.3	0.17	8.5	42.9
苯酚	C_6H_5OH	1050(50℃)	181.8(熔点40.9℃)	511	—	3.4(50℃)	—	—	—
萘	$C_{10}H_8$	1145(固体)	217.9(熔点80.2℃)	314	1.80(100℃)	0.59(100℃)	—	—	—
甲醇	CH_3OH	791	64.7	1101	2.48	0.6	0.212	12.2	22.6
乙醇	C_2H_5OH	789	78.3	846	2.39	1.15	0.172	11.6	22.8
乙醇(95%)	—	804	78.2	—	—	1.4	—	—	—
乙二醇	$C_2H_4(OH)_2$	1113	197.6	780	2.35	23	—	—	47.7
甘油	$C_3H_5(OH)_3$	1261	290(分解)	—	—	1499	0.59	5.3	63
乙醚	$(C_2H_5)_2O$	714	34.6	360	2.34	0.24	0.140	16.3	18
乙醛	CH_3CHO	783(18℃)	20.2	574	1.9	1.3(18℃)	—	—	21.2
糠醛	$C_5H_4O_2$	1168	161.7	452	1.6	1.15(50℃)	—	—	43.5
丙酮	CH_3COCH_3	792	56.2	523	2.35	0.32	0.17	—	23.7
甲酸	HCOOH	1220	100.7	494	2.17	1.9	0.26	—	27.8
乙酸	CH_3COOH	1049	118.1	406	1.99	1.3	0.17	10.7	23.9
乙酸乙酯	$CH_3COOC_2H_5$	901	77.1	368	1.92	0.48	0.14(10℃)	—	—
煤油	—	780～820	—	—	—	3	0.15	10.0	—
汽油	—	680～800	—	—	—	0.7～0.8	0.19(30℃)	12.5	—

5. 常用固体材料的密度和比热容

名　称	密度/(kg/m³)	比热容/[kJ/(kg·℃)]	名　称	密度/(kg/m³)	比热容/[kJ/(kg·℃)]
(1)金属			(3)建筑材料、绝热材料、耐酸材料及其他		
钢	7850	0.461	干砂	1500～1700	0.796
不锈钢	7900	0.502	黏土	1600～1800	0.754(−20～20℃)
铸铁	7220	0.502	锅炉炉渣	700～1100	—
铜	8800	0.406	黏土砖	1600～1900	0.921
青铜	8000	0.381	耐火砖	1840	0.963～1.005
黄铜	8600	0.379	绝热砖(多孔)	600～1400	—
铝	2670	0.921	混凝土	2000～2400	0.837
镍	9000	0.461	软木	100～300	0.963
铅	11400	0.1298	石棉板	770	0.816
(2)塑料			石棉水泥板	1600～1900	—
酚醛	1250～1300	1.26～1.67	玻璃	2500	0.67
脲醛	1400～1500	1.26～1.67	耐酸陶瓷制品	2200～2300	0.75～0.80
聚氯乙烯	1380～1400	1.84	耐酸砖和板	2100～2400	—
聚苯乙烯	1050～1070	1.34	耐酸搪瓷	2300～2700	0.837～1.26
低压聚乙烯	940	2.55	橡胶	1200	1.38
高压聚乙烯	920	2.22	冰	900	2.11
有机玻璃	1180～1190	—			

6. 固体材料的热导率

（1）常用金属材料的热导率

热导率/[W/(m·K)] \ 温度/℃	0	100	200	300	400
铝	228	228	228	228	228
铜	384	379	372	367	363
铁	73.3	67.5	61.6	54.7	48.9
铅	35.1	33.4	31.4	29.8	—
镍	93.0	82.6	73.3	63.97	59.3
银	414	409	373	362	359
碳钢	52.3	48.9	44.2	41.9	34.9
不锈钢	16.3	17.5	17.5	18.5	—

(2) 常用非金属材料的热导率

名　称	温度/℃	热导率 /[W/(m·℃)]	名　称	温度/℃	热导率 /[W/(m·℃)]
棉绳	—	0.10～0.21	泡沫塑料	—	0.0465
石棉板	30	0.10～0.14	泡沫玻璃	−15	0.00489
软木	30	0.0430		−80	0.00349
玻璃棉	—	0.0349～0.0698	木材(横向)	—	0.14～0.175
保温灰	—	0.0698	(纵向)	—	0.384
锯屑	20	0.0465～0.0582	耐火砖	230	0.872
棉花	100	0.0698		1200	1.64
厚纸	20	0.14～0.349	混凝土	—	1.28
玻璃	30	1.09	绒毛毡	—	0.0465
	−20	0.76	85%氧化镁粉	0～100	0.0698
搪瓷	—	0.87～1.16	聚氯乙烯	—	0.116～0.174
云母	50	0.430	酚醛加玻璃纤维	—	0.259
泥土	20	0.698～0.930	酚醛加石棉纤维	—	0.294
冰	0	2.33	聚碳酸酯	—	0.191
膨胀珍珠岩散料	25	0.021～0.062	聚苯乙烯泡沫	25	0.0419
软橡胶	—	0.129～0.159		−150	0.00174
硬橡胶	0	0.150	聚乙烯	—	0.329
聚四氟乙烯	—	0.242	石墨	—	139

7. 某些固体材料的黑度

材料名称	温度/℃	黑度 ε
表面不磨光的铝	26	0.055
表面被磨光的铁	425～1020	0.144～0.377
用金刚砂冷加工后的铁	20	0.242
氧化后的铁	100	0.736
氧化后表面光滑的铁	125～525	0.78～0.82
未经加工处理的铸铁	925～1115	0.87～0.95
表面被磨光的铸铁件	770～1040	0.52～0.56
表面有一层有光泽的氧化物的钢板	25	0.82
经过刮面加工的生铁	830～990	0.60～0.70
氧化铁	500～1200	0.85～0.95
无光泽的黄铜板	50～360	0.22
氧化铜	800～1100	0.66～0.84
铬	100～1000	0.08～0.26
有光泽的镀锌铁板	28	0.228
已经氧化的灰色镀锌铁板	24	0.276
石棉纸板	24	0.96
石棉纸	40～370	0.93～0.945
水	0～100	0.95～0.963
石膏	20	0.903
表面粗糙没有上过釉的硅砖	100	0.80
表面粗糙上过釉的硅砖	1100	0.85
上过釉的黏土耐火砖	1100	0.75
涂在铁板上的有光泽的黑漆	25	0.875
无光泽的黑漆	40～95	0.96～0.98
白漆	40～95	0.80～0.95
平整的玻璃	22	0.937
烟尘、发光的煤尘	95～270	0.952
上过釉的瓷器	22	0.924

8. 某些液体的热导率

液　体	温度/℃	热导率/[W/(m・℃)]	液　体	温度/℃	热导率/[W/(m・℃)]
石油	20	0.180	四氯化碳	0	0.185
汽油	30	0.135		68	0.163
煤油	20	0.149	二硫化碳	30	0.161
	75	0.140		75	0.152
正戊烷	30	0.135	乙苯	30	0.149
	75	0.128		60	0.142
正己烷	30	0.138	氯苯	10	0.144
	60	0.137	硝基苯	30	0.164
正庚烷	30	0.140		100	0.152
	60	0.137	硝基甲苯	30	0.216
正辛烷	60	0.140		60	0.208
丁醇(100%)	20	0.182	橄榄油	100	0.164
丁醇(80%)	20	0.237	松节油	15	0.128
正丙醇	30	0.171	氯化钙盐水(30%)	30	0.550
	75	0.164	氯化钙盐水(15%)	30	0.590
正戊醇	30	0.163	氯化钠盐水(25%)	30	0.570
	100	0.154	氯化钠盐水(12.5%)	30	0.590
异戊醇	30	0.152	硫酸(90%)	30	0.360
	75	0.151	硫酸(60%)	30	0.430
正己醇	30	0.163	硫酸(30%)	30	0.520
	75	0.156	盐酸(12.5%)	32	0.520
正庚醇	30	0.163	盐酸(25%)	32	0.480
	75	0.157	盐酸(38%)	32	0.440
丙烯醇	25～30	0.180	氢氧化钾(21%)	32	0.580
乙醚	30	0.138	氢氧化钾(42%)	32	0.550
	75	0.135	氨	25～30	0.180
乙酸乙酯	20	0.175	氯水溶液	20	0.450
氯甲烷	−15	0.192		60	0.500
	30	0.154	水银	28	0.360
三氯甲烷	30	0.138			

9. 干空气的重要物理性质（101.3kPa）

温度/℃	密度/(kg/m³)	比热容/[kJ/(kg・℃)]	热导率 λ/[$\times10^{-2}$W/(m・℃)]	黏度 μ/$\times10^{-5}$Pa・s	普兰德数 Pr
−50	1.584	1.013	2.035	1.46	0.728
−40	1.515	1.013	2.117	2.52	0.728
−30	1.453	1.013	2.198	1.57	0.723
−20	1.395	1.009	2.279	1.62	0.716
−10	1.342	1.009	2.360	1.67	0.712
0	1.293	1.005	2.442	1.72	0.707
10	1.247	1.005	2.512	1.77	0.705
20	1.205	1.005	2.591	1.81	0.703
30	1.165	1.005	2.673	1.86	0.701
40	1.128	1.005	2.756	1.91	0.699
50	1.093	1.005	2.826	1.96	0.698
60	1.060	1.005	2.896	2.01	0.696
70	1.029	1.009	2.966	2.06	0.694
80	1.000	1.009	3.047	2.11	0.692
90	0.972	1.009	3.128	2.15	0.690
100	0.946	1.009	3.210	2.19	0.688

续表

温度/℃	密度/(kg/m³)	比热容/[kJ/(kg·℃)]	热导率λ/[×10⁻²W/(m·℃)]	黏度μ/×10⁻⁵Pa·s	普兰德数Pr
120	0.898	1.009	3.338	2.29	0.686
140	0.854	1.013	3.489	2.37	0.684
160	0.815	1.017	3.640	2.45	0.682
180	0.779	1.022	3.780	2.53	0.681
200	0.746	1.026	3.931	2.60	0.680
250	0.674	1.038	4.268	2.74	0.677
300	0.615	1.047	4.605	2.97	0.674
350	0.566	1.059	4.908	3.14	0.676
400	0.524	1.068	5.210	3.30	0.678
500	0.456	1.093	5.745	3.62	0.687
600	0.404	1.114	6.222	3.91	0.699
700	0.362	1.135	6.711	4.18	0.706
800	0.329	1.156	7.176	4.43	0.713
900	0.301	1.172	7.630	4.67	0.717
1000	0.277	1.185	8.071	4.90	0.719
1100	0.257	1.197	8.502	5.12	0.722
1200	0.239	1.206	9.153	5.35	0.724

10. 水的重要物理性质

温度/℃	饱和蒸气压/kPa	密度/(kg/m³)	焓/(kJ/kg)	比热容/[kJ/(kg·℃)]	热导率λ/[×10⁻²W/(m·℃)]	黏度μ/(×10⁻⁵Pa·s)	体积膨胀系数β/(×10⁻⁴/℃)	表面张力σ/×10⁻³(N/m)	普兰德数Pr
0	0.608	999.9	0	4.212	55.13	179.2	−0.63	75.6	13.67
10	1.226	999.7	42.04	4.191	57.45	130.8	0.70	74.1	9.52
20	2.335	998.2	83.90	4.183	59.89	100.5	1.82	72.6	7.02
30	4.247	995.7	125.7	4.174	61.76	80.07	3.21	71.2	5.42
40	7.377	992.2	167.5	4.174	63.38	65.60	3.87	69.6	4.31
50	12.31	988.1	209.3	4.174	64.78	54.94	4.49	67.7	3.54
60	19.92	983.2	251.1	4.178	65.94	46.88	5.11	66.2	2.98
70	31.16	977.8	293	4.178	66.76	40.61	5.70	64.3	2.55
80	47.38	971.8	334.9	4.195	67.45	35.65	6.32	62.6	2.21
90	70.14	965.3	377	4.208	68.04	31.65	6.95	60.7	1.95
100	101.3	958.4	419.1	4.220	68.27	28.38	7.52	58.8	1.75
110	143.3	951.0	461.3	4.238	68.50	25.89	8.08	56.9	1.60
120	198.6	943.1	503.7	4.250	68.62	23.73	8.64	54.8	1.47
130	270.3	934.8	546.4	4.266	68.62	21.77	9.19	52.8	1.36
140	361.5	926.1	589.1	4.287	68.50	20.10	9.72	50.7	1.26
150	476.2	917.0	632.2	4.312	68.38	18.63	10.3	48.6	1.17
160	618.3	907.4	675.3	4.346	68.27	17.36	10.7	46.6	1.10
170	792.6	897.3	719.3	4.379	67.92	16.28	11.3	45.3	1.05
180	1003.5	886.9	763.3	4.417	67.45	15.30	11.9	42.3	1.00
190	1225.6	876.0	807.6	4.460	66.99	14.42	12.6	40.8	0.96
200	1554.8	863.0	852.4	4.505	66.29	13.63	13.3	38.4	0.93
210	1917.7	852.8	897.7	4.555	65.48	13.04	14.1	36.1	0.91
220	2320.9	840.3	943.7	4.614	64.55	12.46	14.8	33.8	0.89
230	2798.6	827.3	990.2	4.681	63.73	11.97	15.9	31.6	0.88
240	3347.9	813.6	1037.5	4.756	62.80	11.47	16.8	29.1	0.87
250	3977.7	799.0	1085.6	4.844	61.76	10.98	18.1	26.7	0.86
260	4698.3	784.0	1135.0	4.949	60.43	10.59	19.7	24.2	0.87
270	5504.0	767.9	1185.3	5.070	59.96	10.20	21.6	21.9	0.88
280	6417.2	750.7	1236.3	5.229	57.45	9.81	23.7	19.5	0.90
290	7443.3	732.3	1289.9	5.485	55.82	9.42	26.2	17.2	0.93
300	8592.9	712.5	1344.8	5.736	53.96	9.12	29.2	14.7	0.97

11. 水在不同温度下的黏度

温度/℃	黏度/cP(或 mPa·s)	温度/℃	黏度/cP(或 mPa·s)	温度/℃	黏度/cP(或 mPa·s)
0	1.7921	33	0.7523	67	0.4233
1	1.7313	34	0.7371	68	0.4174
2	1.6728	35	0.7225	69	0.4117
3	1.6191	36	0.7085	70	0.4061
4	1.5674	37	0.6947	71	0.4006
5	1.5188	38	0.6814	72	0.3952
6	1.4728	39	0.6685	73	0.3900
7	1.4284	40	0.6560	74	0.3849
8	1.3860	41	0.6439	75	0.3799
9	1.3462	42	0.6321	76	0.3750
10	1.3077	43	0.6207	77	0.3702
11	1.2713	44	0.6097	78	0.3655
12	1.2363	45	0.5988	79	0.3610
13	1.2028	46	0.5883	80	0.3565
14	1.1709	47	0.5782	81	0.3521
15	1.1404	48	0.5683	82	0.3478
16	1.1111	49	0.5588	83	0.3436
17	1.0828	50	0.5494	84	0.3395
18	1.0559	51	0.5404	85	0.3355
19	1.0299	52	0.5315	86	0.3315
20	1.0050	53	0.5229	87	0.3276
20.2	1.0000	54	0.5146	88	0.3239
21	0.9810	55	0.5064	89	0.3202
22	0.9579	56	0.4985	90	0.3165
23	0.9359	57	0.4907	91	0.3130
24	0.9142	58	0.4832	92	0.3095
25	0.8937	59	0.4759	93	0.3060
26	0.8737	60	0.4688	94	0.3027
27	0.8545	61	0.4618	95	0.2994
28	0.8360	62	0.4550	96	0.2962
29	0.8180	63	0.4483	97	0.2930
30	0.8007	64	0.4418	98	0.2899
31	0.7840	65	0.4355	99	0.2868
32	0.7679	66	0.4293	100	0.2838

12. 饱和水蒸气表（一）（按温度排列）

温度/℃	绝对压强/kPa	蒸汽密度/(kg/m^3)	焓/(kJ/kg)		汽化热/(kJ/kg)
			液体	蒸汽	
0	0.6082	0.00484	0	2491	2491
5	0.8730	0.00680	20.9	2500.8	2480
10	1.226	0.00940	41.9	2510.4	2469
15	1.707	0.01283	62.8	2520.5	2458
20	2.335	0.01719	83.7	2530.1	2446
25	3.168	0.02304	104.7	2539.7	2435
30	4.247	0.03036	125.6	2549.3	2424
35	5.621	0.03960	146.5	2559.0	2412
40	7.377	0.05114	167.5	2568.6	2401
45	9.584	0.06543	188.4	2577.8	2389
50	12.34	0.0830	209.3	2587.4	2378

续表

温度/℃	绝对压强/kPa	蒸汽密度/(kg/m³)	焓/(kJ/kg)		汽化热/(kJ/kg)
			液体	蒸汽	
55	15.74	0.1043	230.3	2596.7	2366
60	19.92	0.1301	251.2	2606.3	2355
65	25.01	0.1611	272.1	2615.5	2343
70	31.16	0.1979	293.1	2624.3	2331
75	38.55	0.2416	314.0	2633.5	2320
80	47.38	0.2929	334.9	2642.3	2307
85	57.88	0.3531	355.9	2651.1	2295
90	70.14	0.4229	376.8	2659.9	2283
95	84.56	0.5039	397.8	2668.7	2271
100	101.33	0.5970	418.7	2677.0	2258
105	120.85	0.7036	440.0	2685.0	2245
110	143.31	0.8254	461.0	2693.4	2232
115	169.11	0.9635	482.3	2701.3	2219
120	198.64	1.1199	503.7	2708.9	2205
125	232.19	1.296	525.0	2716.4	2191
130	270.25	1.494	546.4	2723.9	2178
135	313.11	1.715	567.7	2731.0	2163
140	361.47	1.962	589.1	2737.7	2149
145	415.72	2.238	610.9	2744.4	2134
150	476.24	2.543	632.2	2750.7	2119
160	618.28	3.252	675.8	2762.9	2087
170	792.59	4.113	719.3	2773.3	2054
180	1003.5	5.145	763.3	2782.5	2019
190	1255.6	6.378	807.6	2790.1	1982
200	1554.8	7.840	852.0	2795.5	1944
210	1917.7	9.567	897.2	2799.3	1902
220	2320.9	11.60	942.4	2801.0	1859
230	2798.6	13.98	988.5	2800.1	1812
240	3347.9	16.76	1034.6	2796.8	1762
250	3977.7	20.01	1081.4	2790.1	1709
260	4693.8	23.82	1128.8	2780.9	1652
270	5504.0	28.27	1176.9	2768.3	1591
280	6417.2	33.47	1225.5	2752.0	1526
290	7443.3	39.60	1274.5	2732.3	1457
300	8592.9	46.93	1325.5	2708.0	1382

13. 饱和水蒸气表（二）（按压强排列）

绝对压强/kPa	温度/℃	蒸汽密度/(kg/m³)	焓/(kJ/kg)		汽化热/(kJ/kg)
			液体	蒸汽	
1.0	6.3	0.00773	26.5	2503.1	2477
1.5	12.5	0.01133	52.3	2515.3	2463
2.0	17.0	0.01486	71.2	2524.2	2453
2.5	20.9	0.01836	87.5	2531.8	2444
3.0	23.5	0.02179	98.4	2536.8	2438
3.5	26.1	0.02523	109.3	2541.8	2433
4.0	28.7	0.02867	120.2	2546.8	2427
4.5	30.8	0.03205	129.0	2550.9	2422

续表

绝对压强/kPa	温度/℃	蒸汽密度/(kg/m^3)	焓/(kJ/kg)		汽化热/(kJ/kg)
			液体	蒸汽	
5.0	32.4	0.03537	135.7	2554.0	2418
6.0	35.6	0.04200	149.1	2560.1	2411
7.0	38.8	0.04864	162.4	2566.3	2404
8.0	41.3	0.05514	172.7	2571.0	2398
9.0	43.3	0.06156	181.2	2574.8	2394
10.0	45.3	0.06798	189.6	2578.5	2389
15.0	53.5	0.09956	224.0	2594.0	2370
20.0	60.1	0.1307	251.5	2606.4	2355
30.0	66.5	0.1909	288.8	2622.4	2334
40.0	75.0	0.2498	315.9	2634.1	2312
50.0	81.2	0.3080	339.8	2644.3	2304
60.0	85.6	0.3651	358.2	2652.1	2294
70.0	89.9	0.4223	376.6	2659.8	2283
80.0	93.2	0.4781	390.1	2665.3	2275
90.0	96.4	0.5338	403.5	2670.8	2267
100.0	99.6	0.5896	416.9	2676.3	2259
120.0	104.5	0.6987	437.5	2684.3	2247
140.0	109.2	0.8076	457.7	2692.1	2234
160.0	113.0	0.8298	473.9	2698.1	2224
180.0	116.6	1.021	489.3	2703.7	2214
200.0	120.2	1.127	493.7	2709.2	2205
250.0	127.2	1.390	534.4	2719.7	2185
300.0	133.3	1.650	560.4	2728.5	2168
350.0	138.8	1.907	583.8	2736.1	2152
400.0	143.4	2.162	603.6	2742.1	2138
450.0	147.7	2.415	622.4	2747.8	2125
500.0	151.7	2.667	639.6	2752.8	2113
600.0	158.7	3.169	676.2	2761.4	2091
700.0	164.7	3.666	696.3	2767.8	2072
800	170.4	4.161	721.0	2773.7	2053
900	175.1	4.652	741.8	2778.1	2036
1×10^3	179.9	5.143	762.7	2782.5	2020
1.1×10^3	180.2	5.633	780.3	2785.5	2005
1.2×10^3	187.8	6.124	797.9	2788.5	1991
1.3×10^3	191.5	6.614	814.2	2790.9	1977
1.4×10^3	194.8	7.103	829.1	2792.4	1964
1.5×10^3	198.2	7.594	843.9	2794.5	1951
1.6×10^3	201.3	8.081	857.8	2796.0	1938
1.7×10^3	204.1	8.567	870.6	2797.1	1926
1.8×10^3	206.9	9.053	883.4	2798.1	1915
1.9×10^3	209.8	9.539	896.2	2799.2	1903
2×10^3	212.2	10.03	907.3	2799.7	1892
3×10^3	233.7	15.01	1005.4	2798.9	1794
4×10^3	250.3	20.10	1082.9	2789.8	1707
5×10^3	263.8	25.37	1146.9	2776.2	1629
6×10^3	275.4	30.85	1203.2	2759.5	1556
7×10^3	285.7	36.57	1253.2	2740.8	1488
8×10^3	294.8	42.58	1299.2	2720.5	1404
9×10^3	303.2	48.89	1343.5	2699.1	1357

14. 液体黏度共线图

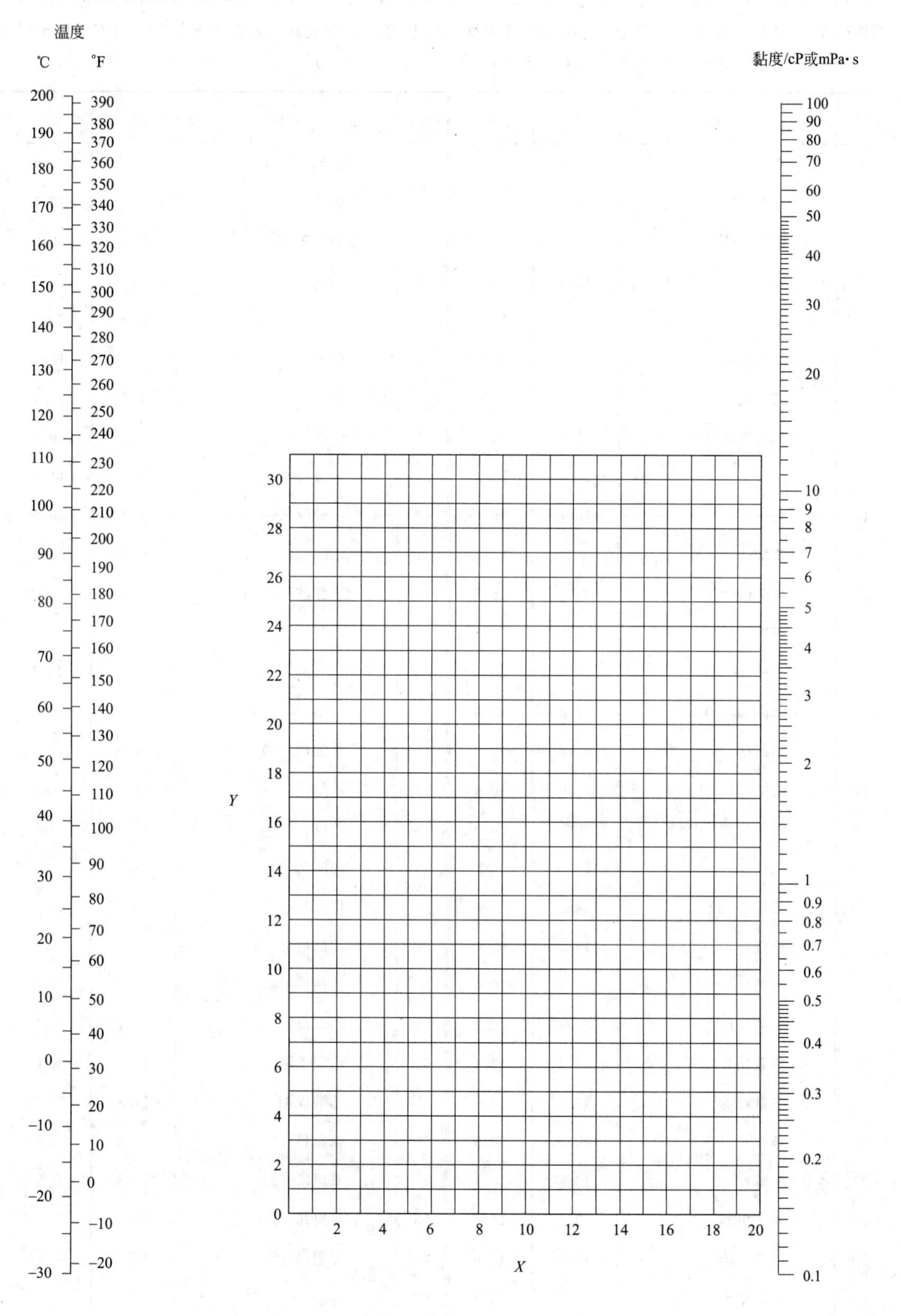
温度
℃
°F
黏度/cP或mPa·s
200
190
180
170
160
150
140
130
120
110
100
90
80
70
60
50
40
30
20
10
0
−10
−20
−30
390
380
370
360
350
340
330
320
310
300
290
280
270
260
250
240
230
220
210
200
190
180
170
160
150
140
130
120
110
100
90
80
70
60
50
40
30
20
10
0
−10
−20
Y
30
28
26
24
22
20
18
16
14
12
10
8
6
4
2
0
2
4
6
8
10
12
14
16
18
20
X
100
90
80
70
60
50
40
30
20
10
9
8
7
6
5
4
3
2
1
0.9
0.8
0.7
0.6
0.5
0.4
0.3
0.2
0.1

液体黏度共线图坐标值

用法举例：求苯在50℃时的黏度，从本表序号26查的$X=12.5$，$Y=10.9$。把这两个数值标在前页共线图的Y-X坐标上的一点，把这点与图中左方温度标尺上的50℃点连成一直线，延长，与右方黏度标尺相交，由此交点定出50℃苯的黏度为0.44mPa·s。

序号	名称	X	Y	序号	名称	X	Y
1	水	10.2	13.0	31	乙苯	13.2	11.5
2	盐水（25%NaCl）	10.2	16.6	32	氯苯	12.3	12.4
3	盐水（25%$CaCl_2$）	6.6	15.9	33	硝基苯	10.6	16.2
4	氨	12.6	2.0	34	苯胺	8.1	18.7
5	氨水（26%）	10.1	13.9	35	酚	6.9	20.8
6	二氧化碳	11.6	0.3	36	联苯	12.0	18.3
7	二氧化硫	15.2	7.1	37	萘	7.9	18.1
8	二硫化碳	16.1	7.5	38	甲醇（100%）	12.4	10.5
9	溴	14.2	13.2	39	甲醇（90%）	12.3	11.8
10	汞	18.4	16.4	40	甲醇（40%）	7.8	15.5
11	硫酸（110%）	7.2	27.4	41	乙醇（100%）	10.5	13.8
12	硫酸（100%）	8.0	25.1	42	乙醇（95%）	9.8	14.3
13	硫酸（98%）	7.0	24.8	43	乙醇（40%）	6.5	16.6
14	硫酸（60%）	10.2	21.3	44	乙二醇	6.0	23.6
15	硝酸（95%）	12.8	13.8	45	甘油（100%）	2.0	30.0
16	硝酸（60%）	10.8	17.0	46	甘油（50%）	6.9	19.6
17	盐酸（31.5%）	13.0	16.6	47	乙醚	14.5	5.3
18	氢氧化钠（50%）	3.2	25.8	48	乙醛	15.2	14.8
19	戊烷	14.9	5.2	49	丙酮	14.5	7.2
20	乙烷	14.7	7.0	50	甲酸	10.7	15.8
21	庚烷	14.1	8.4	51	乙酸（100%）	12.1	14.2
22	辛烷	13.7	10.0	52	乙酸（70%）	9.5	17.0
23	三氯甲烷	14.4	10.2	53	乙酸酐	12.7	12.8
24	四氯化碳	12.7	13.1	54	乙酸乙酯	13.7	9.1
25	二氯乙烷	13.2	12.2	55	乙酸戊酯	11.8	12.5
26	苯	12.5	10.9	56	氟里昂-11	14.4	9.0
27	甲苯	13.7	10.4	57	氟里昂-12	16.8	5.6
28	邻二甲苯	13.5	12.1	58	氟里昂-21	15.7	7.5
29	间二甲苯	13.9	10.6	59	氟里昂-22	17.2	4.7
30	对二甲苯	13.9	10.9	60	煤油	10.2	16.9

15. 气体黏度共线图（常压）

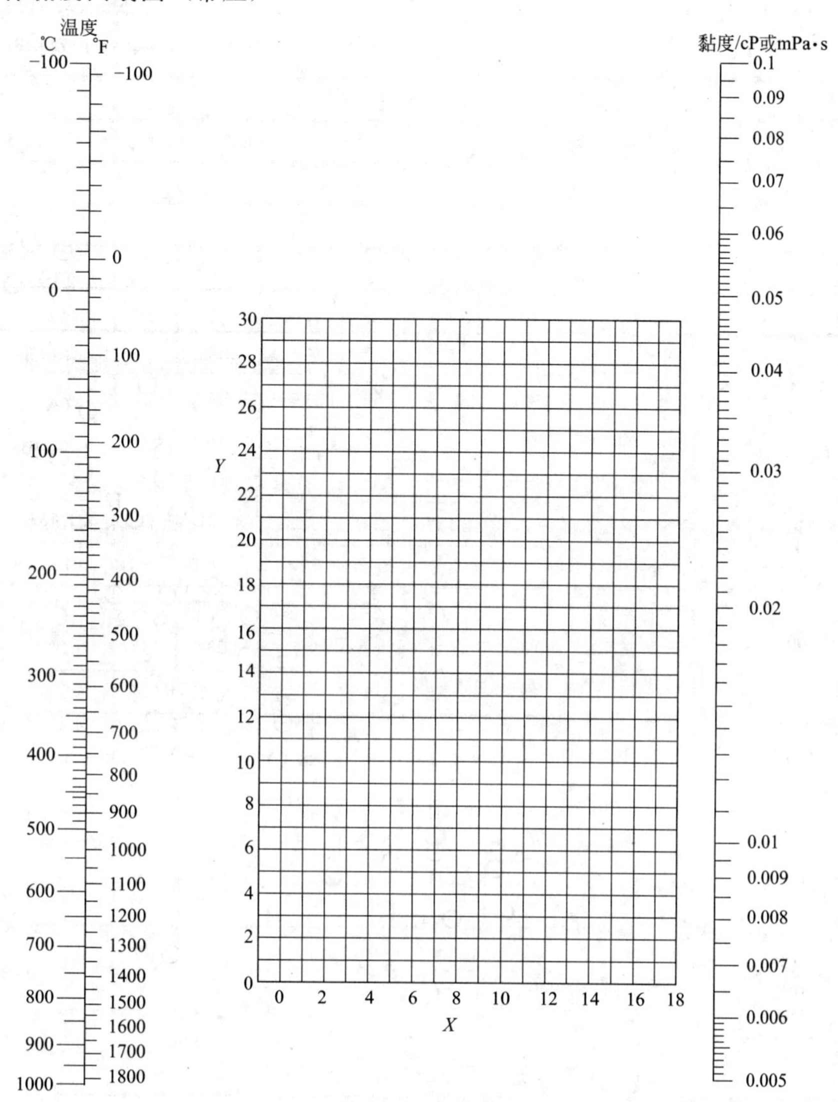

气体黏度共线图坐标值

序号	名称	X	Y	序号	名称	X	Y	序号	名称	X	Y
1	空气	11.0	20.0	15	氟	7.3	23.8	29	甲苯	8.6	12.4
2	氯气	11.0	21.3	16	氯	9.0	18.4	30	甲醇	8.5	15.6
3	氮气	10.6	20.0	17	氯化氢	8.8	18.7	31	乙醇	9.2	14.2
4	氢气	11.2	12.4	18	甲烷	9.9	15.5	32	丙醇	8.4	13.4
5	$3H_2+1N_2$	11.2	17.2	19	乙烷	9.1	14.5	33	乙酸	7.7	14.3
6	水蒸气	8.0	16.0	20	乙烯	9.5	15.1	34	丙酮	8.9	13.0
7	二氧化碳	9.5	18.7	21	乙炔	9.8	14.9	35	乙醚	8.9	13.0
8	一氧化碳	11.0	20.0	22	丙烷	9.7	12.9	36	乙酸乙酯	8.5	13.2
9	氨气	8.4	16.0	23	丙烯	9.0	13.8	37	氟里昂-11	10.6	15.1
10	硫化氢	8.6	18.0	24	丁烯	9.2	13.7	38	氟里昂-12	11.1	16.0
11	二氧化硫	9.6	17.0	25	戊烷	7.0	12.8	39	氟里昂-21	10.8	15.3
12	二硫化碳	8.0	16.0	26	己烷	8.6	11.8	40	氟里昂-22	10.1	17.0
13	一氧化二氮	8.8	19.0	27	三氯甲烷	8.9	15.7				
14	一氧化氮	10.9	20.5	28	苯	8.5	13.2				

16. 液体比热容共线图

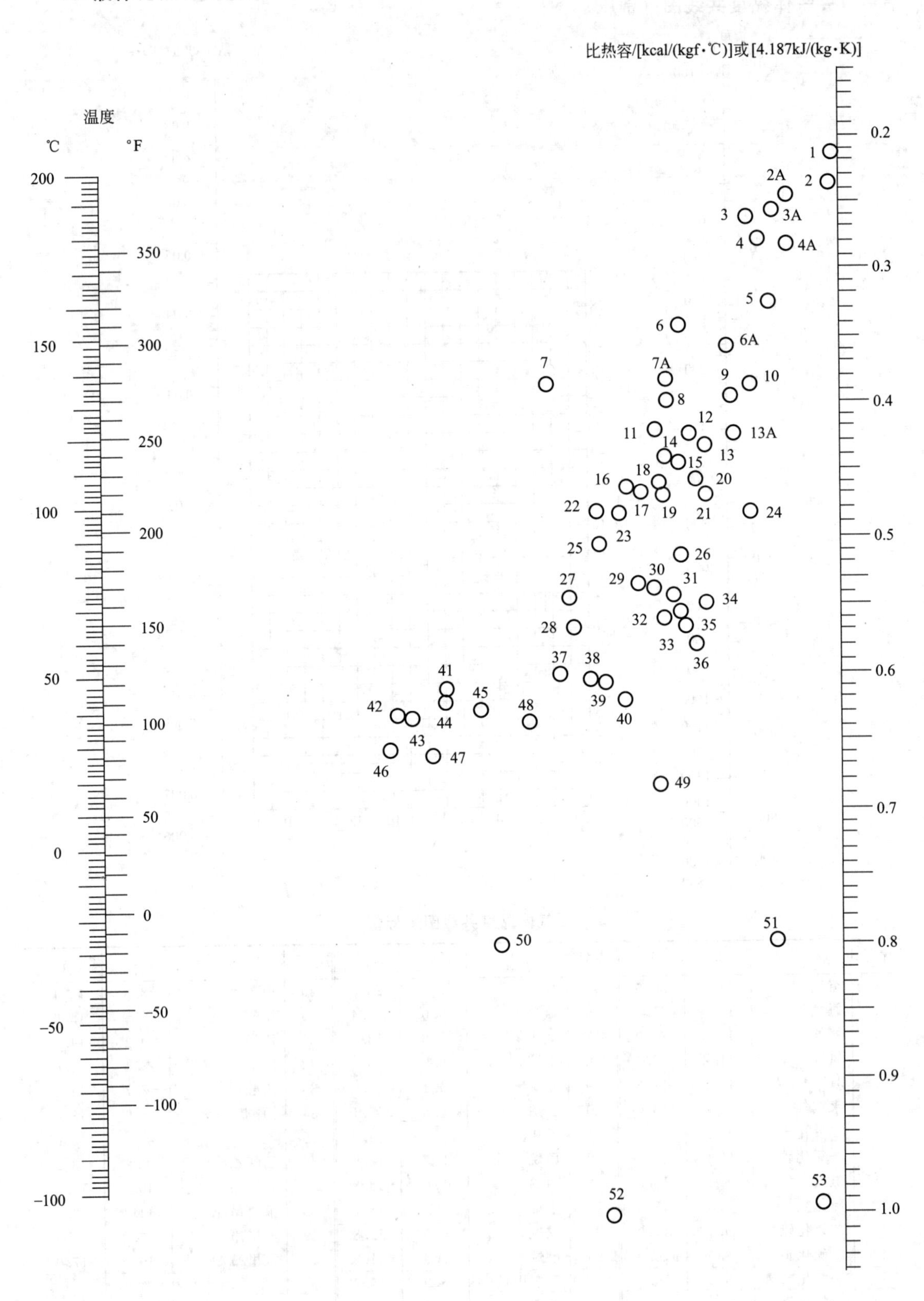

比热容/[kcal/(kgf·℃)]或[4.187kJ/(kg·K)]
温度
℃
°F
200
150
100
50
0
-50
-100
350
300
250
200
150
100
50
0
-50
-100
0.2
0.3
0.4
0.5
0.6
0.7
0.8
0.9
1.0
1
2
2A
3
3A
4
4A
5
6
6A
7
7A
8
9
10
11
12
13
13A
14
15
16
17
18
19
20
21
22
23
24
25
26
27
28
29
30
31
32
33
34
35
36
37
38
39
40
41
42
43
44
45
46
47
48
49
50
51
52
53

液体比热容共线图中的编号

编号	名称	温度范围/℃	编号	名称	温度范围/℃
53	水	10～200	25	乙苯	0～100
51	盐水(25%NaCl)	−40～20	15	联苯	80～120
49	盐水(25%$CaCl_2$)	−40～20	16	联苯醚	0～200
52	氨	−70～50	16	联苯-联苯醚	0～200
11	二氧化硫	−20～100	14	萘	90～200
2	二硫化碳	−100～25	40	甲醇	−40～20
9	硫酸(98%)	10～45	42	乙醇(100%)	30～80
48	盐酸(30%)	20～100	46	乙醇(95%)	20～80
35	己烷	−80～20	50	乙醇(50%)	20～80
28	庚烷	0～60	45	丙醇	−20～100
33	辛烷	−50～25	47	异丙醇	−20～50
34	壬烷	−50～25	44	丁醇	0～100
21	癸烷	−80～25	43	异丁醇	0～100
13A	氯甲烷	−80～20	37	戊醇	−50～25
5	二氯甲烷	−40～50	41	异戊醇	10～100
4	三氯甲烷	0～50	39	乙二醇	−40～200
22	二苯基甲烷	30～100	38	甘油	−40～20
3	四氯化碳	10～60	27	苯甲醇	−20～30
13	氯乙烷	−30～40	36	乙醚	−100～25
1	溴乙烷	5～25	31	异丙醚	−80～200
7	碘乙烷	0～100	32	丙酮	20～50
6A	二氯乙烷	−30～60	29	乙酸	0～80
3	过氯乙烯	−30～40	24	乙酸乙酯	−50～25
23	苯	10～80	26	乙酸戊酯	0～100
23	甲苯	0～60	20	吡啶	−50～25
17	对二甲苯	0～100	2A	氟里昂-11	−20～70
18	间二甲苯	0～100	6	氟里昂-12	−40～15
19	邻二甲苯	0～100	4A	氟里昂-21	−20～70
8	氯苯	0～100	7A	氟里昂-22	−20～60
12	硝基苯	0～100	3A	氟里昂-113	−20～70
30	苯胺	0～130			
10	苯甲基氯	−20～30			

17. 气体比热容共线图（常压）

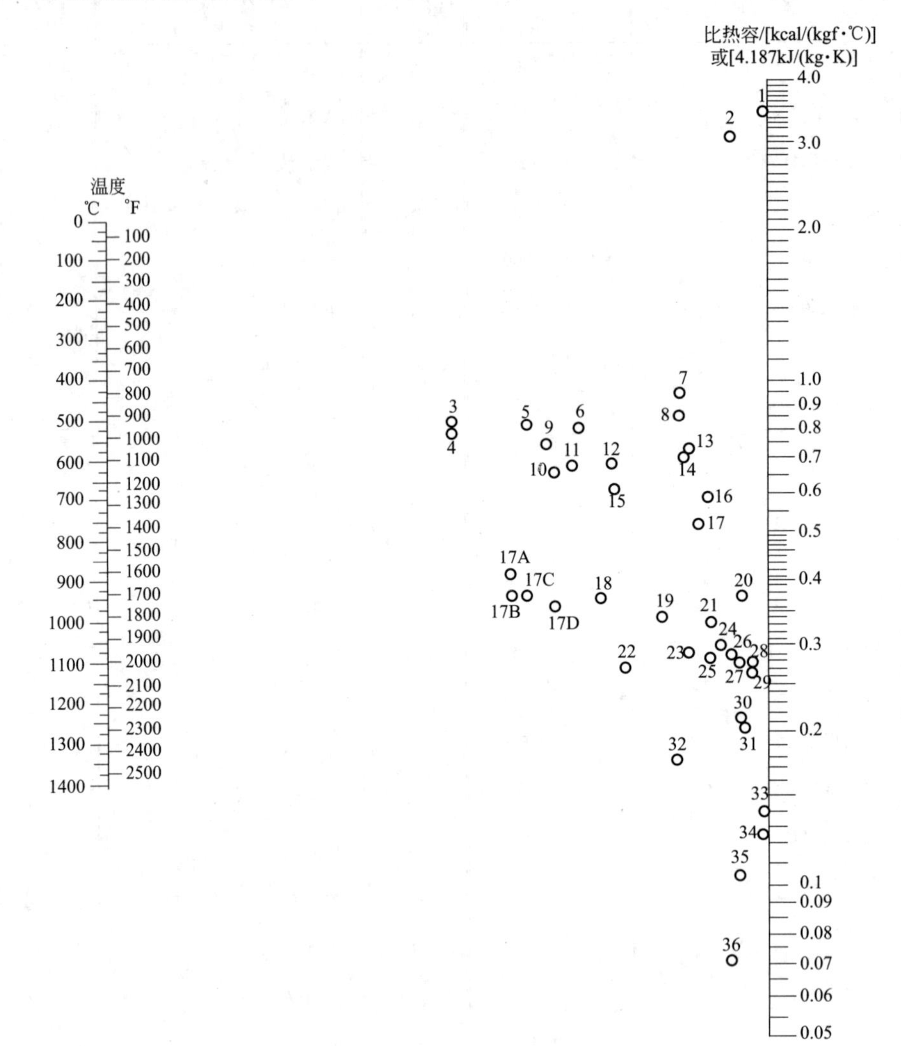

气体比热容共线图中的编号

编号	名称	温度范围/℃	编号	名称	温度范围/℃	编号	名称	温度范围/℃
27	空气	0～1400	24	二氧化碳	400～1400	9	乙烷	200～600
23	氧	0～500	22	二氧化硫	0～400	8	乙烷	600～1400
29	氧	500～1400	31	二氧化硫	400～1400	4	乙烯	0～200
26	氮	0～1400	17	水蒸气	0～1400	11	乙烯	200～600
1	氢	0～600	19	硫化氢	0～700	13	乙烯	600～1400
2	氢	600～1400	21	硫化氢	700～1400	10	乙炔	0～200
32	氯	0～200	20	氟化氢	0～1400	15	乙炔	200～400
34	氯	200～1400	30	氯化氢	0～1400	16	乙炔	400～1400
33	硫	300～1400	35	溴化氢	0～1400	17B	氟里昂-11	0～500
12	氨	0～600	36	碘化氢	0～1400	17C	氟里昂-21	0～500
14	氨	600～1400	5	甲烷	0～300	19A	氟里昂-22	0～500
25	一氧化氮	0～700	6	甲烷	300～700	17D	氟里昂-113	0～500
28	一氧化氮	700～1400	7	甲烷	700～1400			
18	二氧化碳	0～400	3	乙烷	0～200			

18. 液体汽化潜热共线图

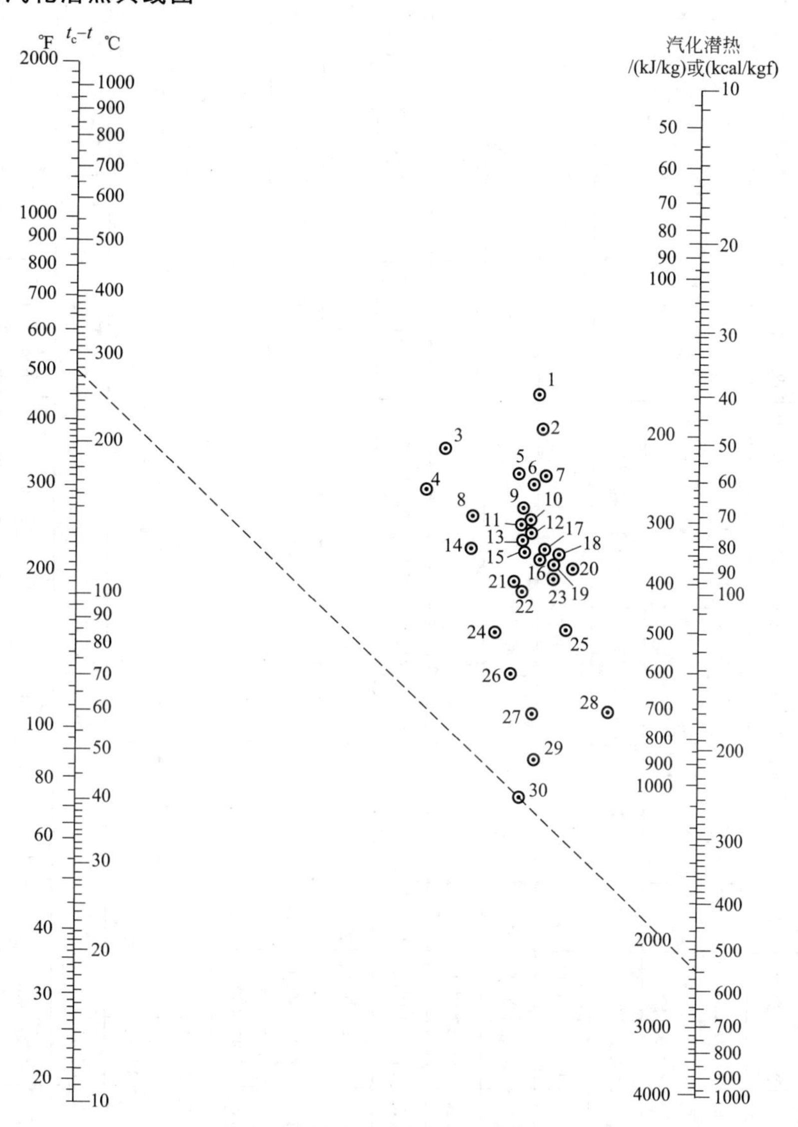

液体汽化潜热共线图中的编号

用法举例：求水在 $t=100$℃时的汽化潜热，从下表中查得水的编号为 30，又查得水的 $t_c=374$℃，故得 $t_c-t=374-100=274$℃，在前页共线图的 t_c-t 标尺上定出 274℃的点，与图中编号为 30 的圆圈中心点连一直线，延长到汽化潜热的标尺上，读出交点读数为 540kcal/kgf 或 2260kJ/kg。

编号	名称	t_c/℃	t_c-t 范围/℃	编号	名称	t_c/℃	t_c-t 范围/℃	编号	名称	t_c/℃	t_c-t 范围/℃
30	水	374	100～500	11	己烷	235	50～225	26	乙醇	243	20～140
29	氨	133	50～200	10	庚烷	267	20～300	24	丙醇	264	20～200
19	一氧化氮	36	25～150	9	辛烷	296	30～300	13	乙醚	194	10～400
21	二氧化碳	31	10～100	20	一氯甲烷	143	70～250	22	丙酮	235	120～210
4	二硫化碳	273	140～275	8	二氯甲烷	216	150～250	18	乙酸	321	100～225
14	二氧化硫	157	90～160	7	三氯甲烷	263	140～270	2	氟里昂-11	198	70～225
25	乙烷	32	25～150	2	四氯化碳	283	30～250	2	氟里昂-12	111	40～200
23	丙烷	96	40～200	17	氯乙烷	187	100～250	5	氟里昂-21	178	70～250
16	丁烷	153	90～200	13	苯	289	10～400	6	氟里昂-22	96	50～170
15	异丁烷	134	80～200	3	联苯	527	175～400	1	氟里昂-113	214	90～250
12	戊烷	197	20～200	27	甲醇	240	40～250				

19. 无机物水溶液在大气压下的沸点

溶液 \ 温度/℃	101	102	103	104	105	107	110	115	120	125	140	160	180	200	220	240	260	280	300	340
	溶液浓度(质量分数)/%																			
$CaCl_2$	5.66	10.31	14.16	17.36	20.00	24.24	29.33	35.68	40.83	54.80	57.89	68.94	75.85	64.91	68.73	72.64	75.76	78.95	81.63	86.18
KOH	4.49	8.51	11.96	14.82	17.01	20.88	25.65	31.97	36.51	40.23	48.05	54.89	60.41							
KCl	8.42	14.31	18.96	23.02	26.57	32.62	36.47	（近于 108.5℃）												
K_2CO_3	10.31	18.37	24.20	28.57	32.24	37.69	43.97	50.86	56.04	60.40	66.94	（近于 133.5℃）								
KNO_3	13.19	23.66	32.23	39.20	45.10	54.65	65.34	79.53												
$MgCl_2$	4.67	8.42	11.66	14.31	16.59	20.23	24.41	29.48	33.07	36.02	38.61									
$MgSO_4$	14.31	22.78	28.31	32.23	35.32	42.86	（近于 108℃）													
NaOH	4.12	7.40	10.15	12.51	14.53	18.32	23.08	26.21	33.77	37.58	48.32	60.13	69.97	77.53	84.03	88.89	93.02	95.92	98.47	（近于 314℃）
NaCl	6.19	11.03	14.67	17.69	20.32	25.09	28.92	（近于 108℃）												
$NaNO_3$	8.26	15.61	21.87	17.53	32.45	40.47	49.87	60.94	68.94											
Na_2SO_4	15.26	24.81	30.73	31.83	（近于 103.2℃）															
Na_2CO_3	9.42	17.22	23.72	29.18	33.66															
$CuSO_4$	26.95	39.98	40.83	44.47	45.12	（近于 104.2℃）														
$ZnSO_2$	20.00	31.22	37.89	42.92	46.15															
NH_4NO_3	9.09	16.66	23.08	29.08	34.21	42.52	51.92	63.24	71.26	77.11	87.09	93.20	69.00	97.61	98.84	100				
NH_4Cl	6.10	11.35	15.96	19.80	22.89	28.37	35.98	46.94												
$(NH_4)_2SO_4$	13.34	23.41	30.65	36.71	41.79	49.73	49.77	53.55	（近于 108.2℃）											

注：表中括号内的数字指饱和溶液的沸点。

20. 某些气体溶于水的亨利系数

气体	温度/℃															
	0	5	10	15	20	25	30	35	40	45	50	60	70	80	90	100
	$E\times10^{-6}$/kPa															
H_2	5.87	6.16	6.44	6.70	6.92	7.16	7.39	7.52	7.61	7.70	7.75	7.75	7.71	7.65	7.61	7.55
N_2	5.35	6.05	6.77	7.48	8.15	8.76	9.36	9.98	10.5	11.0	11.4	12.2	12.7	12.8	12.8	12.8
空气	4.38	4.94	5.56	6.15	6.73	7.30	7.81	8.34	8.82	9.23	9.59	10.2	10.6	10.8	10.9	10.8
CO	3.57	4.01	4.48	4.95	5.43	5.88	6.28	6.68	7.05	7.39	7.71	8.32	8.57	8.57	8.57	8.57
O_2	2.58	2.95	3.31	3.69	4.06	4.44	4.81	5.14	5.42	5.70	5.96	6.37	6.72	6.96	7.08	7.10
CH_4	2.27	2.62	3.01	3.41	3.81	4.18	4.55	4.92	5.27	5.58	5.85	6.34	6.67	6.91	7.01	7.10
NO	1.71	1.96	2.21	2.45	2.67	2.91	3.14	3.35	3.57	3.77	3.95	4.24	4.44	4.54	4.58	4.60
C_2H_6	1.28	1.57	1.92	2.90	2.66	3.06	3.47	3.88	4.29	4.69	5.07	5.72	6.31	6.70	6.96	7.01
	$E\times10^{-5}$/kPa															
C_2H_4	5.59	6.62	7.78	9.07	10.3	11.6	12.9	—	—	—	—	—	—	—	—	—
N_2O	—	1.19	1.43	1.68	2.01	2.28	2.62	3.06	—	—	—	—	—	—	—	—
CO_2	0.738	0.888	1.05	1.24	1.44	1.66	1.88	2.12	2.36	2.60	2.87	3.46	—	—	—	—
C_2H_2	0.73	0.85	0.97	1.09	1.23	1.35	1.48	—	—	—	—	—	—	—	—	—
Cl_2	0.272	0.334	0.399	0.461	0.537	0.604	0.669	0.74	0.80	0.86	0.90	0.97	0.99	0.97	0.96	—
H_2S	0.272	0.319	0.372	0.418	0.489	0.552	0.617	0.686	0.755	0.825	0.689	1.04	1.21	1.37	1.46	1.50
	$E\times10^{-4}$/kPa															
SO_2	0.167	0.203	0.245	0.294	0.355	0.413	0.485	0.567	0.661	0.763	0.871	1.11	1.39	1.70	2.01	—

21. 某些二元物系的气液平衡组成

(1) 乙醇-水 (101.3kPa)

乙醇(摩尔分数)/%		温度/℃	乙醇(摩尔分数)/%		温度/℃
液相中	气相中		液相中	气相中	
0.00	0.00	100	32.73	58.26	81.5
1.90	17.00	95.5	39.65	61.22	80.7
7.21	38.91	89.0	50.79	65.64	79.8
9.66	43.75	86.7	51.98	65.99	79.7
12.38	47.04	85.3	57.32	68.41	79.3
16.61	50.89	84.1	67.63	73.85	78.74
23.37	54.45	82.7	74.72	78.15	78.41
26.08	55.80	82.3	89.43	89.43	78.15

(2) 苯-甲苯 (101.3kPa)

苯(摩尔分数)/%		温度/℃	苯(摩尔分数)/%		温度/℃
液相中	气相中		液相中	气相中	
0.0	0.0	110.6	59.2	78.9	89.4
8.8	21.2	106.1	70.0	85.3	86.8
20.0	37.0	102.2	80.3	91.4	84.4
30.0	50.0	98.6	90.3	95.7	82.3
39.7	61.8	95.2	95.0	97.9	81.2
48.9	71.0	92.1	100.0	100.0	80.2

（3）氯仿-苯（101.3kPa）

氯仿（质量分数）/%		温度/℃	氯仿（质量分数）/%		温度/℃
液相中	气相中		液相中	气相中	
10	13.6	79.9	60	75.0	74.6
20	27.2	79.0	70	83.0	72.8
30	40.6	78.1	80	90.0	70.5
40	53.0	77.2	90	96.1	67.0
50	65.0	76.0			

（4）水-乙酸（101.3kPa）

水（摩尔分数）/%		温度/℃	水（摩尔分数）/%		温度/℃
液相中	气相中		液相中	气相中	
0.0	0.0	118.2	83.3	88.6	101.3
27.0	39.4	108.2	88.6	91.9	100.9
45.5	56.5	105.3	93.0	95.0	100.5
58.8	70.7	103.8	96.8	97.7	100.2
69.0	79.0	102.8	100.0	100.0	100.0
76.9	84.5	101.9			

（5）甲醇-水（101.3kPa）

甲醇（摩尔分数）/%		温度/℃	甲醇（摩尔分数）/%		温度/℃
液相中	气相中		液相中	气相中	
5.31	28.34	92.9	29.09	68.01	77.8
7.67	40.01	90.3	33.33	69.18	76.7
9.26	43.53	88.9	35.13	73.47	76.2
12.57	48.31	86.6	46.20	77.56	73.8
13.15	54.55	85.0	52.92	79.71	72.7
16.74	55.85	83.2	59.37	81.83	71.3
18.18	57.75	82.3	68.49	84.92	70.0
20.83	62.73	81.6	77.01	89.62	68.0
23.19	64.85	80.2	87.41	91.94	66.9
28.18	67.75	78.0			

22. 管子规格

（1）水煤气输送钢管（摘自 YB 234—63）

公称口径		外径	普通管壁厚	加厚管壁厚	公称口径		外径	普通管壁厚	加厚管壁厚
/mm	/in	/mm	/mm	/mm	/mm	/in	/mm	/mm	/mm
6	$\frac{1}{8}$	10	2	2.5	40	$1\frac{1}{2}$	48	3.5	4.25
8	$\frac{1}{4}$	13.5	2.25	2.75	50	2①	60	3.5	4.5
10	$\frac{3}{8}$	17	2.25	2.75	70	$2\frac{1}{2}$	75.5	3.75	4.5
15	$\frac{1}{2}$①	21.25	2.75	3.25	80	3①	88.5	4	4.75
20	$\frac{3}{4}$①	26.75	2.75	3.5	100	4①	114	4	5
25	1①	33.5	3.25	4	125	5	140	4.5	5.5
32	$1\frac{1}{4}$①	42.25	3.25	4	150	6	165	4.5	5.5

① 表示常用规格。

（2）无缝钢管规格简表

冷拔无缝钢管（摘自 YB 231—64）

外径/mm	壁厚/mm		外径/mm	壁厚/mm	
	从	到		从	到
6	1.0	2.0	24	1.0	7.0
8	1.0	2.5	25	1.0	7.0
10	1.0	3.5	27	1.0	7.0
12	1.0	4.0	28	1.0	7.0
14	1.0	4.0	32	1.0	8.0
15	1.0	5.0	34	1.0	8.0
16	1.0	5.0	35	1.0	8.0
17	1.0	5.0	36	1.0	8.0
18	1.0	5.0	38	1.0	8.0
19	1.0	6.0	48	1.0	8.0
22	1.0	6.0	51	1.0	8.0

注：壁厚有 1.0mm、1.2mm、1.5mm、2.0mm、2.5mm、3.0mm、3.5mm、4.0mm、4.5mm、5.0mm、5.5mm、6.0mm、7.0mm、8.0mm。

热轧无缝钢管（摘自 YB 231—64）

外径/mm	壁厚/mm		外径/mm	壁厚/mm	
	从	到		从	到
32	2.5	8	127	4.0	32
38	2.5	8	133	4.0	32
45	2.5	10	140	4.5	35
57	3.0	13	152	4.5	35
60	3.0	14	159	4.5	35
68	3.0	16	168	5.0	35
70	3.0	16	180	5.0	35
73	3.0	19	194	5.0	35
76	3.0	19	219	6.0	35
83	3.5	24	245	7.0	35
89	3.5	24	273	7.0	35
102	3.5	28	325	8.0	35
108	4.0	28	377	9.0	35
114	4.0	28	426	9.0	35
121	4.0	32			

23. IS型离心泵性能表

泵型号	流量/(m^3/h)	扬程/m	转速/(r/min)	汽蚀余量/m	泵效率/%	功率/kW		泵质量/kg	参考价格/元	泵外形尺寸(长×宽×高)/mm	泵口径/mm	
						轴功率	配带功率				吸入	排出
IS50-32-125	7.5		2900				2.2		570	465×190×252	50	32
	12.5	20	2900	2.0	60	1.13	2.2					
	15		2900				2.2					
	3.75		1450				0.55					
	6.3	5	1450	2.0	54	0.16	0.55					
	7.5		1450				0.55					
IS50-32-160	7.5		2900				3		610	465×240×292	50	32
	12.5	32	2900	2.0	54	2.02	3					
	15		2900				3					
	3.75		1450				0.55					
	6.3	8	1450	2.0	48	0.28	0.55					
	7.5		1450				0.55					
IS50-32-200	7.5	525	2900	2.0	35	2.62	5.5		690	465×240×340	50	32
	12.5	50	2900	2.0	48	3.54	5.5					
	15	48	2900	2.5	51	3.84	5.5					
	3.75	13.1	1450	2.0	33	0.41	0.75					
	6.3	12.5	1450	2.0	42	0.51	0.75					
	7.5	12	1450	2.5	44	0.56	0.75					
IS50-32-250	7.5	82	2900	2.0	28.5	5.67	11		850	600×320×405	50	32
	12.5	80	2900	2.0	38	7.16	11					
	15	78.5	2900	2.5	41	7.83	11					
	3.75	20.5	1450	2.0	23	0.91	15					
	6.3	20	1450	2.0	32	1.07	15					
	7.5	20	1450	2.5	35	1.14	15					
IS65-50-125	15		2900				3			465×210×252	65	50
	25	20	2900	2.0	69	1.97	3					
	30		2900				3					
	7.5		1450				0.55					
	12.5	5	1450	2.0	64	0.27	0.55					
	15		1450				0.55					
IS65-50-160	15	35	2900	2.0	54	2.65	5.5		670	465×240×292	65	50
	25	32	2900	2.0	65	3.35	5.5					
	30	30	2900	2.5	66	3.71	5.5					
	7.5	8.8	1450	2.0	50	0.36	0.75					
	12.5	8.0	1450	2.0	60	0.45	0.75					
	15	7.2	1450	2.5	60	0.49	0.75					

续表

泵型号	流量/(m³/h)	扬程/m	转速/(r/min)	汽蚀余量/m	泵效率/%	功率/kW		泵质量/kg	参考价格/元	泵外形尺寸(长×宽×高)/mm	泵口径/mm	
						轴功率	配带功率				吸入	排出
IS65-40-200	15	53	2900	2.0	49	4.42	0.75		730	485×265×340	65	40
	25	50	2900	2.0	60	5.67	0.75					
	30	47	2900	2.5	61	6.29	0.75					
	7.5	13.2	1450	2.0	43	0.63	1.1					
	12.5	12.5	1450	2.0	55	0.77	1.1					
	15	11.8	1450	2.5	57	0.85	1.1					
IS65-40-250	15		2900				15		760	600×320×405	65	40
	25	80	2900	2.0	53	10.3	15					
	30		2900				15					
	7.5		1450				2.2					
	12.5	20	1450	2.0	48	1.42	2.2					
	15		1450									
IS65-40-315	15	127	2900	2.5	28	18.5	30		1060	625×345×450	65	40
	25	125	2900	2.5	40	21.3	30					
	30	123	2900	3.0	44	22.8	30					
	7.5	32.0	1450	2.5	25	2.63	4					
	12.5	32.0	1450	2.5	37	2.94	4					
	15	31.7	1450	3.0	41	3.16	4					
IS80-65-125	30	22.5	2900	3.0	64	2.87	5.5			485×240×292	80	65
	50	20	2900	3.0	75	3.63	5.5					
	60	18	2900	3.5	74	3.93	5.5					
	15	5.6	1450	2.5	55	0.42	0.75					
	25	5	1450	2.5	71	0.48	0.75					
	30	4.5	1450	3.0	72	0.51	0.75					
IS80-65-160	30	36	25900	2.5	61	4.82	7.5		740	485×265×340	80	65
	50	32	2900	2.5	73	5.97	7.5					
	60	29	2900	3.0	72	6.59	7.5					
	15	9	1450	2.5	55	0.67	1.5					
	25	8	1450	2.5	69	0.75	1.5					
	30	7.2	1450	3.0	68	0.86	1.5					
IS80-50-200	30	53	2900	2.5	55	7.87	15		820	485×265×360	80	50
	50	50	2900	2.5	69	9.87	15					
	60	47	2900	3.0	71	10.8	15					
	15	13.2	1450	2.5	51	1.06	2.2					
	25	12.5	1450	2.5	65	1.31	2.2					
	30	11.8	1450	3.0	67	1.44	2.2					

续表

泵型号	流量/(m³/h)	扬程/m	转速/(r/min)	汽蚀余量/m	泵效率/%	功率/kW		泵质量/kg	参考价格/元	泵外形尺寸（长×宽×高）/mm	泵口径/mm	
						轴功率	配带功率				吸入	排出
IS80-50-160	30	84	2900	2.5	52	13.2	22	358	2750	1370×540×565	80	50
	50	80	2900	2.5	63	17.3						
	60	75	2900	3.0	64	19.2						
IS80-50-250	30	84	2900	2.5	52	13.2	22			625×320×405	80	50
	50	80	2900	2.5	63	17.3	22					
	60	75	2900	3.0	64	19.2	22					
	15	21	1450	2.5	49	1.75	3					
	25	20	1450	2.5	60	2.27	3					
	30	18.8	1450	3.0	61	2.52	3					
IS80-50-315	30	128	2900	2.5	41	25.5	37			625×345×505	80	50
	50	125	2900	2.5	54	31.5	37					
	60	123	2900	3.0	57	35.3	37					
	15	32.5	1450	2.5	39	3.4	5.5					
	25	32	1450	2.5	52	4.19	5.5					
	30	31.5	1450	3.0	56	4.6	5.5					
IS100-80-125	60	24	2900	4.0	67	5.86	11			485×280×340	100	80
	100	20	2900	4.5	78	7.00	11					
	120	16.5	2900	5.0	74	7.28	11					
	30	6	1450	2.5	64	0.77	1.5					
	50	5	1450	2.5	75	0.91	1.5					
	60	4	1450	3.0	71	0.92	1.5					
IS100-80-160	60	36	2900	3.5	70	8.42	15		940	600×280×360	100	80
	100	32	2900	4.0	78	11.2	15					
	120	28	2900	5.0	75	12.2	15					
	30	9.2	1450	2.0	67	1.12	2.2					
	50	8.0	1450	2.5	75	1.45	2.2					
	60	6.8	1450	3.5	71	1.57	2.2					
IS100-65-200	60	54	2900	3.0	65	13.6	22		1020	600×320×405	102	65
	100	50	2900	3.6	76	17.9	22					
	120	47	2900	4.8	77	19.9	22					
	30	13.5	1450	2.0	60	1.84	4					
	50	12.5	1450	2.0	73	2.33	4					
	60	11.8	1450	2.5	74	2.61	4					

续表

泵型号	流量/(m^3/h)	扬程/m	转速/(r/min)	汽蚀余量/m	泵效率/%	功率/kW 轴功率	功率/kW 配带功率	泵质量/kg	参考价格/元	泵外形尺寸(长×宽×高)/mm	泵口径/mm 吸入	泵口径/mm 排出
IS100-65-250	60	87	2900	3.5	61	23.4	37		1120	625×360×450	100	65
	100	80	2900	3.8	72	30.3	37					
	120	74.5	2900	4.8	73	33.3	37					
	30	21.3	1450	2.0	55	3.16	5.5					
	50	20	1450	2.0	68	4.00	5.5					
	60	19	1450	2.5	70	4.44	5.5					
IS100-65-315	60	133	2900	3.0	55	39.6	75		1280	655×400×505	100	65
	100	125	2900	3.6	66	51.6	75					
	120	118	2900	4.2	67	57.5	75					
	30	34	1450	2.0	51	5.44	11					
	50	32	1450	2.0	63	6.92	11					
	60	30	1450	2.5	64	7.67	11					
IS125-100-200	120	57.5	2900	4.5	67	28.0	45		1150	625×360×480	125	100
	200	50	2900	4.5	81	33.6	45					
	240	44.5	2900	5.0	80	36.4	45					
	60	14.5	1450	2.5	62	38.3	7.5					
	100	12.5	1450	2.5	76	4.48	7.5					
	120	11.0	1450	3.0	75	4.79	7.5					
IS125-100-250	120	87	2900	3.8	66	43.0	75		1380	670×400×505	125	100
	200	80	2900	4.2	78	55.9	75					
	240	72	2900	5.0	75	62.8	75					
	60	21.5	1450	2.5	63	5.59	11					
	100	20	1450	2.5	76	7.17	11					
	120	18.5	1450	3.0	77	7.84	11					
IS125-100-315	120	132.5	2900	4.0	60	72.1	11		1420	670×400×565	125	100
	200	125	290	4.5	75	90.8	11					
	240	120	2900	5.0	77	101.9	11					
	60	33.5	1450	2.5	56	9.4	15					
	100	32	1450	2.5	73	11.9	15					
	120	30.5	1450	3.0	74	13.5	15					
IS125-100-400	60	52	1450	2.5	53	16.1	30		1570	670×500×635	125	100
	100	50		2.5	65	21.0						
	200	48.5		3.0	67	23.6						
IS150-125-250	120	22.5	1450	3.0	71	10.4	18.5		1440	670×400×605	150	125
	200	20		3.0	81	13.5						
	240	17.5		3.5	78	14.7						

续表

泵型号	流量/(m³/h)	扬程/m	转速/(r/min)	汽蚀余量/m	泵效率/%	功率/kW		泵质量/kg	参考价格/元	泵外形尺寸(长×宽×高)/mm	泵口径/mm	
						轴功率	配带功率				吸入	排出
IS150-125-315	120	32	1450		78		30		1700	670×500×630	150	125
	200											
	240											
IS150-125-400	120	53	1450	2.0	62	27.9	45		1800	670×500×715	150	125
	200	50		2.6	75	36.3						
	240	46		3.5	74	40.6						
IS200-150-250	240	20	1450		82	26.6	37		1960	690×500×655	200	150
	400											
	460											
IS200-150-315	240	37	1450	3.0	70	34.6	55		2050	830×550×715	200	150
	400	32		3.5	82	42.5						
	460	28.5		4.0	80	44.6						
IS200-150-400	240	55	1450	3.0	74	48.6	90		2140	830×550×765	200	150
	400	50		3.8	81	67.2						
	460	45		4.5	76	74.2						

24. Y型离心油泵（摘录）

泵型号	流量/(m³/h)	扬程/m	转速/(r/min)	允许汽蚀余量/m	泵效率/%	功率/kW	
						轴功率	电机功率
50Y60	13.0	67	2950	2.9	38	6.24	7.5
50Y60A	11.2	53	2950	3.0	35	4.68	7.5
50Y60B	9.9	39	2950	2.8	33	3.18	4
50Y60×2	12.5	120	2950	2.4	34.5	11.8	15
50Y60×2A	12	105	2950	2.3	35	9.8	15
50Y60×2B	11	89	2950	2.25	32	8.35	11
65Y60	25	60	2950	3.05	50	8.18	11
65Y60A	22.5	49	2950	3.0	49	6.13	7.5
65Y60B	20	37.5	2950	2.7	47	4.35	5.5
65Y100	25	110	2950	3.2	40	18.8	22
65Y100A	23	92	2950	3.1	39	14.75	18.5
65Y100B	21	73	2950	3.05	40	10.45	15
65Y100×2	25	200	2950	2.85	42	35.8	45
65Y100×2A	23	175	2950	2.8	41	26.7	37
65Y100×2B	22	150	2950	2.75	42	21.4	30
80Y60	50	58	2950	3.2	56	14.1	18.5
80Y100	50	100	2950	3.1	51	26.6	37
80Y100A	45	85	2950	3.1	52.5	19.9	30
80Y100×2	50	200	2950	3.6	53.5	51	75
80Y100×2A	47	175	2950	3.5	50	44.8	55
80Y100×2B	43	153	2950	3.35	51	35.2	45
80Y100×2C	40	125	2950	3.3	49	27.8	37

25. F型耐腐蚀泵

泵型号	流量 /(m^3/h)	扬程 /m	转速 /(r/min)	汽蚀余量 /m	泵效率 /%	功率/kW	
						轴功率	电机功率
25F-16	3.60	16.00	2960	4.30	30.00	0.523	0.75
25F-16A	3.27	12.50	2960	4.30	29.00	0.39	0.55
40F-26	7.20	25.50	2960	4.30	44.00	1.14	1.50
40F-26A	6.55	20.00	2960	4.30	42.00	0.87	1.1
50F-40	14.4	40	2900	4	44.00	3.57	7.5
50F-40A	13.1	32.5	2900	4	44.00	2.64	7.5
50F-16	14.4	15.7	2900		62.00	0.99	1.5
50F-16A	13.1	12	2900			0.69	1.1
65F-16	28.8	15.7	2900			0.69	
65F-16A	26.2	12	2900			1.65	2.2
100F-92	94.3	92	2900	6	64.00	39.5	55.0
100F-92A	88.6	80				32.1	40.0
100F-92B	100.8	70.5				26.6	40.0
150F-56	190.8	55.5	2900	6	67.00	43.0	55.0
150F-56A	170.2	170.2				34.8	45.0
150F-56B	167.8	167.8				29.0	40.0
150F-22	190.8	190.8	2900	6	75.00	15.3	30.0
150F-22A	173.5	173.5				11.3	17.0

注：电机功率应根据液体的密度确定，表中数据仅供参考。

26. 4-72-11型离心泵通风规格（摘录）

机号	转速/(r/min)	全压/Pa	流量/(m^3/h)	效率/%	所需功率/kW
6C	2240	2432.1	15800	91	14.1
	2000	1941.8	14100	91	10.0
	1800	1569.1	12700	91	7.3
	1250	755.1	8800	91	2.53
	1000	480.5	7030	91	1.39
	800	294.2	5610	91	0.73
8C	1800	2795.0	29900	91	30.8
	1250	1343.6	20800	91	10.3
	1000	863.0	16600	91	5.52
	630	343.2	10480	91	1.51
10C	1250	2226.2	41300	94.3	32.7
	1000	1422.0	32700	94.3	16.5
	800	912.1	26130	94.3	8.5
	500	353.1	16390	94.3	2.3
6D	1450	1961.4	20130	89.5	14.2
	960	441.3	6720	91	1.32
8D	1450	1961.4	20130	89.5	14.2
	730	490.4	10150	89.5	2.06
16B	900	2942.1	121000	94.3	127
20B	710	2844.0	186300	94.3	190

注：传动方式中A为电动机直联；B，C，E为皮带轮传动；D为联轴器传动。

27. 管壳式热交换器系列标准（摘录）

外壳直径/mm	159			273					400				600		800			
公称压强/(kgf/cm^2)	25			25					16、25				10、16、25		6、10、16、25			
公称面积/m^2	1	2	3	3	4		5	7	10		20	40	60	120	100		200	230
管子排列方法[①]	△	△	△	△	△		△	△	△		△	△	△	△	△		△	△
管长/m	1.5	2	3	1.5	1.5	2	2	3	1.5		3	6	3	6	3		6	6
管子外径/mm	25	25	25	25	25	25	25	25	25		25	25	25	25	25		25	25
管子总数	13	13	13	32	38	32	38	32	102	86	86	86	269	254	456	444	444	501
管程数	1	1	1	2	1	2	1	2	2	4	4	4	1	2	4	6	6	1
壳程数	1	1	1	1	1	1	1	1	1	1	1	1	1	1	1	1	1	1
管程通道截面积/m^2	0.00408	0.00408	0.00408	0.00503	0.01196	0.00503	0.01196	0.00503	0.01605	0.00692	0.00692	0.00692	0.0845	0.0399	0.0358	0.02325	0.02325	0.1574
壳程通道截面积/m^2 折流板间距/mm 150 a型	—	—	—	0.0156	0.01435	0.017	0.0144	0.01705	0.0214	0.0231	0.0208	0.0196	—	—	—	—	—	—
150 b型	—	—	—	0.0165	0.0161	0.0181	0.0176	0.0181	0.0286	0.0296	0.0276	0.0137	—	—	—	—	—	—
300 a型	—	—	—	—	—	—	—	—	—	—	—	—	0.0377	0.0378	0.0662	0.0806	0.0724	0.0594
300 b型	—	—	—	—	—	—	—	—	—	—	—	—	0.053	0.0534	0.097	0.0977	0.0898	0.0836
600 a型	0.01024	0.01295	0.01223	0.0273	0.0232	0.0312	0.0266	0.0197	0.0308	0.0332	0.0363	0.036	0.0504	0.0553	0.0718	0.0875	0.094	0.0774
600 b型	0.01325	0.015	0.0143	0.029	0.0282	0.0332	0.0323	0.0316	0.013	0.0427	0.0466	0.05	0.0707	0.0782	0.105	0.0344	0.14	0.1092
折流板切去弓形缺口高度/mm a型	50.5	50.5	50.5	85.5	80.5	85.0	80.5	85.5	93.5	104.5	104.5	104.5	132.5	138.5	166	188	188	177
折流板切去弓形缺口高度/mm b型	46.5	46.5	46.5	71.5	71.5	71.5	71.5	71.5	86.5	86.5	86.5	86.5	122.5	122.5	158	152	152	158

① △表示管子为正三角形排列。

注：型折流板缺口上下排列，b型折流板缺口左右排列。

参考文献

[1] 夏清，陈常贵，姚玉英.化工原理：上下册.天津：天津大学出版社，2010.
[2] 陈敏恒，丛德滋，方图南，等.化工原理：上下册.北京：化学工业出版社，2007.
[3] 钟理，伍钦，马四朋.化工原理：上下册.北京：化学工业出版社，2008.
[4] 管国锋，赵汝溥.化工原理：上下册.北京：化学工业出版社，2008.
[5] 柴诚敬.化工原理：上下册.北京：高等教育出版社，2009.
[6] 丁忠伟.化工原理学习指导.北京：化学工业出版社，2006.
[7] 王晓红，田文德，王英龙.化工过程计算机应用基础.北京：化学工业出版社，2009.
[8] 王晓红，田文德，王英龙.化工原理.北京：化学工业出版社，2009.
[9] 丁惠华.化工原理的教学与实践.北京：化学工业出版社，1992.
[10] 时均.化学工程手册.北京：化学工业出版社，2002.
[11] 余国琮.化学工程词典.北京：化学工业出版社，2008.
[12] 范文元.化工单元操作节能技术.合肥：安徽科学技术出版社，2000.
[13] 林爱光，阴金香.化学工程基础.北京：清华大学出版社，2008.
[14] 王志魁.化工原理.北京：化学工业出版社，2005.
[15] [美]博德RB，斯图沃特WE，莱特富特EN.传递现象.戴干策，荣顺熙，石炎福译.北京：化学工业出版社，2004.
[16] [美]麦凯布，史密斯，哈里奥特.化学工程单元操作.伍钦，钟理，夏清，熊丹柳改编，北京：化学工业出版社，2008.
[17] 廖应琪.水污染控制技术.南京：东南大学出版社，2002.
[18] 惠绍棠，阮国岭，于开录.海水淡化与循环经济.天津：天津人民出版社，2005.
[19] 路忠兴，周元培.氯碱化工生产工艺：氯碱分册.北京：化学工业出版社，1995.
[20] 贾绍义，柴诚敬.化工原理课程设计（化工传递与单元操作课程设计）.天津：天津大学出版社，2002.
[21] 金涌，祝京旭，汪展文.流态化工程原理.北京：清华大学出版社，2001.
[22] 陈雪梅.化工原理学习辅导与习题解答.武汉：华中科技大学出版社，2007.
[23] 匡国柱.化工原理学习指导.大连：大连理工大学出版社，2002.
[24] 王湛，王志，高学理等.膜分离技术基础：第3版.北京：化学工业出版社，2019.
[25] 刘茉娥，等.膜分离技术应用手册.北京：化学工业出版社，2001.